ASM ENGINEERED MATERIALS REFERENCE BOOK

ASM ENGINEERED MATERIALS REFERENCE BOOK

Compiled by
The Editorial Staff
Reference Publications
ASM INTERNATIONAL

ASM INTERNATIONAL
METALS PARK, OHIO 44073

Library of Congress Catalog Card Number: 88-071241

ISBN: 0-87170-350-5

SAN: 204-7586

PRINTED IN THE UNITED STATES OF AMERICA

Preface

When the American Society for Metals became ASM International in 1986, its responsibility to report materials data and information increased. This book is one result of that effort. Designed as a companion volume to the **ASM Metals Reference Book, Second Edition**, the scope of the **ASM Engineered Materials Reference Book** is composites, ceramics, engineering or high-performance plastics, and electronic materials. Although both volumes can stand alone, and there is no overlap of material between them, together they cover the entire technical scope of the Society as it presently stands.

This book contains material from a wide variety of sources. The Editors make no claim as to the completeness of the information, as a great deal exists, and more is being generated with every new study of engineered materials. This volume should be used as a first-hand reference for engineered materials data; every piece of information is sourced, and a listing of organizations, professional societies and other data and information sources is included. We believe the book represents the best compilation of engineered materials property data currently available

The organization and scope of this volume was carefully devised. For example, plastics (a very large field in its own right) was limited to engineering plastics — high performance materials that comprise less than ten percent of the total plastics picture. On the other hand, for electronic materials, as much basic information as possible was included, for this is a rapidly evolving field and anything other than basic information would probably be obsolete before the book was printed. Composites information, which makes up a good portion of the book, was organized by constituent materials, and by matrix/filler combinations, to make it easier to look up specific composite types. Phase diagrams were selected for both the electronic materials and the ceramics sections based on commercially used materials.

Just as the **ASM Metals Reference Book Second Edition** was designed as a first-stop source for metals data, the **ASM Engineered Materials Reference Book** is the first place to look for information on engineered materials. We hope that the materials scientists, engineers and technicians find this book to be of as much use as the **Metals Reference Book**.

Sunniva K. Refsnes
Metals Park, OH
August 1988

CONTENTS

Composites

COMPOSITES

Fibers

FIBERS

Fibers

General

Classification of some fibers

Category	Materials
Metal	Beryllium, molybdenum, steel, tungsten
Glass	Vitreous silica, E-glass, S-glass
Carbonaceous(a)	Polyacrylonitrile (high strength), polyacrylonitrile (high modulus), pitch, rayon (very high modulus), rayon (high modulus)
Polymer	Aramid, olefin, nylon, rayon
Inorganic	Alumina (monocrystal), alumina (polycrystal), alumina (whisker), alumina silicates, asbestoes, boron (tungsten core), boron nitride, silicon carbide (carbon core), silicon carbide (polycrystal), silicon carbide (whisker), silicon nitride (whisker), zirconia (polycrystal)

(a) Material refers to starting process.

Source: Ref 16, p 5.1

Properties of fiber materials

Material	Density, g/cm^3	Longitudinal Young's modulus		Tensile strength	
		GPa	10^6 psi	MPa	ksi
Polyester	1.36	13.8	2.0	1100	160
E-glass	2.52	72.3	10.5	3450	500
S-glass	2.49	85.4	12.4	4130	600
Kevlar 49	1.44	124	18.0	2760	400
T-300	1.72	218	31.6	2240	325
VSB-32	1.99	379	55.0	1210	175
FP	3.96	379	55.0	1380	200
Boron	2.35	455	66.0	2070	300
Silicon Carbide	3.19	483	70.0	1520	220
GY-70	1.97	531	77.0	1720	250

Source: Ref 6, p 175

Properties of natural fibers and typical synthetic fibers

Fiber	Density, g/cm^3	Tensile strength		Tensile modulus		Elongation at failure, %
		GPa	10^6 psi	GPa	10^6 psi	
Hemp	1.52	0.92	0.13	70	10	1.7
Jute	1.52	0.86	0.12	60	9	2.0
Flax	1.52	0.84	0.12	100	15	1.8
Cotton	1.52	0.2-0.8	0.03-0.12	27	4	6-12
Silk	1.34	0.6	0.09	10	2	18-20
S-glass	2.50	4.6	0.67	84	12	2-5
Carbon (type I)	1.90	2.0	0.29	380	55	1-2
Aramid	1.44	2.8	0.41	133	20	2-4

Source: Ref 6, p 117

Thermal properties of selected fibers

Fiber	Diameter, μm	Heat capacity, kJ/(kg · K)[a]	Thermal conductivity, W/(m · K)[b]	Coefficient of thermal expansion, 10^{-6}/°C
Graphite				
PAN HM	7	0.7	1003	−1.1
PAN HTS (T300)	8	0.7	1003	−1.1
Rayon (T50)	6	0.7	1003	−1.1
Thornel 75 (T75)	5	0.7	1003	−1.1
Pitch (type P)	5–10	0.7	1003	−1.1
Pitch UHM	11	0.7	1003	−1.1
Boron on tungsten	102–203	1.3	38	5.0
Borsic	102–203	1.3	38	5.0
Boron on carbon	102–203	1.3	38	5.0
Silicon carbide on tungsten	102–203	1.2	16	4.3
Silicon carbide on carbon	102	1.2	16	4.3
Beryllium	127	1.9	150	11.5
Alumina (FP)	20	...	...	8.3
S glass	9	0.7	13	5.0
E glass	9	0.7	13	5.0
Molybdenum	127	0.3	145	4.9
Steel	127	0.5	29	13.3
Tantalum	508	0.2	55	6.5
Tungsten	381	0.1	168	4.5
Whisker ceramic				
Al_2O_3	10–25	0.6	24	7.7
Metallic (Fe)	127	0.5	29	13.3

[a]To convert kJ/(kg · K) to Btu/(lb · °F), divide by 4.184. [b]To convert W/(m · K) to Btu · ft/(ft² · h · °F), divide by 1.729.

Source: Ref 16, p 5.34

Differences between bulk and whisker strengths of various materials

Material	Tensile strength, GPa, Bulk	Tensile strength, GPa, Whisker	Tensile strength, 10^6 psi, Bulk	Tensile strength, 10^6 psi, Whisker
Iron	0.028	13.1	0.004	1.9
Copper	0.0014	2.8	0.0002	0.4
Silicon	0.034	3.79	0.005	0.55
Graphite	0.28	20.7	0.04	3.0
Boron carbide	0.1551	6.653	0.0225	0.965
Alumina	0.55	42.7	0.08	6.2
Silicon carbide	0.21	20.7[a]	0.03	3.0[a]

[a]Bend test (all other values are tensile data).

Source: Ref 16, p 5.37

Effect of fiber type on selected ultimate properties

Fiber type	Specific gravity	Tensile strength MPa	Tensile strength ksi	Tensile modulus GPa	Tensile modulus 10^6 psi	Compressive strength MPa	Compressive strength ksi	Thermal conductivity W/m · k	Thermal conductivity Btu · in./h · ft^2 · °F
Glass(a)	2.0	690	100	40	6	410	60	0.30	2
Carbon(b)	1.65	1000-1500	150-220	100-140	15-20	620-970	90-140	0.85-1.4	6-10
Aramid(c)	1.28	1400	200	80	12	280	40	0.15	1

(a) E-glass unidirectional rovings. (b) Type AS graphite fibers. (c) DuPont Kevlar 49 fibers.

Source: Ref 6, p 540

Comparison of long and short fibers

Property	Polycarbonate matrix reinforced with 40% E-glass: Short	Polycarbonate matrix reinforced with 40% E-glass: Long	Nylon 6/6 matrix reinforced with 40% E-glass: Short	Nylon 6/6 matrix reinforced with 40% E-glass: Long
Izod impact strength, J/cm (ft·lb/in.):				
Notched	1.2 (2.2)	4.8 (8.9)	1.4 (2.6)	5.94 (11.13)
Unnotched	6.4 (12)	12 (23)	8.0 (15)	17 (31)
Flexural strength, MPa (ksi)	180 (26)	255 (37)	275 (40)	340 (49)
Flexural modulus, GPa (10^6 psi)	9.65 (1.40)	13.4 (1.95)	11.7 (1.70)	11.7 (1.70)
Tensile strength, MPa (ksi)	145 (21)	150 (22)	220 (32)	180 (26)
Tensile modulus, GPa (10^6 psi)	11.7 (1.70)	17.9 (2.60)	13.1 (1.90)	17.9 (2.60)
Compressive strength, MPa (ksi)	150 (22)	220 (32)	160 (23)	255 (37)
Specific gravity	1.50	1.50	1.45	1.45

Reported value(a) for the columns above.

(a) Short = length-to-diameter aspect ratio of 50. Long = length-to-diameter aspect ratio of 300.

Source: Ref 27, p 82

Fiber bundle dimensions

Material	Yield/tow m/kg	Yield/tow yd/lb	Filament size μm	Filament size μin.
Graphite (1000 to 12000 filaments per tow)	300-1200	150-600	5-10	200-390
Fiberglass (2450-12240 filaments per tow)	490-2400	245-1200	4-13	160-510
Aramid (800-3200 filaments per tow)	2000-7850	980-3900	12	470

Source: Ref 6, p 143

Some filament properties at room and elevated temperatures

Property	Type of filament				
	Borsic	SiC	Mo (TZM)	Sapphire	Be
Diameter:					
μm	107–145	102	102	254	127–1525
mils	4.2–5.7	4	4	10	5–60
Density:					
g/cm^3	2.768	3.460	10.186	3.958	1.855
$lb/in.^3$	0.100	0.125	0.368	0.143	0.067
Modulus, GPa:					
Room temperature	415	450	295	470	295
205 °C (400 °F)	...	450	...	...	...
315 °C (600 °F)	...	450	...	...	...
425 °C (800 °F)	345	450	...	...	290
540 °C (1000 °F)	275	450	...	460	255
Modulus, 10^6 psi:					
Room temperature	60	65	43	68	43
205 °C (400 °F)	...	65	...	...	...
315 °C (600 °F)	...	65	...	...	...
425 °C (800 °F)	50	65	...	...	42
540 °C (1000 °F)	40	65	...	67	37
Tensile strength, MPa:					
Room temperature	2930	2275	2550	2550	690–965
205 °C (400 °F)	2860	2070	...	2035	550
315 °C (600 °F)	2655	2040	...	1895	345
425 °C (800 °F)	1795	2025	...	1795	240
540 °C (1000 °F)	1550	2015	...	1795	170
Tensile strength, ksi:					
Room temperature	425	330	370	370	100–140
205 °C (400 °F)	415	300	...	295	80
315 °C (600 °F)	385	296	...	275	50
425 °C (800 °F)	260	294	...	260	35
540 °C (1000 °F)	225	292	...	260	25
Coefficient of thermal expansion, room temperature to 540 °C (1000 °F):					
$10^{-6}/°C$	5.22	3.24	5.94	7.92	15.3
$10^{-6}/°F$	2.90	1.80	3.30	4.40	8.50

Source: Ref 79, p 5.31

Mechanical properties and maximum service temperatures of whiskers

	Whisker diameter		Density		Tensile strength		Tensile modulus		Maximum service temperature	
	μm	mils	kg/m³	lb/in.³	MPa	ksi	GPa	10⁶ lb/in.²	°C	°F
Al_2O_3	3–10	0.11–0.39	3958	0.143	20 685	3000	427	62	1650	3000
BeO	10–30	0.39–1.17	2851	0.103	13 100	1900	345	50	1925	3500
B_4C	...	...	2519	0.091	13 790	2000	483	70	1095	2000
SiC	1–3	0.03–0.11	3211	0.116	20 685	3000	483	70	600	1110
Si_3N_4	...	...	3183	0.115	13 790	2000	379	55	600	1110
Quartz	...	...	...	...	4 135	600	76	11	...	...
Sapphire	...	...	...	...	11 720	1700	510	74	...	...
Cr	...	...	7197	0.260	8 895	1290	241	35	540	1000
Cu	...	...	8913	0.322	3 275	475	124	18	260	500
Fe	...	...	7833	0.283	13 100	1900	200	29	540	1000
Ni	...	...	8968	0.324	3 860	560	200	29	540	1000
Graphite	...	...	1661	0.060	19 615	2845	703	102	(a)	(a)
Au	...	...	...	...	1 655	240	76	11	...	...
Zirconia	...	...	...	...	4 135	600	427	62	...	...
Silicon	...	...	...	...	3 790	550	158	23	...	...

(a) 600 °F in air; 5000 °F in inert atmosphere.

Source: Ref 16, p 5.35

Some fibers used in plastic

Fiber[a]	Fiber diameter μm	Fiber diameter mil	Density g/cm³	Density lb/in.³	Tensile strength MPa	Tensile strength ksi	Modulus GPa	Modulus 10⁶ psi	Use limit °C	Use limit °F	Price/lb (1985)
E glass (C)	3–20	0.12–0.79	2.49	0.090	3450	500	72.4	10.5	425	800	\$0.80–\$1.20
S glass (C)	10–20	0.39–0.79	2.49	0.090	4590	665	86.9	12.6	425	800	\$4
Kevlar (C)	12	0.47	1.44	0.052	2760	400	125	18	425	800	\$16
Carbon/graphite-PAN (C)	7.0	0.28	1.72–1.80	0.062–0.065	2410–4830	350–700	230–395	33–57	>1650[b]	>3000[b]	\$17–\$450
Carbon/graphite-pitch (C)	5.1–12.7	0.20–0.50	1.99–2.16	0.072–0.078	2070	300	380–690	55–100	>1650[b]	>3000[b]	\$26–\$1250
Processed mineral fiber (DC)	4–6	0.16–0.24	2.68	0.097	830	120	105	15	760	1400	\$0.30–\$0.50
Fiberfrax (DC)	2–5	0.08–0.20	2.60	0.094	1030	150	105	15	1150	2100	\$1
Fibermax (DC)	3–6	0.12–0.24	2.99	0.108	860	125	150	22	1760	3200	\$16.65

[a]C = continuous; DC = discontinuous. [b]Oxidation begins at lower temperatures.

Source: Ref 16, p 5.29

Textile fibers comparison

		Glass		Polyester			Nylon			
		Fiberglas		Dacron (Du Pont)			Nylon, 6/6			Nomex (Du Pont)
Fiber type		E glass	S Glass	Regular tenacity	High tenacity	Staple and tow	Regular tenacity	High tenacity	Staple and tow	Staple, tow and filament
		Single filament		Filament			Monofilament filament	Filament		
Breaking tenacity (gram per denier)	**Std.**	15.3	19.9	2.8 to 5.6	6.0 to 9.5 6.0 to 9.5	2.2 to 6.0 2.2 to 6.0	2.3 to 6.0	5.9 to 9.5	3.5 to 7.2 3.0 to 6.1	4.0 to 5.3
	Wet	15.3	19.9	2.8 to 5.6		2.0 to 5.5	2.0 to 5.2	5.1 to 8.0	3.7 to 5.9	3.04 to 4.1
	Std. loop			2.5 to 5.2		2.0 to 5.5	2.0 to 5.1	5.0 to 7.6	3.7 to 5.9	4.0 to 5.0
	Std. knot						2.0 to 5.1	5.0 to 7.6		
Tensile strength (psi)		450 to 550	650 to 700	50 to 99	106 to 168	39 to 106	40 to 106	86 to 134		90,000
Breaking elongation (%)	**Std.**	4.8	5.7	19 to 34	12 to 16	12 to 55	25 to 65	15 to 28	16 to 66	22 to 32
	Wet	4.8	5.7	19 to 34	12 to 16	12 to 55	30 to 70	18 to 32	18 to 68	20 to 30
Elastic recovery (%)		100	100	97 at 2% 80 at 8%	100 at 1%	100 at 1%	100 at 5% 98 to 100 at 10%	100 at 4% 96 at 5% 95 at 10%		
Average stiffness (gram per denier)		320	380	12 to 27	46 to 82	12 to 17	5 to 24	21 to 58	10 to 45	
Average toughness		0.37	0.53	0.40 to 0.80	0.50 to 0.70	0.20 to 1.10	0.50 to 1.00	0.74 to 0.84	0.58 to 0.84	0.85
Specific gravity		2.54	2.48	1.38			1.13 to 1.14			1.38
Water absorbency, %	**70 °F, 65% r.h.**	None	None	0.4 or 0.8 (depends on type)			4.0 to 4.5			6.5
	70 °F, 95% r.h.	Up to 0.3	Up to 0.3				6.1 to 8.0			12.5
Effect of heat		Will not burn. Retains 95% tensile at 650 °F. Softens at 1,350 °F.	Will not burn. Retains 80% tensile at 650 °F. Softens at 1,560 °F.	Sticks at 445 °F. Melts at 482 °F.			Sticks at 445 °F. Melts at 480° to 500 °F. Yellows slightly at 300 °F, when held for 5 hrs.			Does not melt. Decomposes at 700 °F.
Effect of acids and alkalis		Resists most acids and alkalis.		Good resistance to most mineral acids. Dissolves with partial decomposition in concentrated solutions of sulfuric acids. Good resistance to weak alkalis. Moderate resistance to strong alkalis at room temperatures. Disintegrates in strong alkalis at boil.			Unaffected by most mineral acids, except hot mineral acids. Dissolves with partial decomposition in concentrated solutions of hydrochloric, sulfuric, and nitric acids. Soluble in formic acid. Substantially inert in alkalis.			Unaffected by most acids, except some strength loss after long exposure to hydrochloric, nitric, and sulfuric. Generally good resistance to alkalis
Effect of bleaches and solvents		Unaffected		Excellent resistance to bleaches and other oxidizing agents. Generally insoluble except in some phenolic compounds.			Can be bleached in most bleaching solutions. Generally insoluble in most organic solvents. Soluble in some phenolic compounds.			Unaffected by most bleaches and solvents except for slight strength loss from exposure to sodium chlorite.
Dyes used		Resin-bonded pigment systems. Vat, acid, or chrome dyes will tint.		Disperse, developed, and cationic (for some types), with carrier or at high temperature.			Disperse, acid, and premetalized are usually preferred, but most other classes are also used.			Industrial yarn is nondyeable. Staple is dyeable with cationic dyes.
Resistance to mildew, aging, sunlight, abrasion		Excellent resistance to sunlight and aging. Not attacked by mildew. (Binder may be affected by mildew.)		Not weakened by mildew. Excellent resistance to aging and abrasion. Prolonged exposure to sunlight causes some strength loss but no discoloration.			Excellent resistance to mildew, aging, and abrasion. Prolonged exposure to sunlight causes some deterioration.			Excellent resistance to mildew and aging. A 50% strength loss after 60 weeks' exposure to sunlight. Good abrasion resistance

(continued)

Textile fibers comparison (continued)

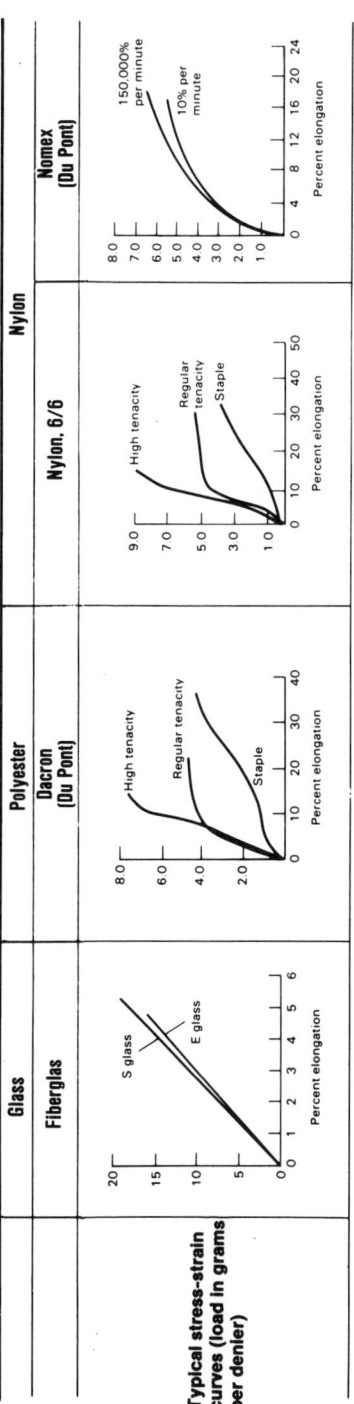

	Glass	Polyester	Nylon	
	Fiberglas	Dacron (Du Pont)	Nylon, 6/6	Nomex (Du Pont)
Typical stress-strain curves (load in grams per denier)	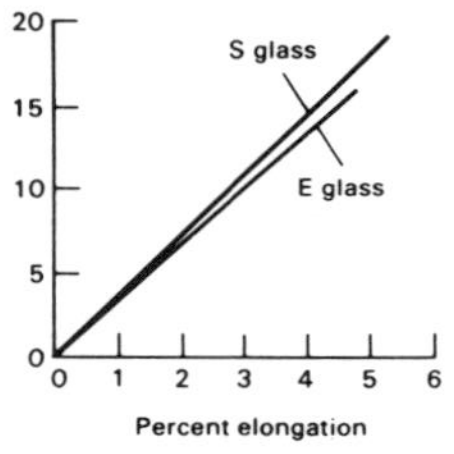	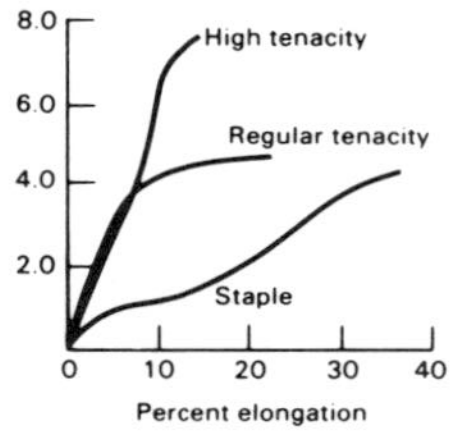	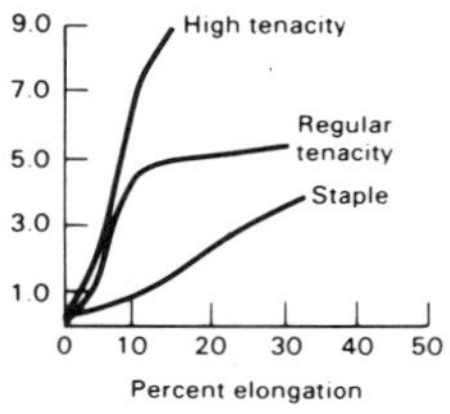	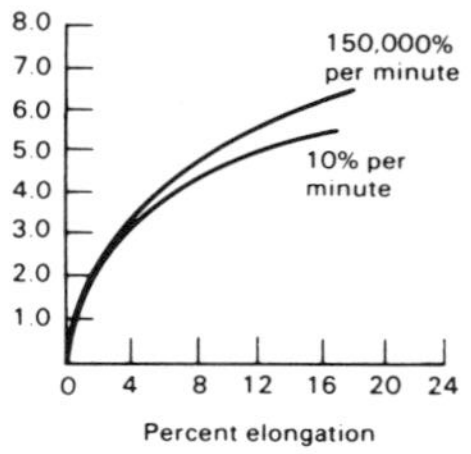

Source: Ref 16, p 5.32

Some fibers used in metal-matrix composites

Fiber[a]	Fiber diameter μm	Fiber diameter mils	Tensile strength MPa	Tensile strength ksi	Modulus GPa	Modulus 10^6 psi	Use limit °C	Use limit °F	Price/lb (1985)
Boron (C)	102–103	4.0–8.0	3450	500	400	58	540	1000	$262
Carbon/graphite-PAN (C)	7.0	0.28	2410–4830	350–700	230–395	33–57	>1650[b]	>3000[b]	$17–$450
Carbon/graphite-pitch (C)	5.1–12.7	0.20–0.50	2070	300	380–690	55–100	>1650[b]	>3000[b]	$26–$1250
SiC monofilament (C)	140	5.51	4140	600	425	62	930	1700	$800
SiC (W)	6	0.24	3340	485	485–825	70–120	930	1700	$95
FP alumina (C)	20	0.79	1380	200	380	55	>1650	>3000	$200
Fiberfrax (DC)	2–5	0.08–0.20	1030	150	105	15	1150	2100	$1
Fibermax (DC)	3–6	0.12–0.24	830	120	150	22	1760	3200	$16.65

[a]C = continuous; W = whisker; DC = discontinuous. [b]Oxidation begins at lower temperatures.

Source: Ref 16, p 5.29

Critical lengths (L_C) and critical aspect ratios (L_C/d)

Fiber	Matrix	L_c mm	L_c in.	L_c/d mm	L_c/d in.
E-glass(a)	Polypropylene	1.78	0.0700	140	6
E-glass(a)	Epoxy	0.43	0.017(b)	34(b)	1.3
E-glass(a)	Polyester	1.27	0.0500(c)	100(c)	4
Carbon (AS1)(d)	Epoxy	0.33	0.013	47	1.9
Carbon (AS4)(d)	Epoxy	0.42	0.017	60	2.4
Carbon (AS4)(d)	Polycarbonate	0.74	0.030	106	4.17
Carbon (XAS)(e)	Polycarbonate	0.35	0.014	50	2
Carbon (XAS)(e)	Epoxy	0.36	0.014	51	2

(a) Proprietary sizing. (b) At 40 °C (105 °F). (c) At 50 °C (120 °F). (d) Hercules Inc. (e) Hysol-Grafil Ltd.

Source: Ref 6, p 120

Fiber tow characteristics before impregnation

Before impregnation

Material	Yield/tow m/kg	Yield/tow yd/lb	Filament size µm	Filament size µin.
Graphite (1000-12 000 filaments/tow)	300-1200	150-600	5-10	200-390
Fiberglass (2450-12 240 filaments/tow)	490-2400	245-1200	4-13	160-510
Aramid (800-3200 filaments/tow)	2000-7850	980-3900	12	470

Source: Ref 6, p 151

Filament diameter nomenclature

Alphabet	Filament diameter µm	Filament diameter 10^{-4} in.
AA	0.8 - 1.2	0.3 - 0.5
A	1.2 - 2.5	0.5 - 1.0
B	2.5 - 3.8	1.0 - 1.5
C	3.8 - 5.0	1.5 - 2.0
D	5.0 - 6.4	2.0 - 2.5
E	6.4 - 7.6	2.5 - 3.0
F	7.6 - 9.0	3.0 - 3.5
G	9.0 - 10.2	3.5 - 4.0
H	10.2 - 11.4	4.0 - 4.5
J	11.4 - 12.7	4.5 - 5.0
K	12.7 - 14.0	5.0 - 5.5
L	14.0 - 15.2	5.5 - 6.0
M	15.2 - 16.5	6.0 - 6.5
N	16.5 - 17.8	6.5 - 7.0
P	17.8 - 19.0	7.0 - 7.5
Q	19.0 - 20.3	7.5 - 8.0
R	20.3 - 21.6	8.0 - 8.5
S	21.6 - 22.9	8.5 - 9.0
T	22.9 - 24.1	9.0 - 9.5
U	24.1 - 25.4	9.5 - 10

Source: Ref 6, p 109

Towpreg form parameters

Parameter	Typical range
Strand weight per length, g/m (lb/yd)	0.74-1.48 (0.0015-0.0030)
Resin content, %	28-45
Tow width, cm (in.)	0.16-0.64 (0.06-0.25)
Package size, kg (lb)	0.25-4.5 (0.5-10)

Source: Ref 6, p 151

Polymer Fibers

Properties of aramid fibers

Property	Kevlar 49	Kevlar 149
Physical properties		
Specific gravity	1.44	1.38
Density, g/cm^3 (lb/in.^{3})	1.44 (0.052)	1.42 (0.050)
Filament diameter, μm (mil)	12 (0.47)	...
Mechanical properties		
Tensile strength, MPa (ksi)	...	1400 (203), 0°, Dir. 1
Tensile modulus, GPA (10^6 psi)	...	105 (15.4) 0°, Dir. 1
Elongation at RT, 5	2.5	1.3, 0°, Dir. 1
Thermal properties		
Coefficient of expansion, axial and lateral, 10^{-6}/K	...	-268, 0°; 120, 90°
Electrical properties		
Dielectric constant	...	At 1 kHz, 4.14 dry; at 1 MHz, 3.90 dry
Dissipation factor	...	At 1 kHz, 0.0103 dry; at 1 MHz, 0.0142 dry

Source: Ref 6, p 361

Organic fibers for filament winding (in order of ascending modulus of strand, normalized to 100% fiber volume)

Type	Strand nominal tensile modulus		Strand nominal tensile strength		Maximum number of filaments/strand	Fiber density, g/cm^3
	GPa	10^6 psi	MPa	ksi		
Aramid (medium modulus)	62	9.0	2758	400	1000	1.44
Oriented polyethylene(a)	117	17	2585	375	118	0.97
Aramid(a)	121	17.5	4067	590	...	1.44
Aramid	124	18	3792	550	5000	1.44
Oriented polyethylene(b)	170	24.8	3274	471	...	0.97

(a) Development status. (b) Research and development status

Source: Ref 6, p 504

Thermal, physical, and chemical properties of a typical aramid

Heat-deflection temperature at 1.82 MPa, °C	260
Maximum resistance to continuous heat, °C	150
Coefficient of linear thermal expansion, 10^{-5}/°C	2.6
Tensile strength, MPa	120
Elongation, %	5
Specific gravity	1.2
Dielectric constant	3.0
Resistance to chemicals at 25 °C(a):	
Nonoxidizing acids (10% H_2SO_4)	Q
Oxidizing acids (10% HNO_3)	U
Aqueous salt solutions (NaCl)	S
Aqueous alkalis (NaOH)	S
Polar solvents (C_2H_5OH)	Q
Nonpolar solvents (C_6H_6)	S
Water	S

(a) S, satisfactory; Q, questionable; U, unsatisfactory.

Source: Ref 23, p 71

Mechanical properties and maximum use temperature of Kevlar 49

Tensile strength:	
At room temperature (16 months)	No loss in strength
At 50 °C (120 °F) in air (2 months)	No loss in strength
At 100 °C (210 °F) in air, MPa (ksi)	3170 (460)
At 200 °C (390 °F) in air, MPa (ksi)	2720 (395)
Tensile modulus:	
At room temperature (16 months)	No loss in modulus
At 50 °C (120 °F) in air (2 months)	No loss in modulus
At 100 °C (210 °F) in air, GPa (10^6 psi)	114 (16.5)
At 200 °C (390 °F) in air, GPa (10^6 psi)	110 (16.0)
Long-term use temperature in air, °C (°F)	160 (320)
Decomposition temperature, °C (°F)	500 (930)

Source: Ref 16, p 5.25

Isotropic properties of 20% aramid-fiber-reinforced nylon, 6/6 composites

Property	Flow direction	Transverse direction	Difference %
Flexural strength, MPa (ksi)	107 (15.5)	101 (14.7)	5
Flexural modulus, GPa (10^6 psi)	3.6 (0.52)	3.3 (0.48)	8
Izod impact strength, J/cm (ft·lb/in.):			
Notched	0.64 (1.2)	0.64 (1.2)	0
Unnotched	1.6 (3.0)	1.4 (2.7)	11
Coefficient of linear thermal expansion, 10^{-5}/°C (10^{-5}/°F)	4.3 (2.4)	4.7 (2.6)	8

Source: Ref 19, p 84

Properties of pultruded unidirectional IPN-Kevlar (58 wt %)

	Reported value at:		
Property	Room temperature	99 °C (210 °F)	150 °C (300 °F)
Flexural strenth, MPa (ksi)	605.8 (87.86)	491. 4 (71.28)	331.4 (48.07)
Flexural modulus, GPa (10^6 psi)	62 (9.0)	...	...
Short-beam shear, MPa (ksi)	37.9 (5.49)	35.2 (5.10)	22.1 (3.20)

Source: Ref 22, p 248

Properties of Spectra polyethylene fibers

	Spectra 900	Spectra 1000
Density		
g/cm³	0.97	0.97
Filament diameter		
μm (μin.)	38 (1500)	27 (1060)
Tensile modulus		
GPa (10^6 psi)	117 (17)	172 (25)
Tensile strength		
GPa (10^6 psi)	2.6 (0.380)	2.9-3.3 (0.430-0.480)
Tensile elongation		
%	3.5	0.7
Available yarn counts		
No. filaments	60-120	60-120

Source: Ref 6, p 56

Effect of electron radiation on Kevlar 49

Using resonant transformer and filament wrapped in aluminum foil over dry ice, conditions of exposure were 1 Mrad every 13.4 s, 0.5 mA, 2 MV, and 30-cm (10-in.) distance.

	Single-filament properties				
	Tenacity		Tensile modulus		Elongation,
Mrad exposure	MPa	ksi	GPa	10^6 psi	%
0	2860	415	128	18.6	2.4
100	2940	426	130	18.8	2.4
200	3010	436	133	19.3	2.5

Source: Ref 6, p 56

Kevlar 49 yarn and roving sizes

	Yield		
Denier	m/kg	yd/lb	Number of filaments
55	163636	81175	25
195	46155	22895	134
380	23684	11749	267
1140	7895	3916	768
1420	6388	3144	1000
2130	4225	2097	1000
4560	1973	980	3072
7100	1268	630	5000

Note: All Kevlar 49 yarns have approximately 1.5 denier units/filament, with the exception of the 2130-denier product and the 55-denier product, which have a denier/filament ratio of approximately 2.1 and 2.25, respectively.

Source: Ref 6, p 114

Kevlar 29 yarn and roving sizes

	Yield		
Denier	m/kg	yd/lb	Number of filaments
200	45000	22320	134
400	22500	11160	267
1000	9000	4464	1000
1500	6000	2976	4000
9000	1000	497	4000
15000	600	298	10000

Source: Ref 6, p 114

Typical properties of Compet RPC fibers

Property	Type 1W69	Type 1W71	Type IR69
Generic name	Polyester	Polyester	Nylon 6
Tensile strength, MPa (ksi)	1105 (160)	1035 (150)	965 (140)
Extension at break, %	14.0	22.0	20.0
Modulus of elasticity, GPa (10^6 psi)	14 (2.0)	9.7 (1.4)	5.2 (0.75)
Shrinkage at 175 °C (350 °F), %	10	2.5	12.0
Diameter, μm (mil)	23 (0.91)	23 (0.91)	23 (0.91)
Specific gravity	1.38	1.38	1.16
Melting point, °C (°F)	253 (488)	253 (488)	220 (428)

Source: Ref 27, p 73

Properties of hybrid yarns

Property	PEEK hybrid	PPS hybrid
Density, g/cm^3	1.60	1.61
Total Denier, g/10,000 m	3000	2950
Number of filaments(a)	3100	3100
Fiber area, cm^2	0.0019	0.0019
Yield, m/kg	3370	3390

(a) Fabricated with Celion G30-500 3K carbon fibers. Available with up to 61 vol % carbon, with other high-performance carbon fibers, and filament counts.

Source: Ref 28, p 272

Effect of reinforcement on mechanical properties

Material	Tensile strength, MPa	Tensile strength, ksi	Tensile modulus, GPa	Tensile modulus, 10^6 psi	Elongation, %
E-glass-ortho polyester					
Ambient temperature	157	22.8	11.0	1.59	1.7
50 °C (125 °F)	148	21.5	8.41	1.22	2.4
65 °C (150 °F)	140	20.3	7.31	1.06	2.6
S-glass-ortho polyester					
Ambient temperature	159	23.0	10.3	1.49	1.9
50 °C (125 °F)	165	24.0	8.55	1.24	2.6
65 °C (150 °F)	157	22.8	7.38	1.07	2.6
Aramid-ortho polyester					
Ambient temperature	212	30.7	12.5	1.82	2.0
50 °C (125 °F)	208	30.2	11.5	1.67	2.1
65 °C (150 °F)	200	29.0	9.52	1.38	2.4
E-glass-vinyl ester					
Ambient temperature	206	29.9	12.6	1.83	2.1
50 °C (125 °F)	192	27.9	11.4	1.66	2.2
65 °C (150 °F)	201	29.2	11.6	1.68	2.4
S-glass-vinyl ester					
Ambient temperature	198	28.7	11.4	1.66	2.2
50 °C (125 °F)	172	25.0	9.6	1.39	2.3
65 °C (150 °F)	199	28.8	10.3	1.49	2.5
Aramid-vinyl ester					
Ambient temperature	189	27.4	12.1	1.76	1.8
50 °C (125 °F)	221	32.0	11.7	1.70	2.2
65 °C (150 °F)	218	31.6	11.7	1.69	2.2

Source: Ref 6, p 92

Mechanical properties of aramid, polyamide, polyester, and nylon fibers

Fiber	Density Mg/m³	Density lb/in.³	Tensile strength MPa	Tensile strength ksi	Tensile modulus GPa	Tensile modulus 10^6 psi	Ultimate elongation, %
Aramid-Kevlar 29	1.44	0.052	3620	525	83	12	4.4
Aramid-Kevlar 49	1.44	0.052	3620	525	124	18	2.9
Polyamide	1.13	0.041	830	120	2.8	0.4	...
Polyester-Dacron Type 68	1.38	0.050	1120	162	4.1	0.6	14.5
Nylon–Du Pont 728[a]	1.13	0.041	990	143	5.5	0.8	18.3
Spectra-900	0.97	0.035	2590	375	117	17	...

[a]Unimpregnated twisted yarn test–ASTM D2256.

Source: Ref 16, p 5.26

Kevlar 49 fabric and woven roving specifications

Style no.	Weave	Basis weight g/m²	Basis weight oz/yd²	Fabric construction ends/cm	Fabric construction ends/in.	Yarn denier	Fabric thickness mm	Fabric thickness 10^{-3} in.
Light weight								
166(a)	Plain	30.6	0.9	37 × 37	94 × 94	55	0.04	1.5
199(a)	Plain	61.13	1.8	24 × 24	60 × 60	55	0.05	2
120	Plain	61.1	1.8	13 × 13	34 × 34	195	0.11	4.5
220	Plain	74.7	2.2	9 × 9	22 × 22	380	0.11	4.5
Medium weight								
181	8-harness satin	169.8	5.0	20 × 20	50 × 50	380	0.23	9
281	Plain	169.8	5.0	7 × 7	17 × 17	1140	0.25	10
285	Crowfoot	169.8	5.0	7 × 7	17 × 17	1140	0.25	10
328	Plain	230.9	6.8	7 × 7	17 × 17	1420	0.33	13
335	Crowfoot	230.9	6.8	7 × 7	17 × 17	1420	0.30	12
500	Plain	169.8	5.0	5 × 5	13 × 13	1420	0.28	11
Unidirectional								
143	Crowfoot	190.2	5.6	39 × 8	100 × 20	380 × 195	0.25	10
243	Crowfoot	227.5	6.7	15 × 7	38 × 18	1140 × 380	0.33	13
Woven roving								
1050	4 × 4 basket	356.6	10.5	11 × 11	28 × 28	1420	0.46	18
1033	8 × 8 basket	509.4	15.0	16 × 16	40 × 40	1420	0.66	26
1350	4 × 4 basket	458.5	13.5	10 × 9	26 × 22	2130	0.64	25

(a) Only available on special order; custom fabric will be woven to specifications.

Source: Ref 6, p 114

Metal Fibers, Whiskers and Wires

Composite wire properties

	6061-Al (%ROM)(a)	AZ91-Mg (%ROM)
Fiber vol%	50.2 ± 2.0(b)	54.3 ± 2.0
Mean tensile strength, MPa (ksi)	1385 ± 110 (201.0 ± 16)(b)(c)	1240 ± 90 (180.0 ± 13.0)(d)(e)
Mean elastic modulus, GPa (10^6 psi)	420 ± 25 (60.7 ± 3.4)(c)(f)	>390 (>57)

(a) Based on the mean properties of virgin yarn. (b) Average of 3200 tests. (c) 100% ROM. (d) Average of 360 tests. (e) >95% ROM. (f) Average of 105 tests

Source: Ref 6, p 869

Representative properties of refractory alloy wires

Alloys	Density, g/cm^3	Wire diam mm	in.	Ultimate tensile strength MPa	ksi	Stress for 100-h rupture MPa	ksi	Stress/density for 100-h rupture cm × 10^3	in. × 10^3
Tungsten alloys, 1093 °C (2000 °F) data									
218CS	19.1	0.20	0.008	869	126	434	63	234	92
W-1%ThO_2	19.1	0.20	0.008	979	142	531	77	282	111
W-2%ThO_2	18.9	0.38	0.015	1193	173	655	95	356	140
W-3%Re	19.4	0.20	0.008	1475	214	476	69	249	98
W-5%Re-2%ThO_2	19.1	0.20	0.008	1213	176	483	70	254	100
W-24%Re-2%ThO_2	19.4	0.20	0.008	1455	211	345	50	183	72
W-Hf-C	19.4	0.38	0.015	1427	207	1110	161	584	230
W-Re-Hf-C	19.4	0.38	0.015	2165	314	1413	205	744	293
Tungsten alloys, 1204 °C (2200 °F) data									
218CS	19.1	0.20	0.008	745	108	317	46	170	67
W-1%ThO_2	19.1	0.20	0.008	841	122	372	54	198	78
W-2%ThO_2	18.9	0.38	0.015	1034	150	483	70	257	101
W-3%Re	19.4	0.20	0.008	1082	157	317	46	168	66
W-5%Re-2%ThO_2	19.1	0.20	0.008	1020	148	303	44	160	63
W-24%Re-2%ThO_2	19.4	0.20	0.008	1014	147	193	28	102	40
W-Hf-C	19.4	0.38	0.015	1386	201	765	111	404	159
W-Re-Hf-C	19.4	0.38	0.015	1937	281	910	132	480	189

Source: Ref 6, p 880

Mechanical properties and melting points of stainless steel and other metallic fibers

Fiber	Fiber diameter μm	mils	Density g/cm^3	lb/in.^{3}	Tensile strength MPa	ksi	Tensile modulus GPa	10^6 psi	Melting point °C	°F
AFC-77	150-1270	6-50	7.75	0.280	3540-4135	528-600	207	30.0	1370	2500
Rene 41	25-50	1-2	8.25	0.298	2000-2345	290-340	220	31.9	1370	2500
Udimet 700	255	10	7.92	0.286	1515-2330	220-338	221	32.0	1404	2559
Ribtec-HT(a)	205-510	8-20	...	...	57	8.3	82.7	12.0	1480-1530	2700-2790
Ribtec-LR 430(a)	205-510	8-20	...	...	47	6.8	82.7	12.0	1480-1530	2700-2790
Ribtec-GR 304(a)	205-510	8-20	...	...	124	18	124	18.0	1400-1455	2550-2650
Ribtec-OS 446(a)	205-510	8-20	...	...	52	7.6	96.5	14.0	1425-1510	2600-2750
Ribtec-310(a)	205-510	8-20	...	...	151	22	124	18.0	1400-1455	2550-2650
Ribtec-OC 330 (a)	205-510	8-20	...	...	193	28	134	19.5	1345-1425	2450-2600
Steel (wire)	25	1	...	...	3445	500	207	30.0	...	...
Aluminum	...	...	2.68	0.097	620	90	73.1	10.6	...	...

(a) Modulus of elasticity computed at 315 °C (600 °F); tensile strength computed at 870 °C (1600 °F).

Source: Ref 16, p 5.20

Tensile properties of refractory-metal wires

Wire material	Wire diameter cm	Wire diameter in.	Test temperature °C	Test temperature °F	Tensile strength MPa	Tensile strength ksi	Elongation in 25 mm (1 in.), %	Reduction in area, %
W-Hf-C, in-process annealed	0.038	0.015	21	70	2700	392	5.4	21.1
			1095	2000	1430	207	...	67.8
			1205	2200	1390	201	...	70.9
W-Hf-C, hard drawn	0.038	0.015	21	70	2250	326	2.8	1.9
			1095	2000	1740	253	...	44.2
			1205	2200	1540	224	...	46.9
W-Re-Hf-C, hard drawn	0.038	0.015	21	70	3160	458	4.8	27.5
			1095	2000	2160	314	...	24.7
			1205	2200	1940	281	...	37.6
ASTAR 811C	0.051	0.020	21	70	1700	247	6.9	51.0
			1095	2000	744	108	...	80.8
			1205	2200	490	71	...	89.8
	0.038	0.015	21	70	1740	253	5.3	42.9
			1095	2000	779	113	...	66.4
			1205	2200	550	80	...	66.9
B-88	0.051	0.020	21	70	1480	215	4.8	26.5
			1095	2000	530	77	...	87.4
			1205	2200	350	50	...	97.9
	0.038	0.015	21	70	1620	235	7.7	54.8
			1095	2000	490	71	...	94.5
			1205	2200	310	45	...	95.7
W-2ThO_2	0.038	0.015	21	70	2650	384	5.5	14.2
			1095	2000	1190	173	...	50.2
			1205	2200	1030	150	...	51.0

Source: Ref 29, p 116

Mechanical properties and melting points of refractory metallic fibers/wires

Type of filament or wire	Nominal composition	Melting point °C	Melting point °F	Density, ρ g/cm^3	Density, ρ lb/in.^{3}	Tensile strength, σ MPa	Tensile strength, σ ksi
Cr	Cr	1865	3390	7.2	0.26	1586	230
Nb-Su16	Nb-11W-3Mo-2Hf-0.08C	~2590	~4700	9.27	0.335	889-979	129-142
Nb-Su31	Nb-17W-3.5Hf-0.12C	~2590	~4700	9.46	0.342	1034-1503	150-218
Nb, FS85	Nb-28Ta-10W-1Zr-0.005C	2590	4695	10.6	0.383	1510	219
Nb, AS30	Nb-20W-1Zr	~2590	~4700	9.6	0.347	1758	255
Nb, B88	Nb-28W-2Hf-0.06C	~2590	~4700	10.3	0.372	1620	235
Mo	Mo	2610	4730	10.2	0.369	2206	320
Mo + 0.5 Ti	Mo-0.5Ti	2610	4730	10.1	0.367	1793	260
Mo, TZM	Mo-0.5Ti-0.08Zr-0.015C	~2610	~4730	10.2	0.369	1965	285
Mo, TZC	Mo-1.25Ti-0.3Zr-0.15C	~2610	~4730	10.1	0.367	2268	329
Ta, ASTAR 811C	Ta-8W-1Re-0.9Hf-0.03C	~2990	~5400	16.9	0.610	1703	247
W	W	3410	6170	19.3	0.697	1648-3268	239-474
W, 218CS	W	3410	6170	19.2	0.695	2386-2661	346-386
W + 1 ThO_2 (NF)	W-1ThO_2	~3410	~6170	19.1	0.691	2255-2310	327-335
W + 2 ThO_2	W-2ThO_2	~3410	~6170	18.9	0.683	2647-2751	384-399
W + 3 Re (3D)	W-3Re	~3410	~6170	19.4	0.70	2785	404
W + 5 Re	W-5Re	~3410	~6170	19.4	0.70	1689-2647	245-384
W + Hf + C	W-0.03Hf-0.036C	~3410	~6170	19.4	0.70	2248-2358	326-342
W + Re + Hf + C	W-4Re-0.38Hf-0.02C	~3410	~6170	19.4	0.70	3158	458

(continued)

Recrystallization temperature of tungsten wires in various matrices

System	Ni content of matrix, wt%	Recrystallization temperature(a) °C	°F
W-Ni	100	1150-1200	2100-2190
W-NiCr	20	1300	2370
W-2%ThO_2 NiCr	20	1250	2280
W-2%ThO_2 Ni	100	1080-1130	1980-2065
W-2%ThO_2 Inconel 718	52	1175	2150
W-2%ThO_2 Hastelloy X	48	1200	2190
W-2%ThO_2 Kovar	29.5	1250-1300	2280-2370
W-2%ThO_2 Stainless steel	10	1435	2615
W-2%ThO_2 Stainless steel	10	1465	2670

(a) 50 μm (1950 μin.) in 1 h.

Source: Ref 6, p 879

Physical properties of metallic wires

Material	Specific gravity	Melting point °C	°F	Tensile strength MPa	ksi	Modulus of elasticity GPa	10^6 psi	Coefficient of thermal expansion, 10^{-6}/K
Aluminum	2.71	660	1220	290	40	68.9	10.0	23.6
Beryllium	1.85	1350	2460	1100	160	310	45.0	11.6
Copper	8.90	1083	1980	413	60	124	18.0	16.5
Tungsten	19.3	3410	6170	2890	130	345	50.0	4.6
Austentic stainless steel	7.9	1539	2800	2390	350	200	29.0	8.5
Molybdenum	10.2	2625	4750	2200	320	331	48.0	...

Source: Ref 6, p 118

Mechanical properties and melting points of refractory metallic fibers/wires (continued)

Type of filament or wire	Specific strength, σ/ρ 10^6 cm	10^6 in.	Modulus of elasticity, E GPa	10^6 psi	Specific modulus, E/ρ 10^6 cm	10^6 in.	Typical cross section (diameter) 10^{-3} cm	10^{-3} in.
Cr	2.2	0.89	290	42	411	162	25	10
Nb-Su16	0.98-1.1	0.39-0.42	121-134	17.6-19.5	~159-177	~63-70	25-89	10-35
Nb-Su31	1.1-1.6	0.44-0.64	122-135	17.7-19.6	~160-178	~63-70	61-102	24-40
Nb, FS85	1.5	0.57	138	20	132	52	13	5
Nb, AS30	1.9	0.74	~138	~20	~147	~58	13	5
Nb, B88	1.6	0.63	~138	~20	~137	~54	38	15
Mo	2.2	0.87	358	52	358	141	15	6
Mo + 0.5 Ti	1.8	0.71	317	46	318	125	13	5
Mo, TZM	2.0	0.77	317	46	318	125	20-25	8-10
Mo, TZC	2.3	0.89	317	46	318	125	13	5
Ta, ASTAR 811C	1.0	0.41	200	29	122	48	38-51	15-20
W	0.9-1.7	0.34-0.68	406	59	216	85	5.1-127	2-50
W, 218CS	1.3-1.4	0.5-0.56	406	59	216	85	20-38	8-15
W + 1 ThO_2 (NF)	1.2-1.2	0.47-0.48	406	59	216	85	20-51	8-20
W + 2 ThO_2	1.3-1.5	0.56-0.58	406	59	218	86	20-51	8-20
W + 3 Re (3D)	1.5	0.58	406	59	213	84	7.6-20	3-8
W + 5 Re	0.9-1.4	0.35-0.55	406	59	213	84	25-127	10-50
W + Hf + C	1.2-1.4	0.47-0.56	406	59	213	84	38	15
W + Re + Hf + C	1.7	0.65	406	59	213	84	38	15

Source: Ref 16, p 5.21

Stress-rupture properties of wire materials

Wire material	Wire diameter cm	Wire diameter in.	Test temperature °C	Test temperature °F	Stress MPa	Stress ksi		Rupture time, h	Reduction in area, %
Tungsten-base alloys									
W-Hf-C, thermally annealed during drawing	0.038	0.015	1095	2000	1300	189	....	4.4	44.2
					1290	187	....	10.3	58.4
					1230	178	....	21.1	23.2
					1210	175	....	19.1	35.0
					1150	167	....	61.5	44.5
					1110	161	....	108.5	18.0
	0.038	0.015	1205	2200	918	133	....	28.3	15.3
					841	122	....	42.9	21.9
					765	111	....	104.3	11.5
					689	100	....	188.4	28.5
W-Hf-C, hard drawn	0.038	0.015	1095	2000	1310	190	....	17.7	44.2
					1240	180	....	139.4	37.0
					1230	178	....	88.6	22.6
					1210	175	....	262.0	57.3
					1100	160	....	(a)	...
	0.038	0.015	1205	2200	1170	170	....	6.0	29.4
					1140	165	....	4.3	30.6
					1100	160	....	11.1	20.2
					1040	150	....	22.5	24.9
					965	140	....	18.6	20.2
					896	130	....	37.4	16.6
					827	120	....	63.0	22.6
					793	115	....	74.3	17.8
					758	110	....	334.1	50.2
					689	100	....	329.6	65.8
W-Re-Hf-C, hard drawn	0.038	0.015	1095	2000	1590	230	....	15.6	15.7
					1520	220	....	36.2	19.5
					1480	215	....	42.8	34.0
					1450	210	....	72.1	27.1
					1380	200	....	442.6	34.7
					1310	190	....	104.2	16.7
					1240	180	....	522.3	37.3
	0.038	0.015	1205	2200	1140	165	....	14.4	32.0
					1040	150	....	18.4	43.2
					965	140	....	39.7	23.0
					896	130	....	49.8	32.8
					862	125	....	365.5	33.7
					827	120	....	345.5	44.3
					793	115	....	342.2	43.2
Tantalum-base alloy									
ASTAR 811C	0.051	0.020	1095	2000	690	100	....	7.3	17.5
					620	90	....	68.5	8.7
					590	85	....	43.0	7.0
					520	75	....	338.2	3.8
	0.038	0.015	1095	2000	620	90	....	14.6	29.8
					590	85	....	94.6	4.2
					570	82	....	19.1	19.3
					570	82	....	162.8	6.6
					550	80	....	338.2	7.4
					550	80	....	(b)	...
	0.051	0.020	1205	2200	350	50	....	10.8	8.8
					310	45	....	28.5	9.7
					280	40	....	78.3	8.3
					240	35	....	166.7	7.3
	0.038	0.015	1205	2200	520	75	....	10.2	15.3
					480	70	....	14.7	10.3
					410	60	....	45.4	5.2
					380	55	....	20.1	6.5
					350	50	....	62.7	2.8
					310	45	....	391.9	<1.0

(continued)

Stress-rupture properties of wire materials (continued)

Wire material	Wire diameter cm	Wire diameter in.	Test temperature °C	Test temperature °F	Stress MPa	Stress ksi		Rupture time, h	Reduction in area, %
Niobium-base alloy									
B-88	0.051	0.020	1095	2000	380	55	....	4.1	32.8
					370	53	....	37.1	15.8
					350	50	....	101.3	18.6
					310	45	....	102.1	16.6
	0.038	0.015	1095	2000	350	50	....	44.8	22.1
					310	45	....	55.4	23.3
					280	40	....	199.3	20.7
	0.051	0.020	1205	2200	280	40	....	2.6	34.7
					240	35	....	14.4	39.9
					210	30	....	78.5	20.9
	0.038	0.015	1205	2200	240	35	....	25.4	32.0
					210	30	....	86.8	28.6
					170	25	....	224.1	26.1

(a) Test stopped at 233.4 h. (b) Test stopped at 348.9 h.

Source: Ref 29, p 117-118

Carbon and Graphite Fibers

Classification of carbon fibers by modulus and strength

Fiber type	State-of-the-art Modulus GPa	State-of-the-art Modulus 10^6 psi	State-of-the-art Strength GPa	State-of-the-art Strength ksi	Newer Fibers Modulus GPa	Newer Fibers Modulus 10^6 psi	Newer Fibers Strength GPa	Newer Fibers Strength ksi
High strength	228	33	3.45	500	241	35	4.14	600
					255	37	5.17	750
Intermediate modulus					283	41	4.83	700
High modulus	379	55	2.41	350				
Very high modulus	517	75	2.07	300				
Ultrahigh modulus					690	100	2.24	325
					827	120	2.41	350

Source: Ref 16, p 5.7

Carbon and graphite fibers (in order of ascending modulus of strand, normalized to 100% fiber volume)

Class of fiber	Strand nominal tensile modulus MPa	Strand nominal tensile modulus 10^6 psi	Strand nominal tensile strength MPa	Strand nominal tensile strength ksi	Maximum number of filaments/strand	Fiber density, g/cm^3
High tensile strength	227	33	3102	450	12 000	1.75
High strain	234	34	4100	594	6 000	1.79
Intermediate modulus	275	40	4295	623	12 000	1.74
High modulus	358	52	2482	360	3000	1.81
High modulus pitch	379	55	2068	300	4000	2.0
Ultra-high modulus	517	75	1816	270	384	1.96
Ultra-high modulus, pitch	517	75	2068	300	2000	2.0
Extreme-high modulus, pitch	689	100	2240	325	2000	2.15

Source: Ref 6, p 504

Properties of carbon fibers: low-modulus rayon and isotropic pitch precursor fibers

Properties	Rayon precursor		Isotropic pitch precursor	
Axial				
Tensile modulus, GPa (10^6 psi)	41	6	55	8
Tensile strength, GPa (10^6 psi)	1.0	(0.15)	0.7	0.10
Elongation at break, %	2.5	...	1.4	...
Electrical resistivity, μΩ · m (μΩ · cm)	20	(2000)	30	(3000)
Bulk				
Density, g/cm^3	1.6	...	1.6	...
Filament diameter, μm (μin.)	8.5	(330)	10	(390)
Carbon assay, %	99	...	98	...

Source: Ref 6, p 52

Thermal, physical, and chemical properties of typical graphite fibers

Tensile strength, MPa	2000
Elongation, %	0.6
Specific gravity	1.63
Modulus of elasticity, GPa	550

Source: Ref 23, p 78

Graphite fiber properties

Fiber	Bulk dc resistivity ohm·m × 10^{-4}	Tensile modulus GPa	Tensile modulus 10^6 psi
Carbon			
AS-4	15	231	33.5
T300	18	230	33.4
Graphite			
HMS-4	7.5	341	49.9
P-75	1.8	517	75.0

Source: Ref 6, p 113

Properties of carbon fiber types

Fiber type	Density, g/cm^3	Modulus of elasticity GPa	Modulus of elasticity 10^6 psi	Tensile strength GPa	Tensile strength 10^6 psi	Electric resistivity Ω·m	Thermal conductivity W/m·K	Thermal conductivity Btu·in./ft^2·h·°F
High-strength (PAN)	1.7-1.8	230-250	33-36	2.8-4.0	0.41-0.58	12-30	7-10	50-70
Ultrahigh strength (PAN)	1.7-1.8	260-290	38-42	4.1-5.7	0.59-0.83	14-20	7-9	50-60
High-modulus (PAN, mesophase pitch)	1.8-2.0	350-550	50-80	1.7-3.5	0.25-0.50	5-10	60-200	420-1400
Ultrahigh-modulus (mesophase pitch)	2.0-2.2	600-900	90-130	2.1-2.5	0.30-0.36	1-4	400-2500	2800-17,300
Low-modulus (rayon, pitch)	1.3-1.7	40-60	6-9	0.6-1.0	0.085-0.145	30-100	7-28	50-190

Source: Ref 6, p 867

Properties of carbon fibers

Properties	Amoco T300	Amoco	Toray	Hercules	FMI
Physical					
Specific gravity	1.77	2.18	1.82	1.88	1.80
Density, g/cm^3 (lb/in.^{3})	1.77 (0.064)	2.18 (0.079)	1.82 (0.066)		
Filament diameter, μm (mils)	7.01 (0.276)	...	...	...	...
Mechanical					
Tensile strength at RT, MPa (ksi)	3650 (530)	2240 (325)	6900 (1000)	3790 (550)	5170 (750)
Tensile modulus at RT, GPa (10^6 psi)	230 (33.5)	830 (120)	290 (42)	425 (62)	260 (38)
Elongation, %	1.4	...	2.4	0.75	1.97
Thermal					
CTE, 10^{-6}/K	−0.54	...	...	...	...
Thermal conductivity, W/m · K (Btu/ft · h · °F)	8.7 (5.0)	...	...	...	...
Specific heat					
At RT, kJ/kg · K (Btu/lb · °F)	0.92 (0.22)	...	...	...	...
From 20-1480 °C mean (70-2700 °F mean)	1.7 (0.4)	...	...	...	...
Electrical					
Electrical resistivity, Ω · m	18.0×10^{-6}	...	...	...	...

Source: Ref 6, p 361

Typical mechanical property values of commercially available carbon fibers

Product name	Manufacturer	Precursor type	Density, g/cm^3	Tensile strength GPa	Tensile strength 10^6 psi	Tensile modulus GPa	Tensile modulus 10^6 psi
AS-4	Hercules, Inc.	PAN	1.78	4.0	0.580	231	33.5
AS-6	Hercules, Inc.	PAN	1.82	4.5	0.652	245	35.5
IM-6	Hercules, Inc.	PAN	1.74	4.8	0.696	296	42.9
T300	Union Carbide/Toray	PAN	1.75	3.31	0.480	228	32.1
T500	Union Carbide/Toray	PAN	1.78	3.65	0.530	234	33.6
T700	Toray	PAN	1.80	4.48	0.650	248	36.0
T-40	Toray	PAN	1.74	4.50	0.652	296	42.9
Celion	Celanese/ToHo	PAN	1.77	3.55	0.515	234	34.0
Celion ST	Celanese/ToHo	PAN	1.78	4.34	0.630	234	39.0
XAS	Grafil/Hysol	PAN	1.84	3.45	0.500	234	34.0
HMS-4	Hercules, Inc.	PAN	1.78	3.10	0.450	338	49.0
PAN 50	Toray	PAN	1.81	2.41	0.355	393	57.0
HMS	Grafil/Hysol	PAN	1.91	1.52	0.220	341	49.4
G-50	Celanese/ToHo	PAN	1.78	2.48	0.360	359	52.0
GY-70	Celanese	PAN	1.96	1.52	0.220	483	70.0
P-55	Union Carbide	Pitch	2.0	1.73	0.250	379	55.0
P-75	Union Carbide	Pitch	2.0	2.07	0.300	517	75.0
P-100	Union Carbide	Pitch	2.15	2.24	0.325	724	100
HMG-50	Hitco/OCF	Rayon	1.9	2.07	0.300	345	50.0
Thornel 75	Union Carbide	Rayon	1.9	2.52	0.365	517	75.0

Source: Ref 6, p 113

Properties of pultruded unidirectional graphite

Property	Reported value, at temperature of: Room temperature	99 °C (210 °F)	150 °C (300 °F)
IPN-graphite (70 wt %)			
Flexural strength, MPa (ksi)	1610 (233)	1300 (188)	855 (124)
Flexural modulus, GPa 910[6] psi)	122 (17.7)	...	...
Short-beam shear, MPa (ksi)	97.9 (14.2)	69 (10)	37.0 (5.37)
Epoxy graphite			
Flexural strength, MPa (ksi)	1790 (260)	...	1140 (165)
Flexural modulus, GPa (10^6 psi)	128 (18.5)	...	103 (15.0)
Short-beam shear, MPa (ksi)	97 (14)	...	52 (7.5)

Source: Ref 22, p 248

Specific tensile strength and modulus of carbon fiber relative to other reinforcements

Fiber	Tensile strength/density 10^7 cm	10^6 in.	Tensile modulus/density 10^9 cm	10^9 in.
AS-4	2.25	8.86	1.29	0.508
IM-6	2.76	10.9	1.70	0.669
E-glass	1.33	5.24	0.28	0.11
S-glass	1.73	6.81	0.32	0.13
Kevlar 49	2.50	9.84	0.90	0.35
Boron	1.50	5.91	1.60	0.63
P-75	1.04	4.09	2.58	1.02

(a) Data derived in epoxy matrix

Source: Ref 6, p 113

Properties of graphite fibers

Property	Hercules HMS PAN	Amoco WYB rayon	Amoco P-120 pitch	Amoco-rayon T50	Amoco-rayon T-75
Physical properties					
Specific gravity	1.83	1.32	2.18	1.67	1.80
Density, g/cm^3 (lb/in.^{3})	1.83 (0.066)	1.32 (0.048)	2.18 (0.079)	1.67 (0.060)	1.80 (0.065)
Filament diameter, μm (mil)	8.00 (0.315)	9.4 (0.37)	10 (0.4)	6.6 (0.26)	6.0 (0.24)
Mechanical properties					
Tensile strength, MPa (ksi)	2200 (320)	620 (90)	2240 (325)	2170 (315)	2620 (380)
Tensile modulus, GPa (10^6 psi)	340 (50)	40 (6)	825 (120)	395 (57)	540 (78)
Elongation, %	0.58	1.5	0.27	0.60	0.50
Thermal properties					
Coefficient of thermal expansion, 10^{-5}/K:					
Axial	-0.99	...	-1.62	...	...
Lateral	16.8	...	...	...	...
Thermal conductivity, axial, W/m·K (Btu/ft·h·°F)	104 (60.0)	...	609 (352)	118 (68)	156 (90)
Specific heat, kJ/kg·K (Btu/lb·°F)	0.71 (0.17)	...	...	0.71 (0.17)	0.71 (0.17)
Electrical properties					
Electrical resistivity, 10^{-6} Ω·m:					
Axial	13	...	2.2	...	...
Lateral	10	...	...	...	...

Source: Ref 6, p 361

Mechanical properties of selected carbon/graphite commercial fibers

Fiber	Fiber diameter µm	Fiber diameter mils	Density Mg/m³	Density lb/in.³	Tensile strength MPa	Tensile strength ksi	Tensile modulus GPa	Tensile modulus 10^6 psi	Ultimate elongation, %
Magnamite HMS	8.00	0.315	1.83	0.066	2210	320	345	50	0.58
Magnamite HMU	8.00	0.315	1.85	0.067	2760	400	380	55	0.70
Magnamite chopped fiber	8.00	0.315	1.77	0.064	2480	360	205	30	1.2
Magnamite AS1	8.00	0.315	1.80	0.065	3100	450	230	33	1.32
Magnamite AS2	8.00	0.315	1.80	0.065	2760	400	230	33	1.3[a]
Magnamite AS4	8.00	0.315	1.80	0.065	3590	520	235	34	1.53
Magnamite AS6	...	...	1.82	0.0657	4140	600	243	35.3	1.65[a]
Magnamite IM6	...	...	1.74	0.0627	4380	635	279	40.4	1.50
Microfil 55	4.32	0.170	1.77	0.064	3620	525	380	55	1.00
Microfil 40	4.32	0.170	1.69	0.061	4480	650	275	40	1.65
Celion C-6S	7.11	0.280	1.77	0.064	3790	550	231	33.5	1.64
Celion G-50	6.60	0.260	1.77	0.064	2480	360	360	52	0.7
Celion GY-70	8.38	0.330	1.91–1.97	0.069–0.071	1520	220	485	70	0.38
Celion GY-80	...	...	1.91–1.97	0.069–0.071	1520	220	550	80	...
Celion 1000	7.11	0.280	1.77	0.064	3240	470	235	34	1.4
Celion 3000	7.11	0.280	1.77	0.064	3790	550	231	33.5	1.64
Panex $^1/_4$CF-30	7.92	0.312	1.74	0.063	2410	350	205	30	1.2
Panex 30	7.92	0.312	1.74	0.063	2590	375	220	32	1.3
Fortafil 3(C)	7.37	0.290	1.77	0.064	3100	450	230	33	1.4
Fortafil 3(O)	5.33–13.97	0.210–0.550	1.77	0.064	2760	400	230	33	1.2
Fortafil 5(O)	4.32–11.94	1.170–0.470	1.77	0.064	3100	450	345	50	0.9
Grafil XA-S (standard)	...	...	1.79	0.0646	3100	450	234	34.0	1.31
Grafil XA-S (high performance)	...	...	1.79	0.0646	3450	500	234	34.0	1.45
Grafil XA-S (high strain)	...	...	1.79	0.0648	3860	560	234	34.0	1.65
Grafil IM-S	...	...	1.76	0.0635	3100	450	290	42.0	1.07
Grafil HM-S/10/K	...	...	1.85	0.067	2480	360	345	50.0	0.73
Grafil HM-S/16/K	...	...	1.85	0.067	2760	400	372	54.0	0.74
Hi-Tex	...	...	1.80	0.065	3100–3240	450–470	228	33.0	...
Hi-Tex HS	...	...	1.80	0.065	3620–3690	525–535	234	34.0	...
Thornel P-25W 4K	10.92	0.430	1.91	0.069	1380	200	160	23	0.90
Thornel T-40 12K	5.94	0.234	1.80	0.065	5650	820	275–290	40–42	2.0
Thornel T-50 3K	6.45	0.254	1.80	0.065	2410	350	395	57	0.70
Thornel 75	5.56	0.219	1.83	0.066	2620	380	545	79	...
Thornel T-300 6K	6.93	0.273	1.77	0.064	3240	470	231	33.5	...
Thornel 400	...	...	1.77	0.064	3100	450	205	30	...
Thornel T-500	6.93	0.273	1.80	0.065	3860	560	241	35.0	1.5
Thornel T-600	...	...	1.80	0.065	4140	600	241	35.0	1.7
Thornel T-700	...	...	1.80	0.065	4480	650	250	36	1.8
Modmor I	7.75	0.305	1.97	0.071	1380	200	380	55	...
Modmor II	8.03	0.316	1.74	0.063	2410	350	240	35	...

[a]Minimum elongation.

Source: Ref 16, p 5.9

Properties of carbon fibers: polyacrylonitrile precursor fibers

Properties	Standard grades: Low modulus		Standard grades: High modulus		New grades: Low modulus		New grades: Intermediate modulus		New grades: High modulus	
Axial										
Tensile modulus, GPa (10^6 psi)	230	(30)	390	(55)	230	(35)	270	(40)	320	(45)
Tensile strength, GPa (10^6 psi)	3.3	(0.48)	2.4	(0.35)	4.5	(0.65)	5.3-6.8	(0.77-0.99)	5.5	(0.80)
Elongation at break, %	1.4	...	0.6	...	2.0	...	2.0-2.5	...	1.8	...
Thermal conductivity, W/m · K (Btu · in./h · ft^2 · °F)	8.5	(59)	70	(490)	7	(50)				
Electrical resistivity, μΩ · m (μΩ · cm)	18	(1800)	9.5	(950)	18	(1800)				
Coefficient of thermal expansion at 21 °C (70 °F), 10^{-6}/K	−0.7	...	−0.5	...						
Transverse										
Tensile modulus, GPa (10^6 psi)	40	(6)	21	(3)						
Coefficient of thermal expansion at 50 °C (120 °F), 10^{-6}/K	10	(2)	7	(1)						
Bulk										
Density, g/cm^3	1.76	...	1.9	...	1.8	...	1.8	...	1.8	...
Filament diameter, μm (μin.)	7-8	(280-310)	7	(280)	5-6	(200-240)	6	(240)	4	(160)
Carbon assay, %	92-97	...	100	...	92-97	...	96	...	96	...

Source: Ref 6, p 49

Properties of carbon fibers: mesophase pitch precursor fibers

Properties	Standard grades: Low modulus		Standard grades: High modulus		Standard grades: Very high modulus		New grades: Low modulus		New grades: High modulus	
Axial										
Tensile modulus, GPa (10^6 psi)	160	(25)	380	(55)	725	(110)	225	(35)	380	(55)
Tensile strength, GPa (10^6 psi)	1.4	(0.20)	1.7	(0.25)	2.2	(0.32)	3.1	(0.45)	3.1	(0.45)
Elongation at break, %	0.9	...	0.4	...	0.3	...	...	...	...	...
Thermal conductivity, W/m · K (Btu · in./h · ft^2 · °F)	...	...	100	(690)	520	(3600)				
Electrical resistivity, μΩ · m (μΩ · cm)	13	(1300)	7.5	(750)	2.5	(250)	...	...	...	...
Coefficient of thermal expansion at 21 °C (70 °F), 10^{-6}/K	...	...	−0.9	...	−1.6	...	...	...	...	...
Transverse										
Tensile modulus, GPa (10^6 psi)	...	...	21	(3)	...	...	...	...	...	...
Coefficient of thermal expansion at 50 °C (120 °F), 10^{-6}/K	...	...	7.8	...	...	...	...	...	...	...
Bulk										
Density, g/cm^3	1.9	...	2.0	...	2.15	...	2.05	...	2.15	...
Filament diameter, μm (μin.)	11	(430)	10	(390)	10	(390)	8	(310)	8	(310)
Carbon assay, %	97+	...	99+	...	99+	...	99+	...	99+	...

Note: The data in this table was obtained from technical data sheets from Great Lakes Carbon Corp., BASF Structural Materials Inc., Amoco Corp., Hercules Corp., Fiber Materials Inc., Stackpole Fibers Co., Polycarbon Inc., and Ashland Petroleum Co.

Source: Ref 6, p 51

Classification of carbon-base fibers

Classification	Carbon content, %	Maximum processing temperature °C	°F	X-ray diffraction crystal structure	Crystallite orientation	Treatment	Approximate modulus of elasticity GPa	10^6 psi	Approximate tensile strength MPa	ksi
Carbon	>80	<1000	<1830	Crystallites too small to be detected	"Amorphous"	Carbonization	34	5	690	100
Graphite	>99	>2500	>4530	Crystallites large enough to be detected	Similar to precursor "random"	Graphitization	97	14	1035	150
Structural carbon or structural graphite ..	>99	>2500	>4530	Crystallite number and size greater than in graphite fiber	Preferred orientation of graphite crystallites in a carbon matrix ("turbostratic")	Combination thermomechanical treatments	>172	>25	>1240	>180

Source: Ref 16, p 5.6

Ceramic Fibers

Maximum use temperatures of some refractory fibers in oxidizing and nonoxidizing atmospheres

Fiber type	Maximum use temperature: Oxidizing atmosphere °C	°F	Nonoxidizing atmosphere °C	°F
Al_2O_3	1540	2805	1600	2910
ZrO_2	1650	3000	1650	3000
SiO_2	1060	1940	1060	1940
Al_2O_3-SiO_2	1300	2370	1300	2370
Al_2O_3-SiO_2-Cr_2O_3	1427	2600	1427	2600
Al_2O_3-SiO_2-B_2O_3	1427	2600	1427	2600
C	400	750	2500	4530
B	560	1040	1200	2190
BN	700	1290	1650	3000
SiC	1800	3270	1800	3270
Si_3N_4	1300	2370	1800	3270

Source: Ref 16, p 5.1.2

Characteristics of discontinuous fibers

Material	Form	Diameter µm	Diameter µin.	Density, g/cm³	Ultimate tensile strength MPa	Ultimate tensile strength ksi	Tensile modulus GPa	Tensile modulus 10^6 psi	Coefficient of thermal expansion, 10^{-6}/K
Silicon carbide	Crystalline	0.2	8	3.2	800	120	500	75	4.3
Silicon carbide	Crystalline	120	4700	3.2	...	...	...	...	...
Alumina	Crystalline	3	120	3.3	2000	290	300	45	8.1
Mullite	Crystalline	3	120	3.2	690	100	152	20	5.1
Aluminosilicate	Amorphous	2	80	2.7	1730	250	104	15	...
Zirconia	Crystalline	5	200	5.7	...	...	206	30	10.5

Source: Ref 6, p 903

Properties of short fibers and whiskers

Material	Density g/cm³	Tensile strength GPa	Tensile strength 10^6 psi	Tensile modulus GPa	Tensile modulus 10^6 psi
Alumina					
Whiskers	4.0	10-20	1-3	700-1500	100-220
Sintered fibers	<4.0	0.2-0.7	0.030-0.10	140-300	20-40
Boron, thermally formed fibers	2.3	2.75	0.400	400	60
Boron nitride, fibers	1.8-2.0	0.3-1.4	0.045-0.20	28-80	4-10
Carbon					
Whiskers	>2.0	...	...	700	100
Fibers	1.8-2.0	2-3	0.30-0.45	230-550	35-80
Silicon nitride, whiskers	3.2	5-7	0.75-1.0	350-380	50-55

Source: Ref 6, p 119

Mechanical properties and maximum use temperatures of refractory fibers

Fiber	Fiber diameter μm	Fiber diameter mils	Density g/cm³	Density lb/in.³	Tensile strength MPa	Tensile strength ksi	Tensile modulus GPa	Tensile modulus 10⁶ psi	Maximum use temperature °C	Maximum use temperature °F
Avco boron	102	4.0	2.57	0.093	3520	510	400	58	...	...
	142	5.6	2.49	0.090	3520	510	400	58	...	...
	203	8.0	2.46	0.089	3520	510	400	58	...	...
Boron (tungsten core)	51	2.0	3.38	0.122	2760	400	400-415	58-60	...	...
Boron on carbon	102	4.0	2.24	0.081	3280	475	365	53	315	600
	142	5.6	2.27	0.082	3280	475	380	55	315	600
	203	8.0	2.30	0.083	3170	460	345	50	...	...
SiC-coated boron	107	4.2	2.66	0.096	2410	350	400-415	58-60	...	...
	145	5.7	2.57	0.093	2410	350	400-415	58-60	...	...
Boron carbide	102	4.0	2.35	0.085	2690	390	425	62	315	600
Boron nitride	6.9	0.27	1.91	0.069	1380	200	90	13	1095	2000
Titanium boride (TiB_2)	...	...	4.48	0.162	105	15	510	74	2205	4000
TiC	...	...	4.90	0.177	1540	224	450	65	...	...
Zirconium oxide	...	...	4.84	0.175	2070	300	345	50	1925	3500
Nextel 312	9.9-11.9	0.39-0.47	2.71	0.098	1380-1720	200-250	150	22	1205	2200
Nextel 440	9.9-11.9	0.39-0.47	3.10	0.112	1380-2070	200-300	205-240	30-35	...	...
Fiberfrax	2-12	0.08-0.47	2.60	0.094	1030-1720	150-250	105	15	1795(a)	3260(a)
Al_2O_3 (polycrystalline)	...	...	3.15	0.114	2070	300	170	25	1650	3000
FP Al_2O_3	15.2-25.4	0.6-1.0	3.71	0.134	1380(b)	200(b)	345	50	2045(a)	3710(a)
SiO_2-coated alumina	...	...	3.71	0.134	1900(b)	275(b)	380	55	...	...
Al_2O_3 monocrystal (sapphire)	...	...	3.96	0.143	2550	370	470	68	2040(a)	3700(a)
Nicalon	...	...	...	...	>2410	>350	180	26	1095	2000
Avco SiC	142	5.6	3.04	0.110	3450	500	425	62	...	...
SiC	102	4.0	3.46	0.125	2280	330	450	65	...	...
SiC (carbon core)	142	5.6	3.29	0.119	3790(b)	550(b)	345	50	...	...
SiC (tungsten core)	142	5.6	3.29	0.119	3790(b)	550(b)	415	60	600	1110
$Al_2O_3 \cdot Cr_2O_3$ monocrystal (ruby)	...	...	3.99	0.144	3450-4140	500-600	470	68	2040(a)	3700(a)
Borsic (SiC/B/W)	107-145	4.2-5.7	2.77	0.100	2930	425	415	60	2300(a)	4170(a)
Borsic/C	107-145	4.2-5.7	2.30	0.083	3170	460	350-365	51-53	...	...
Saffil alumina, RF grade	...	...	3.29	0.119	2000	290	295	43	1600	2910

(a) Melting or softening point. (b) Ultimate tensile strength.

Source: Ref 16, p 5.5

Influence of alumina and aluminosilicate fiber reinforcement on coefficient of thermal expansion

Matrix	Fiber vol%	Coefficient of thermal expansion, 10^{-6}/K 0°		90°
332.0 T5	...	...	24.5	...
	Alumina, 5	23.9	...	23.6
	Mullite, 5	22.3	...	23.8
	Alumina, 15	18.9	...	22.3
339.0 T5	...	...	20.4	...
	Alumina, 10	18.0	...	...
	Alumina, 20	16.4	...	...

Source: Ref 6, p 905

Typical properties of Nicalon SiC fiber

Production method .... Si- and C-containing polymer spun, cured, and pyrolyzed

Diameter .... 10 to 20 μm, or 0.39 to 0.79 mil (500 per yarn)

Modulus .... 180 GPa, or 26.1 × 10^6 psi (420 GPa, or 60.9 × 10^6 psi, for β-SiC)

Strength at 20 °C (68 °F):

As-produced .... 2 GPa (0.29 × 10^6 psi)

After treatment at 1400 °C or 2550 °F (argon) .... <1 GPa (<0.15 × 10^6 psi)

Strength at 1400 °C or 2550 °F (oxygen) .... <0.5 GPa (<0.07 × 10^6 psi)

Creep strain at 1300 °C (2370 °F), 0.6 GPa (0.09 × 10^6 psi), 20 h 4.5%

Source: Ref 16, p 903

Characteristics of ceramic whiskers

Material	Whisker morphology and size	Crystal structure	Properties
SiC	Rod or needle, 3-10 μm (120-390 μin.) diam, 10-1000 μm (40-3950 μin.) long	Alpha and/or beta phases	>500 GPa (>70 × 10^6 psi) modulus of elasticity; 2-7 GPa (0.3-1.0 × 10^6 psi) tensile strength
Si_3N_4	Rod or needle, 0.2-0.5 μm (8-20 μin.) diam, 50-300 μm long	Alpha plus beta phase	390 GPa (60 × 10^6 psi) modulus of elasticity; up to 1.5 GPa (0.2 × 10^6 psi) tensile strength

Note: Whiskers are commercially produced by several sources, including Arco Chemical Company; Versar Manufacturing, Inc.; Tateho Chemical Company, Ltd., and Tokai Carbon Company, Ltd.

Source: Ref 6, p 941

Ultimate tensile strengths of tungsten-coated and uncoated SiC filaments at various temperatures

Temperature °C	Temperature °F	UTS[a] (coated[b]) GPa	UTS[a] (coated[b]) ksi	UTS (uncoated) GPa	UTS (uncoated) ksi	σ/σ_0
R.T.	R.T.	1.835	266	2.828	410	0.649
800	1470	1.648	239	2.359	342	0.699
1000	1830	1.531	222	2.083	302	0.735
1200	2190	1.414	205	1.883	273	0.751
1400	2550	1.083	157	1.359	197	0.797
1600	2910	0.855	124	1.062	154	0.805

[a]Mean ultimate tensile strength. [b]Coating thickness, 12.7 μm (0.0005 in.).

Source: Ref 16, p 5.12

Thermal shock response of SiC whisker reinforced alumina as indicated by flexural strength retained after quenching from elevated temperature into boiling water

	Temperature quenched from °C	Temperature quenched from °F	Retained fracture strength MPa	Retained fracture strength ksi
Alumina-20 vol% SiC	No thermal shock		620	89.9
Single thermal shock cycle	400	750	630	91.4
	600	1110	685	99.3
	800	1470	615	89.2
	1000	1830	710	103.0
Ten thermal shock cycles	400	750	610	88.5
	500	930	570	82.7
	800	1470	540	78.3
	1000	1830	545	79.0
Alumina	No thermal shock		315	45.7
Single thermal shock cycle	300	570	250	36.3
	400	750	225	32.6
	500	930	125	18.1

Source: Ref 6, p 943

Influence of silicon carbide fiber reinforcement on coefficient of thermal expansion

Matrix	Fiber, vol%	Coefficient of thermal expansion, 10^{-6}/K 0°		Coefficient of thermal expansion, 10^{-6}/K 90°	
339.0 T5	...	...	20.4	...	
	Silicon carbide, 15	16.9	...	...	
356.0 T5	...	...	21.4	...	
	Silicon carbide, 14	16.5	...	...	
2024 T6	...	...	21.1	...	
	Silicon carbide, 25	14.9	...	16.4	
	Silicon carbide, 40	13.0	...	...	
6061 T7	...	...	21.6	...	
(Extruded)	Silicon carbide, 25	12.1	...	...	

Source: Ref 6, p 906

Influence of fiber reinforcement on thermal conductivity at 200 °C (390 °F)

Matrix	Fiber, vol%	Thermal conductivity, W/m · K (Btu · in./h · ft^2 · °F)					
		0 °C				90°	
332.0 T5	...	...	...	176	1220	...	...
	Alumina, 5	158	1100	...	...	140	970
	Mullite, 5	152	1050	...	...	128	890
	Alumina, 15	134	930	...	...	98	680
	Zirconia, 19	116	800	...	...	87	600
339.0 T5	...	...	...	144	1000	...	...
	Alumina, 10	112	780	...	...	...	...
	Alumina, 15	99	690	...	...	...	...
	SiC, 15	136	940	...	...	...	...
2024 T6	...	...	...	151	1050	...	...
	SiC, 25	102	710	...	...	...	...

Source: Ref 6, p 908

Properties of boron and alumina fibers

Fiber	Density, g/cm^3	Diameter, μm	Diameter, μin.	Tensile strength, GPa	Tensile strength, 10^6 psi	Tensile modulus, GPa	Tensile modulus, 10^6 psi
Boron-tungsten	2.6	100-200	3950-7850	5.5-7.0	0.80-1.0	400	58
Boron-carbon	2.3	100-200	3950-7850	5.0	0.73	400	58
α-alumina(a)	3.95	20	790	14(b)	2.0	390	57
				19(c)	2.8	390	57

(a) Slurry-spun continuous fiber. (b) Uncoated. (c) Silicon carbide-coated.

Source: Ref 6, p 118

Commercially available continuous oxide fibers

Composition, wt%	Identification	Company	Forms(a)
Al_2O_3, > 99	Fiber FP	E.I. Du Pont de Nemours & Co., Inc.	C,Y,F
Al_2O_3, 85; SiO_2, 15	High performance alumina fiber	Sumitomo Chemical Co. Ltd., Japan; distributed by Avco Specialty Materials, Textron, Inc.	C,Y,F
Al_2O_3, 80; SiO_2, 20	Long alumina fiber	Denki Kagaku Kogyo K.K., with Nichibi Co. Ltd.	C,Y
Al_2O_3, 70; B_2O_3, 2; SiO_2, 28	Nextel 440 and Nextel 480	Minnesota Mining & Manufacturing Co.	C,R,Y,F
Al_2O_3, 62; B_2O_3, 14; SiO_2, 24	Nextel 312	Minnesota Mining & Manufacturing Co.	C,R,Y,F
SiO_2, 99.95	Astroquartz II	Distributed by J.P. Stevens & Co. Inc.	C,R,Y,F
SiO_2, 98; rem. 2	Refrasil	Hitco Materials Div., Armco Inc.	Y,F
SiO_2, 98; rem. 2	Siltemp	Ametek, Inc.	F
ZrO2, 68; SiO_2, 32	Nextel Z-11	Minnesota Mining & Manufacturing Co.	C,R,Y,F

(a) C, continuous; Y, yarn; F, fabric; R, roving

Source: Ref 6, p 60

Commercially available discontinuous oxide fibers

Composition, wt%	Identification	Company	Forms(a)
Al_2O_3, 95; SiO_2, 5	Saffil	Imperial Chemical Industries, PLC Ltd., England; distributed by Babcock & Wilcox	D,B,M,Ch
Al_2O_3, 72; SiO_2, 28	Fibermax	Sohio Engineered Materials (formerly Carborundum)	D,B
Al_2O_3, 70; B_2O_3, 2; SiO_2, 28	Nextel 440 and Nextel 480 Ultrafiber	Minnesota Mining & Manufacturing Co.	D,M,Ch
Al_2O_3, 60-68; B_2O_3, 4-9; SiO_2, 23-32	Staple Fiber	Nichias Corp.	D,B
Al_2O_3, 62; B_2O_3, 14; SiO_2, 24	Nextel 312 Ultrafiber	Minnesota Mining & Manufacturing Co.	D,M,Ch
Al_2O_3, 52; SiO_2, 48	Fiberfrax	Sohio Engineered Materials	D,B,M,F
Al_2O_3, 49-50; SiO_2, 50-51	Innswool	A.P. Green Refractories	D,B,M
Al_2O_3, 52-55; SiO_2, 41-44	Cer-wool	Combustion Engineering	D,B,M
Al_2O_3, 47; SiO_2, 53	Cerafiber	Manville Corp.	D,B,M
Al_2O_3, 42.5; Cr_2O_3, 2.5; SiO_2, 55	Cerachrome	Manville Corp.	D,B,M
Al_2O_3, 40; SiO_2, 50; CaO, 5; MgO, 3.5; TiO_2, 1.5	Cerawool	Manville Corp.	D,B,M
ZrO_2, 92; Y_2O_3, 8	Zircar	Zircar Products, Inc.	D,B,M,F

(a) D, discontinuous; B, bulk; M, mat or blanket; Ch, chopped; F, fabric

Source: Ref 6, p 61

Commercially available carbide and nitride fibers

Composition	Identification	Company	Forms(a)
SiC	Nicalon	Nippon Carbon Co. Ltd.	C,Y,Ch,F,M,R
Si-Ti-C	Tyranno	Ube Industries, Ltd.	C
Si-Zr-C	Tyranno	Ube Industries Ltd.	C
SiC on C Core	CVD SiC	Tokai Carbon Co., Ltd.; distributed by Avco Specialty Materials, Textron, Inc.	C
SiC	Tokawhisker	Tokai Carbon Co., Ltd.; distributed by Avco Specialty Materials, Textron, Inc.	D
SiC	Silar	Arco Metals Co.	D
SiC	Tateho	Tateho Chemical Industry Co., Ltd.; distributed by ICD Group Inc.	D
Si_3N_4	Tateho	Tateho Chemical Industry Co., Ltd.; distributed by ICD	D

(a) C, continuous; Y, yarn; Ch, chopped; F, fabric; M, mat or blanket; R, roving; D, discontinuous

Source: Ref 6, p 61

Properties of continuous oxide fibers

Fiber	Composition, wt %					Density,	Average diameter		Tensile strength		Tensile modulus of elasticity		Use temperature	
	Al_2O_3	B_2O_3	SiO_2	ZrO_2	Rem	g/cm^3	μm	μin.	GPa	10^6 psi	GPa	10^6 psi	°C	°F
Fiber FP(a)	>99	...	...	...	...	3.95	20	790	1.38	0.200	379	55	1320	2410
Sumitomo Alumina	85	...	15	...	...	3.2	17	670	1.45	0.210	193	28	1250	2280
Nextel 440	70	2	28	...	...	3.05	11	430	2.07	0.300	193	28	1430	2605
Nextel 480	70	2	28	...	...	3.1	11	430	2.24	0.325	207-241	30-35	1430	2605
Nextel 312	62	14	24	...	...	2.7	11	430	1.72	0.250	155	22	1200	2190
Nextel Z-11	...	...	32	68	...	3.7	14	550	1.31	0.190	76	11	1000	1830
Astroquartz II	...	...	99.95	...	...	2.2	9	350	3.45	0.500	69	10	1050	1920
Refrasil	...	...	97.9	...	2.1	2.1	...	...	0.21-0.41	0.03-0.06	72	10.5	1095	2000
Siltemp	...	...	98	...	2	2.2	...	...	...	...	...	...	...	...
Denki	80	...	20	...	...	...	10	390	...	...	...	...	...	...

Fiber	Melt or liquidus temperature		Coefficient of thermal expansion	Dielectric constant	Resistivity		Refractive index
	°C	°F	10^{-6}/K		Ω·m at 20 °C	Ω·cm at 68 °F	
Fiber FP(a)	2045	3710	...	...	...	...	...
Sumitomo Alumina	...	...	8.8(b)	...	10^{11}	10^{13}	1.65
Nextel 440	>1800	>3270	5	5.6	...	...	1.617
Nextel 480	>1800	>3270	...	5.8	...	...	>1.617
Nextel 312	1800	3270	3.5	5	...	...	1.572
Nextel Z-11	2000	3630	...	...	...	...	1.75
Astroquartz II	1700	3090	0.54	3.8	10^{16}	10^{18}	1.459
Refrasil	>1760	>3200	...	...	...	...	...
Siltemp	>1760	>3200	...	...	...	...	...
Denki	...	...	...	...	...	...	...

(a) Compressive strength is 6.9 GPa (10^6 psi). (b) At 200 to 400 °C (390 to 750 °F).

Source: Ref 6, p 62

Stress-rupture strengths (100-hour) of SiC/C and SiC/W filaments

Filament	1093 °C (2000 °F)		1204 °C (2200 °F)		1316 °C (2400 °F)	
	GPa (ksi)	$S/\rho \times 10^6$ in.	GPa (ksi)	$S/\rho \times 10^6$ in.	GPa (ksi)	$S/\rho \times 10^6$ in.
SiC/C[a]	1.93 (280)	2.3	1.034 (150)	1.4	0.69 (100)	0.96
SiC/W[b]	1.07 (155)	1.3	0.863 (126)	1.1	0.34 (50)	0.43

[a]C is used as a substrate. [b]W is used as a substrate.

Source: Ref 16, p 5.12

Properties of discontinuous silicon carbide and silicon nitride whiskers

Fiber	Composition	Max free carbon, wt %	Crystalline species	Whisker ocntent, %	Particle content, %	Density g/cm^3	Average diameter μm	Average diameter μin.	Predominant length μm	Predominant length μin.
Silar SC-9	SiC	0.10	α	80-90	10-20	3.2	0.6	24	10-80	390-3150
Silar SC-10	SiC	0.20	α	70-80	20-30	3.2	6.6	260	10-80	390-3150
Tokawhisker	SiC	Negligible	β	...	<1	3.19	0.1-0.5	4-20	30-100	1200-3950
Tateho SiC	SiC	...	β	...	...	3.18	0.05-1.5	2-59	20-200	790-7900
Taheto Si_3N_4	Si_3N_4	...	α	...	...	3.18	0.1-1.6	4-63	20-200	790-7900

Fiber	Surface area m^2/kg	Surface area ft^2/lb	Tensile strength GPa	Tensile strength 10^6 psi	Tensile modulus of elasticity GPa	Tensile modulus of elasticity 10^6 psi	Use temperature: stability to °C	Use temperature: stability to °F	Electrical conductivity
Silar SC-9	3000	14,600	6.9(a)	1.0	690	100	1760(b)	3200	...
Silar SC-10	3000	14,600	6.9(a)	1.0	690	100	1760(b)	3200	...
Tokawhisker	...	...	3-14	0.44-2.03	400-700	58-101	1600 (in air)	2910	...
Tateho SiC	...	...	...	...	...	...	...	...	Conductive
Tateho Si_3N_4	...	...	...	...	...	...	...	...	Nonconductive

(a) Estimated. (b) Atmospheric environment not stated.

Source: Ref 6, p 63

Properties of continuous silicon carbide fibers

Fiber	Composition Si	O	C	Ti	H	N	Rem	Crystalline species	Density, g/cm^3	Average diameter μm	Average diameter μin.	Tensile strength GPa	Tensile strength 10^6 psi	Tensile modulus of elasticity GPa	Tensile modulus of elasticity 10^6 psi
Nicalon SiC	54.3	11.8	30.0	...	...	...	3.9	β-SiC	2.55	10-15(a)	390-590	2.5-3.2	0.36-0.47	180-200	26-29
Tyranno Si-Ti-C	44.2	12.3	24.5	11.0	0.6	3.4	4.0	β-SiC + TiC	2.3	10-15	390-590	1.99	0.286	117	17
Avco CVD SiC	...	...	...	...	..	...	...	β on carbon core	...	~140, with 33-μm core	~5510, with 1300μin. core	>3.4	>0.50	428	62

Fiber	Use temperature °C	Use temperature °F	Coefficient of thermal expansion, 10^{-6}/K	Dielectric constant	Resistivity Ω·m at 20 °C	Resistivity Ω·m at 68 °F	Chemical resistance, % wt reduction in 24 h at 80 °C (175 °F): 6N HCl	18N H_2SO_4	7N $HClO_3$	30% NaOH aqueous
Nicalon SiC	1200(b) (c)	2190	3.1(d)	6-8	10^3	10^1	<1%	<1%	<1%	<1%
Tyranno	1300	2370	...	...	...	...	...	...	...	...
Avco CVD SiC	1100	2010	4.9	...	...	...	...	...	...	...
	1200	2190	...	...	...	...	...	...	...	...
	>1400	2550	...	...	...	...	...	...	...	...

(a) Round cross section. (b) Substantial loss of strength. (c) Loss in strength above ~ 1000 °C (1830 °F). (d) Along fiber axis, 0 to 200 °C (32 to 390 °F).

Source: Ref 6, p 63

Properties of discontinuous oxide fibers

Fiber	Composition, wt %											Density, g/cm³	Average diameter	
	Al_2O_3	B_2O_3	CaO	Cr_2O_3	Fe_2O_3	MgO	SiO_2	TiO_2	Y_2O_3	ZrO_2	Rem		μm	μin.
Cerachrome	42.5	...	...	2.5	...	...	55	...	...	...	...	...	3.5	138
Cerafiber	47	...	...	...	...	...	52.8	...	...	...	0.2	2.65	3.5	138
Cerawool	40	...	5	...	...	3.5	50	1.5	...	...	...	2.54	3.5	138
Cer-wool	52-55	...	...	...	0.1-0.2	...	41-44	0.1-0.2	...	...	1-2	...	3.0	118
Fiberfrax	51.9	...	...	...	...	...	47.9	...	...	...	0.1	2.73	2.3	79-118
Fibermax	72	...	...	...	...	...	28	...	...	...	...	3.0	2-3.5	79-138
Innswool	49-50	...	...	...	...	...	50-51	...	...	...	<0.5	...	3-5	118-197
Kaowool	45	...	...	...	...	...	53	...	...	...	2	2.56	2.8	110
Nextel 312 Ultrafiber	62	14	...	...	...	...	24	...	...	...	...	2.75	3.5	138
Nextel 440 Ultrafiber	70	2	...	...	...	...	28	...	...	...	...	3.1	3.3	130
Nichias	60-68	4-9	...	...	...	...	23-32	...	...	...	...	...	10.5	413
Saffil RF Grade	96-97	...	...	...	...	...	3-4	...	...	...	...	3.3	3.0	118
Saffil RG Grade	96-97	...	...	...	...	...	3-4	...	...	...	...	3.3-3.5	3.0	118
Zircar	...	...	...	...	...	...	...	...	8	92	...	5.6-5.9	4-6	157-236

Fiber	Tensile strength		Tensile modulus of elasticity		Use temperature		Melt or liquidus temperature		Specific heat	
	GPa	10^6 psi	GPa	10^6 psi	°C	°F	°C	°F	J/kg·K at 1000 °C	Btu/lb·°F at 1830 °F
Cerachrome	...	...	...	...	1425	2600	>1760	>3200	1148	0.2741
Cerafiber	...	...	...	...	1315	2400	>1760	>3200	1148	0.2741
Cerawool	...	...	...	...	875	1610	>1648	>3000	1148	0.2741
Cer-wool	...	...	...	...	...	...	...	...	...	...
Fiberfrax	1.90	0.276	100	14.6	1260	2300	1790	3255	1130	0.2698
Fibermax	1.03	0.150	150	22	1650	3000	1890	3435	...	...
Innswool	...	...	...	...	1235	2255	1760	3200	...	...
Kaowool	1.13	0.165	84	12.2	1260	2300	1760	3200	1088	0.2598
Nextel 312 Ultrafiber	1.72	0.250	152	22	1200	2190	1800	3270	...	...
Nextel 440 Ultrafiber	1.31	0.190	207-241	30-35	1430	2605	>1800	>3270	...	...
Nichias	1.79	0.260	...	...	...	...	...	...	...	...
Saffil RF Grade	2.0	0.290	310	45	1600	2910	>2000	>3630	...	...
Saffil RG Grade	1.0-2.0	0.145-0.290	297	43	1600	2910	>2000	>3630	...	...
Zircar	...	...	...	...	2200	3990	2600	4710	...	...

Source: Ref 6, p 62

Glass Fibers

Glass fibers for filament winding (in order of ascending modulus of strand, normalized to 100% fiber volume)

Type	Strand nominal tensile modulus GPa	Strand nominal tensile modulus 10^6 psi	Strand nominal tensile strength MPa	Strand nominal tensile strength ksi	Maximum number of filaments/strand	Fiber density, g/cm^3
E	72.4	10.5	3447	500	4000	2.60
R	86.2	12.5	2068	300	60	2.49
S	86.9	12.6	4585	665	...	2.55

Source: Ref 6, p 504

Inherent properties of glass fibers

	Specific gravity	Tensile strength MPa	Tensile strength ksi	Tensile modulus GPa	Tensile modulus 10^6 psi	Coefficient of thermal expansion, 10^{-6} K	Dielectric constant(a)	Liquidus temperature °C	Liquidus temperature °F
E-glass	2.58	3450	500	72.5	10.5	5.0	6.3	1065	1950
A-glass	2.50	3040	440	69.0	10.0	8.6	6.9	996	1825
ECR-glass	2.62	3625	525	72.5	10.5	5.0	6.5	1204	2200
S-glass	2.48	4590	665	86.0	12.5	5.6	5.1	1454	2650

(a) At 20 °C (72 °F) and 1 MHz.

Source: Ref 6, p 107

Properties of quartz fibers

Properties	Astroquartz II
Physical	
Specific gravity	2.20
Density, g/cm^3 (lb/in.^{3})	2.19 (0.079)
Filament diameter, μm (mils in.)	8.9 (0.35)
Mechanical	
Tensile strength at RT, MPa (ksi)	3450 (500)
Tensile modulus at RT, GPa (10^6 psi)	69.0 (10.0)
Elongation, %	5
Thermal (pure silica block)	
Specific heat from −20 to 500 °C (0° to 932 °F), kJ/kg · K (Btu/lb · °F)	0.96 (0.23)
Electrical (pure fused silica block)	
Electrical resistivity at RT, Ω · m	10^{16}
Dielectric constant at RT, 1 MHz	3.78

Source: Ref 6, p 360

Glass-fiber compositions (Wt %)

Component	Grade of glass: A (high alkali)	C (chemical)	S (high strength)	E (electrical)
Silicon oxide	72.0	64.6	64.2	54.3
Aluminum oxide	0.6	4.1	24.8	15.2
Ferrous oxide	...	...	0.21	...
Calcium oxide	10.0	13.2	0.01	17.2
Magnesium oxide	2.5	3.3	10.27	4.7
Sodium oxide	14.2	7.7	0.27	0.6
Potassium oxide	...	1.7	...	...
Boron oxide	...	4.7	0.01	8.0
Barium oxide	...	0.9	0.2	...
Miscellaneous	0.7	...	...	...

Source: Ref 16, p 5.17

Commercial forms of glass-fiber reinforcements

Nominal form	General description	Process	Nominal glass content of typical laminates, %	Typical applications
Rovings	Continuous strands of glass fibers	Filament winding, continuous panel, preforming (matched-die molding), spray-up, pultrusion	25–80	Pipe, automobile bodies, rod stock, rocket-motor cases, ordnance
Chopped strands	Strands cut to lengths of 0.125–2 in. (3.2–50.8 mm)	Premix molding, wet slurry preforming	15–40	Electrical and appliance parts, ordnance components
Reinforcing mats	Continuous or chopped strands in random matting	Matched-die molding, hand lay-up, centrifugal casting	20–45	Translucent sheets, truck and auto body panels
Surfacing and overlaying mats	Nonreinforcing random mat	Matched-die molding, hand lay-up, filament winding	5–15	Where smooth surfaces are required (automobile bodies, some housings)
Yarns	Twisted strands	Weaving, filament winding	60–80	Aircraft, marine, electrical laminates
Woven fabrics	Woven cloths from glass fiber yarns	Hand lay-up, vacuum bag, autoclave, high-pressure laminating	45–65	Aircraft structures, marine, ordnance hardware, electrical flat sheet and tubing
Woven roving	Woven glass fiber strands (coarser and heavier than fabrics)	Hand lay-up	40–70	Marine, large containers
Nonwoven fabrics	Unidirectional and parallel rovings in sheet form	Hand lay-up, filament winding	60–80	Aircraft structures

Source: Ref 16, p 5.16

The effect of glass form and amount on mechanical properties

Type of glass fiber reinforcement	Glass content, wt%	Density, g/cm^3	Tensile strength		Tensile modulus		Elongation, %	Flexural strength		Flexural modulus		Compressive strength	
			MPa	ksi	10^{-3} Pa	10^{-6} psi		MPa	ksi	10^{-3} Pa	10^{-6} psi	MPa	ksi
Neat cured resin	0	1.22	59	8.6	5.40	0.783	2.0	88	12.8	3.90	0.565	156	22.6
Chopped strand mat	30	1.50	117	17.0	10.80	1.566	3.5	197	28.6	9.784	1.419	147	21.3
Chopped strand mat	50	1.70	288	41.8	16.70	2.422	3.5	197	28.6	14.49	2.102	160	23.2
Roving fabric	60	1.76	314	45.5	19.50	2.828	3.6	317	46.0	15.00	2.175	192	27.8
Woven glass fabric	70	1.88	331	48.0	25.86	3.750	3.4	403	58.4	17.38	2.520	280	40.6
Unidirectional roving fabric	70	1.96	611	88.6	32.54	4.720	2.8	403	58.4	29.44	4.270	216	31.3

Source: Ref 6, p 91

Typical properties of glass fiber types

Material	Density, bulk annealed g/cm³	Tensile strength at -190 °C (-310 °F) MPa	ksi	at 23 °C (72 °F) MPa	ksi	at 371 °C (700 °F) MPa	ksi	at 538 °C (1000 °F) MPa	ksi	Modulus of elasticity at 538 °C (1000 °F) GPa	10^6 psi	Elongation, %
E-glass	2.62	5310	770	3445	500	2620	380	1725	250	81.3	11.8	4.88
S-glass	2.50	8725	1200	4585	665	4445	645	2415	350	88.9	12.9	5.7
C-glass	2.56	5380	780	3310	480	...	...	...	...	...	...	4.8

Material	Chemical resistance (percent weight loss), in: H_2O 24 h	H_2O 168 h	10% HCl 24 h	10% HCl 168 h	10% H_2SO_4 24 h	10% H_2SO_4 168 h	1% Na_2CO_3 24 h	1% Na_2CO_3 168 h	10% NaOH 168 h	Relative permittivity At 1 MHz	Relative permittivity At 60 Hz	Dissipation factor At 1 MHz	Dissipation factor At 60 Hz
E-glass	0.7	0.9	42	43	39	42	2.1	2.1	20	6.6	6.7	0.0025	0.0034
S-glass	0.5	0.7	3.8	5.1	4.1	5.7	2.0	2.1	66	5.3	5.4	0.0034	0.0129
C-glass	1.1	2.9	4.1	7.5	2.2	4.9	24	31	...	6.9	...	0.0085	...

Material	Volume resistivity, Ω·m	Surface resistivity, Ω	Dielectric strength kV/cm	V/mil	Viscosity softening point °C	°F	Viscosity annealing point °C	°F	Viscosity strain point °C	°F	Thermal expansion(a), 10^{-6}/K	Specific heat At 23 °C (72 °F) kJ/kg·K (Btu/lb·°F)	At 200 °C (392 °F) kJ/kg·K (Btu/lb·°F)	Refractive index, bulk annealed
E-glass	0.402×10^{15}	0.42×10^{16}	103	262	846	1555	657	1215	615	1140	5.4	0.810 (0.193)	1.03 (0.247)	1.562
S-glass	0.905×10^{13}	0.886×10^{13}	130	330	970	1778	810	1490	760	1400	1.6	0.737 (0.176)	...	1.525
C-glass	...	...	...	...	750	1382	588	1090	552	1025	6.3	0.787 (0.188)	0.90 (0.215)	1.537

(a) From -30 °C (-20 °F) to 250 °C (480 °F)

Source: Ref 6, p 46

Effect of glass content on mechanical properties

Material	Glass content, wt%	Flexural strength MPa	ksi	Flexural modulus 10^{-2} Pa	10^{-5} psi	Tensile strength MPa	ksi	Tensile modulus 10^{-2} Pa	10^{-5} psi	Compressive strength MPa	ksi
Orthophthalic	30	170	25	55	8.0	140	20	48	7.0	...	...
	40	220	32	69.0	10.0	150	22	55	8.0	...	...
Isophthalic	30	190	28	55	8.0	150	22	82.7	12.0	...	...
	40	240	35	75.8	11.0	190	28	117	17.0	210	30
BPA fumarate	25	120	17	51	7.4	80	12	75.8	11.0	170	24
	35	150	22	82.7	12.0	100	14	103	15.0	170	24
	40	160	23	89.6	13.0	120	18	110	16.0	180	26
Chlorendic	24	120	17	59	8.5	80	11	75.8	11.0	140	21
	34	160	23	68.9	10.0	120	18	96.5	14.0	120	18
	40	190	28	96.5	14.0	140	20	96.5	14.0	120	18
Vinyl ester	25	110	16	54	7.9	86.2	12.5	69.6	10.1	180	26.5
	35	260	37.3	95.2	13.8	153.4	22.25	108	15.6	230	34
	40	220	32	88.9	12.9	160	23	110	15.9	210	30

Source: Ref 6, p 91

Compositional ranges for glass fibers used in composite materials

	E-glass range, %	S-glass range, %	C-glass range, %
Silicon dioxide	52-56	65	64-68
Aluminum oxide	12-16	25	3-5
Boric oxide	5-10	...	4-6
Sodium oxide and potassium oxide	0-2	...	7-10
Magnesium oxide	0-5	10	2-4
Calcium oxide	16-25	...	11-15
Barium oxide	...	...	0-1
Zinc oxide	...	...	...
Titanium dioxide	0-1.5	...	64-68
Zirconium oxide	...	...	...
Iron oxide	0-0.8	...	0-0.8
Iron	0-1	...	...

Source: Ref 6, p 46

Polymer Matrix Composites

POLYMER MATRIX COMPOSITES

Polymer Matrix Composites

General

Fiber-resin composite properties

Physical	Mechanical	Thermal	Electrical	In-service conditions
Specific gravity Density	Tensile strength Tensile modulus Poisson's ratio Compressive strength Compressive modulus Poisson's ratio Shear strength Shear modulus	Coefficient of thermal expansion Thermal conductivity Specific heat	Dielectric constant Dielectric strength Dissipation factor Volume resistivity	Service temperature TGA Temperature allowed on all standard loads Flammability EMI/RFI protection

Source: Ref 6, p 356

Thermoset and thermoplastic trade-offs for commercial aircraft composites

Property	Thermosets	Thermoplastics
Resin cost	Low	Low to high
Prepregability	Excellent	Poor (new methods such as emulsions could change this)
Prepreg tack/drape	Excellent	None (revised lay-up techniques are required)
Volatile-free prepreg	Good	Good to excellent
Prepreg shelf life and out-time	Poor	Excellent
Prepreg quality assurance	Fair	Excellent
Prepreg cost	Good	High (new methods needed)
Composite processing	Slow	Slow (unless automated processes are developed)
Shrinkage	Moderate	Low
Composite mechanical properties	Good (room for improvement in damage tolerance)	Good (more data and experience needed)
Interlaminar fracture toughness	Low	High
Resistance to fluids/solvents	Good	Poor to good
Resistance to creep	Good	Currently not known
Crystallinity problems	None	Yes

Source: Ref 6, p 100

Comparison of engineering properties of fiberglass reinforced plastics and competitive materials(a)

Material	Glass fiber content, wt %	Mechanical and physical properties: Flexural strength, ksi(b)	Flexural modulus, 10^5 psi(c)	Tensile strength at yield, ksi(b)	Tensile modulus, 10^5 psi(c)	Compressive strength, ksi(b)	Ultimate tensile elongation, %	Izod impact strength, ft·lb/in. of notch(d)	Thermal conductivity, Btu·in./ft^2·h·°F (K value) (e)	Specific heat, Btu/lb·°F	Flammability (UL94) (g)
Glass-fiber-reinforced thermosets											
Sheet molding compound (SMC)	15-30	18-30	14-20	8-20	16-25	15-30	0.3-1.5	8-22	1.3-1.7	0.30-0.35	5V
Bulk molding compound (BMC)	15-35	10-20	14-20	4-10	16-25	20-30	0.3-5	2-10	1.3-1.7	0.30-0.35	5V
Preform/mat (compression molded)	25-50	10-40	13-18	25-30	9-20	15-30	1-2	10-20	1.3-1.8	0.30-0.33	VO
Cold press molding-polyester	20-30	22-37	13-19	12-20	...	...	1-2	9-12	1.3-1.8	0.30-0.33	VO
Spray-up-polyester	30-50	16-28	10-12	9-18	8-18	15-25	1.0-1.2	4-12	1.2-1.6	0.30-0.34	VO
Filament wound-epoxy	30-80	100-270	50-70	80-250	40-90	45-70	1.6-2.8	40-60	1.92-2.28	0.23-0.25	VO
Rod stock-polyester	40-80	100-180	40-60	60-180	40-60	30-70	1.6-2.5	45-60	1.92-2.28	0.22-0.25	VO
Molding compound-phenolic	5-25	18-24	30	7-17	26-29	14-35	0.25-0.6	1-8	1.1-2.0	0.20-0.30	VO
Glass-fiber-reinforced thermoplastics											
Acetal	20-40	15-28	8-13	9-18	8-15	11-17	2	0.8-2.8	...	...	HB
Nylon	6-60	7-50	2-28	13-33	2-20	13-24	2-10	0.8-4.5	...	0.30-0.35	VO
Polycarbonate	20-40	17-30	7.5-15	12-25	7.5-17	14-24	2	1.5-3.5	...	...	VO
Polyethylene	10-40	7-12	2.1-6	6.5-11	4-9	4-8	1.5-3.5	1.2-4.0	...	...	VO
Polypropylene	20-40	7-11	3.5-8.2	5.5-10.5	4.5-9	6-8	1-3	1-4	...	...	VO
Polystyrene	20-35	10-17	8-12	10-15	8.4-12.1	13.5-19	1.0-1.4	0.4-4.5	...	0.23-0.35	VO
Polysulfone	20-40	21-27	8-15	13-20	15	21-26	2-3	1.3-2.5	...	...	VO
ABS (acrylonitrile butadiene styrene)	20-40	23-26	9.2-15	11-16	6-10	12-22	3-3.4	1-2.4	...	...	VO
PVC (polyvinyl choride)	15-35	20-25	9-16	14-18	10-18	13.4-16.8	2-4	0.8-1.6	...	...	VO
Polyphenylene oxide (modified)	20-40	17-31	8-15	15-22	9.5-15	18-20	1.7-5	1.6-2.2	...	...	VO
SAN (styrene acrylonitrile)	20-40	15-21	8.0-18	13-18	9-18.5	12-23	1.1-1.6	0.4-2.4	...	...	VO
Thermoplastic polyester	20-35	19-29	8.7-15	14-19	13-15.5	16-18	1-5	1.0-2.7	1.3	...	VO
Unreinforced thermoplastics											
Acetal	N.Ap.	13-14	4	8-10	4-5	5	25-60	1.2-2.3	...	0.35	HB
Nylon	N.Ap.	5-18	2-4	9	2-5	7-10	29	1-4	...	0.40	VO
Polycarbonate	N.Ap.	13	3	9-11	3.5	12	100-130	16	...	0.30	VO
Polyethylene (high density)	N.Ap.	...	0.7-2.6	4	0.6-1.5	2.7-3.6	30-900	0.6-20.0	...	...	VO
Polypropylene	N.Ap.	5-8	1.2-2.7	3-5	1.2	3.7-8	200-700	0.5-20.0	...	...	VO
Polystyrene (high impact)	N.Ap.	3-10	1-5	3-5	2-4	4-9	15-30	0.7-3.6	...	...	VO
Polysulfone	N.Ap.	1.5	4	10	3.6	14	50-100	1.3	...	...	VO
ABS (high heat)	N.Ap.	9	3-4	6-6	2.8-4.1	6.8-12.5	10-20	2.5	...	...	VO
PVC	N.Ap.	13-16	4	6-7	4	...	...	2-20	...	...	VO
Polyphenylene oxide (modified)	N.Ap.	15	4	10	3.7	15	50-100	1.5-1.9	...	...	VO
SAN	N.Ap.	9.7-17.5	5	9-11	5	14-17	2.5-3.7	0.4	...	...	VO
Metals											
Gray cast iron	N.Ap.	10	N.Av.	15-30	120	25	1	4-4.4	288-408	0.13-0.19	N.Ap.
Low-carbon steel (cold rolled)	N.Ap.	28	300	29-33	300	28	38-39	N.Ap.	260-460	0.10-0.11	N.Ap.
Stainless steel	N.Ap.	30-35	280	30-35	280	30	50-60	8.5-11.0	96-185	0.12	N.Ap.
Aluminum, wrought	N.Ap.	20	100	6-27	100	N.Av.	30-40	N.Ap.	810-1620	0.22-0.23	N.Ap.
Alumnium, die cast	N.Ap.	8-26	100	8-26	100	9	6-8	N.Ap.	610-1100	0.22-0.23	N.Ap.
Magnesium, die cast	N.Ap.	14	65	8-30	65	10-14	4-6	3	288-960	0.245-0.25	N.Ap.
Zinc, die cast	N.Ap.	N.Av.	N.Av.	10-25	N.Av.	N.Av.	10	4.3	764-792	0.10	N.Ap.
Brass, plain yellow wrought	N.Ap.	14	150	14	150	N.Av.	60-65	N.Ap.	804	0.09	N.Ap.

Comparison of engineering properties of fiberglass reinforced plastics and competitive materials (continued)

Material	Glass fiber content, %	Hardness, Rockwell	Dielectric strength V/mil(h)	Specific gravity	Density, lb/in.^{3} (j)	Heat distortion at 284 psi, °F(k)	Continuous heat resistance, °F(k)	Coefficient of thermal expansion 10^{-6}/°F(m)	Weak acids	Strong acids	Weak alkalis	Strong alkalis	Organic solvents
	Mechanical and physical properties								Chemical resistance(n)				
Glass-fiber-reinforced thermosets													
Sheet molding compound (SMC)	15-30	H50-H112	300-450	1.7-2.1	0.061-0.075	400-500	300-400	8-12	G to E	F	F	P	G to E
Bulk molding compound (BMC)	15-35	H80-H112	300-450	1.8-2.1	0.065-0.075	400-500	300-400	8-12	G to E	F	F	P	G to E
Preform/mat (compression molded)	25-50	H40-H105	300-600	1.5-1.7	0.054-0.061	350-400	150-400	10-18	G to E	F	F	P	G to E
Cold press molding-polyester	20-30	H40-H105	300-600	1.5-1.7	0.054-0.061	350-400	150-400	10-18	G to E	F	F	P	G to E
Spray-up polyester	30-50	0.30-0.34	V0	H40-H105	200-400	1.4-1.6	0.050-0.058	350-400	150-350	12-20	G to E	F	F
Filament wound-epoxy	30-80	M98-120	300-400	1.7-2.2	0.061-0.079	350-400	500	2-6	E	F	E	G	E
Rod stock-polyester	40-80	H80-112	200-400	1.6-2.0	0.058-0.072	325-375	150-500	3-8	G to E	F	F	F	G to E
Molding compound-phenolic	5-25	M90-99	150-370	1.7-1.9	0.061-0.069	400-500	325-350	4.5-9	F	P	F	P	F
Glass-fiber-reinforced thermoplastics													
Acetal	20-40	M78-M94	500-600	1.55-1.69	...	315-335	185-220	19-35	F	P	F	P	E
Nylon	6-60	...	400-500	1.47-1.7	0.049	300-500	300-400	11-21	G	P	E	F	G
Polycarbonate	20-40	M75-M100	450	1.24-1.52	...	285-300	275	12-18	E	G(p)	G	F	P(q)
Polyethylene	10-40	...	450-500	1.16-1.28	...	150-200	280-300	17-27	E	G(p)	E	E	G(r)
Polypropylene	20-40	R95-R115	500-600	1.04-1.22	...	230-300	300-320	16-24	E	G(p)	E	E	G(r)
Polystyrene	20-35	M70-M95	350-425	1.20-1.29	0.045-0.048	200-220	180-200	17-22	E	G(p)	G	G	P(q)
Polysulfone	20-40	M85-M92	...	1.38-1.55	...	330-350	...	12-17	E	E(p)	E	E	G
ABS (acrylonitrile butadiene styrene)	20-40	M75-M102	...	1.23-1.38	...	215-240	200-230	16-20	E	G	E	E	P(s)
PVC (polyvinyl chloride)	15-35	M80-M88	500-550	1.45-1.62	...	155-165	...	12	E	G	E	E	P(s)
Polyphenylene oxide (modified)	20-40	M95	...	1.20-1.38	...	220-315	240-265	10-20	E	F	E	E	G(t)
SAN (styrene acrylonitrile)	20-40	M77-M103	...	1.22-1.40	...	210-230	200-220	16-21	G	G(u)	G	G	P(s)
Thermoplastic polyester	20-35	R118-M70	560-750	1.45-1.61	...	380-470	275-375	24-33	F	P	P	P	E
Unreinforced thermoplastics													
Acetal	N.Ap.	M78-M94	465-500	1.42	0.052	230-255	185-220	45	F	P	F	P	E
Nylon	N.Ap.	R108-118	300-470	1.12-1.14	0.039-0.041	120-150	250-300	55-63	G	P	E	F	G
Polycarbonate	N.Ap.	M70	400-425	1.20	0.043	265-290	275	39	E	G(p)	G	F	P(q)
Polyethylene (high density)	N.Ap.	...	450-500	0.95	...	100-130	180-230	6	E	G(p)	E	E	P(q)
Polypropylene	N.Ap.	R50-R110	500-600	0.9	...	125-140	190-240	38	E	G(p)	E	E	P(q)
Polystyrene (high impact)	N.Ap.	M12-M45	300-600	1.05	0.039	175-205	150-180	22-56	E	G(p)	G	G	P(q)
Polysulfone	N.Ap.	M69-R12.0	425	1.24	...	345	300-345	31	E	E	E	E	G
ABS (high heat)	N.Ap.	R113	350-500	1.05	...	215-245	190-230	41-52	G	G(u)	G	G	P(s)
PVC	N.Ap.	D80	...	1.4	...	155-165	...	...	E	E	E	E	P(s)
Polyphenylene oxide (modified)	N.Ap.	M75	...	1.06	...	375	...	30	E	E	E	E	G(v)
SAN	N.Ap.	M80	400-500	1.08	...	190-220	140-205	36	G	G(u)	G	G	P(s)

(continued)

Comparison of engineering properties of fiberglass reinforced plastics and competitive materials (continued)

Material	Glass fiber content, %	Hardness, Rockwell	Dielectric strength V/mil(h)	Specific gravity	Density, lb/in.^{3} (j)	Heat distortion at 284 psi, °F(k)	Continuous heat resistance, °F(k)	Coefficient of thermal expansion 10^{-6} °F(m)	Chemical resistance(n): Weak acids	Strong acids	Weak alkalis	Strong alkalis	Organic sol-vents
Metals													
Gray cast iron	N.Ap.	B93	C	7.19	0.26	N.Ap.	N.Av.	6			(w)		
Low-carbon steel (cold rolled)	N.Ap.	B72	C	7.8	0.28	N.Ap.	N.Av.	6-8			(w)		
Stainless steel	N.Ap.	B90	C	7.92	0.29	N.Ap.	N.Av.	9-10			(x)		
Aluminum, wrought	N.Ap.	B1-85	C	2.6-2.8	0.10	N.Ap.	N.Av.	12-13			(y)		
Aluminum, die cast	N.Ap.	E59	C	2.57-2.96	0.09	N.Ap.	N.Av.	12-13			(y)		
Magnesium, die cast	N.Ap.	E50-E59	C	1.81	0.07	N.Ap.	N.Av.	14-16			(z)		
Zinc, die cast	N.Ap.	B44	C	6.6	0.24	N.Ap.	N.Av.	15-16			(aa)		
Brass, plain yellow wrought	N.Ap.	F58-F64	C	8.5	0.31	N.Ap.	N.Av.	11-12			(bb)		

(a) Data from Owens-Corning Fiberglas. N.Av. = not available; N.Ap. = not applicable. (b) To convert to MPa, multiply listed values by 6.894757. (c) To convert to GPa, multiply listed values by 0.6894757. (d) To convert to J/cm of notch, multiply listed values by 0.5338. (e) To convert to W/m·K, multiply listed values by 0.14423. (f) To convert to kJ/kg·K, multiply listed values by 4.1868. (g) Classification shown is highest obtainable rating. Less-critical applications may permit use of materials with lower classifications. (h) To convert to kV/mm, multiply listed values by 0.03937. (j) To convert to g/cm^3, multiply listed value by 27.68. (k) To convert to °C, subtract 32 from listed values, then divide by 1.8. (m) To convert to 10^{-6}/°C, multiply listed values by 1.8. (n) E = excellent (outstanding); G = good (acceptable); F = fair (test before using); P = poor (not recommended). (p) Attacked by oxidizing acids. (q) Soluble in aromatic and chlorinated hydrocarbons. (r) Below 80° C (176 ° F). (s) Soluble in ketones and esters, and in aromatic and chlorinated hydrocarbons. (t) Soften in some aromatic and chlorinated aliphatics. Resistant to alcohol. (u) Disintegrates in sulfuric acid. (v) Dissolves or swells in some aromatic and chlorinated aliphatics. Resistant to alcohol.

Source: Ref 16, p 6-56

Resin-dependent properties of graphite composites

Resin type	Continuous use temperature °C	°F	Maximum use temperature °C	°F	Interlaminar shear strength MPa	ksi	Comments
Thermosets							
Epoxy, 120 °C (250 °F) cure	70	160	105	225	55–103	8–15	690 kPa (100 psi) mold pressure
Epoxy, 175 °C (350 °F) cure	120	250	150	300	103	15	690 kPa (100 psi) mold pressure
Polyimide, 315 °C (600 °F) cure	290	550	370	700	103	15	690 kPa (100 psi) mold pressure with postcure
Phenolic	260	500	315	600	...	...	...
Thermoplastics							
Polysulfone	150	300	175	350	97	14	Moisture resistant
Polyphenylsulfone	180	360	205	400	97	14	Moisture resistant

Source: Ref 16, p 6-55

Selection of resins used with glass fiber reinforcements

Selection of the resin system to be used is based on satisfying chemical, electrical, and thermal performance requirements of the finished product. There are two general classes of resins: *thermosets,* which become hard when heated; and *thermoplastics,* which are soft when heated and harden upon cooling.

Type of resin	Properties	Processes
Thermosets		
Polyesters[b]	Simplest, most versatile, most economical, and most widely used family of resins, having good electrical properties and good chemical resistance (especially to acids)	Compression molding; filament winding; hand lay-up; mat molding; pressure bag molding; continuous pultrusion; injection molding; spray-up; centrifugal casting; cold molding; comoform; encapsulation
Epoxies	Excellent mechanical properties, dimensional stability, and chemical resistance (especially to alkalis); low water absorption; self-extinguishing (when halogenated); low shrinkage; good abrasion resistance; very good adhesion properties	Compression molding; filament winding; hand lay-up; continuous pultrusion; encapsulation; centrifugal casting
Phenolics	Good acid resistance; good electrical properties (except arc resistance); high heat resistance	Compression molding; continuous laminating
Silicones	Highest heat resistance; low water absorption; excellent dielectric properties; high arc resistance	Compression molding; injection molding; encapsulation
Melamines	Good heat resistance; high impact strength	Compression molding
Diallyl phthalate	Good electrical insulation; low water absorption	Compression molding
Thermoplastics		
Polystyrene	Low cost; moderate heat distortion; good dimensional stability; good stiffness; good impact strength	Injection molding; continuous laminating
Nylon	High heat distortion; low water absorption; low elongation, good impact strength; good tensile and flexural strength	Injection molding; blow molding; rotational molding
Polycarbonate	Self-extinguishing; high dielectric strength; high mechanical properties	Injection molding
Styrene-acrylo-nitrile	Good solvent resistance; good long-term strength; good appearance	Injection molding
Acrylics	Good gloss, weather resistance, optical clarity, and color; excellent electrical properties	Injection molding; vacuum forming; compression molding; continuous laminating
Vinyls	Excellent weatherability; superior electrical properties; excellent moisture and chemical resistance; self-extinguishing	Injection molding; continuous laminating; rotational molding
Acetals	Very high tensile strength and stiffness; exceptional dimensional stability; high chemical and abrasion resistance; no known room-temperature solvent	Injection molding
Polyethylene	Good toughness; light weight; low cost; good flexibility; good chemical resistance; can be "welded"	Injection molding; rotational molding; blow molding
Fluorocarbons	Very high heat and chemical resistance; nonburning; lowest coefficient of friction; high dimensional stability	Injection molding; encapsulation; continuous pultrusion
Polyphenylene oxide modified	Very tough engineering plastic; superior dimensional stability; low moisture absorption; excellent chemical resistance	Injection molding
Polypropylene	Excellent resistance to stress or flex cracking; very light weight; hard, scratch-resistant surface; can be electroplated; good chemical and heat resistance; exceptional impact strength; good optical qualities	Injection molding; continuous laminating; rotational molding
Polysulfone	Good transparency; high mechanical properties, heat resistance, and electrical properties at high temperatures; can be electroplated	Injection molding

[a]From Owens-Corning Fiberglas.
[b]Properties shown also apply to some polyesters formulated for thermoplastic processing by injection molding.

Source: Ref 16, p 6-6

Process comparisons for glass fiber composites

	Resin transfer molding	Injection molding	Pultrusion	RRIM	Hand lay-up spray-up
Factor limiting maximum size of part	Machine size	Machine size	Materials	Metering equipment	Mold size; part transport
Maximum size to date, m^2 (ft^2)	9.3 (100)	9.3 (100)	1.6 mm to 1.22 m × 2.44 m (1/16 to 48 × 96 in.)	4.6 (50)	279 (3 000)
Shape limitations	Moldable	Moldable	Round, rectangular	Moldable	None
Usual useful production	Medium	High	Medium	Medium-high	Low-medium
Volume, No. parts/year	1000–20 000	50 000–1 000 000	3050 m (10 000 ft)	15 000–100 000	0–1 000
Production cycle time	10–20 min	15 s to 15 min	10 to 30 min	1–2 min	3 min to 24 h
Typical glass content, %	15–25	20–40	30–75	5–25	20–35
Strength orientation	Random	Random	Highly oriented	With flow	Random (usually)
Strength category	Low to medium	Low	High	Low	Medium
Wall thickness:					
Minimum, mm (in.)	0.76 (0.03)	0.76 (0.03)	1.6 (1/16)	2.03 (0.080)	0.76 (0.030)
Maximum, mm (in.)	25.4 (1.0)	12.7–25.4 (1/2–1)	12.7 (1/2)	12.7 (0.500)	38–up (1.5–up)
Tolerance, mm (in.)	. . .	. . .	+0.25 ± 25.4 (+0.01 ± 1.0)	±0.05 (±0.002)	+0.51 (+0.02)
Variations	Uniform	Uniform	Uniform	Uniform	As desired
Minimum draft:					
To depth 150-mm (6-in.)	1°	1°	0–2°	1°–3°	0–2°
Over depth 150-mm (6-in.)	1°	1°+	0–2°	3°+	0–2°
Minimum inside radius, mm (in.)	1/2 part depth	1/2 part depth	1.5 (0.06)	1/2 part depth	6.4 (0.25)
Ribs	Yes	Yes	No	Yes	Yes
Bosses	Yes	Yes	No	Yes	Yes
Undercuts	Possible	Possible	No	Yes, with proper	Avoid
Holes:					
Parallel	Yes	Yes	No	Yes	Yes
Perpendicular	Yes	Yes	No	Yes	Yes
Built-in cores	Yes	Yes	No	Yes	Possible
Metal inserts	Yes	Yes	No	Yes	Yes
Metal edge stiffeners	Yes	Yes	No	Yes	Yes
Surface finish:					
Number of finished surfaces	All	2	2	2	1
Quality of surface	Excellent	Excellent	Fair to good	Excellent	Excellent
Gel-coat surface	. . .	Yes	No	No	Yes
Surfacing mat	Yes	Yes	No	No	Yes
Combination with thermoplastic liner	Yes	Yes	No	No	Yes
Trim in mold	No	No	No	Yes	No
Molded-in labels	Yes	Yes	No	No	Yes
Raised numbers	Yes	Yes	No	Yes	Yes
Translucency	Yes	Yes	No	No	Yes
Tool cost	High	High	Low	Low-medium	Low
Capital equipment cost	High	High	Low	Low-medium	Low

*Data from Owens-Corning Fiberglas.

Source: Ref 16, p 6-5

Process comparisons for glass fiber composites (continued)

	Filament winding	Compression: sheet molding compound	Compression: bulk molding compound	Preform molding
Factor limiting maximum size of part	Winding machine	Press rating and size	Press rating and size	Press rating and size
Maximum size to date, m^2 (ft^2)	93 (1 000)	4.6 (50)	4.6 (50)	18.6 (200)
Shape limitations	Surface of revolution	Moldable	Moldable	Moldable
Usual useful production	Low-medium	High	High	High
Volume, No. parts/year	0–1 000	1 000–1 000 000	1 000–1 000 000	1 000–1 000 000
Production cycle time	5 min	1 1/2 to 5 min	1 1/2 to 5 min	1 1/2 to 5 min
Typical glass content, %	65–90	15–35	15–35	24–45
Strength orientation	Highly oriented	Random	Random	Random
Strength category	Very high	Low-medium	Low-medium	Medium-high
Wall thickness:				
Minimum, mm (in.)	0.25 (0.01)	0.76 (0.03)	1.5 (0.06)	0.76 (0.03)
Maximum, mm (in.)	51 (2.0)	6.4 (0.25)	25.4 (1.0)	6.4 (0.25)
Tolerance, mm (in.)	+0.25 (+0.01)	+0.20 (+0.008)	+0.13 (+0.005)	+0.13 (+0.005)
Variations	As desired	Uniform desirable <3:1	As desired	Uniform desirable <2:1
Minimum draft:				
To depth 150-mm (6-in.)	3°	1°–3°	1°–3°	1°–3°
Over depth 150-mm (6-in.)	3°+	3°+	3°+	3°+
Minimum inside radius, mm (in.)	3.18 (0.125)	1.5 (0.06)	1.5 (0.06)	3.18 (0.125)
Ribs	No	Yes	Yes	Not recommended
Bosses	No	Yes	Yes	Not recommended
Undercuts	No	Avoid	No	No
Holes:				
Parallel	Yes	Yes	Yes	Not recommended
Perpendicular	—	Undesirable	Undesirable	Undesirable
Built-in cores	Possible	Possible	Possible	Possible
Metal inserts	Yes	Yes	Yes	Not recommended
Metal edge stiffeners	No	No	No	Yes
Surface finish:				
Number of finished surfaces	1	2	All	2
Quality of surface	Excellent	Very good	Excellent	Very good
Gel-coat surface	Yes	No	No	Yes
Surfacing mat	Yes	No	No	Yes
Combination with thermoplastic liner	Yes	No	No	No
Trim in mold	Yes	Yes	Yes	Yes
Molded-in labels	Yes	Difficult	Difficult	Difficult
Raised numbers	—	Yes	Yes	Yes
Translucency	Yes	No	No	Yes
Tool cost	Low	High	High	Medium
Capital equipment cost	Low	High	High	High

*Data from Owens-Corning Fiberglas.

Comparing conductive polymers and composites

	Advantages	Disadvantages
Naturally conductive polymers by heat cure	Ease of processing, void-free, stable in oxygen	High cost (at present), weight loss up to 17%, long time for cure
Doped polymers	Can increase and control degree of conductivity	Unstable in air, mechanical properties reduced, temp. rise causes loss of dopant
Coatings (thin metal coating on plastic)	Moderate cost, good shielding, even with thin coatings	Extra mfg step, coating may scratch or peel, more equipment required
Particulate composites	Particles are low in cost, can be used as a filler	Need much larger vol/fraction than fibers, doesn't provide tensile strength
Flake composites	Low cost	May require painting, alignment is difficult
Long-fiber composites	High aspect ratio, low loading required, high strength	Higher cost than short fibers
Short-fiber composites	Low cost, ease of processing	Lower aspect ratio, higher loading required
Nickel-coated graphite-fiber composites	Moderate cost, can use with wide variety of resins, complex shapes	High cost
Thermoplastic matrix composites	Easy to process	Lose aspect ratio and hence, conductivity during processing
Thermoset matrix composites	Retain high aspect ratio during processing	Requires longer time for cure
Steel-fiber composites	Low cost, and low loading required	Fibers not lightweight

Source: Ref 19, p 223

Thermoplastics

General

Properties of long-glass-fiber-reinforced thermoplastic composites

Matrix	Reinforcement	Composite efficiency	Tensile strength, MPa	Modulus of elasticity, GPa	Impact strength, J/cm
PET	Twill weave	0.37	220.0	13.4	9.7
PP	Twill weave	0.45	270.0	16.2	11.1
PP	Noncrimp fabric	0.24	275.1	7.6	...
PP	Long chopped	0.19	120.0	6.3	3.0
PP(a)	Swirl mat	0.19	120.0	5.0	7.8

(a) Azdel.

Source: Ref 28, p 258

Applications for glass fiber reinforced plastics

Industry	Advantages of FRP	Applications
Automotive	High-volume production; fine finishes, reduced costs	Automobile body components; fender extenders; front ends; headlamp and taillamp housings; hoods; spoilers, instrument panels; shift consoles; under-the-hood components; truck hoods; fenders; cab and body components; insulated tanks; engine covers; housings; fender liners
Agricultural	Ruggedness; corrosion resistance	Farm tractor hoods, grilles, instrument housings, seating, fenders; garden tractor and lawn mower bodies and housings; fertilizer and pesticide tanks and sprayers; feed troughs
Appliances	Ability to produce complex molded parts without fasteners or welds	Room air conditioner cases, base pans, bulkheads; condenser and compressor fans; humidifier cases and blower wheels; dishwasher pump bodies; dryer ducts; home laundry tubs; water softener tanks, controls, piping; fan housings; gears; vacuum cleaner housings; iron handles; soap dispensers; microwave oven cook trays; television swivel stands; sump pump bases
Aviation/aerospace	High strength with light weight	Aircraft interior components for passengers and cargo; wing tips; antenna components; radomes; wing fuel tanks; ducting; rocket-motor cases; nozzles; nose cones; pressure vessels; instrument housings; launch tubes
Business machines	Excellent surface finish and dimensional stability at elevated temperatures; high strength	Machine covers and housings; access panels; keyboard caps; keys; printer heads; gears, cams, and levers; frames; mounting panels; printed circuit boards; fans and blowers
Chemical processing	Corrosion resistance	Chemical and fuel tanks, pipes, and ducting; storage tanks and hoppers; process pump and valve bodies, casings, impellers; pressure vessels; filters; fume-collection hoods and duct systems; scrubbing towers; electroplating racks and handling equipment; photographic processing equipment
Construction	Ruggedness; moderate cost; good appearance	Structural shapes; paneling; siding; skylighting; curtain wall components; glazing panels; patio covers; concrete pouring forms
Electrical/electronic	High dielectric strength with low moisture absorption	Electrical pole line hardware, crossarms, strain insulators, standoffs, brackets; shatterproof street lighting globes; switch control rods; hot sticks; electronic components; housings and backboards; utility line maintenance equipment
House and home	Beauty; low maintenance; low cost	Architectural components; appliance and equipment components; furniture–chairs, tables, lawn furnishings; sinks; bathroom tub/shower units; skylights
Marine	Ease of repair; low maintenance; high performance	Pleasure, commercial, and military boat hulls and superstructures; barge covers; lighters; fuel tanks; water tanks; masts and spars; bulkheads; duct work; ventilation cowls; marker buoys; floating docks; outboard engine shrouds
Materials handling	High strength with light weight	Tote trays and bins; food-processing and delivery trays, boxes, bins; tanks and pipes; conveyor-system components; pallets and skids; cargo-handling equipment
Recreational	Low maintenance; good appearance	Motor homes; travel trailers; truck campers; camping trailers; pickup covers; water and snow skis; surfboards; golf clubs; hockey sticks; lacrosse sticks; archery bows; fishing rods; vaulting poles; recreational water craft, canoes; snowmobile and all-terrain vehicle bodies; golf carts; protective helmets; swimming pools; diving boards; playground equipment
Transportation	Toughness; lightness	Railway passenger and freight car components; transport seating; freight car roofs; hopper car covers; refrigerator car liners; air cargo "igloos"; motor truck and bus components; rapid transit car ends; third-rail covers; barges; truck trailer panels; refrigerated truck bodies

*From Owens-Corning Fiberglas.

Source: Ref 16, p 6-1

Candidate matrix resins for thermoplastic advanced composites

Name	Company	Polymer	Glass transition temperature °C	Glass transition temperature °F	Melt temperature °C	Melt temperature °F
Semicrystalline						
Victrex PEEK	ICI	Polyether etherketone	143	290	343	650
Ryton	Phillips	Polyphenylene sulfide	88	190	290	555
HTX	ICI	Poly aromatic ketone	205	400	358	675
Ultra Pek	BASF	Polyether ketone Etherketone ketone	172	340	372	700
Liquid crystal						
Xydar	Dartco	Polyester			415	780
Vectra	Celanese	Polyester			415	780
Amorphous thermoplastic						
Udel	Amoco	Polysulfone	190	375	...	...
Victrex PES	ICI	Polyether sulfone	230	445	...	...
Ultem	GE	Polyetherimide	215	420	...	...
Various	Various	Polycarbonate	140-150	285-300	...	...
PASII	Phillips	Poly aromatic sulfide	204	400	...	...
Pseudothermoplastics						
Avimid KIII	DuPont	Polyimide	255	490	...	...
Torlon	AMOCO	Polyamideimide	275	525	...	...
LARC-TPI	NASA	Polyimide	255	490	...	...

Source: Ref 6, p 545

Properties of selected resins filled with mica(a)

Property	Polypropylene homopolymer(b) Unfilled	Polypropylene homopolymer(b) 40% mica(c)	Nylon 6/6(d) Unfilled	Nylon 6/6(d) 40% mica(e)	PBT(f) Unfilled	PBT(f) 40% mica(g)	PET(h) Unfilled	PET(h) 40% mica(j)
Tensile strength:								
MPa	33.1	46.9	77.9(k)	86.2(k)	50.3	68.3	71.7	79.3
ksi	4.8	6.8	11.3(k)	12.5(k)	7.3	9.9	10.4	11.5
Flexural strength:								
MPa	25.5	71.0	58.6(k)	137.9(k)	62.7	110.3	90.3	117.9
ksi	3.7	10.3	8.5(k)	20.0(k)	9.1	16.0	13.1	17.1
Flexural modulus:								
GPa	1.241	7.584	3.054(k)	9.377(k)	2.392	11.58	3.116	17.31
10^6 psi	0.180	1.100	0.443(k)	1.360(k)	0.347	1.680	0.452	2.510
Heat-distortion temperature at 1.82 MPa (284 psi):								
°C	56	116	67	201	53	169	80	218
°F	133	241	153	394	127	336	176	424
Izod impact resistance at 23 °C (73 °F):								
Unnotched								
J/cm	No break	2.6	...	...	No break	1.7	3.9	1.0
ft·lb/in.	No break	4.9	...	...	No break	3.1	7.3	1.9
Notched								
J/cm	0.24	0.26	0.43(k)	0.28(k)	...	...	0.28	0.24
ft·lb/in.	0.45	0.49	0.80(k)	0.53(k)	...	...	0.52	0.45

(a) Data from Marietta Resources International. Micas are all phlogopite grades from Marietta Resources. (b) Profax 6523, Hercules, Inc. (c) Suzorex 200 QX. (d) Zytel 101, Du Pont. (e) Suzorex 325-PO. (f) Valox 310, General Electric. (g) Suzorite 325 HK. (h) Eastman 7352. (j) Suzorite 605. (k) Dry.

Source: Ref 19, p 112

Comparative properties of carbon- and glass-reinforced engineering thermoplastics

Resin type	Tensile strength MPa	Tensile strength ksi	Flexural modulus GPa	Flexural modulus 10^6 psi	Impact strength, notched/unnotched J/cm	Impact strength, notched/unnotched ft·lb/in.	Heat-deflection temperature °C	Heat-deflection temperature °F
Amorphous resins								
Acrylonitrile-butadiene-styrene (ABS):								
30% glass fiber	100	14.5	7.6	1.1	0.75/3.5	1.4/6.5	105	220
30% carbon fiber	130	18.8	12.4	1.8	0.59/2.4	1.1/4.5	105	220
Nylon:								
30% glass fiber	148	21.5	7.9	1.15	0.64/3.7	1.2/7.0	140	285
30% carbon fiber	207	30	15.2	2.2	0.64/4.3	1.2/8.0	145	290
Polycarbonate:								
30% glass fiber	128	18.5	8.3	1.2	2.0/9.34	3.7/17.5	150	300
30% carbon fiber	165	24	13.1	1.9	0.96/5.34	1.8/10.0	150	300
Polyetherimide:								
30% glass fiber	197	28.5	8.6	1.25	0.75/5.60	1.4/10.5	215	420
30% carbon fiber	234	34	17.2	2.5	0.75/6.67	1.4/12.5	215	420
Polyphenylene oxide (PPO):								
30% glass fiber	145	21	9.0	1.3	1.2/5.1	2.3/5.1	155	310
30% carbon fiber	159	23	11.7	1.7	0.53/3.0	1.0/5.6	155	310
Polysulfone:								
30% glass fiber	124	18	8.3	1.2	0.96/7.5	1.8/14	185	365
30% carbon fiber	159	23	14.5	2.1	0.64/3.5	1.2/6.5	185	365
Styrene-maleic-anhydride (SMA):								
30% glass fiber	103	15	9.0	1.3	0.59/2.4	1.1/4.5	120	250
Thermoplastic polyurethane:								
30% glass fiber	57	8.2	1.3	0.19	5.1/15	9.5/28	170	340
Crystalline resins								
Acetal:								
30% glass fiber	134	19.5	9.7	1.4	0.96/4.8	1.8/9.0	165	325
20% carbon fiber	81	11.8	9.3	1.35	0.53/1.6	1.0/3.0	160	320
Nylon 66:								
30% glass fiber	179	26	9.0	1.3	1.5/11	2.9/20	255	490
30% carbon fiber	241	35	20.0	2.9	0.80/6.4	1.5/12	257	495
Polybutylene terphthalate (PBT):								
30% glass fiber	134	19.5	9.7	1.4	1.4/9.1	2.6/17	210	410
30% carbon fiber	152	22	15.9	2.3	0.64/3.5	1.2/6.5	210	410
Polythylene terephthalate (PET):								
30% glass fiber	159	23	9.0	1.3	1.0/...	1.9/...	225	435
Polyphenylene sulfide (PPS):								
30% glass fiber	138	20	11.0	1.6	0.75/4.5	1.4/8.5	260	500
30% carbon fiber	186	27	16.9	2.45	0.59/2.9	1.1/5.5	265	505

Source: Ref 16, p 6-34

Interlaminar fracture toughness of thermoplastic composites as determined by the double-cantilever-beam test

Material	Process conditions °C/kPa	Process conditions °F/psi	Interlaminar fracture toughness, G_{Ic} J/m²	Interlaminar fracture toughness, G_{Ic} ft·lbf/ft²
Polysulfone (Udel)	340/4150	650/600	1175	80
Polyetherimide (Ultem)	400/4850	750/700	950	65
Polyamideimide (Torlon)	340/1400	650/200	1050	70
Polyphenylene sulfide (Ryton)	340/1400	650/200	720	50
Polyether etherketone (PEEK)	400/700	750/100	1600	110

Source: Ref 6, p 100

Design properties of carbon-fiber-reinforced thermoplastic composites(a)

Base resin	PAN carbon fiber content, %	Physical properties: Specific gravity(b)	Physical properties: Water absorption,(e) %	Physical properties: Mold shrinkage(d), in./in.	Mechanical properties: Tensile strength ksi(e)	Mechanical properties: Flexural strength ksi(g)	Mechanical properties: Flexural modulus 10^6 psi(h)	Mechanical properties: Shear strength ksi(f)	Izod impact strength(k), ft·lb/in.(m): Notched	Izod impact strength(k), ft·lb/in.(m): Unnotched
Nylon 6/6	10	1.18	0.80	0.0040	20.0	30.0	1.00	...	1.0	6.0
	20	1.23	0.60	0.0025	28.0	42.0	2.40	12.0	1.1	8.0
	30(x)	1.28	0.50	0.0020	35.0	51.0	2.90	13.0	1.5	12.0
	40	1.34	0.40	0.0020	40.0	60.0	3.40	14.0	1.6	13.0
	30(y)	1.38	0.48	0.0025	30.5	43.0	2.30	12.5	1.4	10.0
	30(z)	1.36	0.45	0.0025	27.0	36.0	1.75	11.5	1.4	11.0
	30(aa)	1.27	0.80	0.0020	32.0	46.5	2.50	12.5	2.1	14.0
Nylon 6	30	1.28	0.80	0.0020	32.0	46.0	2.40	12.5	1.8	13.0
Nylon 6/12	30	1.22	0.15	0.0020	29.0	42.0	2.30	12.0	1.8	13.0
Nylon 6/10	30	1.23	0.12	0.0020	28.0	40.5	2.20	12.0	1.8	14.0
Supertough nylon	30	1.22	0.35	0.0025	24.0	34.0	2.00	11.0	3.0	19.0
Amorphous nylon	30	1.27	0.12	0.0015	30.0	47.5	2.20	12.5	1.2	8.0
Polycarbonate	30	1.33	0.08	0.0015	24.0	36.0	1.90	10.0	1.8	10.0
Polysulfone	30	1.37	0.15	0.0015	23.0	32.0	2.05	9.5	1.2	6.5
Polyethersulfone	30	1.48	0.30	0.0015	26.0	37.0	2.05	10.5	1.0	5.5
	30(y)	1.57	0.20	0.0020	23.5	33.5	1.90	9.3	1.0	4.8
PEEK	15(y)	1.46	0.10	0.0020	19.5	24.0	1.50	7.8	0.9	9.0
	30	1.39	0.10	0.0010	31.0	36.0	2.20	12.4	1.2	12.0
Acetal	20	1.46	0.50	0.0050	11.8	13.7	1.35	8.2	1.0	3.0
Polyester (PBT)	30	1.41	0.04	0.0020	22.0	29.0	2.30	8.0	1.2	6.5
Polyphenylene sulfide	30	1.45	0.04	0.0010	27.0	34.0	2.45	9.5	1.1	5.5
	30(y)	1.57	0.03	0.0015	25.5	28.0	2.40	8.0	0.8	4.5
ETFE(bb)	20	1.72	0.02	0.0025	12.0	16.0	1.20	6.5	5.5	7.5
PVDF	15	1.77	0.02	0.0060	13.5	18.0	1.15	7.5	1.0	7.0

Design properties of carbon-fiber-reinforced thermoplastic composites (continued)

Base resin	PAN carbon fiber content,	Thermal properties: Heat deflection temperature °F(p)	Thermal properties: Thermal conductivity Btu·in./ft²·h·°F(r)	Thermal properties: Coefficient of linear thermal expansion 10^{-5}/°F(t)	Flammability (UL94)
Nylon 6/6	10	485	4.0	2.0	HB
	20	495	5.5	1.4	HB
	30(x)	495	7.0	1.1	HB
	40	500	8.5	0.8	HB
	30(y)	490	7.0	1.1	HB
	30(z)	485	7.0	1.1	HB
	30(aa)	495	6.8	1.1	HB
Nylon 6	30	425	7.0	1.0	HB
Nylon 6/12	30	420	6.5	0.9	HB
Nylon 6/10	30	425	6.5	0.9	HB
Supertough nylon	30	410	5.8	1.2	HB
Amorphous nylon	30	290	7.0	1.1	HB
Polycarbonate	30	300	4.9	0.9	VI
Polysulfone	30	365	5.5	0.6	VO
Polyethersulfone	30	415	6.0	0.8	VO
	30(y)	410	5.7	0.8	VO
PEEK	15(y)	600	4.5	1.8	VO
	30	600	7.1	0.7	VO
Acetal	20	320	4.6	2.2	HB
Polyester(PBT)	30	430	4.6	0.5	HB
Polyphenylene sulfide	30	505	5.2	0.6	VO
	30(y)	500	5.4	0.8	VO
ETFE(bb)	20	435	6.0	1.0	VO
PVDF	15	300	2.2	2.5	VO

(continued)

Design properties of carbon-fiber reinforced thermoplastic composites (continued)

Base resin	PAN carbon fiber content, %	Tribological properties: Coefficient of friction, Static	Coefficient of friction, Dynamic	Wear factor 10^{-10} in.^{3}·min/ft·lb·h	Limiting PV, 10 fpm	Limiting PV, 100 fpm	Limiting PV, 1000 fpm	Electrical: surface resistivity(u) Ω/in.^{2}	Chemical resistance(v, w): Acids	Bases	Solvents
Nylon 6/6	10	0.17	0.21	60	10000	16000	5000	10^7	F	G	E
	20	0.16	0.20	40	19000	25000	7000	1300	F	G	E
	30(x)	0.16	0.20	20	21000	27000	8000	150	F	G	E
	40	0.13	0.18	14	22000	27500	8500	75	F	G	E
	30(y)	0.11	0.15	10	29000	42000	19000	200	F	G	E
	30(z)	0.10	0.11	6	29000	43000	20000	1000	F	G	E
	30(aa)	0.18	0.22	22	20000	26500	7000	150	F	G	E
Nylon 6	30	0.18	0.21	30	18000	22000	7500	150	F	G	E
Nylon 6/12	30	0.19	0.23	25	18000	20000	17000	250	F	G	E
Nylon 6/10	30	0.20	0.25	25	18000	21000	7500	250	F	G	E
Supertough nylon	30	0.20	0.25	25	20000	26000	7500	1100	F	G	E
Amorphous nylon	30	0.19	0.24	90	10000	11000	6000	150	F	F	F
Polycarbonate	30	0.18	0.17	85	8000	8500	5500	3500	F	F	P
Polysulfone	30	0.17	0.14	75	8500	8500	6000	200	G	E	P
Polyethersulfone	30	0.17	0.15	80	10000	10000	7000	100	G	E	P
	30(y)	0.13	0.17	40	35000	33000	16000	100	G	E	P
PEEK	15(y)	0.18	0.20	60	42000	40000	22000	5000	E	E	G
	30	0.19	0.13	60	...	...	...	120	E	E	G
Acetal	20	0.11	0.14	40	13000	20000	15000	2000	F	G	E
Polyester (PBT)	30	0.12	0.15	24	18000	22000	10000	500	F-G	P-F	E
Polyphenylene sulfide	30(y)	0.23	0.20	160	12000	20000	10000	250	E	E	E
	30(y)	0.13	0.15	75	27000	35000	30000	150	E	E	E
ETFE(bb)	20	0.16	0.18	28	...	...	...	1200	E	E	E
PVDF	15	0.25	0.25	14	15000	11000	<5000	700	E	E	E

(a) Data from LNP Corp. This information is based on experience and is intended for use as a guide only. (b) ASTM D792. (c) ASTM D570. (d) ASTM D955. (e) ASTM D638. (f) To convert to MPa, multiply listed values by 6.894757. (g) ASTM D790. (h) To convert to GPa, multiply listed values by 6.894757. (j) ASTM D732. (k) ASTM D256. (m) To convert to J/cm, multiply listed values by 0.5338. (n) ASTM D648. (p) To convert to °C, subtract 32 from listed values, then divide by 1.8. (q) ASTM C177. (r) To convert to W.m·K, multiply listed values by 0.14423. (s) ASTM D696. (t) To convert to 10^{-5}/°C, multiply listed values by 1.8. (u) To convert to Ω/m^2, multiply listed values by 1550. (v) ASTM D257. (w) E = excellent; G = good; F = fair; P = poor. (x) Grade RC 1006. (y) Plus 15% PTFE lubricant. (z) Plus 15% PTFE/silcone lubricant. (aa) Grade RD 1006 HI. (bb) Based on Du Pont Tefzel fluoropolymer.

Source: Ref 16, p 6-48 to 6-49

Properties of nylon 6/6, nylon 6, and PBT reinforced with glass/mica mixture

Property	Nylon 6/6	Nylon 6	PBT
Resin content, wt %	60	60	60
Mica content, wt %	25	25	25
Glass-fiber content, wt %	15	15	15
Tensile strength, MPa (ksi)	141 (20.5)	141 (20.4)	100 (14.5)
Tensile elongation, %	3.4	3.8	3.1
Flexural strength, MPa (ksi)	197 (28.5)	252 (36.6)	148 (21.5)
Flexural modulus, GPa (10^6 psi)	9.997 (1.450)	11.34 (1.645)	9.653 (1.400)
Impact strength, J/cm (ft·lb/in.):			
Unnotched	0.43 (0.8)	0.48 (0.9)	0.43 (0.8)
Notched	8.01 (15.0)	6.14 (11.5)	5.87 (11.0)
Heat-distortion temperature at 1.82 MPa (264 psi), °C, (°F)	225 (490)	200 (395)	215 (420)
Mold shrinkage, mm/mm (in./in.)	0.006	0.0055	0.0065
Coefficient of thermal expansion, 10^{-5}/°C (10^{-5}/°F)	3.8 (2.1)	3.6 (2.0)	3.4 (1.9)

Source: Ref 19, p 112

Properties of pitch carbon fiber reinforced composites

ASTM test	Property	Nylon 20 wt % chopped fiber	Nylon 30 wt % chopped fiber	Nylon 40 wt % chopped fiber	Nylon 30 wt % chopped fiber[f]	Polycarbonate 25 wt % chopped fiber	Polypropylene 40 wt % chopped fiber
Mechanical Properties							
D638	Tensile strength:						
	MPa	89.6	107	121	138	72.3	31.7
	ksi	13.0	15.5	17.5	20.0	10.5	4.60
D638	Elongation, %	2.5	2.0	1.5	1.9	2.0	1.1
D638	Tensile modulus:						
	GPa	10.3	13.8	17.9	17.2	11.4	...
	10^5 psi	15.0	20.0	26.0	25.0	16.5	...
D790	Flexural strength:						
	MPa	152	179	193	227	68.9	23.4
	ksi	22.0	26.0	28.0	33.0	10.0	3.4
D790	Flexural modulus:						
	GPa	6.9	10.3	13.8	13.1	...	7.6
	10^5 psi	10.0	15.0	20.0	19.0	...	11.0
D256	Izod impact strength, notched:[b]						
	J/m	32.0	37.4	42.7	64.1	58.7	37.4
	ft · lb/in.	0.6	0.7	0.8	1.2	1.1	0.7
D256	Izod impact strength, unnotched:[b]						
	J/m	...	694	...	...	320	214
	ft · lb/in.	...	13.0	...	...	6.0	4.0
D695	Compressive strength:						
	MPa	138	138	158	172	...	...
	ksi	20.0	20.0	23.0	25.0	...	...
D732	Shear strength:						
	MPa	65.5	65.5	68.9	68.9	...	...
	ksi	9.5	9.5	10.0	10.0	...	...
D785	Rockwell hardness	E44	E53	E54	E55	...	...
D621	Deformation under load,[c] %	0.54	0.38	0.2	...	...	...
Physical Properties							
D792	Specific gravity	1.24	1.30	1.36	1.40	1.35	1.15
D570	Water absorption in 24 h, %	0.75	0.6	0.5	0.5	...	0.03
...	Linear mold shrinkage in 3 mm ($^1/_8$ in.), %	0.4	0.3	0.2	0.2	0.1	0.1
Thermal Properties							
D696	Coefficient of linear thermal expansion:						
	10^{-5} m/m · °C	2.3	1.6	0.9	1.3	...	...
	10^{-5} in./in. · °F	1.3	0.9	0.5	0.7	...	...
D648	Deflection temperature under load:[d]						
	°C	232	241	246	243	143	127
	°F	450	465	475	470	290	260
D648	Deflection temperature under load:[e]						
	°C	254	254	259	254	...	...
	°F	490	490	498	490	...	...
...	Flammability at minimum thickness (UL 94)	HB	HB	HB	HB	V-O	HB

(continued)

Properties of pitch carbon fiber reinforced composites (continued)

ASTM test	Property	Nylon 20 wt % chopped fiber	Nylon 30 wt % chopped fiber	Nylon 40 wt % chopped fiber	Nylon 30 wt % chopped fiber[f]	Polycarbonate 25 wt % chopped fiber	Polypropylene 40 wt % chopped fiber
Electrical Properties							
D257	Volume resistivity, Ω · cm	10^6	10^4	10^3	10^4	10^5	10^2
D257	Surface resistivity, Ω/sq	10^6	10^4	10^3	10^4	10^5	10^2

[a]Data from Wilson-Fiberfil International. All tests conducted at 23 °C (73 °F) unless otherwise noted. [b]Izod impact test bars 6.35 × 12.7 mm (1/4 × 1/2 in.). [c]Deformation at 27.6 MPa and 50 °C (4 ksi and 122 °F). [d]At 1.82 MPa (264 psi). [e]At 0.45 MPa (66 psi). [f]Contains 15 wt % glass fiber.

Source: Ref 16, p 6-51

Properties of thermoplastic matrix composites

Properties	Aromatic copolyester Resin	Aromatic copolyester 40% glass fiber-resin	PBT Resin	PBT 15-40% glass fiber-resin	PET Resin	PET 30-45% glass fiber-resin
Physical						
Heat deflection temperature at 1800 kPa (264 psi), °C (°F)	355 (671)	...	55 (130)	205 (400)	...	225 (435)
U.L. in-service temperature rating, °C (°F)	240 (464)	...	120 (248)	140 (284)	140 (284)	150-180 (302-356)
Processing melt temperature, °C (°F)	400-450 (750-840)	...	270 (520)	...	290 (550)	...
Specific gravity	1.35	1.70	1.31	1.53	...	1.56-1.69
Density, g/cm^3 (lb/in.^{3})	1.35 (0.049)	1.70 (0.061)	1.31 (0.047)	1.53 (0.055)	...	1.56-1.69 (0.056-0.061)
U.L. flammability	94 V-0	94 V-0	94 HB / 94 V-0	94 HB / 94 V-0	94 HB / 94 V-0	94 HB / 94 V-0

Source: Ref 6, p 363

Tensile strengths of filled and unfilled heat-resistant resins after thermal aging at 260 °C (500 °F)

Base resin	Fiberglass content, wt %	0 h	100 h	250 h	500 h	750 h	1000 h	1500 h
		Tensile strength, ksi (MPa), after aging for:						
ETFE	20	11.3 (77.9)	11.5 (79.3)	10.0 (69.0)	7.0 (48.3)	5.0 (34.5)	3.8 (26.2)	2.3 (15.9)
FEP	20	5.0 (34.5)	5.1 (35.2)	4.8 (33.1)	4.7 (32.4)	4.7 (32.4)	4.6 (31.7)	4.5 (31.0)
Polyphenylene sulfide	40	23.2 (160)	16.4 (113)	16.0 (110)	15.5 (107)	15.0 (103)	14.5 (100)	13.8 (95.2)
Polyethersulfone	40	22.7 (157)	15.6 (108)	14.8 (102)	14.3 (98.6)	13.7 (94.5)	12.2 (84.1)	10.5 (72.4)
Polyimide	30	13.0 (89.6)	15.0 (103)	14.3 (98.6)	13.4 (92.4)	12.8 (88.3)	12.0 (82.7)	11.2 (77.2)
Polyamide-imide	0	27.4 (189)	27.2 (188)	26.6 (183)	26.0 (179)	24.5 (169)	23.5 (162)	22.0 (152)
Polyarylsulfone	0	13.1 (90.3)	11.5 (79.3)	10.5 (72.4)	10.0 (69.0)	9.5 (65.5)	8.4 (57.9)	7.6 (52.4)
Poly-*p*-oxybenzoate	0	23.0 (159)	18.0 (124)	16.3 (112)	16.0 (110)	15.5 (107)	15.1 (104)	13.0 (89.6)
Nylon 66	50	31.0 (214)	17.6 (121)	10.3 (11.0)	9.4 (64.8)	...	...	...
Polyester	40	22.1 (152)	Melted	...	...	...	...	...
Polysulfone	40	20.3 (140)	Melted	...	...	...	...	...

[a]Tested at 23 °C (73 °F).

Source: Ref 16, p 6-64

Comparison of fatigue behavior of various resins

Base resin	Fiber type(a)	Cyclic failure stress MPa	ksi
SAN	Glass	45	6.5
Styrene	Glass	41	6.0
Polycarbonate	Glass	38	5.5
ETFE copolymer	Glass	24	3.5
	Carbon	42	6.1
Polysulfone	Glass	34	5.0
Polyethersulfone	Carbon	55	8.0
Acetal copolymer	Glass	48	7.0
Polypropylene	Glass	31	4.5
Polyphenylene sulfide	Carbon	66	9.5
Nylon 6/6c(b)	Glass	41	6.0
	Carbon	55	8.0
	Bead	27	3.9
Polyester (PBT)	Glass	39	5.6
	Carbon	51	7.4
Modified PPO	Glass	34	4.9
PEEK	Carbon	121	17.5

(a) Fiber content, 30%. (b) The "c" indicates moisture conditioned at 50% RH.

Source: Ref 19, p 227

Specific

Comparison of injection-moldable polycarbonates for EMI shielding

Property	Un-filled	Propri-etary system[b]	25% carbon fiber	25% metal-lized glass	30% Al flake
Tensile strength:					
MPa	62	55	138	82.7	41
ksi	9.0	8.0	20.0	12.0	6.0
Elongation, %	6–8	7	1.1	2.0	2.9
Izod impact:					
J/cm	1.3	0.9	1.1	0.9	1.1
ft · lb/in.	2.4	1.7	2.0	1.6	2.0
Flexural strength:					
MPa	89.6	93.1	193	145	75.8
ksi	13.0	13.5	28.0	21.0	11.0
Flexural modulus:					
GPa	2.3	2.3	13.8	6.8	4.4
10^6 psi	0.33	0.33	2.00	0.98	0.64
DTUL[c]:					
°C	132	142	146	143	141
°F	270	288	295	290	285
Specific gravity	1.20	1.27	1.31	1.40	1.44
Attenuation at 1.0 GHz, dB	0	40	30	20	30

[a]Data from Wilson-Fiberfil International. [b]Conductive additives include stainless steel fibers. [c]At 1.82 MPa (264 psi).

Source: Ref 16, p 6-42

Properties of fiberglass-reinforced polyester resins

ASTM test method	Property	Woven cloth	Chopped roving	Sheet molding compound
Mechanical properties				
D792	Specific gravity	1.5-2.1	1.35-2.30	1.65-2.60
D790	Flexural yield strength:			
	MPa	276-552	68.9-276	68.9-248
	ksi	40-80	10-40	10-36
D790	Flexural modulus of elasticity:			
	GPa	...	6.9-20.7	6.9-15.2
	10^5 psi	...	10-30	10-22
D695	Compressive strength:			
	MPa	172-345	103-207	103-207
	ksi	25-50	15-30	15-30
D638	Tensile strength:			
	MPa	207-345	103-207	55.2-138
	ksi	30-50	15-30	8-20
D638	Elongation, %	0.5-2.0	0.5-5.0	...
D638	Tensile modulus of elasticity:			
	GPa	10.3-31.0	5.5-13.8	...
	10^5 psi	15.0-45.0	8.0-20.0	...
D256	Izod impact strength, notched(a):			
	J/m	267-1600	107-1070	374-1175
	ft·lb/in.	5.0-30.0	2.0-20.0	7.0-22.0
...	Barcol hardness	60-80	50-80	50-70
Electrical properties				
D257	Volume resistivity(b), Ω·cm	10^{14}	10^{14}	10^{14}-10^{15}
D149	Dielectric strength(c), V/mil:			
	Short-time test	350-500	350-500	380-450
	Step-by-step test	300-400	300-450	350-400
D150	Dielectric constant:			
	At 60 Hz	4.1-5.5	3.8-6.0	4.4-6.3
	At 10^3 Hz	4.2-6.0	4.0-6.0	4.4-6.1
	At 10^6 Hz	4.0-5.5	3.5-5.5	4.2-5.8
D150	Dissipation (power) factor:			
	At 60 Hz	0.01-0.04	0.01-0.04	0.007-0.021
	At 10^3 Hz	0.01-0.06	0.01-0.05	0.007-0.015
	AT 10^6 Hz	0.01-0.03	0.01-0.03	0.016-0.024
D495	Arc resistance, s	60-120	120-180	120-200
Resistance characteristics				
...	Heat resistance (continuous):			
	°C	149-177	149-177	149-204
	°F	300-350	300-350	300-400
D570	Water absorption(d), in 24 h, %	0.05-0.50	0.1-1.0	0.10-0.15
...	Effect of sunlight	Slight	Slight	Slight

(a) Test bar 13 × 13 mm (0.5 × 0.5 in.). (b) At 50% RH and 23 °C (73 °F). (c) Material thickness, 3.2 mm (0.125 in.). (d) Material thickness, 3.1 mm (0.125 in.).

Source: Ref 16, p 6-65

Properties of injection-molding composites(a)

Property	Reported value for glass/PET ratio of: 100/0	75/25	50/50	0/100
Notched impact, J/cm (ft·lb/in.)	0.59 (1.1)	1.4 (2.7)	2.3 (4.3)	3.2 (6.0)
Unnotched impact, MPa (ksi)	12 (1.7)	27 (3.9)	34 (5.0)	38 (5.5)
Tensile strength, MPa (ksi)	30 (4.4)	32 (4.6)	30 (4.4)	26 (3.7)
Tensile modulus, GPa (10^6 psi)	4.1 (0.59)	4.1 (0.60)	3.8 (0.55)	3.8 (0.55)
Flexural strength, MPa (ksi)	72 (10.4)	70 (10.2)	66 (9.5)	50 (7.2)
Flexural modulus, GPa (10^6 psi)	10 (1.5)	10 (1.5)	9.7 (1.4)	9.0 (1.3)
Abrasion (1000 cycles), g	3.3	3.2	2.5	2.4

(a) System is typical commercial polyester BMC formulation: 20 wt % glass loading; glass replaced by PET on equal-volume basis; fiber length, 6.4 mm (0.25 in.).

Source: Ref 19, p 74

Thermochemical properties of nylon-composite samples(a) 3.2-mm (1/8-in.) laminates with various fillers

Composite(b)	Linear mold shrinkage								Coefficient of linear thermal expansion							
	Longitudinal				Transverse at 12.7-mm (0.5-in.) width				Longitudinal				Transverse at 12.7-mm (0.5-in.) width			
	3.2-mm (0.125-in.) sample		6.4-mm (0.25-in.) sample		3.2-mm (0.125-in.) sample		6.4-mm (0.25-in.) sample		3.2-mm (0.125-in.) sample		6.4-mm (0.25-in.) sample		3.2-mm (0.125-in.) sample		6.4-mm (0.25-in.) sample	
	mm	in.	mm	in.	mm	in.	mm	in.	$10^{-5}/°C$	$10^{-5}/°F$	$10^{-5}/°C$	$10^{-5}/°F$	$10^{-5}/°C$	$10^{-5}/°F$	$10^{-5}/°C$	$10^{-5}/°F$
100% nylon 6/6	0.3683	0.0145	0.6706	0.0264	0.7010	0.0276	0.8661	0.0341	7.2	4.0	11.2	6.2	14	7.8	14	7.8
10% GF/90% nylon	0.1270	0.0050	0.2616	0.0103	0.3759	0.0148	0.6299	0.0248	5.7	3.2	6.1	3.4	20	11.1	19	10.6
10% CF/90% nylon	0.0559	0.0022	0.1168	0.0046	0.4064	0.0160	0.6655	0.0262	3.5	1.9	2.5	1.4	19	10.6	19	10.6
20% GF/80% nylon	0.0686	0.0027	0.1321	0.0052	0.4013	0.0158	0.6200	0.0244	2.8	1.6	2.9	1.6	21	11.7	19	10.6
20% CF/80% nylon	0.0076	0.0003	0.0610	0.0024	0.3912	0.0154	0.5588	0.0220	1.6	0.9	1.6	0.9	19	10.6	17	9.4
30% GF/70% nylon	0.0356	0.0014	0.0940	0.0037	0.4064	0.0160	0.5385	0.0212	2.4	1.3	2.1	1.2	18	10.0	18	10.0
30% CF/70% nylon	0.0076	0.0003	0.1780	0.0007	0.3810	0.0150	0.5334	0.0210	1.2	0.7	1.1	0.6	22	12.2	15	8.3

(a) Measurement vs. flow direction. (b) GF, glass fibers; CF, carbon fibers.

Source: Ref 19, p 78

Corrosion resistance of glass fiber-polyester resin composites

Resin	75% H_2SO_4 80 °C (175 °F)	15% NaOH 65 °C (150 °F)	5¼% NaOCl 65 °C (150 °F)	Xylene ambient	Deionized water 100 °C (210 °F)	Seawater 80 °C (180 °F)
Isophthalic	−	−	−	+	−	−
Chlorendic	+	−	−	+	+	+
BPA fumarate	−	+	−	+	−	+
Vinyl ester	−	−	+	−	−	+

Source: Ref 6, p 94

Electrical properties of isophthalic polyester 3.2-mm (1/8-in.) laminates with various fillers

Material	Dielectric strength short time V/µm	Dielectric strength short time V/mil	Volume resistivity, 10^{-13} Ω·m	Dielectric constant, 1 MHz	Dissipation factor, 1 MHz	Dielectric constant, 1 kHz	Dissipation factor, 1 kHz	Dielectric constant, 60 Hz	Dissipation factor, 60 Hz	Arc resistance Avg	Arc resistance Max	Arc resistance Min	Track resistance, V	Dielectric breakdown short time, kV	Dielectric breakdown step-by-step, kV
Calcium carbonate	15.0	380	7.8	4.10	0.007	4.18	0.005	4.19	0.003	157	181	140	840	58	61
Gypsum CaSO4	14.4	365	2.1	3.69	0.011	4.04	0.023	4.19	0.027	153	184	141	840	70	55
Alumina trihydrate	15.4	390	2.6	3.67	0.009	3.81	0.010	3.89	0.011	183.5	184	183	860	67	51
Clay	14.4	365	6.4	4.08	0.018	4.61	0.040	5.10	0.057	182.5	183	182	840	59	57

(a) Vinyl toluene monomer

Source: Ref 6, p 94

Electrical properties of BPA fumarate polyester 3.2-mm (1/8 in.) laminates with various fillers

Material	Dielectric strength short time V/μm	Dielectric strength short time V/mil	Volume resistivity, 10^{-13} Ω·m	Dielectric constant, 1 MHz	Dissipation factor, 1 MHz	Dielectric constant, 1 kHz	Dissipation factor, 1 kHz	Dielectric constant, 60 Hz	Dissipation factor, 60 Hz	Arc resistance Avg	Arc resistance Max	Arc resistance Min	Track resistance, V	Dielectric breakdown short time, kV	Dielectric breakdown step-by-step, kV
Calcium carbonate	6.1	155	1.6	3.94	0.005	4.00	0.004	4.03	0.004	140	143	133	840	58	52
Gypsum CaSO4	5.9	150	3.3	3.72	0.009	4.03	0.024	4.24	0.029	144	151	137	820	50	40
Alumina trihydrate	11.8	300	3.3	3.64	0.008	3.81	0.015	3.93	0.025	182	184	181	820	55	52
Clay	12.6	320	3.5	4.08	0.023	4.68	0.043	5.11	0.053	183	184	181	840	61	43

(a) Vinyl toluene monomer

Source: Ref 6, p 94

Selected process factors for carbon-PEEK composite 3.2-mm (1/8-in.) laminates with various fibers

	Area/width limits	Weight/thickness	Properties	Economics	Advantages/disadvantages
Preimpregnated					
Woven fabric (APC-2 woven)	460 mm (18 in.) only. Wider very difficult	216 g/m^2 max	~APC-2	Difficult to prepreg	Thin skins, good props, limited widths, no cold drape
Woven prepreg tow (APC-2 tow 12K)	1780 mm (70 in.)	300 g/m^2	APC-2	Simple weaving	Large area, thin skins, good props, easy consolidation, no cold drape
Postimpregnated					
Film stacked (PEEK film/woven carbon fiber fabric)	Large area limited only by weaving loom and consolidation press size (applies to film stacked, slit film, cowoven, and comingled and woven)		Poorer ↑	Simple weaving and film	Straightforward, large area, difficult impregnation, flat sheet only, no drape
Slit film (cowoven slit PEEK film)		200-300 g/m^2 (applies to slit film and cowoven)	90% push and pull, 70% bend, 50% impact ↑	Slow weaving. Expensive slit tape	Large area, limited drape, difficult impregnation
Cowoven (PEEK monofill or multifill)				Simple weaving	Large area, limited drape with monofill, difficult impregnation
Comingled and woven (PEEK multifill)				Four-step process	Large area, good drape, improved impregnation
Powder impregnated (fine PEEK powder trapped in tow or fabric		200-500 g/m^2 (0.7-1.6 oz/ft^2)	Poor		Better impregnation; poor polymer weight, function, and control
3D woven preform (knitted or Z-stitched thick fabric or true 3D)		Very thick or true 3D		Expensive	May be only way of producing, for example, pultrusion

Source: Ref 6, p 546

Properties of Kynol-fiber reinforced polypropylene

Property	Fiber, wt % 4	18(a)	18(b)	33	Powder, 20 wt %	Poly-propylene(c)
Density, g/cm³	0.91	0.94	0.94	0.98	0.94	0.903
Izod impact strength, J/cm (ft·lb/in.):						
Unnotched	7.10 (13.3)	5.50 (10.3)	4.8 (9)	4.3 (8)	2.0 (3.8)	12 (22)
Notched	0.37 (0.7)	0.62 (1.17)	0.37 (0.7)	0.64 (1.2)	0.32 (0.6)	0.27 (0.5)
Elongation, %	>50	25	17	7	6	>100
Tensile strength, MPa (ksi)	30 (4.3)	28 (4.1)	35 (5.1)	30 (4.4)	26 (3.7)	34 (4.9)
Heat-deflection temperature, °C (°F):						
At 1.82 MPa (264 ksi)	65 (149)	77 (171)	84 (183)	100 (212)	67 (153)	58 (136)
At 455 kPa (66 psi)	...	131 (268)	136 (277)	...	...	98 (208)
Tensile modulus, GPa (10^6 psi)	1.59 (0.23)	1.86 (0.27)	1.86 (0.27)	1.93 (0.28)	1.59 (0.23)	1.38 (0.20)

(a) Without coupling agent. (b) With coupling agent. (c) Profax 6523, Hercules.

Source: Ref 19, p 76

Powder blending vs. melt compounding of mica/polypropylene composites

Property	Unfilled	40 wt % fine mica(a) Powder blend	40 wt % fine mica(a) Melt compounded	50 wt % coarser mica(b) Powder blend	50 wt % coarser mica(b) Melt compound
Tensile strength, MPa (ksi)	28 (4.1)	34 (4.9)	35 (5.1)	36 (5.2)	29 (4.2)
Flexural modulus, GPa (10^6 psi)	1.372 (0.199)	7.033 (1.020)	6.640 (0.963)	11.38 (1.650)	9.928 (1.440)
Flexural strength, MPa (ksi)	29 (4.2)	57 (8.3)	59 (8.5)	63 (9.2)	52 (7.5)

(a) Suzorite 325-H, phlogopite. (b) Suzorite 60-S, phlogopite.

Source: Ref 19, p 112

Performance of selected polyester composites in fire tests

	System I	II	III
Material			
Resin	100(a)	100(b)	100(c)
Alumina trihydrate	100	100	100
Antimony oxide	...	5	...
Ferrous oxide	...	...	5
Test method and property			
ASTM E 162			
Flame-spread index	75	7	7
ASTM E 84			
Flame spread	64	23	25
Smoke emission	608	270	268
NBS chamber			
Flaming mode			
Max density	203	433	264
90-s density	2.5	18	11
240-s density	162	245	128
Nonflaming mode			
Max density	481	400	350
90-s density	1	1	5
240-s density	16	45	50

(a) Orthophthalic resin. (b) HET acid resin A, 26% Cl. (c) HET acid resin B, 26% Cl.

Source: Ref 6, p 95

Properties of polysulfone resin and fiber-resin composites

Properties	Neat resin	30% glass fiber-resin	30% carbon fiber-resin
Physical			
Specific gravity	1.24	1.45	1.37
Density, g/cm³ (lb/in.³)	1.24 (0.045)	1.45 (0.052)	1.37 (0.049)
U.L. electrical rating, °C (°F)	160 (320)	...	...
Deflection temperature at 1800 kPa (264 psi), °C (°F)	175 (345)	185 (365)	185 (365)
In-service temperature, °C (°F)	150-205 (300/400)	...	...
U.L. flammability rating	94 V-0	94 V-0	94 V-0

Source: Ref 6, p 546

Electrical properties of glass-polymer composites

Volume resistivity, Ω·m		
50% relative humidity	10^{10}-10^{12}	
Dielectric strength, V/μm (V/mil)		
Short-time, 3.2 mm (1/8 in.)	13.6-16.5	345-420
Step-by-step, 3.2-mm (1/8-in.) increments	10.8-15.4	275-390
Dielectric constant		
60 Hz	5.3-7.3	
1 kHz	4.68	
1 MHz	5.2-6.4	
Dissipation factor		
60 Hz	0.011-0.041	
1 MHz	0.008-0.022	
Arc resistance	120-200	

Source: Ref 6, p 94

Thermosets

General

Interlaminar fracture toughness of carbon fiber-thermoset resins as determined by double-cantilever-beam specimen

Material(a)	Interlaminar fracture toughness, G_{Ic} J/m²	ft·lbf/ft²
5208	80-90	5-6
3502	120-150	8-10
3501-6	150-214	10-15
1504	95-123	7-8
2220-1	256	18
914	220-250	15-17
BP907	324-397	22-27
HST-7	543	37
V378A(b)	72-86	5-6
5245	134-141	9-10

(a) Composites fabricated at 175 °C (350 °F). (b) Postcured at 205 °C (400 °F)

Source: Ref 6, p 98

Properties of unidirectional advanced composites

Property	Boron/ epoxy	Boron/ polymide	S-glass/ epoxy	High-modulus graphite/ epoxy	High-modulus graphite/ polyimide	High-strength graphite/ epoxy(a)	Aramid/ epoxy(b)	High-strength graphite/ epoxy(c)
Reinforcement content, vol %	50	49	72	45	45	70	54	60
Density, g/cm^3 (lb/in.^{3})	2.02 (0.073)	1.99 (0.072)	2.13 (0.077)	1.55 (0.056)	1.55 (0.056)	1.61 (0.058)	1.36 (0.049)	1.58 (0.057)
Tensile strength, MPa (ksi):								
Longitudinal	1370 (199)	1040 (151)	1290 (187)	840 (122)	807 (117)	1500 (218)	1190 (172)	1520 (220)
Transverse	56 (8.1)	11 (1.6)	46 (6.7)	42 (6.1)	15 (2.2)	40 (5.8)	11 (1.6)	55 (8.0)
Tensile modulus, GPa (10^6 psi):								
Longitudinal	201 (29.2)	221 (32.1)	61 (8.8)	190 (27.5)	216 (31.3)	145 (21.0)	84 (12.2)	110 (16.0)
Transverse	22 (3.2)	14 (2.1)	25 (3.6)	6.9 (1.0)	5.0 (0.72)	10 (1.5)	4.8 (0.70)	15 (2.2)
Compressive strength, MPa (ksi):								
Longitudinal	1600 (232)	1090 (158)	820 (119)	883 (128)	652 (94.5)	1700 (247)	290 (42)	1240 (180)
Transverse	123 (17.9)	63 (9.1)	162 (23.5)	197 (28.5)	70 (10.2)	246 (35.7)	65 (9.4)	248 (36.0)
Shear modulus, GPa (10^6 psi)	5.38 (0.78)	7.65 (1.11)	12.0 (1.74)	6.2 (0.9)	4.48 (0.65)	6.9 (1.0)	2.83 (0.41)	4.96 (0.72)
Intralaminar shear strength, MPa (ksi)	63 (9.1)	26 (3.8)	45 (6.5)	61 (8.9)	22 (3.2)	68 (9.8)	28 (4.0)	69 (10.0)
Poisson's ratio:								
Major	0.17	0.16	0.23	0.10	0.25	0.28	0.32	0.25
Minor	0.02	0.02	0.09	...	0.02	0.01	0.02	0.034
Moisture coefficient, 10^{-2}mm (10^{-2} in.):								
Longitudinal	0.0762 (0.003)	0.0762 (0.003)	0.3556 (0.014)	0.0762 (0.003)	0.0762 (0.003)	0.1524 (0.006)	0.2032 (0.008)	0.1524 (0.006)
Transverse	4.267 (0.168)	4.267 (0.168)	3.251 (0.128)	3.277 (0.129)	3.277 (0.129)	3.277 (0.129)	3.835 (0.151)	3.277 (0.129)
Coefficient of thermal expansion, $10^{-6}/°C$ ($10^{-6}/°F$):								
Longitudinal	6.1 (3.4)	4.9 (2.7)	3.8 (2.1)	...	0.0 (0.0)	0.02 (0.01)	-2.88 (-1.60)	0.72 (0.40)
Transverse	30.4 (16.9)	28.4 (15.8)	16.7 (9.3)	33.3 (18.5)	25.4 (14.1)	22.5 (12.5)	56.3 (31.3)	29.5 (16.4)

(a) Union Carbide Thornel 300 fibers. (b) Du Pont Kevlar 49. (c) Hercules AS fibers.

Source: Ref 4, p 36

Properties of low-temperature thermoset matrix composites

Properties	Neat resin	10-40 wt% glass fiber-resin
Physical		
Specific gravity	1.2-1.4	1.6-1.9
Density, g/cm^3(lb/in^3)	1.2-1.4 (0.043-0.051)	1.6-1.9 (0.058-0.069)
In-service temperature °C (°F)	120-150 (250-300)	120-205 (250-400)
Heat deflection temperature at 1820 kPa (264 psi) °C (°F)	50-205 (120-400)	190-205 (375-400)
U.L. rated in-service temperature °C (°F)	180 (356)	...
U.L. flammability	...	...

Source: Ref 6, p 392

Thermosets

Specific

Thermal properties of fiber-epoxy composite materials

Type of material	Coefficient of thermal expansion, 10^{-6}/K: 0°	Coefficient of thermal expansion, 10^{-6}/K: Quasi-isotropic	Thermal conductivity, W/m·K (Btu·ft/h·ft^{2}°F): 0°	Thermal conductivity, W/m·K (Btu·ft/h·ft^{2}°F): Quasi-isotropic
AS graphite-epoxy	-0.1	0.6	7-9 (4-5)	5-7 (3-4)
HMS graphite-epoxy	-0.2	0.3	10-55 (30-32)	28-30 (16-18)
Kevlar 49-epoxy	-1.1	-0.63	0.17 (0.101)	...
S-glass-epoxy	1.9	3.3	3.5 (2)	0.35 (0-2)

Source: Ref 6, p 818

Behavior of carbon-epoxy versus metals under various conditions

Condition	Carbon-epoxy behavior relative to metals
Stress-strain relationship	More linear strain to failure
Notch sensitivity	
Static	Greater sensitivity
Fatigue	Less sensitivity
Transverse properties	Weaker
Mechanical properties variability	Higher
Sensitivity to aircraft hygrothermal environment	Greater
Damage growth mechanism	In-plane delamination instead of through-the-thickness cracks

Source: Ref 6, p 348

Epoxy resin and fiber-resin properties

Property	Epoxy resin systems				Continuous fiber-fabric epoxy resin systems					
	DGEBA	Phenolic-novolac	Cyclo-aliphatic	Glass-resin fiber	Chopped glass fabric resin	S-2 filament winding glass fiber-resin	Kevlar-resin	Carbon-resin	Graphite fiber-resin	Lay-up quartz fabric-resin
Physical										
Specific gravity	1.15	1.24	1.22	1.72	1.79	1.86	1.25	1.46	1.58	1.75
Density, g/cm^3 (lb/in.^{3})	1.15 (0.042)	1.24 (0.045)	1.22 (0.044)	1.72 (0.062)	1.79 (0.065)	1.86 (0.067)	1.25 (0.045)	1.46 (0.053)	1.58 (0.057)	1.75 (0.063)
Service temperature, °C (°F)	80-88 (175-190)	230 (450)	230-260 (450-500)	150 (300)	150 (300)	150 (300)	150 (300) for 1-2 min	150 (300) (300)	150 (300)	150 (300)
Heat-deflection temperature at 1.82 MPa (264 psi), °C (°F)	110 (230)	150-205 (300-400)	150-275 (300-525)	NA ...	NA ...	NA ...	NA ...	NA ...	NA ...	NA ...
TGA, stability temperature, °C (°F) 0% wt loss	200 (390)	...	...	NA	...	...	...	...	..	...
Reinforcement volume	NA	NA	NA	35	49.3	60	62.9	56	54	65 wt %

Source: Ref 6, p 399

Typical properties of molded epoxy composite materials

Fiber		Specific gravity,	Fiber content		Tensile strength		Flexural strength		Flexural modulus		Compressive strength		Impact strength	
Type	Product code	g/cm^3	wt %	vol %	MPa	ksi	MPa	ksi	GPa	10^6	MPa	ksi	J/mm	ft·lbf/in.
Fiberglass	Type E	1.88	63	46	190	27	470	68	28	4.1	290	42	1.6	30
Fiberglass	Type S-2	1.85	63	46	210	30	430	62	30	4.3	260	38	1.7	32
PAN-based carbon	High-strength	1.48	58	49	140	20	330	48	38	5.5	190	28	0.55	10
PAN-based carbon	High-modulus	1.51	58	48	170	25	340	50	55	8.0	210	30	0.70	13
Aramid	Kevlar 49	1.34	53	49	160	23	290	42	21	3.0	150	22	1.8	34
Aramid	Kevlar 29	1.33	53	49	110	16	270	39	19	2.8	130	19	2.1	40

Source: Ref 6, p 161

Properties of typical fiber-epoxy composites and structural metals

Material	Density, g/cm²	Modulus of elasticity, Longitudinal, GPa	Modulus of elasticity, Longitudinal, 10^6psi	Modulus of elasticity, Transverse, GPa	Modulus of elasticity, Transverse, 10^6psi	Shear modulus, G_{LT}, GPa	Shear modulus, G_{LT}, 10^6psi	Poisson's ratio, ν_{LT}	Tensile strength, Longitudinal, MPa	Tensile strength, Longitudinal, ksi	Tensile strength, Transverse, MPa	Tensile strength, Transverse, ksi	Compressive strength, Longitudinal, MPa	Compressive strength, Longitudinal, ksi	Compressive strength, Transverse, MPa	Compressive strength, Transverse, ksi	Shear strength τ^{su}_{LT}, MPa	Shear strength τ^{su}_{LT}, ksi
Unidirectional composites (V_f = 0.6)																		
E-glass	1.94	45	6.5	12	1.7	4.4	0.64	0.25	1000	150	34	5	550	80	140	20	40	6
Kevlar-49	1.30	76	11.0	5.5	0.8	2.1	0.3	0.34	1380	200	28	4	280	40	140	20	55	8
T-300	1.47	132	19.2	10.3	1.5	6.5	0.95	0.25	1240	180	45	6.5	830	120	140	20	62	9
VSB-32	1.63	229	33.2	6.9	1.0	5.5	0.8	0.25	1170	170	41	6	690	100	140	20	75	11
Boron	1.86	274	39.8	15	2.2	52	7.5	0.25	1310	190	34	5	2480	360	310	45	100	15
GY-70	1.61	320	46.4	5.5	0.8	4.1	0.6	0.25	690	100	41	6	620	90	140	20	96	14
Metals																		
2024-T3	2.77	72.3	10.5	72.3	10.5	27.6	4.0	0.31	462	67	455	66	345	50	345	50	276	40
7075-T6	2.80	71.0	10.3	71.0	10.3	27.6	4.0	0.31	544	79	530	77	475	69	475	69	324	47
4130	7.84	207	30.0	207	30.0	82.7	12.0	0.25	655	95	655	95	1100	160	1100	160	380	55

Source: Ref 6, p 178

Typical in-plane stiffness properties of epoxy matrix composite materials

Material	E_{11} GPa	E_{11} 10^6psi	E_{22} GPa	E_{22} 10^6psi	ν_{12}	G_{12} GPa	G_{12} 10^6psi
Carbon-epoxy prepreg (AS4/3501–6)	131.0	19.0	11.2	1.63	0.28	6.52	0.945
Carbon-epoxy prepreg (T300/5208)	153	22.1	11.2	1.63	0.33	7.10	1.03
Boron-epoxy	204	29.6	18.5	2.68	0.23	5.59	0.810
Kevlar 49-epoxy	76.0	11.0	5.50	0.80	0.34	2.30	0.33
E-glass-epoxy	38.6	5.60	8.27	1.20	0.26	4.14	0.60

Source: Ref 6, p 459

Typical mechanical properties of graphite/epoxy composites with various fiber orientations

Fiber type	Ultimate tensile strength, MPa	Ultimate tensile strength, ksi	Tensile modulus (*E*), GPa	Tensile modulus (*E*), 10^6 psi	Ultimate compressive strength, MPa	Ultimate compressive strength, ksi	Compressive modulus (*E*), GPa	Compressive modulus (*E*), 10^6 psi
Chopped fiber:								
Molding compound	352	51	108	15.7	469	68	...	...
Unidirectional (nonwoven):								
High-strength fiber (0°)	1625	236	138	20.0	993	144	113	16.4
High-strength fiber (0°/±45°)	496	72	57	8.3	503	73	50	7.3
Medium-strength fiber (0°)	1455	211	147	21.3	1405	204	131	19.0
Medium-strength fiber (0°/±45°)	503	73	65	9.4	503	73	64	9.3
High-modulus fiber (0°)	1240	180	215	31.2	758	110	177	25.6
Fabric (woven):								
Medium-strength fiber	510	74	70	10.2	510	74	63	9.2

Source: Ref 16, p 6-55

Comparison of 2D and 3D models of Kevlar/epoxy composites in Charpy impact test

	Impact energy - Y			Impact energy - Z		
Model	J	ft·lb	Standard deviation	J	ft·lb	Standard deviation
2D	14.57	10.75	1.06	17.83	13.15	0.92
3D	15.25	11.25	0.35	19.0	14.0	0.71

Source: Ref 22, p 189

Prepreg characteristics, cure conditions, and mechanical properties of SiC/epoxy laminates(a)

Prepreg characteristics	
Resin content	26 wt %
Volatile content	1.0%
Tack (at 23 °C; 73 °F)	Good
Width tolerance	± 0.51 mm (± 0.02 in.)
Filament count (nominal)	5.51/mm (140 in.)
Storage temperature	-18 °C (0 °F)
Typical cure conditions	
Temperature	177 °C (350 °F)
Pressure	345-585 kPa (50-85 psi)
Time	90 min(b)
Mechanical properties	
Tensile strength:	
At room temperature	1579 MPa (229 ksi)
At 127 °C (260 °F)	1309 MPa (190 ksi)
Tensile modulus:	
At room temperature	227 GPa (33 x 10^6 psi)
At 127 °C (260 °F)	227 GPa (33 x 10^6 psi)
Compressive strength:	
At room temperature	2248 MPa (326 ksi)
At 127 °C (260 °F)	1600 MPa (232 ksi)
Horizontal shear strength:	
At room temperature	103 MPa (15.0 ksi)
At 127 °C (260 °F)	62 MPa (9.0 ksi)
Flexural strength:	
At room temperature	2165 MPa (314 ksi)
At 127 °C (260 °F)	2179 MPa (316 ksi)
Flexural modulus:	
At room temperature	221 GPa (32.0 x 10^6 psi)
At 127 °C (260 °F)	207 GPa (30.0 x 10^6 psi)
Density	2.325 g/cm^3 (0.084 lb/in.^{3})

(a) Laminates were fabricated and tested by the University of Dayton under AFML Contract (F33615-78-C-5172). (b) Postcure, 4 h at 190 °C (375 °F).

Source: Ref 16, p 6-45

Comparison of 2D (with and without a circular hole) and 3D models of Kevlar/epoxy composites in four-point bending test

	Maximum flexural strength		Fracture toughness	
Model	MPa	ksi	J/cm^3	in.·lb/in.^{3}
2D without hole	30.9	4.48	506	273
2D with hole	26.1	3.79	136	73.4
3D with hole	17.8	2.58	95.4	51.5

Source: Ref 22, p 189

Typical properties of boron/epoxy laminates(a)

Temperature °C	°F	0° tensile strength MPa	ksi	0° tensile modulus GPa	10^6 psi	0° failure strain %	90° tensile strength MPa	ksi	0° compressive strength MPa	ksi
Unidirectional construction (0°)										
-55	-67	1240-1380	180-200	205-220	30-32	0.59-0.64	68-90	10-13	3105-3310	450-480
24	75	1275-1450	185-210	205-215	30-31	0.62-0.69	55-83	8-12	2585-3070	375-445
125	260	1275-1345	185-195	200-205	29-30	0.63-0.66	41-55	6-8	1860-2070	270-300
190	375	1105-1170	160-170	195-200	28-29	0.56-0.61	21-28	3-4	585-1000	85-145
Cross-plied construction (0°/90°)										
-55	-67	690-760	100-110	130	18.9	0.54-0.56	690-760	100-110	...	...
24	75	690-760	100-110	125	18.0	0.57-0.61	690-760	100-110	...	...
125	260	655-690	95-100	120	17.5	0.58	655-690	95-100	...	...
190	375	580-655	84-95	105-115	15.5-16.5	0.56-0.58	580-655	84-95	...	...

(a) Nominal fiber content, 55 vol %.

Source: Ref 16, p 6-46

Phenolic glass fiber-resin and fiber-resin properties

Physical and service properties	Phenolic resin	Fiber-resin composites: Molding compound glass fiber-resin	Tape wrapping fabric-resin: E-glass fabric-resin	Carbon fabric-resin	Graphite fabric-resin
Physical					
Specific gravity	1.28	1.95	1.90	1.45	1.42
Density, g/cm^3(lb/in.^{3})	1.28 (0.046)	1.95 (0.070)	1.90 (0.069)	1.45 (0.052)	1.42 (0.051)
Service temperature, °C (°F)	150-230 (300-450)	150 (300)	538 (1000)	3038 (5500)	3038 (5500)
Heat deflection temperature at 1820 MPa (264 psi), °C (°F)	120-315 (250-600)	na	na	na	na
Flammability (U.L.)	94 V-1	94 V-0	94 V-0	...	...
Thermogravimetric analysis—stability temperature at 5-10% wt loss, °C (°F)	230 (450)	na	...	...	...
Reinforcement, wt%		35-45	63	55	55

Source: Ref 6, p 381

Properties of polyimide thermoset resin and fiber-resin composites

Properties	Resin	50% glass fiber-resin
Physical		
Heat deflection temperature at 1820 kPa (264 psi), °C (°F)	305–360 (582–680)	350 (660)
UL rating, °C (°F)	... ...	... ...
In-service temperature, °C (°F)	260–370 (500–700)	260 (500)
Specific gravity	1.43	1.65
Density, g/cm^3 (lb/in.^{3})	1.43 (0.052)	1.65 (0.060)

Source: Ref 6, p 373

Properties of a high-modulus graphite-epoxy lamina

E_1 = 170 GPa (25.0 × 10^6 psi)
E_2 = 12 GPa (1.7 × 10^6 psi)
G_{12} = 4.5 GPa (0.65 × 10^6 psi)
ν_{12} = 0.30
ρ = 0.056
α_1 = −0.54 × 10^{-6}/K
α_2 = 35.1 × 10^{-6}/K
σ_L^{tu} = 758 MPa (110.0 ksi)
σ_T^{tu} = 28 MPa (4.0 ksi)
σ_L^{u} = 62 MPa (9.0 ksi)
σ_L^{cu} = 758 MPa (110.0 ksi)
σ_T^{cu} = 138 MPa (20.0 ksi)
V_f = 0.6

Note: Ply thickness, 0.13 mm (0.0052 in.)

Source: Ref 6, p 223

Metal Matrix Composites

METAL MATRIX COMPOSITES

Metal Matrix Composites

General

Classification of most composite system interfaces

Class I[a]	Class II[b]	Class III[c]
Copper/tungsten	Copper (chromium)/ tungsten	Copper (titanium)/ tungsten
Copper/alumina	Eutectics	Aluminum/carbon (<700 °C)
Silver/alumina	Niobium/tungsten	Titanium/alumina
Aluminum/BN-coated boron	Nickel/carbon	Titanium/boron
Magnesium/boron	Nickel/tungsten[e]	Titanium/silicon carbide
Aluminum/boron[d]		Aluminum/silica
Aluminum/stainless steel[d]		
Aluminum/silicon carbide[d]		

[a]Filament and matrix do not react and are mutually insoluble. [b]Filament and matrix do not react, but exhibit some solubility. [c]Filament and matrix react to form surface coating. [d]Pseudo–Class I system. [e]Becomes reactive at lower temperatures with formation of Ni_4W.

Source: Ref 16, p 7-6

Coefficients of thermal expansion of matrices and reinforcing fibers

Matrix	Expansion, $10^{-6}/°C$	Filament	Expansion, $10^{-6}/°C$
Aluminum	23.9	Boron	6.3
Titanium	8.4	Borsic	6.3
Iron	11.7	SiC	4.0
Nickel	13.3	Alumina	8.3
Silcon	2.5	Carbon(a)	~0
Copper	17	SiO_2	~0
Silver	19	Tungsten	4.5
Magnesium	26	Al_2O_3	8
		Beryllium	12

(a) Parallel to basal plane.

Source: Ref 16, p 7-11

Some fiber/metal combinations

Matrix	C	B	Al_2O_3	SiC	Steel	Be	W	Ni_3Nb
Ti		+	+	+		+		
Al	+	+	+	+	+	+		
Ni	+		+	+			+	+

Source: Ref 16, p 7-7

Comparison of fiber-matrix reactions for various matrix materials

Annealing temperature °C	Annealing temperature °F	Matrix	No. of compositions investigated	Relative number of cases, %: Recrystal-lization	Intermetallic compound	Diffusion penetration	No recrystal-lization	No reaction
1200	2190	Nickel-based	27	93	55	...	7	4
		Cobalt-based	29	10	83	12	90	10
		Iron-based	30	3	30	...	97	70
1300	2370	Nickel-based	27	96	63	...	4	4
		Cobalt-based	19	21	84	...	79	10
		Iron-based	30	20	80	3	80	13

Source: Ref 6, p 878

Strength properties of several types of fiber-reinforced metals

Filament	Matrix	Fabrication technique(a)	Test temperature, °C	V_f	L/d_f	....	Strength properties(b): S_c, MPa	S_c, ksi	$\bar{S}_f$, MPa	$\bar{S}_f$, ksi	$\bar{S}_f^*$, MPa	$\bar{S}_f^*$, ksi	β
Metallic filaments													
Stainless steel	Al	HPF	RT	0.11	250	....	179	26.9	1515	220	1405	204.0	0.93
W	Cu	VI	RT	0.63	∞	....	772	112.0	1240	180	1215	176.0	0.98
W	Cu	VI	250	0.69	∞	....	855	124.0	1240	180	1225	178.0	0.99
W	Cu	VI	250	0.56	40	....	558	81.0	1240	180	972	141.0	0.78
W	Cu	VI	250	0.51	20	....	490	71.0	1240	180	938	136.0	0.76
W	Cu	VI	250	0.57	10	....	386	56.0	1240	180	655	95.0	0.53
W	Cu	VI	RT	0.768	∞	....	1755	254.9	2255	327	2250	326.0	~1.00
W	Cu	VI	RT	0.357	75	....	827	120.0	2255	327	2220	322.0	0.99
Oxide filaments													
Al-coated E-glass	Al	HP	RT	0.50	∞	....	310	45.0	896	130	593	86.0	0.66
Al-coated E-glass	Al	HP	480	0.50	∞	....	178	28.85	731	106	342	49.6	0.47
Al-coated E-glass	Al	HP	540	0.50	∞	....	135	19.63	710	103	263	38.1	0.37
Al-coated E-glass	Al	VI	RT	0.50	∞	....	121	17.5	896	130	230	33.4	0.26
Al-coated E-glass	Al	VI	480	0.50	∞	....	96	13.89	731	106	189	27.48	0.26
Al-coated SiO_2	Al	HP	RT	0.48	∞	....	871	126.3	3035	440	1725	250.0	0.57
Al-coated SiO_2	Al	HP	100	0.48	∞	....	958	139.0	3035	440	1935	281.0	0.65
Al-coated SiO_2	Al	HP	500	0.48	∞	....	288	41.8	1795	260	593	86.0	0.33
Pb-coated E-glass	Pb	VI	RT	0.08	∞	....	103	15.0	1515	220	1055	153.0	0.70

(a) HPF = hot pressed filaments on powdered matrix; VI = vacuum infiltration; HP = hot pressed metal-coated filaments. (b) S_c = composite tensile strength; $\bar{S}_f$ = average strength of fibers tested individually; $\bar{S}_f^*$ = average strength of fibers in the matrix during fracture; β = effective fiber strength factor = $\bar{S}_f^*/\bar{S}_f$.

Source: Ref 16, p 7-2

Room-temperature tensile and specific strengths of various fiber-reinforced composites

Composition		Density, lb/in.^{3}		Fiber content,[b]	Composite		
Matrix	Fiber	Matrix	Fiber	vol %	Tensile strength,[c] ksi	Strength/density, 10^3 in.	Source
Al	B/W	0.097	0.095	50	110	1150	United Aircraft Research Laboratory
	B	···	···	10	43	453	GE-AETD
	Steel	0.097	0.282	25	173	1210	Harvey Aluminum Co.
	Be	0.097	0.067	40	80	830	North American
	SiO_2	0.097	0.079	48	118	1340	Rolls-Royce Ltd.
	Al_2O_3	0.097	0.143	35	161	1425	GE-SSL
	B_4C	0.097	0.091	10	29	302	GE-SSL
	$CuAl_2$	0.097	0.157	50	39	307	United Aircraft Research Laboratory
	Al_3Ni	0.097	0.143	10	48	470	United Aircraft Research Laboratory
Nb	Nb_2C	0.299	0.286	31	172	570	United Aircraft Research Laboratory
Ni	B/W	0.320	0.095	75	384	1470	General Technology, Inc.
	W	0.320	0.695	9.4	61.4	173	Battelle N.W.
	W	···	···	40	161	344	NGTE (England)
	Al_2O_3	0.320	0.143	19	171	600	GE-SSL
	C	0.320	0.054	56	80	467	Union Carbide Corp.
Al-10.2Si	Al_2O_3	0.096	0.143	15[d]	40.7	395	Melpar, Inc.
Ni-20Cr	Al_2O_3	0.308	0.143	9	255	870	Horizons, Inc.
NiCr	W	···	···	22	73	185	Clevite Corp.
Fe	Al_2O_3	0.284	0.143	36	237	1017	Horizons, Inc.
Ta	Ta_2C	0.598	0.544	29	155	267	United Aircraft Research Laboratory, twice solidified
	Ta_2C	0.598	0.544	29	118	203	United Aircraft Research Laboratory, cold swaged, 67% reduction
Ag	Si_3N_4	0.378	0.115	15[d]	40	119	Explosive Research Development Est. (England)
	Al_2O_3	0.378	0.143	24	232	720	GE-SSL
	Steel	···	···	44	65	191	MIT
Ti	Mo[e]	···	···	20	96	457	Clevite Corp.
Cu	W	···	···	77	255	420	NASA-Lewis
Co	W	···	···	30	107	244	Clevite Corp.
	Mo	···	···	17	52	156	Clevite Corp.
316 stainless steel	W	···	···	18	58.6	175	Clevite Corp.

[a]Early and developmental metal-matrix composite work. [b]Unidirectional fiber orientation except where otherwise noted. [c]Highest reported values. [d]Random fiber orientation. [e]Short, discontinuous fibers and whiskers.

Source: Ref 16, p 7-10

Typical mechanical properties of some metal-matrix composites

Fiber	Matrix	Reinforcement, vol %	Density, g/cm^3 [a]	Longitudinal tensile strength, MPa[b]	Longitudinal modulus, GPa[b]	Transverse tensile strength, MPa[b]	Transverse modulus, GPa[b]
G T50	201 Al	30	2.380	620	170	50	30
G T50	201 Al	49		1120	160	...	...
G GY70	201 Al	34	2.380	660	210	30	30
G GY70	201 Al	30	2.436	550	160	70	40
G HM pitch	6061 Al	41	2.436	620	320	...	...
G HM pitch	AZ31 Mg	38	1.827	510	300	...	...
B on W, 142-μm fiber	6061 Al	50	2.491	1380	230	140	160
Borsic	Ti	45	3.681	1270	220	460	190
G T75	Pb	41	7.474	720	200	...	...
G T75	Cu	39	6.090	290	240	...	...
FP	201 Al	50	3.598	1170	210	(140)	140
SiC	6061 Al	50	2.934	1480	230	(140)	140
SiC	Ti	35	3.931	1210	260	520	210
SiC whisker	Al	20	2.796	340	100	340	100
B_4C on B	Ti	38	3.737	1480	230	>340	>140
G T75	Mg	42	1.799	450	190	...	...
G HM	Pb	35	7.750	500	120	...	...
G T75	Al-7% Zn	38	2.408	870	190	...	...
G T75	Zinc	35	5.287	770	120	...	...
G T50	Nickel	50	5.295	790	240	...	...
G T75	Nickel	50	5.342	828	310	30	40
G (81.3 μm)	2024 Al	50	2.436	760	140	...	...
G (142 μm)	2024 Al	60	2.436	1100	180	...	...
Superhybrid	Grafitic	60	2.048	860	120	220	60
Superhybrid	S-glass	60	2.159	740	80	190	30
Superhybrid	Kevlar	60	1.799	700	80	190	10

[a]To convert g/cm^3 to $lb/in.^3$, divide by 27.68. [b]To convert MPa to psi, multiply by 145; to convert GPa to psi, multiply by 145 000.

Source: Ref 16, p 7-2

Factors affecting the realizable strengths of fiber-reinforced metals

Factor	Ideal case
Fiber parameter-strength	
Initial strength	Should be at a maximum.
Scatter in strength	Variation in strength between individual fibers should be minimized.
Strength retention	Should be maximized during handling, fabrication, etc.
Fracture statistics	All fibers should be loaded evenly. Also depends on points discussed above for initial strength and scatter in strength.
Fiber parameter-shape and size	
Taper	Should be minimized
Diameter	Should be minimized. Fiber strength usually increases with decreasing diameter. Statistically, more fibers can be packed in per unit volume for small diameters.
Length	Should be optimized in terms of length greater than critical length. Note: strength also increases with decreasing length.
Variation in fiber diameter	Should be as uniform as possible
Cross section	Not all fibers are circular, hence cross-sectional shape will affect maximum packing density and stress distribution.
Fiber parameters-relation to matrix	
Orientation	Fibers should be parallel to tensile axis. However, properties can be tailor-made by varying orientation.
Distribution	Fibers should be uniformly distributed in the matrix.
Overlap	Short fibers must have a minimum overlap distance; otherwise, crack in matrix will propagate through regions between fiber ends.
Fiber-fiber	Fibers must be separated by matrix; otherwise, contact points serve as stress concentrators.
Matrix parameters	
Mechanical	Rate of work hardening should be optimized, properties relative to fibers—i.e., E_m, E_f.
Chemical	Should wet fibers, but not react to weaken fibers, particularly at high temperatures; oxidation and corrosion resistance are important.
Thermal	Coefficient of expansion should be similar to that of fiber, recrystallization, temperature.
Fabrication parameters	
Fiber production	Factors include cost to obtain optimum values cited under "Fiber parameters."
Fiber handling	Technique depends on length and length-to-diameter ratio, and on properties desired in composite (see "Fiber parameters-relation to matrix"); surface treatment (metallizing).
Incorporation into matrix	Many techniques—powder metallurgy, infiltration, impact forming, etc. Fibers must be wetted and bonded to matrix. Must form composite of theoretical density.
Composite shaping	Extrusion, forging, pressing—depends on fiber brittleness (and size).
Joining	Chemical, pressure, electron welding, and diffusion bonding techniques are more effective than mechanical joining.

Source: Ref 16, p 7-2

Methods used for fabrication of metallic-matrix whisker composites

Method used to combine constituents	Matrix	Whisker(a) Composition	Alignment	Coating	Consolidation process(b)
Deposition (molecular)					
Chemical vapor deposition	Ni or NiCr	Al_2O_3, SiC	R	C	H.P. or L.P.H.P.
Electrodeposition	Ni	Al_2O_3	R	C	H.P.
	Ni	Al_2O_3, SiC	R	C, U	...
	Ni	Al_2O_3	A	C, U	As deposited or hot forged
	Ni	Al_2O_3	A	C	H.P.
Electroplate	Ni	Al_2O_3, SiC	A, R	C	Cold and hot rolling, H.P.
Electroform	Ni	Al_2O_3	A	C	...
Liquid (matrix)					
Alloy (eutectic)	Al, Nb	$NiAl_3$, Nb_2C	A	U	Unidirectional solidification
	Resin	Al_2O_3	A	C	Polymerize matrix
	Ag	Al_2O_3	A	C	Solidification of matrix
	Cu, Ni-alloys	Al_2O_3	A	C	Solidification of matrix
Infiltration	Al	B_4C	A	C	Solidification of matrix
	Al-alloys	Al_2O_3	A	C	Solidification of matrix
Melting of powdered matrix	Fe, NiCr	Al_2O_3	R	C, U	L.P.H.P.
Casting of melt and whiskers	Co-alloys	Al_2O_3, SiC	R	C, U	Solidification of melt
	Ag, Fe, Ni	Si_3N_4	A	U	Dry, burn off organic carrier, H.P.
Spin, extrude, draw slurry of whisker matrix powder, and carrier solution	Al	SiC	A	U	Dry, burn off organic carrier, L.P.H.P.
	Al-alloy	SiC	A	U	Dry, burn off organic carrier, H.P.
	Cu, Al, Mg	SiC	A	U	Dry, burn off, sinter, H.P.
	Al-alloy	SiC	A	U	Dry, burn off, H.P.
	Ag-alloy	Si_3N_4	R	U	L.P.H.P.
Filter slurry or settle out whiskers and matrix	Cu, Al-alloy	Al_2O_3, SiC	R	U	H.P.
	NiCr, Al-alloy	SiC	A	C, U	L.P.H.P.
	NiCr, Al-alloy	Al_2O_3, SiC	A, R	C, U	L.P.H.P.
	Mg-alloy	Al_2O_3, SiC	A, R	C, U	L.P.H.P.
Impregnation into whisker strands and tapes	Resin	SiC	A	U	H.P. and squeeze out excess resin
	Resin	Al_2O_3	A	U	H.P. and squeeze out excess resin
	Resin	Si_3N_4	A	U	H.P. and squeeze out excess resin
Solid state					
Powder matrix and whiskers	Ni	Al_2O_3	R,	C	H.P.
	Al	Si_3N_4	A	U	Extrusion
	Ni, Ti	Al_2O_3, SiC	R	U	High-energy-rate forming
Deposit whiskers on Al-foil	Al	SiC	A	U	Diffusion bonding

(continued)

Methods used for fabrication of metallic-matrix whisker composites (continued)

Method used to combine constituents	Matrix	Forming or shaping process	Remarks
Deposition (molecular)			
Chemical vapor deposition	Ni or NiCr	Hot roll	SiC whiskers reacted chemically with matrix
Electrodeposition	Ni	Hot roll	Problem of voids and low whisker content
	Ni	Hot roll	Problem of voids and low whisker content
	Ni	Hot roll	Align whiskers via flow in electroplating bath
	Ni	Hot roll	Problem of consolidation and whisker breakage
Electroplate	Ni	Hot roll	SiC whiskers reacted with matrix, also whisker breakage
Electroform	Ni	Hot roll	Excellent properties, limited to very small specimens
Liquid (matrix)			
Alloy (eutectic)	Al, Nb	Extrude, roll	Whisker content fixed by alloy composition
	Resin	Extrude, roll	Achieved high-strength composites
	Ag	Extrude, roll	Achieved high-strength composites
	Cu, Ni-alloys	Extrude, roll	Achieved varying degress of success; coating stability a problem
Infiltration	Al	Extrude, roll	Achieved varying degress of success; coating stability a problem
	Al-alloys	Extrude, roll	Achieved varying degress of success; coating stability a problem
Melting of powdered matrix	Fe, NiCr	Extrude, roll	Problems of interfacial reactions and whisker segregation
Casting of melt and whiskers	Co-alloys	Extrude, roll	Whiskers concentrated at grain boundaries, low whisker content
	Ag, Fe, Ni	Extrude, roll	Excellent whisker alignment and packing
Spin, extrude, draw	Al	Extrude, roll	Excellent whisker alignment and packing
slurry of whisker	Al-alloy	Extrude, roll	Excellent whisker alignment, fabrication of complex shapes
matrix powder, and	Cu, Al, Mg	Extrude, roll	Problem of matrix porosity and wetting
carrier solution	Al-alloy	Extrude, roll	Excellent whisker alignment
	Ag-alloy	Extrude, roll	Excellent whisker alignment, problem carbon burn-off
Filter slurry or settle	Cu, Al-alloy	Hot roll	Little whisker alighnment, much whisker breakage
out whiskers and	NiCr, Al-alloy	Clad and roll	Ni-coated whiskers were aligned magnetically during settling in slurry
matrix	NiCr, Al-alloy	Clad and roll	Ni-coated whiskers were aligned magnetically during settling in slurry
	Mg-alloy	Hot extrusion	Successfully rolled composites with aligned whiskers
Impregnation into	Resin	Hot extrusion	Achieved excellent alignment, little whisker breakage
whisker strands	Resin	Hot extrusion	Achieved excellent alignment, little whisker breakage
and tapes	Resin	Hot extrusion	Achieved excellent alignment, little whisker breakage
Solid state			
Powder matrix and whiskers	Ni	Hot extrusion	Many whiskers broken
	Al	Hot extrusion	Extended long rods, but many whiskers broken
	Ni, Ti	Hot extrusion	Chemical reactions minimized, but whiskers still broken
Deposit whiskers on Al-foil	Al	Hot extrusion	Achieved excellent alignment of whiskers

(a) R = random whisker alignment; A = whiskers unidirectonally aligned; C = coated whiskers; U = uncoated whiskers. (b) H.P. = hot pressing; L.P.H.P. = liquid phase hot pressing.

Source: Ref 16, p 7-8

Carbon/Metal

Typical properties of 55MSI graphite/6061 aluminum composites

Density, g/cm³	Density, lb/in.³	Reinforcement content, vol %	Fiber orientation	Modulus of elasticity, GPa	Modulus of elasticity, 10^6 psi
2.35	0.085	34	0°	182.2 ± 6.6	26.43 ± 0.95
2.35	0.085	34	90°	33	4.8

Source: Ref 27, p 97

Comparative properties of various graphite fiber reinforced metals

Composite	Fiber content, vol %	Strength, ksi	Modulus of elasticity, 10^6 psi	Density, lb/in.^{3}	Strength/density, 10^6 in.	Modulus/density, 10^6 in.
Graphite[a]/lead	41	104	29.0	0.270	0.385	107.0
Graphite[b]/lead	35	72	17.4	0.280	0.260	62.3
Graphite[a]/zinc	35	110.9	16.9	0.191	0.580	88.5
Graphite[a]/magnesium	42	65	26.6	0.064	1.016	393.7

[a]Thornel 75 fiber. [b]Courtaulds HM fiber.

Source: Ref 16, p 7-11

Typical properties of graphite-magnesium castings(a)

Fiber type	Fiber content/orientation	Casting	Fiber preform method	Tensile strength, 0° GPa	Tensile strength, 0° 10^6 psi	Modulus of elasticity, 0° GPa	Modulus of elasticity, 0° 10^6 psi	Tensile strength, 90° GPa	Tensile strength, 90° 10^6 psi	Modulus of elasticity, 90° GPa	Modulus of elasticity, 90° 10^6 psi	CTE 10^{-6}/K
P55	40%/0°	Rod	Filament wound	0.72	0.105	172	25	...	...	...	...	...
P100	35%/0°	Rod	Filament wound	0.72	0.105	248	36	...	...	...	...	...
P75	40%/± 16° plus 9%/90°	Hollow cylinder	Filament wound	0.45	0.065	179	26	0.061	0.089	86	12.5	1.3
P100	40%/± 16°	Hollow cylinder	Filament wound	0.56	0.081	228	33	0.38	0.055	30	4.4	-0.07
P55	40%/0°	Plate	Prepreg	0.48(b)	0.070(b)	159	23	0.02	0.003	21	3	2.3
P55	30%/0° plus 10%/90°	Plate	Prepreg	0.28	0.04	83	12	0.10	0.015	34	5	4.5
P55	20%/0° plus 20%/90°	Plate	Prepreg	8.45(b)	1.225(b)	90	13	0.24	0.035	90	13	...

(a) All materials contain pitch-base fibers. (b) Equivalent 0° tensile strength at 400 °C (750 °F)

Source: Ref 6, p 871

Tensile strengths and moduli for two graphite/Al composites

Composite	Fiber loading, vol %	Tensile strength MPa	Tensile strength ksi	Tensile modulus GPa	Tensile modulus 10^6 psi	Wire diameter mm	Wire diameter in.
VS0054/201 Al	48–52	1035–1070	150–155	345	50	0.64[a]	0.025[a]
GY70SE/201 Al	37–38	793–827	115–120	207	30	0.71[b]	0.028[b]

[a]Two strand. [b]Eight strand.

Source: Ref 16, p 7-18

Property data for a commercial graphite/aluminum composite

	Modulus of elasticity				Tensile strength			
	Longitudinal		Transverse		Longitudinal		Transverse	
Fiber(a)	GPa	10^6 psi	GPa	10^6 psi	MPa	ksi	MPa	ksi
P55	207-221	30-32	28-41	4-6	517-621	75-90	28-48	4-7
P75	276-296	40-43	28-41	4-6	621-724	90-105	28-48	4-7
P100	379-414	55-60	28-41	4-6	552-834	80-121	28-48(b)	4-7(b)

(a) Union Carbide Thornel fibers. (b) Estimated.

Ref 16, p 7-18

Silicon Carbide/Metal

Tensile strength of SCS-2-Al

Fiber orientation	No. of plies	Tensile strength MPa	Tensile strength ksi	Tensile modulus GPa	Tensile modulus 10^6 psi	Total strain	Poisson's ratio	Coefficient of thermal expansion, 10^{-6}/K
0°	6,8,12	1462	212	204.1	29.6	0.89	0.268	6.6
90°	6,12,40	86.2	12.5	118.0	17.1	0.08	0.124	21.3
$[0°/90°/0°/90°]_s$	8	673	97.6	136.5	19.8	0.90	...	...
$[0°_2/90°/0°]_s$	8	1144	166.0	180.0	26.1	0.92	...	...
$[90°_2/0°/90°]_s$	8	341.3	49.5	96.5	14.0	1.01	...	...
±45°	8,12,40	309.5	44.9	94.5	13.7	10.6	0.395	...
$[0°/\pm45°/0°]_{s+2s}$	8,16	800.0	116	146.2	21.2	0.86	...	...
$[0°/\pm45°/90°]_s$	8	572.3	83.0	127.0	18.4	1.0	...	...

Source: Ref 6, p 862

Data on investment-cast SCS-Al

Fiber orientation	Fiber, vol%	Ultimate tensile strength MPa	Ultimate tensile strength ksi	ROM, %	Tensile modulus GPa	Tensile modulus 10^6 psi	ROM, %	Ultimate compressive strength MPa	Ultimate compressive strength ksi	Compressive modulus GPa	Compressive modulus 10^6 psi
$0°_3/90°_6/0°_3$	33	458.5	66.5	75	122.0	17.7	107	1378.9	200	...	...
$90°_3/0°_6/90°_3$	33	584.0	84.7	95	124.8	18.1	110	1378.9	200	...	...
0°	34	1034.2	150	85	172.4	25	100	1896.1	275	186.2	27.0

Source: Ref 6, p 864

Room-temperature tensile strength of silicon carbide-aluminum alloy composites

		Ultimate tensile strength			
		Base		Reinforced	
Material	Fiber, vol%	MPa	ksi	MPa	ksi
Pure aluminum	11	59	8.6	235	34.1
6061-T6	16	300	43.5	441	64.0
2024-T4	20	470	68.2	565	81.9

Source: Ref 6, p 908

Reactivity of metals with SiC fibers

Metal	Melting temperature, °C	Reactivity[a] in 1 h in H_2 gas at temperature, °C, of: 450	530	620	650	750	850	950	1000	1100	1200
Al	660	○	○	○							
Ag	961				○	○	○				
Cu	1083					○	○	○			
Ni	1453								○	x	x
Co	1495								○	△	x
Fe	1537								○	○	○
Ti	1668								△	△	△
Cr	1875								○	○	△
Mo	2610								○	○	○

[a]Key: ○ = no reaction ; △ = slight reaction; x = significant reaction.

Source: Ref 16, p 7-7

Tensile data for SiC-whisker-reinforced aluminum alloy

Fiber content, vol %	Yield strength (0.2%) MPa	ksi	Standard deviation	Range of measurement	Tensile strength MPa	ksi	Standard deviation	Range of measurement	Modulus of elasticity GPa	10^6 psi	Standard deviation	Range of measurement
0	210	30.5	3.8	9.5	297	43.1	1.8	3.5	71.9	10.4	4.5	13
0.12	266.5	38.7	4.2	10.6	359	52.1	33.6	85.6	95.3	13.8	1.6	6
0.16	264.5	38.4	0.6	1.6	374	54.2	8.0	23.0	90.0	13.1	3.7	9
0.20	298	43.2	4.0	10.2	383.6	55.6	15.2	38.8	111.0	16.1	5.0	13

Source: Ref 6, p 890

Yield strength and ultimate tensile strength of aluminum alloy reinforced with SiC whiskers at different temperatures

Fiber, vol%	350 °C (660 °F) Yield strength MPa	ksi	Ultimate tensile strength MPa	ksi	300 °C (570 °F) Yield strength MPa	ksi	Ultimate tensile strength MPa	ksi	250 °C (480 °F) Yield strength MPa	ksi	Ultimate tensile strength MPa	ksi
Polycrystalline alumina												
0	35	5.1	55	8.0	...	...	70	10.2	70	10.2	115	16.7
0.05	54	7.8	63	9.1	79	11.5	88	12.8	112	16.2	134	19.4
0.12	68	9.9	74	10.7	...	...	...	...	...	...	...	...
0.20	110	16.0	112	16.2	154	22.3	155	22.5	186	27.0	198	28.7
SiC whiskers												
0	35	5.1	55	8.0	...	...	70	10.2	70	10.2	115	16.7
0.12	94	13.6	124	18.0	153	22.2	180	26.1	197	28.6	226	32.8
0.16	120	17.4	147	21.3	...	...	...	...	...	...	...	...
0.20	163	23.6	184	26.3	207	30.0	235	34.1	268	38.9	284	41.2

Source: Ref 6, p 891

Compression strength of SCS-2-Al

Direction	Plies	Load N	Load lb	Stress MPa	Stress ksi	Tensile modulus GPa	Tensile modulus 10^6 psi	Poisson's ratio
0°	12	36 000	8 100	2 647	383.9	...	...	...
		38 250	8 600	2 708	392.7	...	...	...
		38 700	8 700	2 739	397.3	...	...	...
		40 500	9 100	2 878	417.4	...	...	...
		48 900	11 000	3 296	478.0	212.4	30.8	0.241
		53 100	11 940	3 689	535.0	222.7	32.3	...
90°	12	4 220	948	294.4	42.7	104.8	15.2	...
		4 380	985	300.6	43.6	116.5	16.9	0.174
		4 270	960	294.4	42.7	...	...	...
		4 230	950	292.3	42.4	113.1	16.4	0.173
		3 960	890	273.0	39.6	115.8	16.8	...
		3 780	850	259.2	37.6	124.1	18.0	...
90°	40	13 480	3 030	293.7	42.6	...	...	...
		14 610	3 285	294.4	42.7	131.7	19.1	0.136
		13 280	2 985	290.0	42.0	102.7	14.9	...
		13 430	3 020	287.5	41.7	108.9	15.8	...
		13 520	3 040	294.4	42.7	115.1	16.7	...
		13 680	3 075	297.2	43.1	142.0	20.6	0.158

Source: Ref 6, p 863

Shear strength of SCS-2-Al

Test temperature °C (°F)	Failure stress MPa	Failure stress ksi	Shear strength MPa	Shear strength ksi	Shear modulus MPa	Shear modulus ksi
Room temperature	455.7	66.1	113.8	16.5	42.5	6.17
	452.3	65.6	113.1	16.4	39.5	5.73
	479.2	69.5	120.0	17.4	39.8	5.77
	422.6	61.3	105.5	15.3	40.3	5.85
Average	**452.5**	**65.6**	**113.1**	**16.4**	**40.5**	**5.88**
75 (165)	437.1	63.4	109.6	15.9	40.2	5.83
	434.4	63.0	108.9	15.8	43.2	6.27
	424.7	61.6	106.2	15.4	41.7	6.05
Average	**432.1**	**62.6**	**108.2**	**15.7**	**41.7**	**6.05**
−55 (−65)	501.2	72.7	125.5	18.2	44.5	6.46
	482.6	70.0	120.7	17.5	39.6	5.75
	453.0	65.7	113.1	16.4	39.6	5.75
Average	**479.0**	**69.4**	**119.8**	**17.3**	**41.3**	**5.98**

Source: Ref 6, p 863

Typical properties of SiC whisker-reinforced aluminum alloy sheet

Sheet thickness mm	Sheet thickness in.	Test specimen orientation	Ultimate tensile strength MPa	Ultimate tensile strength ksi	Yield strength(a) MPa	Yield strength(a) ksi	Elongation (e), %	Young's modulus, E GPa	Young's modulus, E 10^6 psi	Fracture toughness, K_c MPa $\sqrt{m}$	Fracture toughness, K_c ksi $\sqrt{in.}$
2.54	0.100	Longitudinal (Along roll direction)	718	104	573	83.1	5.3	114	16.5	55	50
2.54	0.100	Transverse (90° to roll direction)	559	81.0	386	56.4	8.5	95	14	59	54

(a) 0.2% offset

Source: Ref 6, p 901

Notched strengths of SCS-2-Al

Specimen	Average gross stress, RT MPa	ksi	Average net stress, RT MPa	ksi	Notch factor RT	75°C (165°F)	−55°C (−65°F)
Double-edge notch, 0°	814.8	118.2	1269.5	184.1	0.92	0.85	0.80
Center hole, 1.6-mm (1/16-in.) diam, 0°	1125.9	163.3	1163.8	168.8	0.84	...	...
Center hole, 3.2-mm (1/8-in.) diam, 0°	991.7	143.8	1061.6	154.0	0.77	0.75	0.72
	898.0(a)	130.2	956.1(a)	138.7	0.70(a)	...	...
Center hole, 6.4-mm (1/4-in.) diam, 0°	842.1	122.1	966.4	140.2	0.70	...	...
	800.9(a)	116.2	911.3(a)	132.2	0.66(a)	...	...
Center hole, 3.2-mm (1/8-in.) diam, 0°/±45°	728.1	105.6	777.4	112.8	0.90	...	...
Center hole, 6.4-mm (1/4-in.) diam, 0°/±45°	620.5	90.0	710.8	103.1	0.83	...	...
Center hole, 3.2-mm (1/8-in.) diam, 0°/±45°/90°	437.1	63.4	467.5	67.8	0.90	...	...
Center hole, 6.4-mm (1/4-in.) diam, 0°/±45°/90°	400.6	58.1	460.6	66.8	0.89	...	...
Center hole, 2.4-mm (3/32-in.) diam, ±45°	244.7	35.5	256.6	37.2	0.85	...	...
Center crack, 6.4-mm (1/4-in.) EDM slot, 0°	822.0	119.2	944.2	137.0	0.68	...	...
Center crack, 12.7-mm (1/2-in.) EDM slot, 0°	659.9	95.7	886.9	128.6	0.64	...	...
	621.6(a)	90.20	819.8(a)	118.9	0.60(a)	...	...

RT, room temperature. (a) 40-ply material

Source: Ref 6, p 863

Typical properties of SiC/2024-T6 Al billet and extruded plate showing the effects of SiC whisker alignment

MMC material form	Test specimen orientation	Ultimate tensile strength MPa	ksi	Yield strength(a) MPa	ksi	Coefficient of thermal expansion (α), 10^{-6}/K	Density (ρ), g/cm³
12-in.-diam cylindrical billet	Longitudinal (axial)	496	71.9	351	50.9	16.1	2.86
1/2-in. by 5-in. extrusion	Longitudinal	737	107	448	64.9	13.0	2.86
12-in.-diam cylindrical billet	Transverse	503	72.9	358	51.9	16.4	2.86
1/2-in. by 5-in. extrusion	Transverse (long)	462	67.0	379	54.9	19.6	2.86

(a) 0.2% offset

Source: Ref 6, p 900

Tensile properties of a forged SiC/6061-T6 Al turbine wheel(a)

Test-specimen orientation	Tensile strength MPa	ksi	Yield strength(b) MPa	ksi	Elongation, %	Modulus of elasticity GPa	10^6 psi
Radial	518	75.1	386	55.9	4.7	111	16.1
Transverse (circumferential)	490	71.0	379	54.9	6.1	108	15.7

(a) SiC whisker content, 20 vol %. (b) At 0.2% offset.

Source: Ref 6, p 901

Properties of aluminum/SiC$_p$ composites(a)

Property	Al 2124-T6/ 30 vol % SiC$_p$	Al 2124-T6/ 40 vol % SiC$_p$
Microyield strength, MPa (ksi)	177 (17)	...
Coefficient of thermal expansion, 10^{-6}/K (10^{-6}/°F)	12.4 (6.9)	10.8 (6.0)
Modulus of elasticity, GPa (10^6 psi)	117 (17)	145 (21)
Density, g/cm^3 (lb/in.^{3})	2.91 (0.105)	2.96 (0.107)
Specific modulus, 10^6 m (10^6 in.)	4.11 (162)	4.98 (196)
Thermal conductivity, W/m·K (Btu/ft·h·°F)	125 (72)	116 (67)
Specific heat, kJ/kg·K (Btu/lb·°F)	0.80 (0.19)	0.67 (0.16)

(a) SiC$_p$ stands for particulate SiC.

Source: Ref 16, p 7-51

Typical mechanical properties of SiC particulate reinforced aluminum alloy composites

Alloy and vol%	Modulus of elasticity GPa	Modulus of elasticity 10^6 psi	Yield strength MPa	Yield strength ksi	Ultimate tensile strength MPa	Ultimate tensile strength ksi	Ductility, %
6061							
Wrought	68.9	10	275.8	40	310.3	45	12
15	96.5	14	400.0	58	455.1	66	7.5
20	103.4	15	413.7	60	496.4	72	5.5
25	113.8	16.5	427.5	62	517.1	75	4.5
30	120.7	17.5	434.3	63	551.6	80	3.0
35	134.5	19.5	455.1	66	551.6	80	2.7
40	144.8	21	448.2	65	586.1	85	2.0
2124							
Wrought	71.0	10.3	420.6	61	455.1	66	9
15	...	...	...	...	...	...	...
20	103.4	15	400.0	58	551.6	80	7.0
25	113.8	16.5	413.7	60	565.4	82	5.6
30	120.7	17.5	441.3	64	593.0	86	4.5
35	...	...	...	...	...	...	...
40	151.7	22	517.1	75	689.5	100	1.1
7090							
Wrought	72.4	10.5	586.1	85	634.3	92	8
15	...	...	...	...	...	...	...
20	103.4	15	655.0	95	724.0	105	2.5
25	115.1	16.7	675.7	98	792.9	115	2.0
30	127.6	18.5	703.3	102	772.2	112	1.2
35	131.0	19	710.2	103	724.0	105	0.90
40	144.8	21	689.5	100	710.2	103	0.90
7091							
Wrought	72.4	10.5	537.8	78	586.1	85	10
15	96.5	14	579.2	84	689.5	100	5.0
20	103.4	15	620.6	90	724.0	105	4.5
25	113.8	16.5	620.6	90	724.0	105	3.0
30	127.6	18.5	675.7	98	765.3	111	2.0
35	...	...	...	...	...	...	...
40	139.3	20.2	620.6	90	655.0	95	1.2

Source: Ref 6, p 893

Pin bearing strengths of SiC particulate reinforced aluminum

Composite	Edge distance (multiplied by pin diameter) from edge	Bearing yield strength MPa	ksi	Bearing ultimate strength MPa	ksi
20 vol% SiC-7091 (T-6) (Ref 17)	1.5	689.5	100	724.0	105
	2.0	1000.0	145	1310.0	190
	3.0	1000.0	145	1448.0	210
25 vol% SiC-2124 (Ref 19)	2.0	827.4	120	...	...

Source: Ref 6, p 893

Property comparison of aluminum/SiC$_p$ composites

Property	Al 2124-T6	Al 2124-T6/ 30 vol % SiC$_p$(a)	Al 2124-T6/ 40 vol % SiC$_p$(a)	Al 6061-T6/ 35 vol % SiC$_p$(a)
Coefficient of thermal expansion(b), 10^{-6}/K (10^{-6}/°F)	23.4 (13.0)	12.4 (6.9)	10.8 (6.0)	12.1 (6.7)
Microyield strength(c), MPa (ksi)	117 (17)	117 (17)	124 (18)	124 (18)
Modulus of elasticity, GPa (10^6 psi)	72 (10.5)	117 (17)	145 (21)	131 (19)
Density, g/cm^3 (lb/in.^{3})	2.78 (0.100)	2.91 (0.105)	2.96 (0.107)	2.88 (0.104)

(a) SiC mean particle size of 3.5 μm (1.38×10^{-4} in.) used in instrument-grade MMC materials. (b) Average value over a temperature range of 222 to 366 K (-60 to +200 °F). (c) Microyield strength, also known as precision elastic limit, is the stress required to cause 10^{-6} m/m (1 μin./in.) of residual (plastic) strain.

Source: Ref 22, p 71

Mechanical data on SCS-Mg cast rod

Sample No.	Exposure time, min	Ultimate tensile strength MPa	ksi	Strain to failure, %	Elastic modulus GPa	10^6 psi	Fiber, vol%
VIR 67	5	1000	145	0.83	169.6	24.6	34
VIR 69	10	1524	221	0.88	209.6	30.4	46
VIR 72	10	1331	193	0.78	230.3	33.4	50
VIR 77	10	1379	200	0.95	180.6	26.2	37

Source: Ref 6, p 865

Shear strength of SiC particulate reinforced aluminum composites

Composite	Shear strength MPa	ksi
25 vol% SiC-6061 (T-6)	277.9	40.3
30 vol% SiC-6061 (T-6)	289.6	42.0
25 vol% SiC-7091 (T-6)	279.2	55.0
30 vol% SiC-7090	430.9	62.5
25 vol% SiC-2124 (T-4)	344.8	50.0

Source: Ref 6, p 893

Mechanical data on SCS-Cu

Panel	Fiber, vol%	Axial ultimate tensile stress MPa	ksi	Axial modulus GPa	10^6 psi
84-014	0.23	690	100	172.4	25.0
84-153	0.33	965	140	202	29.3
84-377	0.33	900	130	187.5	27.2

Source: Ref 6, p 865

Abrasive wear testing of 20% silicon carbide reinforced 2024-20-T6 Al alloy

	Unreinforced 2024	Silicon carbide-2024
6000 cycles		
Wt loss, g (oz)	0.12 (0.0040)	zero
Scar depth, μm (mil)	15-25 (0.6-1.0)	zero
20 000 cycles		
Wt loss, g (oz)	0.34 (0.0120)	0.06 (0.0020)
Scar depth, μm (mil)	35-60 (1.4-2.4)	1-15 (0.4-0.6)

Source: Ref 6, p 909

Data on SCS-6-Ti (sample size, 62 panels)

	Tensile strength, MPa	Tensile strength, ksi	Elastic modulus, GPa	Elastic modulus, 10^6 psi	Strain to failure, %
Mechanical properties of SiC-Ti-6Al-4V (35 vol %)					
As fabricated:					
Mean	1690	245	186.2	27.0	0.96
Standard deviation	119.3	17.3	7.58	1.1	0.091
After heating 7 h at 905 °C (1660 °F)					
Mean	1434	208	190.3	27.6	0.86
Standard deviation	108.9	15.8	8.3	1.2	0.087
Mechanical properties of SiC-Ti-15V-3Sn-3Cr-3Al (38 to 41 vol %)					
As fabricated:					
Mean	1572	228	197.9	28.7	...
Standard deviation	138	20	6.21	0.9	...
After heat treating 16 h at 480 °C (900 °F), 13 samples:					
Mean	1951	283	213.0	30.9	...
Standard deviation	96.5	14	4.83	0.7	...

Source: Ref 6, p 865

Alumina/Metal

Adhesion of metals to polycrystalline Al_2O_3

Metal	Sintering temperature, °C	Sintering atmosphere	Adhesion, erg/cm^2
Cu	450	H_2	815
Ni	1000	H_2	435
Au	1000	Air	530
Fe	1000	H_2	810
Ni	1400	Ar	518
Ag	700	H_2	435

Source: Ref 16, p 8-28

Coefficients of linear thermal expansion of polycrystalline alumina and alumina/chromium composites

Material	Method	Temperature, °C	Temperature, °F	Coefficient of linear thermal expansion, 10^{-6}/K	Coefficient of linear thermal expansion, 10^{-6}/°F
Al_2O_3	(a)	27-799	80-1470	7.90	4.39
		27-1315	80-2400	9.40	5.22
$70Al_2O_3/30Cr$	Telemicroscope	25-800	77-1472	8.6	4.8
		22-1315	72-2400	9.45	5.25
$34Al_2O_3/66(Cr, Mo)$	Telemicroscope	24-802	75-1475	7.96	4.42
		24-1315	75-2400	10.5	5.82
$28Al_2O_3/72Cr$	Telemicroscope	25-800	77-1472	8.62	4.79
		22-1315	72-2400	10.4	5.75
$23Al_2O_3/77Cr$ (LT-1)	...	25-1000	77-1832	8.91	4.95

(a) Average of many reported values obtained by a variety of methods.

Source: Ref 16, p 8-28

Yield strength and ultimate tensile strength of aluminum alloy reinforced with polycrystalline alumina at different temperatures

Fiber, vol%	350 °C (660 °F) Yield strength MPa	ksi	350 °C (660 °F) Ultimate tensile strength MPa	ksi	300 °C (570 °F) Yield strength MPa	ksi	300 °C (570 °F) Ultimate tensile strength MPa	ksi	250 °C (480 °F) Yield strength MPa	ksi	250 °C (480 °F) Ultimate tensile strength MPa	ksi
Polycrystalline alumina												
0	35	5.1	55	8.0	...	...	70	10.2	70	10.2	115	16.7
0.05	54	7.8	63	9.1	79	11.5	88	12.8	112	16.2	134	19.4
0.12	68	9.9	74	10.7	...	...	...	...	...	...	...	...
0.20	110	16.0	112	16.2	154	22.3	155	22.5	186	27.0	198	28.7
SiC whiskers												
0	35	5.1	55	8.0	...	...	70	10.2	70	10.2	115	16.7
0.12	94	13.6	124	18.0	153	22.2	180	26.1	197	28.6	226	32.8
0.16	120	17.4	147	21.3	...	...	...	...	...	...	...	...
0.20	163	23.6	184	26.3	207	30.0	235	34.1	268	38.9	284	41.2

Source: Ref 6, p 891

Tensile data for polycrystalline-alumina-reinforced aluminum alloy

Fiber content, vol %	Yield strength (0.2%) MPa	ksi	Standard deviation	Range of measurement	Tensile strength MPa	ksi	Standard deviation	Range of measurement	Modulus of elasticity GPa	10^6 psi	Standard deviation	Range of measurement
0	210	30.5	3.8	9.5	297	43.1	1.8	3.5	71.9	10.4	4.5	13
0.05	232	33.6	4.2	10.4	282	40.9	6.5	15.1	78.4	11.4	2.3	6
0.12	251.5	36.5	14.6	38.3	273	40.0	19.6	49.6	83.0	12.0	7.8	21
0.20	282.5	41.0	11.3	25.2	312	45.3	16.0	42.3	95.2	13.8	2.7	7

Source: Ref 6, p 890

Typical compositions and physical properties of alumina/chromium composites

Typical composition wt %	Approximate vol %	Density, g/cm^3	Melting point °C	°F
$70Al_2O_3/30Cr$	$81Al_2O_3/19Cr$	4.60–4.65	>1705[a]	>3100[a]
$34Al_2O_3/66(Cr,Mo)$[b]	$49.9Al_2O_3/50.1(Cr,Mo)$	5.82	>1730[a]	>3150[a]
$28Al_2O_3/72Cr$	$42Al_2O_3/58Cr$	5.9	>1730[a]	>3150[a]
$23Al_2O_3/77Cr$	$35Al_2O_3/65Cr$	5.9–6.0	1850[c]	3360[c]
Al_2O_3		3.98	2050	3720

[a]Temperature given is sintering temperature at which composite was prepared. [b]Second phase added as 80Cr-20Mo alloy. [c]Maximum service temperature: 1315 °C (2400 °F) for long term; 1650 °C (3000 °F) for short term.

Source: Ref 16, p 8-28

Wetting angles in various alumina/liquid metal systems

System	Wetting angle	Temperature, °C, atmosphere
Al_2O_3/Ni	>90°	1450, air
Al_2O_3/Cu	138°	1200, vacuum
Al_2O_3/Fe-Cr	~40°	1650, reducing
Al_2O_3/Cr	1–10°	1950, reducing
Al_2O_3/Fe	139°	1550, N_2

Source: Ref 16, p 8-30

Hardness at 25 °C of Al-9Si-3Cu alloy reinforced with Al_2O_3 fibers

Volume fraction (V_f)	Vickers hardness number, HV 10
0	131
0.12	179
0.18	190
0.24	212

(a) "Saffil" fiber, RF grade.

Source: Ref 16, p 7-18

Coefficients of thermal expansion for Al_2O_3 fiber/Al alloy composites

	Coefficient of thermal expansion (a), 10^{-5}/°C	
Volume fraction (V_f)	In-plane	Normal
0	2.03	2.03
0.12	1.66	1.76
0.18	1.54	1.66
0.24	1.55	1.57

(a) Coefficients (at 20 to 200 °C) of composites containing "Saffil" fiber, RF grade, in Al-9Si-3Cu alloy measured parallel and normal to planes of fiber orientation.

Source: Ref 16, p 7-18

Boron/Metal

Reactivity of boron with various metals

	Definite reaction			Little or no reaction		
	Temperature		Time[a],	Temperature		Time[a],
Metal	°C	°F	h	°C	°F	h
Fe	700	1292	50	600	1112	(b)
	900	1652	1	...	...	...
Co	700	1292	50	600	1112	(b)
	900	1652	1	...	...	...
Al	700[c]	1292[c]	0.2	600	1112	1
Mg	...	...	...	600	1112	100
	...	...	...	700[c]	1292[c]	0.2
Be	...	...	...	1000	1832	24
Ti	...	...	...	600	1112	100
Cr	900	1652	1	...	...	...
Ag	...	...	...	900	1652	2
Ni	600	1112	100	...	...	...
	700	1292	100	...	...	...
	900	1652	1	...	...	...

[a]Time required for consolidation of metal powders and fibers by hot pressing. [b]Reaction time so short that the reaction is finished as soon as fibers have been hot pressed. [c]Molten.

Source: Ref 16, p 7-6

Wetting angles of borides by iron-group melts

Boride	Iron: Wetting angle, Vacuum	Iron: Wetting angle, Argon	Iron: Work of adhesion, erg/cm²	Nickel: Wetting angle, Vacuum	Nickel: Wetting angle, Argon	Nickel: Work of adhesion, erg/cm²	Cobalt: Wetting angle, Vacuum	Cobalt: Wetting angle, Argon	Cobalt: Work of adhesion, erg/cm²
TiB_2	39	94	1650	25	72	2220	...	64	2590
ZrB_2	97	102	1410	72	78	2020	...	81	2100
HfB_2	...	98	1530	...	99	1430	...	...	...
NbB_2	...	0	...	...	24	...	...	22	3460
TaB_2	...	0	...	...	...	...	...	...	...
CrB_2	...	0	...	20	21	3310	...	0	...
Mo_2B_5	...	0	...	...	0	...	...	22	3460
W_2B_5	...	...	...	...	...	...	...	94	1680

Source: Ref 16, p 8-33

Typical properties of eutectoid bonded boron-aluminum(a)

Type of loading	Tensile strength: Room temperature, MPa	Tensile strength: Room temperature, ksi	Tensile strength: 315 °C (600 °F), MPa	Tensile strength: 315 °C (600 °F), ksi	Ultimate strain 10^{-6} m/m (10^{-6} in./in.): Room temperature	Ultimate strain: 315 °C (600 °F)	Initial modulus: Room temperature, GPa	Initial modulus: Room temperature, 10^6 psi	Initial modulus: 315 °C (600 °F), GPa	Initial modulus: 315 °C (600 °F), 10^6 psi
Longitudinal tension:										
Coupon	1100	160	1070	155	6,500	5,800	205	29.8	194	28.1
Beam	1340	194	...	...	7,200	...	198	28.7	...	...
Longitudinal compression, beam	2360(b)	343(b)	...	...	10,500	...	254	36.8	...	...
Transverse tension, coupon	115	16.7	27	3.9	4,200	6,500	134	19.5	97	14.0
Transverse compression, beam	259	37.5	66	9.6	2,400	10,700	138	20.0	120	17.4
In-plane shear, rail shear	69	10	26	3.7	16,000	13,000	69	10.0	54	7.8

(a) All tests conducted on 42 to 45 vol % B/Al. (b) Failure occurred in honeycomb core — no failure in B/Al face sheet.

Source: Ref 29, p 214

Wetting angles of refractory borides with various metal melts

Boride	Cu 1130 °C	Al 900 °C	Ga 800 °C	In 250 °C	Si 1500 °C	Ge 1000 °C	Sn 300 °C	Pb 400 °C	Bi 320 °C
	Wetting angle for metal melt and temperature of:								
TiB_2	143	98	115	124	15	...	114	106	141
ZrB_2	135	106	117	114	...	102	110	...	...
HfB_2	...	134	...	114	...	140	...	...	...
NbB_2	109	125	101	133	0	60	102	125	110
TaB_2	...	138	...	117	...	...	...	...	...
CrB_2	26	107	123	97	...	126	100	124	128
Mo_2B_5	0	134	...	...	...	28	100	...	...
W_2B_5	104	...	...	130	22	128	100	...	...

Source: Ref 16, p 8-33

Mechanical properties of boron/aluminum composites(a)

Matrix	Fiber orientation	Tensile strength MPa	Tensile strength ksi	Modulus of elasticity GPa	Modulus of elasticity 10^6 psi
Al-6061	0°	1515	220	207	30
	90°	138	20	138	20
Al-2024	0°	1550	225	207	30
	90°	214	31	145	21

(a) These samples contain 48% Avco 5.6 mil (142-μm) boron. The longitudinal tensile specimens are 6 in. (152 mm) by 0.3125 in. (7.9 mm) by 6 ply, and the transverse tensile bars are 6 in. (152 mm) by 0.5 in. (12.7 mm) by 6 ply.

Source: Ref 16, p 7-11

Room-temperature properties of boron-aluminum (0°) with 50 vol % filament

Property	Reported value
Tensile strength, MPa (ksi):	
Longitudinal	1100 (160.0)
Transverse	110 (16.0)
Compressive strength, MPa (ksi):	
Longitudinal	1215 (176.0)
Transverse	159 (23.0)
Shear strength, MPa (ksi):	
In-plane	69 (10.0)
Interlaminar	126 (18.3)
Strain, μm/m (μin./in.):	
Longitudinal	5000-6000
Transverse	6000-12,000
Tensile modulus, GPa (10^6 psi):	
Longitudinal	235 (34.0)
Transverse	138 (20.0)
Compressive modulus, GPa (10^6 psi):	
Longitudinal	207 (30.0)
Transverse	131 (19.0)
In-plane shear modulus, GPa (10^6 psi)	66 (9.5)
Poisson's ratio:	
Longitudinal	0.23
Transverse	0.17
Density, g/cm^3	2.7
Coefficient of thermal expansion, 10^{-6}/°C (10^{-6}/°F):	
Longitudinal	5.8 (3.2)
Transverse	19.1 (10.6)

Source: Ref 6, p 854

Average tensile and shear strengths of soldered boron/aluminum tee sections

Test mode	Test temperature K	Test temperature °F	Failure stress MPa	Failure stress ksi
Tension	294	70	39	5.70
	366	200	38	5.50
Shear	294	70	73	10.60
	366	200	75	10.70

Source: Ref 29, p 220

Typical mechanical properties of 50 vol % unidirectional reinforced B/Al 6061 composites

Tensile strength:	
Longitudinal	1490 MPa (216 ksi)
Transverse	138 MPa (20 ksi)
Tensile modulus:	
Longitudinal	214 GPa (31×10^6 psi)
Transverse	138 GPa (20×10^6 psi)
Poisson's ratio:	
Longitudinal	0.23
Transverse	0.13
Compressive strength:	
Longitudinal	1725 MPa (250 ksi)
Transverse	207 MPa (30 ksi)
Compressive modulus:	
Longitudinal	221 GPa (32×10^6 psi)
Transverse	138 GPa (20×10^6 psi)
Longitudinal shear strength	159 MPa (23 ksi)
Longitudinal shear modulus	41 GPa (6×10^6 psi)
Longitudinal bearing strength	827 MPa (120 ksi)
Unnotched fatigue strength at runout (10^7 cycles):	
Longitudinal	1035 MPa (150 ksi)
Transverse	41 MPa (6 ksi),
Creep at 370 °C (700 °F)	At 1105 MPa (160 ksi) total strain averages 0.06% in 100 h.

Source: Ref 16, p 7-14

Room-temperature properties of cross-ply boron-aluminum (0°/90°) with 50 vol % filament

Property	Reported value
Tensile strength, MPa (ksi):	
Longitudinal	483 (70)
Transverse	483 (70)
Compressive strength, MPa (ksi):	
Longitudinal	607 (88)
Transverse	607 (88)
Shear strength, MPa (ksi):	
In-plane	103 (15)
Interlaminar	69 (10)
Strain, longitudinal, μm/m (μin./in.)	6700
Tensile modulus, GPa (10^6 psi):	
Longitudinal	145 (21)
Transverse	145 (21)
Compressive modulus, GPa (10^6 psi):	
Longitudinal	145 (21)
Transverse	145 (21)
Density, g/cm^3	2.7

Source: Ref 6, p 854

Axial tensile strength of 140-μm (5.6-mil) boron-aluminum at various fiber levels

Matrix	Boron content, %	Tensile strength MPa	Tensile strength ksi	Modulus of elasticity GPa	Modulus of elasticity 10^6 psi	Strain to failure, %
2024, as fabricated	45	1287.5	186.7	202.1	29.3	0.775
	47	1420.7	206.0	222.1	32.2	0.795
	52	1721.0	249.6	...	...	...
	54	1798.6	260.8	...	...	...
	64	1527.6	221.5	275.9	40.0	0.72
	66	1739.2	251.6	...	...	...
	70	1927.6	279.5	...	...	...
2024-T6	46	1458.7	211.5	220.7	32.0	0.810
	64	1924.1	279.0	275.9	40.0	0.755
6061, as fabricated	48	1489.7	216.0	...	...	...
	50	1343.4	194.8	217.2	31.5	0.695
6061-T6	51	1417.2	205.5	231.7	33.6	0.735

Source: Ref 6, p 855

Room-temperature longitudinal tensile strengths of thermally exposed and unexposed boron-aluminum

Exposure condition	Average monolayer strength MPa	Average monolayer strength ksi	Average filament-bundle strength MPa	Average filament-bundle strength ksi	Degradation, %
Unexposed controls	1275	185	3040	441	...
30 min at 550 °C (1020 °F)	1035	150	2510	364	17.6
7 min at 555 °C (1030 °F)	1150	167	2680	389	11.9
15 min at 555 °C (1030 °F)	1095	159	2470	358	19.0
7 min at 570 °C (1060 °F)	1080	157	2580	374	15.5
15 min at 570 °C (1060 °F)	915	133	2185	317	28.3
35 min at 570 °C (1060 °F)	850	123	1995	289	34.6
7 min at 595 °C (1100 °F)	930	135	2180	316	28.6

Source: Ref 29, p 211

Impact energy of full-size notched Charpy specimens containing 55% boron fibers

Fiber diameter, μm	Matrix	Total impact energy, J
200	Aluminum 1100	95
145	Aluminum 1100	55
145	Al alloy 6061	35
Unreinforced	Ti-6Al-4V	25
Unreinforced	Al alloy 6061-T6	15

Source: Ref 16, p 7-11

Refractory Metal Fiber/Metal Matrix Composites

Elevated-temperature tensile properties of W/Fe-Cr-Al-Y

Test temperature °C	Test temperature °F	Specimen	Filament orien-tation	Ultimate tensile strength MPa	Ultimate tensile strength ksi	Total elonga-tion, %
648	1200	P34-15-2	±15°	776	112.6	25.2
		P36-15-2	±15°	717	104.0	12.8
		P34-45-2	±45°	564	81.8	35.5
		P36-45-2	±45°	539	78.2	24.6
		P34-90-2	90°	189	27.4	3.5
		P36-90-2	90°	180	26.2	3.2
760	1400	P34-15-1	±15°	571	82.8	12.0
		P36-15-1	±15°	534	77.4	14.2
		P34-45-1	±45°	185	26.9	23+[a]
		P36-45-1	±45°	163	23.6	24.6
		P36-90-1	90°	111	16.1	6.5
648	1200	P21-1-1[b]	0°	737	106.9	3.0
		P21-1-2[c]	0°	768	111.4	2.9

[a]Test terminated before failure. [b]Creep specimen after 1077-h creep test at 1037 °C (1900 °F). [c]Creep specimen after 990-h creep test at 1093 °C (2000 °F).

Source: Ref 79, p 7-32

Room-temperature tensile properties of tungsten fiber/copper alloy composites

Binder material	Maximum solubility of alloying element in tungsten	Alloying element content wt %	Alloying element content at. %	Specimen	Fiber content, vol %	Tensile strength MPa	Tensile strength ksi	Reduction in area, %	Type of fracture
Pure copper	Insoluble in tungsten	0	0	...	65	1556	225.7	...	Ductile
				...	70.2	1641	238.0	...	Ductile
				...	75.4	1722	249.8	...	Ductile
Copper-nickel	0.3	5	5.4	1	79	1700	246.6	34	Ductile
				2	78.4	1724	250.0	37	Ductile
				3	76	1509	218.9	32	Ductile
		10	10.9	4	74.1	908	131.7	Nil	Brittle
				5	75.5	750	108.8	Nil	Brittle
				6	79.5	354	51.3	Nil	Brittle
Copper-cobalt	0.3	1	1.1	7	77.3	1513	219.4	...	Semiductile
		5	5.4	8	76	1470	213.2	1.5	Semiductile
				9	74.8	1581	229.3	2.3	Ductile
				10	74.7	1015	147.2	...	Brittle
				11	74.9	1187	172.1	...	Brittle
Copper-aluminum	2.6	5	11.3	12	63.4	682	98.9	Nil	Brittle
				13	72.4	1060	153.8	Nil	Semiductile
				14	76.1	1065	154.5	Nil	Semiductile
		10	20.8	15	76.7	955	138.5	...	Brittle
Copper-titanium	8	10	12.8	16	78.2	1542	223.7	...	Semiductile
				17	71.7	1518	220.1	10	Semiductile
		25	30.7	18	76.3	1287	186.7	...	Brittle
Copper-zirconium	3	10	7.2	19	72.8	1489	216.0	Nil	Brittle
				20	78.5	1760	255.3	Nil	Ductile
				21	75.6	1564	226.9	Nil	Semiductile
				22	64.7	1190	172.6	Nil	Brittle
				23	64.3	1349	195.7	Nil	Semiductile
		33	25.5	24	75.9	736	106.7	Nil	Brittle
Copper-chromium	Complete solid solubility (miscibility gap)	1	1.2	25	78.7	1541	223.5	7.4	Semiductile
				26	77.5	1572	228.0	25.8	Ductile
				27	77.2	1558	225.9	7.5	Semiductile
		2	2.4	28	76.4	1666	241.7	16.4	Ductile
Copper-niobium	Complete solid solubility	1	0.6	29	75.4	1635	237.1	20.6	Ductile
				30	75.1	1538	223.1	24.7	Ductile

Source: Ref 16, p 7-26

Rupture strengths for tungsten alloy wire reinforced composites

Alloy, wt%	Wire	Wire diam mm	Wire diam in.	Vol%	Density, g/cm^3	100-h rupture strength MPa	100-h rupture strength ksi	Strength to density for 100-h rupture cm × 10^3	Strength to density for 100-h rupture in. × 10^3
100-h rupture strength at 1100 °C (2010 °F)									
Ni-12.5Cr-7W-4.8Mo-5Al-2.5Ti(ZhS6)	VRN Tungsten	0.3-0.5	0.012-0.020	40	12.5	138	20	112.5	44.3
Ni-11W-6Al-6Cr-2Mo-1.5Nb(EPD-16)	No reinforcement	...	...	0	8.3	51	7.4	63.5	25.0
	Tungsten	0.25	0.010	40	12.7	131	19	104	40.9
Ni-12.5Cr-2.5Fe-2Nb-4Mo-6Al-1Ti(Nimocast 713C)	No reinforcement	...	...	0	8.0	48	7	61.3	24.1
	Tungsten	1.27	0.050	20	10.3	93	13.5	92.7	36.5
Co-21.5Cr-25W-10Ni-3.5Ta-0.8Ti(Mar-M322E)	No reinforcement	...	...	0	...	48	7	...	...
	W-2%ThO_2	0.08	0.003	40	...	207	30	...	...
Ni-25W-15Cr-2Al-2Ti	No reinforcement	...	...	0	9.15	23	3.3	25.4	10
	218CS (Tungsten)	0.38	0.015	40	13.3	138	20	105.8	41.7
	W-2%ThO_2	0.38	0.015	40	13.0	193	28	151.3	59.6
	W-Hf-C	0.38	0.015	40	13.3	324	47	249.1	98.0
Fe-24Cr-5Al-1Y	W-1%ThO_2	0.38	0.015	56	12.5	242(a)	35	195.7	76.8
Fe-24Cr-5Al-1Y	W-Hf-C	0.38	0.015	35	11.3	242	35	214.7	84.5

(a) 831-h rupture strength.

Source: Ref 6, p 881

Effect of thermal cycling on residual room-temperature tensile properties of W-1ThO_2/Fe-Cr-Al-Y

Exposure	Observations	Ultimate tensile strength MPa	Ultimate tensile strength ksi	Modulus GPa	Modulus 10^6 psi	Strain to failure µin./in.
As-fabricated	...	655	95.0	179	26.0	3400
As-fabricated	...	581	84.2	201	29.2	3700
100 cycles, 29–1095 °C (85–2000 °F)	No visual change	563	81.7	259	37.6	2400
100 cycles, 29–1095 °C (85–2000 °F)	No visual change	618	89.6	219	31.7	4300
1000 cycles, 29–1095 °C (85–2000 °F)	Surface roughening	590	85.5	258	37.4	3200
1000 cycles, 29–1095 °C (2000 °F)	Surface roughening	557	80.8	177	25.6	3300
100 cycles, 29–1205 °C (85–2200 °F)	No visual change	624	90.5	250	36.3	3100
100 cycles, 29–1205 °C (85–2200 °F)	No visual change	587	85.2	228	33.1	3600
1000 cycles, 29–1205 °C (85–2200 °F)	Surface roughening	503	72.9	158	22.9	3200
1000 cycles, 29–1205 °C (85–2200 °F)	Surface roughening	487	70.6	170	24.6	3000

Source: Ref 16, p 7-31

Cermets

Properties of simultaneously pressed and sintered Al_2O_3/TiC cermet

Mechanical Properties	
Transverse rupture strength[b], MPa (ksi)	760 (110)
Fracture toughness, MPa · $m^{1/2}$ (ksi · $in.^{1/2}$)	5.9 (5.3)
Elastic constants:	
Elastic modulus, GPa (10^6 psi)	
At 25 °C (77 °F)	376 (54.5)
At 500 °C (930 °F)	353 (51.2)
At 1000 °C (1830 °F)	324 (47.1)
Shear modulus, GPa (10^6 psi)	
At 25 °C (77 °F)	154 (22.4)
At 500 °C (930 °F)	144 (21.0)
At 1000 °C (1830 °F)	134 (19.4)
Poisson's ratio	
At 25 °C (77 °F)	0.22
At 500 °C (930 °F)	0.22
At 1000 °C (1830 °F)	0.21
Microhardness, Vickers DPH, GPa (kg/mm^2)	
100-g load	23.0 (2350)
500-g load	18.6 (1900)
Thermal Properties	
Linear thermal expansion, 10^{-6}/°C	
From 25 to 300 °C (77 to 570 °F)	7.0
From 25 to 500 °C (77 to 930 °F)	7.6
From 25 to 800 °C (77 to 1470 °F)	8.3
Thermal conductivity, W/cm · K (cal/cm · s · °C)	
At 25 °C (77 °F)	0.13 (0.030)
At 500 °C (930 °F)	0.13 (0.030)
At 1000 °C (1830 °F)	0.11 (0.027)
Thermal diffusivity, cm^2/s	
At 25 °C (77 °F)	0.042
At 500 °C (930 °F)	0.029
At 1000 °C (1830 °F)	0.025
Specific heat (C_p), J/g · K (cal/g · °C)	
At 25 °C (77 °F)	0.58 (0.14)
At 500 °C (930 °F)	0.84 (0.20)
At 1000 °C (1830 °F)	0.88 (0.21)
Electrical Property	
Volume resistivity, dc, Ω · cm	
At 25 °C (77 °F)	2×10^{-2}
At 100 °C (212 °F)	2×10^{-2}
At 300 °C (570 °F)	3×10^{-2}

[a]Composite is electrically conductive. Density, 5.16 g/cm^3. Porosity, vacuum tight. Water absorption, none. Grain size, ~2 μm. Color, black. Data from Greenleaf Corp. [b]In four-point bending.

Source: Ref 16, p 8-32

General property data for titanium carbide cermets

Property	Typical values reported at 24 °C (75 °F)[a] TiC/metal composite	TiC
Density, g/cm^3	5.5–6.8	4.65–4.92
Melting point:		
°C	1455[b]	3065
°F	2650[b]	5550
Specific heat:		
kJ/kg · K	0.46	0.50
Btu/lb · °F	0.11	0.12
Thermal conductivity:		
W/m · K	28–35	26–35
Btu/ft · h · °F	16–20	15–20
Coefficient of linear thermal expansion (21–980 °C or 70–1800 °F):		
10^{-6}/°C	2.8–3.6	2.4
10^{-6}/°F	5.0–6.5	4.3
Bend strength:		
MPa	689–1380	414–827
ksi	100–200	60–120
Compressive strength:		
MPa	2760–3450[c]	1380–2760
ksi	400–500[c]	200–400
Tensile strength:		
MPa	689–965[d]	241–276
ksi	100–140[d]	35–40
Impact strength:		
J	5.42–21.47	<1.36
in. · lb	48–190	<12
Young's modulus:		
GPa	310–379	276–448
10^6 psi	45–55	40–65
Hardness, Rockwell A	84–89	93
Oxidation resistance (20 h at 870–980 °C or 1600–1800 °F), % weight gain	<1	<1

[a]Data are for composite materials containing up to 50 wt % metal; values were obtained from various sources. [b]Melting point for nickel or cobalt metal component. [c]Compressive yield stress. [d]Yield of 0.2 to 0.3% for indicated stresses.

Source: Ref 30, p 8-36

Components of typical carbide/metal cermets

Carbide components[a] Major	Minor	Metal components[a] Major	Minor
TiC	...	Ni or Co	...
TiC	...	Ni or Co	Cr, Mo, W, Al, or Fe
TiC (NbC, TaC)[b]	...	Ni or Co	(c)
TiC	Cr_3C_2	Ni or Co	...
TiC	(d)	Ni or Co	(c)
Cr_xC_y[e]	(d)	Ni	(c)
WC	(d)	Co	(c)

[a]Composite range of about 30 to 70 wt % of either major component. [b]A solid-solution component. [c]With or without other metal additions, such as Cr, Mo, W, Al, or Fe. [d]With or without other additions of transition-metal carbides. [e]As Cr_3C_2 or mixed carbides of chromium.

Source: Ref 30, p 8-36

Ceramic Matrix Composites

CERAMIC MATRIX COMPOSITES

Ceramic Matrix Composites

General

Some ceramic/metal combinations

Oxide matrix	Metal	Oxide matrix	Metal
Binary			
Cr_2O_3	Cr, Mo, W, Re	$Nd_2O_3(CeO)$	Nb
$(Cr,Al)_2O_3$	Cr, Mo, W	TiO_2	Cr, Nb, Ta
Gd_2O_3	Mo, W	UO_2	Mo, Ta, W
$Gd_2O_3(CeO_2)$	Mo	$UO_2(ThO_2)$	W
$HfO_2(CaO)$	Mo, W	$Y_2O_3(CeO_2)$	Mo, W
$HfO_2(Y_2O_3)$	W	ZrO_2	Ta
La_2O_3	Mo, W	$ZrO_2(CaO)$	W
Nd_2O_3	Mo, W	$ZrO_2(Y_2O_3)$	W
$LaCrO_3$	Cr, Mo, W	SiO_2	Cr
$YCrO_3$	Cr, Mo, W	TiC	Mo, Fe, Ni, Co
SiC	Ag, Co, Cr	Al_2O_3	Al, Co, Fe, Cr
WC	Co		

Oxide matrix	Metal	Oxide matrix	Metal
Ternary			
UO_2-MgO	W	La_2O_3-$LaCrO_3$	W
MgO-ZrO_2	W	CaO-$CaCr_2O_4$	W
MgO(Cr_2O_3)-ZrO_2	Mo	Cr_2O_3-ZrO_2	W, Mo
Cr_2O_3-$LaCrO_3$	W	Cr_2O_3-HfO_2	Mo

Source: Ref 16, p 8-1

Fabrication techniques for ceramic composites

I Architectures	II Matrix densification
Filament winding	Infiltration
Chopped fiber	Glass
Braiding	Polymer precursor
Fabric lay-up	Sol gel
1D, 2D, 3D	Si
Whiskers	CVD
Particle dispersion	Hot pressing
	Sintering
	Reaction sintering
	Plasma spraying

Source: Ref 16, p 8-1

Flaw size particle or fiber spacings, and strengths of ceramic composites

Material	Flaw size[a], μm	Particle or fiber: Material	Particle or fiber: Diameter, μm	Spacing, μm, for volume fraction of: 10%	20%	30%
Al_2O_3	65	Random carbon fibers	8	32	21	14
Glass	23	Al_2O_3 particles	60	45	23	12
			15	11	5.7	3.1
Glass	30	Aligned carbon fibers	8	14	8	5
Glass	30	Random carbon fibers	8	32	21	14
Glass	72	Al_2O_3 particles	3.5	2.6	1.4	0.7
			11	8	4.2	2.2
			44	33	17	9
MgO	82[b]	Random Ni fibers	89	360	240	160
	G[c]	Co, Fe, or Ni particles	~40	30	15	8
Si_3N_4	62	SiC	5	3.7	1.9	1.0
			9	6.7	3.4	1.8
			32	24	12	6.6

[a]Flaw size calculated from *S*, *E*, and γ of matrix. [b]Assuming uniform spacing of equal spherical or cylindrical particles, for random fibers, only one-third were assumed to be oriented to significantly affect crack propagation. [c]Grain size (*G*), 10 μm or less and $0.03 < P < 0.07$.

Source: Ref 16, p 8-7

Ceramic matrix composite toughening concepts

Concept	Basic requirements	Status of verification and modeling
1. Modulus transfer of load from matrix to fibers	$E_f > E_m$, preferably by a factor of >2	Verified, reasonable modeling
2. Prestressing of fibers and matrix	$\alpha_f > \alpha_m$ so axial tensile stresses in fibers < their fracture stress to give reasonable compressive axial stress in matrix	Not verified; basic modeling not expected to be difficult
3. Crack-impeding second phases	Fracture toughness of fibers (or particles) > local matrix so crack is either arrested or bows out, i.e., gives line tension effects between fibers or particles	Arrest impractical; line tension modeling, but uncertain verification
4. Fiber pull-out	Fiber (or elongated particles) have high enough transverse fracture toughness so failure occurs along fiber matrix interface	Limited verification and modeling
5. Crack deflection or multiplication	Sufficiently weak fiber (or particle) matrix interfaces, or appropriate mismatch of properties, especially thermal expansions between matrix and particles (fibers) and use of appropriate particles (fiber sizes)	Limited verification: no modeling. Some verification: possible modeling developing
6. Phase-transformation toughening	Second-phase particles (fibers) increase one or more dimensions.	Verified with ZrO_2 particles; modeling developing

Source: Ref 16, p 8-2

Theoretical maximum temperatures for some fiber-reinforced ceramics (based on melting or softening points)

System	Maximum temperature, °C
Carbon–Pyrex glass	700–800
Carbon–glass ceramic	1300
Silicon carbide–glass	>650
FP alumina–glass	>650
Silicon carbide–silicon	1410
Silicon carbide–silicon nitride	1900
Carbon–carbon	3550

[a]Carbon oxidizes in air above 400 °C, and SiC oxidizes rapidly between 980 and 1150 °C but is stable between 1150 and 1400 °C.

Source: Ref 16, p 8-6

Fabrication methods for metal fiber/ceramic matrix composites

Fibers	Matrix	Fabrication method
Cr Mo	Al_2O_3-Cr_2O_3 CeO_2-doped Gd_2O_3	Hot pressing of grains of previously directionally solidified eutectic composites
V, Nb, Ta	Cr_2O_3	Directional solidification
Cr, Nb, Ta	TiO_2	Hot pressing of grains of previously grown eutectic
Ta	ZrO_2	Directional solidification
Cr	Fe_3O_4, Al_2O_3, Cr_2O_3, and mixtures	Directional solidification
Ta	Unstabilized HfO_2	Hot pressing
W, Mo	Stabilized HfO_2	Hot pressing
Ni, Fe, Co	MgO	Hot pressing
W	Fused SiO_2	Directional solidification
Ta, Mo, Nb	UO_2	Hot pressing
Stainless steel	Wustite	Hot pressing, plasma spraying, or vapor deposition
Cu, Cu-Be, Be	Be_4B, Be_2B	Whisker formation *in situ*
Ti, Cr	SiC	Hot pressing
Ta, W	Si_3N_4	Flame spraying of silicon and heating in nitriding atmosphere
W, Mo	Si_3N_4	Hot pressing
Mo, Ta, W	Sialon, Si_3N_4, Si_3N_4-C, TaC	Hot pressing
Ta	TaC	Hot pressing
W, W-Re	TaC	Hot pressing
Nb	$MoSi_2$	Hot pressing
Nb	Borosilicate glass	Hot pressing
Ni	Glass-ceramic	Fusing of glass-coated fibers together using pressure
W, Mo, stainless steel, or carbon steel	Glass, glass-ceramic	Hot pressing, vacuum injection, or pultrusion
Stainless steel	PbO glass	

Source: Ref 16, p 8-3

Cyclic fatigue results for some ceramic composites

Material	Environment	Mean strength, MPa	Maximum stress, MPa	Number of cycles to failure
30 vol % BN/mullite	Liquid N_2	...	320–360	2–100
30 vol % BN/alumina	Liquid N_2	...	410	3
30 vol % BN/Si_3N_4	Silicone oil	330	275	10–100
Woven carbon/carbon	Silicone oil	97	80–110	2–70

Source: Ref 16, p 8-28

Some potential matrix, fiber, and dispersion material options for ceramic composites

I Matrix Materials			
• Si_3N_4	• ZrO_2, HfO_2	• Glass	• Mullite
• SiC	• Al_2O_3	• Glass ceramic	• Cordierite
II Fiber Materials			
• SiC	• Si_3N_4	• Graphite	• Mullite
• BN	• $Al_2O_3 \cdot B_2O_3 \cdot SiO_2$	• Coatings	• Al_2O_3
III Dispersion Materials			
• SiC	• ZrO_2	• TiC	• BN

Source: Ref 16, p 8-1

Ceramic

Calculated abrasion wear resistance parameters for Si_3N_4-base composites compared with some Al_2O_3 systems

Material	Abrasion wear resistance parameter, $H^{1/2} \cdot K_{Ic}^{3/4}$	Fracture toughness (K_{Ic}), $MPa\sqrt{m}$
Si_3N_4 + 6 wt % Y_2O_3	11.87	4.8
Si_3N_4 + 10 wt % TiC	11.91	4.8
Si_3N_4 + 30 wt % TiC	11.84	4.4
Si_3N_4 + 50 wt % TiC	8.48	2.7
Si_3N_4 + 30 wt % (W,Ti)C	9.57	3.5
Si_3N_4 + 30 wt % WC	13.06	5.2
Si_3N_4 + 30 wt % TaC	11.15	4.6
Si_3N_4 + 30 wt % HfC	9.81	3.6
Si_3N_4 + 30 wt % SiC	9.74	3.6
Al_2O_3	7.38	2.3
Al_2O_3 + ZrO_2	9.32	3.2
Al_2O_3 + TiC	9.92	3.2
Sialon	9.87	4.0

Source: Ref 16, p 8-38

Hardness and fracture toughness of Al_2O_3 ceramic cutting-tool materials compared with those of Sialon and Si_3N_4 + Y_2O_3

Material	Hardness (KHN), GPa	Fracture toughness (K_{Ic}), $MPa\sqrt{m}$
Al_2O_3	15.6	2.3
Al_2O_3 + TiC	17.2	3.2
Al_2O_3 + ZrO_2	15.2	3.2
Sialon	12.2	4.0
Si_3N_4 + Y_2O_3	13.4	4.8

Source: Ref 16, p 8-30

Fabrication methods for ceramic fiber/cermic matrix composites

Fibers	Matrix	Fabrication method
Al_2O_3	Al_2O_3	Sintering
AlN	Al_2O_3	Tape casting, aligning of AlN needles, and sintering
AlN, Si_3N_4	AlN, Si_3N_4	Hot pressing
Al_2O_3, C, ZrO_2	$Mg_3(PO_4)_2$	Hot pressing
Al_2O_3, C, B, SiC, SiO_2	Al_2O_3, $3Al_2O_3 \cdot 2SiO_2$	Hot pressing
Al_2O_3, C, B, BN, SiC	Si_3N_4	Reaction sintering
$3Al_2O_3 \cdot 2SiO_2$	$3Al_2O_3 \cdot 2SiO_2$-Al_2O_3	Slip casting and firing
BN	BN	Hot pressing
BN	BN	Chemical vapor deposition
BN	BN	Firing of B_2O_3-containing composite in nitriding atmosphere
C	Al_2O_3, $3Al_2O_3 \cdot 2SiO_2$	Coating of fibers with LiC; sintering
C	Al_2O_3	Hot pressing
C	Carbides, borides, silicides, oxides	Hot pressing
C	Pyrolytic materials	Chemical vapor deposition
C	C-SiC, TiC	Chemical vapor deposition
C	Si_3N_4	Hot pressing, reaction sintering; coating of fibers — e.g., with SiC — to improve compatibility
C	Sialon, Si_3N_4, Si_3N_4-C, TaC	Hot pressing
C	TaC	Vacuum impregnation with precursor solution and pyrolyzing
C	C-TaC	Hot pressing of Ta-coated fibers
C	ZrB_2-Si-C	Hot pressing
C, fused SiO_2	Powdered ceramic	Application of aqueous slurry and drying
MgO	Cubic ZrO_2	Directional solidification
SiC	Si	Impregnation of carbon fiber preform with molten silicon
SiC	Si	Heating of a mixture of carbon fibers and silicon powder
SiC	Si	Infiltration of SiC fibers with molten silicon
SiC	SiC	Chemical vapor deposition
SiC	SiC, Si_3N_4, AlN, BN	Hot pressing or sintering
SiC	Si_3N_4	Reaction sintering
Si_3N_4	Si_3N_4	Hot pressing
ZrO_2	Al_2O_3	Directional solidification
ZrO_2	CaO-ZrO_2	Directional solidification
ZrO_2	MgO	Hot pressing
ZrO_2	ZrO_2	Hot pressing
ZrO_2	ZrO_2	Impregnation

Source: Ref 16, p 8-4

Manufactured fiber-reinforced ceramic composites and their corresponding processes

Processing	Composite (fiber-matrix)	Comments
Hot pressing	W-glass, Ni-glass, Mo-thoria, Mo-alumina, W-ceramic, stainless steel-alumina, C-glass, C-glass ceramic, C-MgO, C-Al_2O_3, ZrO_2-MgO, ZrO_2-ZrO_2, SiC-glass, SiC-glass-ceramics, Al_2O_3-glass, C-Si_3N_4 Ta-Si_3N_4	Fibers and matrix powder are mixed together and hot pressed to produce low-porosity composites, with uncracked matrices, if thermal expansion coefficients are matched. Aligned continuous fiber composites can have very high strengths.
Cold pressing and sintering	C-glass, metal fiber-ceramic	Fibers and matrix are mixed, cold pressed, and sintered. Disappointing results because the large shrinkage of the matrix during sintering produces cracked composites.
Devitrification	C-glass-ceramic, SiC-glass-ceramic	Fibers and glass powder are hot pressed at relatively low temperatures to give a reinforced glass. Further high-temperature heat treatment is used to devitrify the glass to a glass-ceramic. The C-glass system gives disappointingly low strengths, probably because of volume changes during devitrification. The SiC glass-ceramic system is reported to have good properties.
Reaction bonding	Reinforced Si_3N_4	Fibers are incorporated into flame-sprayed silicon that is subsequently reaction-sintered in nitrogen.
Slip-casting	Ceramic fiber-fused silica	Ceramic fibers are incorporated into slips of finely divided fused silica and fired. Increased porosity that is due to the presence of fiber usually results in a degradation of properties.
Plasma-spraying ...	Mo-Al_2O_3, W-Al_2O_3	Alumina powder is plasma-sprayed. Other unpublished work employing plasma-spraying is reported in France and Japan (1983). Processing is believed to be slow.
Chemical vapor infiltration and deposition	SiC fibers in SiC, C fibers in SiC	Fiber integrity can be preserved by lack of mechanical movement and relatively low process temperatures.

Source: Ref 6, p 927

Izod impact strength after elevated-temperature exposure for glass-reinforced ceramic composite(a)

Temperature (2-h exposure) °C	°F	Izod impact strength J/cm	ft·lb/in.
315	600	17.1	32.0
425	800	9.0	16.9

(a) S-944, style 181 fiberglass fabric in matrix of a modified aluminum phosphate.

Source: Ref 16, p 8-21

Fabrication methods for some ceramic whisker/ceramic matrix composites

Whiskers	Matrix	Fabrication method
$3Al_2O_3 \cdot 2SiO_2$, α-Al_2O_3, ZrO_2	Oxides and nitrides	Hot pressing
$3Al_2O_3 \cdot 2SiO_2$, α-Al_2O_3, SiC, Si_3N_4, ZnO	TiO_2	Hot pressing
$3Al_2O_3 \cdot SiO_2$	Al_2O_3, Al_2O_3-Mo, Cr_2O_3, ZrO_2, Al_2O_3-Cr, AlN, BN, Si_3N_4, V_2O_3, TiN, SiO_2	Hot pressing
α-Al_2O_3, AlN, SiC	$3Al_2O_3 \cdot 2SiO_2$-Al_2O_3	Hot pressing
Si_3N_4	Si_3N_4	Sintering
SiC, BN, C	Si_3N_4, AlN	Sintering or hot pressing
ZrO_2	Stabilized ZrO_2	Hot pressing
ZrO_2	MgO	Hot pressing
Ground whiskers	Several oxides	Powder metallurgy techniques

Source: Ref 16, p 8-4

Formulation and fabrication parameters of multidirectional continuous fiber reinforced ceramic-ceramic composites

Property/composite	Three-directional silica-silica	Four-directional silica-silica	Three-directional alumina-alumina	Three-directional alumina-silica	Three-directional boron nitride/boron nitride
Preform reinforcement type	Three-directional orthogonal	Four-directional cubic braided	Three-directional orthogonal	Three-directional orthogonal	Three-directional orthogonal
Fiber type	Fused quartz continuous	Fused quartz continuous	Polycrystalline alumina staple	Polycrystalline alumina staple	Continuous polycrystalline boron nitride
Fiber electrical resistivity, ohm · m	1×10^{16} (at 293 K) 2×10^{5} (at 1073 K)	1×10^{16} (at 293 K) 2×10^{5} (at 1073 K)	1×10^{10} (estimated)	1×10^{10} (estimated)	1×10^{8} (at 1075 K)
Fiber strength, GPa (10^6 psi)	0.7 (0.10)	0.7 (0.10)	1.4 (0.20)	1.4 (0.20)	0.35-0.7 (0.05-0.10)
Fiber volume fraction	50%	50%	30%	30%	40%
Matrix/densification	Colloidal silica	Colloidal silica	Colloidal alumina	Collodial silica	Nitrided boric oxide
Process or stabilization temperature, K	923-1013	923	110	923	2073
Composite bulk density, g/cm^3	1.6	1.6	1.9	2.0	1.6

Source: Ref 6, p 935

Performance properties of multidirectional continuous fiber-reinforced ceramic-ceramic composite materials at 300 K

Property	Three-directional silica-silica	Four-directional silica-silica	Three-directional alumina-alumina	Three-directional alumina-silica	Three-directional boron nitride-boron nitride
Tensile					
Modulus of elasticity, GPa (10^6 psi)	15.6 (2.26)	9.7-13.1 (1.4-1.90) (a)	36.3-5.26	33.8 (4.90)	15.4 (2.23)
Tensile strength, MPa (ksi)	26.7 (3.87)	20.4-26.6 (2.96-3.86) (a)	71.1 (10.3)	74.8 (10.8)	24.8 (3.60)
Tensile strain, %	0.2	1.2	0.2	0.2	0.2
Compressive					
Compressive modulus, GPa (10^6 psi)	21.9 (3.18)	8.6 (1.3)	31.4 (4.55)	...	29.2 (4.23)
Compressive strength, MPa (ksi)	144.8 (21.0)	70.6 (10.2)	224.7 (32.59)	...	36.5 (5.29)
Compressive strain, %	1.6	1.5	>0.6	...	0.2
Shear					
Shear modulus, GPa (10^6 psi)	1.5 (0.22)	4.4 (0.64)	3.4 (0.49)	1.7 (0.25)	...
Poisson's ratio	0.09	...	...	...	...
Coefficient of thermal expansion, mean, to 600 K, 10^{-6}/K	0.54	0.47	6.4	6.4	2.7
Thermal conductivity at 300 K, W/m·K (Btu·in./h·ft^2·°F)	0.66 (4.6)	0.65 (4.5)	1.62 (11.2)	0.68 (4.7)	9.0 (62.4)
Dielectric constant(b)	3.5	2.9	3.7	3.8(c)	3.0
Loss tangent(b)	0.0009	0.001	0.0045	0.004(c)	0.002

(a) Lower value for "water-desensitized" version corresponding to dielectric properties. (b) At 8.5 to 9.4 GHz, or as indicated. (c) At 35 GHz.

Source: Ref 6, p 937

Mechanical properties of Si_3N_4 composites containing 30 vol % of metal carbide dispersoid (2 μm average particle diameter)

Matrix	Dispersed phase	Density, g/cm^3	Hardness (KHN), GPa	Fracture toughness (K_{Ic}), MPa $\sqrt{m}$	Modulus of rupture, MPa: RT	Modulus of rupture, MPa: 1000°C	Modulus of rupture, MPa: 1200°C
Si_3N_4 + 6 wt % Y_2O_3	None	3.26	13.4 ± 0.3	4.8 ± 0.3	110.9 ± 1.6	88.3 ± 3.5	49.2 ± 5.0
Si_3N_4 + 6 wt % Y_2O_3	TiC	3.81	15.21 ± 0.3	4.4 ± 0.5	80.6 ± 5.9	120.4 ± 12.2	64.4 ± 2.9
	(Ti,W)C	4.55	14.06 ± 0.3	3.5 ± 0.3	75.5 ± 3.2	86 ± 0	52.9 ± 0.5
	WC	7.70	14.4 ± 0.4	5.2 ± 0.4	89.1 ± 31.8	136.4 ± 1.6	55.7 ± 0.5
	TaC	6.87	12.6 ± 0.2	4.6 ± 0.4	86.2 ± 7.3	124.5 ± 16.0	43.2 ± 2.0
	HfC	5.74	14.1 ± 0.4	3.6 ± 0.2	86 ± 0.8	...	68.6 ± 0.5
	SiC	3.24	13.6 ± 0.2	3.65 ± 0.5	97.6 ± 8.5	94.0 ± 4.9	52.3 ± 3.2
Al_2O_3	TiC	4.28	17.2 ± 0.2	3.2 ± 0.4	72.2 ± 13.0	69.4 ± 4.3	57.0 ± 4.1

Source: Ref 16, p 8-38

Selected properties of BN/Al_2O_3 and ZrO_2/Al_2O_3 composites

Material	Bend strength, MPa	Modulus of elasticity, (*E*), GPa	Impact energy, (γ_{Ic}), J/m^2	Critical temperature, (ΔT_c), °C	Fracture toughness: (K_{Ic}) MPa$\sqrt{m}$	Fracture toughness: Fracture strength, σ_f, MPa	Specific strength, σ_T/σ_i
30 vol % BN-1/Al_2O_3, ⊥ HPA	170	120	60	700-850(b)	3.8	...	...
30 vol % BN-2/Al_2O_3, ⊥ HPA	400	190	40->100	450(b)	3.9-9.0	...	...
30 vol % BN-2/Al_2O_3, ‖ HPA	160	140	25	...	2.6	...	...
30 vol % BN/Al_2O_3, ⊥ HPA(c)	200	200	3	400-800(b)	1.1	...	...
0.5 vol % ZrO_2/Al_2O_3	...	400	20	600	...	410	0.35
4.0 vol % ZrO_2/Al_2O_3	...	410	25	600	...	420	0.4
9.0 vol % ZrO_2/Al_2O_3	...	380	55	950	...	700	...
11.5 vol % ZrO_2/Al_2O_3	...	380	60	1150	...	780	0.6-0.75
14.0 vol % ZrO_2/Al_2O_3	...	375	35	800	...	680	0.65
19.0 vol % ZrO_2/Al_2O_3	...	350	35	800	...	350	0.65

(a) ⊥ HPA = stress direction perpendicular to hot pressing axis. ‖ HPA = stress direction parallel to hot pressing axis. (b) Determined by 22 °C (72 °F) water quench test using 3-mm-(0.19-in.-)square bars. (c) Calculated treating BN as pseudoporosity.

Source: Ref 16, p 8-24

Thermal and physical properties of glass-reinforced ceramic composites(a)

Specific gravity	1.8-1.9
Specific heat:	
kJ/kg·K	0.84
Btu/lb·°F	0.2
Coefficient of thermal expansion:	
Parallel to reinforcement	
10^{-6}/K	3.6
10^{-6}/°F	2.0
Perpendicular to reinforcement	
10^{-6}/K	1.1
10^{-6}/°F	0.6
Thermal conductivity:	
W/m·K	0.58
Btu·in./ft^2·h·°F	4.0
Total normal emissivity:	
At 425 °C (800 °F)	0.73
At 650 °C (1200 °F)	0.76

(a) S-944, style 181 fiberglass fabric in matrix of a modified aluminum phosphate.

Source: Ref 16, p 8-19

SiC/Ceramic

Types and properties of Si/SiC

		Properties at 25 °C (77 °F) (a)					
	SiC content,	Bond strength		Elastic modulus		Density	
Designation	vol %	MPa	ksi	GPa	10^6 psi	g/cm^3	lb/in.^{3}
G.E.-reported data							
Type TH(b)	80-85	483	70	393	57	3.30	0.012
Type THL	38-40	276	40	303	44	2.70	0.098
Type F(c)	20-25	172	25	200	29	2.60	0.094

Fansteel-reported data

Tensile strength at room temperature:	
MPa	827
ksi	120
Charpy impact strength:	
J	13.6
ft·lb	10
Oxidation resistance	Excellent
Thermal shock resistance	Excellent

(a) Measured on 2.54 × 2.54 × 15.8 mm (0.1 × 0.1 × 0.625 in.) specimens tested in three-point bending. (b) Unidirectional orientation. (c) Omnidirectional orientation.

Source: Ref 16, p 8-10

Physical properties of fiber-grain CVI SiC matrix composites

	Density		Modulus of rupture		Modulus of elasticity		Strain,
Material system	g/cm^3	lb/in.^{3}	MPa	ksi	GPa	10^6 psi	%
Carbon fibers/ SiC grain	2.06	0.074	40.0	5.8	20	2.9	0.57
Mullite fibers/ SiC grain	1.80	0.065	35.9	5.2	15.9	2.3	0.56
Silica fibers/ SiC grain	1.95	0.070	33.1	4.8	22.1	3.2	0.46

Source: Ref 16, p 8-6

Physical, thermal, and mechanical property data for silicon carbide composites

	Reported property value(a)			
Property	Silicon nitride-bonded SiC(b)	Silicon oxynitride-bonded SiC(c)	SiC bonded graphite aggregate(d)	Self-bonded SiC(e)
Density, g/cm^3	2.5-2.8	~2.7	2.3-2.8	3.10
Porosity, %	13-15	~18	8-26	~5
Maximum recommended working temperature:		1650		
Oxidizing environment				
°C	1650		815-1540	1650
°F	3000	3000	1500-2800	3000
Neutral environment		2205		
°C	2205		2205	2315
°F	4000	4000	4000	4200
Specific heat (27-1370 °C; 80-2500 °F):		0.84-1.55		
kJ/kg·K	0.84-1.55		0.71-1.47	0.71-1.38
Btu/lb·°F	0.2-0.37	0.2-0.37	0.17-0.35	0.17-0.33
Thermal conductivity (980 °C; 1800 °F):		16		
W/m·K	16		48-59	42
Btu/ft·h·°F	9.5	9.5	28-34	24
Coefficient of linear thermal expansion (21-1370 °C; 70-2500 °F):		1.4		
10^{-6}/°C	1.4		1.5(f)	1.51
10^{-6}/°F	2.6	2.6	2.7(f)	2.72
Bend strength:		41-55		
24 °C (75 °F)				
MPa	41-69		34-69	165
ksi	6-10	6-8	5-10	24
1095 °C (2000 °F)		55-62		
MPa	48-69		34-69	172
ksi	7-10	8-9	5-10	25
1500 °C (2730 °F)		~21		
MPa	21		48-83	124
ksi	3	~3	7-12	18
Compressive strength (24 °C; 75 °F):		~138		
MPa	138		434(f)	1380
ksi	20	~20	63(f)	200
Impact strength (21 °C; 70 °F):		...		
J	<0.113		<0.113	<0.113
in.·lb	<1	...	<1	<1
Modulus of elasticity:		117		
24 °C (75 °F)				
GPa	117		~241(f)	379
10^6 psi	17	17	~35(f)	55
1000 °C (1830 °F)		69		
GPa	69		207(f)	365
10^6 psi	10	10	30(f)	53
Bend-creep resistance—conditions to promote a tensile creep strain of 0.125%:				
Time, h	>300 (1095° C; 2000° F)		>630 (1205° C; 2200° F)	>1000 (1205° C; 2200° F)
Temperature (300 h)		...		
°C	>1095		>1205	>1300
°F	>2000	...	>2200	>2370
Bend stress (100 h)		...		
MPa	28-41 (1000 °C)		28-41 (1000 °C)	159 (1205 °C)
ksi	4-6 (1830 °F)	...	4-6 (1830 °F)	23 (2200 °F)

(a) Data were obtained primarily from company literature. (b) Composition is approximately 80 wt % SiC and 20 wt % Si_3N_4. (c) Composition is approximately 80 wt % SiC and 20 wt % Si_2ON_2 (silicon oxynitride). (d) Variable graphite content; compositions reportedly range from 78 to 50 wt % SiC (dense bond phase), 20 to 46 wt % graphite aggregate, and 1 to 4 wt % silicon. (e) Fine-grain, high-density material. (f) Value for low-graphite-content material.

Source: Ref 16, p 8-37

Mean matrix crack spacing and first cracking stress for SiC/RBSN composites

Fiber fraction, %	Matrix crack spacing, mm	Composite stress at which matrix first cracked, MPa
23 ± 3	2.0 ± 0.3[a]	237 ± 25[b]
40 ± 2	0.9 ± 0.2	293 + 15

[a]Standard deviation for 30 cracks on five bend specimens. [b]Standard deviation for 5 specimens measured in 3-point bend.

Source: Ref 16, p 8-8

Density and porosity data for SiC/RBSN composites

Volume fraction of fibers, %	Before nitriding: Density, g/cm^3	Before nitriding: Matrix porosity(a) %	After nitriding: Density, g/cm^3	After nitriding: Matrix porosity(a) %
0	1.56	35	1.98	37
23 ± 3	1.70	54(a)	2.19	39(a)
40 ± 2	1.90	51(a)	2.36	40(a)

(a) Matrix porosity calculated from composite density and from theoretical density for CVD SiC fiber (3.0 g/cm^3) and from density for silicon (2.4 g/cm^3) or for Si_3N_4 (3.2 g/cm^3).

Source: Ref 16, p 8-8

Room-temperature strengths of RBSN and SiC/RBSN composites

Test	Axial strength, MPa: 0% fiber	23 ± 3% fiber	40 ± 2% fiber
4-point bend (L/h[a] = 15)	107 ± 26[b]	539 ± 48[b]	616 ± 36[b]
4-point bend (L/h[a] = 45)	...	675 ± 42	868 ± 32
3-point bend (L/h[a] = 35)	...	717 ± 80	958 ± 45
Tensile[c]	...	352 ± 73	536 ± 20

[a]L/h refers to span-to-height ratio of test specimen (h ≈ 1.2 mm). [b]Tested at 50-mm gauge length. [c]Standard deviation for five tests.

Source: Ref 16, p 8-8

Physical properties of various ceramic- and graphite-reinforced CVI SiC composites

Material system	Density g/cm^3	Density lb/in.^{3}	Modulus of rupture MPa	Modulus of rupture ksi	Modulus of elasticity GPa	Modulus of elasticity 10^6 psi	Strain, %
PAN 8HS/SWB	1.74	0.063	229	33.2	40.7	5.9	3.0
PAN knit/KFB	1.95	0.070	215	31.2	43.4	6.3	3.2
Nicalon/8HS	2.09	0.076	262(a)	38.0(a)	53.1	7.7	0.64
Saffil/Al_2O_3 paper	...	...	296	43.0	208.3	30.2	1.0
Saffil/Al_2O_3 paper	...	...	125	18.2	94.5	13.7	1.0
Saffil/Al_2O_3 paper	...	...	101	14.6	17.9	2.6	0.8

(a) Ceramic-grade fiber; results for nonceramic-grade fiber were 108 MPa (15.6 ksi) high and 37 MPa (5.4 ksi) low.

Source: Ref 16, p 8-6

Thermal properties of SiC whisker reinforced ceramics

Composite	Thermal conductivity at 22 °C (70 °F), W/m · K	Thermal conductivity at 22 °C (70 °F), Btu · in./h · ft² · °F	Thermal conductivity at 600 °C (1110 °F), W/m · K	Thermal conductivity at 600 °C (1110 °F), Btu · in./h · ft² · °F	Linear coefficient of thermal expansion at 22-1100 °C (70-2010 °F), 10^{-6}/K
Alumina	36 ± 5	250 ± 35	12 ± 3	85 ± 20	7.8-8.2
Alumina-20 vol% SiC whiskers	32	220	16	110	7.35
30 vol% SiC whiskers	...	...	...	...	6.70
60 vol% SiC whiskers	...	...	...	...	5.82
SiC	95	660	50	350	4.8
20 vol% SiC whiskers-mullite	7.2	50			5.60

Source: Ref 6, p 943

Shock strengths of SiC-whisker-reinforced ceramics

Matrix	Whisker content, vol %	Aid/wt %	Temperature °C	Temperature °F	Pressure MPa	Pressure ksi	Time, min	Density g/cm³	Density lb/in.³	Charpy shock strength, J (in.·lb) Room temperature	Charpy shock strength, J (in.·lb) 1095 °C (2000 °F)	Charpy shock strength, J (in.·lb) 1315 °C (2400 °F)
Si_3N_4	5	MgO/1	1700	3092	28	4	65	3.15	0.114	0.150 (1.33)	0.160 (1.40)	0.15 (1.31)
	10	MgO/4	1700	3092	28	4	65	3.12	0.113	0.195 (1.73)	0.165 (1.46)	0.11 (0.99)
	38	MgO/1	1700	3092	28	4	75	3.00	0.108	0.094 (0.83)	0.080 (0.73)	0.21 (1.85)
	10	MgO/1	1700	3092	28	4	95	2.79	0.101	0.069 (0.61)	0.230 (2.03)	0.146 (1.29)
SiC	38	Al_2O_3/3; C/2	2140	3884	28	4	180	3.12	0.113	0.094 (0.83)	0.076 (0.67)	0.069 (0.61)
	50	Al_2O_3/3	2170	3938	28	4	180	3.11	0.112	0.072 (0.64)	0.079 (0.70)	0.069 (0.61)
	65	Al_2O_3/3	2140	3884	28	4	180	2.93	0.106	0.067 (0.59)	0.060 (0.54)	0.11 (0.96)

Source: Ref 16, p 8-12

Graphite Ceramic

Thermal-strain parameters (at 20 °C) and measured works of fracture for 20 vol % carbon-fiber composites

Material	Work of fracture, J/m²	$(\alpha_m - \alpha_f)\Delta T$
MgO	10	...
CM	110	6.6×10^{-3}
Al_2O_3	38-66	...
CA	40	0.4×10^{-3}
Pyrex	4	...
CP	344	-2.3×10^{-3}
Glass-ceramic	4	...
CGC	100	-6.0×10^{-3}

Source: Ref 16, p 8-13

Properties of HMS graphite/$Li_2O \cdot Al_2O_3 \cdot 8SiO_2$ composites(a) tested perpendicular and parallel to hot pressing direction

Test direction	Modulus of rupture, MPa	Modulus of rupture, ksi	Average modulus of rupture, MPa	Average modulus of rupture, ksi	Standard deviation, MPa	Standard deviation, ksi	Variance	Modulus of elasticity, GPa	Modulus of elasticity, 10^6 psi	Average modulus of elasticity, GPa	Average modulus of elasticity, 10^6 psi	Standard deviation, GPa	Standard deviation, 10^6 psi	Variance
Perpendicular to hot pressing direction	804.6	116.7	793.2	115.0	57	8.2	7.1	146	21.2	144	20.9	11.7	1.7	8.1
	756.4	109.7						149	21.6					
	777.7	112.8						143	20.7					
	766.7	111.2						123	17.9					
	721.9	104.7						134	19.5					
	892.9	129.5						153	22.2					
	832.2	120.7						158	22.9					
Parallel to hot pressing direction	837.7	121.5	860.7	124.8	41	5.9	4.7	159	23.0	159	23.0	4.8	0.7	3.1
	897.7	130.2						155	22.5					
	815.0	118.2						152	22.0					
	877.0	127.2						160	23.2					
	912.2	132.3						166	24.1					
	824.6	119.6						159	23.0					

(a) Fiber content, 39.6 vol . Bulk density, 2.06 g/cm^3.

Source: Ref 16, p 8-18

Properties of HMS graphite fiber/$Li_2O \cdot Al_2O_3 \cdot 8SiO_2$ composites(a)

Fiber content, vol %	Bulk density, g/cm^3	Modulus of rupture, MPa	Modulus of rupture, ksi	Average modulus of rupture, MPa	Average modulus of rupture, ksi	Standard deviation, MPa	Standard deviation, ksi	Variance
29.0	2.19	637	92.4	633	91.8	37	5.4	5.9
		641	93.0					
		689	99.9					
		601	87.1					
		597	86.6					
33.1	2.16	698.4	101.3	782.1	113.4	48	7.0	6.2
		799.1	115.9					
		821.2	119.1					
		806.7	117.0					
		785.3	113.9					
35.1	2.17	845.3	122.6	834.4	121.0	58	8.4	7.0
		878.4	127.4					
		830.8	120.5					
		737.7	107.0					
		879.8	127.6					
36.3	2.15	804.6	116.7	878.4	127.4	51	7.4	5.8
		839.1	121.7					
		896.3	130.0					
		951.5	138.0					
		892.9	129.5					
		886.0	128.5					

(a) Hot pressed for 5 min at 1375 °C (2510 °F) and 6.9 MPa (1.0 ksi).

Source: Ref 16, p 8-19

Mechanical properties of graphite fiber (MODMOR II) reinforced ceramic composites

Property[b]	RT	650 °C (1200 °F)
Flexural strength:		
MPa	153	194
ksi	22.2	28.2
Flexural modulus:		
GPa	86.2	80.7
10^6 psi	12.5	11.7
Compressive strength:		
MPa	205	247
ksi	29.8	35.8
Compressive modulus:		
GPa	106	...
10^6 psi	15.4	...
Tensile strength[c]:		
MPa	303	...
ksi	44.0	...
Tensile modulus:		
GPa	68.7	...
10^6 psi	9.96	...
Interlaminar shear strength:		
MPa	13.0	13.2
ksi	1.89	1.92

[a]Data from Acurex Corp. Ceramic matrix is Chemceram, a modified aluminum phosphate. [b]Flexural specimens failed in shear. Specific gravity, 1.77 g/cm^3 (0.064 lb/in.^{3}). Load applied once temperature achieved on specimen. [c]228 MPa (33.0 ksi) at 1260 °C (2300 °F); 210 MPa (30.4 ksi) at 1650 °C (3000 °F).

Source: Ref 16, p 8-13

Izod impact energy of HMS graphite fiber/$Li_2O \cdot Al_2O_3 \cdot 8SiO_2$ composites, Udimet 700, MAR-M200, and alumina

Composition	Test bar No.	Density, g/cm^3	Fiber content, vol %	Notched	Izod impact energy J/cm^2	Izod impact energy ft · lb/in.^{2}
HMS graphite fiber/	2	2.19	31.6	No	15.9	75.5
$Li_2O \cdot Al_2O_3 \cdot 8SiO_2$	1	2.19	31.6	Yes	5.94	28.3
composites	3	2.18	34.2	Yes	4.20	20.0
	3	2.18	34.4	Yes	4.68	22.3
Udimet 700	2	...	...	Yes	34.31	163.4
MAR-M200	3	...	...	Yes	18.9	90.0
Al_2O_3 AP 35	3	...	...	Yes	0.315	1.50
	2	...	...	No	0.701	3.34

Source: Ref 16, p 8-18

Modulus and strength fractions of rule-of-mixture values for some carbon-fiber-reinforced ceramics

Matrix	Modulus fraction	Strength fraction
Pyrex glass	0.86	0.81
Glass-ceramic	0.71–0.80	0.81
Carbon	1.00	0.50

[a]Fibers are continuous and aligned, with V_f between 0.4 and 0.5.

Source: Ref 16, p 8-13

Modulus of rupture vs. thermal shock cycles between room temperature and 1200° C (2190° F) for graphite fiber/$Li_2 \cdot Al_2O_3 \cdot 8SiO_2$ composites

Thermal shock cycles	Modulus of rupture MPa	Modulus of rupture ksi	Average modulus of rupture MPa	Average modulus of rupture ksi
0	951.5	138.0	857.7	124.4
	839.8	121.8		
	781.9	113.4		
1	897.7	130.2	893.6	129.6
	889.4	129.0		
5	838.4	121.6	861.8	125.0
	885.3	128.4		

Source: Ref 16, p 8-18

Glass Matrix

Coefficients of thermal expansion for borosilicate 7740 glass matrix composites

Filament	Orientation	Coefficient of thermal expansion(a), $10^{-6}/°C$
35 vol % SiC monofilament	0°	4.20
	90°	4.60
40 vol % SiC yarn	0°	3.25
	90°	2.70

(a) From 22 to 500 °C

Source: Ref 16, p 8-9

Coefficients of thermal expansion for graphite fiber unidirectionally reinforced borosilicate matrix composites

Fiber type	Fiber elastic modulus GPa	Fiber content, vol %	Coefficient of thermal expansion(a) $10^{-6}/°C$ 0°	90°
Thornel 300	234	54	-0.10	+4.6
HM	350	70	-0.50	+6.5
P-100	654	50	-1.0	+4.4
Chopped Cel 6000	234	30	1.7	1.7(4.2)
Borosilicate glass	...	0	3.25	3.25

(a) From 22 to 500 °C

Source: Ref 16, p 8-15

Properties of SiC-fiber-reinforced borosilicate 7740 glass

Property	Monofilament		Yarn
Fiber content, vol %	35	65	40
Density, g/cm^3	2.6	2.9	2.4
Axial flexural strength, MPa:			
At 22 °C	650	830	290
At 350 °C	...	930	360
At 600 °C	825	1240	520
Axial elastic modulus, GPa, at 22 °C	185	290	120
Axial fracture toughness, MN·m$^{3/2}$:			
At 22 °C	18.8	...	11.5
At 600 °C	14.3	...	7.0

Source: Ref 16, p 8-9

Three-point flexural strengths and elastic moduli of SiC-yarn-reinforced 7930 glass composites

Test temperature, °C	Specimens fabricated at 1500 °C: Flexural strength, MPa	Specimens fabricated at 1500 °C: Flexural modulus, GPa	Specimens fabricated at 1600 °C: Flexural strength, MPa	Specimens fabricated at 1600 °C: Flexural modulus, GPa
22	467	111	509	109
	413	107	482	97.1
	447	112	527	101
950	643	108	724	101
	565	88	657	102
1050	...	...	742	83
	...	...	668	87
1150	541	64	...	...
	498	65	...	...
1200	419	60	...	...
1250	243	48	...	...
	280	52	...	...

Source: Ref 16, p 8-10

Fracture toughness of HM-fiber reinforced borosilicate glass matrix composites

Test temperature, °C	Test speed, cm/s	Fracture toughness MPa$\sqrt{m}$
22	330.0	21.4
22	0.002	22.4
600	330.0	15.8
650	330.0	19.0

Source: Ref 16, p 8-14

Coefficients of thermal expansion for some graphite-reinforced borosilicate glass composites

Composite type	Coefficient of thermal expansion, 10^{-6}/K(a): Longitudinal	Transverse
60 vol % GY-70/7740	-0.29	7.6
60 vol % HMS/7740	-1.0	3.6
60 vol % T-300/7740	0.38	4.3

(a) From 295 to 423 K

Source: Ref 16, p 8-14

Design, Tooling and Manufacturing

DESIGN, TOOLING AND MANUFACTURING

Design, Tooling and Manufacturing

Composite Design Considerations

Physical properties relating to reinforced plastic design considerations

Mechanical properties
- Tensile properties
- Compressive properties
- Flexural properties
- Shear properties
- Impact strength
- Properties at high rates of loading (dynamic properties)
- Bearing strength
- Surface hardness
- Creep properties (creep-rupture and stress-relaxation)
- Fatigue (cyclic properties)
- Poisson's ratio
- Notch sensitivity
- Shatterproofness
- Shockproofness
- Tear resistance

Thermal properties
- Thermal conductivity
- Thermal expansion
- Specific heat
- Flow temperature
- Flammability (flame resistance)
- Heat-distortion (deflection temperature under load)
- Thermal shrinkage
- Maximum safe operating temperature
- Ignition properties
- Brittleness temperature

Electrical properties
- Arc resistance
- Electrical resistance (insulation resistance—volume and surface)
- Dielectric strength and dielectric breakdown voltage
- Dielectric constant and power factor

Optical properties
- Index of refraction (refractive index)
- Light diffusion
- Crazing resistance
- Spectral transmission (haze)
- Internal stress (transparent plastics)
- Surface stability, optical
- Optical uniformity and distortion

Chemical and permanence properties
- Water absorption
- Water vapor permeability (diffusion)—gas transmission rate
- Sunlight and weather exposure (aging)
- Resistance to chemical reagents
- Effects of radiation
- Toxicity
- Volatile loss (outgassing)
- Stress-crazing
- Impact sensitivity (LOX)
- Accelerated service (temperature and humidity)

Source: Ref 16, p 6-9

Characteristics of three-directional woven preforms

Material	Bulk density, g/cm³	No. of yarn bundles X	Y	Z	Center-to-center bundle spacing X, Y mm	X, Y in.	Z mm	Z in.	Fiber, vol%(a) X	Y	Z
Thornel 50(b)	0.64	1	1	1	0.56	0.022	0.58	0.023	0.14	0.14	0.13
	0.75	1	1	2	0.71	0.028	0.58	0.023	0.11	0.11	0.23
	0.68	2	2	1	1.02	0.040	0.58	0.023	0.14	0.14	0.12
	0.80	2	2	6	0.69	0.027	1.02	0.040	0.12	0.12	0.24
Thornel 75(b)	0.70	1	1	2	0.56	0.022	0.58	0.023	0.09	0.09	0.17
	0.65	2	2	1	0.84	0.033	0.58	0.023	0.12	0.12	0.09
	0.72	2	2	2	1.07	0.042	0.58	0.023	0.09	0.09	0.18

(a) Volume fraction of total preform volume occupied by fiber in each orthogonal direction. (b) Center-to-center bundle spacing.

Source: Ref 6, p 879

Design guides for fiberglass composites

Characteristic		Compression molding: Sheet molding compound	Compression molding: Bulk molding compound	Compression molding: Pre-form molding	Injection molding (thermo-plastics)	Cold press molding	Spray-up and hand lay-up
Minimum inside radius, mm (in.)		1.59 (1/16)	1.59 (1/16)	3.18 (1/8)	1.59 (1/16)	6.35 (1/4)	6.35 (1/4)
Molded-in holes	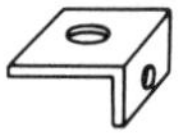	Yes[a]	Yes[a]	Yes[a]	Yes[a]	No	Large
Trimmed in mold		Yes	Yes	Yes	No	Yes	No
Core pull and slides		Yes	Yes	No	Yes	No	No
Undercuts		Yes	Yes[b]	No	Yes	No	Yes[b]
Minimum recommended draft, °/in.	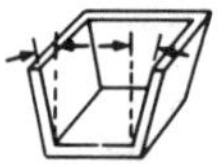	Depths of 6.35–152 mm (1/4–6 in.), 1–3°; depths of 152 mm (6 in.) and over, 3° or as required				2° 3°	0°
Minimum practical thickness, mm (in.)	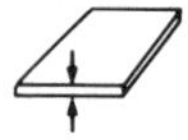	1.27 (0.50)	1.52 (0.060)	0.76 (0.030)	0.89 (0.035)	2.03 (0.080)	1.52 (0.060)
Maximum practical thickness, mm (in.)	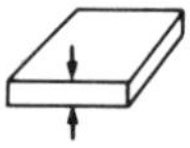	25.4 (1)	25.4 (1)	6.35 (0.250)	12.7 (0.500)	12.7 (0.500)	No limit
Normal thickness variation, mm (in.)	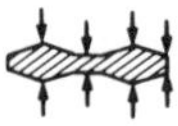	±0.13 (±0.005)	±0.13 (±0.005)	±0.20 (±0.008)	±0.13 (±0.005)	±0.25 (±0.010)	±0.50 (±0.020)
Maximum thickness build-up, heavy build-up and increased cycle	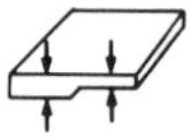	As required	As required	2-to-1 maximum	As required	2-to-1 maximum	As required
Corrugated sections		Yes	Yes	Yes	Yes	Yes	Yes
Metal inserts		Yes	Yes	NR[c]	Yes	No	Yes
Bosses		Yes	Yes	Yes	Yes	NR[c]	Yes
Ribs		As required	Yes	NR[c]	Yes	NR[c]	Yes
Molded-in labels		Yes	Yes	Yes	No	Yes	Yes
Raised numbers		Yes	Yes	Yes	Yes	Yes	Yes
Finished surfaces (reproduces mold surface)	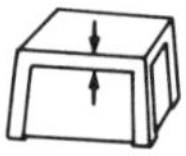	Two	Two	Two	Two	Two	One

[a]Parallel or perpendicular to ram action only. [b]With slides in tooling, or split mold. [c]Not recommended. [d]Owens-Corning Fiberglas Corp.

Source: Ref 16, p 6-25

Reinforcement efficiencies of selected composites

Matrix	Reinforcement	Composite efficiency (Φ)	Tensile strength (σ), MPa	Strain, (ε)
Epoxy	Twill weave	0.39	251.3	0.26
Epoxy	Noncrimp fabric	0.42	427.2	0.41
Epoxy	Random mat	0.43	250.4	0.22
PP(a)	Short chopped	0.19	103.4	0.17
PP(b)	Swirl mat	0.19	120.0	0.20
SMC	Long chopped	0.35	158.6	0.13

(a) Injection molded. (b) Azdel.

Source: Ref 28, p 256

Reinforcement efficiencies of thermoplastic composites

Matrix	Reinforcement	Composite efficiency (Φ)	σ, MPa	ε
PP	Long chopped	0.19	120.0	0.20
PP	Noncrimp fabric	0.24	275.1	0.44
PP	Twill weave	0.33	236.5	0.27
PP	Twill weave	0.45	270.0	0.23
PET	Twill weave	0.37	237.1	0.21
PBT	Twill weave	0.28	208.4	0.24
PBT	Long chopped	0.16	120.6	0.20
Nylon 12	Atochem weave	0.33	209.9	0.24
PEEK	Random mat	0.31	227.5	0.21

Source: Ref 28, p 256

Typical fabric styles and composite properties

Weave	Yarns/in. Warp × Fill	Weight kg/m^2	Weight oz/yd^2	Thickness at 25 kPa (3.4 psi) mm	Thickness at 25 kPa (3.4 psi) in.	Yarn (carbon)
Eight-harness satin	24 × 23	0.370	10.9	0.46	0.018	Thornel 300 3K
Eight-harness satin	24 × 23	0.370	10.9	0.48	0.019	Celion 3K
Plain	12½ × 12½	0.190	5.6	0.30	0.012	Thornel 300 3K, Kevlar aramid tracers
Five-harness satin	24 × 24	0.125	3.7	0.20	0.008	Thornel 300 1K
CFS	24 × 12	0.20	6.0	0.23	0.009	Celion 3K warp 150 l/o glass fill
Plain	11½ × 11½	0.19	5.7	0.25	0.010	Magnamite AS-4 3K
Five-harness satin	11 × 11	0.370	10.9	0.50	0.020	Magnamite AS-4 6K
Plain	8 × 8	0.525	15.5	0.81	0.032	Celion 12K
Eight-harness satin	10½ × 10½	0.755	22.2	1.0	0.040	Thornel 300 15K
Plain	10 × 10	0.345	10.2	0.48	0.019	75 l/o glass warp, Grafil E/XA-S 12K fill
8HS	21 × 21	0.393	11.6	0.38	0.015	HITEX 3K

Typical composite properties (balanced weave)

Tensile strength, MPa (ksi)	620-690 (90-100)
Tensile modulus, GPa (10^6 psi)	69-76 (10-11)
Flexural strength, MPa (ksi)	690-900 (100-130)
Flexural modulus, GPa (10^6 psi)	62-69 (9-10)
Compressive strength, MPa (ksi)	620-690 (90-100)
Compressive modulus, GPa (10^6 psi)	62-69 (9-10)
Short beam shear strength, kPa (psi)	55-69 (8-10)
Specific gravity	1.6

Source: Ref 6, p 126

Methods for determining composite density

Given	Unknown	Equation
% Fiber by volume	% Fiber by weight	$\dfrac{va}{va + (1 - v)b}$
Fiber density = a Fiber volume = v	% Resin by weight	$\dfrac{(1 - v)b}{va + (1 - v)b}$
Fiber weight = $v \cdot a$ Resin density = b	Composite density	$va + (1 - v)b$
Resin volume = $1 - v$ Resin weight = $(1 - v) \cdot b$	Bulk factor	$1/v$
% Resin by volume	% Fiber by volume	$\dfrac{(1 - w)/a}{(1 - w)/a + w/b}$
Fiber density = a Fiber weight = $1 - w$	% Resin by volume	$\dfrac{w/b}{(1 - w)/a + w/b}$
Fiber volume = $(1 - w)/a$ Resin density = b Resin weight = w	Composite density	$\dfrac{1}{(1 - w)/a + w/b}$
Resin volume = w/b	Bulk factor	$1 + \dfrac{wa}{(1 - w)b}$
Bulk factor = BF	% Fiber by weight	$\dfrac{a}{a + (BF - 1)b}$
Fiber density = a Fiber volume = 1	% Resin by weight	$\dfrac{(BF - 1)b}{a + (BF - 1)b}$
Fiber weight = a Resin density = b	% Fiber by volume	$1/BF$
Resin volume = $BF - 1$ Resin weight = $(BF - 1)b$	% Resin by volume	$(BF - 1)/BF$
	Composite density	$\dfrac{a + (BF - 1)b}{BF}$

Source: Ref 6, p 509

Properties of woven cloth in epoxy

Property	100% "S2" glass		100% Kevlar		100% carbon	
Tensile strength (0°), MPa (ksi)	495	(72)	565	(82)	565	(82)
Tensile modulus (0°), GPa (10^6 psi)	30	(4.3)	42	(6.1)	71.0	(10.3)
Flexural strength, (0°), MPa (ksi)	670	(97)	475	(69)	600	(87)
Flexural modulus (0°), GPa (10^6 psi)	26	(3.8)	33	(4.8)	61	(8.9)
Short-beam shear strength, MPa (ksi)	52	(7.5)	28	(4.0)	59	(8.5)

Source: Ref 19, p 58

Effects of weave pattern on fiberglass composite mechanical properties

Fabric style(a)	Plies, number	Resin content, wt%	Thickness, mm	Thickness, in.	Flexural strength, MPa	Flexural strength, ksi	Flexural modulus, GPa	Flexural modulus, 10^6 psi	Compressive strength, MPa	Compressive strength, ksi	Tensile strength, MPa	Tensile strength, ksi
7628	18	37.1	3.15	0.124	371	53.8	26.8	3.89	177	25.7	317	45.9
76281(b)	18	36.7	3.07	0.121	584	84.7	23.5	3.41	393	57.0	408	59.2
16-149	12	36.5	3.05	0.120	436	63.2	26.3	3.81	331	48.0	405	58.7
7781(c)	12	37.6	3.05	0.120	600	87.0	22.3	3.24	443	64.3	414	60.0

(a) Each material is polyester compatible. (b) 5-end satin weave version of style 7781. (c) 8-end satin weave.

Source: Ref 6, p 150

Typical composite efficiencies attained in reinforced plastics

Fiber configuration		Fiber length	Total fiber content (V_f), vol %	F_{long}(a), ksi (MPa) F_{theor}(b)	F_{test}(c)	Composite efficiency(d) %
Filament-wound	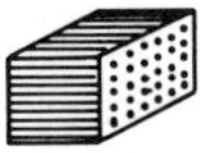	Continuous	0.77	310 (2140)	180 (1240)	58.0
Cross-laminated fibers	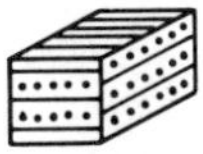	Continuous	0.48	197 (1360)	72.5 (500)	36.8
Cloth-laminated fibers		Continuous	0.48	197 (1360)	43.0 (296)	21.8
Mat-laminated fibers	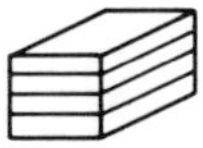	Continuous	0.48	197 (1360)	57.2 (394)	29.0
Chopped fiber systems (random)		Noncontinuous	0.13	60.7 (418)	15.0 (103)	24.7
Glass flake composites	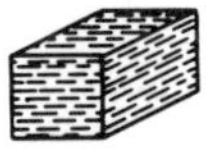	Noncontinuous	0.70	165.5 (1141)	20.0 (138)	12.1

(a) F_{long} = tensile strength in direction of greatest fiber content (longitudinal), if there is one. (b) Theoretical strength based on "rule of mixtures": $F_{theor} = V_f S_f + (1 - V_f) S_m$ = 400 ksi (2758 MPa)—typical boron or carbon fiber strength, and S_m = 10 ksi (69 MPa)—typical resin strength. (c) F_{test} = typical experimental strength values. (d) Composite efficiency = $(F_{test}/F_{theor}) \times 100$.

Source: Ref 79, p 6-8

Composite Tooling and Machining

Properties of typical composite tooling materials

Material	Coefficient of thermal expansion, 10^{-6}/K	Thermal conductivity, W/m·K	Btu·in./h·ft$^{2\circ}$ F
Fiberglass-epoxy	7.9		
Graphite-epoxy	-0.9	0.222	0.15
Aluminum	23.0	0.221	1.53
Steel	13.9	0.048	0.33
Electroless nickel	13.3	0.035	0.24

(a) Includes base or substructure.

Source: Ref 6, p 703

Parameters for composite trimming

Operation	Equipment	Cutter type	Speed m/s	Speed ft/min
Straight-line cuts	Pneumatic saw	Diamond-coated circular saw(a)	60-90	12 000-18 000
	Hand router	Diamond router(b) Carbide router(c)	10-15	2000-3000
Irregular outline	Hand router	Diamond router(b) Carbide router(c)	10-15	2000-3000
Chamfer, deburr	Hand	Abrasive drum(d)	NA	NA
Finish operations	Hand drill motor	Abrasive disc(d)	5-60	1000-12 000
	Hand	Abrasive cloth(d)	NA	NA

(a) Diamond circular saw 0.050 kerf, 36/44 grit. (b) Diamond routers, 36/44 grit roughing, 80/100 grit finishing. (c) Carbide diamond-shaped, chisel-cut routers. (d) 80 grit (rough), 220 grit (finish)

Source: Ref 6, p 668

Parameters for peck drilling

Material	Speed, rpm	Feed rate per revolution mm	Feed rate per revolution in.	Peck cycle, in.
Titanium	550	0.050	0.002	60 min
Graphite	550	0.10	0.004	30 min
Aluminum	550	0.10	0.004	30 min

Source: Ref 6, p 671

Performance of tooling materials

Material	Coefficient of thermal expansion 10^{-6}/K	Thermal correction required 0.9-m (3-ft) part mm	0.9-m (3-ft) part in.	9-m (30-ft) part mm	9-m (30-ft) part in.	Springback of square corners, degrees (b)
Graphite-epoxy(a)	3.6	0.28	0.011	2.79	0.110	1°15′
Glass-epoxy(a)	11.7-13.1	0.76	0.030	9.14	0.360	...
Steel	12.1	0.76	0.030	9.14	0.360	1°30′
Nickel electroform	13.3	0.76	0.030	8.89	0.350	...
Fiberglass wet lay-up	14.4-18.0	...	...	...	...	...
Aluminum	22.5	1.52	0.060	13.97	0.550	2°0′

(a) MXG-7620 resin. (b) Angle on tool must be larger than engineering call-out.

Source: Ref 6, p 586

Thermal characteristics of tooling materials

Tool material	Tool thickness mm	Tool thickness in.	Tool rise time(a) (ambient to 110 °C, or 230 °F) min	Overshoot (degrees over ambient) °C	Overshoot (degrees over ambient) °F
Aluminum	6.4	0.25	41.6	13	23
Steel	6.4	0.25	(b)	9	15
Carbon-epoxy	6.4	0.25	45.8	14	24
Aluminum	25.4	1.00	48.0	(b)	(b)
Steel	12.7	0.50	48.6	7	12
Carbon-epoxy	12.7	0.50	51.7	13	23
Steel	38.1	1.50	73.8	2	3
Aluminum	101.6	4.00	86.3	0	0

(a) In response to 2 °C/min (4 °F/min) ramp from ambient to 240 °F. (b) Testing difficulties prevented accurate determination of values.

Source: Ref 6, p 578

Approved fastener materials

Materials being joined	Fastener material: Preferred	Fastener material: Recommended with barrier coatings	Fastener material: Not recommended
Graphite-epoxy(a) to aluminum(b)	Titanium pin with aluminum collars or nuts bearing on aluminum	Multiphase alloys, Inco 718, or austenitic stainless steel, with aluminum or stainless steel collars or nuts bearing on aluminum, or titanium-columbium systems	Copper, brass, aluminum, low-alloy steel, martensitic stainless steels, or cadmium-plated fastener systems
Graphite-epoxy to titanium, A286, austenitic stainless steel, or graphite-epoxy	Multiphase alloys, Monel fastening systems, Inconel, or titanium, with stainless steel, Inconel, or titanium collars or nuts	Stainless steel pin or screw with aluminum collars or nuts bearing on stainless steel or titanium. Stainless steel collars or nuts bearing on graphite-epoxy, titanium or stainless steel, or titanium-columbium systems	Aluminum collars or nuts bearing on graphite-epoxy, copper, brass, aluminum, or low-alloy steel elements, or cadmium-plated fastener systems

(a) Graphite is used to mean either graphite or carbon reinforcement. (b) These materials are incompatible at the faying surface without a barrier coating.

Source: Ref 6, p 709

Parameters for graphite drilling

Hole diam, max		Speed, rpm	Feed rate per revolution	
mm	in.		mm	in.
3.967	0.1562	2800	0.025-0.040	0.0010-0.0015
4.763	0.1875	2800	0.025-0.040	0.0010-0.0015
6.350	0.2500	2800	0.025-0.040	0.0010-0.0015
7.938	0.3125	1800	0.045-0.055	0.0017-0.0022
9.525	0.3750	1800	0.045-0.055	0.0017-0.0022

Source: Ref 6, p 669

Beam press versus roller press

	Beam	Roller
Tonnage	70 - 200	Unlimited
Cutting speed	125 mm/s (5 in./s)	430 mm/s (17 in./s)
Material ply height	12.7 mm (½ in.) or more, depending on capacity of die and press stroke	3.2-4.8 mm (⅛ - 3/16 in.) max
Die maintenance	Up and down hydraulic action, easy on dies	Mechanical pincer action can distort die knives. More die repairs
Press maintenance	Minimal	Minimal
Purchase price	\$50 000-\$200 000	\$10 000-\$40 000
Cutting pad	Fixed-in head	Rides on top of die and tends to curl because of the roller action. Disposable pads required
Cuttable materials	All composite unidirectional tapes and fabrics	Same

Source: Ref 6, p 611

Faying-surface sealant categories

Material type	Use	Specifications(a)
Polysulfide(b)	General-purpose corrosion-inhibiting sealant	MIL-S-81733C(c)
	Fuel tank sealant (also for general-purpose nonfuel areas)	MIL-S-8802E
	General-purpose low-adhesion sealant	MIL-S-8784B
	Low-adhesion sealant for fuel tank areas	AMS 3267(d)
	High-temperature sealant (to 180 °C, or 360 °F, peak)	MIL-S-83430A
Silicone(e)	High temperature sealing applications (-60 to 205 °C, or -80 to 400 °F) (f)	AMS 3373

(a) Some of these specifications have several types, grades, and classes. A careful review of these categories and the recommended use for each must be made before sealant selection. (b) Manufacturers of polysulfide sealant include Products Research and Chemical Corporation, Goal Chemical Sealants Corporation, and Chem-Seal Corporation. (c) Material conforming to this specification is recommended for permanent graphite composite assemblies because of the corrosion-inhibiting formulation and also because of the long pot life, which permits a long assembly time for complex structures. (d) Material conforming to this specification is recommended for removable graphite composite assemblies because of the corrosion-inhibiting formulation. (e) Manufacturers of silcone sealant include General Electric Company, Silicone Division, and Dow-Corning Corporation. (f) A two-part silicone sealant, with catalytic curing agents, must be used. One-part silcones require moisture (from exposure to air) for cure and thus cannot be used on a faying surface with more than 25-mm (1-in.) width.

Source: Ref 6, p 719

Cutting speeds versus materials and thicknesses

Material	Thickness mm	Thickness in.	Cutting speed m/min	Cutting speed in./min
Abrasive cutting at 240 MPa (35 ksi), 100-grit garnet, 20 hp				
Resin-impregnated graphite, aramid, glass fibers	3.2	0.125	1.60	63
	6.4	0.250	0.75	30
	12.7	0.500	0.46	18
	19.1	0.750	0.30	12
	25.4	1.000	0.13	5
Ferrous metal, stainless, Hastelloy, Inconel, mild steel, high carbon, 60 HRC	1.6	0.063	0.50	20
	3.2	0.125	0.28	11
	6.4	0.250	0.18	7
	12.7	0.500	0.075	3
	25.4	1.000	0.019	0.75
Aluminum, magnesium titanium, brass alloys	1.6	0.063	1.47	58
	3.2	0.125	0.74	29
	6.4	0.250	0.41	16
	12.7	0.500	0.25	10
	25.4	1.000	0.075	3
Nonabrasive cutting at 360 MPa (52 ksi), 20 hp				
Balsa wood, light foam rubber styrofoam	1.6	0.063	15	600+
	12.7	0.500	12	450
	25.4	1.000	7	275
Polyurethane, rubber compounds, polyethylene(a) (30 + Durometer)	6.4	0.250	6	225
	12.7	0.500	3	100
	25.4	1.000	1	40
Paper, fabric, corrugated board, rubber(a) (30 - Durometer)	0.13	0.005	15	600+
	0.38	0.015	8	300+
	0.81	0.032	8	300+
	1.6	0.063	8	300+
	3.2	0.125	8	300+
	6.4	0.250	8	300+

(a) These materials may require additional support on a table.

Source: Ref 6, p 674

Factors for drilling of composites

Material	Speed sfm	Feed (equivalent) mm/rev	ipr	Tool recommendation	Drilling time, min	Tool life, holes
Boron-epoxy, 2.03 mm (0.08 in.) ...	300-600	0.013	0.0005	Core with end-set diamonds	0.1	300-400
Boron-epoxy, 25.4 mm (1.00 in.) ...	300-600	0.013	0.0005	Core with end-set diamonds	1.0	75-100
Boron-epoxy/titanium multilayer, 12.7 mm (0.50 in.) total	200-500	0.0013	0.00005	Core	4.0	30-50

Source: Ref 29, p 389

Composite Manufacturing Methods

Factors affecting prepreg form selection

10 = highest cost or worst case; 1 = lowest cost or best case

Material form and fabrication process	Facility cost	Production rate	Importance of operator's skill	Part complexity possible	Part reproducibility	Material cost	Material use efficiency
Unidirectional tape							
Hand lay-up...........	1	10	10	5	10	3	7
Machine-cut, hand lay-up	5	5	5	5	5	3	4
Machine lay-down	10	1	1	5	1	6	2
Multidirectional tape							
Hand lay-up...........	1	5	9	7	8	8	5
Machine-cut, hand lay-up	5	3	7	7	4	8	4
Fabrics							
Hand lay-up...........	1	10	8	1	8	5	7
Machine-cut, hand lay-up	5	5	4	1	4	5	7
Towpreg	8	5	7	3	5	2	3

Source: Ref 6, p 145

Sizing classifications and functions

Type	Purpose	Example	Remarks
Film-forming organics and polymers	To protect the reinforcement during processing	Polyvinyl alcohol (PVA), polyvinyl acetate (PVAc)	The polymer is formulated to wet-spread to form a uniform coating that is applied to aid procesing but later may be removed by washing or heat cleaning (e.g., fugitive sizing).
Adhesion promotors	To improve composite mechanical properties and/or moisture resistance	Silane coupling agents	Principally used on inorganic reinforcement, (e.g., glass fiber).
Interlayer	To enhance composite properties by creating an interphase between matrix and reinforcement	Elastomeric coating	Not in commercial use
Chemical modifiers	React to form protective coating	Silicon carbide on boron fibers	

Source: Ref 6, p 122

Commercial designations and sources of processing materials

Material	Purpose	Commercial designation	Source	Comments
Peel ply	Provides a bondable surface	Burlease 51789 (nylon-6)	Burlington Mills	Not suitable for phenolics
		Burlease 60 000 (polyester)	Burlington Mills	Suitable for phenolics
		Bleeder-lease E (fiberglass with release agent)	Airtech International	
Separator (release)	Separates cured laminates from other process materials without damage	A5000 Teflon FEP film	Richmond Technology	
		A4000 halogen release film	Airtech International	
		Wrightlon 4500 (halocarbon film)	Airtech International	
		104 TFE Teflon-coated fiberglass-style fabric	Various	
Bleeder	Absorbs excess resin	120-style fiberglass fabric	Commercial	
		181-style fiberglass fabric	Commercial	
		Organic fiber felts	Commercial	
Barrier	Limits resin flow to only bleeder plies	See Separator		
		Unperforated TFE film		
Breather	Evacuates the vacuum bag so that the desired autoclave pressure is applied	Airweave N-10, Airweave N-4	Airtech International	Organic fiber, stretchable felt
		A-3000	Richmond Technology	Organic fiber, stretchable felt
		Airweave HP	Airtech International	1.4-mm (0.055-in.) thick fiberglass fabric
		Burflo 4819	Burlington Mills	Polyester nonwoven fabric
Dam	Prevents resin flow from edges	Rubber neoprene cork, Rubatex 886	Groendyke	Tape with pressure-sensitive adhesive
Vacuum bag	Applies autoclave pressure	Vac-Pak	Richmond Technology	Various types for different cure temperatures
		Wrightlon, IPPLON, Wrightcast, Vacalloy	Airtech International	Various types for different cure temperatures
		Silicone rubbers	Various	Semipermanent

Source: Ref 6, p 643

Effect of resin matrices on properties of composite materials

Material			Typical values												
Fiber type	Fiber product code	Resin type	Specific gravity, g/cm^3	Fiber wt%	Fiber vol%	Tensile strength MPa	Tensile strength ksi	Flexural strength MPa	Flexural strength ksi	Flexural modulus GPa	Flexural modulus 10^6 psi	Compressive strength MPa	Compressive strength ksi	Impact strength J/mm	Impact strength ft · lbf/in.
Fiberglass	Type E	Epoxy	1.88	63	46	190	27	470	68	28	4.1	290	42	1.6	30
Fiberglass	Type E	Polyimide	1.95	63	47	140	21	260	37	21	3.1	220	32	1.2	22
Fiberglass	Type E	Phenolic	1.78	56	34	110	16	240	35	21	3.0	340	35	1.1	20
Fiberglass	Type E	Polyester	1.98	55	39	80	12	170	25	17	2.5	180	26	0.8	15
Fiberglass	Type E	Silicone	2.02	46	34	30	4	70	10	14	2.0	80	11	0.25	5

Source: Ref 16, p 6-4

Manufacturing processes for polymer matrix composites

Process	Tooling material	Tool life, number of parts: Low	Tool life, number of parts: High	Relative dollars	Typical materials used	Advantages of process	Limitations of process	Applications
Structural foam (thermoplastic and RIM)	Steel	200 000	1 million	100	Polyurethanes, polycarbonates, polyphenylene oxide, phenylene ether copolymers, polybutylene terephthalate, ABS, polyethylene polystyrene, polypropylene	Large detailed parts, low production costs, rigidity, complex shapes, parts consolidation. High strength-to-weight ratio. Low density. Molded-in inserts. Low pressure molding due to low viscosity. Wide material selection. Minimizing or eliminating of sink marks. Low molded-in stresses. Improved chemical resistance.	Surfaces usually need secondary finishing and/or painting. Some sacrifice of physical properties relative to base resin. Longer cycle times than injection-molded parts.	Business machine housings, automotive fascias, medical and electronic cabinetry, furniture, materials-handling equipment.
	Machined aluminum	5 000	250 000	60 to 80				
	Cast aluminum; Kirksite	500	50 000	50 to 70				
	Cast aluminum-filled epoxy	10	500	20 to 30				
	Cast epoxy (100%)	2	25	5 to 20				
Injection molding	Steel	200 000	1 million	100	Broad range of thermoplastics and thermosets	High-volume production runs, close tolerances, molded-in color, low part cost, large material selection, parts consolidation.	High tooling investment. Long tooling lead times.	Wide range of applications in all industries.
	Machined aluminum	5 000	250 000	60 to 80				
Compression molding	Steel	200 000	1 million	100	Thermosets (polyesters, sheet molding compound, alkyds, ureas, phenolics, epoxies dialyl phthalates)	High-strength, heat-resistant parts. High modulus. Complex shapes. Parts consolidation. Excellent surface finish. Used for molding large parts, such as with polyester, as well as full range of sizes. Excellent fatigue resistance.	High tooling costs. Deflashing needed. Labor intensive. Parts generally need painting.	Automotive parts, electrical connectors, business-machine parts, power-tool housings.
Vacuum forming	Machined aluminum	5 000	250 000	60 to 80	ABS and ABS alloys, PVC, PPO, acrylic, polystyrene, polyethylene, polycarbonate, polypropylene	Excellent for complex contours, with minimum internal details. Large or small parts.	Often limited by large radii, shallow depths, large draft angles, loose tolerances. Exposed edges must be trimmed and buffed or milled.	Signage, business-machine housings, furniture, medical cabinetry, recreational products (boats, campers), transportation (interior and exterior parts), packaging (cups, plates).
	Cast aluminum; Kirksite	500	50 000	50 to 70				
	Cast aluminum-filled epoxy	10	500	20 to 30				
	Cast epoxy (100%)	2	25	5 to 20				
Hand layup sprayed glass fiber	Machined aluminum	5 000	250 000	60 to 80	Thermoset polyesters	Large parts, basically shells, can be molded with complex curves, excellent surface finish, and high rigidity.	Internal details (bosses, ribs) must be manually layed into inside wall and then overlayed with glass fiber. Labor intensive.	Recreational boating, materials handling, furniture, construction, transportation (truck hoods, bus seats).
	Cast aluminum; Kirksite	500	50 000	50 to 70				
	Wooden pattern	10	1 000	3 to 10				
Die casting	Steel	200 000	1 million	100	Zinc, magnesium, aluminum alloys	Complex shapes, intricate details, tight tolerances, excellent surface finishes direct from die. High-strength fatigue-resistant parts.	Usually requires costly secondary operations, such as deburring, deflashing, tapping, painting.	Small appliances, hardware, automotive parts, motors, power tools.

(continued)

Manufacturing processes for polymer matrix composites (continued)

Process	Tooling material	Tool life, number of parts Low	High	Relative dollars	Typical materials used	Advantages of process	Limitations of process	Applications
Matched metal die forming	Steel	200 000	1 million	100	Mild steel, aluminum	Effective for production of compound-surfaced parts which require high strength, rigidity, heat resistance.	High tooling costs due to need for progressive dies. Deburring required. Painting necessary.	Automotive body panels, major appliances, housings, large containers.
Sandcasting	Wooden pattern	10	1 000	3 to 10	Iron, bronze, brass, copper, aluminum	Complex shapes with low capital investment. Very rigid parts. High heat resistance and strength. Good surface finish.	Limited to bulky parts. Frequently require machining. Labor intensive. Secondary finishing operations can be extensive.	Wide range of industrial uses. Transportation components, materials-handling equipment, large generator parts.
Sheetmetal/brake formed	Fixtures				Steel, aluminum, brass	Excellent for punching and bending sheetmetal.	Compound curves, internal ribs should be avoided. Assembly intensive. Painting.	Electronic cabinetry, ducting, furniture, construction.

Source: Ref 16, p 6-3

Hand, mechanized, and fully automated lay-up processes

Process	Ply generation	Placement on tool	Forming of prepreg to tool shape
Hand lay-up	Lay tape on tool or on lay-up templates Cut fabric to shape using templates	Place manually using tooling to coordinate the partial plies to the tool: Lay-up templates Bank rails at edges	Use vacuum bag and localized heating to soften the prepreg
Mechanized	Cut from wide goods with the Gerber cutter Cut from wide goods with Clicker press and steel rule dies	Stack on tool manually as above, but use an optical/layer system to locate plies on the mold	Use vacuum bag plus localized heat as above. Use mechanical devices to seat plies on the webs of sine wave spars (rollers) Use devices to apply vacuum/pressure to seat plies on male tools by means of a diaphragm Use devices to heat the prepreg so it can be formed easily by hand
	Lay-up plies in the flat with a tape layer	Use robotics or other mechanized system to transfer plies to the tool	
Fully automated	For wing and tail skins and parts of similar gentle contour, use contour tape layer to lay up the plies directly on the mold. Applications limited by the capability of the tape layer For automated laminating cell, cut the plies with a Gerber cutter Use a robot to pick up the plies from the Gerber table and place them on the tool Braiding of structural shapes, filament winding of surface of revolution parts, and pultrusion of structural shapes are described elsewhere.		

Source: Ref 6, p 577

Comparison of filament-winding impregnation methods

	Prepreg	Wet winding	Wet rerolled
Cleanliness	Best	Worst	Almost equal to prepreg, mess is away from winder
Fiber availability	Poor. Not all fibers are available; many necessitate special order	Best. Any fiber that system will handle	Best. All fibers
Control of resin content	Best. Constant speed and viscosity	Poor. Speed of mandrel varies, viscosity of resin may vary	Better. Process is away from winding and is faster; little viscosity change
Quality assurance	Highest. Can be done far ahead	Worst. Imposes quality control procedures onto factory floor and can lead to errors	Good. Can be done ahead
Ability to use complex resin systems	Yes. Hot melts available.	Very difficult. Requires complex impregnators to remove solvents or liquify hot melts	Difficult. Still requires complex impregnators
Large data base resin systems	Yes	Commercial resins generally not available as liquids; the wet systems with large data bases may be proprietary	Same as wet winding
Graphite fibers encapsulated (to prevent electronic shortouts)	Yes	No	Graphite fibers not released at winder
Storage	Must be refrigerated and storage records maintained	Easy mix at winder; dry fibers have long shelf life	Must be stored like prepreg, but shorter storage life records must be kept
Fiber damage	Depends on impregnator; fiber is handled twice	May require special equipment; less damage potential because of less handling	All handling of fiber is under control of user
Cost	Highest	Lowest	Slightly above wet but also requires capital investment for impregnation equipment
Large roving package	Depends on impregnator	Whatever is available dry from fiber manufacturers	Whatever is available dry from fiber manufacturers
Room-temperature cure	Not possible	Possible	Possible
Simple resin formulation	Possible	Necessary	Necessary
Winding speed	Can be highest. Resin throw from fiber is minimized	Lowest speed	Intermediate. Resin can be staged to lower resin throw
Stability on nongeodesic path	Highest possible	Lowest. Wet resin may cause slippage	Intermediate. Resin can be staged to increase tack

Source: Ref 6, p 506

Property comparison by process(a)

Process	Reinforcement, wt %	Tensile strength MPa	Tensile strength ksi	Tensile modulus GPa	Tensile modulus 10^6ksi	Flexural strength MPa	Flexural strength ksi
Spray	30-50 glass-polyester	60-120	9-18	5.5-12	0.8-1.8	110-190	16-28
Compression	15-30 glass-SMC	55-140	8-20	11-17	1.6-2.5	120-210	18-30
Compression	25-50 glass mat-polyester	170-210	25-30	6.2-14	0.9-2.0	70-280	10-40
Filament winding	30-80 glass-epoxy	550-1700	80-250	28-62	4.0-9.0	690-1850	100-270
Pultrusion	40-80 glass mat-polyester	410-1050	60-150	28-41	4.0-6.0	690-1050	100-150
Pultrusion	30-50 glass mat-polyester	80-120	12-30	6.9-17	1.0-2.5	170-210	25-30
Pultrusion	30-55 glass mat and roving vinyl ester resin	70-280	10-40	6.9-21	1.0-3.0	100-280	15-40
Pultrusion	30-55 glass mat and roving vinyl ester resin	50-240	7-35	5.5-17	0.8-2.5	70-210	10-30

Property comparison by process (continued)

Process	Compressive strength MPa	Compressive strength ksi	Impact strength J/m	Impact strength ft·lbf/ft	Thermal conductivity W/m·K	Thermal conductivity Btu·in./h·ft^2·°F
Spray	100-170	15-25	210-640	48-144	0.17-0.23	1.2-1.6
Compression	100-210	15-30	430-1150	96-264	0.19-0.25	1.3-1.7
Compression	100-210	15-30	530-1050	120-240	0.19-0.26	1.3-1.8
Filament winding	310-480	45-70	2150-3200	480-720	0.27-0.33	1.9-2.3
Pultrusion	210-480	30-70	2400-3200	540-720	0.27-0.33	1.9-2.3
Pultrusion	210-340	30-50	530-1350	120-300	0.22-0.27	1.5-1.85
Pultrusion	140-340	20-50	270-1600	60-360	0.22-0.33	1.5-2.3
Pultrusion	100-280	15-40	210-1350	48-300	0.22-0.33	1.5-2.3

Property comparison by process (continued)

Process	Heat distortion at 1.8 MPa °C	Heat distortion at 1.8 MPa °F	Dielectric strength kV/cm	Dielectric strength kV/in.
Spray	175-205	350-400	80-160	200-400
Compression	205-260	400-500	120-180	300-450
Compression	175-205	350-400	120-240	300-600
Filament winding	175-205	350-400	120-180	300-400
Pultrusion	205-260	400-500	80-160	200-400
Pultrusion	95-150	200-300	80-120	200-300
Pultrusion	175-230	350-450	80-130	200-325
Pultrusion	175-205	350-400	80-120	200-300

(a) Range of values reflects transverse and axial testing directions as well as percent reinforcement.

Source: Ref 6, p 540

Comparison of tape and fabric prepregs for manual lay-up

Advantages	Disadvantages
Tape	
Better strength/ stiffness control	More plies required
Lower resin content	Longer time to cut patterns and lay up
Can be spliced parallel to fibers	...
Lower coefficient of thermal expansion	...
Fabric	
Fewer plies required	Lower mechanical strength (due to higher resin content)
Less time to lay up; easier to form over large curved areas	Difficulties in splicing large parts
Lower cost	Higher void content

Source: Ref 6, p 602

CVD densification of a multidirectional preform (typical)

Process conditions

Temperature, °C (°F)	1100 (2012)
Pressure, Pa (torr)	1350 (10)
Hydrocarbon	natural gas

Process steps

1. Process three-directional preform to 1.2 g/cm^3 density
2. Machine preform surfaces
3. Process for 650 h to a density of 1.6 g/cm^3

Source: Ref 6, p 916

Effect of matrix precursor on composite modulus

	Heat treatment	
	at 1000 °C (1830 °F)	at 2600 °C (4710 °F)
Phenolic resin	110%	140%
Pitch	130%	210%

Note: Fiber stiffening factor assuming all stiffness comes from HM fiber

Source: Ref 6, p 913

Ceramics

CERAMICS

GENERAL

SINGLE OXIDES

MIXED OXIDES

CARBIDES

NITRIDES

BORIDES

BINARY PHASE DIAGRAMS

Ceramics

General

Strong ceramics for engineering

Main compound	Fabrication method	Minor components, %	Typical mean strength(a) MPa	Property advantages
Al_2O_3	Sintering	~0.05 MgO	200-400	Refractory, high-purity types, can be translucent
		0-2 MgO, 0-4 SiO_2	200-250	Acid-resistant, debased type
		0-3 CaO, 0-10 SiO_2	200-300	General-purpose electrical and mechanical types
		10-20 ZrO_2	200-500	Transformation-toughened type
	Hot pressing	20-40 TiC	200-500	High-strength machine-tool tips
ZrO_2	Sintering	4-8 CaO, MgO, Y_2O_3	200	Stabilized type, refractory, heat-insulating but not strong or thermal shock resistant
	Sintering plus aging	2-4 CaO, MgO, Y_2O_3	500	Partially stabilized type, transformation-toughened but not refractory
	Sintering	1-2 Y_2O_3	1000	All-tetragonal (TZP) type, transformation-toughened but not refractory
SiC	Reaction bonding	8-20 Si	400	Hard, wear-resistant, quite strong, thermal shock resistant, refractory, acid-resistant
	Hot pressing	1-2 Al_2O_3	500	Stronger but less dimensional flexibility
	Sintering	~1 B	300	Acid- and alkali-resistant
Si_3N_4	Reaction bonding	...	200	Thermal shock resistant, moderate strength, refractory, but porous
	Hot pressing	1-5 MgO, Y_2O_3	800	Harder, less refractory, impermeable
	Sintering	5-15 MgO, Y_2O_3	600	High strength, more shape flexibility
Sialons	Sintering	5-15 MgO, Y_2O_3	600	Strong, hard, wear-resistant, and some are quite refractory
B_4C	Hot pressing	~1 B	400	Very hard and wear-resistant, but oxidizes above 800° C

(a) Typical flexural strength as determined from small test bars tested in bending. Strength figures are functions of material composition and microstructure, specimen-preparation methods, test geometry, and loading rate. Fracture-stress levels for real components may be rather different.

Source: Ref 1, p 6

Property requirements for ceramic engine materials

Thermal-property requirements	Low thermal conductivity, low specific heat, high thermal shock resistance
Mechanical-property requirements	High strength, high fracture toughness, good wear resistance, low coefficient of friction
Chemical-property requirement	Chemical inertness for high resistance to corrosion and erosion

Source: Ref 1, p 16

Properties of selected high-temperature ceramics

Property	Alumina (Al_2O_3) 96%	99%	99.5%	Silicon nitride (Si_3N_4) Reaction bonded	Hot pressed	Titanium diboride (TiB_2)	Boron nitride (BN), 96%
General properties							
Color	White	White	Ivory	Gray	Black	Gray	White
Particle size, μm	9	9	13	...	...	...	...
Maximum continuous temperature (no load):							
°F	3000	3092	3092	2372	2372	2192	5027
°C	1650	1700	1700	1300	1300	1200	2775
Physical properties							
Thermal conductivity:							
$Btu\cdot in./ft^2\cdot h\cdot °F$	220	205	225	95	173.6	175.79	13.3
$W/m\cdot K$	31	29	32	13.7	25	25.35	1.9
Coefficient of thermal expansion:							
$10^{-6}/°F$	3.5	4.1	3.5	1.4 to 1.8	2	4.5	(a)
$10^{-6}/°C$	6.4	7.4	6.4	2.6 t0 3.2	3.6	8.1	(b)
Specific heat:							
Btu/lb·F	0.19	0.20	0.20	0.17	...	0.23	...
J/kg·K	795	837	837	712	...	963	...
Specific gravity	3.76	3.85	3.94	0.75 to 2.7	3.19	4.52	2.08
Water absorption (porosity), %	0.00	0.00	0.00	(76 to 15)	(0.00)	...	1.1
Mechanical properties							
Compressive strength:							
ksi	340	375	380	77 to 112	500	192	(c)
MPa	2344	2585	2620	531 to 772	3448	1324	(d)
Tensile strength:							
ksi	25	30	30	...	...	...	...
MPa	172	207	207	...	...	...	...
Flexural strength:							
ksi	46	50	50	30	125	35	(e)
MPa	317	345	345	206.9	861.9	241	(f)
Modulus of elasticity:							
10^6psi	45	50	55	24	45	62	(g)
GPa	310	345	379	165	310	427.5	(h)
Impact resistance:							
in.·lb Charpy	7	7	6	...	...	...	...
N·m	0.79	0.79	0.68	...	...	...	...
Hardness:							
Mohs scale	9	9	9	9	...	...	...
Vickers	...	...	...	...	1800	...	...
Electrical properties							
T_e value:							
°F	1832	2012	1472	...	...	...	...
°C	1000	1100	800	...	...	...	...
Dielectric strength:							
V/mil	225	230	230	...	...	...	950
kV/mm	8.9	9.0	9.0	...	...	...	374
Electrical resistivity, Ω·cm	$>10^{15}$	$>10^{14}$	$>10^{14}$	6.6×10^{10}	10^{12}	$9\text{ to }15\times10^{16}$	$>2\times10^{14}$
Dissipation factor, MHz	0.00019	0.0001	0.0001	...	...	...	0.00034
Loss factor, MHz	0.0018	0.0009	0.0009	...	...	...	0.00139
Dielectric constant, MHz	9.3	9.3	9.5	...	...	...	4.08

(a) WG, 2.3; AG, 1.1. (b) WG, 4.1; AG, 2. (c) WG, 45; AG, 34. (d) WG, 310; AG, 234. (e) WG, 11.7; AG, 13.9. (f) WG, 80.7; AG, 95.8. (g) WG, 9.4: AG, 6. (h) WG, 64.8; AG, 41.4.

Source: Ref 2

Comparison of properties important for structural use of advanced ceramics

Material(a)	Strength, MPa	Toughness, MPa$\sqrt{m}$	Thermal expansion, $10^{-6}/°C$
Silicon nitrides			
RBSN	300	3.6	3.3
HPSN	1100	6.6	3.5
GPSSN	440	2.9	3.5
Silicon carbides			
Alpha SSC	420	2 to 3	4.1
Beta SSC	533	2.4	4.1
HPSC	800	3.9	4.2
Transformation-toughened ceramics			
PSZ	700+	8+	10.2
TTA	900	8	7
Ceramic-ceramic composite			
SIC-LAS	620	15	1 to 4

(a) Key: RBSN—reaction-bonded silicon nitride; HPSN—hot pressed silicon nitride; GPSSN—gas-pressure-sintered silicon nitride; Alpha SSC—alpha-phase-sintered silicon carbide; Beta SSC—beta-phase-sintered silicon carbide; HPSC—hot pressed silicon carbide; PSZ—partially stabilized zirconia; TTA—transformation-toughened alumina; SIC-LAS—silicon carbide fibers in lithium aluminosilicate glass.

Source: Ref 1, p 144

Properties of ceramic materials for heat engines

Material	Density (ρ), g/cm^3	Properties
Zirconia	5.5	+ Low coefficient of friction, good wear and corrosion resistance, low thermal conductivity, relatively high thermal expansion, good thermal shock resistance, high fracture toughness - Some uncertainty about high-temperature performance
Sintered silicon nitride	3.3	+ Good thermal shock resistance, good mechanical properties, significantly higher strength and lower thermal conductivity than SiC at high temperatures - Poor thermal insulator
Sintered silicon carbide	3.1	+ Low coefficient of friction, good wear resistance, good corrosion resistance, good thermal shock resistance, retains strength at high temperatures
Reaction-bonded silicon nitride	2.8	+ <1 % shrinkage during firing, retains accurate dimensions - Lower density than sintered Si_3N_4
Lithium aluminum silicate	2.3	+ Very low thermal expansion, very low thermal conductivity, good thermal shock resistance, good thermal insulator - Very poor strength and fatigue life

Source: Ref 1, p 17

Properties of aluminas and other ceramic materials(a)

Property	Test	Units	Test or material condition	Refractory (alumina-mullite)
Specific gravity	ASTM C20-70	...	....	2.84
Hardness	ASTM E18-67	Rockwell 45N	....	...
	(1000-g load)	Knoop (GPa)	....	...
Surface finish	Profilometer	μm (μin.)	As-fired	...
	(0.75-m,m cutoff)		Ground	...
			Polished	...
Crystal size	...	μm (μin.)	Range	...
			Average	...
Water absorption	ASTM C373-72	...	....	...
Gas permeability(d)	...	...	....	...
Color	...	...	....	White
Compressive strength	ASTM C773-74	MPa (ksi)	25 °C	83 (12)
			1000 °C	...
Flexural strength	ASTM F417-75T	MPa (ksi)	25 °C: typ	...
			min (e)	...
			1000 °C: typ	...
			min(e)	...
Tensile strength	ACMA Test #4	MPa (ksi)	25 °C	...
			1000 °C	...
Modulus of elasticity	ASTM C623-71	GPa (10^6 psi)	....	...
Shear modulus	ASTM C623-71	GPa (10^6 psi)	....	...
Bulk modulus	ASTM C623-71	GPa (10^6 psi)	....	...
Transverse sonic velocity	ASTM C623-71	m/s (ft/s)	....	...
Poisson's ratio	ASTM C623-71	...	....	...
Maximum-use temperature	...	°C (°F)	No load	1750 (3180)
Coefficient of linear thermal expansion	ASTM C372-56	10^{-6}/°C (10^{-6}/°F)	-200 to 25 °C	...
			25 to 200 °C	...
			25 to 500 °C	...
			25 to 800 °C	...
			25 to 1000 °C	...
			25 to 1200 °C	6.0 (3.4)
Thermal conductivity	ASTM C408-58	W/m·K (cal/cm·s·°C)	20 °C	...
			100 °C	...
			400 °C	...
			800 °C	2.1 (0.005)
Specific heat	ASTM C351-61	J/kg·K (cal/g·°C)	100 °C	630 (0.15)
Dielectric strength	ASTM D116-69	ac-kV/mm (ac-V/mil)	6.35 mm (f)	...
			3.18 mm(f)	...
			1.27 mm(f)	...
			0.64 mm(f)	...
			0.25 mm(f)	...
Dielectric constant (at 25 °C)	ASTM D150-70, D2520-70	...	1 kHz	...
			1 MHz	...
			100 MHz	...
Dissipation factor	ASTM D150-70, D2520-70	...	1 kHz	...
			1 MHz	...
			100 MHz	...
Loss index	ASTM D150-70, D2520-70	...	1 kHz	...
			1 MHz	...
			100 MHz	...
Volume resistivity	ASTM D1829-66	Ω·cm²/cm	25 °C	...
			300 °C	...
			500 °C	...
			700 °C	...
			1000 °C	...
T_e value(g)	...	°C (°F)	....	...

(a) All data measurements are typical and made at room temperature unless otherwise noted. Ceramic property values vary somewhat with method of manufacture, size, and shape of part. Ceramic compositions are controlled using modern chemical, spectrographic, and x-ray fluorescent methods. (b) Partially stabilized with MgO. (c) Nominal. (d) No helium leak through a plate 25.4 mm in diameter by 0.25 mm thick

Source: Coors Porcelain

Properties of aluminas and other ceramic materials (continued)

			Alumina		
Zirconia(b)	90%(c) Al_2O_3 (opaque)	94%(c) Al_2O_3	96%(c) Al_2O_3	99.5%(c) Al_2O_3	99.9%(c) Al_2O_3
5.4	3.69	3.62	3.72	3.89	3.96
68	75	78	78	83	90
...	10.8	11.1	11.1	14.7	15.2
...	1.6 (63)	1.6 (63)	1.6 (63)	0.9 (35)	0.5 (20)
...	1.0 (39)	1.3 (51)	1.3 (51)	0.5 (20)	0.9 (35)
0.4 (14)	0.1 (3.9)	0.3 (12)	0.3 (12)	0.1 (3.9)	$<$0.03 ($<$1.2)
...	2-40 (79-1575)	2-25 (79-985)	2-20 (79-788)	5-50 (197-1970)	1-6 (39-236)
55 (2170)	6 (2336)	12 (473)	11 (433)	17 (670)	3 (118)
None	None	None	None	None	None
None	None	None	None	None	None
Ivory	Black	White	White	Ivory	Ivory
...	2413 (350)	2103 (305)	2068 (300)	2620 (380)	3792 (550)
...	...	345 (50)	...	...	1930 (280)
414 (60)	365 (53)	352 (51)	358 (52)	379 (55)	552 (80)
...	324 (47)	317 (46)	324 (47)	...	517 (75)
...	...	138 (20)	172 (25)	...	414 (60)
...	...	117 (17)	138 (20)	...	379 (55)
...	228 (33)	193 (28)	193 (28)	262 (38)	310 (45)
...	...	103 (15)	96 (14)	...	221 (32)
...	308 (45)	283 (41)	303 (44)	372 (54)	386 (56)
...	124 (18)	117 (17)	124 (18)	152 (22)	158 (23)
...	...	165 (24)	172 (25)	228 (33)	228 (33)
...	9.3 (31) $\times 10^3$	8.9 (29) $\times 10^3$	9.1 (30) $\times 10^3$	9.8 (32) $\times 10^3$	9.9 (32) $\times 10^3$
...	0.24	0.21	0.21	0.22	0.22
...	1500 (2730)	1700 (3100)	1700 (3100)	1750 (3180)	1900 (3450)
...	4.4 (2.4)	3.4 (1.9)	3.4 (1.9)	3.4 (1.9)	3.4 (1.9)
...	6.4 (3.6)	6.3 (3.5)	6.0 (3.4)	7.1 (4.0)	6.5 (3.6)
...	7.3 (4.0)	7.1 (4.0)	7.4 (4.1)	7.6 (4.3)	7.4 (4.1)
...	8.0 (4.4)	7.6 (4.3)	8.0 (4.5)	8.0 (4.5)	7.8 (4.4)
4.9 (2.7)	8.4 (4.7)	7.9 (4.4)	8.2 (4.6)	8.3 (4.6)	8.0 (4.5)
...	...	8.1 (4.5)	8.4 (4.7)	...	8.3 (4.6)
...	12.6 (0.030)	18.0 (0.043)	24.7 (0.059)	35.6 (0.085)	38.9 (0.093)
...	11.3 (0.027)	14.2 (0.035)	18.8 (0.045)	25.9 (0.062)	27.6 (0.066)
...	7.5 (0.018)	7.9 (0.017)	10.0 (0.024)	12.1 (0.028)	13.4 (0.032)
...	...	5.0 (0.010)	5.4 (0.013)	6.3 (0.015)	6.3 (0.015)
...	1045 (0.25)	880 (0.21)	880 (0.21)	880 (0.21)	880 (0.21)
...	5.3 (135)	8.7 (220)	8.3 (210)	8.7 (220)	9.4 (240)
...	8.8 (225)	11.8 (300)	10.8 (275)	11.4 (290)	12.8 (325)
...	16.3 (415)	16.7 (425)	14.6 (370)	16.9 (430)	18.1 (460)
...	21.2 (540)	21.6 (550)	17.7 (450)	22.8 (580)	23.2 (590)
...	28.3 (720)	28.3 (720)	22.8 (580)	33.1 (840)	31.5 (800)
...	22.0	8.9	9.0	9.8	9.9
...	9.8	8.9	9.0	9.7	9.8
...	...	8.9	9.0	...	...
...	0.3000	0.0002	0.0011	0.0002	0.0020
...	0.0200	0.0001	0.0001	0.0003	0.0002
...	...	0.0005	0.0002	...	...
...	6.6	0.002	0.010	0.002	0.020
...	0.200	0.001	0.001	0.003	0.002
...	...	0.004	0.002	...	...
...	$>10^{14}$	$>10^{14}$	$>10^{14}$	$>10^{14}$	$>10^{15}$
...	5.5×10^{10}	9.0×10^{11}	3.1×10^{11}	...	1.0×10^{15}
...	5.5×10^{7}	2.5×10^{9}	4.0×10^{9}	...	3.3×10^{12}
...	1.7×10^{6}	5.0×10^{7}	1.0×10^{8}	...	9.0×10^{9}
...	4.0×10^{4}	5.0×10^{5}	1.0×10^{6}	...	1.1×10^{7}
...	...	950 (1742)	1000 (1832)	...	1170 (2138)

measured at a vacuum of 3×10^{-7} torr vs. approximately one atmosphere of helium pressure for 15 s at room temperature. (e) Minimum flexural strength is a minimum mean for a sample of ten specimens. (f) Specimen thickness. (g) Temperature at which resistivity is 1 MΩ·cm.

Properties of some ceramic materials

Material	Melting point, °C	Limit of application, °C	Hardness, Mohs scale	Density, g/cm³	Specific heat (mean), J/kg·°C, at 25 to 1000 °C	Linear coefficient of expansion, 10^{-6}/°C, at 25 to 800 °C	Thermal conductivity, W/m·°C, at temperature, °C	Electrical resistivity, Ω·cm, at temperature °C
Alumina (Al_2O_3)	2050	1950	9	3.96	1050	8.0	4 at 1315	10^6 at 1100
Beryllia (BeO)	2550	2400	9	3.0	2180	7.5	29 at 1000	4×10^8 at 600; 8×10^{12} at 2100
Magnesia (MgO)	2850	2400	6	3.60	1170	13.5	59 at 1100	2×10^8 at 850
Thoria (ThO_2)	3220(a)	2700	7	9.5-9.9	290	9.5	3 at 1000	2.6×10^7 at 550 1.5×10^4 at 1200
Zirconia (ZrO_2)	2700	2400	6.5	5.5-5.8	590	7.5	3 at 1315	10^6 at 385 3.6×10^2 at 1200
Zircon ($ZrO_2 \cdot SiO_2$)	2500(a)	1870	7.5	4.5-4.7	630	4.5	4 at 1200	High
Spinel ($MgO \cdot Al_2O_3$)	2130	1900	8	3.60	1050	8.5	2 at 1315	2.8×10^7 at 500 2.0×10^5 at 1100
Mullite ($3Al_2O_3 \cdot 2SiO_2$)	1850	1800	...	2.8	840	5.0	4 at 1200	10^5-10^3 at 815-1370
Sillimanite ($Al_2O_3 \cdot SiO_2$)	1800(a)	1800	6.5	3.2	840	5.0	2 at 1300	10^4-10^5 at 815-1370
Silicon carbide (SiC)	2200-2700(b)	1400-1700(c)	9	3.2	840	4.5	13 at 1100	7420-745 at 1000-1500
Silicon nitride (Si_3N_4)	1900(d)	1400 in air; 1850 in inert atmosphere	9	3.18	1050	α=2.9; β=2.3	9.5 at 1200	10^{13} at 25; 10^{10} at 480
Carbon graphite (C)(e)	3600(d)	...	0.5-1.0	2.2	1600	2.2	147 at 50; 63 at 900	10^{-3}
Quartzite (SiO_2)	1400	1090	7	2.65	1170	8.6	2.06 at 1200	10^{14} at 20 5×10^3 at 1300
Boron carbide (B_4C)	2350	540 in air; 2260 in inert atmosphere	9.3	2.5	2090	5.7	17.3 at 800	...
Titanium carbide (TiC)	3140	1500(c)	9-10	6.5	1050	6.9	40 at 1100	...
Tungsten carbide (WC)	2780	...	9-10	14.3	300	6.3	43.3 at 1100	...
Boron nitride (BN)	2721	650(c)	2	2.1	1570	7.5(‖); 0.77(⊥)	26 at 900	1.7×10^{13} at 25 (‖); 2.3×10^{10} at 480 (‖)

(a) Approximate. (b) Decomposes. (c) Oxidizes. (d) Sublimes. (e) Properties vary with type.

Source: Cotronics Corp.

Thermal conductivity of selected ceramics

Material	Thermal conductivity, 100 °C	cal/cm·s·° C, at 1000 °C
Al_2O_3	0.072	0.015
BeO	0.525	0.049
MgO	0.090	0.017
$MgAl_2O_4$	0.036	0.014
ThO_2	0.025	0.007
Mullite	0.014	0.009
$UO_{2.00}$	0.024	0.008
Graphite	0.43	0.15
ZrO_2 (stabilized)	0.0047	0.0055
Fused silica glass	0.0048	0.006
Soda-lime-silica glass	0.004	...
TiC	0.060	0.014
Porcelain	0.004	0.0045
Fire-clay refractory	0.0027	0.0037
TiC cermet	0.08	0.02

Source: Ref 3, p 642

Moduli of elasticity for selected ceramics

Material	Modulus of elasticity (E) GPa	10^6 psi
Aluminum oxide crystals	380	55
Sintered alumina(a)	365	53
Alumina porcelain(b)	365	53
Sintered beryllia(a)	310	45
Hot pressed boron nitride(a)	83	12
Hot pressed boron carbide(a)	290	42
Graphite(c)	9.0	1.3
Sintered magnesia(a)	210	30.5
Sintered molybdenum silicide(a)	405	59
Sintered spinel(a)	238	34.5
Dense silicon carbide(a)	470	68
Sintered titanium carbide(a)	310	45
Sintered stabilized zirconia(a)	150	22
Silica glass	72.4	10.5
Vycor glass	72.4	10.5
Pyrex glass	69	10
Mullite porcelain	69	10
Steatite porcelain	69	10
Superduty fire-clay brick	97	14
Magnesite brick	170	25
Bonded silicon carbide(c)	345	50

(a) Approximately 5% porosity. (b) 90 to 95% Al_2O_3. (c) Approximately 20% porosity.

Source: Ref 3, p 777

Physical properties of ceramic materials

Material	Bending strength, MPa, at: 800 K	1400 K	Density g/cm^3	Modulus of elasticity GPa at 1260 K	Thermal expansion, 10^{-6}/K at 300-1260 K	Thermal conductivity, W/m·K, at 1260 K
S-Si_3N_4	530	300	3.1	300	3.2	12
RB-Si_3N_4	300	300	2.6	180	3.0	9
S-SiC	450	450	3.15	400	4.5	40
MAS	70	20	2.2	12	0.6	1
ZrO_2	600	300	5.7	200	9.8	2.5
Al_2O_3-TiO_2	40	20	3.2	23	3.0	2
Inco 713 C	900	200	7.9	170	15.0	25

Source: Ref 1, p 48

Average strengths of ceramics(a)

Material	Compressive strength MPa	Compressive strength ksi	Tensile strength MPa	Tensile strength ksi	Flexural strength MPa	Flexural strength ksi
Alumina: 85%	1620	235	125	18	293	42.5
90%	2410	350	140	20	315	46
95%	2410	350	195	28	340	49
99%	2590	375	205	30	345	50
Alumina silicate	275	40	17	2.5	62	9
$ZrO_2 \cdot Al_2O_3$	2410	350	...	...	...	...
3% ½O_3 PSZ(b)	2960	430	...	...	1170	170
TTZ(c)	1760	255	350	51	635	92
9% MgO PSZ(b)	1860	270	...	...	690	100
Slip cast Si_3N_4	140	20	24	3.5	69	10
Reaction-bonded SiC	690	100	140	20	255	37
Pressureless-sintered SiC	3860	560	170	25	550	80
Sintered SiC with free silicon	1030	150	165	24	325	47
Sintered SiC with graphite	415	60	35	5	55	8
Reaction-bonded Si_3N_4	770	112	...	...	205	30
Hot pressed Si_3N_4	3450	500	...	...	860	125

(a) Strength is dependent on test method, sample preparation, and sample size. Data are from a variety of sources. (b) Partially stabilized zirconia. (c) Transformation-toughened zirconia.

Source: Ref 4, p 42

Moduli of rupture for selected ceramics

Material	Modulus of rupture MPa	Modulus of rupture ksi
Aluminum oxide crystals	345 to 1034	50 to 150
Sintered alumina(a)	207 to 345	30 to 50
Alumina porcelain(b)	345	50
Sintered beryllia(a)	138 to 276	20 to 40
Hot pressed boron nitride(a)	48 to 103	7 to 15
Hot pressed boron carbide(a)	345	50
Sintered magnesia(a)	103	15
Sintered molybdenum silicide(a)	690	100
Sintered spinel(a)	90	13
Dense silicon carbide(a)	172	25
Sintered titanium carbide(a)	1100	160
Sintered stabilized zirconia(a)	83	12
Silica glass	107	15.5
Vycor glass	69	10
Pyrex glass	69	10
Mullite porcelain	69	10
Steatite porcelain	138	20
Superduty fire-clay brick	5.2	0.75
Magnesite brick	27.6	4.0
Bonded silicon carbide(c)	13.8	2.0
1090 °C insulating firebrick(d)	0.28	0.04
1430 °C insulating firebrick(e)	1.17	0.17
1650 °C insulating firebrick(f)	2.0	0.3

(a) Approximately 5% porosity. (b) 90 to 95% Al_2O_3. (c) Approximately 20% porosity. (d) 80 to 85% porosity. (e) Approximately 75% porosity. (f) Approximately 60% porosity.

Source: Ref 5, p 319

Mechanical properties of SiC-whisker-reinforced ceramics

Composite	Whisker control, vol %	Toughness MPa$\sqrt{m}$	Toughness ksi$\sqrt{in.}$	Flexural strength MPa	Flexural strength ksi
Corning 1723 glass	0	<1	<0.9		
	25	2.1-3.4	1.9-3.1	200-340	30-50
Barium osumilite	25	4.5	4.1	360-400	50-60
Si_3N_4	0	5-7	4.6-6.4	400-650	60-95
	10	6.5-9.5	5.9-8.6	400-500	60-75
	30	7.5-10	6.8-9.1	350-450	50-65
Spinel	30			415	60
Mullite	0	1.8-2.2	1.6-2.0	250	40
	20	4.6	4.2	440	65
ZrO_2	0	6.2	5.6	1080	160
Toughened alumina	20	8.5-13.5	7.7-12.3	700-880	100-130
Cordierite	0	2.2	2.0	180	25
	20	3.7	3.4	260	40
$MoSi_2$	0	5.3	4.8	150	20
	20	8.2	7.5	310	45

Source: Ref 6, p 942

Comparison of basic properties of selected compounds used in advanced ceramics

Compound	Density, g/cm^3	Hardness, kg/mm^2	Melting point, °C	Thermal conductivity, cal/cm·s·°C
Aluminum oxide	3.98	2100	2050	0.07
Zirconium oxide(a)	6.27	1200	2715	0.005
Silicon carbide	3.22	2500	2220(b)	0.16
Silicon nitride	3.17	2400	1900	0.04
Silica glass	2.20	...	...	0.002

(a) Stabilized in cubic form. (b) Decomposes.

Source: Ref 1, p 144

Fracture toughness of ceramics

Material	Comments	Fracture toughness (K_{Ic}) (a), MPa$\sqrt{m}$
Soda-lime glass(b)	Amorphous	0.74 DCB
Aluminosilicate glass	Amorphous	0.91 DCB
ZnSe	Vapor deposited	0.9
WC	Co-bonded	13.0
ZnS	Vapor deposited	1.0
Si_3N_4	Hot pressed	5.0
Al_2O_3	MgO-doped	4.0
Al_2O_3(sapphire)	Monocrystal	2.1
SiC	Hot pressed	4.0
SiC-ZrO_2	Hot pressed(c)	5.0
MgF_2	Hot pressed	0.9
MgO	Hot pressed	1.2
B_4C	Hot pressed	6.0
Si	Monocrystal	0.6
ZrO_2	Ca-stabilized	7.6 DCB

(a) Double-torsion measurement technique, except where double-cantilever-beam test (DCB) indicated. (b) Commercial sheet glass. (c) 20% ZrO_2, 14 wt % mullite. ZrO_2 present in monoclinic form; no transformation toughening.

Source: Ref 4, p 83

Properties of wear-resistant ceramics(a)

Property	Alumina 85%	90%	92%	95%	96%	99%	99.5%	99.8%	Alumina silicate
Modulus of elasticity:									
GPa	220	270	290	295-315	310	345	380	385	55
10^6 psi	32	39	42	43-46	45	50	56	8	
Impact resistance:									
J	0.71-0.73	0.73	0.73-0.77	0.73-0.86	0.8	0.8	0.7	0.8	0.37
in.·lb Charpy	6.3-6.5	6.5	6.5-6.8	6.5-7.6	7	7	6	7	3.3
Hardness:									
Vickers	...	...	...	...	...	...	...	...	...
Mohs scale	9	9	9	9	9	9	9	9	9
Knoop	...	...	...	...	...	...	...	...	...
Rockwell	...	...	...	...	...	...	...	93.5(d)	...

(continued)

Properties of wear-resistant ceramics (continued)

Property	ZrO_2·Al_2O_3	3% ½O_3 PSZ(b)	TTZ(c)	9% MgO PSZ(b)	SiC composite graphite	Cast Si_3N_4 bonded SiC	Reaction-bonded SiC	Pres-sureless-sintered SiC
Modulus of elasticity								
GPa	260	...	200	205	14	115	385	405
10^6 psi	38	...	29	30	2	17	56	59
Impact resistance								
J	...	...	...	...	...	Up to 0.09	...	...
in.·lb Charpy	...	...	...	...	...	Up to 0.8	...	...
Hardness								
Vickers	...	...	...	...	...	...	...	...
Mohs scale		...	...	...	8+	...	...	...
Knoop	1470	...	...	1080-	...	...	2700	2800
			1520					
Rockwell	...	91.5(d)	74-79(e)	...	...	...	...	...

(continued)

Properties of wear-resistant ceramics (continued)

Property	Sintered SiC with free silicon	Sintered SiC with graphite	Reaction-bonded Si_3N_4	Hot pressed Si_3N_4
Modulus of elasticity				
GPa	380	205	165	310
10^6 psi	55	30	24	45
Impact resistance				
J	...	Similar to silicon carbide		
in.· lb Charpy	...			
Hardness				
Vickers	...	...	...	1800
Mohs scale	...	...	>9	...
Knoop	1900	1000	...	...
Rockwell	...	...	...	75-80(e)

(a) Data are from a variety of commercial sources. (b) Partially stabilized zirconia. (c) Transformation-toughened zirconia. (d) Rockwell A scale. (e) Rockwell C scale.

Source: Ref 4, p 96-97

Results of corrosion tests of selected ceramics in liquids

Test environ-ment(a), wt %	Temperature °C	°F	Corrosive weight loss(b), mg/cm²·yr Si/SiC composites (12% Si)	Tungsten carbide (6% Co)	Aluminum oxide (99%)	Silicon carbide (no free Si)
98% H_2SO_4	100	212	55.0	>1000	65.0	1.8
50% NaOH	100	212	>1000	5.0	75.0	2.5
53% HF	25	77	7.9	8.0	20.0	<0.2
85% H_3PO_4	100	212	8.8	55.0	>1000	<0.2
70% HNO_3	100	212	0.5	>1000	7.0	<0.2
45% KOH	100	212	>1000	3.0	60.0	<0.2
25% HCl	70	158	0.9	85.0	72.0	<0.2
10% HF + 57% HNO_3	25	77	>1000	>1000	16.0	<0.2

(a) Test time, 125 to 300 h of submersive testing (with continuous stirring). (b) Corrosion weight loss guide: >1000 mg/cm²·yr - specimen completely destroyed within days; 100 to 999 - not recommended for service longer than one month; 50 to 100—not recommended for service longer than one year; 10 to 49 - caution recommended, based on the specific application; 0.3 to 9.9—recommended for long-term service; <0.2 - recommended for long-term service (no corrosion, other than as a result of surface cleaning, was evidenced).

Source: Ref 4, p 114

Applications of hot isostatic pressed ceramics

Materials			Applications	
Al_2O_3	ThO_2	ZrB_2	Letdown valves for coal liquefaction	Drill-bit inserts
Al_2O_3 + ZrO_2	TiC	HfB_2	Turbine disks	Nuclear waste consolidation
Al_2O_3 + TiC	B_4C	$BaTiO_3$	Turbine blades	Nuclear waste containers
Al_2O_3 + TiN	SiC	$BaFeO_4$	Turbine vanes	Bearings
BeO	SiC + Si_3N_4	$MnZnFeO_4$	Stirling heater heads	Radomes and infrared domes
UO_2	Si_3N_4	$NiZnFeO_4$	Cutting tools	β-alumina tubes
ZrO_2	Sialon	$Pb(ZrTi)O_3$	Orthopedic implants and dental ceramics	Nuclear reactor core insulators
ZnO_2	TiN	$(PbBi)(ZrTi)O_3$	Magnetic tape heads	Fusion reaction insulators
SiO_2	BN	$(PbLa)(ZrTi)O_3$	Transducers	Special dies
T_2O_3	TiB_2		Laser windows	Multilayer capacitors
			Sputtering targets	

Source: Ref 7, p 26

Fricton behavior of glasses and glass-ceramics

Materials	Static coefficient of friction
Glass on brass	0.18
Glass on chromium	0.16
Glass on tin plate	0.29
Brass on glass	0.24
Steel on glass	0.18
Glass-ceramic on glass-ceramic:	
Lithium-zinc silicate (low ZnO)	0.19
Lithium-zinc silicate (high ZnO)	0.09
Lithium aluminosilicate	0.16

Source: Ref 4, p 99

Advantages and disadvantages of ceramics as engine components

+ Abundant raw materials	— More efficient fuel burn
	— Reduced emissions
+ Higher operating temperatures	— Brittleness
	— High-temperature lubrication
+ Lower density/lighter weight	— Lower inertia for moving parts
	— Lower overall engine weight

Source: Ref 1, p 16

Typical bulk properties of heat-engine ceramics

Material	Modulus of rupture, MPa, at: Room temperature	1000 °C	1375 °C	Modulus of elasticity(a), GPa	Linear coefficient of expansion, $10^{-6}/°C$	Thermal conductivity, W/m·°C
Aluminosilicates:						
Lithium aluminosilicate (LAS)	83	69	...	...	450 ppm(b)	...
Magnesium aluminosilicate (MAS)	140	140	...	160	2.2	3.5
Aluminosilicate (AS)(c)	...	...	...	...	...	...
Silicon nitrides:						
Hot pressed (MgO additive)	690	620	330	315	3.0	30-15
Sintered (Y_2O_3 additive)	655	585	275	275	3.2	28-12
Reaction bonded (2.45 g/cm^3)	205	345	380	165	2.8	6-3
Silicon carbides:						
Hot pressed (Al_2O_3 additive)	655	585	515	450	4.5	85-35
Sintered (α phase)	310	310	310	405	4.8	100-50
Reaction sintered (20 vol % free Si)	380	415	275	345	4.4	100-50
CVD (lower values)	415	550	550	415	...	...

(a) At room temperature. (b) Total thermal excursion in ppm from 0 to 1000 °C. (c) Properties available only in matrix form.

Source: Ref 1, p 62

Applications of ceramics in internal-combustion engines

Component	Material properties needed(a)	Material types being considered(b)	Reasons for substituting for metals
Turbines			
Flame can	1, 2, 3	A, B	Metal parts limited to 1100 °C surface temperature; ceramic parts may not need parasitic losses of air cooling to cope with higher temperatures
Volute	1, 2	A, C	
Stator, nozzle rings	1, 2, 3	C, D, E, F	
Rotor	1, 2, 3	D, E, F	Lower mass; no cooling required
Bearings	1, 4	D, E, F	Bearings running hot are needed to cope with higher temperatures
Heat exchanger	2	H	Required to improve efficiency of land-vehicle engines
Diesels			
Cylinder liners	2, 4, 5	A, G, G′	Reduce heat losses through cylinder wall (adiabatic diesel); may no longer need cooling system, saving weight and space; may reduce emissions
Cylinder block, head	2, 4, 5	A, C	
Valve heads	1, 2	G′	Reduce oxidation at higher running temperatures
Valve seats	1, 2, 3, 4	A, B, D, E, F	More corrosion resistant in hot exhaust stream
Pistons or piston caps, swirl chambers	2, 3, 5	C, D, G, I, G'	Reduce heat losses through piston
Precombustion chambers	1, 2, 3	C, E	Cheaper than nickel alloy types
Exhaust ports	1, 2, 5	C, G′	Maintain high temperature to maximize turbocharger efficiency
Turbocharger rotors	1, 2, 3	D, E, F	Cheaper and lower inertia than nickel alloy types
Wear parts:			
Cam followers	3, 4	E, F, G	Reduce inertia and friction giving less wear, less frequent adjustment, and smaller parasitic losses
Valve guides	3, 4	E, F, G	
Tappet faces	3, 4	E, F, G, G'	
Gudgeon pins	3, 4	D, E, F	Lower thermal expansion to match that of ceramic piston
Gasoline engines			
For gasoline engines which have temperature limitations on fuel injection, consideration is mostly being given to wear parts such as those listed above for diesels.			

(a) 1, high temperature capability (>1000 °C); 2, thermal shock resistance; 3, high strength; 4, hard or wear-resistant; 5, low thermal conductivity. (b) A, reaction-bonded silicon carbide; B, sintered silicon carbide; C, reaction-bonded silicon nitride; D, hot pressed silicon nitride; E, sintered silicon nitride; F, sintered sialons; G, zirconia ceramics; H, aluminous keatite; I, aluminum titanate; ′, coating on a metal.

Source: Ref 1, p 9

Defects in crystalline ceramics

Defect	Cause	Relative scale
Impurity	Foreign atom or ion introduced by substitution or interstitially; vacancies may be created to balance valance charges	Atomic in three dimensions
Vacancy	Missing positive or negative ion, or atom	Atomic in three dimensions
Vacancy pair	Missing positive ion coupled with missing negative ion	Atomic in three dimensions
Vacancy cluster	Aggregation of vacancies	Microscopic (large clusters can be resolved optically)
Color center	Anion vacancy plus electron or cation vacancy plus electron hole	Atomic in two dimensions, microscopic in length
Dislocation	Structural misfit, linear character	Atomic in two dimensions, microscopic in length
Subgrain boundary	Structural misfit (at small angles), surface character	Atomic in one dimension, microscopic in two dimensions
Grain boundary	Structural misfit (at large angles), surface character	Atomic to microscopic in one dimension, microscopic to macroscopic in two dimensions
Phase boundary	Compositional or crystallographic discontinuity, surface character	Atomic to microscopic in one dimension, microscopic to macroscopic in two dimensions

Source: Ref 5, p 302

Refractive indices of selected crystalline materials

	Average refractive index	Birefringence
Silicon chloride, $SiCl_4$	1.412	...
Lithium fluoride, LiF	1.392	...
Sodium fluoride, NaF	1.326	...
Calcium fluoride, CaF_2	1.434	...
Corundum, Al_2O_3	1.76	0.008
Periclase, MgO	1.74	...
Quartz, SiO_2	1.55	0.009
Spinel, $MgAl_2O_4$	1.72	...
Zircon, $ZiSiO_4$	1.95	0.055
Orthoclase, $KAlSi_3O_8$	1.525	0.007
Albite, $NaAlSi_3O_8$	1.529	0.008
Anorthite, $CaAl_2Si_2O_8$	1.585	0.008
Sillimanite, $Al_2O_3{\cdot}SiO_2$	1.65	0.021
Mullite, $3Al_2O_3{\cdot}2SiO_2$	1.64	0.010
Rutile, TiO_2	2.71	0.287
Silicon carbide, SiC	2.68	0.043
Litharge, PbO	2.61	...
Galena, PbS	3.912	...
Calcite, $CaCO_3$	1.65	0.17
Silicon, Si	3.49	...
Cadmium telluride, CdTe	2.74	...
Cadmium sulfide, CdS	2.50	...
Strontium titanate, $SrTiO_3$	2.49	...
Lithium niobate, $LiNbO_3$	2.31	...
Yttrium oxide, Y_2O_3	1.92	...
Zinc selenide, ZnSe	2.62	...
Barium titanate, $BaTiO_3$	2.40	...

Source: Ref 3, p 662

Typical properties of cemented carbide, sintered alumina, and Sialon cutting-tool materials

Property	Cemented carbide	Sintered alumina	Alumina-titanium carbide composite	Sialon
Hardness:				
GPa	12.3-15.1	15.3-15.9	17.0-17.4	12.2-15.2
HRA	91-92	93.2-93.6	94.0-94.4	90.8-93.5
Melting point:				
K	1673	2273	3413 (TiC)	Decomposes
°F	2552	3632	5684 (TiC)	at 2200 K
Coefficient of thermal expansion (α), $10^{-6}/K$	4.5-7.2	7.5	7.6	3.2
Modulus of elasticity:				
GPa	520-660	440	420	300
10^6 psi	75-96	64	61	44
Transverse rupture strength at 25 °C:				
MPa	1000-2400	700-840	840-940	830
ksi	145-350	100-122	122-136	120
Fracture toughness (K_H), $MPa\sqrt{m}$	...	2.2-2.5	3.1-3.5	3.6-5.2
Density, g/cm^3	12.0-15.1	3.80-3.90	4.20-4.30	3.35

Source: Ref 1, p 190

Comparative cutting performances of hard metal, alumina, and Sialon

Cutting parameter	Cast iron	Hardened steel En31	Incoloy 901
Hard metal			
Cutting speed:			
m/min	245	4.6	21
ft/min	800	15	70
Depth of cut:			
mm	6.4	...	...
in.	0.25	...	...
Feed rate:			
mm/rev	0.51	...	...
in/rev.	0.020	...	...
Alumina			
Cutting speed:			
m/min	610	(a)	90
ft/min	2000	(a)	300
Depth of cut:			
mm	6.4	(a)	(b)
in.	0.25	(a)	(b)
Feed rate:			
mm/rev	0.25	(a)	(b)
in./rev	0.010	(a)	(b)
Sialon			
Cutting speed:			
m/min	1070	120	305
ft/min	3500	400	1000
Depth of cut:			
mm	9.6	0.51	2.03
in.	0.38	0.020	0.080
Feed rate:			
mm/rev	0.51	0.25	0.25
in./rev	0.020	0.010	0.010

(a) Impossible to cut. (b) No second entry.

Source: Ref 1, p 170

Single Oxides

Property-value comparison of the principal single oxides

Property	Typical range of values at: Room Temperature	1095° C (2000° F)	Exceptions (room-temperature values)
Physical properties			
Crystal system	(a)	...	...
Theoretical density, g/cm^3	3-11.5	...	...
Melting point, °C (°F)	2040-2870 (3700-5200) (b)	...	ThO_2 (3220 °C); TiO_2 (1840 °C) (b)
Thermal properties			
Specific heat, kJ/kg·K (Btu/lb·°F)	0.4-0.8 (0.1-0.2)	0.4-1.3 (0.1-0.3)	BaO; ThO_2; UO_2 (all 0.25 kJ/kg·K)
Thermal conductivity, W/m·K (Btu/ft·h·°F)	1.7-17 (1-10)	1.7-6.9 (1-4)	BeO (24 W/m·K); MgO (52 W/m·K); Al_2O_3 (35 W/m·K)
Linear thermal expansion, %	0.2-0.3 (c)	0.8-1.5	SrO (0.4); CaO (0.4); Cr_2O_3 (0.1) (c)
Mechanical properties			
Bend strength, MPa (ksi)	140-275 (20-40)	69-140 (10-20)	Al_2O_3 (450 MPa); CaO (69 MPa)
Compressive strength, MPa (ksi)	2070 (300)	550-895 (80-130)	MgO (825 MPa); UO_2 (690 MPa)
Tensile strength, MPa (ksi)	69-140 (10-20)	34-105 (5-15)	Al_2O_3 (275 MPa)
Impact strength, J (in.·lb)	0.09-0.12 (0.8-1.1)	...	(d)
Modulus of elasticity, GPa (10^6 psi)	205-415 (30-60)	205-345 (30-50)	...
Shear modulus, GPa (10^6 psi)	69-140 (10-20)	69-140 (10-20)	...
Bulk modulus, GPa (10^6 psi)	0-170 (0-25)	...	...
Poisson's ratio	0.2-0.5	(e)	...
Creep rate, 1/h	...	(f)	...
Hardness, Mohs	6-9	...	BaO; CaO; SrO (all 3-4)
Microhardness, kg/mm^2	600-1000	...	Al_2O_3 (3000)
Thermal-stress resistance	...	(f)	...
Oxidation and corrosion resistance	...	(f)	...

(a) The most common or useful phases of the majority of the principal single oxides crystallize in the cubic or hexigonal systems. (b) "Room temperature" heading not applicable. (c) At 315 °C (600 °F); "room temperature" heading not applicable. (d) Data available ony for Al_2O_3 and MgO. (e) Values generally higher at elevated temperatures. (f) Varies widely; direct comparison difficult.

Source: Ref 1, p 249

Electrical properties of high-alumina ceramics

Property	Frequency, Hz	Vitreous 85% 25 °C (77 °F)	Vitreous 85% 500 °C (930 °F)	Vitreous 95% 25 °C (77 °F)	Vitreous 95% 500 °C (930 °F)	Vitreous 99.5=%	Porous 99.5=%
Dielectric constant	60	8.4	...	9.2	...	...	...
	1×10^3 (1 kHz)	7.65-8.75	13.86	8.84-10.51	13.3	10	...
	1×10^6 (1 MHz)	7.4-8.95	8.87	8.81-9.60	9.03	...	5.5
	1×10^8	8.10-8.95	...	8.80-9.60	...	...	5.3
	1×10^9	...	...	8.60	...	...	...
	3×10^9	8.14	...	8.80	...	...	...
	1×10^{10}	8.08-8.77	8.26	8.40-9.36	9.03	...	7.07
Power factor (tan δ)	60	0.0013-0.0015	...	0.0005	...	...	...
	1×10^3	0.0002-0.0014	0.580	0.00007-0.0006	1.1	...	...
	1×10^6	0.0007-0.0012	0.024	0.00035-0.0035	0.012	...	0.0005
	1×10^8	0.0009	...	0.00035-0.00040	...	...	0.0005
	1×10^9	...	...	0.0006	...	...	...
	3×10^9	0.0014	...	0.0010	...	...	...
	1×10^{10}	0.0027	0.0033	0.0008-0.0015	0.0021	...	...
Loss factor	60	0.011-0.013	...	...	...	...	...
	1×10^3	0.00175-0.0115	8.0	0.0008-0.0053	14.6	...	...
	1×10^6	0.0018-0.0078	0.21	0.0014-0.0035	0.108	...	0.003
	1×10^8	0.006-0.0074	...	0.0031-0.0040	...	...	0.003
	1×10^9	0.0076	...	0.0038	...	...	...
	3×10^9	0.0114	...	0.0038	...	...	...
	1×10^{10}	0.013-0.218	0.027	0.0067-0.0140	0.019	...	0.00075

Source: Ref 8

Physical properties of beryllium oxide

Modification	Crystal system	Theoretical density, g/cm³	Melting point (a) K	Melting point (a) °F
α-BeO	Hexagonal	3.008	2843	4658
β-BeO(b)	Tetragonal	2.69	2843	4658

(a) Most frequently reported value. Other values range from 2793 to 2923 K (4568 to 4801 °F). (b) α-β transition at 2322 ± 50 K (3720 ± 90 °F).

Source: Ref 9, p 5.4.2-1

Values of bulk modulus and Poisson's ratio for polycrystalline beryllia

Property	Temperature K	Temperature °F	Reported values
Bulk modulus, GPa (10^6 psi)	294	70	375 (54.5)
			291 (42.1)
Poisson's ratio	294-1273	70-1830	0.34
	>1273	>1830	>0.34
	294	70	0.38
			0.30 ± 0.05

Source: Ref 9, p 5.4.2-16

Mechanical properties of SiC-whisker-reinforced alumina ceramics

Whisker control, vol %(a)	Temperature °C	Temperature °F		Fracture strength MPa	Fracture strength ksi	Fracture toughness MPa $\sqrt{m}$	Fracture toughness ksi $\sqrt{in.}$
0	22	70	...	...	...	4.5	4.1
	700	1290	...	...	..	4.0	3.6
	1000	1830	...	...	...	3.8	3.5
10	22	70	...	455 ± 55	65 ± 8	7.1	6.5
	1000	1830	...	320 ± 36	45 ± 5		
20	22	70	...	655 ± 135	95 ± 20	7.5-9.0	6.8-8.2
	700	1290	...	535 ± 35	80 ± 5		
	1000	1830	...	570 ± 20	85 ± 3	7.0-8.0	6.4-7.3
40	22	70	...	850 ± 130	120 ± 20	6.0	5.5
	700	1290	...	740 ± 61	110 ± 9		
	1000	1830	...	665 ± 88	96 ± 13	6.2	5.6

(a) Hot pressed mixture of alumina powder and SiC whiskers.

Source: Ref 6, p 942

Dielectric constants and loss tangents for microwave-absorbing ceramics

Composition, wt %	Dielectric constant (K′) 8.6 GHz	Dielectric constant (K′) 10.5 GHz	Dielectric constant (K′) 12 GHz	Loss tangent (tan δ) 8.6 GHz	Loss tangent (tan δ) 10.5 GHz	Loss tangent (tan δ) 12 GHz
98% MgO - 2% SiC	11.22	11.13	10.94	0.15	0.06	0.03
95% MgO - 5% SiC	12.80	12.72	12.64	0.20	0.12	0.10
90% MgO - 10% SiC	17.12	16.84	16.61	0.30	0.19	0.19
80% MgO - 20% SiC	26.58	26.08	25.73	0.40	0.28	0.30
60% MgO - 40% SiC	54.24	53.81	50.46	0.92	0.70	0.73
98% BeO - 2% SiC	8.17	7.40	7.13	0.001	0.039	0.017
95% BeO - 5% SiC	10.60	8.95	8.54	0.032	0.075	0.043
90% BeO - 10% SiC	13.65	11.81	11.53	0.21	0.15	0.14
80% BeO - 20% SiC	20.26	18.70	17.81	0.33	0.24	0.22
60% BeO - 40% SiC	67.68	54.55	49.54	0.73	0.73	0.72

Source: Ref 10, p 200

Properties of microwave-absorbing ceramics

Composition	Fracture energy (Υ_c) J/m^2	ft·lb./ft^2	Modulus of rupture (σ) MPa	ksi	Modulus of elasticity (E) GPa	10^6 psi
90% BeO - 10% SiC	36	2.47	525	76	385	56
80% BeO - 20% SiC	24	1.64	450	65	395	57
60% BeO - 40% SiC	14	0.96	505	73	405	59
40% BeO - 60% SiC	30	2.05	745	108	415	60
99% MgO - 1% SiC	...	..	230	33	285	41
95% MgO - 5% SiC	28	1.92	...	...	290	42
90% MgO - 10% SiC	9	0.62	125	18	295	43
80% MgO - 20% SiC	13	0.89	195	28	315	46
70% MgO - 30% SiC	16	1.10	215	31	330	48
60% MgO - 40% SiC	15	1.03	205	30	345	50
40% MgO - 60% SiC	44	3.01	310	45	380	55

Source: Ref 10, p 200

Physical properties of zirconia and hafnia

Oxide	Crystal system	Theoretical density, g/cm^3	Melting temperature K	°F
ZrO_2	Monoclinic	5.56	(a)	(a)
	Tetragonal	6.10	3037 ± 83(b)	5008 ± 150(b)
	Cubic (stabilized ZrO_2)	(c)	(d)	(d)
HfO_2	Monoclinic	9.68	(e)	(e)
	Tetragonal	10.01	3117 ± 55(f)	5152 ± 10(f)
	Cubic (stabilized HfO_2)	(g)	(g)	(g)

(a) Monoclinic ZrO_2 transforms to tetragonal ZrO_2 at 1222 to 1494 K (1740 to 2230 °F), the exact temperature depending on purity. (b) Range of most reliable values reported. (c) The theoretical density of stabilized, cubic ZrO_2 will depend on the amount and kind of stabilizer used; approximate theoretical densities for 7, 5, and 3 wt % CaO-stabilized ZrO_2 are reported as 5.5, 5.75, and 5.9 g/cm^3, respectively. For the same percentage additives, MgO-stabilized ZrO_2 will have nearly identical densities, whereas Y_2O_3-stabilized ZrO_2 will have slightly higher densities than those of CaO-stabilized ZrO_2. (d) Stabilization also leads to lower melting points. For 7 and 5 wt % CaO-stabilized ZrO_2, respective melting points are reported as 2772 and 2872 K (4530 and 4710 °F). (e) Monoclinic HfO_2 transforms to tetragonal HfO_2 at 1883 to 2072 K (2930 to 3270 °F). (f) Range of reported values. (g) For cubic-stabilized HfO_2, values of theoretical density and melting point will depend on the amount and kind of stabilizer used.

Source: Ref 9, p 5.4.5-1

Representative strength values of fused silica and quartz at room temperature

Type of material	Percent of theoretical density	Bend strength MPa	Bend strength ksi	Compressive strength MPa	Compressive strength ksi	Tensile strength MPa	Tensile strength ksi
Bulk fused silica							
Clear	100	100(a)	16(a)	690-1380	100-200	69(a)	10(a)
	...	55(b)	8(b)	...	...	...	...
	100	48	7	1035	150	...	...
Translucent	100	45(a)	6.5(a)	275	40	21	3
Polygranular fused silica							
Hot pressed	~95	...	...	690-1380	100-200	...	...
Slip cast, sintered(c)	~34-98	3.4-120	0.5-17.5	17-490	2.5-7.1	4.8-32	0.7-4.7
Foamed (open pores)	13.6-35	0.69-2.76	0.10-0.40	2.76-8.62	0.40-1.25	...	...
Foamed (closed pores)	7-10	0.69-1.03	0.10-0.15	0.90-1.38	0.13-0.20	...	...
Single-crystal quartz(d)							
Parallel to C-axis	100	...	...	~2070	~300	110-275	16-40
Perpendicular to C-axis	...	...	...	<2070	<300	...	...

(a) Polished surfaces. (b) Abraded surfaces. (c) Fracture energy of slip cast fused silica is reported to be about 1 kg/cm (6 lb/in.). (d) Natural crystal.

Source: Ref 9, p 5.4.6-12

Properties of dense alumina

Property	Temperature °C	°F	94% Al_2O_3	96% Al_2O_3	97.6% Al_2O_3	99.5% Al_2O_3
Flexural strength, MPa (ksi)	Room temperature		345 (50)	365 (53)	296 (43)	310 (45)
Compressive strength, MPa (ksi)	Room temperature		>2070 (>300)	>2070 (>300)	>1720 (>250)	>2070 (>300)
Density, g/cm^3 (lb/in.^{3})	Room temperature		3.67 (0.132)	3.72 (0.134)	3.76 (0.134)	3.86 (0.139)
Porosity, % water absorption	...		0.00(a)	0.00(a)	0.00(a)	0.00(a)
Color	...		White	White	White	White
Hardness, HR45N	...		78	79	75	81
Thermal conductivity, W/m·K (Btu/ft·h·°F)	Room temperature		20.5 (11.9)	25.6 (14.8)	26.8 (15.5)	29.3 (16.9)
Coefficient of linear thermal expansion, 10^{-6}/°C (10^{-6}/°F)	25- 200	77- 390	6.3 (3.5)	6.4 (3.6)	6.9 (3.8)	6.9 (3.8)
	200- 400	390- 750	7.5 (4.2)	7.6 (4.2)	7.8 (4.3)	7.8 (4.3)
	400- 600	750-1110	8.0 (4.4)	8.2 (4.6)	8.5 (4.7)	8.3 (4.6)
	600- 800	1110-1470	8.6 (4.8)	8.7 (4.8)	8.8 (4.9)	9.0 (5.0)
	800-1000	1470-1830	9.1 (5.1)	9.0 (5.0)	9.0 (5.0)	9.4 (5.2)
Maximum working temperature, °C (°F)	...		1600 (2910)	1620 (2950)	1650 (3000)	1725 (3150)
Dielectric strength (b), dc-kV/mm (dc-V/mil)	Room temperature		25.6 (650)	26.6 (675)	43.3 (1100)	31.5 (800)
T_e value, °C (°F)	...		>950 (>1740)	>950 (>1740)	>1000 (>1800)	>975 (>1790)
Volume resistivity, Ω·cm	25	77	>10^{14}	>10^{14}	>10^{14}	>10^{14}
	300	570	2.0 × 10^{12}	2.0 × 10^{12}	2.0 × 10^{12}	2.0 × 10^{11}
	600	1110	4.6 × 10^{8}	5.2 × 10^{8}	2.3 × 10^{10}	6.0 × 10^{8}
	900	1650	3.5 × 10^{6}	4.1 × 10^{6}	5.0 × 10^{8}	2.5 × 10^{6}
Dielectric constant (K′)						
10 MHz	25	77	9.07	9.30	9.53	9.58
	300	570	9.53	9.65	9.91	9.92
	500	930	9.91	10.10	10.14	10.20
1000 MHz	25	77	9.04	9.20	9.00	9.30
	300	570	...	...	...	...
	500	930	...	...	...	...
8500 MHz	25	77	8.98	9.16	9.04	9.37
	300	570	9.26	9.30	9.32	9.61
	500	930	9.40	9.45	9.54	9.82
Dissipation factor (tan δ):						
10 MHz	25	77	0.00026	0.00030	0.00004	0.00003
	300	570	0.00028	0.00061	0.00016	0.00009
	500	930	0.00341	0.00330	0.00052	0.00040
1000 MHz	25	77	0.00062	0.00044	0.00030	0.00014
	300	570	...	...	...	...
	500	930	...	...	...	...
8500 MHz	25	77	0.00078	0.00062	0.00045	0.00009
	300	570	0.00155	0.00085	0.00040	0.00014
	500	930	0.00155	0.00121	0.00072	0.00025

(continued)

Properties of dense alumina (continued)

Property	Temperature °C	°F	94% Al_2O_3	96% Al_2O_3	97.6% Al_2O_3	99.5% Al_2O_3
Loss factor (K′ tan δ):						
10 MHz	25	77	0.00236	0.00279	0.00038	0.00029
	300	570	0.00267	0.00588	0.00158	0.00089
	500	930	0.03369	0.03333	0.00527	0.00408
1000 MHz	25	77	0.00560	0.00405	0.00270	0.00130
	300	570	...	...	...	...
	500	930	...	...	...	...
8500 MHz	25	77	0.00700	0.00568	0.00407	0.00084
	300	570	0.01165	0.00719	0.00373	0.000135
	500	930	0.01457	0.01143	0.00687	0.00245

(a) Vacuum tight. (b) For material 2.54 mm (0.100 in.) thick under oil.

Source: GTE Wesco Div. Al_2O_3 Brochure

Properties of high-alumina ceramics

Property	Temperature °C	Temperature °F	85% Al_2O_3 vitreous body	95% Al_2O_3 vitreous body	99.5% Al_2O_3 vitreous body	99.5% Al_2O_3 porous body
Tensile strength, MPa (ksi)	...		115-160 (17-23)	170-240 (25-35)	259-262 (37.5-38)	...
Compressive strength, MPa (ksi)	...		965-2760 (140-400)	1720-2760 (250-400)	2940 (427)	69-860 (10-125)
Flexural strength, MPa (ksi)	...		205-310 (30-45)	310-345 (45-50)	295-325 (42.7-47)	69-150 (10-22)
Modulus of elasticity GPa (10^6 psi)	...		215-240 (31-35)	270-295 (39-43)	360 (52)	...
Impact resistance, J (in.·lb Charpy)	...		0.66-0.79 (5.8-7.0)	0.70-0.86 (6.2-7.6)	...	0.34 (3.0)
Specific gravity	...		3.40-3.53	3.61-3.75	3.7-3.97	2.4-3.40
Water absorption, %	...		0.00-0.02	0.00	0.00	7-8
Porosity, %	...		<1(a)	<1(a)	<1	7.25
Hardness:						
Mohs scale	...		8.5-9	9	9	...
Knoop	...		1450	1720	...	...
Maximum working temperature, °C (°F)	...		1200-1400 (2200-2550)	1600-1700 (2910-3100)	1950 (3542)	1400-1800 (2550-3270)
Pore size, μm (mil)	...		...	...	...	2-3 (0.08-0.12)
Heat capacity, kJ/kg (Btu/lb)	...		0.419 (0.180)	0.437-0.442 (0.188-0.190)	0.51 (0.22)	...
Thermal conductivity, W/m·K (Btu·in./ft²·h·°F)	38	100	13-17 (90-116)	19-22 (130-150)	19 (135)	17 (116)
	425	800	26 (180)	35 (240)	...	...
	870	1600	33 (230)	43 (300)	...	...
Coefficient of thermal expansion, 10^{-6}/°C (10^{-6}/°F)	25-200	77-390	5.47-5.68 (3.04-3.16)	5.7-6.67 (3.2-3.70)	...	5.1 (2.8)
	25- 600	77-1110	6.55-6.96 (3.64-3.87)	6.7-7.65 (3.7-4.25)	...	...
	25-700	77-1290	7.6-7.9 (4.2-4.4)	8.07 (4.48)	7.7 (4.3)	...
	25-800	77-1470	7.33 (4.07)	7.6 (4.2)	...	...
	25-1000	77-1830	7.67-7.89 (4.26-4.38)	8.45-9.14 (4.69-5.08)	8.4 (4.7)	...
Thermal shock resistance	...		Fair	Good	Good	Good
Dielectric strength, ac-kV/mm ac-V/mil	25	77	8.07-13.8 (205-350)	9.84-15.7 (250-400)	15.0 (380)	1.97 (2.8)
	500	930	...	3.94-4.72 (100-120)	...	...
	1000	1830	...	0.79-1.18 (20-30)	...	...
Volume resistivity, Ω·cm²/cm	25	77	$1\text{-}3.6 \times 10^{14}$	10^{16}	...	10^{14}
	100	212	$2\text{-}7.5 \times 10^{13}$	9.0×10^{14}	...	$8.5 \times 10^{13} - 1 \times 10^{14}$
	200	390	...	10×10^{13}	...	...
	300	570	$1\text{-}5.0 \times 10^{10}$	5.3×10^{12}	1.2×10^{13}	$1 \times 10^{10} - 1.5 \times 10^{11}$
	400	750	...	10×10^{10}	...	...
	500	930	$1 \times 10^{8} - 7.5 \times 10^{9}$	$1.2\text{-}4.5 \times 10^{10}$	1.3×10^{11}	$7.5 \times 10^{7} - 1.0 \times 10^{9}$
	600	1110	...	10^{8}	...	...
	700	1290	$3\text{-}7.0 \times 10^{6}$	6.0×10^{8}	...	$3.6 \times 10^{6} - 3.0 \times 10^{7}$
	800	1470	...	...	3.5×10^{8}	...
	900	1650	$4\text{-}5.0 \times 10^{5}$	...	...	5.6×10^{5}
T_e value, °C (°F)	...		750-1000 (1380-1830)	800-1100 (1470-2010)	1100 (2010)	835-1100 (1535-2010)

(a) Gas tight (helium mass spectrometer test on 0.254-mm, or 0.010-in., sections).

Source: Ref 8

Physical properties of barium, calcium, and strontium oxides

Oxide	Crystal system(a)	Theoretical density, g/cm^3	Melting temperature K	°F
BaO	Cubic	5.72	2196	3493
CaO	Cubic	3.32(b)	2887(c)	4737(c)
SrO	Cubic	4.7	2727 ± 22	4450 ± 40

(a) Hexagonal forms have also been reported, particularly as sublimation products. Amorphous forms also exist, which convert to the cubic crystalline form at temperatures of about 672 to 700 K (750 to 800 °F). (b) Other reported values are 3.37 and 3.40 g/cm^3. (c) Most reliable value; other reported values range from about 2805 to 2894 K (4590 to 4750 °F).

Source: Ref 9, p 5.4.4-1

Physical properties of titanium and chromium oxides

Oxide	Crystal systems	Theoretical density, g/cm^3	Melting point K	°F
α-TiO	Monoclinic	4.93	(b)	(b)
β-TiO	Cubic	...	2023	3180
Ti_2O_3	Hexagonal	4.6	2128(c)	3770(c)
Ti_3O_5	Monoclinic	4.24	(d)	(d)
TiO_2 (anatase)	Tetragonal	3.84	(e)	(e)
TiO_2 (rutile)	Tetragonal	4.25	(f)	(f)
TiO_2 (brookite)	Orthorhombic	4.17	2109 ± 17(g)	3340 ± 30(g)
Cr_2O_3(h)	Hexagonal	5.21	2299 ± 122(g)	4170 ± 220(g)

(a) Some values may be experimental rather than theoretical. (b) Transforms to a β-phase at 1263 K (1815 °F). (c) Transforms to a β-phase at 472 K (390 °F). (d) Transforms to a β-phase at 450 K (350 °F). (e) Reported to exist in three forms. Transformation temperatures are given as 915 K (1188 °F) for anatase I → anatase II, 1188 K (1679 °F) for anatase II → rutile, and 1323 K (1922 °F) for anatase III → rutile. (f) Transforms to brookite at 1572 K (2370 °F). (g) Range of values most frequently reported. (h) Other oxides of chromium exist but are not stable at elevated temperatures

Source: Ref 9, p 5.4.8-1

Hardnesses of thoria, urania, and plutonia

Oxide	Theoretical density, %	Load, g	Knoop hardness, kg/mm^2
ThO_2	...	500	640
ThO_2 + ½% CaO	98.3	500	700
UO_2	~100	100	662(a)
UO_2	...	...	666
UO_2	93.3	10	355
UO_2	93.3	50	585
UO_2	93.3	100	625
UO_2	93.3	500	600
UO_2	93.3	2000	520
UO_2	97	...	600 Vickers
	80	...	~260 Vickers
	60	...	~45 Vickers
$UO_{2.00}$	...	500	640
$UO_{2.02}$	...	500	787
$UO_{2.145}$	...	500	880
UO_2	...	...	6-7 (Mohs)
PuO_2	~90	...	~440 Vickers

(a) Average value for annealed single crystal, independent of orientation.

Source: Ref 9, p 5.4.10-14

Mechanical properties of various zirconia materials

Property	Temperature K	Temperature °F	Reported values MPa	Reported values ksi
Yttria-stabilized ZrO_2				
Bend strength	294	70	364-577(a)	52.9-83.8(a)
	294	70	334-527(b)	48.6-76.6(b)
	294	70	654-663(c)	95.1-96.3(c)
	294	70	151	22
	773	932	138	20
	1273	1832	117	17
	1473	2192	69	10
Magnesia-stabilized ZrO_2				
Bend strength	293	68	290	42.2
	1323	1922	217	31.6
Compressive strength	294	70	2064	300
	773	932	1569	228
	1273	1830	1176	171
	1673	2552	127	18.5
	1773	2732	19	2.8
	293	68	450	65.4
	1323	1922	137	19.9
Calcia-stabilized ZrO_2				
Bend strength	294	70	239	34.7
	1273	1830	169	24.5
	1573	2372	113	16.4
Compressive strength	294	70	2085	303
	1473	2192	689	100
Tensile strength	294	70	30.0-42.2	4.36-6.13
	294	70	79.7	11.6
	294	70	144.5	21
	1472	2190	82.5	12
	1811	2800	~13.8	~2

(a) Dry pressed and fired to a density of about 6.0 g/cm^3. (b) Wet pressed and fired to a density of about 6.0 g/cm^3. (c) Pressure cast and fired to a density of about 6.0 g/cm^3.

Source: Ref 9, p 5.4.5-12

Typical properties of transformation-toughened zirconia

Property	Test	Temperature °C	Temperature °F	TTZ
Stabilizer	...	...	...	MgO
Density	ASTM C20-70	...	...	5.75
Water absorption	ASTM C373-72	...	...	None
Hardness, HR45N	Rockwell R45N	...	...	74-79
Flexural strength, MPa (ksi)	ASTM F417-75T	25	77	634 (92)
		500	930	414 (60)
		1000	1830	290 (42)
Tensile strength, MPa (ksi)	ALMA Test #4	25	77	352 (51)
Compressive strength, MPa (ksi)	ASTM C773-74	25	77	1758 (255)
Modulus of elasticity, GPa (10^6 psi)	ASTM C623-71	...	...	200 (29)
Shear modulus, GPa (10^6 psi)	ASTM C623-71	...	...	69 (10)
Bulk modulus, GPa (10^6 psi)	ASTM C623-71	...	...	373 (54)
Poisson's ratio	...	...	...	0.22
Coefficient of thermal expansion, 10^{-6}/°C (10^{-6}/°F)	ASTM C372-56	25-1000	77-1830	10.1 (5.6)
Fracture toughness (K_{Ic}), MPa$\sqrt{m}$ (ksi$\sqrt{in.}$	SENB	...	...	8-12 (7-11)
Weibull modulus, m (ft)	Four-point bend	...	...	20 (66)

Source: Coors Ceramics

Physical properties of vanadium, niobium, and tantalum oxides

Oxide	Crystal systems	Theoretical density, g/cm^3	Melting point K	Melting point °F
VO	Cubic	5.23	2322	3720
V_2O_3	Hexagonal	4.87	2243-2273	3578-3632
VO_2	Tetragonal	4.65	1818-1911	2813-2980
V_2O_5	Orthorhombic	3.35	943	1238
NbO	Cubic	7.30	2218	3533
Nb_2O_3	...	...	2045-2050	3222-3230
NbO_2	Tetragonal	5.90	2188	3479
Nb_2O_5	Orthorhombic(a)	4.46	1761-1783	2710-2750
Ta_2O_5	Orthorhombic	8.02	2155	3420

(a) Nb_2O_5 transforms to a monoclinic form at about 1373 K (2012 °F).

Source: Ref 9, p 5.4.7-1

Physical properties of oxides of rare-earth metals

Oxide	Crystal system	Theoretical density g/cm^3	Melting point K	Melting point °F
Sc_2O_3	Cubic(a)	3.841	2660	4330 ± 158(b)
Y_2O_3	Cubic	5.03	~2700	~4400
La_2O_3	Hexagonal	6.57	2539	4110 ± 90
Ce_2O_3	Hexagonal	6.87	2448	3949 ± 61(b)
CeO_2	Cubic	7.28 ± 0.1	2614	4245 ± 740(b)
PrO_2	Cubic	...	...	...
Pr_6O_{11}	Cubic	5.47	2315	3710 ± 54(b)
Pr_2O_3	Cubic	6.32	2485	4013 ± 153(b)
	Hexagonal(c)	...	...	...
Nd_2O_3	Hexagonal(d)	7.28	2378	3821 ± 369(b)
Pm_2O_3	(e)	...	...	...
SmO	Cubic	...	...	...
Sm_2O_3	Monoclinic	7.43	...	(f)
Sm_2O_3	Cubic(g)	7.62	2576	4177 ± 84(b)
EuO	Cubic	8.16	2131	3376 ± 284(b)
Eu_2O_3	Cubic	7.3 ± 0.04	...	(f)
Eu_2O_3	Monoclinic(h)	7.95	2439	3931 ± 295(b)
Eu_3O_4	Orthorhombic	8.07	2273	3632 ± 180(i)
$Eu_{16}O_{21}$	Orthorhombic	6.74	...	...
Gd_2O_3	Cubic	7.65 ± 0.02	...	(f)
β-Gd_2O_3	Monoclinic	...	2629	4273 ± 70(b)
Tb_2O_3	Cubic(j)	7.74 ± 0.06	2612	4242 ± 94(b)
Tb_4O_7	Cubic	...	...	...
Dy_2O_3	Cubic(k)	8.17	2583	4190 ± 148(b)
Ho_2O_3	Cubic	8.41 ± 0.6	2636	4285 ± 59(b)
Er_2O_3	Cubic	8.64	2618	4253 ± 99(b)
Tm_2O_3	Cubic	8.83 ± 0.13	2619	4255 ± 81(b)
Yb_2O_3	Cubic	9.25 ± 0.05	2608	4235 ± 135(b)
Lu_2O_3	Cubic	9.423	2740	4473(l)

(a) Cubic form transforms to monoclinic at an elevated temperature. (b) Range of reported values. (c) Cubic form reportedly transforms to hexagonal at about 1153 K (1616 °F). (d) Cubic form reportedly transforms to hexagonal at about 1113 K (1544 °F). (e) Three types: hexagonal, monoclinic, cubic. (f) Transforms to β modification irreversibly, the rate being a function of time and temperature. (g) Cubic form reportedly transforms to monoclinic at about 1253 K (1796 °F). Monoclinic form reportedly begins transformation to hexagonal at 2573 K (4142 °F). (h) An apparently irreversible transformation from cubic to monoclinic takes place at approximately 1348 K (1967 °F). At higher temperatures, transformations to three other polymorphic forms occur, two of which are hexagonal. (i) Estimated from experimental data. (j) Cubic form reportedly transforms to monoclinic at about 2123 K (3362 °F), and monoclinic form becomes hexagonal at a higher temperature. (k) Cubic form reportedly transforms to monoclinic at about 2413 K (3884 °F). (l) Value most often reported.

Source: Ref 9, p 5.4.9-1

Mechanical properties of rare-earth oxides

Oxide	Temperature K	Temperature °F		Property values	
Bend strength				**MPa**	**ksi**
Y_2O_3	293	68	....	123	17.92
	1273	1830	....	133	19.34
	1573	2370	....	122	17.78
	293	68	....	104	15.07
	1273	1830	....	139	20.19
	1773	2730	....	125	18.20
	2023	3180	....	129	18.77
Sc_2O_3	293	68	....	213	31.00
	293	68	....	176	25.60
	973	1290	....	186	27.00
	1273	1830	....	181	26.30
	1623	2460	....	127	18.50
CeO_2	293	68	....	139	20.19
	1073	1470	....	95	13.79
	1273	1830	....	71	10.38
	1473	2190	....	51	7.39
	1623	2460	....	23	3.41
Gd_2O_3	293-1673	68-2550	....	108-137	15.64-19.91
Er_2O_3	293-1673	68-2550	....	98-127	14.22-18.49
Dy_2O_3	293	68	....	78	11.38
	1073	1470	....	88	12.80
	1473	2190	....	137	19.91
Yb_2O_3	293	68	....	73	10.66
	1073	1470	....	73	10.66
	1473	2190	....	78	11.38
Compressive strength				**MPa**	**ksi**
Y_2O_3	293	68	....	392	57.00
CeO_2	293	68	....	588	85.40
Nd_2O_3	293	68	....	107-134	15.57-19.42
Gd_2O_3	293	68	....	166-215	24.17-31.28
Er_2O_3	293	68	....	420-430	61.15-62.57
Dy_2O_3	293	68	....	381	55.46
Yb_2O_3	293	68	....	225	32.71
Bulk modulus				**GPa**	**10^6 psi**
Tm_2O_3	293	68	....	130	18.86
	791	964	....	123	17.91
	1269	1832	....	114	16.56
Lu_2O_3	293	68	....	139	20.25
	790	962	....	132	19.24
	1269	1825	....	125	18.24
Y_2O_3	293	68	....	135	19.69
Dy_2O_3	293	68	....	150	21.84
Ho_2O_3	293	68	....	134	19.53
Er_2O_3	293	68	....	140	20.40
Poisson's ratio					
Yb_2O_3	293	68	....	0.284	
Gd_2O_3	293	68	....	0.276	
	1273	1832	....	0.267	
Sm_2O_3	293	68	....	0.32	
Tm_2O_3	293	68	....	0.292	
	791	964	....	0.289	
	1269	1832	....	0.285	
Lu_2O_3	293	68	....	0.287	
	790	962	....	0.288	
	1269	1825	....	0.289	
Y_2O_3	293	68	....	0.298	
Dy_2O_3	293	68	....	0.313	
Ho_2O_3	293	68	....	0.290	
Er_2O_3	293	68	....	0.292	

(continued)

Mechanical properties of rare-earth oxides (continued)

Oxide	Temperature K	°F		Property values
Y_2O_3	293	68	....	0.295
CeO_3	293	68	....	0.275-0.315
Hardness				**kg/mm²**
Sc_2O_3	293	68	....	790-910
La_2O_3	293	68	....	300-380
CeO_3	293	68	....	5-6 Mohs
Pr_2O_3	293	68	....	370-380
Nd_2O_3	293	68	....	~380-650
Sm_2O_3	293	68	....	380-480
Eu_2O_3	293	68	....	140-500
Gd_2O_3	293	68	....	380-550
Tb_2O_3	293	68	....	380
Dy_2O_3	293	68	....	400-700
Ho_2O_3	293	68	....	380
Er_2O_3	293	68	....	380-700
Tm_2O_3	293	68	....	380
Yb_2O_3	293	68	....	650
Lu_2O_3	293	68	....	650-830

Source: Ref 9, p 5.5.9-2

Bend and compressive strengths of thoria

Theoretical density, %	Average grain size µm	mils	Temperature K	°F	Strength values MPa	ksi
Bend strength						
80.2	~4	~0.16	297	75	89.4	13.0
91.4	16	0.63	297	75	128.6	18.7
68.9	43	1.70	297	75	31.6	4.6
Compressive strength						
80.2	~4	~0.16	297	75	1444.8	210
91.4	16	0.63	297	75	1561.8	227
68.9	43	1.70	297	75	217.4	31.6
92	...	...	300	80	1472.3	214
			672	750	1073.3	156
			1072	1470	488.5	71
			1472	2190	195.4	28
			1772	2730	<10	<2

Source: Ref 9, p 5.4.10-8

Hardness of zirconia

Stabilizer(a)	Hardness, Mohs	Micro-hardness, kg/mm²
8% CaO	...	1180
10% CaO	>6	1470
12% CaO	...	1130
8% MgO	...	1080
10% MgO	...	1520
12% MgO	...	1430
15% Y_2O_3(b)	>6	1360

(a) Mole percent. (b) Single crystal: Knoop, 500-g load.

Source: Ref 9, p 5.4.5-10

Physical properties of thorium, uranium, and plutonium oxides

Oxide	Crystal system	Theoretical density, g/cm³	Melting point K	°F
ThO_2	Cubic	10.00 ± 0.01	3493	5830 ± 180(a)
UO_2	Cubic	10.96 ± 0.01	3113	5144 ± 36
U_3O_8	Orthorhombic(b)	8.39	...	...
UO_3	Orthorhombic(c)	8.34	...	(d)
PuO	Cubic	13.9	...	...
Pu_2O_3	Cubic	10.2	2484	4020 ± 80
	Hexagonal	11.2	2484	4020 ± 80
PuO_2	Cubic	11.46	2633	4334 ± 36

(a) The value 5970 °F (3572 K) is most frequently reported. (b) Becomes "hexagonal" above 673 K (752 °F). (c) As many as five allotropic forms may exist, including an amorphous form. (d) Decomposes to U_3O_8 in the range 772 to 972 K (930 to 1290 °F).

Source: Ref 9, p 5.4.10-1

Moduli of elasticity for niobium oxides

Oxide	Temperature K	Temperature °F	Modulus of elasticity GPa	Modulus of elasticity 10^6 psi
Nb_2O_5				
Sintered	293	68	~40	~5.8
	1273	1830	~100	~14.5
Hot pressed	293	68	~150	~21.8
	1273	1830	~125	~18.2
Nb_2O_3				
Hot pressed	293	68	~161	~23.4

Source: Ref 9, p 5.4.7-5

Mixed Oxides

Typical strengths of commercial silicate ceramics at room temperature

Major phase	Bend strength MPa	Bend strength ksi	Tensile strength MPa	Tensile strength ksi	Compressive strength MPa	Compressive strength ksi
$3Al_2O_3{\cdot}2SiO_2$ (mullite)	~170	~25	~110	~16	690-1310	100-190
$ZrO_2{\cdot}SiO_2$ (zircon)	~140	~20	~75	~11	480-690	70-100
$2MgO{\cdot}SiO_2$ (forsterite)	~140	~20	~65	~9.5	~550	~80
$MgO{\cdot}SiO_2$ (steatite)	~130	~19	~64	~9.3	~550	~80
$CaO{\cdot}SiO_2$ (wollastonite)	~130	~19	~55	~8	...	...
$2MgO{\cdot}2Al_2O_3{\cdot}5SiO_2$ (cordierite)	~110	~16	~54	~7.8	~345	~50

Source: Ref 9, p 5.5.1-11

Thermal conductivities of titanates

Titanate	Thermal conductivity(a) W/m·K	Btu/ft·h·°F
$SrO \cdot TiO_2$	5.9 to 5.5 (43 to 140 °C)	3.4 to 3.2 (110 to 280 °F)
	6.2 to 4.8 (49 to 150 °C)	3.6 to 2.8 (120 to 300 °F)
$CaO \cdot TiO_2$	4.7 to 4.3 (43 to 130 °C)	2.7 to 2.5 (110 to 270 °F)
$BaO \cdot TiO_2$	3.3 to 2.6 (43 to 230 °C)	1.9 to 1.5 (110 to 440 °F)
	2.9 to 2.6 (49 to 230 °C)	1.7 to 1.5 (120 to 390 °F)

(a) Reported values; no material or test details were available.

Source: Ref 9, p 5.5.4-2

Physical properties of titanates

Titanate	Crystal system	Density, g/cm^3	Melting point K	°F
$\alpha Al_2O_3 \cdot TiO_2$	Orthorhombic	3.68	2135	3380
$\beta Al_2O_3 \cdot TiO_2$		...	(a)	(a)
$2BaO \cdot TiO_2$		5.3(b)	2135	3380
$BaO \cdot TiO_2$	Hexagonal (c)	5.9(b)	1890 ± 5.5	2940 ± 10
$BaO \cdot 2TiO_2$		...	1595(d)	2410(d)
$BaO \cdot 3TiO_2$		4.7(b)	1630	2475
$BaO \cdot 4TiO_2$		4.6(b)	1700	2600
$CaO \cdot TiO_2$	Cubic	4.10	2245	3580
$3CaO \cdot 2TiO_2$		...	2020(d)	3180(d)
$HfO_2 \cdot TiO_2$	Orthorhombic	7.21	...	...
$2MgO \cdot TiO_2$	Cubic	3.52	2005	3150
$MgO \cdot TiO_2$	Hexagonal	~4.00	1905	2970
$MgO \cdot 2TiO_2$		3.66	1920	3000
$2MnO \cdot TiO_2$	Cubic	4.49	1725	2650
$MnO \cdot TiO_2$	Hexagonal	4.54	1635(c)	2480(c)
$SrO \cdot TiO_2$	Cubic	5.11	2310	3700
$2ZnO \cdot TiO_2$	Cubic	5.12	1820	2820

(a) Unstable between 1020 and 1570 K (1380 and 2370 °F); converts to alpha phase at 2090 K (3300 °F). (b) Pycnometric measurements. (c) Three polymorphic forms reported: tetragonal (295 to 395 K, or 70 to 250 °F); cubic (395 to 1735 K, or 250 to 2660 °F); and hexagonal (above 1735 K, or 2660 °F). (d) Incongruent melting.

Source: Ref 9, p 5.5.4-1

Mechanical properties of niobates

Niobate	Temperature °C	°F	Bend strength MPa	ksi	Modulus of elasticity GPa	10^6 psi	Material and test conditions
$MgO \cdot Nb_2O_5$	24	75	37	5.4	...	...	Pressed and sintered; 91 to 92% dense.
	800	1470	78	11.3	...	...	
	1000	1830	90	13.0	...	...	
	1200	2190	51	7.4	...	...	
$TiO_2 \cdot Nb_2O_5$	24	75	29	4.2	41	6	Pressed and sintered; 91% pure; 86 to 88% dense; Modulus of elasticity determined sonically.
	800	1470	34	4.9	41	6	
	1000	1830	48	6.9	41	6	
	1200	2190	36	5.3	...	...	

Source: Ref 9, p 5.5.5-8

Physical properties of silicates

Silicate	Crystal system	Density(a), g/cm^3	Melting point K	Melting point °F
$3Al_3O_3\cdot 2SiO_2$	Orthorhombic	3.13-3.36	2120	3360
$2BaO\cdot SiO_2$	Orthorhombic	5.20	>2025	>3190
$BaO\cdot SiO_2$	Monoclinic	4.40	1875	2920
$2BaO\cdot 3SiO_2$	...	3.93	1720	2640
$BaO\cdot 2SiO_2$	Orthorhombic	3.73	1690	2585
$BaO\cdot Al_2O_3\cdot 2SiO_2$	Hexagonal	3.21-3.30	1990	3120
$BaO\cdot 2CaO\cdot 3SiO_2$	Hexagonal	...	1595(d)	2410(d)
$BaO\cdot TiO_2\cdot SiO_2$	...	...	1670	2550
$BaO\cdot TiO_2\cdot 2SiO_2$	...	...	1520	2280
$2BeO\cdot SiO_2$	Hexagonal	2.99	1835(e)	2840(e)
$3CaO\cdot SiO_2$	Monoclinic	...	2170(e)	3450(e)
$2CaO\cdot SiO_2$	Monoclinic	3.28	2400	3865
$3CaO\cdot 2SiO_2$	Orthorhombic	...	1740(d)	2670(d)
$CaO\cdot SiO_2$	Triclinic	2.9	1810	2800
$CaO\cdot Al_2O_3\cdot 2SiO_2$	Triclinic	2.77	1820	2820
$2CaO\cdot Al_2O_3\cdot 2SiO_2$	Tetragonal	3.04	1865	2895
$CaO\cdot MgO\cdot SiO_2$	Orthorhombic	3.2	1770(d)	2730(d)
$CaO\cdot MgO\cdot 2SiO_2$	Monoclinic	3.28	1665	2535
$2CaO\cdot MgO\cdot 2SiO_2$	Tetragonal	2.94	1735	2660
$3CaO\cdot MgO\cdot 2SiO_2$	Monoclinic	3.15	1845(d)	2865(d)
$CaO\cdot K_2O\cdot SiO_2$	...	...	1905	2970
$CaO\cdot TiO_2\cdot SiO_2$	Monoclinic	3.5	1655	2520
$2CaO\cdot ZnO\cdot 2SiO_2$	Tetragonal	...	1700	2600
$CaO\cdot ZrO_2\cdot SiO_2$	...	...	1855	2880
$2CoO\cdot SiO_2$	Orthorhombic	4.68	1695	2590
$Dy_2O_3\cdot SiO_2$	...	...	2200	3505
$2Dy_2O_3\cdot 3SiO_2$	...	...	2195	3490
$Dy_2O_3\cdot 2SiO_2$	...	...	1995(d)	3130(c)
$Er_2O_3\cdot SiO_2$	...	6.80	2250	3595
$2Er_2O_3\cdot 3SiO_2$	...	6.22	2170	3450
$Er_2O_3\cdot 2SiO_2$	...	6.10	2070	3270
$Gd_2O_3\cdot SiO_2$	...	6.55	2170	3450
$2Gd_2O_3\cdot 3SiO_2$	Hexagonal	6.29	2220	3450
$Gd_2O_3\cdot 2SiO_2$	...	5.34	1993	3128
$HfO_2\cdot SiO_2$	Tetragonal	...	...	...
$2FeO\cdot SiO_2$	Orthorhombic	4.24	1470	2190
$La_2O_3\cdot SiO_2$	Monoclinic	5.72	2200	3505
$2La_2O_3\cdot 3SiO_2$	Hexagonal	5.30	2245	3580
$La_2O_3\cdot 2SiO_2$	Monoclinic	4.85	2020(d)	3180(d)
$Li_2O\cdot SiO_2$	Orthorhombic	2.48	1475	2200
$2Li_2O\cdot SiO_2$	Orthorhombic	2.33	1525	2290(d)
$Li_2O\cdot Al_2O_3\cdot 2SiO_2$	Orthorhombic	2.36	1670(d)	2550(d)
$Li_2O\cdot Al_2O_3\cdot 4SiO_2$	Monoclinic	...	1700	2600
$Li_2O\cdot Al_2O_3\cdot 6SiO_2$	...	2.41	1455	2160
$4Li_2O\cdot 3ZrO_2\cdot 5SiO_2$	...	4.02	1425	2110
$MgO\cdot SiO_2$(f)	...	...	1825	2850
$2MgO\cdot SiO_2$	Orthorhombic	3.22	2185	3470
$2MgO\cdot 2Al_2O_3\cdot 5SiO_2$	Orthorhombic	2.51	1745	2680(d)
$4MgO\cdot 5Al_2O_3\cdot 2SiO_2$	...	...	~1725(d)	~2650(d)
$2MnO\cdot SiO_2$	Orthorhombic	4.05	1615(d)	2445(d)
$MnO\cdot SiO_2$	Triclinic	3.71	1570(d)	2370(d)
$3MnO\cdot Al_2O_3\cdot 3SiO_2$	Cubic	4.18	1470	2190
$2MnO\cdot 2Al_2O_3\cdot 5SiO_2$	...	...	1470(d)	2190(d)
$Nd_2O_3\cdot SiO_2$	Hexagonal	...	...	...
$Nd_2O_3\cdot 2SiO_2$	Monoclinic	...	...	...
$K_2O\cdot Al_2O_3\cdot SiO_2$	Cubic	...	...	...
$K_2O\cdot Al_2O_3\cdot 2SiO_2$	Hexagonal	2.6	2020	3180
$K_2O\cdot Al_2O_3\cdot 4SiO_2$	Cubic	2.47	1960	3070
$K_2O\cdot Al_2O_3\cdot 6SiO_2$	Triclinic	2.56	1420	2100
$K_2O\cdot ZnO\cdot SiO_2$	Cubic	...	1570	2370(d)
$Pr_2O_3\cdot SiO_2$	...	...	1670(e)	2550(e)
$Sm_2O_3\cdot SiO_2$	Monoclinic	6.36	2215	3525

(continued)

Physical properties of silicates (continued)

Silicate	Crystal system	Density(a), g/cm^3	Melting point K	°F
$2Sm_2O_3 \cdot 3SiO_2$	...	5.77	2195	3490
$Sm_2O_3 \cdot 2SiO_2$	Monoclinic	5.20	2050(d)	32309(d)
$Sc_2O_3 \cdot SiO_2$	...	3.49	2223	3542
$S_2O_3 \cdot 2SiO_2$	...	3.39	2133	3380
$2SrO \cdot SiO_2$	Monoclinic	3.84	>1975	>3100
$SrO \cdot SiO_2$	Monoclinic	3.65	1850	2975
$SrO \cdot Al_2O_3 \cdot 2SiO_2$	Triclinic	3.12	1935	3020
$ThO \cdot SiO_2$	Monoclinic	5.3	2250(d)	3595(d)
$Yb_2O_3 \cdot SiO_2$	...	...	2220	3540
$2Yb_2O_3 \cdot 3SiO_2$	...	...	2195	3490
$Yb_2O_3 \cdot 2SiO_2$	...	...	2120	3360
$Y_2O_3 \cdot SiO_2$	Monoclinic	4.49	2250	3595
$2Y_2O_3 \cdot 3SiO_2$	Hexagonal	4.39	2200	3540
$Y_2O_3 \cdot 2SiO_2$	Monoclinic	4.06	2050(b)	3230(b)
$ZnO \cdot SiO_2$	Hexagonal	4.1	1785	2750
$ZrO_2 \cdot SiO_2$	Tetragonal	4.68	1815(e)	2805(e)

(a) Values either are calculated from lattice parameters or are pycnometric measurements. (b) Approximate values. (c) Transformation from monoclinic structure to hexagonal structure occurs at 570 K (570 °F) with significant volume change resulting. (d) Incongruent melting. (e) Decomposition. (f) Phase inversions at about 1535 K (about 2300 °F). (g) Transforms to orthorhombic structure at 1820 K (2820 °F) on cooling.

Source: Ref 9, 5.5.5-1

Mechanical properties of titanates

Titanate	Test temperature °C	°F	Bend strength MPa	ksi	Compressive strength MPa	ksi	Modulus of elasticity GPa	10^6 psi
$Al_2O_3 \cdot TiO_2$	21	70	21-76(a)	3-11(a)	...	...	...	...
$BaO \cdot TiO_2$	21	70	...	...	...	...	110(b)	16(b)
	21	70	66-108(c)	9.6-15.6(c)	...	...	...	...
	21	70	83-163(d)	12.0 -23.6(d)	...	...	...	...
$CaO \cdot TiO_2$	21	70	...	...	48-131(e)	7-19(e)	...	...
$3CaO \cdot 2TiO_2$	21	70	...	...	69(f)	10(f)	...	...
$MgO \cdot 2TiO_2$	21	70	12-14(g)	1.8-2.0(g)	...	...	7.6-10(g)	1.1-1.5(g)
	400	750	14-17	2.0-2.5	...	...	9.0-23	1.3-3.3
	600	1110	18-24	2.6-3.5	...	...	12-41	1.8-6.0
	800	1470	25-41	3.6-5.9	...	...	26-94	3.8-13.7
	1000	1830	44-58	6.4-8.4	...	...	43	6.3

(a) Pressed and sintered; 88 to 90% dense. (b) Pressed and sintered; 95% dense; 95% pure; dynamic method. (c) Commercial material, as received; average value, 80.7 MPa (11.7 ksi). (d) Same material described in footnote (c), thermally conditioned at 1290 °C (2350 °F); average value, 101 MPa (14.6 ksi), (e) Pressed and sintered; 75 to 91% dense. (f) Pressed and sintered; 89% dense. (g) Pressed and sintered; 95.5% dense; strength and modulus determined in flexure (three-point loading).

Source: Ref 9, p 5.5.4-5

Specific heats of aluminates

Aluminate	Specific heat (a) kJ/kg·K	Specific heat (a) Btu/lb·°F	Material and test conditions
$CaO \cdot 2Al_2O_3$	0.75	0.18	Powder samples; "high-purity", essentially stoichiometric compositions; 135-to-150-g samples; Nernst method, adiabatic copper calorimeter; 51 to 298 K
$CaO \cdot Al_2O_3$	0.75	0.18	
$3CaO \cdot Al_2O_3$	0.75	0.18	
$FeO \cdot Al_2O_3$	0.71	0.17	Powder samples; "high-purity", essentially stoichiometric compositions; 217-g sample; Nernst method, adiabatic copper calorimeter; 51 to 298 K;

(a) Room-temperature values.

Source: Ref 9, 5.5.2-5

Physical properties of zirconates

Zirconate	Crystal system	Theoretical density, g/cm^3	Melting point K	Melting point °F
$CaO \cdot ZrO_2$	Monoclinic(a)	4.76 ± 0.02	~2615	~4250
$SrO \cdot ZrO_2$	Cubic(b)	5.48(c)	>2970(d)	>4890(d)
$BaO \cdot ZrO_2$	Cubic	6.26	~2920	~4800

(a) Pseudocubic also reported. (b) Also reported as othorhombic or distorted cubic. (c) Also reported as 5.52 g/cm^3. (d) Melting point of about 3115 K (about 5150 °F) also reported.

Source: Ref 9, p 5.5.3-1

Physical properties of aluminates

Aluminate	Crystal system	Theoretical density, g/cm^3	Melting point(a) K	Melting point(a) °F
$BaO \cdot Al_2O_3$	Cubic/hexagonal	3.99(b)	2270	3630
$BaO \cdot 6Al_2O_3$	Hexagonal	3.64(b)	2135	3380
$BeO \cdot Al_2O_3$	Orthorhombic	3.76	2155	3420
$CaO \cdot 2Al_2O_3$	Monoclinic	2.90	2035(c)	3200(c)
$3CaO \cdot 5Al_2O_3$	Orthorhombic	...	1995	3130
$CaO \cdot Al_2O_3$	...	2.98(b)	1870	2910
$3CaO \cdot Al_2O_3$	Cubic	3.00	1810(c)	2800(c)
$CeO \cdot Al_2O_3$	Cubic	6.17	2345 ± 28	3765 ± 50
$CoO \cdot Al_2O_3$	Cubic	4.37(b)	2235	3560
$Dy_2O_3 \cdot 2Al_2O_3$	Cubic	6.05	2090	3300
$Gd_2O_3 \cdot Al_2O_3$	Cubic	...	2255	3600
$FeO \cdot Al_2O_3$	Cubic	4.35	1710(c)	2620(c)
$K_2O \cdot Al_2O_3$	Cubic	...	>1920	>3000
$Li_2O \cdot Al_2O_3$	...	2.55(b)	2170-2270	3450-3630
$Li_2O \cdot 5Al_2O_3$	Cubic(d)	3.60(b)	...	...
$MgO \cdot Al_2O_3$	Cubic	3.59 ± 0.01	2410	3875
$MnO \cdot Al_2O_3$	Cubic	4.12	1835(c)	2840(c)
$Na_2O \cdot Al_2O_3$	...	...	>1970	>3090
$NiO \cdot Al_2O_3$	Cubic	4.45	2300	3685
$Sm_2O_3 \cdot Al_2O_3$	Cubic	...	2255	3600
$SrO \cdot Al_2O_3$	...	...	2285	3650
$SrO \cdot 2Al_2O_3$	Monoclinic	3.03	2045	3220
$2Y_2O_3 \cdot Al_2O_3$	Cubic	...	2310	3700
$3YrO_3 \cdot 5Al_2O_3$	Cubic	...	2255	3600
$Y_2O_3 \cdot Al_2O_3$	Cubic	5.50	...	...
$ZnO \cdot Al_2O_3$	Cubic	4.58(b)	2220	3540

(a) Approximate values. (b) Measured density. (c) Incongruent melting. (d) Reportedly exists in many forms. Other stable spinel phase formed above 1560 K (2350 °F).

Source: Ref 9, p 5.5.2-1

Carbides

Mechanical properties of silicon carbide materials

Property	Temperature K	°F	Reported value
CVD SiC			
Compressive strength, MPa (ksi)	293	68	>345 (>50)
Modulus of elasticity, GPa (10^{-6} psi)	1570	2370	>200 (>29)
	273	32	480 (70)
	297	75	420 (61)
	1210	1725	370 (54)
	1490	2220	340 (49)
	1670	2550	270 (39)
Microhardness (Knoop), kg/mm^2	297	75	3000 (a)
	1070	1470	790 (b)
	1670	2550	410 (b)
Recrystallized SiC			
Compressive strength, MPa (ksi)	297	75	689 (100)
Modulus of elasticity, GPa (10^6 psi)	297	75	210 (30)
Hot pressed SiC			
Compressive strength, MPa (ksi)	297	75	1380-3450 (200-500)
	810	1000	1720-2750 (250-400)
	1144	1600	2060-3100 (300-450)
	293	68	~3900 (~567)
	813	1000	~4100 (~596)
Tensile strength, MPa (ksi)	810	1000	280 (40)
	1255	1800	240 (35)
	1420	2100	190 (28)
	1530	2300	270 (39)
	1640	2500	190 (28)
Modulus of elasticity, GPa (10^6 psi)	297	75	430-448 (62-65)
	1640	2500	380 (55)
Poisson's ratio	293	68	0.17
Microhardness (Knoop), kg/mm^2	297	75	2500(a)
SiC whiskers			
Tensile strength, MPa (ksi)	297	75	690-11,380 (100-1650)
Modulus of elasticity, GPa (10^6 psi)	297	75	190 (28)
	297	75	90-855 (13-124)
	297	75	380-655 (55-95)
Sintered SiC			
Modulus of elasticity, GPa (10^6 psi)	295	72	406 (59)
	810	1000	392 (57)
	1255	1800	385 (56)
	1754	2700	372 (54)
Reaction-sintered SiC			
Modulus of elasticity, GPa (10^6 psi)	293	68	430 (62)
	293	68	370 (53)
	770	930	345 (50)
	1270	1830	340 (49)
	1370	2010	330 (48)
	1570	2370	320 (47)
Bulk modulus, GPa (10^6 psi)	297	75	96 (14)
Torsional modulus, GPa (10^6 psi)	293	68	155 (22.5)
Shear modulus, GPa (10^6 psi)	297	75	165-190 (24-27.5)
Poisson's ratio	293	68	0.13-0.24
Single-crystal SiC			
Microhardness (Knoop), kg/mm^2	297	75	2200-2950(a)

(a) 100-g load. (b) 950-g load.

Source: Ref 9, p 5.2.3-1

General thermal properties of beryllium carbide (Be_2C)

Property	Temperature K	Temperature °F	Reported value(a)
Specific heat, kJ/kg·K (Btu/lb·°F)	293	68	1.632 (0.39)
	302-372	85-210	1.397 (0.33)
Thermal conductivity, W/m·K	298	77	42.56 (24.6)
Btu/ft·h·°F)	422	300	23.35 (13.5)
Coefficient of linear thermal expansion, 10^{-6}/K (10^{-6}/°F)	298-872	77-1110	10.1 (5.6)

(a) These data are for Be_2C bodies which contain 2 to 12 wt % of other materials as impurities or additions for facilitating fabrication and which are greater than 90% dense.

Source: Ref 9, p 5.2.1-2

Properties of silicon-carbide-base high-temperature structural ceramic materials

Type of silicon carbide	Bulk density, g/cm³	Porosity, %	Flexural strength, MPa, at: 20 °C	1000 °C	1400 °C	Modulus of elasticity GPa, at 20 °C	Thermal expansion, 10^{-6}/K at 20-1400 °C
Reaction bonded	2.7	16	250	250	250	280	4.5
Silicon impregnated	3.1	0	400	500	250	380	4.3
Recrystallized	2.6	20	100	100	100	200	4.5
Sintered	3.0	5	500	450	400	400	4.6
Hot pressed	3.2	0	550	550	450	420	4.6

Source: Ref 1, p 126

Physical properties of sintered silicon carbide materials

Property	Reported value at temperature of: Room temperature	600 °C (1110 °F)	1000 °C (1830 °F)	1200 °C (2190 °F)
Reaction-sintered SiC				
Modulus of elasticity, GPa (10^6 psi)	365 (53)	344 (50)	338 (49)	331 (48)
Thermal conductivity, W/m·K (cal/cm·s·°C)	82.5 (0.197)	25 (0.060)	...	...
Thermal expansion, %	...	0.3	0.45	0.55
Sintered alpha SiC				
Modulus of elasticity, GPa (10^6 psi)	410 (59.5)	...	...	...
Thermal conductivity, W/m·K (cal/cm·s·°C)	90.4 (0.216)	50 (0.12)	...	...
Thermal expansion, %	...	0.28	0.44	0.54

Source: Ref 10, p 198

Properties of representative grades of cemented carbide

Cemented carbide	Room-temperature hardness, HV	Modulus of elasticity GPa	Transverse rupture strength, MPa	Coefficient of thermal expansion, 10^{-6}/K	Thermal conductivity, W/m·K	Density, g/cm³
Iron-bonded TiC	1000	305	2070	7.83	17	6.60
WC-20 wt % Co	1050	490	2850	6.4	100	13.55
WC-10 wt % Co	1625	580	2280	5.5	110	14.50
WC-3 wt % Co	1900	673	1600	5.0	110	15.25
WC-10 wt % Co-22 wt % (Ti, Ta, Nb)C	1500	510	2000	6.1	40	11.40
(Ti, Mo)C-Ni	1900	460	1100	7.5	17	5.50

Source: Ref 1, p 151

Physical properties of chromium, molybdenum, and tungsten carbides

Carbide	Crystal system	Theoretical density g/cm^3	Melting point K	Melting point °F
$Cr_{23}C_6$	Cubic	6.99	~1794	~2770
Cr_7C_3	Hexagonal	6.92	~2053	~3235
Cr_3C_2	Orthorhombic	6.68	~2166	~3440
Mo_2C	Hexagonal(a)	9.12	2761-2794(b)	4510-4570(b)
MoC_{1-x}	Hexagonal(c)	...	~2825(d)	~4625(d)
MoC	Hexagonal	~8.8	~2866(e)	~4700(e)
W_2C	Hexagonal	~17.3	~3050(f)	~5030(f)
WC_{1-x}	Cubic	...	~3016(g)	~4970(g)
WC	Hexagonal	~15.8	~3050(h)	~5030(h)

(a) An orthorhombic form (α Mo_2C) reportedly exists below 1473 K (2192 °F). (b) Alpha and beta phases indicated for C:Mo atomic ratios of 0.30 to 0.34; α Mo_2C melts congruently; β Mo_2C melts congruently but is not stable below 1746 K (2685 °F). (c) Pseudocubic lattice; see also footnote (d). (d) Phase indicated for C:Mo atomic ratios of 0.38 and 0.39; congruent melting; not stable below 1927 K (3010° F). (e) Phase indicated for C:Mo atomic ratios of 0.41 to 0.50; congruent melting; not stable below 2283 K (3650 °F). (f) Alpha and beta phases indicated for C:W atomic ratios of 0.29 to 0.35; α W_2C melts congruently but is not stable below 1522 K (2280 °F); β W_2C decomposes at about 2794 K (4570 °F) and is not stable below 2722 K (4440 °F). (g) Phase indicated for C:W atomic ratios of 0.375 to 0.40; congruent melting; not stable below 2802 K (4585 °F). (h) Phase at stoichiometric composition; decomposes into melt and graphite.

Source: Ref 9, p 5.2.6-1

Properties of sintered alpha silicon carbide

Property	Room temp.	1000 °C (1830 °F)	1200 °C (2190 °F)	1400 °C (2550 °F)	1500 °C (2730 °F)	1650 °C (3000 °F)
Hardness (Knoop), kg/mm^2	2800	...	...	...	...	...
Wet abrasion (Riley-Stoker)	3.4	...	...	...	...	...
Density, g/cm^3	3.14-3.18	...	...	...	...	...
Modulus of elasticity, GPa (10^6 psi)	406 (58.9)	378 (54.9)	...	...	...	...
Shear modulus, GPa (10^6 psi)	178 (25.8)	169 (24.5)	...	...	...	...
Poisson's ratio	0.142	0.118	...	...	...	...
Flexural strength(a), MPa (ksi):						
In air	459 (66.6)	442 (64.1)	450 (65.3)	432 (62.7)	404 (58.6)	...
In argon	446 (64.8)	...	...	...	...	494 (71.6)
Weibull modulus(b)	12.3	...	...	...	...	...
Fracture toughness(c), MPa$\sqrt{m}$ (ksi$\sqrt{in.}$)	4.6 (4.2)	6.4 (5.8)	...	...	...	...
Oxidation	...	...	...	Not detectable		

Property	Room temp.	200 °C (390 °F)	400 °C (750 °F)	600 °C (1110 °F)	1500 °C (2730 °F)
Thermal diffusivity(d), cm^2/s	0.413	0.230	0.185	0.140	...
Specific heat(e), J/kg·K (cal/g·°C)	670 (0.160)	921 (0.220)	1060 (0.252)	1120 (0.268)	1400 (0.334)
Thermal conductivity, W/m·K (cal/cm·s·°C)	87.1 (0.208)	67.0 (0.160)	61.5 (0.147)	49.4 (0.118)	...

Property	RT-700 °C (RT-1290 °F)	700-2000 °C (1290-3630 °F)
Coefficient of thermal expansion, 10^{-4}/°C (10^{-4}/°F)	4.02 (2.23)	5.32 (2.96)

(a) Four-point. (b) Two-parameter. (c) Double torsion and SENB. (d) Laser flash. (e) Drop calorimeter.

Source: Ref 1, p 20

Applications of silicon carbide

Industries	Environment conditions	Applications	Primary benefits
Oil and gas	High temperature, abrasive, high fluid pressures	Nozzles, bearings, seal faces, valve seats, choke inserts	Abrasion and wear resistance
Chemical processing	Strong acids (conc. HNO_3; H_2SO_4; HCl; HF) and strong bases (NaOH)	Seal faces, bearings, pump sleeves, pump components, heat exchangers	Wear and corrosion resistance, impermeability
	High-temperature oxidation	Gasifier tubes, thermocouple tubes	Corrosion resistance at high temperatures
Automotive and truck, aircraft, aerospace	Engine combustion	Combustion components, turbocharger rotors, gas-tubine vanes, blades, rocket nozzles	Low friction and high strength, low inertial loads thermal shock, resistance
Automotive and truck	Engine oils	Valve-train components	Low friction wear resistance
Sand blasting	High velocity abrasive	Nozzles	Abrasion resistance
Paper	Pulp-black liquor (50% NaOH)	Seal faces, sleeves, bearings	Corrosion resistance
	Pulp/paper slurry	Suction-box covers foils, forming boards	Abrasion resistance, low friction
Heat treating, furnacing steel	High-temperature	Thermocouple tubes, radiant tubes, recuperators burner components	Temperature and corrosion resistance impermeability
Mining and Mineral processing	Abrasive	Linings, pump components	Abrasion resistance
Nuclear	Boronated high-temperature water	Seal faces, bushings	Radiation resistance
All	In-process forming	Wire dies, can dies, textile guides, molds	Abrasion, wear and corrosion resistance

Source: Ref 1, p 78

Mechanical properites of beryllium carbide (Be_2C)

Property	Reported value(a)
Bend strength, MPa (ksi)	89.4 (13)
Compressive strength, MPa (ksi)	722.4 (105)
Modulus of elasticity, GPa (10^6 psi):	
Compression	316 (46)(b)
	241 (35)(c)
	206 (30)(d)
Flexure	344 (50)
Poisson's ratio	0.10
Microhardness (Knoop), kg/mm^2	2700

(a) These data are for hot pressed Be_2C that is 90 to 96% pure and about 92% dense. Tested at room temperature unless otherwise noted. (b) At 811 K (1000 °F). (c) At 1089 K (1500° F). (d) At 1366 K (2000 °F).

Source: Ref 9, p 5.2.1-3

Physical properties of aluminum and boron carbides

Carbide	Crystal system	Theoretical density g/cm^3	Melting point K	Melting point °F
Al_4C_3	Rhombohedral	2.99	2977 ± 55(a)	~4900 ± 100(a)
B_4C	Rhombohedral	2.52	2700(b)	4400(b)

(a) Decomposition obscures melting-point determinations; estimated value. (b) Congruent melting.

Source: Ref 9, p 5.2.2-1

Mechanical properties of chromium, molybdenum, and tungsten carbides

Property	Reported values(a) for: Cr_3C_2	Mo_2C	WC
Bend strength, MPa (ksi)	241-323 (35-47)	...	...
	~571 (~83) (b)	...	...
	~138 (~20) (c)	...	...
	...	...	482-832 (70-121)
	...	...	78-195 (11.4-28.4)(d)
Compressive strength, MPa (ksi)	1039 (151)	...	2683 (390)
	929 (135) (e)	...	1404 (204) (e)
	564 (82) (f)	...	...
	413 (60) (g)	...	...
	1101 (160)	901 (131)	...
	...	...	2958 (430)
	...	...	3523 (512)
Tensile strength, MPa (ksi)	...	...	344 (50)
Modulus of elasticity, GPa (10^6 psi)	385 (56)	...	...
	...	227 (33)	~688 (~100)
	...	...	667 (97)
	...	533 (77.5)	...
Hardness:			
Microhardness(h), kg/mm^2	1350-2280	1500-1800	1700-2400
Rockwell A	~91	~89	~90

(a) Tested at room temperature unless otherwise noted. Data from a variety of sources. (b) At 1255 to 1366 K (1800 to 2000 °F). (c) At 1589 K (2400 °F). (d) At 1589 to 2366 K (2400 to 3800 °F). (e) At 1273 K (1830 °F). (f) At 1477 L (2200 °F). (g) At 1672 K (2550 °F). (h) Essentially consistent ranges of microhardness values have also been reported from various sources for $Cr_{23}C_6$, Cr_7C_3, and W_2C. Representative ranges for these compounds, in kg/mm^2, are 970 to 1650, 1336 to 2200, and 1500 to 2060 respectively.

Source: Ref 9, p 5.2.6-6

Shaping methods for silicon carbide ceramics

Type of silicon carbide	Shaping methods: Dry pressing Axial	Dry pressing Isostatic	Injection molding	Warm molding	Extruding	Slip casting	Hot pressing
Reaction bonded	X	X	X	X	X	X	
Silicon infiltrated	X	X	X	X	X	X	
Recrystallized	X	X				X	
Sintered	X	X				X	
Hot pressed							X

Source: Ref 1, p 126

Room-temperature mechanical properties of uranium carbides

Property	Reported values(a) for: UC	U_2C_3	UC_2
Bend strength, MPa (ksi)	55-83 (8-12)	89-110 (13-16)	55-69 (8-10)
	103 (15)	...	...
Compressive strength, MPa (ksi)	351 (51)	454 (66)	...
	296 (43) (a)	...	...
	124 (18) (b)	...	...
Modulus of elasticity(c), GPa (10^6 psi)	179-220 (26-32)	179-220 (26-32)	...
Microhardness (Knoop), kg/mm^2	500-800	650-800	~500
	935	...	620

(a) Stress applied parallel to pressing direction of specimen. (b) Stress applied perpendicular to pressing direction of specimen. (c) Static method.

Source: Ref 9, p 5.2.8-8

Mechanical properties of boron carbide

Property	Reported values(a)
Bend strength, MPa (ksi)	323-346 (46.9-50.3)
	242-243 (35.2-35.3) (b)
	205-238 (29.8-34.6) (c)
	192-205 (27.9-29.8) (d)
	199 (28.9) (e)
	241 (35.0)
Compressive strength, MPa (ksi)	2752 (400)
Impact strength, J (in.·lb):	
Unnotched	0.026-0.031 (0.23-0.28)
Notched	0.0028-0..003 (0.025-0.028)
Modulus of elasticity, GPa (10^6 psi)	289-454 (42-66)
	362-400 (52.6-58.2)
	322-384 (46.8-55.8) (f)
	277-356 (40.2-51.8) (g)
	228-265 (33.2-38.5) (h)
Shear modulus, GPa (10^6 psi)	165-206 (24-30)
Bulk modulus, GPa (10^6 psi)	193-255 (28-37)
Poisson's ratio	0.19
Microhardness, kg/mm^2:	
Knoop (100-g load)	2800
Vickers (100-g load)	3200
	4200 ± 280
	4750 ± 250
	5000

(a) Tested at room temperature unless otherwise noted. Data from a variety of sources. (b) At 1144 K (1600 °F). (c) At 1366 K (2000 °F). (d) At 1589 K (2400 °F). (e) At 1700 K (2600 °F). (f) At 672 K (750 °F). (g) At 1072 K (1470 °F). (h) At 1472 K (2190 °F).

Source: Ref 9, p 5.2.2-4

Physical properties of titanium, zirconium, and hafnium carbides

Carbide	Crystal system	Theoretical density g/cm^3	Melting point K	Melting point °F
TiC	Face-centered cubic	4.92	3340 ± 15(b)	5550 ± 27(b)
ZrC	Face-centered cubic	6.56	3693 ± 20(c)	6188 ± 36(c)
HfC	Face-centered cubic	12.67	4203 ± 20(d)	7106 ± 36(d)

(a) Representative values; congruent melting. (b) Maximum melting point at about 44 at. % carbon. (c) Maximum melting point at about 45 at. % carbon. (d) Maximum melting point at about 18.5 at. % carbon.

Source: Ref 9, p 5.2.4-1

Room-temperature mechanical properties of hafnium carbide

Property	Reported values
Bend strength, MPa (ksi)	234-241 (34-35)
Modulus of elasticity, GPa (10^6 psi)	316-461 (46-67)
Shear modulus, GPa (10^6 psi)	179-193 (26-28)
Bulk modulus, GPa (10^6 psi)	241 (35)
Poisson's ratio	0.17
	0.18
Microhardness, kg/mm^2:	
Knoop (50-g load)	2000-2500
Knoop (100-g load)	2260-3050
	1800-2500(a)
Vickers (200-g load)	1900-2100(a)

(a) Single crystal.

Source: Ref 9, p 5.2.4-17

Mechanical properties of titanium carbide

Property	Reported values(a)
Bend strength, MPa (ksi)	282-667 (41-97)
	688-5504 (100-800) (b)
Compressive strength, MPa (ksi)	757-2958 (110-430)
	227 (33) (c)
	89 (13) (d)
Tensile strength, MPa (ksi)	241-275 (35-40)
	110-117 (16-17) (e)
	55-62 (8-9) (f)
Modulus of elasticity, GPa (10^6 psi)	447 (65) (b)
	268-461 (39-67)
Shear modulus, GPa (10^6 psi)	186 (27) (b)
	110-193 (16-28)
Bulk modulus, GPa (10^6 psi)	241 (35) (b)
	227 (33)
Poisson's ratio	0.19(b)
	0.18-0.19
Hardness:	
Knoop (50-g load)	2000-2750(b)
Knoop (100-g load)	2000-2400(b)
	1800-5900
Vickers (100-g load)	~3200
Rockwell A	93

(a) Tested at room temperature unless otherwise noted. Data from a variety of sources. (b) Single crystal. (c) At 1872 K (2910 °F). (d) At 2199 K (3990 °F). (e) At 1255 K (1800 °F). (f) At 1477 K (2200 °F).

Source: Ref 9, p 5.2.4-2

Mechanical properties of zirconium carbide

Property	Reported values(a)
Bend strength, MPa (ksi)	103-206 (15-30)
Compressive strength, MPa (ksi)	826-2958 (120-430)
	1637 (238)
	488 (71) (b)
	255 (37) (c)
Tensile strength, MPa (ksi)	103 (15)
	83-96 (12-14) (d)
	89-110 (13-16) (e)
	69-110 (10-16) (f)
	14-48 (2-7) (g)
Modulus of elasticity, GPa (10^6 psi)	358 (52)
	550 (80)
	385-406 (56-59) (h)
	345-400 (50-58) (i)
Shear modulus, GPa (10^6 psi)	172 (25) (h)
	165-200 (24-29)
	162 (23.5)
Bulk modulus, GPa (10^6 psi)	220 (32) (h)
	206 (30)
Poisson's ratio	0.19
	0.20 (h)
Hardness:	
Microhardness, kg/mm^2:	
Knoop (100-g load)	1830
Vickers 100-g load)	2600
Vickers (50-g load)	2310-2520
Rockwell A	92

(a) Tested at room temperature unless otherwise noted. Data from a variety of sources. (b) At 1273 K (1830 °F). (c) At 1472 K (2190 °F). (d) At 1255 K (1800 °F). (e) At 1477 K (2200 °F). (f) At 1866 K (2900 °F). (g) At 1999 K (3630 °F). (h) Single crystal. (i) At 294 to 1505 K (70 to 2250 °F).

Source: Ref 9, p 5.2.4-15

Physical properties of vanadium, niobium, and tantalum carbides

Carbide	Crystal system	Theoretical density g/cm^3	Melting point K	°F
VC	Cubic	5.48	2972 ± 50	4890 ± 90
V_2C	Hexagonal	5.75	2438 ± 25	3930 ± 45
NbC	Cubic	7.82	3772 ± 75	6330 ± 135
Nb_2C	Orthorhombic(a)	7.85	3360 ± 50	5590 ± 90
TaC	Cubic	14.50	4152 ± 28	7015 ± 50
Ta_2C	Hexagonal(b)	15.00	3602 ± 100(b)	6025 ± 180(b)

(a) Hexagonal above about 1470 K (2200 °F). (b) Ta_2C reportedly exists as α and β phases; the α phase (C6 structure) is stable up to about 2200 ± 28 K (3500 ± 50 °F), and the high-temperature β phase (L'3 structure) is stable between 2200 and about 3600 K (3500 and about 6000 °F).

Source: Ref 9, p 5.2.5-1

Physical properties of the carbides of thorium, uranium, and plutonium

Carbide	Crystal system	Theoretical density g/cm^3	Melting point K	°F
ThC	Cubic	10.67	2772 ± 33(a)	4530 ± 60(a)
ThC_2	(b)	9.60	2927 ± 28(c)	4810 ± 50(c)
UC	Cubic	13.63	2769 ± 33(d)	4525 ± 60(d)
U_2C_3	Cubic	12.88	2002(e)	3145(e)
UC_2	Tetragonal-cubic(f)	11.68	2722 ± 28	4440 ± 50
PuC	Cubic	13.5-14.1	1927(g)	3010(g)
Pu_2C_3	Cubic	12.70	2322	3720
PuC_2	Tetragonal(h)	10.9	2022-2522	3180-4080

(a) Earlier tests indicated a value of 2898 K (4755 °F). (b) Transitions from monoclinic to tetragonal and from tetragonal to cubic occur at 1672 to 1700 K (2550 to 2600 °F) and 1750 to 1772 K (2690 to 2730 °F), respectively. (c) A value of 2823 K (4620 °F) has also been reported. (d) Several values ranging from 2548 to 2866 K (4127 to 4694 °F) have been reported. (e) U_2C_3 has a body-centered-cubic structure that retains its composition up to about 2002 K (3145 °F). (f) Exists in high-temperature phases above 1772 K (2730 °F). (g) Peritectic decomposition. (h) PuC_2 exists in a high-temperature phase above 1922 K (3000 °F) produced by the decomposition of body-centered-cubic Pu_2C_3. Some researchers report a face-centered-cubic structure for PuC_2.

Source: Ref 9, p 5.2.8-1

Mechanical properties of vanadium, niobium, and tantalum carbides

Property	Reported values(a) for: VC	NbC	TaC
Bend strength, MPa (ksi)	...	~289 (~42)	...
	...	...	275-310 (40-45) (b)
	...	...	~378 (~55) (c)
Compressive strength, MPa (ksi)	607 (88)	~2374 (~345)	...
Tensile strength, MPa (ksi)	...	244 (36)	96-291 (14-42)
Modulus of elasticity, GPa (10^6 psi)	268-420 (39-61)	330-537 (48-78)	241-722 (35-105)
Shear modulus, GPa (10^6 psi)	157 (23)	197-245 (29-36)	215-227 (31-33)
Bulk modulus, GPa (10^6 psi)	389 (57)	296-378 (43-55)	248-343 (36-50)
Poisson's ratio	0.22	0.22	0.24
	...	0.25(d)	...
Hardness:			
Microhardness(e), kg/mm^2	2000-3000	1900-2600	1600-2400
Rockwell A	~91	~90	~88

(a) Tested at room temperature unless otherwise noted. (b) At 294 to 1273 K (70 to 1830 °F). (c) At 2033 K (3200 °F). (d) At 2200 K (3500 °F). (e) 30- to 100-g load.

Source: Ref 9, p 5.2.5-9

Nitrides

Physical properties of mononitrides of rare-earth metals(a)

Metal	Nitride	Crystal system	Theoretical density, g/cm^3	Melting or decomposition temperature K	Melting or decomposition temperature °F
Scandium	ScN	Cubic	4.21	~4800	~2923
Yttrium	YN	Cubic	5.90	~4840	~2493
Lanthanum	LaN	Cubic	6.85	...	...
Cerium	CeN	Cubic	8.09	~4660	~2843
Praseodymium	PrN	Cubic	7.49	...	...
Neodymium	NdN	Cubic	7.69	...	...
Samarium	SmN	Cubic	8.50	...	...
Europium	EuN	Cubic	8.77	...	...
Gadolinium	GdN	Cubic	9.10	...	...
Terbium	TbN	Cubic	9.57	...	...
Dysprosium	DyN	Cubic	9.93	...	...
Holmium	HoN	Cubic	10.26	...	...
Erbium	ErN	Cubic	10.35	...	...
Thulium	TmN	Cubic	10.84	...	...
Ytterbium	YbN	Cubic	11.33	...	...
Lutetium	LuN	Cubic	11.59	...	...

(a) Data as reported by various investigators.

Source: Ref 9, p 5.3.7-1

Density and flexural strength of sintered silicon nitride

Composition	Fabrication method	Density g/cm^3	Room temperature flexural strength(a), MPa
Si_3N_4-15Y_2O_3-3Al_2O_3	Cold isopressed	3.141	432 (3-point)
Si_3N_4-8Y_2O_3-4Al_2O_3	Injection molded	3.04-3.09	380-402 (4-point)
Si_3N_4-6Y_2O_3-2Al_2O_3	Injection molded	3.13	385.8 (4-point)

(a) Unmatched.

Source: Ref 1, p 105

Properties of silicon-nitride-base high-temperature structural ceramic materials

Type of silicon nitride	Bulk density, g/cm^3	Porosity, %	Flexural strength, MPa, at: 20 °C	1000 °C	1400 °C	Modulus of elasticity, GPa, at 20 °C	Thermal expansion 10^{-6}/K, at 20-1400 °C
Reaction bonded(a)	1.9-2.2	...	150	150	150	120	3.0
Reaction bonded(b)	2.4-2.6	...	250	250	300	160	3.0
Hot pressed	3.2	0	700	700	400	320	3.2

(a) Axial dry pressed, extruded. (b) Isopressed, slip cast, injection molded, warm molded.

Source: Ref 1, p 126

Room temperature mechanical properties of reaction-sintered silicon nitride

Property	Reported value(a)
Bend strength, MPa (ksi)	117-241 (17-35)
Compressive strength, MPa (ksi)	345-690 (50-100)
Tensile strength, MPa (ksi)	69-172 (10-25)
Modulus of elasticity GPa (10^6 psi)	96-220 (13.9-32)
Poisson's ratio	0.25-0.26
Microhardness, kg/mm^2:	
100-g load	1700-2200
50-g load	2300-3000
25-g load	2670-3260

(a) Range of values reported from a variety of sources.

Source: Ref 9, p 5.3.3-18

Compatibility of silicon nitride with various media

Compatible	Marginally compatible	Not compatible
Hot concentrated mineral acids	Soda-lime and borosilicate glasses at 1470 K (2190 °F)	Hot hydrofluoric acid
Molten aluminum, lead, tin, zinc, light alloys, silver, gold, brass, and nickel/silver at their normal foundry temperatures	Molten metals and alloys up to 1370 K (2000 °F) except as indicated for specific metals	Hot concentrated caustic solutions and fused caustic salts
		Molten magnesium, copper, nichrome, and stainless steels

Source: Ref 9, p 5.3.3-28

Room-temperature values of bend strength, modulus of elasticity and hardness for Si_2ON_2

Fabrication method	Density, g/cm^3	Bend strength(a) MPa	Bend strength(a) ksi	Modulus of elasticity GPa	Modulus of elasticity 10^6 psi	Hardness (Knoop), kg/mm^2
Cold pressed(b)	1.95	32.4	4.7	70.3	10.2	...
	2.09	37.9	5.5	75.8	11.0	...
	2.31	67.6	9.8	113.8	16.5	...
Hot pressed(c)	2.65	210.3	30.5	191.7	27.8	...
	2.70	...	...	...	...	1580

(a) Tested in three-point loading on a 200-mm (8-in.) span for cold pressed specimens, and on a 50-mm (2-in.) span for hot pressed specimens. (b) Si_2ON_2 content, 90% or higher. (c) Si_2ON_2 content, 80%.

Source: Ref 9, p 5.3.3-33

Oxide additives used in densifying silicon nitride powder and its surface layer of silica

Additive (M_xO_y)	Temperature of liquid formation, °C: Silicate (M_xO_y-SiO_2)	Temperature of liquid formation, °C: Oxynitride (M_xO_y-SiO_2-Si_3N_4)
Li_2O	1030	1030
MgO	1543	1390
Y_2O_3	1650	1480
CeO_2	1560	1460
ZrO_2	1640	1590
CaO	1435	1435
Al_2O_3	1595	1470

Source: Ref 1, p 167

Coefficients of linear thermal expansion for nitrides of vanadium, niobium, and tantalum

Nitride	Temperature range K	Temperature range °F	Coefficient of linear thermal expansion(a) $10^{-6}/K$	Coefficient of linear thermal expansion(a) $10^{-6}/°F$
VN	294-1366	70-2000	8.1	4.5
Nb_2N	294-1272	70-1830	3.2	1.8
TaN	294-977	70-1300	3.6	2.0
Ta_2N	294-1270	70-1830	5.2	2.9
Ta_2N	294-1644	70-2500	4.7	2.6

(a) Values are approximate.

Source: Ref 9, p 5.3.5-3

Typical properties of boron nitride

Property	Test	Temperature, °C	100% BN: Parallel to pressing direction	100% BN: Perpendicular to pressing direction	60% SiO_2 - 40% BN: Parallel to pressing direction	60% SiO_2 - 40% BN: Perpendicular to pressing direction
Compressive strength, MPa	ACMA-1	25	310	234	317	289
Modulus of elasticity (sonic), GPa	...	25	65	41	94	106
Modulus of rupture, GPa	...	25	80.70	96.11	97.90	106.2
		1000	62.71	67.09	...	...
		1350	36.09	36.20	77.22	77.91
Density, g/cm^3	ASTM C20	...	2.08		2.12	
Hardness	...	...	385 (K100)		89.5 (15T)	
Water absorption, % of weight gain(a)	...	25	1.1		0.04	
Thermal expansion $10^{-6}/°C$	...	75-500	2.3	1.1	1.8	0.20
	...	75-1000	2.8	0.9	2.5	0.40
	...	75-1500	6.7	1.8	...	...
	...	75-2000	6.2	2.0	...	...
Thermal conductivity, W/m·°C	...	100	23.1	44.0	9.1	25.2
	...	350	18.1	36.3	7.2	19.8
	...	700	16.3	31.6	6.8	15.9
Maximum use temperature, °C;						
In inert or reducing atmosphere	...	...	2000		1400	
In oxidizing atmosphere	...	...	985		1400	
Dielectric constant (1 MHz)	ASTM D150	...	4.08		3.7	
Dielectric strength, V/μm	...	...	37.4		38.8	
Dissipation factor (1 MHz)	ASTM D150	...	0.00034		0.0015	
Loss factor (1 MHz)	ASTM D150	...	0.00139		0.006	
MIL-I-10A grade	...	...	...		L-542	
Volume resistivity, 10^{12} Ω·cm	...	23	>200		250	
		150	>30		>32	
Surface resistivity, 10^{12} Ω	...	23	>250		>250	
		150	>250		>250	
Insulation resistance, 10^{12} Ω	...	23	>200		>250	
		150	0.25		0.25	
Oxidation rate in air (weight loss), $mg/cm^2·h$	...	712	0.046		0.0059	
		1000	0.622		0.0110	
Typical chemical analysis, %:						
Total boron (B)	...	...	40		18	
Total nitride (N)	...	...	50		22	
Boric oxide (B_2O_3)	...	...	6		0.2	
Calcium (Ca)	...	...	0.2		0.03	
Silica (SiO_2)	...	...	0.2		59	
Other	...	...	3.6		0.77	

(a) In 168 h at 25 °C and 80 to 100% relative humidity.

Source: Standard Oil Engineered Materials

Hardness values for TiN, ZrN, and HfN

Nitride	Mohs scale	Microhardness, kg/mm^2 Knoop, 100-g load	Microhardness, kg/mm^2 Vickers, 50-g load
TiN	9-10	~1800	1800-2100
ZrN	7-8	~1500	1500-1850
HfN	7-8	...	1600-2150

Source: Ref 9, p 5.3.4-8

Physical properties of titanium, zirconium, and hafnium mononitrides

Nitride	Crystal system	Theoretical density g/cm^3	Melting point K	Melting point °F
TiN	Cubic	5.44	3223 ± 50	5340 ± 90
ZrN	Cubic	7.35	3253 ± 55	5400 ± 100
HfN	Cubic	13.94	3660 ± 43	6128 ± 77

Source: Ref 9, p 5.3.4-1

Physical properties of vanadium, niobium, and tantalum nitrides

Nitride	Crystal system	Theoretical density g/cm^3	Melting point(a) K	Melting point(a) °F
VN	Cubic	6.08 ± 0.02	2450 ± 139	3950 ± 250
V_2N	Hexagonal	5.99 ± 0.01	...	...
NbN	Cubic	8.36	~2477	~4000
Nb_2N	Hexagonal	8.31	~2589	~4200
TaN	Hexagonal	14.36	3366 ± 44	5600 ± 80
Ta_2N	Hexagonal	15.86	3223	5342

(a) Rapid loss of nitrogen at elevated temperatures precludes reliability of melting-point values, particularly for the mononitride phases.

Source: Ref 9, p 5.3.5-1

Mechanical properties of hot pressed boron nitride

Property	Reported values(a) Parallel to pressing direction	Reported values(a) Perpendicular to pressing direction
Bend strength, MPa (ksi)	48-97 (7-14)	41-110 (6-16)
	41-62 (6-9) (b)	48-69 (7-10) (b)
	28-76 (4-11) (c)	28-76 (4-11) (c)
	...	83 (12) (d)
	...	117 (17) (e)
Tensile strength, MPa (ksi)	41-55 (6-8)	45-62 (6.5-9)
	4 (0.6) (b)	4 (0.6) (b)
	4 (0.6) (f)	4 (0.6) (f)
	10 (1.5) (e) (g)	~15 (~2.2 (e) (g)
	28 (4.0) (g) (h)	~28 (~4.0 (g) (h)
Compressive strength, MPa (ksi)	41-317 (6-46)	52-290 (7.5-42)
Modulus of elasticity, GPa (10^6 psi)	41-97 (6-14)	41-103 (6-15)
Hardness (Knoop, 100-g load), kg/mm^2	210-390	

(a) Tested at room temperature unless otherwise noted. (b) At 1273 K (1832 °F). (c) At 1623 K (2462 °F). (d) At 2073 K (3272 °F). (e) At 2273 K (3632 °F). (f) At 1772 K (2730 °F). (g) Strength increase might represent carbide formation from test environment. (h) At 2477 K (4000 °F).

Source: Ref 9, p 5.3.2-27

Physical properties of beryllium, magnesium, and calcium nitrides

Nitride	Crystal system	Theoretical density g/cm^3	Melting of decomposition temperature
α-Be_3N_2	Cubic	2.70	(a)
β-Be_3N_2	Hexagonal	2.71	(b)
Mg_3N_2	Cubic	2.71	(c)
Ca_3N_2	Rhombohedral	2.18	(d)
α-Ca_3N_2	Cubic	2.61-2.64	(e)
β-Ca_3N_2	Pseudohexagonal	2.67	(f)
γ-Ca_3N_2	Orthorhombic	2.73	(g)

(a) Transforms to β-Be_3N_2 at about 1866 K (2900 °F). (b) This high-temperature phase vaporizes congruently above 1644 K (2500 °F). Sublimation reportedly occurs at 2255 K (3600 °F) in vacuum. (c) Reportedly exists in three allotropic forms: $\alpha \rightarrow \beta$ transition occurs at about 823 K (1022 °F), and $\beta \rightarrow \gamma$ transition at 1061 K (1450 °F). Vaporizes above 1310 K (1900 °F) by a complex mechanism. (d) Decomposes in air; stable under argon. (e) α-Ca_3N_2 is an intermediate phase which vaporizes congruently above 993 K (1330 °F). (f) A low-temperature phase which is metastable up to 593 K (608 °F) and converts to alpha phase at elevated temperature. (g) A high-temperature phase which forms above 1310 K (1900 °F) but is not stable.

Source: Ref 9, p 5.3.1-1

Mechanical properites of aluminum nitride

Property	Reported value(a)
Bend strength, MPa (ksi)	265-970 (385-140)
	186 (27) (b)
	125 (18.1) (c)
Compressive strength, MPa (ksi)	2070 (300)
Modulus of elasticity, GPa (10^6 psi)	274-344 (39.7-50)
	317 (46) (b)
	276 (40) (c)
Poisson's ratio	0.25
Hardness, kg/mm^2:	
Knoop	1225
Vickers	1400

(a) Tested at room temperature unless otherwise noted. (b) At 1273 K (1832 °F). (c) At 1673 K (2552 °F).

Source: Ref 9, p 5.3.2.1-4

Physical properties of chromium, molybdenum, and tungsten nitrides

Nitride(a)	Crystal system	Theoretical density g/cm^3	Melting point K	Melting point °F
CrN	Cubic	6.14 ± 0.02	~1772(b)	~2730(b)
Cr_2N	Hexagonal	6.51	...	...
MoN	Hexagonal	9.18	~1022(b)	~1380(b)
Mo_2N	Cubic	>8.04	~1172(b)	~1650(b)
Mo_3N	Tetragonal	...	(c)	(c)
WN	Hexagonal	12.1	~873(b)	~1112(b)
W_2N	Cubic	12.2	1072-1144(b)	1470-1600(b)

(a) These nitrides are considered to be relatively unstable compounds; their reactivity with nitrogen decreases in the order Cr→Mo→W. (b) Dissociation. (c) Stable only up to 873 K (1112 °F).

Source: Ref 9, p 5.3.6-1

Thermal properties of nitrides of chromium

Property	Temperature K	Temperature °F	Reported values(a) for: CrN	Reported values(a) for: Cr_2N
Specific heat, J/kg·K (Btu/lb·°F)	294	70	711 (0.17)	586 (0.14)
	533	500	753 (0.18)	669 (0.16)
	811	1000	795 (0.19)	711 (0.17)
Thermal conductivity, W/m·K (Btu/ft·h·°F)	294	70	~12 (~7)	~22 (~13)
	273-1000	32-1340	~3.5 (~2)	...
Coefficient of linear thermal expansion, 10^{-6}/K (10^{-6}/°F)	294-477	70-400	0.7 (0.37)	5.9 (3.3)
	477-700	400-800	1.5 (0.83)	8.1 (4.5)
	700-811	800-1000	3.1 (1.75)	8.6 (4.8)

(a) Data are considered approximate values.

Source: Ref 9, p 5.3.6-2

Thermal properties of dimolybdenum nitride (Mo_2N)

Property	Temperature K	Temperature °F	Reported value(a)
Specific heat, J/kg· (Btu/lb· °F)	294	70	293 (0.07)
	533	500	376 (0.09)
	811	1000	418 (0.10)
Thermal conductivity, W/m·K (Btu/ft·h·°F)	294	70	~17 (~10)
Coefficient of linear thermal expansion, 10^{-6}/K (10^{-6}/°F)	294-477	70-400	1.8 (1.0)
	477-700	400-800	4.1 (2.3)
	700-811	800-1000	5.4 (3.0)

(a) Data are considered approximate values.

Source: Ref 9, p 5.3.6-2

Physical properties of thorium, uranium, and plutonium nitrides

Nitride	Crystal system	Theoretical density g/cm³	Melting point K	Melting point °F
ThN	Cubic	11.60	3093(a)	5108(a)
Th_3N_4	Hexagonal(b)	10.50	(c)	(c)
UN	Cubic	14.32	3123(d)	5160(d)
α-U_2N_3	Cubic(e)	11.24	(f)	(f)
β-U_2N_3	Hexagonal(e)	12.24	(g)	(g)
UN_2	Cubic(h)	11.73	...	...
PuN	Cubic	14.25	(i)	(i)

(a) Congruent melting under nitrogen pressure of about 1 atm; other values of 2903 and 3063 K (4766 and 5040 °F) are also reported. (b) Rhombohedral structure. (c) Decomposes to ThN above 1923 K (3000 °F). (d) Congruent melting under nitrogen pressure greater than 2 atm; decomposition occurs at lower nitrogen pressures. (e) Body-centered-cubic (α) phase transforms slowly to the hexagonal (β) phase at temperatures above 1093 K (2000 °F). The change also is a function of composition and nitrogen pressure. (f) Decomposes at elevated temperatures but is stable to about 1644 K (2500 °F) at 1 atm nitrogen pressure. (g) Converts at α-U_2N_3 below 1373 K (2000 °F). (h) Exists only at very high nitrogen pressures and low temperatures. (i) Reported melting points ranged from 2870 to about 3030 K (4700 to about 5000° F) at 1 atm nitrogen pressure. However, these values were probably decomposition temperatures.

Source: Ref 9, p 5.3.7-1

Room-temperature mechanical properties of hot pressed silicon nitride (HS-130)

Property	Reported value
Bend strength, MPa (ksi)	707 (103) (a)
	680 (98.6) (b)
	860 (125) (c)
	898 (130) (c)
	763 (111) (d)
	790-913 (115-132) (e)
Compressive strength, MPa (ksi)	689-2760 (100-400)
Tensile strength, MPa (ksi)	397-434 (57.5-62.9) (f)
	360 (52.2)
Modulus of elasticity, GPa (10^6 psi)	290-307 (42-44.5)
Impact strength, J (in. · lb)	0.23-0.40 (2-3.5)
Poisson's ratio	0.22

(a) Laboratory material containing 1% MgO, high alpha phase; tested in four-point loading. (b) Tested in four-point loading; maximum stress perpendicular to pressing direction. (c) Tested in three-point loading; maximum stress perpendicular to pressing direction. (d) Tested in three-point loading; maximum stress parallel to pressing direction. (e) Laboratory material containing 1% MgO, high alpha phase; tested in three-point loading. (f) Expanding ring test; maximum stress perpendicular to pressing direction.

Source: Ref 9, p 5.3.3.5

Shaping methods for silicon nitride ceramics

Type of silicon nitride	Shaping methods: Dry pressing, Axial	Dry pressing, Isostatic	Injection molding	Warm molding	Extruding	Slip casting	Hot pressing
Reaction bonded	x	x	x	x	x	x	
Hot pressed							x

Source: Ref 1, p 126

Borides

Physical properties of titanium, zirconium, and hafnium borides

Boride	Crystal system	Theoretical density g/cm^3	Melting point K	Melting point °F
Ti_2B	Tetragonal	...	2477(a)	4000(a)
TiB	Cubic	5.26	2333(b)	3740(b)
TiB_2	Hexagonal	4.52	3144	5200
Ti_2B_5	Hexagonal	...	2366(c)	3800(c)
ZrB	Cubic	6.70	(d)	(d)
ZrB_2	Hexagonal	6.09	3311	5500
ZrB_{12}	Cubic	3.63	2838(e)	4650(e)
HfB	Cubic	12.80	3172	5250
HfB_2	Hexagonal	11.20	3522	5880

(a) Not stable below 2072 K (3270 °F). (b) Not stable below 950 K (1250 °F). (c) Not stable below 1977 K (3100 °F). (d) Melting point and thermal-stability range not resolved. (e) Not stable below 1822 K (2820 °F).

Source: Ref 6, p 5.1.2-1

Elastic moduli and hardness of vanadium, niobium, and tantalum diborides

Property	Reported value for: VB_2	NbB_2	TaB_2
Modulus of elasticity(a), GPa (10^6 psi)	260 (38)	...	250 (36)
Microhardness(a), kg/mm^2:			
Knoop (100-g load)	2800	2500	...
Vickers (50-g load)	...	2900	2000
Hardness, Rockwell A	88	87-88	89

(a) At 21 °C (70 °F).

Source: Ref 6, 5.1.3-7

Physical properties of thorium, uranium, and plutonium borides

Boride	Crystal system	Theoretical density g/cm^3	Melting point K	°F
ThB_4	Tetragonal	8.45	2755	4500
ThB_6	Cubic	6.80	2422	3900
UB_2	Hexagonal	12.73	2644	4300
UB_4	Tetragonal	9.38	2755	4500
UB_{12}	Cubic	5.86	2505	4050
PuB	Cubic	14.10	...	...
PuB_2	Hexagonal	12.81	...	...
PuB_4	Tetragonal	9.36	...	...
PuB_6	Cubic	~7.25	...	...

Source: Ref 6, p 5.1.6-1

Physical properties of vanadium, niobium, and tantalum borides

Boride	Crystal system	Theoretical density g/cm^3	Melting point K	°F
V_3B_2	Tetragonal	...	2338	3750
VB	Orthorhombic	5.4	2522	4080
V_3B_4	Orthorhombic	...	2550	4130
VB_2	Hexagonal	5.10	2700	4400
Nb_3B_2	Tetragonal	...	2088	3300
NbB	Orthorhombic	7.60	2533	4100
Nb_3B_4	Orthorhombic	...	2466	3980
NbB_2	Hexagonal	7.21	3272	5430
Ta_2B	Tetragonal	...	2172	3450
Ta_3B_2	Tetragonal	...	2311	3700
TaB	Orthorhombic	14.29	2700	4400
Ta_3B_4	Orthorhombic	13.60	2922	4800
TaB_2	Hexagonal	12.60	3366	5600

Source: Ref 6, p 5.1-3.1

Physical properties of chromium, molybdenum, and tungsten borides

Boride	Crystal system	Theoretical density g/cm^3	Melting point K	°F
Cr_4B	Orthorhombic	6.24	1922	3000
Cr_2B	Hexagonal	6.53	2088	3300
Cr_5B_3(a)	Tetragonal	6.12	2172	3450
CrB	Orthorhombic	6.11	2272	3630
Cr_3B_4	Orthorhombic	5.76	2200	3500
CrB_2	Hexagonal	5.6	2422	3900
Cr_2B_5	Orthorhombic	...	2272(b)	3630(b)
Mo_2B	Tetragonal	9.31	2522(c)	4135(c)
α-MoB	Tetragonal	8.77	(d)	(d)
β-MoB	Orthorhombic	...	2872(e)	4710(e)
MoB_2	Hexagonal	7.78	2747(f)	4485(f)
Mo_2B_5	Rhombohedral-hexagonal	7.48	2414(g)	3885(g)
W_2B	Tetragonal	16.72	2944(h)	4840(h)
α-WB	Tetragonal	16.00	(i)	(i)
β-WB	Orthorhombic	...	2938(j)	4830(j)
W_2B_5	Hexagonal	13.1	2638(h)	4290(h)

(a) Also reported as Cr_3B_2. (b) Not stable below 1672 K (2550 °F). (c) Decomposes into β-MoB and melt (peritectic reaction). (d) Decomposes into Mo_2B and β-MoB at about 2452 K (3955 °F) (peritectic reaction). (e) Congruent melting at 2872 K (4710 °F); decomposes slowly at about 2072 K (3270 °F) into α-MoB and MoB_2. (f) Incongruent melting; decomposes slowly at 1794 K (2770 °F) into α-MoB and Mo_2B_5. (g) Decomposes into MoB_2 and melt (peritectic reaction). (h) Congruent melting. (i) Decomposes into β-WB and W_2B_5 at about 2444 K (3940 °F) (peritectoid reaction). (j) Congruent melting at 2938 K (4830 °F); decomposes in a rapid eutectoid reaction at about 2383 K (3830 °F) into W_2B and α-WB.

Source: Ref 6, p 5.1.4-1

Binary Phase Diagrams

All the following material has been extracted from Ref 11, Vol. 1 and Vol. 2

Al-N (Aluminum-Nitrogen)

26.98154 14.0067

The established equilibrium solid phases of the Al-N system are (1) the fcc terminal solid solution (Al), and (2) the hexagonal nitride AlN.

The assessed Al-N diagram is primarily qualitative, because of the lack of data on the composition ranges of the solid and liquid phases. It was obtained by thermodynamic calculation and review of the experimental data [Hultgren, E; 26Iwa; 46Eas; 59Vol; 74Sch1; 74Sch2; 78Cha; 79Bor]. The eutectic near the melting point of Al and the monotectic near that of AlN are speculative; the respective displacements in temperature of these reactions from the melting points of pure Al and of stoichiometric AlN are believed to be slight.

The phase (Al) exists stably only at very small nitrogen (N_2) fugacities; at greater N_2 fugacities, AlN is stable. The N concentrations in (Al) that are in equilibrium with AlN and the (Al) solidus have not been measured.

An Al-rich liquid, L_1, exists stably at small N_2 fugacities. No experimental data on the Al-rich liquidus of L_1 or of the location of the probable eutectic reaction at its termination exist. The composition of eutectic liquid is estimated to be $\sim 1 \times 10^{-11}$ at.% N. The N-saturated L_1, with very small N concentrations, is in equilibrium with AlN from the temperature of the eutectic reaction to that of the monotectic reaction; compositions of this liquid have not been measured.

At low hydrostatic pressures, the nitride AlN has a wurtzite-type *hP*4 structure. AlN transforms at elevated hydrostatic pressures from the wurtzite-type structure to a possibly sphalerite-type *cF*8 structure [68Ver] or to NaCl-type *cF*8 structure [82Kon].

Up to the monotectic temperature, where it coexists with Al-rich L_1 and O-rich L_2, AlN is in equilibrium with (Al) or L_1 on its Al-rich side. The value 2800 ± 50 °C for the melting point of AlN was reported by [68Cla], who used high-pressure N_2 to retard evaporation. There are no experimental data for the L_2 liquidus from the melting point of AlN to the monotectic reaction or for the monotectic reaction itself. The compositions of the immiscible liquids L_1 and L_2 coexisting above the monotectic temperature are unknown.

The phase AlN_9 was reported to be formed by reaction of NH_3 with AlH_3 in an ether solution at a temperature slightly above the melting point of the solvent [54Wib1] and by reaction of NaN_3 with $AlCl_3$ in tetrahydrofuran [54Wib2]. This nitride may be a stable phase of the condensed Al-N system.

26Iwa: K. Iwasé, *Sci. Rep. Tohoku Imp. Univ.*, Ser. 1, *15*(4), 531-566 (1926).

46Eas: L.W. Eastwood, *Gases in Light Alloys*, John Wiley & Sons, New York, 26 (1946).

54Wib1: E. Wiberg and H. Michaud, *Z. Naturforsch.*, *9b*, 495-496 (1954) in German.

54Wib2: E. Wiberg and H. Michaud, *Z. Naturforsch.*, *9b*, 496-497 (1954) in German.

59Vol: A.E. Vol, *Handbook of Binary Metallic Systems—Structures and Properties*, Vol. 1, Gos. Izd. Fiz.-Mat. Lit., Moscow (1959).

68Cla: W. Class, NASA Contract Rep. NASA CR-1171 (1968).

68Ver: L.F. Vereschagin, G.A. Adadurov, O.N. Breusov, K.P. Burdina, L.N. Burenkova, A.N. Dremin, E.V. Zubova, and A.I. Rogacheva, *Dokl. Akad. Nauk SSSR*, *182*(2), 301-303 (1968) in Russian.

74Sch1: A. Schweighofer and S. Kúdela, *Kovove Mater.*, *12*(1), 95-107 (1974) in Slovak.

74Sch2: A. Schweighofer and S. Kúdela, *Kovove Mater.*, *12*(2), 268-279 (1974) in German.

78Cha: M.W. Chase, Jr., J.L. Curnutt, R.A. McDonald, and A.N. Syverud, *J. Phys. Chem. Ref. Data*, *7*(3), 793-794 (1978).

79Bor: P.C. Borbe, F. Erdmann-Jesnitzer, and E.-J. Jun, *Metall*, *33*(10), 1054-1060 (1979) in German.

82Kon: K. Kondo, A. Sawaoka, K. Sato, and M. Ando, *Shock Waves in Condensed Matter—1981*, AIP Conf. Proc. No. 78, Am. Inst. Phys., NY, 325-329 (1982).

Complete evaluation contains 1 figure, 4 tables, and 60 references.

Al-N Three-Phase Equilibria and Other Transformations (Condensed System)

Reaction	Composition, at.% N			Temperature, °C	Reaction type
$L \rightleftarrows (Al) + AlN$	$\sim 1 \times 10^{-11}$	$<1 \times 10^{-11}$	50	~660	Probably eutectic
$L_2 \rightleftarrows L_1 + AlN$	Unknown	Unknown	~50	Unknown	Probably monotectic
$L \rightleftarrows Al$		0		660.452	...
$L \rightleftarrows AlN$		50		2800 ± 50	Congruent melting

Al-N Crystal Structure Data

Phase	Composition, at.% N	Pearson symbol	Space group	Strukturbericht designation	Prototype
(Al)	0	*cF*4	*Fm*3*m*	*A*1	Cu
AlN	~50	*hP*4	$P6_3mc$	*B*4	SZn(wurtzite)

Al-O (Aluminum-Oxygen)

26.98154 15.9994

The equilibrium phases of the Al-O system are: (1) the liquid, which exists as two immiscible liquids—liquid (Al) when Al-rich and liquid Al_2O_3 when O-rich; (2) the fcc terminal solid solution, (Al), in which the solubility of oxygen is unknown but small; (3) the hexagonal aluminum oxide, αAl_2O_3, α alumina, or corundum, for which deviations from the exact stoichiometric composition are unknown and small; and (4) the gas.

The assessed diagram is based on review of the work of [34Bau], [66Git], and [68Yan], who prepared diagrams for the condensed phases at compositions ranging from 0 to 60 at.% O. The studies of [78Cha], [81Ots], [81She], and [82Lia] were also taken into account.

No stable condensed phase exists that is richer in oxygen than αAl_2O_3. At all temperatures up to its melting point, αAl_2O_3 has a hexagonal structure, which may also be regarded as rhombohedral. In condensed systems up to the monotectic temperature, αAl_2O_3 can coexist stably only with solid or liquid (Al), that is, oxygen-saturated aluminum. Above the O_2 fugacities for coexistence with (Al), αAl_2O_3 exists alone. The present diagram is primarily qualitative, because of the lack of data for the composition ranges of the solid and liquid phases. The existence of a eutectic near the melting point of Al is speculative; however, the existence of a monotectic near the melting point of Al_2O_3 is fairly well established. Displacements in temperature of these three-phase equilibria from the melting points of pure Al and stoichiometric Al_2O_3 are small and are probably not measurable for the eutectic.

At low hydrostatic pressures, the terminal (Al) solid solution with a very limited range exhibits only the fcc structure up to the melting point. No reliable measurement of the solubility of oxygen in (Al) has been reported. The concentration of dissolved oxygen in solid (Al) saturated with respect to αAl_2O_3 is probably smaller than that in liquid (Al) and not measurable with existing techniques.

Liquid (Al) exists stably only at low oxygen O_2 fugacities. At extremely low O_2 fugacities, a short liquidus terminating at the postulated eutectic is considered to exist for liquid (Al) in equilibrium with solid (Al). No experimental data on this liquidus or the probable eutectic have been reported. Liquid (Al) with very low concentrations of dissolved oxygen is in equilibrium with solid αAl_2O_3 from the temperature of the eutectic reaction to that of the monotectic reaction.

[68Yan] found no measurable lowering of the melting point of αAl_2O_3 with Al addition, but [66Git] found a 15 °C lowering. They agree on the existence of two liquid phases in samples heated slightly below or above the melting point of αAl_2O_3.

In addition to amorphous αAl_2O_3, there are at least five varieties of crystalline Al_2O_3 that are well established, but many more have been reported. All crystalline forms of Al_2O_3 are constructed of stacked, close-packed layers of oxygen ions with Al ions and vacancies distributed on the tetrahedral and octahedral sites among these oxygen ions. Polymorphism arises from the possibilities for different oxygen layer stacking sequences and disorder therein, from variations in the distribution of Al ions on their two types of sites, and from ordering among the Al ions and vacancies on those sites.

Al-O Three-Phase Equilibria and Other Transformations (Condensed System)

Reaction	Composition, at.% O			Temperature, °C	Reaction type
(Al, L) ⇄ (Al) + αAl_2O_3	$\leq 3 \times 10^{-8}$	$< 3 \times 10^{-8}$	60	~660	Probably eutectic
Al_2O_3, L ⇄ (Al, L) + αAl_2O_3	59.5 ± 0.5	~0.1	~60	2046.5 ±7.5	Monotectic
L ⇄ Al		0		660.452	Melting
L ⇄ Al_2O_3		60		2054	Congruent melting

At least eight other solid Al oxides have been reported: Al_2O, Al_4O_3, AlO, Al_2O_2, Al_8O_9, Al_3O_4, Al_4O_6, and AlO_2. It is possible that none of these phases exists even metastably. Oxide species occur in vapors of the Al-O system, as well as single-element oxygen and Al species.

34Bau: E. Baur and R. Brunner, *Z. Elektrochem., 40*(3), 154-158 (1934) in German.

66Git: G. Gitlesen, O. Herstad, and K. Motzfeldt, *Selected Topics in High Temperature Chemistry,* T. Førland, Ed., Universitetsforlaget, Oslo, Norway, 179-196 (1966).

68Yan: H. Yanagida and F.A. Kroger, *J. Am. Ceram. Soc., 51*(12), 700-706 (1968).

78Cha: M.W. Chase, Jr., J.L. Curnutt, R.A. McDonald, and A.N. Syverud, *J. Phys. Chem. Ref. Data, 7*(3), 793-940 (1978).

Al-O Crystal Structure Data

Phase	Composition, at.% O	Pearson symbol	Space group	Strukturbericht designation	Prototype
(Al)	0	*cF*4	*Fm*3*m*	*A*1	Cu
αAl_2O_3	60	*hR*10	$R\bar{3}c$	$D5_1$	αAl_2O_3
Metastable Al_2O_3 polymorphs					
γAl_2O_3	...	*cF*56	*Fd*3*m*	$H1_1$	Al_2MgO_4
δAl_2O_3	...	...	...	...	δAl_2O_3
θAl_2O_3	...	...	*C*2/*m*	...	βGa_2O_3
κAl_2O_3(a)	...	...	...	...	...
χAl_2O_3	...	...	$P6_3/mcm$ or $P6/mmm$ or $P6_3/mmc$	...	...

(a) Called η by [Pearson2]. η usually denotes a form indistinguishable from γ.

81Ots: S. Otsuka and Z. Kozuka, *J. Jpn. Inst. Met., 22*(8), 558-566 (1981).

81She: V.E. Shevtsov, *Izv. Akad. Nauk SSSR, Met.,* (1) 60-65 (1981) in Russian; *Russ. Metall.,* (1), 52-57 (1981).

82Lia: W.W. Liang, *Z. Metallkd., 73,* 369-375 (1982).

Complete evaluation contains 1 figure, 6 tables, and 46 references.

B-C (Boron-Carbon)

10.81 12.011

B_4C: *hR*29 or *hR*15; $R\bar{3}m$.

The B-C phase diagram is redrawn from [Elliott].

B-Cr (Boron-Chromium)

10.81 51.996

The assessed phase diagram for the Cr-B system is based primarily on the results of [69Por], [72Por], and [76Guy], with modifications based on the data of [68And]. The system consists of six intermetallic compounds (Cr_2B, Cr_5B_3, CrB, Cr_3B_4, CrB_2, and CrB_4) and the terminal solid solutions—bcc (Cr) and rhombohedral (βB). The existence of several other reported borides, such as Cr_4B, Cr_3B_2, and CrB_6, have been ruled out and/or remain uncertain, and these are not included in the assessed diagram.

The liquidus temperatures for compositions greater than 80 at.% B on the B-rich side of the system are rather uncertain, and the invariant temperatures are uncertain to the extent of ±30 to 50 K. Table 1 summarizes the type, temperature, and compositions of invariant reactions of this system.

Invariant Points on the Assessed Cr-B Phase Diagram

Reaction	Composition, at.% B			Temperature, °C	Reaction type
L ⇄ bcc (Cr)		0.0		1857	Melting
L ⇄ bcc (Cr) + Cr_2B	13.5	0.0	33.3	1630	Eutectic
L + Cr_5B_3 ⇄ Cr_2B	31.0	37.5	33.3	1870	Peritectic
L + CrB ⇄ Cr_5B_3	34.0	50.0	37.5	1900	Peritectic
L ⇄ CrB		50.0		2100	Congruent melting
L ⇄ CrB + Cr_3B_4	53.5	50.0	57.1	2050	Eutectic
L + CrB_2 ⇄ Cr_3B_4	56.0	57.1	66.7	2070	Peritectic
L ⇄ CrB_2		66.7		2200	Congruent melting
L ⇄ CrB_2 + (βB)	83.0	66.7	100	1830	Eutectic
CrB_2 + (βB) ⇄ CrB_4	66.7	100	80.0	1500	Peritectoid
L ⇄ (βB)		100		2092	Melting

The solubility of B in (Cr) has been reported to be approximately 0.7 at.% B at 1100 °C by [Shunk], which is consistent with the value of 0.6 at.% B reported by [71Bor] for a temperature at ≈1527 °C. [69Por] reported the maximum solubility of Cr in (B) to be approximately 2 at.%, which was confirmed by [70And] using X-ray diffraction data.

68And: S. Andersson and T. Lundstrom, *Acta Chem. Scand., 22*(10), 3103-3110 (1968).
69Por: K.I. Portnoi, V.M. Romashov, and I.V. Romanovich, *Poroshk. Metall., 4*(76), 51-57 (1969).
70And: S. Andersson and T. Lundstrom, *J. Solid State Chem., 2*, 603-611 (1970).
71Bor: M.L. Borlera and G. Pradelli, *Metall. Ital., 63,* 61 (1971).
72Por: K.I. Portnoi and V.M. Romanshov, *Poroshk. Metall., 5*(113), 48-56 (1972).

Cr-B Crystal Structure Data

Phase	Pearson symbol	Space group	Strukturbericht designation	Prototype
(αCr)	*cI*2	*Im*3*m*	*A*2	W
Cr_4B(a)	*oF*40	*Fddd*	*D*1_f	Mn_4B
Cr_2B	*oF*40	*Fddd*	*D*1_f	Mn_4B
Cr_2B(a)	(b)	*Abmm*	...	...
Cr_2B(a)	*tI*12	*I*4/*mcm*	*C*16	Al_2Cu
Cr_5B_3	*tI*32	*I*4/*mcm*	*D*8_1	Cr_5B_3
CrB	*oC*8	*Cmcm*	B_f	CrB
Cr_3B_4	*oI*14	*Immm*	*D*7_b	Ta_3B_4
CrB_2	*hP*3	*P*6/*mmm*	*C*32	AlB_2
CrB_4	(b)	...	...	...
CrB_6(a)	(c)	...	...	...
(βB)	*hR*108	*R*3*m*	...	...

(**a**) Unstable or stability is uncertain. (**b**) Orthorhombic. (**c**) Tetragonal.

76Guy: C.N. Guy and A.A. Uraz, *J. Less-Common Met., 48,* 199-203 (1976).

Complete evaluation contains 2 figures, 3 tables, and 35 references.

B-Hf (Boron-Hafnium)

10.81 178.49

HfB: *B*27. HfB_2: *C*32.
The Hf-B phase diagram is redrawn from [Hafnium].

B-Mo (Boron-Molybdenum)

10.81 95.94

Mo_2B: *C*16. Mo_3B_2(HT): *D*5_a. αMoB: B_g. βMoB: B_f. MoB_2(HT): *C*32. Mo_2B_5: *D*8_i. MoB_4: *D*1_e.

The Mo-B phase diagram is redrawn from [Molybdenum].

B-Nb (Boron-Niobium)

10.81 92.9064

Nb_3B_2: *D*0_c. NbB: B_f. Nb_3B_4: *D*7_b. NbB_2: *C*32.

The Nb-B phase diagram is redrawn from [Elliott].

B-Pu (Boron-Plutonium)

10.81 (244)

PuB_2: *C*32. PuB_4: *D*1_e. PuB_6: *D*2_1. PuB_{12}: *D*2_f. PuB_{66}: YB_{66}.

The Pu-B phase diagram is redrawn from [Spear].

B-Ta (Boron-Tantalum)

10.81 180.9479

Ta_2B: *C*16. Ta_3B_2: *D*5_a. TaB: B_f. Ta_3B_4: *D*7_b. TaB_2: *C*32.

The Ta-B phase diagram is redrawn from [Elliott].

B-Th (Boron-Thorium)

10.81 232.0381

ThB_4: *D*1_e. ThB_6: *D*2_1. ThB_{66}: *cF*1880.

The Th-B phase diagram is redrawn from [Moffatt].

B-Ti (Boron-Titanium)

10.811 47.88

The equilibrium phases of the Ti-B system are: (1) the terminal solid solutions—high-temperature bcc (βTi), low-temperature cph (αTi), and rhombohedral (βB); (2) two intermetallic compounds, TiB and TiB_2, whose structures and melting mechanisms are well established; and (3) Ti_3B_4, which forms from the melt only in a narrow temperature range and remains to be verified by an independent study.

There are several discrepancies in the literature concerning the structures and stability ranges of the equilibrium phases, in part because many of the reported binary phases are derived from studies of ternary systems. The major studies of the binary system were conducted by [54Pal], [64Fen], and [66Rud]. The work of [54Pal] was not used to construct the assessed diagram in the high-temperature range, because of the evident effect of contamination on their specimens.

[64Fen] and [66Rud] exercised considerable caution to avoid contamination; nevertheless, there are also discrepancies between these two studies, primarily concerning the existence of Ti_3B_4 and the temperatures of the invariant reactions. For the invariant temperatures, the data of [66Rud] are preferred, because their technique for determining the incipient melting temperatures has been successfully tested for a number of high-temperature systems and because results were verified by differential thermal analysis for a large number of samples.

[64Fen] and [66Fen] observed Ti_3B_4, a phase not reported by [66Rud]. It was reported to be isomorphous with that of Ta_3B_4. The composition of this phase is 58.1 at.% B, as determined by chemical extraction.

A second phase, thought to be possibly a high-temperature form of Ti_3B_4, was found with a completely different structure from any other phase of the system. During heating, Ti_3B_4 is transformed at about 2010 °C, but the transformation is not reversible. High- and low-temperature forms of Ti_3B_4 are not distinguished in the assessed diagram, because the irreversibility of the transformation strongly suggests that it is a contamination effect.

[64Fen] indicated that Ti_3B_4 forms from the melt by a peritectic reaction at 2020 °C, 20 °C above the peritectic reaction involving TiB. In the assessed diagram, the peritectic reac-

Ti-B Crystal Structure Data

Reaction	Composition, at.% B	Pearson symbol	Space group	Strukturbericht designation	Prototype
(αTi)	0 to <0.2	*hP*2	$P6_3/mmc$	*A*3	Mg
(βTi)	0 to <0.2	*cI*2	*Im*3*m*	*A*2	W
TiB	49 to 50	*oP*8	*Pnma*	*B*27	FeB
TiB_2	65.5 to 66.7	*hP*3	*P*6/*mmm*	*C*32	AlB_2
Ti_3B_4	56.1	*oI*14	*Immm*	*D*7_b	Ta_3B_4
(βB)	~100	*hR*108	$R\bar{3}m$	...	βB

Special Points of the Assessed Ti-B Phase Diagram

Reaction	Composition, at.% B			Temperature, °C	Reaction type
L ⇄ (βTi) + TiB	7 ± 1	<1	~50	1540 ± 10	Eutectic
L + TiB_2 ⇄ Ti_3B_4	42 ± 3	~65.5	58.1	2200 ± 25	Peritectic
L + Ti_3B_4 ⇄ TiB	~39	58.1	50	2180	Peritectic
L ⇄ TiB_2		66.7		3225 ± 25	Congruent
L ⇄ (βB) + TiB_2	~98	~100	~66.7	2080 ± 20	Eutectic
(βTi) + TiB ⇄ (αTi)	~0.1	49	~0.2	884 ± 2	Peritectoid
L ⇄ (βTi)		0		1670	Melting point
(βTi) ⇄ (αTi)		0		882	Allotropic transformation
L ⇄ (βB)		100		2092	Melting point

tions are separated by 20 °C, but positioned at 2180 and 2200 °C, according to the work of [66Rud].

Reported melting points of TiB_2 range between 2790 °C [52Gla] and 3225 °C [66Rud]. [64Fen] noted that large volumes of vapor prevented the accurate determination of the TiB_2 melting point. The major difficulty is reaction with the crucible material, which results in the detection of a ternary eutectic temperature lower than the melting point of the binary compound. The highest reported congruent temperature was therefore used to construct the assessed diagram.

52Gla: F.W. Glaser, *Trans. AIME, 194,* 391-396 (1952).
54Pal: A.E. Palty, H. Margolin, and J.P. Nielsen, *Trans. ASM, 46,* 312-328 (1954).
64Fen: R.G. Fenish, NRM-138, 1-37 (1964).
66Fen: R.G. Fenish, *Trans. AIME, 236,* 804 (1966).
66Rud: E. Rudy and St. Windisch, Tech. Rep. No. AFML-TR-65-2, Part I, Vol. VII (1966).

Complete evaluation contains 1 figure, 3 tables, and 53 references.

B-U (Boron-Uranium)

10.81 238.0289

UB_2: *C*32. UB_4: $D1_e$. UB_{12}: $D2_f$.

The U-B phase diagram is redrawn from [Elliott].

B-V (Boron-Vanadium)

10.81 50.9415

The equilibrium solid phases of the V-B system are: (1) the terminal solid solutions of bcc (V) and rhombohedral (βB); and (2) the intermediate compounds of V_3B_2, VB, V_5B_6, V_3B_4, V_2B_3, and VB_2. The phase diagram is based on the phase equilibria data of [66Rud] and [81Spe]. The liquidus of the assessed diagram was calculated by the present authors from Gibbs energy functions optimized with respect to thermochemical and phase diagram data.

Homogeneity ranges appear to be quite small for all of the solid phases in the V-B system. Boron is slightly soluble in (V), and (B) dissolves only small amounts (<1 at.%) of V.

66Rud: E. Rudy and St. Windisch, Tech. Rep. No. AFML-TR-65-2, Part I, Vol. X, Wright-Patterson Air Force Base, OH (1966).
81Spe: K.E. Spear, J.H. Blanks, and M.S. Wang, *J. Less-Common Met., 82,* 237-243 (1981).

Complete evaluation contains 2 figures, 5 tables, and 34 references.

V-B Crystal Structure Data

Phase	Composition, at.% B	Pearson symbol	Space group	Strukturbericht designation	Prototype
(V)	0	*cI*2	*Im*3*m*	*A*2	W
V_3B_2	40	*tP*10	*P*4/*mbm*	$D5_a$	U_3Si_2
VB	50	*oC*8	*Cmcm*	*B*33	CrB
V_5B_6	54.5	(a)	*Ammm*	...	...
V_3B_4	57	*oI*14	*Immm*	$D7_b$	Ta_3B_4
V_2B_3	60	(a)	*Cmcm*	...	...
VB_2	67	*hP*3	*P*6/*mmm*	*C*32	AlB_2
(βB)	100	*hR*108	$R\bar{3}m$	...	...

(a) Orthorhombic.

V-B Invariant Reactions

Reaction	Composition, at.% B			Temperature, °C	Reaction type
$L \rightleftarrows V + V_3B_2$	15	0	40	1735	Eutectic
$L + VB \rightleftarrows V_3B_2$	25	50	40	1925	Peritectic
$L + V_3B_4 \rightleftarrows VB$	49	57	50	2551	Peritectic
$VB + V_3B_4 \rightleftarrows V_5B_6$	50	57	54.5	1727	Peritectoid
$L + V_2B_3 \rightleftarrows V_3B_4$	56	60	57	2640	Peritectic
$L + VB_2 \rightleftarrows V_2B_3$	57	67	60	2653	Peritectic
$L \rightleftarrows VB_2$		67		2750	Congruent
$L \rightleftarrows VB_2 + B$	98	67	100	2068	Eutectic

B-W (Boron-Tungsten)

10.81 183.85

W_2B: *C*16. WB: B_f above 2623 K, B_g below. W_2B_5: $D8_i$. WB_4: *hP*20.

The W-B phase diagram is redrawn from [Moffatt].

B-Zr (Boron-Zirconium)

10.81 91.22

$B_{12}Zr$: $D2_f$. B_2Zr: *C*32.

The B-Zr phase diagram is redrawn from [Zirconium].

Be-O (Beryllium-Oxygen)

9.01218 15.9994

The equilibrium solid phases of the Be-O system are: (1) the terminal solid solution (αBe), (2) the terminal bcc solid solution (βBe), (3) the hexagonal oxide αBeO, and (4) the tetragonal oxide βBeO.

No published diagram has been found for this system. The types of three-phase equilibria have not been established. A largely schematic diagram is shown for the range 0 to 50 at.% O.

The phases (αBe) and (βBe) exist stably only at very small oxygen (O_2) fugacities; at greater O_2 fugacities, BeO is stable. At 0.1 MPa hydrostatic pressure, αBe is stable below 1270 °C and βBe is stable from 1270 °C to the melting point. The value 1289 °C is adopted for the melting point of βBe. Effects of O on the triple point and melting point have not been detected.

The very small O concentrations in (αBe) or (βBe) that are in equilibrium with αBeO and the (βBe) solidus have not been determined. At 0.1 MPa hydrostatic pressure and temperatures up to 2105 °C, αBeO is stable; from 2105 °C to the melting point, βBeO is stable [62Bak, 64Smi]. Because the undetermined deviations in BeO composition from the stoichiometric are certainly small, the αBeO → βBeO transformation is probably not appreciably affected by variations in O_2 fugacity. Determination of the exact temperature has been hindered by hysteresis. The value adopted for the melting point of βBeO is 2550 °C, essentially in agreement with [78Cha]. There is no information concerning the liquidus temperatures between the compositions 0 and 50 at.% O.

There is no experimental information on the three-phase equilibria involving (αBe), (βBe), and αBeO, or on that involving L, αBeO, and βBeO, but their existence is required. For the former, either a eutectoid or a peritectoid reaction is possible; for the latter either a peritectic or metatectic reaction is possible. The temperatures are very close to those of the transformations in pure Be and in BeO, respectively. L + (βBe) + αBeO

Be-O Crystal Structure Data

Phase	Composition, at.% O	Pearson symbol	Space group	Strukturbericht designation	Prototype
Stable (0.1 MPa)					
(αBe)	~0	*hP*2	$P6_3/mmc$	*A*3	Mg
(βBe)	~0	*cI*2	*Im*3*m*	*A*2	W
αBeO	50	*hP*4	$P6_3mc$	*B*4	SZn(wurtzite)
βBeO	50	*tP*6?	$P4_2/mnm$	*C*4?	O_2Ti(rutile)?
Other phases					
(Be)	~0	...	*C*1*c*1 or *C*12/*c*1	...	...
Be_2O?	33.3	*cF*12	*Fm*3*m*	*C*1	CaF_2

Be-O Three-Phase Equilibria and Other Transformations (Condensed System)

Reaction	Composition, at.% O			Temperature, °C	Reaction type
L ⇄ (βBe) + αBeO	~0	~0	~50	~1289	Eutectic?
L + βBeO + αBeO	Unknown	~50	~50	~2105	Unknown
(αBe) + (βBe) + αBeO	~0	~0	~50	~1270	Unknown
Pure component transformations					
(βBe) ⇄ (αBe)		0		1270 ± 6	Allotropic
L ⇄ (βBe)		0		1289 ± 4	Melting
Other transformations					
βBeO ⇄ αBeO		50		2105 ± 15	Polymorphic
L ⇄ βBeO		50		2550 ± 40	Congruent melting

is the only three-phase equilibrium for which any experimental study exists. [32Slo] suggested that it is a eutectic reaction.

32Slo: H.A. Sloman, *J. Inst. Met., 49,* 365-388 (1932).
62Bak: T.W. Baker and P.J. Baldock, *Nature, 193,* 1172 (1962).
64Smi: D.K. Smith, C.F. Cline, and S.B. Austerman, *J. Nucl. Mater., 14,* 237-238 (1964).
78Cha: M.W. Chase, J.L. Curnutt, R.A. McDonald, and A.N. Syverud, *J. Phys. Chem. Ref. Data,* 7(3), 793-940 (1978).

Complete evaluation contains 1 figure, 3 tables, and 70 references.

C-Cr (Carbon-Chromium)

12.011 51.996

$Cr_{23}C_6$: *D*8_4. Cr_7C_3: *D*10_1. Cr_3C_2: *D*5_{10}.

The Cr-C phase diagram is redrawn from [Hansen].

C-Hf (Carbon-Hafnium)

12.011 178.49

HfC: *B*1.

The Hf-C phase diagram is redrawn fram [Hafnium].

C-Mo (Carbon-Molybdenum)

12.011 95.94

The assessed phase diagram for the Mo-C system is based primarily on the work of [65Rud], [66Rud], [67Rud], and [69Rud], with minor temperature changes based on the work of [71Rea]. The Mo-Mo_2C eutectic was increased to 2205 °C.

65Rud: E. Rudy, S. Windisch, and Y.A. Chang, Air Force Materials Laboratory, Wright-Patterson AFB, OH, Rep. No. AFML-TR-65-2, Part 1, Vol. 1 (Mar 1965).

Mo-C Crystal Structure Data

Phase	Composition, at.% C	Pearson symbol	Space group	Strukturbericht designation	Prototype
(Mo)	0 to 1.1	*cI*2	*Im*3*m*	*A*2	W
αMo_2C	31.7 to 32.7	*oP*12	*Pbcn*	···	αMo_2C
βMo_2C	26 to 36	*hP*3	$P6_3/mmc$	*L'*3	Fe_2N
ηMoC_{1-x}	37 to 39.5	*hP*~10	$P6_3/mmc$	···	Mo_3C_2
αMoC_{1-x}	39.5 to 43	*cF*8	*Fm*3*m*	*B*1	NaCl
γMoC	50	*hP*2	$P\bar{6}m2$	B_h	WC
$\gamma' MoC_{1-x}$	···	*hP*8	$P6_3/mmc$	B_i	TiAs
(C)	100	*hP*4	$P6_3/mmc$	*A*9	C (graphite)

66Rud: E. Rudy, S. Windisch, and J.R. Hoffman, Air Force Materials Laboratory, Wright-Patterson AFB, OH, Rep. No. AFML-TR-65-2, Part I, Vol. 6 (Jan 1966).

67Rud: E. Rudy, S. Windisch, A.J. Stosick, and J.R. Hoffman, Air Force Materials Laboratory, Wright-Patterson AFB, OH, Rep. No. AFML-TR-65-2, Part I, Vol. 11 (Apr 1967).

69Rud: E. Rudy, Compendium of Phase Diagram Data, Air Force Materials Laboratory, Wright-Patterson AFB, OH, Rep. No. AFML-TR-65-2, Part V (Jun 1969).

71Rea: J.G. Reavis, G.R. Brewer, D.B. Court, and J.W. Schulte, Los Alamos Scientific Lab., NM, Rep. LA-DC-12753 (1971).

Complete evaluation contains 1 figure, 1 table, and 12 references.

C-Nb (Carbon-Niobium)

12.011 92.9064

Nb_2C(β): *hP**. β′: *oP*12. NbC: *B*1. ζ: *cP*7 or *hR*20?

The Nb-C phase diagram is redrawn from [Metals].

C-Pu (Carbon-Plutonium)

12.011 (244)

PuC: *B*1. Pu_2C_3: $D5_c$. PuC_2: cubic or $C11_a$.

The Pu-C phase diagram is redrawn from [Plutonium].

C-Si (Carbon-Silicon)

12.011 28.0855

Because of difficult experimental requirements, only a few comprehensive investigations of the Si-C phase diagram have been undertaken, and the results are somewhat conflicting. The present diagram is based primarily on the experimental work of [60Dol], with a review of the work of [59Sca], [67Str], [70Noz], and [73Vol]. The assessed Si-C phase diagram is characterized by a peritectic reaction at 2545 ± 40 °C, involving Si-C and a liquid of 27 at.% C, and a eutectic reaction at 1404 ± 5 °C and 0.75 ± 0.5 at.% C [60Dol]. In the assessed phase diagram, the high-temperature melting points of [67Str] have been used.

The present evaluators conducted thermodynamic calculations to determine the stability of various phases of the Si-C system. They determined that (1) the cubic form of silicon carbide (βSiC) is more stable than the hexagonal form (αSiC) at any temperature below the peritectic point; (2) the solid βSiC can decompose under atmospheric pressure to C and liquid Si at 3076 °C if nonmixing of C and Si phases is assumed, i.e., zero entropy of mixing (the transformation of the silicon carbide to a mixture of Si and C can occur at a lower temperature because of a positive entropy of mixing); and (3) the atmospheric boiling point of Si is 3227 °C, and the atmospheric sublimation point of C is 3798 °C.

The solid solubility for C in Si at 1200 to 1400 °C is $\sim 10^{-3}$ to 10^{-4} at.% C, and C dissolves substitutionally in Si. At the eutectic temperature, the solubility is approximately 0.0007 [70Noz] or 0.00018 [73Vol] at.% C.

Extended solid solubility of C in Si has been reported in films prepared on quartz substrates by the pyrolysis of heptane vapor at 1252 °C. The electron-diffraction determination of the lattice parameter of the Si phase was interpreted as evidence of the formation of a solid solution, containing as much as 10 at.% C; the SiC phase was also observed in the diffraction patterns.

In Si-C films synthesized by a plasma deposition process on Si and quartz substrates, the substrate temperature ranged from 200 to 600 °C, and the reaction gases consisted of CH_4 and SiH_4. The formation of amorphous SiC films with a composition of 39 at.% Si and 61 at.% C was obtained by RF sputtering.

Si-C Crystal Structure Data

Phase	Composition, at.% C	Pearson symbol	Space group	Strukturbericht designation	Prototype
(Si)	0	*cF*8	*Fd*3*m*	*A*4	C (diamond)
SiC or βSiC	50	*cF*8	*F*$\bar{4}$3*m*	*B*3	SZn
(C) or graphite	100	*hP*4	*P*6_3/*mmc*	*A*9	C (graphite)
Metastable					
αSiC(a)	50	(b)	...	...	...
Amorphous	39 to 61	...	...	...	...
High pressure					
SiC II	...	*tI*4	*I*4_1/*amd*	*A*5	βSn

(a) Other SiC polytypes have been reported. (b) Hexagonal.

Silicon carbide is well known for its numerous polytypes, which are mostly modifications of the αSiC (hexagonal) structure.

59Sca: R.I. Scace and G.A. Slack, *J. Chem. Phys., 30,* 1551-1555 (1959).
60Dol: R.T. Dolloff, WADD Technical Report 60-143 Wright Air Development Division (1960).
67Str: H.M. Strong and R.E. Hanneman, *J. Chem. Phys., 46,* 3668-3676 (1967).
70Noz: T. Nozaki, Y. Yatsurugi, and N. Akiyama, *J. Electrochem. Soc., 117,* 1566-1568 (1970).
73Vol: F.W. Voltmer and F.A. Padovani, Semicond. Silicon, Pop. Int. Symp. Silicon Mater. Sci. Technol., 2nd; 75-82 (1973).

Complete evaluation contains 1 figure, 2 tables, and 28 references.

C-Ta (Carbon-Tantalum)

12.011 180.9479

Ta_2C: $L'3$. β': $C6$. TaC: $B1$.
The Ta-C phase diagram is redrawn from [Tantalum].

C-Th (Carbon-Thorium)

12.011 232.0381

ThC: $B1$. αThC_2: C_g. βThC_2: $tP6$. γThC_2:$C2$. Th_2C_3: $D5_c$, stable at $P > 33$ kbar, 850 to 1450 °C.

The Th-C phase diagram is redrawn from [Smith].

C-Ti (Carbon-Titanium)

12.011 47.88

The condensed Ti-C system has been characterized over the composition range 0 to about 70 at.% C. In addition to the bcc (βTi) and cph (αTi) solutions, there is an equilibrium carbide, TiC, with the homogeneity range 32 to 48.8 at.% C. The carbide has another form, designated Ti_2C, in which vacancies are ordered on the carbon sublattice [67Gor]. This ordered phase is not the same as that mentioned by [Hansen].

The most important phase diagram investigations were conducted by [53Cad], [55Ogd], [56Wag], [59Bic], and [65Rud]. At higher temperatures, the assessed diagram is based on [65Rud]; at lower temperatures, it is based on [53Cad] and [56Wag]. A thermodynamic assessment was presented by [84Uhr], in which agreement is good between calculated and experimental diagrams; the assessed TiC liquidus is drawn from the calculations of [84Uhr].

[59Bic] estimated liquidus compositions by chemical analysis of the liquid held at temperature until the melt composition was changing only very slowly. These data imply a eutectic composition of about 4.5 at.% C. [65Rud] estimated the liquidus to be somewhat above the temperature at which the samples collapsed [65Rud] placed the eutectic composition at slightly less than 2 at.% C, based on metallographic examination of as-cast alloys. They noted that the low carbon content of the eutectic and structural changes that occurred dur-

Ti-C Crystal Structure Data

Reaction	Composition, at.% C	Pearson symbol	Space group	Strukturbericht designation	Prototype
(βTi)	0 to 0.6	*cI*2	*Im*3*m*	*A*2	W
(αTi)	0 to 1.6	*hP*2	$P6_3/mmc$	*A*3	Mg
TiC	~32 to 48.8	*cF*8	*Fm*3*m*	*B*1	NaCl
Ti_2C	~32 to 36	*cF*48	*Fd*3*m*	...	...

Special Points of the Assessed Ti-C Phase Diagram

Reaction	Composition, at.% C			Temperature, °C	Reaction type
L ⇄ (βTi) + Ti_2C	1.8	0.6	32	1648 ± 5	Eutectic
L ⇄ TiC		44		3067 ± 15	Congruent
TiC ⇄ Ti_2C?		33.3		~1900	Congruent
L ⇄ TiC + C	63	48.8	~100	2776 ± 6	Eutectic
(βTi) + TiC ⇄ (αTi)	0.6	38	1.6	920 ± 3	Peritectoid
L ⇄ (βTi)		0		1670	Melting point
(βTi) ⇄ (αTi)		0		882	Allotropic transformation

ing quenching made the metallographic work difficult.

The Ti-rich limit of the TiC phase field was determined metallographically by [53Cad] (38 at.% C at 920 °C) and metallographically and from the melting point data by [65Rud] (32 at.% C at 1650 °C). Reported congruent melting points of TiC range between 1940 and 3250 °C. The value 3067 ± 15 °C [65Rud] is chosen for the assessed diagram. The reaction L ⇄ TiC + C (graphite) is placed at 2776 °C, with the eutectic composition about 63 at.% C [65Rud].

No distinction has yet been made between equilibria involving TiC and those involving ordered Ti_2C. [62Bit] found discontinuities in the magnetic susceptibility of TiC as a function of composition. [67Gor] performed neutron diffraction studies of alloys annealed between 1100 and 2000 °C. Superlattice lines were found in the range 32 to 36 at.% C. At 33 at.% C, Ti_2C was estimated to be stable up to about 1900 °C.

53Cad: I. Cadoff and J.P. Nielsen, *Trans. AIME*, *197*, 248-252 (1953).

55Ogd: H.R. Ogden, R.I. Jaffee, and F.C. Holden, *Trans. AIME*, *203*, 73-80 (1955).

56Wag: F.C. Wagner, E.J. Bucur, and M.A. Steinberg, *Trans. ASM*, *48*, 742-761 (1956).

59Bic: R.L. Bickerdike and G. Hughes, *J. Less-Common Met.*, *1*, 42-49 (1959).

62Bit: H. Bittner and H. Goretzki, *Monatsh Chem.*, *93*, 10(?)-1004 (1962) in German.

65Rud: E. Rudy, D.P. Harmon, and C.E. Brukl, AFML-TR-65-2, Part I, Vol. II (1965).

67Gor: H. Goretzki, *Phys. Status Solidi*, *20*, K141-K143 (1967).

84Uhr: B. Uhrenius, *Calphad*, *8*(2), 101-119 (1984).

Complete evaluation contains 4 figures, 4 tables, and 43 references.

C-U (Carbon-Uranium)

12.011 238.0289

δ: *B*1. ϵ: $C11_a$. ζ: $D5_c$.

The U-C phase diagram is redrawn from [Metals].

C-V (Carbon-Vanadium)

12.011 50.9415

The assessed phase diagram for the V-C system is based primarily on the work of [68Rud], with modifications based on the work of [62Sto], [68Ade], and [73Sto]. The liquidus is taken from [68Ade], who equilibrated V or VC with C at different temperatures and chemically analyzed the liquid phase in equilibrium with the solid at each temperature.

At elevated temperatures, only two intermediate phases (V_2C and VC) exist; both exhibit wide ranges of homogeneity. Decreasing temperature results in a peritectoidal reaction between V_2C and VC near 1320 °C to form the ζ phase, which in contrast to V_2C and VC appears to be nearly invariant in stoichiometry.

This invariance exists even though the phase is carbon deficient, with the crystal structure indicating an ideal stoichiometry of V_4C_3, but with an actual stoichiometry near V_3C_2. On this basis, the composition of the ζ phase is designated V_4C_{3-x}.

At still lower temperatures, transformations occur within the phase fields of both V_2C and VC to produce more complicated phase relationships.

V-C Crystal Structure Data

Phase	Composition, at.% C	Pearson symbol	Space group	Strukturbericht designation	Prototype
(V)	0 to 4.3	*cI*2	*Im*3*m*	*A*2	W
αV_2C	31 to 33	*oP*12	*Pbcn*	...	ζFe_2N
βV_2C	27 to 34	*hP*3	$P6_3/mmc$(c)	*L*′3	W_2C
$\beta' V_2C$(a)	27.5 to 31.5	*hP*9	$P\bar{3}1m$(c)	...	$Ni_3N(\epsilon Fe_2N)$
V_4C_{3-x}	~40	*hR*20	$R\bar{3}m$	...	...
VC(a)	37 to 48	*cF*8	*Fm*3*m*	*B*1	NaCl
V_6C_5(b)	43 to 46	*mB*44	*B*2	...	...
V_8C_7	47 to 48	*cP*60	$P4_132$	...	...
			$P4_332$	...	
(C)	100	*hP*4	$P6_3/mmc$	*A*9	C (graphite)

(**a**) High-temperature form. (**b**) Enantiomorphic and twinned forms have been described with other lattice parameters and/or space groups. (**c**) According to the International Tables for Crystallography, 1983 edition, the Landau constraints forbid a second-order transformation between $P6_3/mmc$ and $P\bar{3}1m$. It must therefore be concluded that (*a*) one or the other of these two space groups is in error, or (*b*) alternatively, the postulated transition in the assessed diagram is in error, with the transition being first order, requiring a two-phase region between the ordered and disordered structures.

The phase relationships for the region between 27 and 34 at.% C must be considered tentative. The relationships as drawn in the assessed diagram agree with the majority of the available experimental data, but some uncertainties exist. Crystallographic data indicate that the V-rich portion of the high-temperature form of V_2C is identifiably different, with space group symmetry distinguishable from the C-rich high-temperature form. On this basis, the C-rich portion of the high-temperature region is designated βV_2C, and the V-rich portion is designated $\beta' V_2C$. The β and β' crystal structures are hexagonal, and their V sublattices are identical.

These crystallographic considerations, in combination with the lack of any experimental evidence for a two-phase separation between the β and β' forms, are the bases for suggesting that the $\beta' V_2C \rightleftarrows \beta V_2C$ transformation is higher order, as indicated by the dash-dot line. The temperature-composition contour is speculative.

62Sto: E.K. Storms and R.J. McNeal, *J. Phys. Chem., 66*, 1401-1408 (1962).

68Ade: L.M. Adelsberg and L.H. Caddof, *J. Am. Ceram. Soc., 51*(4), 213-220 (1968).

68Rud: E. Rudy, S. Windisch, and C.E. Brukl, *Planseeber. Pulvermetall., 16*, 3-33 (1968).

73Sto: E. Storms, *The Refractory Carbides*, Chapter 4, Academic Press, New York (1973).

Complete evaluation contains 2 figures, 3 tables, and 77 references.

C-W (Carbon-Tungsten)

12.011 183.85

β: PbO_2. β': *L*'3. β'': *C*6. γ: *B*1. δ: B_h.

The W-C phase diagram is redrawn from [Metals].

C-Zr (Carbon-Zirconium)

12.011 91.22

γ: *B*1.

The Zr-C phase diagram is redrawn from [Metals].

Ca-N (Calcium-Nitrogen)

40.08 14.0067

Ca_3N_2: $D5_3$.

The Ca-N phase diagram is redrawn from [Hansen].

Ca-O (Calcium-Oxygen)

40.008 15.9994

The established equilibrium solid phases of the Ca-O system are (1) the terminal fcc solid solution, (αCa); (2) the terminal bcc solid solution (βCa); (3) the fcc oxide CaO; and (4) the bct peroxide CaO_2. At high oxygen (O_2) fugacity, the oxides CaO_4 (superoxide) and CaO_6 (ozonide) are considered to exist, although they have not been prepared in the pure state. These two compounds are possibly stable phases of the system, but their stability relative to other oxides is not established.

Ca-O Crystal Structure Data

Phase	Composition, at.% O	Pearson symbol	Space group	Strukturbericht designation	Prototype
(αCa)	0	*cF*4	*Fm*3*m*	*A*1	Cu
(βCa)	0	*cI*2	*Im*3*m*	*A*2	W
CaO	50	*cF*8	*Fm*3*m*	*B*1	NaCl
CaO_2	66.7	*tI*6	*I*4/*mmm*	$C11_a$	CaC_2
CaO_4	80	...	...	...	...
CaO_6	83.3	...	...	...	...

Because the phase equilibria at higher O concentrations are not known, the assessed diagram, based primarily on the work of [59Bev], is restricted to the range 0 to 50 at.% O. In this composition range at high temperatures, two immiscible liquids (Ca-rich L_1 and O-rich L_2) may occur.

The solid solutions (αCa) and (βCa) saturate with respect to CaO at very small O concentrations, which have not been determined. The boundaries of the (αCa) and (βCa) fields corresponding to coexistence of these phases, except where they terminate at 443 °C in pure Ca, and the (βCa) solidus are also undetermined. The nature of the (αCa) + (βCa) + CaO equilibrium reaction is unknown.

For Ca-rich L_1 in equilibrium with (βCa), the liquidus extends from the melting point of pure βBa at 842 °C to the eutectic point at 839 °C and 0.3 at.% O. For L_1 in equilibrium with CaO, no experimental data between the eutectic point and 1350 °C or above 1350 °C have been reported. [59Bev] located this liquidus at about 17 at.% O at 1350 °C and speculated that the liquidus might terminate either at the congruent melting point of CaO or at a monotectic reaction where the Ca-rich L_1 is in equilibrium with CaO and O-rich L_2 containing less than 50 at.% O. This question is unresolved. There are no data on the compositions of L_2 in equilibrium with L_1 or with CaO at temperatures above this speculative monotectic point.

The CaO phase is stable at O_2 fugacities intermediate between those at which it is in equilibrium with (αCa), (βCa), or L_1 (or possibly L_2) and those at which it is in equilibrium with CaO_2 or liquid with more than 50 at.% O. At 0.1 MPa hydrostatic pressure, only the fcc form of CaO has been observed.

Ca-O Three-Phase Equilibria and Pure Component Transformations (Condensed System)

Reaction	Composition, at.% O			Temperature, °C	Reaction type
L ⇄ (βCa) + CaO	0.3	~0	50	~839	Eutectic
L_1 ⇄ L_2 + CaO	Unknown	Unknown	50	Unknown	Possible monotectic
(βCa) or (αCa) ⇄ (αCa) or (βCa) + CaO	~0	~0	50	~443	Eutectoid or peritectoid
(βCa) ⇄ (αCa)		0		443 ± 3	Allotropic
L ⇄ (βCa)		0		842 ± 3	Melting
L ⇄ CaO		50		2613 ± 25	Congruent melting

In examination of the Ca-rich boundary of CaO, [59Bev] and [66Bev] detected no measurable deviation from the stoichiometric composition at temperatures up to 1400 °C; the O-rich boundary of this phase has not been investigated. Pure CaO melts congruently at 2613 °C.

A metastable form of CaO was reported in a layer at the interface of stable CaO and decomposing calcite [76Sea].

When it is Ca-rich, the peroxide CaO_2 is in equilibrium with CaO and, when O-rich, probably with CaO_4. The CaO-CaO_2 equilibrium is difficult to establish and the latter equilibrium has not been realized. The range of deviations from the stoichiometric composition in CaO_2 is unknown; the melting point is also undetermined.

The superoxide CaO_4 is probably a stable phase of the condensed system, that is, stable relative to a mixture of CaO_2 and CaO_6. Pure CaO_4 has never been prepared, so that the structure, melting point, and range of compositions are undetermined. Theoretical examination of CaO_4 has suggested that it may have a rutile-type lattice [79Sad].

The ozonide CaO_6, which is unstable at ambient conditions, is the highest known oxide of Ca. Its structure, melting point, and range of compositions are unknown. Its equilibrium coexistence with CaO_4 has not been observed.

59Bev: D.J.M. Bevan and F.D. Richardson, *Proc. Aust. At. Energy Symp. on the Peaceful Uses of Atomic Energy 1958*, Melbourne Univ. Press, 586-587 (1959).
66Bev: D.J.M. Bevan, F.J. Lincoln, and F.D. Richardson, *Aust. J. Chem., 19*, 725-739 (1966).
76Sea: A.W. Searcy and D. Beruto, *J. Phys. Chem., 80*(4), 425-429 (1976).
79Sad: P. Sadhukhan and A.T. Bell, *J. Solid State Chem., 29*, 97-100 (1979).

Complete evaluation contains 1 figure, 3 tables, and 77 references.

Cr-O (Chromium-Oxygen)

51.996 15.9994

Cr_2O_3: $D5_1$, T_c (antiferromagnetic) = 40 °C. Cr_3O_4: Mn_3O_4.

The Cr-O phase diagram is redrawn from [Hansen].

Hf-O (Hafnium-Oxygen)

178.49 15.9994

HfO_2: cubic = $C1$, monoclinic = $mP12$.

The Hf-O phase diagram is redrawn from [Hafnium].

Mg-O (Magnesium-Oxygen)

24.305 15.9994

The established equilibrium solid phases of the Mg-O system are (1) the terminal cph solid solution (Mg); (2) the fcc oxide MgO; and (3) the cubic peroxide MgO_2. At high O_2 fugacity and subambient temperatures, a rhombohedral superoxide MgO_4 was reported. It is possibly a stable phase. The suboxide Mg_2O reported by [66Pet] is not an equilibrium phase.

No diagram for this system has been published. Because the phase equilibria at higher O concentrations are not known, the schematic assessed diagram (which was constructed from thermodynamic considerations, melting points, and continuity requirements) is for the range 0 to 50 at.% O only.

The only stable crystal structure of (Mg) is cph. The solid solution (Mg) saturates with respect to MgO at very small O concentrations, which have not been determined. The (Mg) solidus is also undetermined.

The liquidus compositions along the branch from the 650 °C melting point of pure Mg to the (Mg) + L + MgO equilibrium and those along the branch for MgO saturation are undetermined. Even the nature of this three-phase equilibrium reaction is unknown.

It is conceivable that, under high hydrostatic pressure at high temperature, a monotectic equilibrium L_1 + L_2 + MgO might exist. This region of the diagram is, however, probably inaccessible to experiment.

Mg-O Three-Phase Equilibria and Other Transformations (Condensed System)

Reaction	Composition, at.% O			Temperature, °C	Reaction type
L + (Mg) + MgO	~0	~0	~50	~650	Probably eutectic or peritectic
L ⇄ Mg		0		650 ± 1	Melting
L ⇄ MgO		50		2827 ± 30	Congruent melting

Mg-O Crystal Structure Data

Phase	Composition, at.% O	Pearson symbol	Space group	Strukturbericht designation	Prototype
(Mg)	0	*hP*2	$P6_3/mmc$	*A*3	Mg
MgO	50	*cF*8	*Fm*3*m*	*B*1	NaCl
MgO_2	66.7	*cP*12	*Pa*3	*C*2	FeS_2(pyrite)
MgO_4	80	(a)	...	...	...
Other possible phase					
Mg_2O	33.3	(b)	...	...	...

(a) Rhombohedral or hexagonal. (b) Cubic?

The only stable crystal structure of MgO is fcc. It melts congruently at 2827 °C. Its range of compositions is very narrow and the deviations from stoichiometric MgO are undetermined. At its Mg-rich limit, MgO is in equilibrium with (Mg) or L with less than 50 at.% O. At lower temperatures, MgO at its O-rich limit is in equilibrium with MgO_2; at higher temperatures, the conjugate phase is L with more than 50 at.% O.

The peroxide MgO_2 is stable in a range of O_2 fugacities greater than those at which MgO is stable. Thus, MgO_2 is in equilibrium with MgO at its Mg-rich limit and probably with MgO_4 at its O-rich limit. The latter equilibrium has not been realized.

The superoxide MgO_4 is probably a stable compound at O_2 fugacities greater than those at which MgO_2 is stable. The MgO_4 that has been prepared is unstable at atmospheric pressure and temperatures above about −30 °C.

66Pet: F. Petrů, V. Brožek, and B. Hájek, *Collect. Czech. Chem. Commun., 31,* 921-927 (1966) in German.

Complete evaluation contains 1 figure, 2 tables, and 76 references.

Mo-N (Molybdenum-Nitrogen)

95.94 14.0067

Mo-N Crystal Structure Data

Phase	Pearson symbol	Space group	Strukturbericht designation	Prototype
(Mo)	*cI*2	*Im*3*m*	*A*2	W
~$Mo_{16}N_7$	*tI*11 or 12	$I4_1/amd$	...	...
~Mo_2N (γ)	*cF*	...	...	...
δMoN	*hP*16	$P6_3/mmc$	...	...
δ'MoN	*hP*16	$P\bar{3}m1$	...	...
δ"MoN	*hP*16	$P\bar{3}m1$	...	...

The assessed phase diagram for the Mo-N system characterizes the phase behavior at a pressure of 850 atm. The present diagram is based primarily on the work of [70Ett] and was obtained by thermodynamic modeling.

The low-temperature tetragonal βMo_2N and the high-temperature cubic γMo_2N phases are closely related and become indistinguishable at high N content. A miscibility gap occurs between the β and γ phases for substoichiometric compositions from 850 °C down to a critical point around 400 °C and 34.5 at.% N, which has not been accurately fixed because of the slow rate of equilibration.

The Mo-rich boundary is reported to extend from 28.7 at.% N at 1100 °C to 27 at.% N in equilibrium with the eutectic liquid of 19 at.% N [78Jeh]. The thermodynamic calculations of the present evaluators indicate the eutectic to occur at 1894 ± 25 °C in equilibrium with 563 ± 90 atm of N_2 gas and liquid with $x_N = 0.20 \pm 0.01$. The Mo solidus at the eutectic is calculated to be $x_N = 0.011 \pm 0.002$.

The liquid boundary saturated by gas in the present diagram is given for N gas at a pressure of 850 atm, with the fugacity varying from 980 atm at 2027 °C to 950 atm at 2527 °C. Thermodynamic calculations indicate that MoN is unstable at all temperatures, with a N pressure of 850 atm. Consequently, MoN is not shown in the assessed diagram. To produce MoN, much higher N pressures are required, or ammonia gas can be used, which is thermodynamically quite unstable at higher temperatures.

δMoN, as well as metastable portions of the Mo_2N range up to $MoN_{0.68}$, can be produced by the reaction of ammonia gas with H_2MoO_4 at 700 to 800 °C [75Bli]. δMoN can also be prepared by ammonia with Mo metal at 700 to 1000 °C [78Jeh]. No data are available on the N-rich boundary of Mo_2N in equilibrium with N gas, and the boundary is arbitrarily fixed at 35 at.%.

70Ett: P. Ettmayer, *Monatsh. Chem., 101*, 127-140 (1970).

75Bli: G. Bliznakov, B. Piperov, and I. Tsolovski, *Izv. Khim., 8*(4), 614-620 (1975).

78Jeh: H. Jehn and P. Ettmayer, *High Temp.-High Pressures, 8*, 83-94 (1976); *J. Less-Common Met., 58*, 85-98 (1978).

Complete evaluation contains 1 figure, 2 tables, and 13 references.

N-Nb (Nitrogen-Niobium)

14.0067 92.9064

Nb_2N: NV_2. NbN(HT): *B*1. NbN(LT): B_i.

The Nb-N phase diagram is redrawn from [Elliott].

N-Ta (Nitrogen-Tantalum)

14.0067 180.9479

Ta_2N: fcc films are superconducting. TaN: *B*1.

The Ta-N phase diagram is redrawn from [Elliott].

N-Th (Nitrogen-Thorium)

14.0067 232.0381

ThN: *B*2. Th_3N_4: *mC*4 or *o**18 or $D7_1$.

The Th-N phase diagram is redrawn from [Smith].

N-Ti (Nitrogen-Titanium)

14.0067 47.88

The equilibrium solid phases of the Ti-N system are: (1) the terminal cph solid solution, (αTi); (2) the terminal bcc solid solution, (βTi); (3) the tetragonal nitride, Ti_2N; (4) the fcc nitride, TiN; and (5) the bct nitride, δ'.

The solid solutions (αTi) and (βTi) each have wide ranges of composition. Dissolved N extends the stable range of (αTi) above the temperature of the $\alpha Ti \rightleftarrows \beta Ti$ transformation (882 °C) to a peak at the 2350 °C L + (αTi) + TiN peritectic, considerably above the 2020 °C L + (βTi) + (αTi) peritectic and the melting piont of βTi (1670 °C). The (αTi) field is bounded on its Ti-rich side from 882 to 2020 °C by compositions in equilibrium with (βTi) and from 2020 to 2350 °C by its solidus. On its N-rich side, the (αTi) range is bounded by compositions in equilibrium with Ti_2N below 1050 °C and by those in equilibrium with TiN from 1050 to the 2350 °C peritectic equilibrium. The phase boundaries depicted for (αTi) and also those (βTi) in the assessed phase diagram are based primarily on the work of [54Pal].

The Ti-rich boundaries of (αTi) intersect at the (αTi) + TiN + Ti_2N equilibrium, which is depicted in the assessed diagram as being eutectic at 1050 °C. Controversy has existed not only over the locations of these boundaries, but also over the temperature and type of the three-phase equilibrium. The eutectoid form was proposed by [65Mcc].

Dissolved N also extends the range of (βTi) above the melting point of βTi to a peak at the 2020 °C peritectic. The (βTi) field is bounded on its Ti-rich side, from 1670 to 2020 °C, by its solidus and on its N-rich side, from 882 to 2020 °C, by compositions in equilibrium with (αTi).

Two of the three stable nitrides exhibit quite restricted phase fields. Thus, Ti_2N is stable over a relatively narrow range of compositions [54Pal, 65Mcc, 77Arb] and transforms congruently at 1100 °C to TiN [65Mcc]. The boundaries of the Ti_2N field on the Ti-rich side are at compositions in equilibrium with (αTi) below the 1050 °C eutectoid; other boundaries, Ti-rich or N-rich, are in equilibrium with TiN. The nitride δ', also with a narrow range of compositions, decomposes in a peritectoid reaction at 800 °C [77Arb]. Below this temperature, it is in equilibrium with Ti_2N or with TiN on its Ti-rich and N-rich sides, respectively.

The third nitride, TiN, exists stably over an extensive range of compositions, which includes the stoichiometric compound, and of temperatures up to a congruent melting temperature of ~3290 °C [76Ero]. The N-rich boundary above 50 at.% N is undetermined, but on the Ti-rich side, the phase field is bounded by the solidus above the 2350 °C peritectic equilibrium and by compositions in equilibrium with (αTi) or with δ' between the 2350 °C peritectic and 1050 °C eutectoid and below the 800 °C peritectoid, respectively. The boundary for equilibrium with Ti_2N, which runs from the 1050 °C eutectoid to the 800 °C peritectoid, exhibits a maximum at the congruent polymorphic transformation of Ti_2N (1100 °C).

Except for points at pure βTi (1670 °C), at the 2020 and 2350 °C peritectic equilibria and at the congruent melting of TiN (3290 °C), compositions along the liquidus segments are largely undetermined.

A metastable tetragonal phase, α', with ordered N atoms precipitated from (βTi) when it was transformed by quenching and aging [83Sun]. Elevated hydrostatic pressure transforms (αTi) stably to a hexagonal phase, ω, which persists metastably after restoration of ambient conditions [63Jam].

Ti-N Crystal Structure Data

Phase	Composition, at.% N	Pearson symbol	Space group	Strukturbericht designation	Prototype
Stable phases					
(αTi)	0 to 23	*hP*2	$P6_3/mmc$	*A*3	Mg
(βTi)	0 to 6	*cI*2	$Im3m$	*A*2	W
Ti_2N	~33	*tP*6	$P4_2/mnm$	*C*4	anti-O_2Ti (rutile)
TiN	28 to >50	*cF*8	$Fm3m$	*B*1	NaCl
δ'	~38	*tI*12	$I4_1/amd$	C_c	Si_2Th
ω	~0	(a)			
Other phases					
α'	...	*tP*6	$P4_2/mnm$	*C*4	anti-O_2Ti (rutile)

(a) Hexagonal.

Ti-N Invariant Points

Reaction(a)	Composition, at.% N			Temperature, °C	Reaction type
L + (αTi) $\rightleftarrows$ (βTi)	4.0	12.5	6.2	2020 ± 25	Peritectic
L + TiN $\rightleftarrows$ (αTi)	15.2	28	20.5	2350 ± 25	Peritectic
(αTi) + TiN + Ti_2N	23	30	33	1050 ± 60	Probably eutectoid (possibly peritectoid)
Ti_2N + δ' + TiN	34	37.5	39	800 ± 100	Probably peritectoid
Congruent transformations—pure components					
αTi $\rightleftarrows$ βTi		0		883 ± 3	Polymorphic
βTi $\rightleftarrows$ L		0		1670 ± 6	Fusion
Congruent transformations—other					
TiN $\rightleftarrows$ L(b)		47.4		~3290	Fusion
Ti_2N $\rightleftarrows$ TiN(c)		33.3		~1100	Polymorphic

(a) Stable equilibria involving ω are possible at elevated hydrostatic pressure. (b) Only actual observation under pressures greater than ~1 MPa. (c) Occurrence if (αTi) + TiN + Ti_2N equilibrium is eutectoid.

54Pal: A.E. Palty, H. Margolin, and J.P. Nielsen, *Trans. ASM, 46,* 312-328 (1954).

63Jam: J.C. Jamieson, *Science, 140,* 72-73 (1963).

65Mcc: L.A. McClaine and C.P. Coppel, U.S. Air Force Systems Command, Res. Technol. Div., Tech. Rep. AFML-TR-65-299 (1965) (AD-474087).

76Ero: M.A. Eron'yan, R.G. Avarbé, and T.N. Danisina, *Teplofiz. Vys. Temp., 14*(2), 398-399 (1976) in Russian; TR: *High Temp., 14*(2), 359-360 (1976).

77Arb: M.P. Arbuzov, S.Ya. Golub, and B.V. Khaenko, *Izv. Akad. Nauk SSSR, Neorg. Mater., 13*(10), 1779-1789 (1977) in Russian; TR: *Inorg. Mater., 13*(10), 1434-1437 (1977).

83Sun: D. Sundararaman, A.L.E. Terrance, V. Seetharaman, and V.S. Raghunathan, *Trans. Jpn. Inst. Met., 24*(7), 510-513 (1983).

Complete evaluation contains 5 figures, 6 tables, and 56 references.

N-U (Nitrogen-Uranium)

14.0067 238.0289

UN: *B*1. αU_2N_3: $D5_2$. βU_2N_3: $D5_3$.
The U-N phase diagram is redrawn from [Metals].

N-V (Nitrogen-Vanadium)

14.0067 50.9415

In spite of extensive study of the V-N system, much uncertainty remains over the number and stability of the phases that occur. Two intermediate phases—fcc δVN_{1-x} and cph βV_2N_{1-y} with wide ranges of homogeneity have been well established. An ordered phase with a stoichiometry of $V_{32}N_{26}$ has also been reported. There appears to be substantial evidence for its existence, and it appears on the assessed phase diagram as $\delta' VN_{1-x}$. Other intermediate phases with the stoichiometries of $V_{16}N$, $V_{13}N$, V_9N, V_8N, V_9N_2, and V_4N have also been reported, but it is believed that these phases are metastable and therefore are not shown in the present diagram. The assessed diagram is based on evaluation of the work of [49Hah], [64Bra], [72Koz], [74Ett], [74Hor], [75Arb], [78Kha], and [78Ono].

The data of [72Koz] extrapolated to lower temperatures were used to construct the V-rich boundary. The N-rich boundary is placed at 50 at.% N (VN), in accord with [64Bra], [75Arb], and [78Ono].

A minimum occurs in the solidus at 3 to 4 at.% N. The (V) solvus is described by log C (at.% N) = 1.50 − 831/*T* (500 to 1500 °C) [74Hor].

V-N Crystal Structure Data

Phase	Composition, at.% N	Pearson symbol	Space group	Strukturbericht designation	Prototype
(V)	0 to 7	*cI*2	*Im*3*m*	*A*2	W
βV_2N_{1-y}	29 to 31	*hP*9	*P*$\bar{3}$1*m*	...	...
δVN_{1-x}	~33 to 50	*cF*8	*Fm*3*m*	*B*1	NaCl
$\delta' VN_{1-x}$	~43 to 46	(a)	$P4_2/nmc$	...	...

(a) Tetragonal, pseudocubic.

The existence of a hexagonal subnitride phase is well established, but there is still some uncertainty over its exact stoichiometry or composition range.

A break in the solvus at 550 °C is associated with the formation of a metastable phase, $V_{16}N$. Other metastable phases have been identified, but their exact stoichiometries and structures have not been determined definitively.

49Hah: H. Hahn, *Z. Anorg. Chem., 258,* 58-68 (1949).

64Bra: G. Brauer and W.D. Schnell, *J. Less-Common Met., 6,* 326-332 (1964).

72Koz: V.A. Kozheurov, V.M. Zhikharev, V.I. Shishkov, and G.V. Gritshina, *Izv. V.U.Z., Chernaya Metall.,* (8), 10-13 (1972).

74Ett: P. Ettmayer, R. Kieffer, and F. Hattinger, *Metall, 28,* 1151-1155 (1974).

74Hor: G. Horz, *J. Less-Common Met., 35,* 207-225 (1974).

75Arb: M.P. Arbuzov, B.V. Khaenko, and O.A. Frenkel, *Inorg. Mater., 11,* 236-241 (1975).

78Kha: B.V. Khaenko and V.G. Fak, *Inorg. Mater., 14,* 1011-1016 (1978).

78Ono: T. Onozuka, *J. Appl. Crystallogr., 11,* 132-136 (1978).

Complete evaluation contains 5 figures, 4 tables, and 59 references.

N-Zr (Nitrogen-Zirconium)

14.0067 91.22

ZrN: *B*1.
The Zr-N phase diagram is redrawn from [Zirconium].

Nb-O (Niobium-Oxygen)

92.9064 15.9994

NbO: *cP*6. NbO_2: *C*4 or *tI*96. Nb_2O_5: *mP*99, *mP*98, or *mC*14.

The Nb-O phase diagram is redrawn from [Elliott] and [Shunk].

O-Pu (Oxygen-Plutonium)

15.9994 (244)

αPu_2O_3: $D5_3$. βPu_2O_3: $D5_2$. $\alpha' Pu_2O_3$: defect-*C*1. PuO_2: *C*1.

The Pu-O phase diagram is redrawn from [Plutonium].

O-Si (Oxygen-Silicon)

15.9994 28.0855

The Si-O phase diagram is redrawn from [Shunk].

O-Ta (Oxygen-Tantalum)

15.9994 180.9479

Ta_2O_5: incommensurate structures based on an orthorhombic subcell (below ~1450 °C), and a monoclinic subcell (above ~1450 °C).

The Ta-O phase diagram is redrawn from [72Jeh].

72Jeh: H. Jehn and E. Olzi, *J. Less-Common Met., 27,* 297-309 (1972).

O-Th (Oxygen-Thorium)

15.9994 232.0381

ThO_2: *C*1.
The Th-O phase diagram is redrawn from [Smith].

O-Ti (Oxygen-Titanium)

15.994 47.88

Special Points of the Ti-O Phase Diagram

Phases	Composition, at.% O			Temperature, °C	Reaction type
$L + (\alpha Ti) \rightleftarrows (\beta Ti)$	5	13	8	1720 ± 25	Peritectic
$L \rightleftarrows (\alpha Ti)$		~24		1885 ± 25	Congruent
$(\alpha Ti) + Ti_3O_2 \rightleftarrows Ti_2O$	33.3	40	33.9	~600	Peritectoid
$(\alpha Ti) + Ti_2O \rightleftarrows Ti_3O$	~17	~25	~24.5	~500	Peritectoid
$L \rightleftarrows (\alpha Ti) + L$	~37	~31	~53	~1800	Monotectic (?)
$L + (\alpha Ti) \rightleftarrows \gamma TiO$	~55	31.4	34.5	1770	Peritectic
$\gamma TiO \rightleftarrows \beta TiO$		···		~1250	Unknown
$\beta TiO \rightleftarrows \beta Ti_{1-x}O$		···		···	Unknown
$\beta Ti_{1-x}O \rightleftarrows \alpha Ti_{1-x}O$		···		···	Unknown
$(\alpha Ti) + \beta TiO \rightleftarrows \alpha TiO$	33.3	51	50	940	Peritectoid
$(\alpha Ti) + \alpha TiO \rightleftarrows Ti_3O_2$	32.4	50	40	920	Peritectoid
$\alpha Ti_{1-x}O \rightleftarrows \alpha TiO + \beta Ti_2O_3$	54.5	50	60	460	Eutectoid
$L \rightleftarrows \gamma TiO + \beta Ti_2O_3$	~57	54.5	59.8	1720	Eutectic
$L \rightleftarrows \beta Ti_2O_3$		60		1842	Congruent
$L + \beta Ti_2O_3 \rightleftarrows \beta Ti_3O_5$	63	60.2	62.5	1770	Peritectic
$\beta Ti_2O_3 \rightleftarrows \alpha Ti_2O_3$		60		~180	Unknown
$\beta Ti_3O_5 \rightleftarrows \alpha Ti_3O_5$		62.5		187	Unknown
$\gamma Ti_4O_7 \rightleftarrows \beta Ti_4O_7$		63.64		−123	Unknown
$\beta Ti_4O_7 \rightleftarrows \alpha Ti_4O_7$		63.64		−148	Unknown
$L \rightleftarrows \beta Ti_3O_5$ + ?	~64	62.5	···	~1670	Eutectic
$\beta Ti_3O_5 + \beta Ti_5O_9 \rightleftarrows \gamma Ti_4O_7$	62.5	64.29	63.64	~1500	Peritectoid
$L \rightleftarrows TiO_2$		66.7		1870	Congruent
$L \rightleftarrows (\beta Ti)$		0		1670	Melting point
$(\beta Ti) \rightleftarrows (\alpha Ti)$		0		882	Allotropic transformation

This assessment of the Ti-O system covers the phase equilibria and crystal structures of the condensed phases in the composition range between pure Ti and TiO_2. The thermodynamic properties of the titanium oxides have been studied and assessed extensively by [75Cha]. The temperature range in which one can construct a reasonable equilibrium diagram excludes some phase transitions of the higher oxides. Moreover, it is not possible to include all the observed higher oxide phases in a diagram. The reader is therefore referred to the crystal structure table for a complete listing of the phases.

Oxygen has a large solubility in the low-temperature cph (αTi) and stabilizes (αTi) with respect to the high-temperature bcc form, (βTi). At low-temperature, the ordered cph phases (Ti_2O, Ti_3O, and possibly Ti_6O) are formed. In the ordered hexagonal phases, oxygen resides in octahedral sites in layers alternating with Ti. Ti_2O has the anti-CdI_2 structure with alternate oxygen layers vacant and additional vacancies randomly distributed in the occupied layer [57And, 59Now, 69Yam, 70Yam].

Three equivalent sites are available in the oxygen plane, which are successively filled at Ti_6O, Ti_3O, and Ti_2O. The ordered cph phases have been shown in the literature either as discrete compounds, or as phases of wide homogeneity range produced from the disordered phase by second-order transitions. They have been shown as persisting to the melting point, or as disordering above about 600 °C. Based on crystal structure data, ordering on the oxygen sublattice occurs over a wide composition range, and it is erroneous to show these phases as discrete compounds. The evidence for persistence of the ordered phases to the melting point is indirect; direct observations of disordering at lower temperatures are preferred as the basis for the assessed diagram. The detailed placement of phase boundaries, however, is still uncertain.

Structures of the monoxides are based on the NaCl structure of the high-temperature γTiO form. Four additional structural modifications have been identified, which are here designated βTiO, αTiO, $\beta Ti_{1-x}O$, and $\alpha Ti_{1-x}O$. "TiO" will be used to refer to the monoxides without restriction to a particular variety. The phase boundaries separating these phases, except for the disordering of αTiO, have not been determined. The phase boundaries of the monoxides in equilibrium with (αTi) and with βTi_2O_3 have been determined, but without distinguishing the various monoxide modifications.

Five polymorphs of TiO_2 are known—anatase and brookite, which are low-temperature, low-pressure forms; TiO_2-II and TiO_2-III, which are formed from anatase or brookite under pressure; and rutile, the stable phase at all temperatures and ambient pressure. The polymorphic transformations anatase → rutile and brookite → rutile do not occur reversibly [66Vah, 68Dac]. This fact and the heat of transformation data [79Mit] show that anatase and brookite are not stable phases at any temperature.

Between the monoxides and TiO_2, there exists a series of discrete phases with stoichiometry Ti_nO_{2n-1}, where $n \geqq 2$. They are called Magneli phases. It has been suggested that there exist discrete equilibrium phases for $n \leqq 99$ [72Roy]. The phases Ti_nO_{2n-1} ($4 \leqq n \leqq 10$) have crystal structures derived from the rutile structure by crystallographic shear. Closely related structures have been observed that can be described as coherent intergrowths of Magneli phases [70And], or as families of phases based on different crystallographic shear operations [71Bur1, 71Bur2]. Magneli phases undergo one or more structural, electrical, or magnetic transitions at low temperature.

57And: S. Andersson, B. Collen, U. Kuylenstierna, and A. Magneli, *Acta Chem. Scand.*, *11*, 1641-1652 (1957).

59Now: H. Nowotny and E. Dimakopoulou, *Monatsh. Chem.*, *90*, 620-622 (1959) in German.

66Vah: F.W. Vahldiek, *J. Less-Common Met., 11,* 99-110 (1966).
68Dac: F. Dachille, P.Y. Simons, and R. Roy, *Am. Miner., 53,* 1929-1939 (1968).
69Yam: S. Yamaguchi, *J. Phys. Soc. Jpn., 27*(1), 155-163 (1969).
70And: J.S. Anderson and A.S. Khan, *J. Less-Common Met., 22,* 219-223 (1970).
70Yam: S. Yamaguchi, K. Hiraga, and M. Hirabayashi, *J. Phys. Soc. Jpn., 28*(4), 1014-1023 (1970).
71Bur1: L.A. Bursill and B.G. Hyde, *Acta Crystallogr., B27,* 210-215 (1971).
71Bur2: L.A. Bursill, B.G. Hyde, and D.K. Philip, *Philos. Mag., 23,* 1501-1513 (1971).
72Roy: R. Roy and W.B. White, *J. Crystal Growth, 13/14,* 78-83 (1972).
75Cha: M.W. Chase, J.L. Curnutt, H. Prophet, and R.A. McDonald, *J. Phys. Chem. Data,* (4) (1975).
79Mit: T. Mitsuhashi and O.J. Kleppa, *J. Am. Chem. Soc., 62*(7-8), 356-357 (1979).

Complete evaluation contains 8 figures, 9 tables, and 140 references.

Ti-O Crystal Structure Data

Phase	Composition, at.% O	Pearson symbol	Space group	Strukturbericht designation	Prototype
(βTi)	0 to 8	*cI*2	*Im*3*m*	*A*2	W
(αTi)	0 to 31.9	*hP*2	$P6_3/mmc$	*A*3	Mg
Ti_3O	~20 to ~30	*hP* ~ 16	$P\bar{3}1c$	...	...
Ti_2O	~25 to 33.4	*hP*3	$P\bar{3}m1$	...	Anti-CdI_2
γTiO	34.9 to 55.5	*cF*8	*Fm*3*m*	*B*1	ClNa
Ti_3O_2	~40	*hP* ~5	*P*6/*mmm*	...	...
βTiO	...	(a)	...	...	...
αTiO	~50	*mC*16	*A*2/*m* or *B**/*	...	...
β$Ti_{1-x}O$	~55.5	*oI*12	*I*222	...	...
α$Ti_{1-x}O$	~55.5	*tI*18	*I*4/*m*	...	...
βTi_2O_3	59.8 to 60.2	*hR*30	$R\bar{3}c$	$D5_1$	αAl_2O_3
αTi_2O_3	59.8 to 60.2	*hR*30	$R\bar{3}c$	$D5_1$	αAl_2O_3
βTi_3O_5	62.5	(b)	...	...	Anosovite
αTi_3O_5	62.5	*mC*32	*C*2/*m*	...	...
α'Ti_3O_5	(a)	*mC*32	*Cc*	...	V_3O_5
γTi_4O_7	63.6	*aP*44	$P\bar{1}$	...	...
βTi_4O_7	63.6	*aP*44	$P\bar{1}$	...	...
αTi_4O_7	63.6	*aP*44	$P\bar{1}$	...	...
γTi_5O_9	64.3	*aP*28	$P\bar{1}$	...	...
βTi_6O_{11}	64.7	*aC*68	$A\bar{1}$	...	...
Ti_7O_{13}	65.0	*aP*40	$P\bar{1}$	...	...
Ti_8O_{15}	65.2	*aC*92	$A\bar{1}$	...	...
Ti_9O_{17}	65.4	*aI*52	$P\bar{1}$	...	...
Rutile	~66.7	*tP*6	$P4_2/mnm$	*C*4	Rutile
Metastable phases					
Anatase	...	*tI*12	$I4_1/amd$	*C*5	Anatase
Brookite	...	*oP*24	*Pbca*	*C*21	Brookite
High-pressure phases					
TiO_2-II	...	*oP*12	*Pbcn*	...	αPbO_2
TiO_2-III	...	~*hP*48	(c)	...	...

(a) Cubic. (b) Monoclinic. (c) Hexagonal.

O-U (Oxygen-Uranium)

15.9994 238.0289

UO_2: *C*1. U_4O_9: *cI*832. U_3O_8: incommensurate structures based on an orthorhombic subcell; *m*** (HP). UO_3: *oC*36, *oF*128 (LT), *oP*16 (HP?).

The U-O phase diagram is redrawn from [Elliott].

O-V (Oxygen-Vanadium)

15.9994 50.9415

α'(V_8O?): V_8N or tetragonal? β(V_4O): *t***. γ(V_2O): *m***. VO: *B*1. V_2O_3: *D*51, *mI*20 at 148 K. V_3O_5: *mP*32, *mC*32 above 693 K. $V_{1.75}$(V_4O_7): *aP*22. $VO_{1.80}$(V_5O_9): *aP*28. $VO_{1.86}$(V_7O_{13}): *aP*42. V_6O_{13}: *mC*38, *mP*38. V_2O_5: *oP*14. VO_2: *mP*12 298 to 337 K, *C*4 above 330 K, *oP*12 below 340 K?

The V-O phase diagram is redrawn from [Metals] and [Elliott].

O-Y (Oxygen-Yttrium)

15.9994 88.9059

Y_2O_3: $D5_3$.

The Y-O phase diagram is redrawn from [Shunk].

O-Zr (Oxygen-Zirconium)

15.9994 91.22

The equilibrium phases of the Zr-O system at 1 atm are: (1) the liquid, L; (2) the bcc terminal solid solution, (βZr), with a maximum solubility of O in (βZr) of 10.5 at.% O at 1970 °C, which is the temperature of the peritectic reaction L + (αZr) ⇄ (βZr); (3) the cph terminal solid solution, (αZr), which is strongly stabilized by oxygen; (4) the nonstoichiometric compound ZrO_{2-x} (where $0 \le x \le 0.44$), which exists in three crystallographic forms—a high-temperature cubic CaF_2-type crystal structure, an intermediate-temperature tetragonal structure, and a low-temperature monoclinic structure; and (5) the gas phase, g (within the temperature range of the assessed diagram, g consists of pure oxygen).

Three-Phase Equilibria and Congruent Transformations in the Zr-O System Above 850 °C at 1 bar

Reaction	Composition, at.% O			Temperature, °C	Reaction type
L + (αZr) ⇄ (βZr)	10.0 ± 0.5	19.5 ± 2	10.5 ± 0.5	1970 ± 10	Peritectic
L ⇄ (αZr) + γZrO_{2-x}	40 ± 2	35 ± 1	62 ± 1	2065 ± 5	Eutectic
γZrO_{2-x} ⇄ (αZr) + βZrO_{2-x}	63.6 ± 0.4	31.2 ± 0.5	66.5 ± 0.1	~1525	Eutectoid
βZrO_{2-x} – (αZr) – αZrO_{2-x}	~66.5	29.8 ± 0.5	~66.6	~1205	?
L – γZrO_{2-x} – g	~66.6	~66.6	~100	~2710	?
γZrO_{2-x} – βZrO_{2-x} – g	~66.6	~66.6	~100	~2377	?
βZrO_{2-x} – αZrO_{2-x} – g	~66.6	~66.6	~100	~1205	?
L ⇄ (αZr)		25 ± 1		2130 ± 10	Congruent
L ⇄ γZrO_{2-x}		66.6		2710 ± 15	Congruent
γZrO_{2-x} ⇄ βZrO_{2-x}		66.6		~2377	Congruent
βZrO_{2-x} ⇄ αZrO_{2-x}		66.6		~1205	Congruent
L ⇄ (βZr)		0		1855	Melting point
(βZr) ⇄ (αZr)		0		863	Allotropic transformation

For compositions higher than 19.5 at.% O, (αZr) is stable up to the melting temperature and melts congruently at 25 at.% O and 2130 °C; the maximum solubility of O in (αZr) is 35 at.% O at 2065 °C, the temperature of the eutectic reaction L ⇄ (αZr) + ZrO_{2-x} (cubic). α'Zr and α''Zr are designated O-ordered variants of (αZr).

The assessed phase diagram is based primarily on the work of [77Ack], [78Ack], and [80Rau], with review of the work of [54Dom], [61Geb], [74Hir], [75Ack], and [76Ara]. All reported temperatures have been adjusted to IPTS-68. The melting point of Zr is 1855 °C [Melt].

In practice, the congruent tetragonal to monoclinic transformation in bulk stoichiometric ZrO_2 does not occur at a fixed temperature (≈1205 °C), but takes place over a range of temperatures. During transformation, monoclinic ZrO_2 coexists in metastable equilibrium with tetragonal ZrO_2. The proportion of each phase does not change with time as long as the temperature remains constant. This transformation in ZrO_2 is martensitic in nature.

Zr-O Crystal Structure Data

Phase	Composition, at.% O	Pearson symbol	Space group	Strukturbericht designation	Prototype
(αZr)	0 to 35	*hP*2	*P*6$_3$/*mmc*	*A*3	Mg
(βZr)	0 to 10.5	*cI*2	*Im*3*m*	*A*2	W
γZrO_{2-x}	61 to 66.6	*cF*12	*Fm*3*m*	*C*1	CaF_2
βZrO_{2-x}	66.5 to 66.6	*tP*6	*P*4$_2$/*nmc*	...	HgI_2
αZrO_{2-x}	66.6	*mP*12	*P*2$_1$/*c*	...	...

Under specific conditions, the high-temperature tetragonal and cubic ZrO_2 structures have been found to exist in metastable states at room temperature. Quasi-amorphous ZrO_2 also can be obtained by precipitation techniques. On heating and annealing below 1200 °C, these metastable states transform either directly or through a gradual transformation sequence into the stable monoclinic phase.

54Dom: R.F. Domagala and D.J. McPherson, *Trans. AIME*, *200*, 238-246 (1954).

61Geb: E. Gebhardt, H.D. Seghezzi, and W. Dürrschnabel, *J. Nucl. Mat.*, *4*, 255-268 (1961) in German.

74Hir: M. Hirabayushi, S. Yamaguchi, T. Arai, H. Asano, and S. Hashimoto, *Phys. Status Solidi (a)*, *23*, 331-339 (1974).

75Ack: R.J. Ackermann, E.G. Rauh, and C.A. Alexander, *High Temp. Sci.*, *7*, 304-316 (1975).

76Ara: T. Arai and M. Hirabayashi, *J. Less-Common Met.*, *44*, 291-300 (1976).

77Ack: R.J. Ackermann, S.P. Garg, and E.G. Rauh, *J. Am. Ceram. Soc.*, *60*, 341-345 (1977).

78Ack: R.J. Ackermann, S.P. Garg, and E.G. Rauh, *J. Am. Ceram. Soc.*, *61*, 275-276 (1978).

80Rau: E.G. Rauh and S.P. Garg, *J. Am. Ceram. Soc.*, *63*, 239-240 (1980).

Complete evaluation contains 10 figures, 3 tables, and 59 references.

Assessed Al-N Phase Diagram (Condensed System)

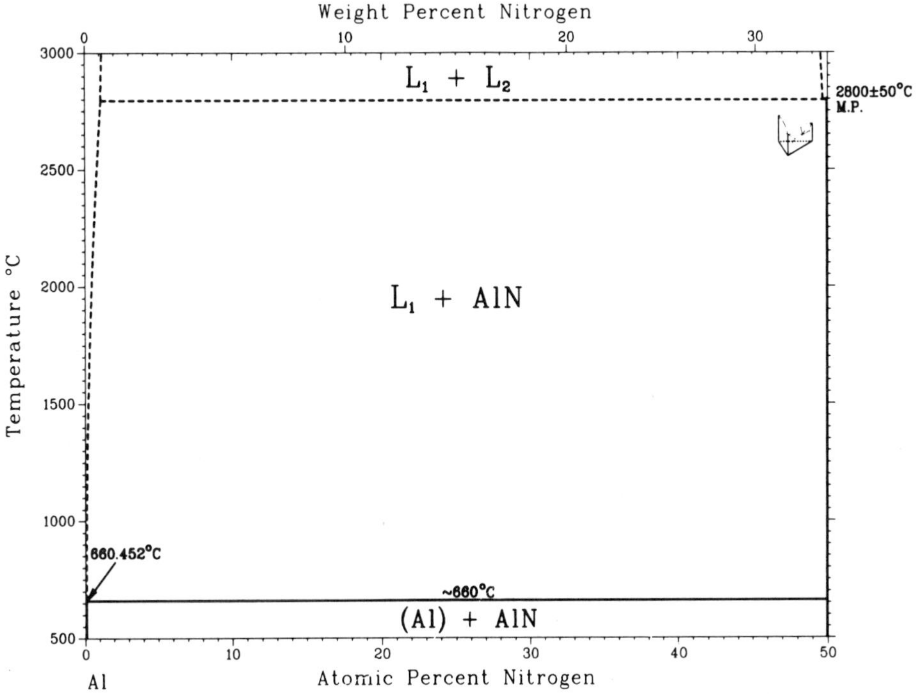

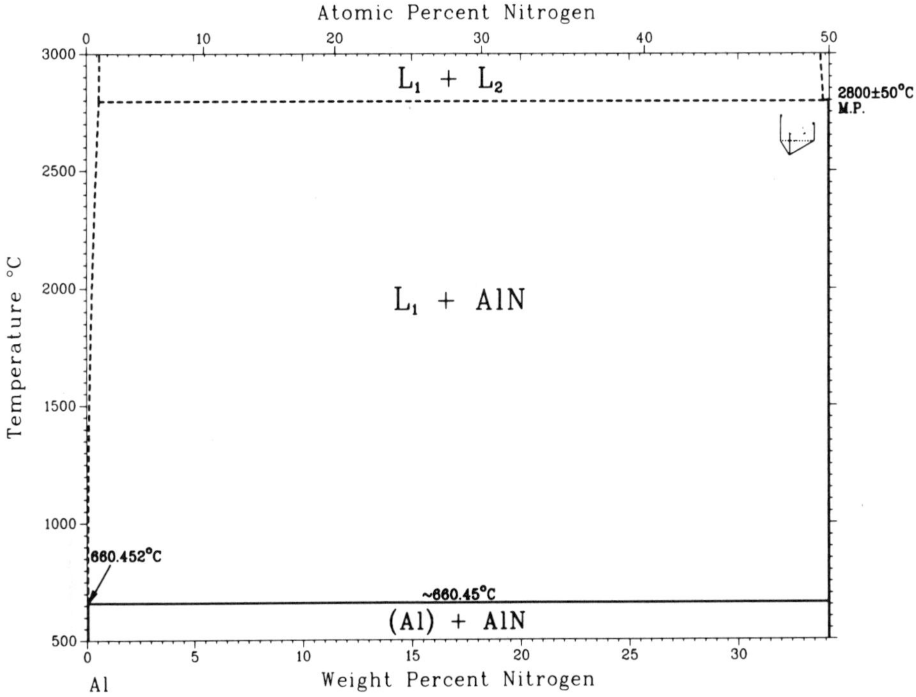

Assessed Al-O Phase Diagram (Condensed System)

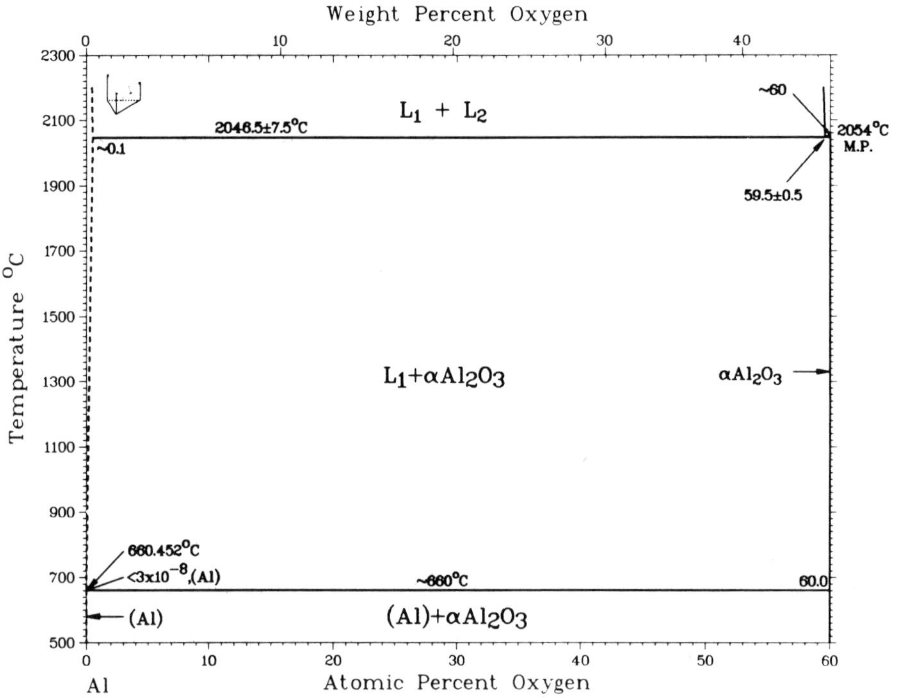

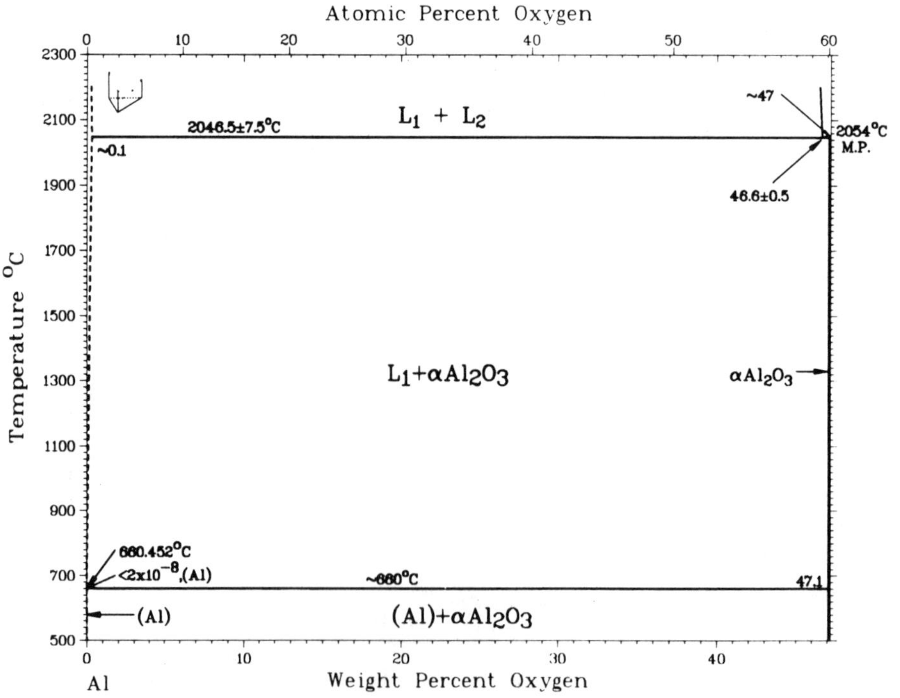

B-C Phase Diagram

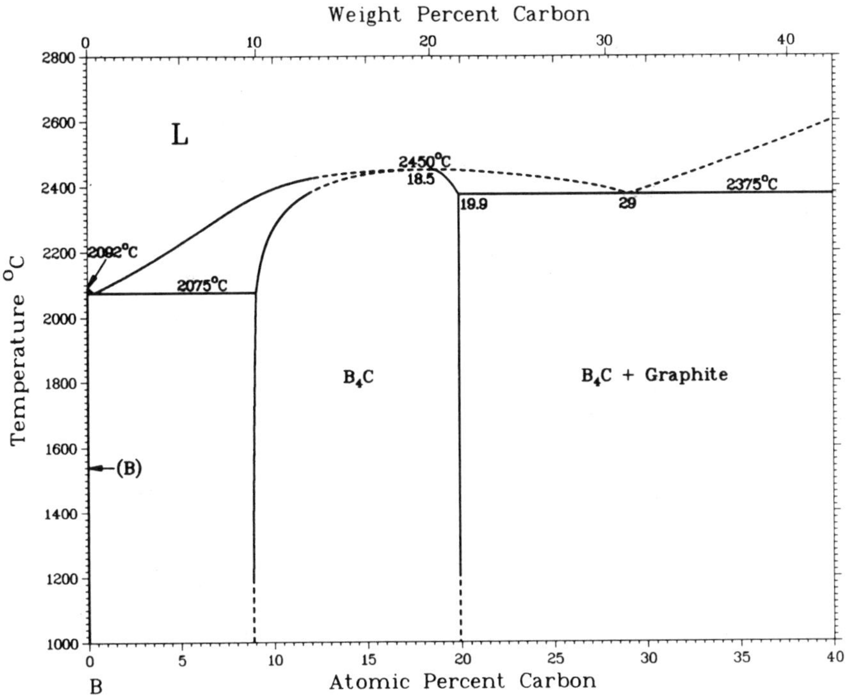

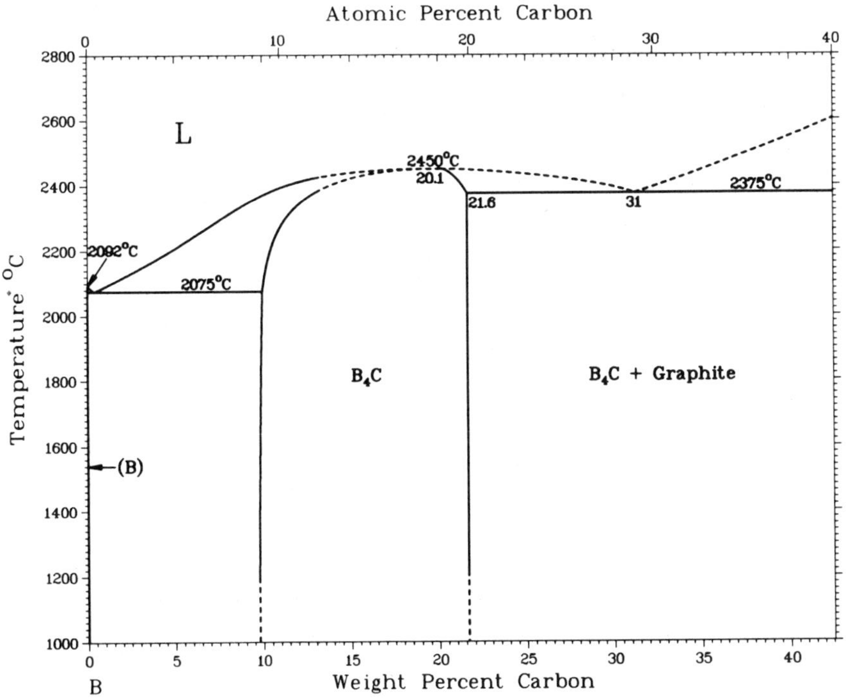

Assessed Cr-B Phase Diagram

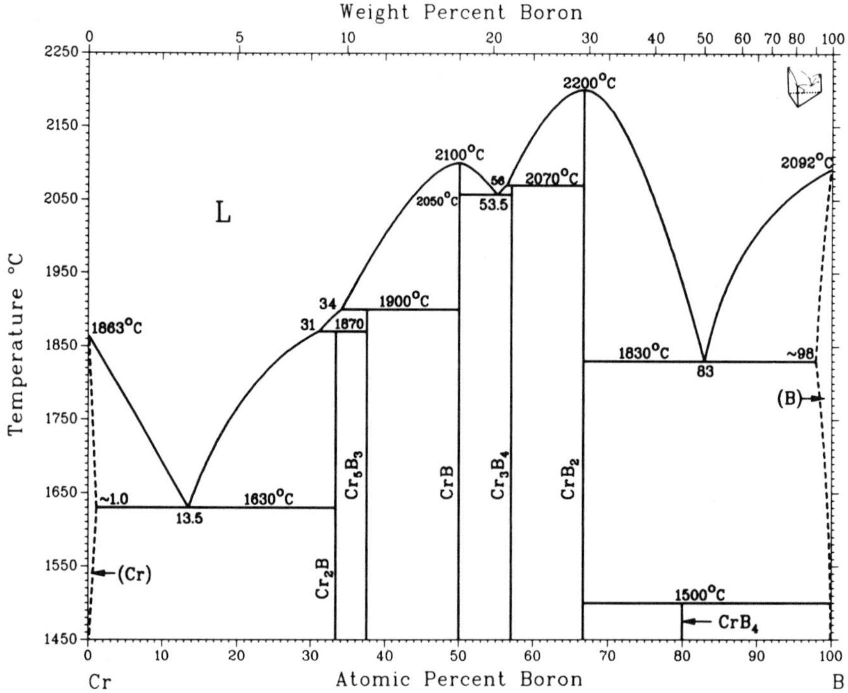

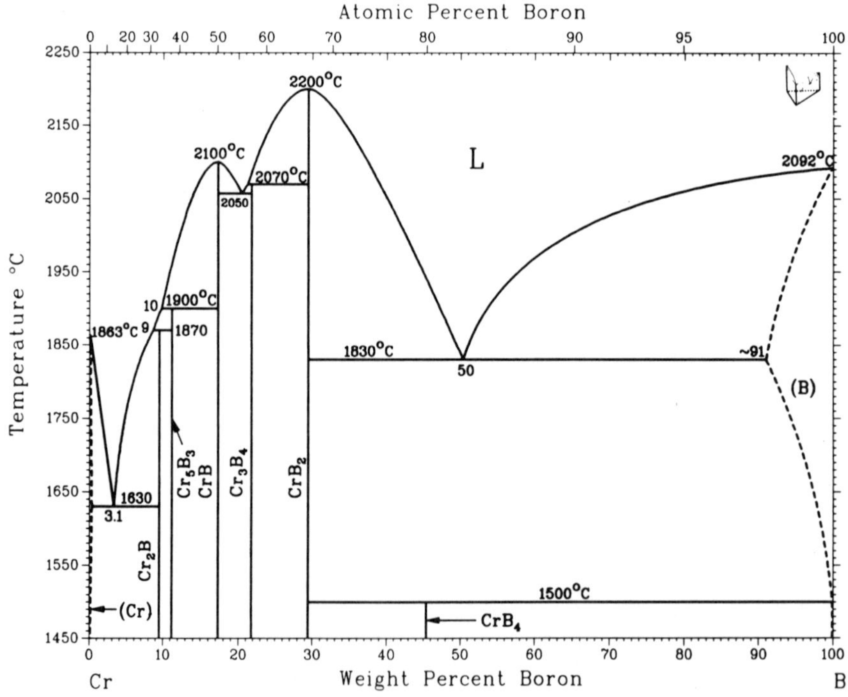

Hf-B Phase Diagram

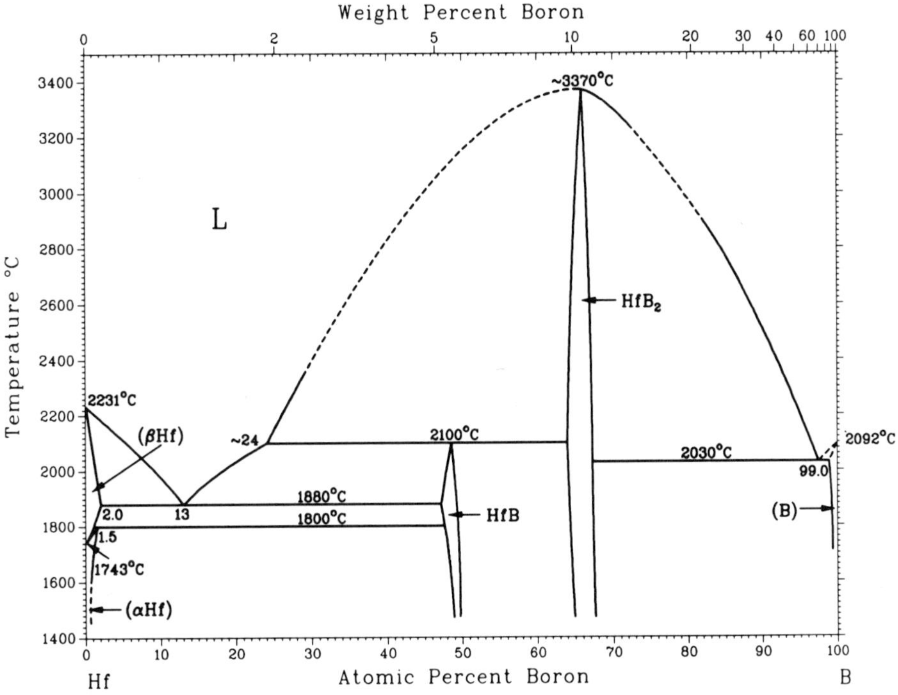

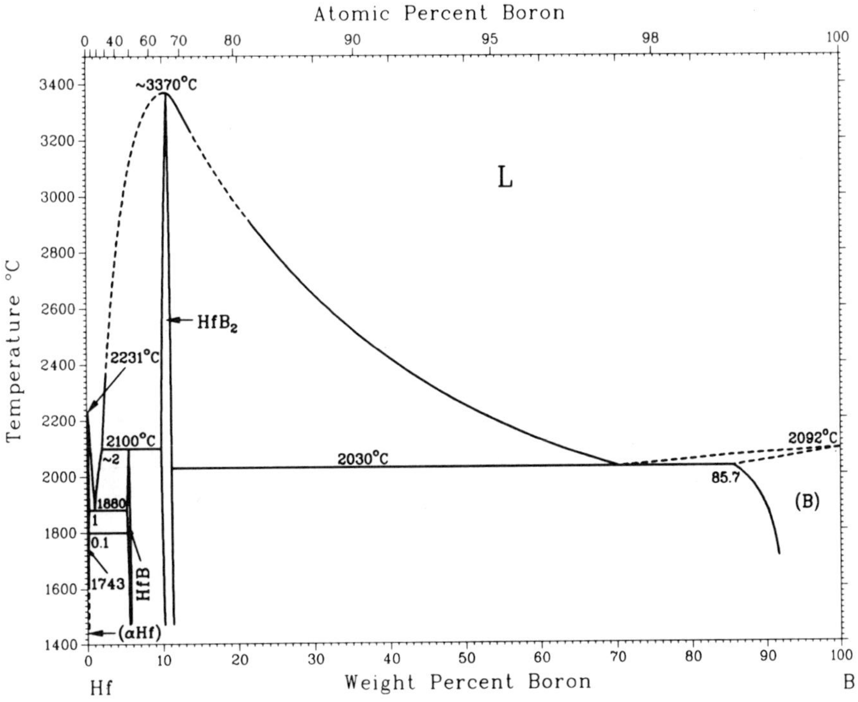

Mo-B Phase Diagram

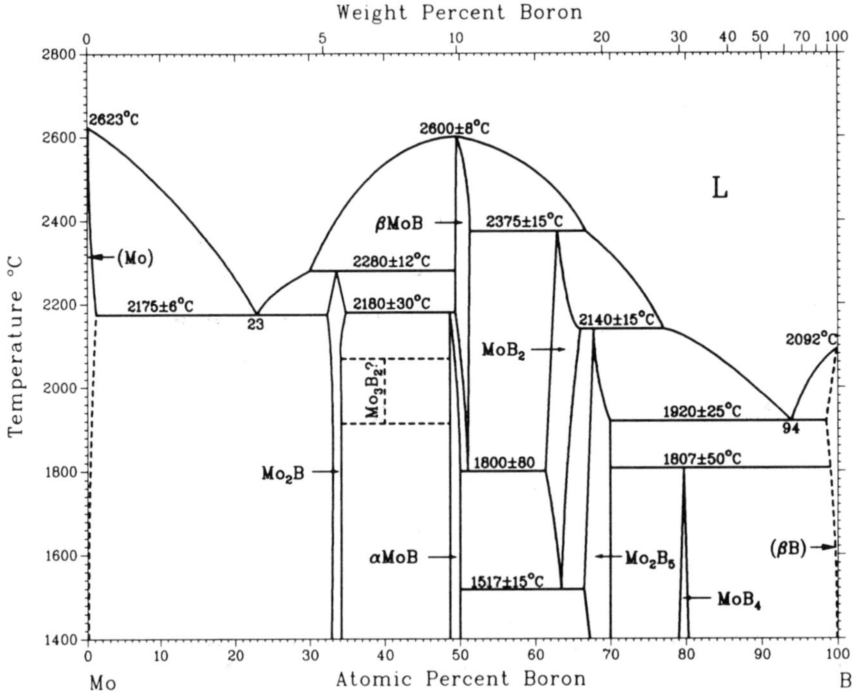

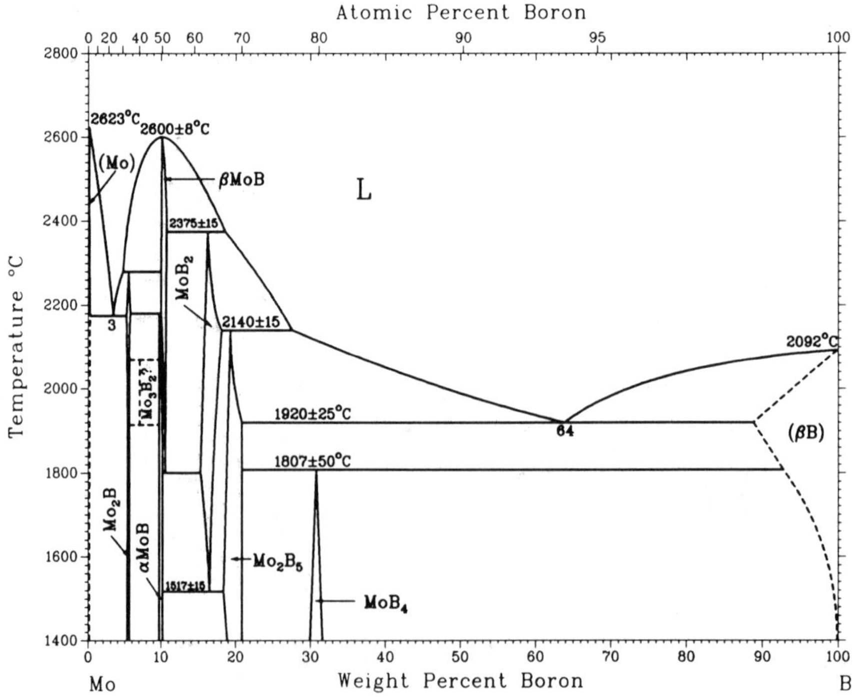

Nb-B Phase Diagram

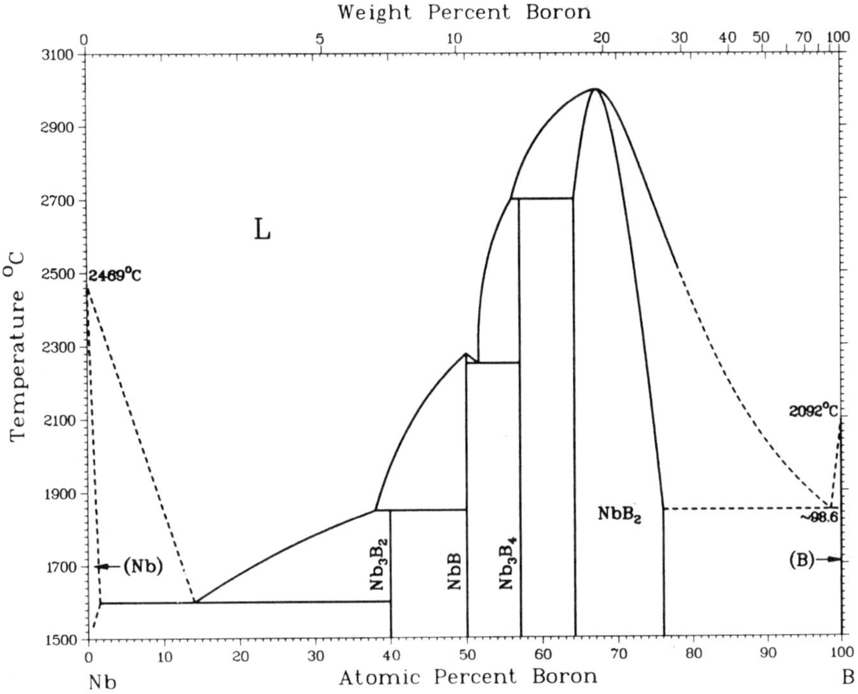

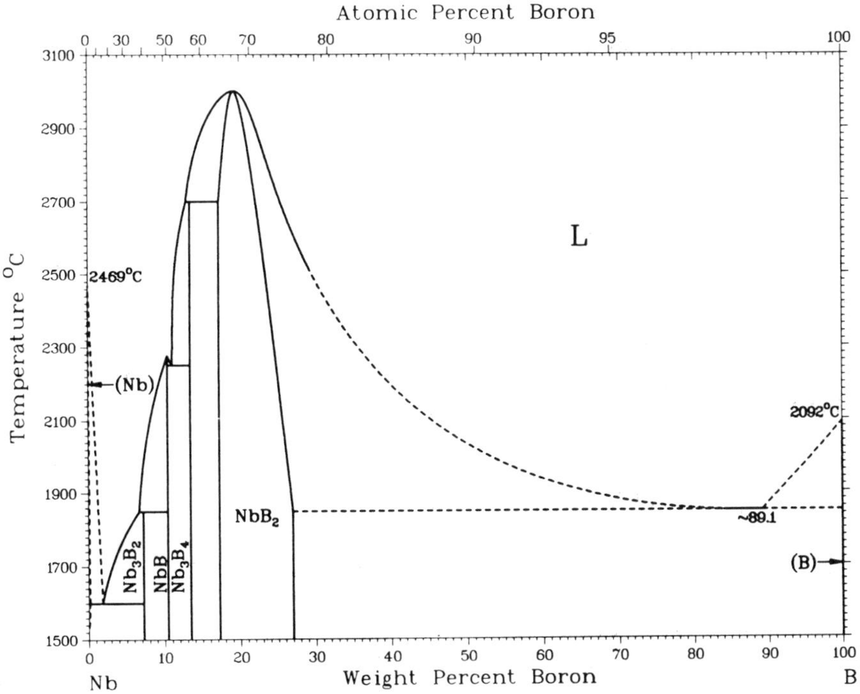

Pu-B Phase Diagram

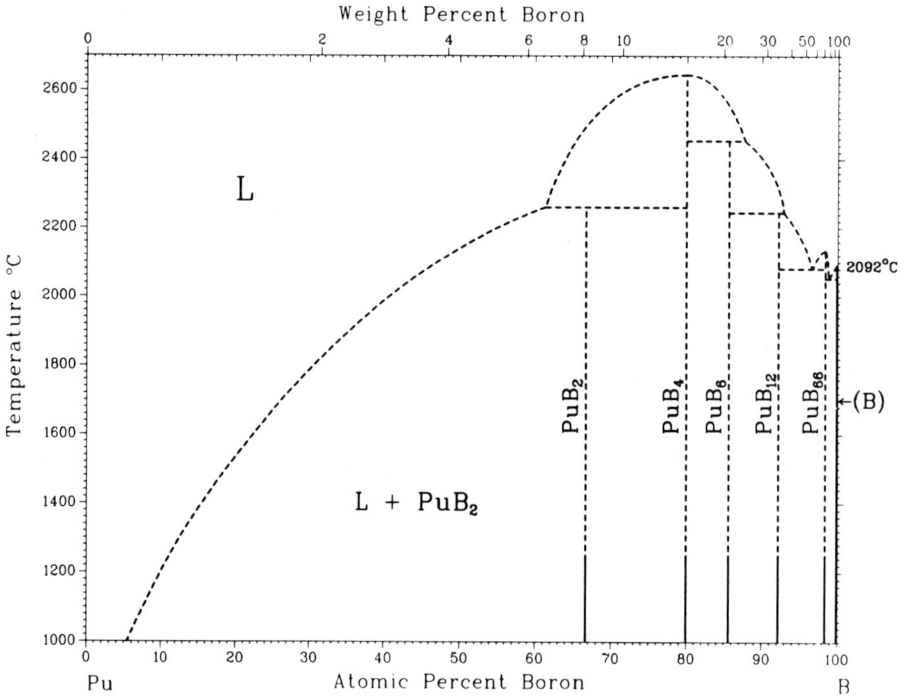

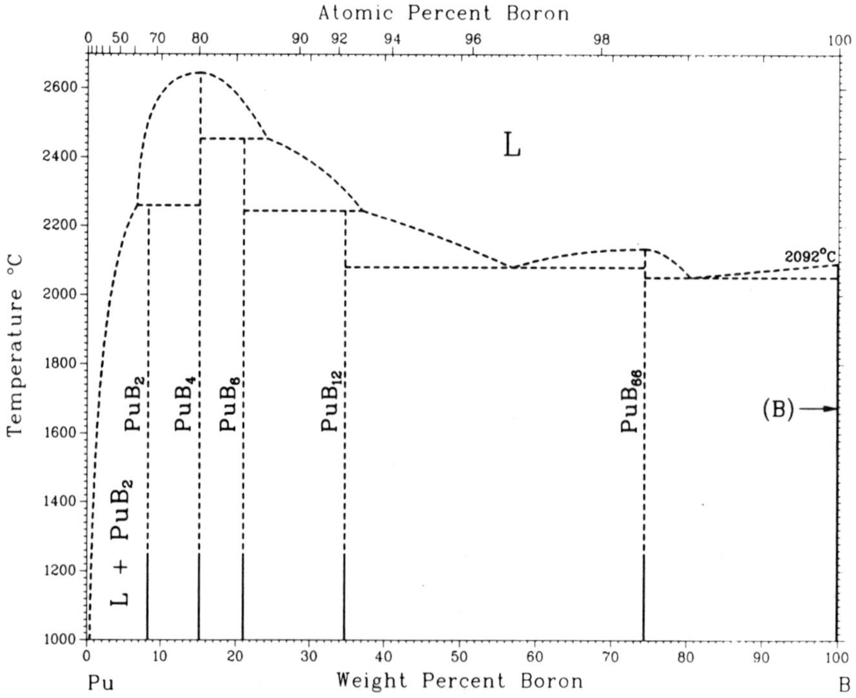

Ta-B Phase Diagram

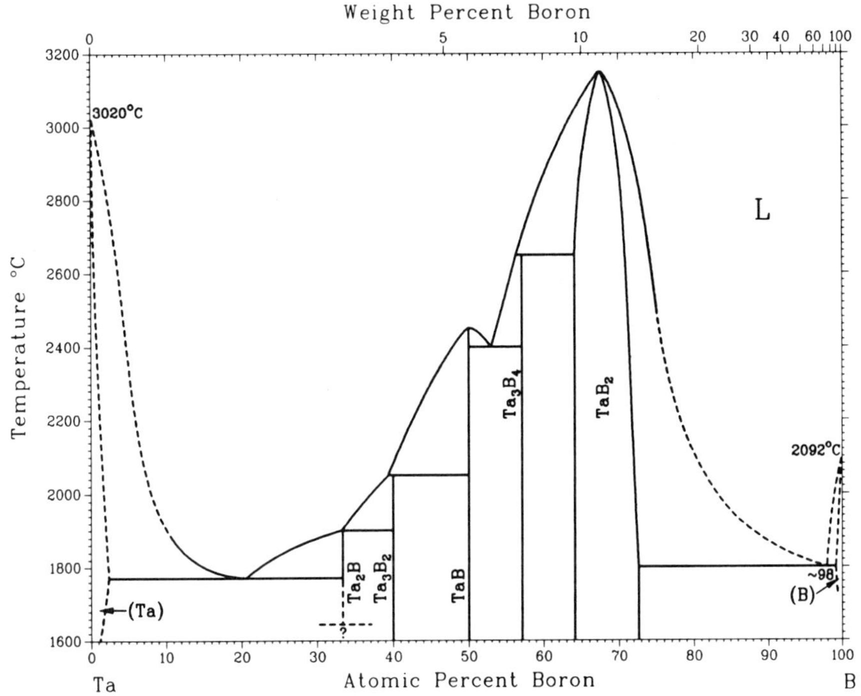

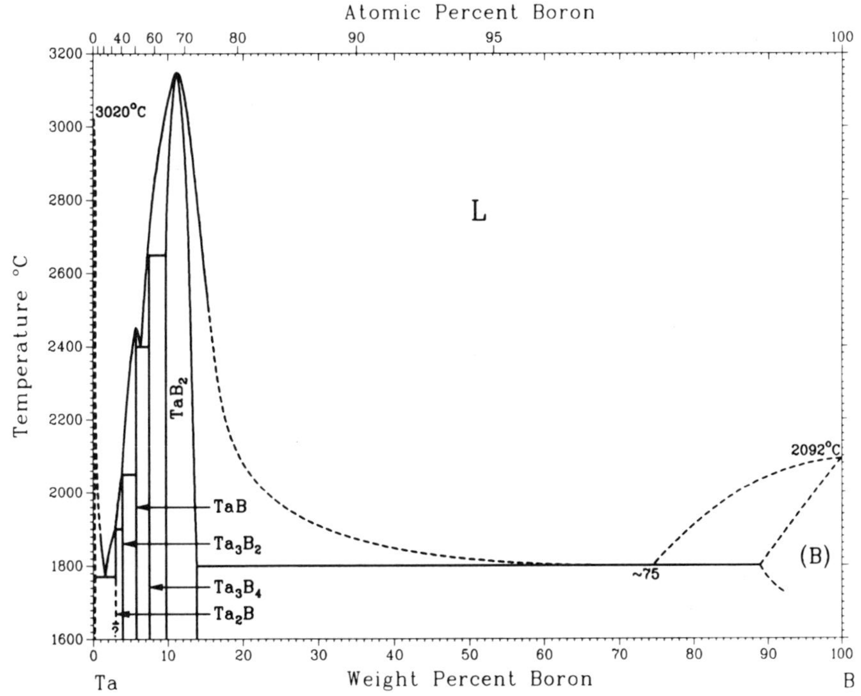

Th-B Phase Diagram

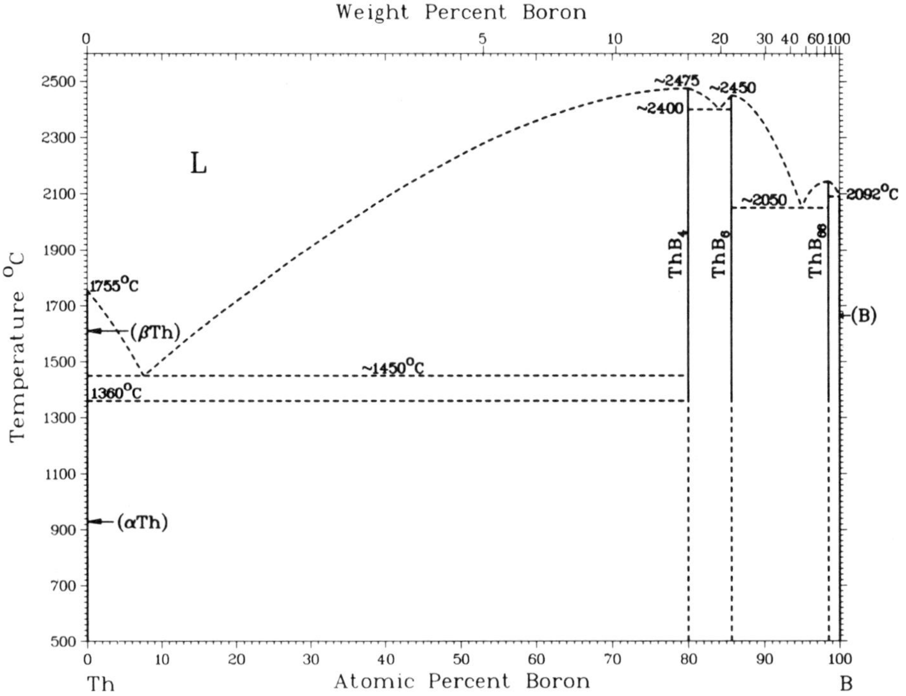

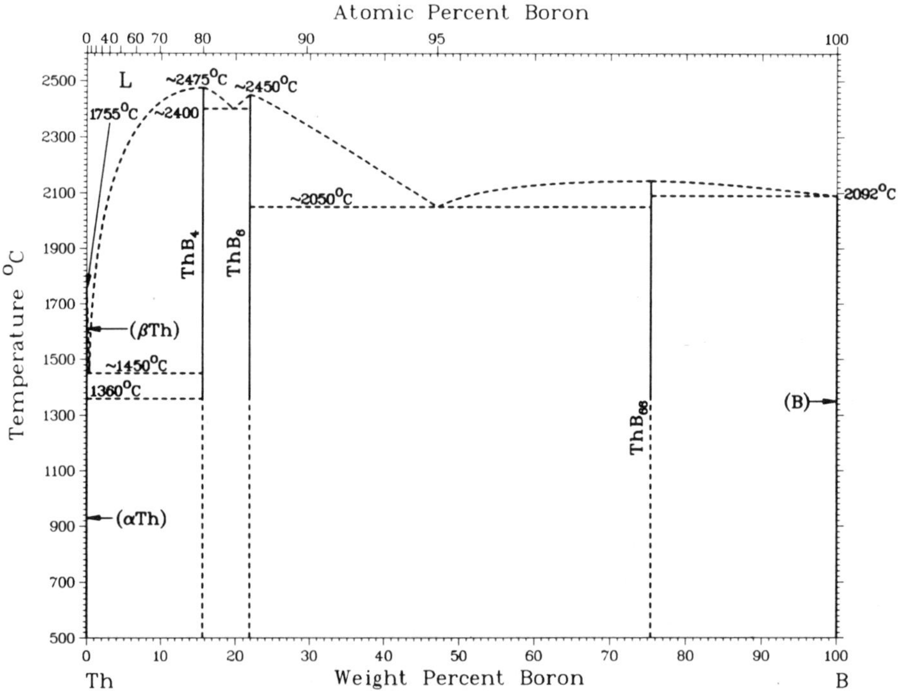

Assessed Ti-B Phase Diagram

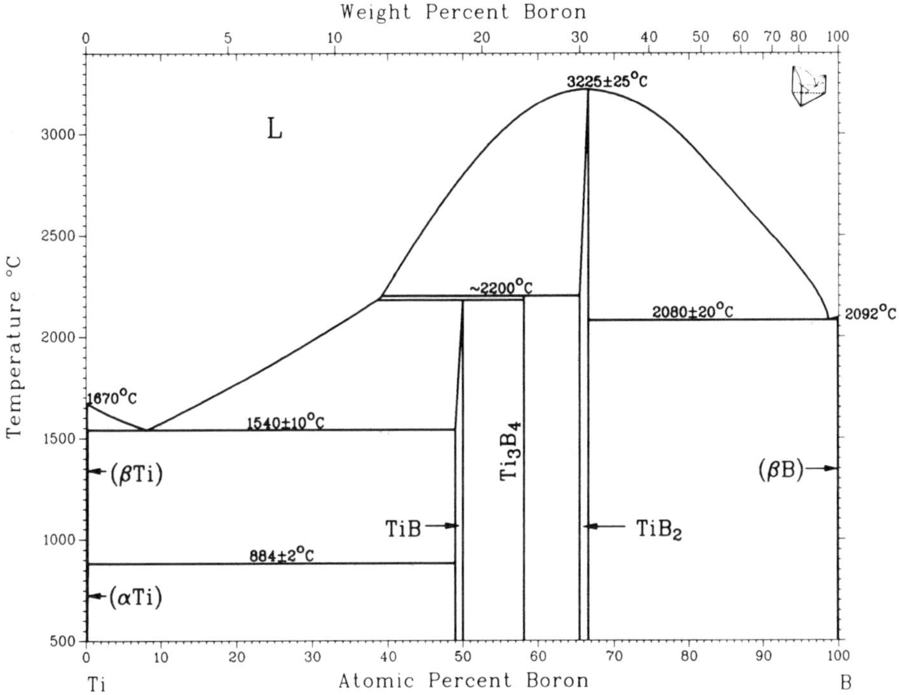

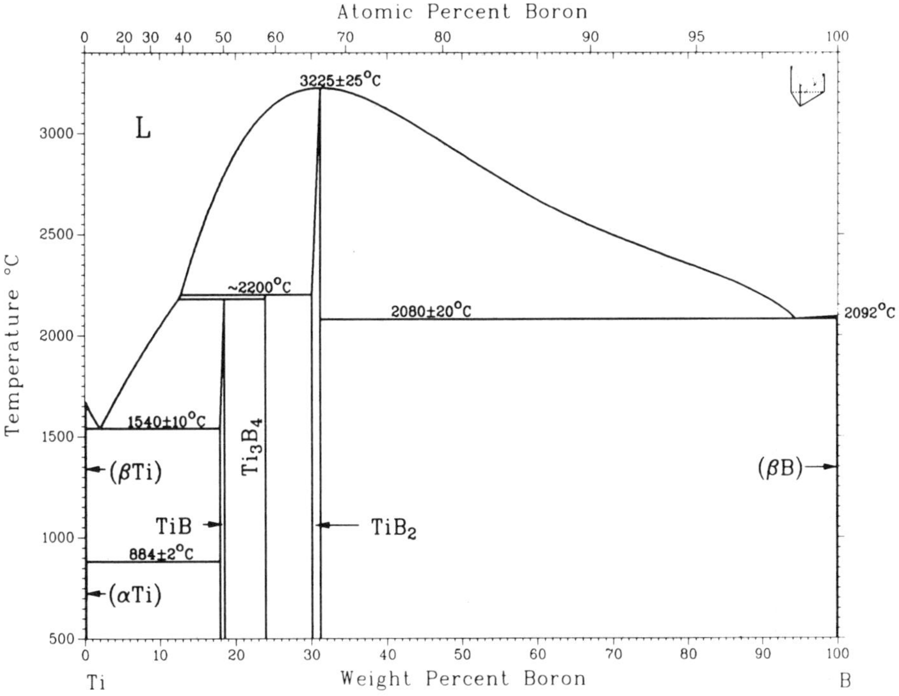

U-B Phase Diagram

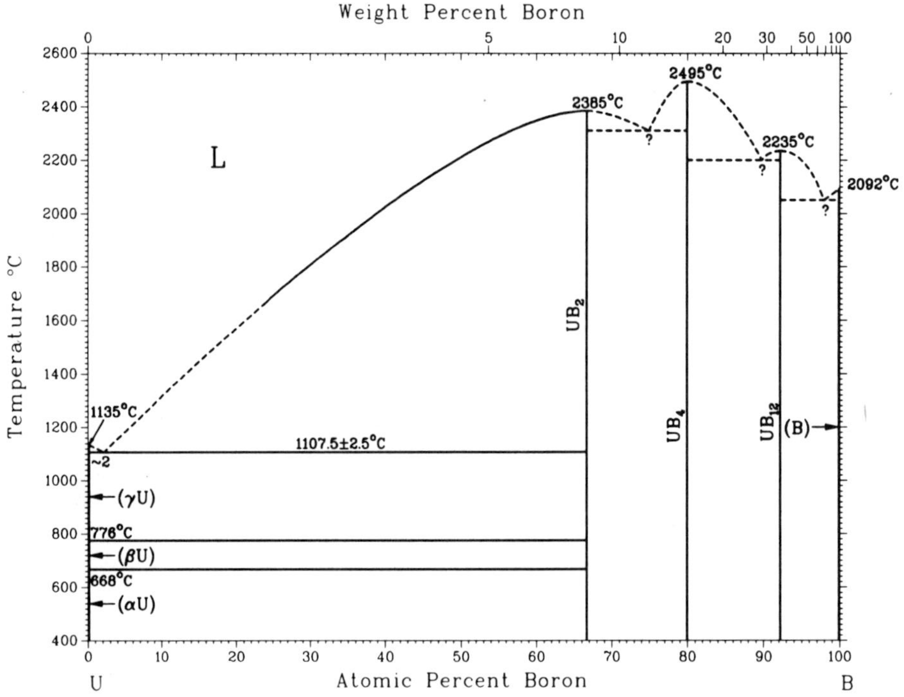

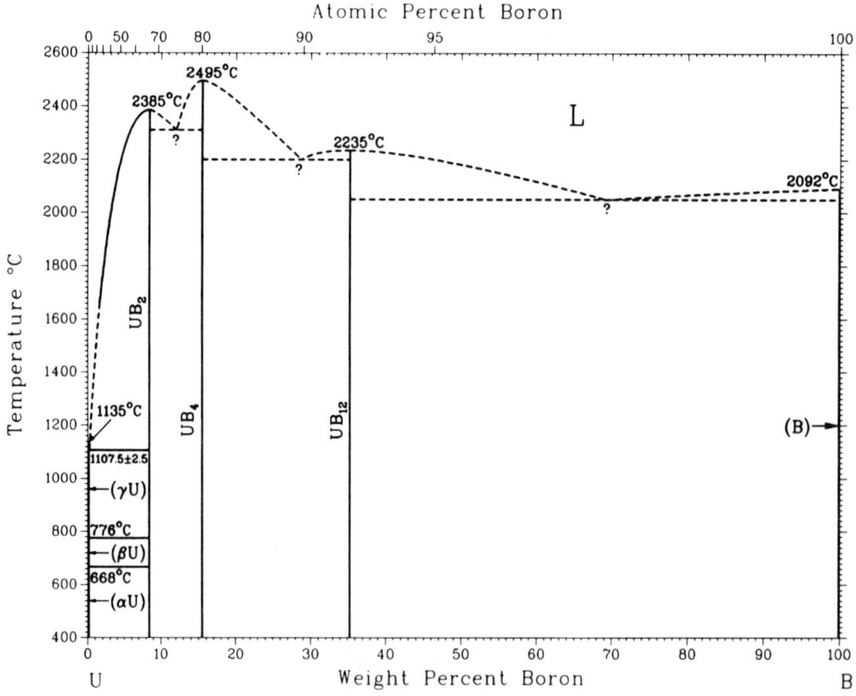

Assessed V-B Phase Diagram

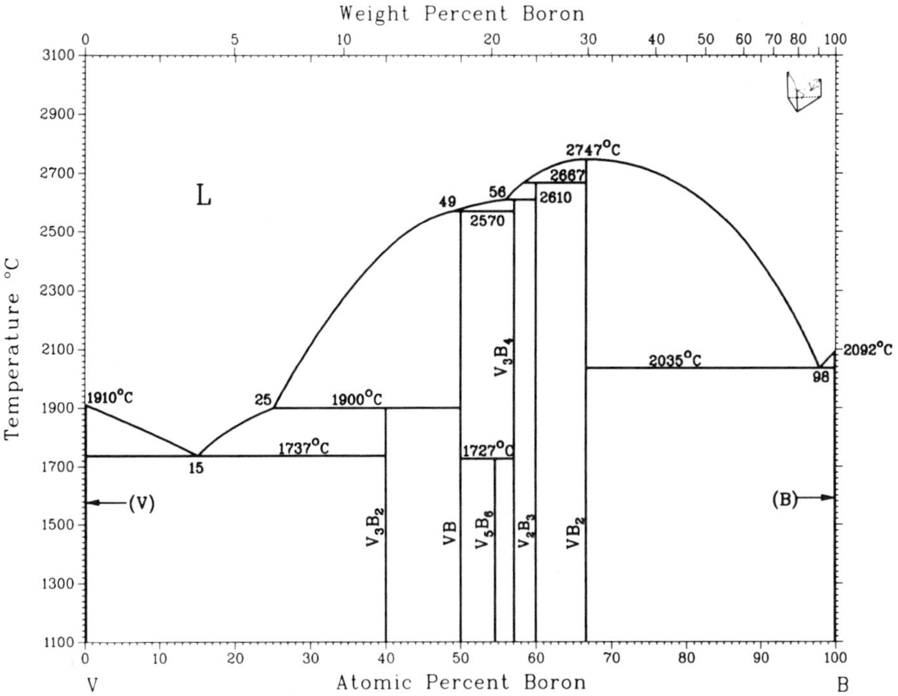

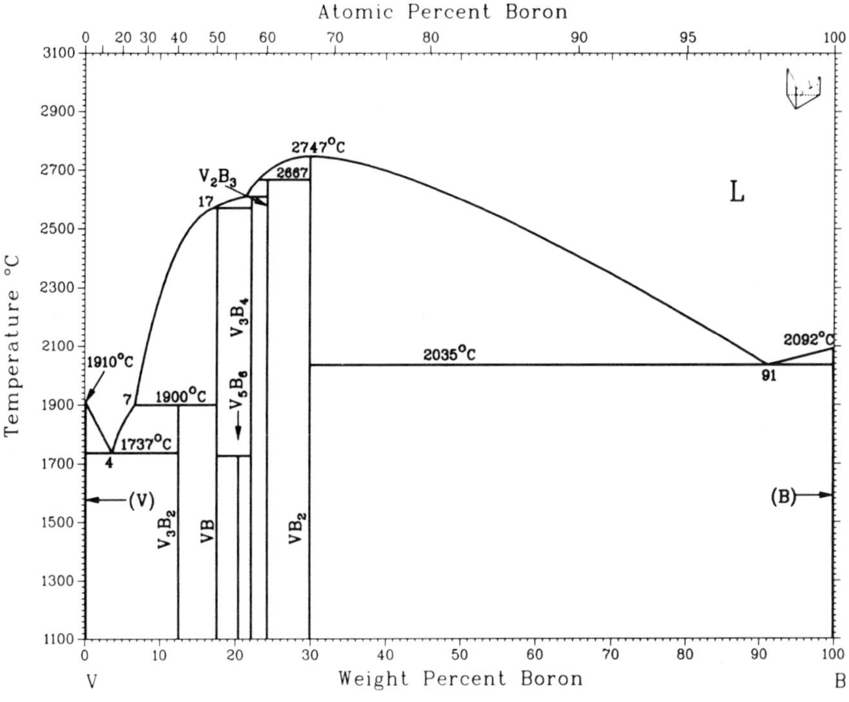

W-B Phase Diagram

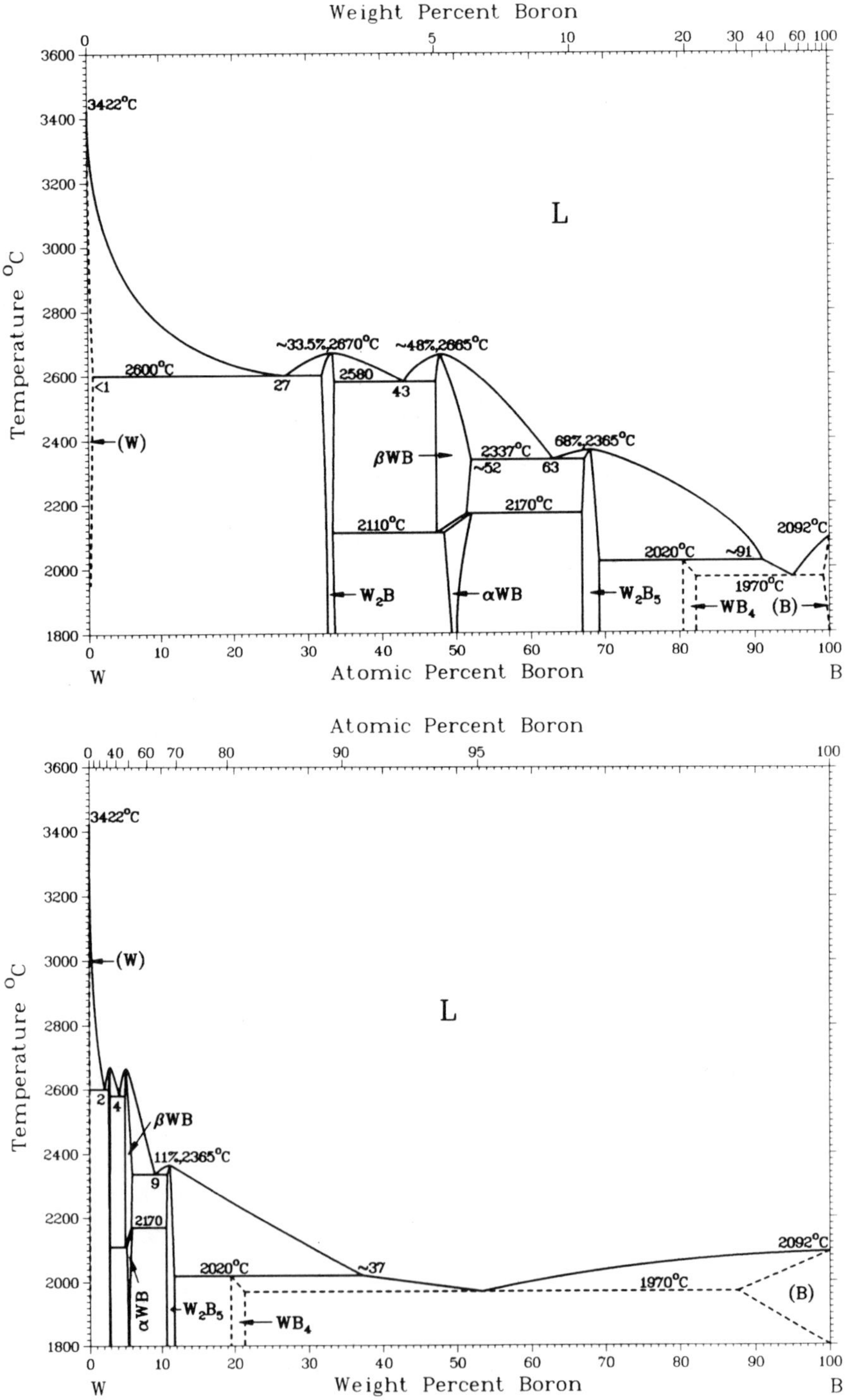

B-Zr Phase Diagram

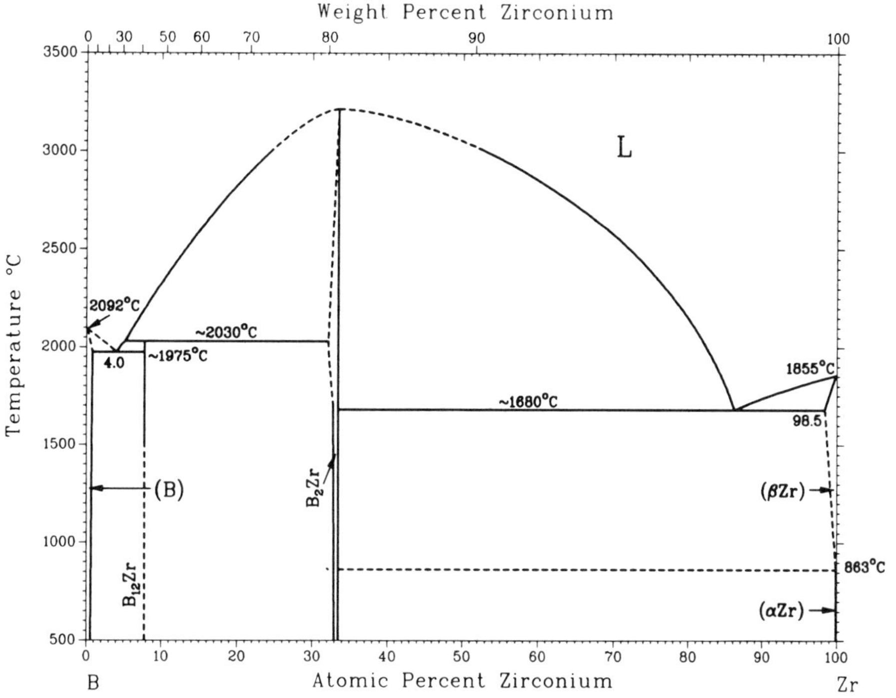

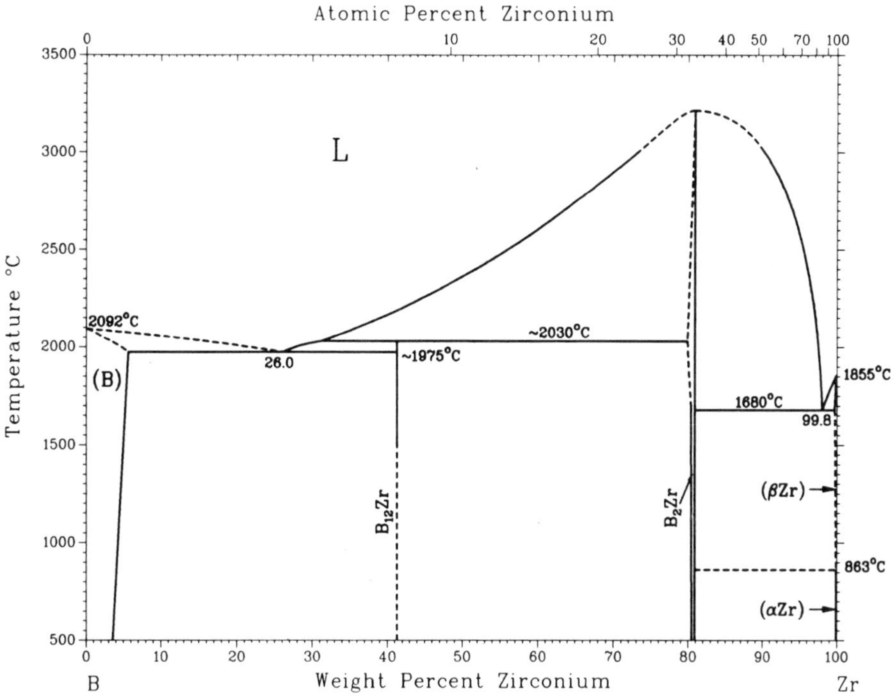

Assessed Be-O Phase Diagram (Condensed System)

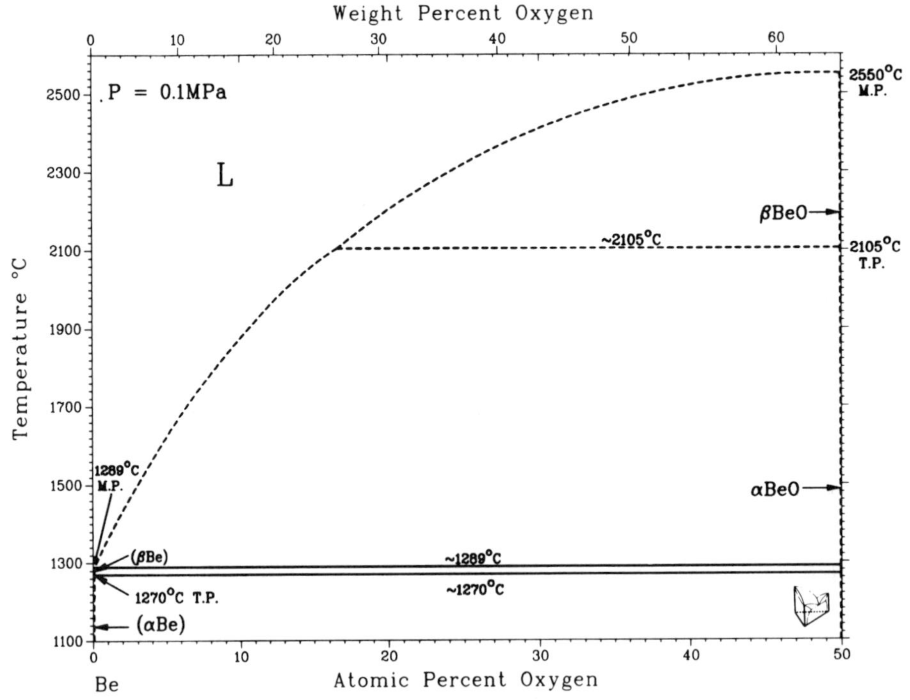

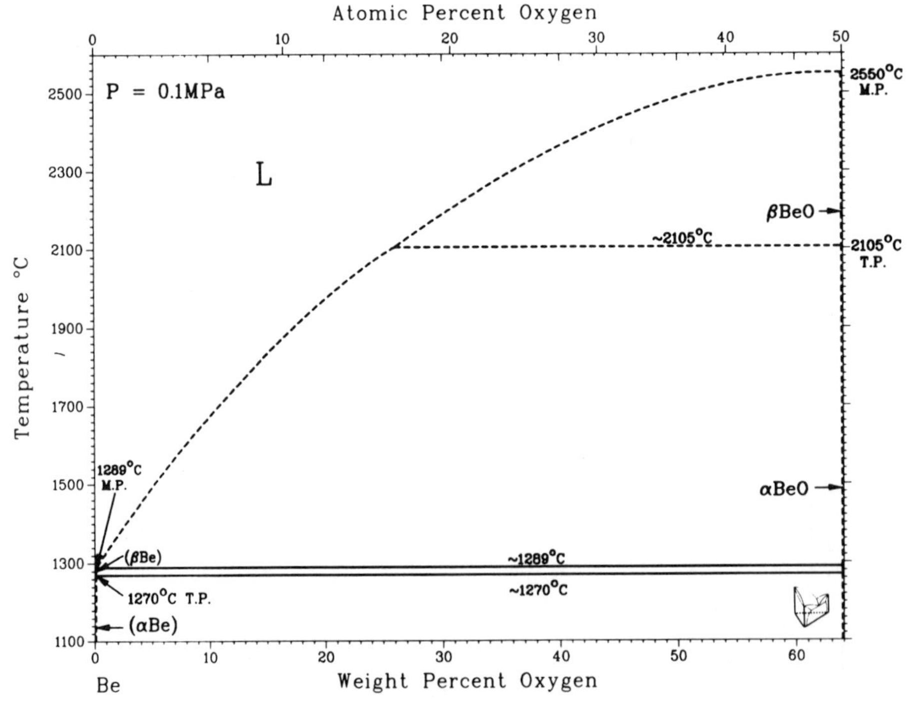

Cr-C Phase Diagram

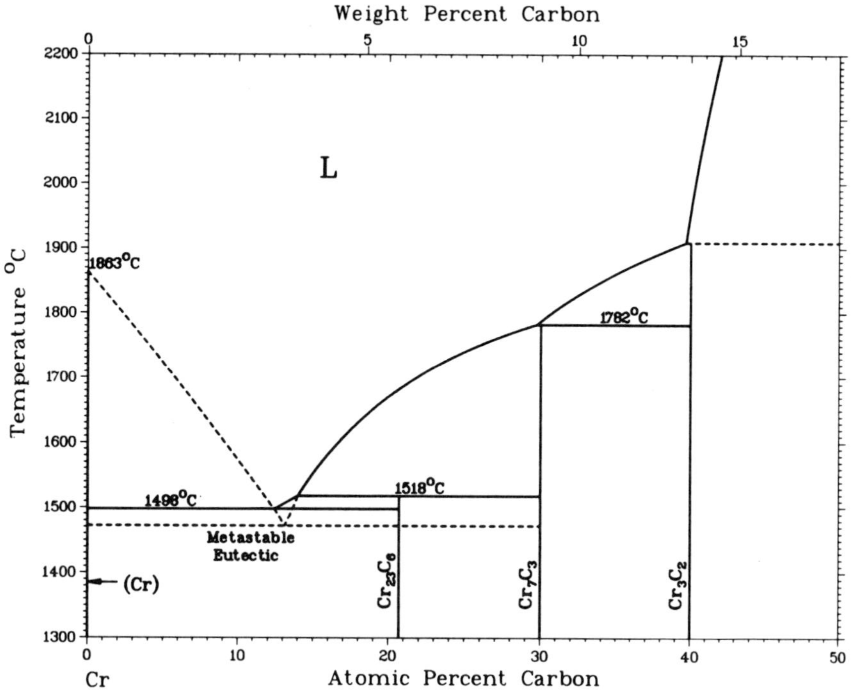

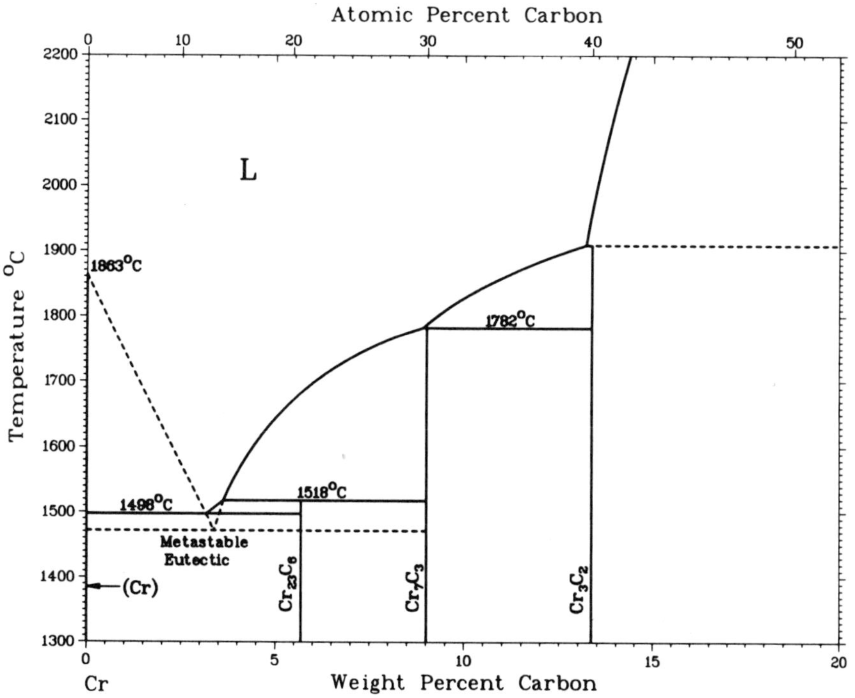

Hf-C Phase Diagram

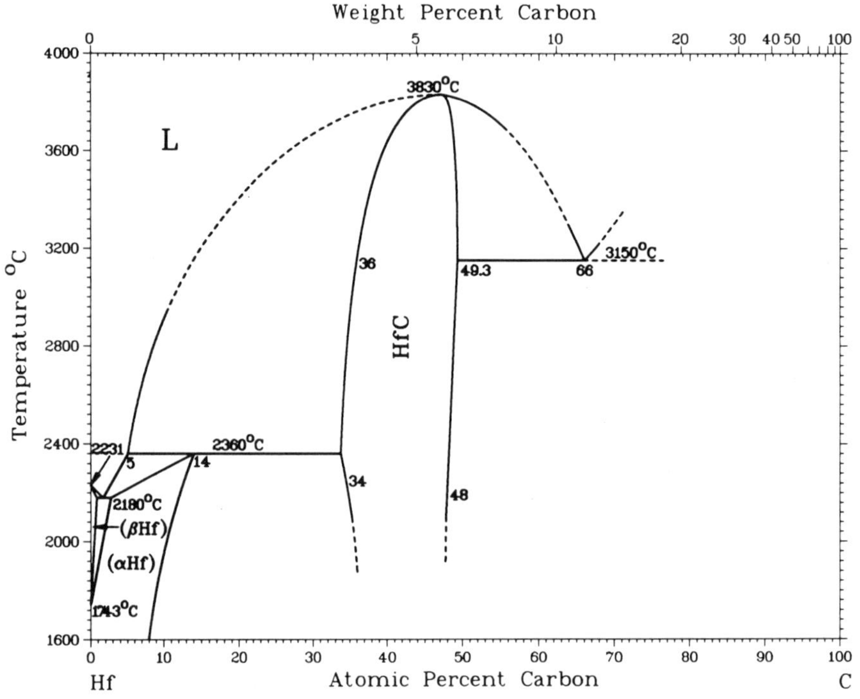

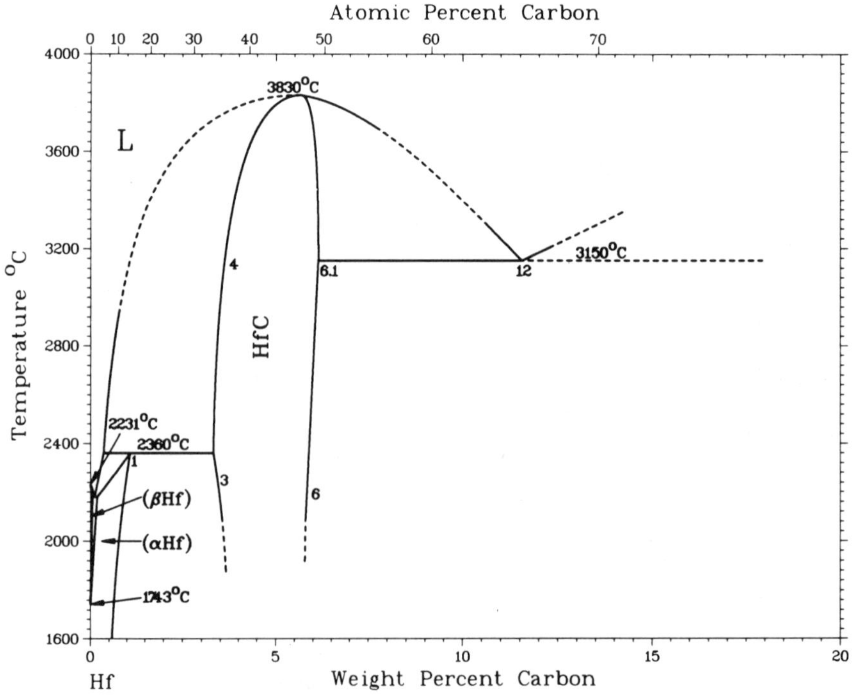

Mo-C Phase Diagram

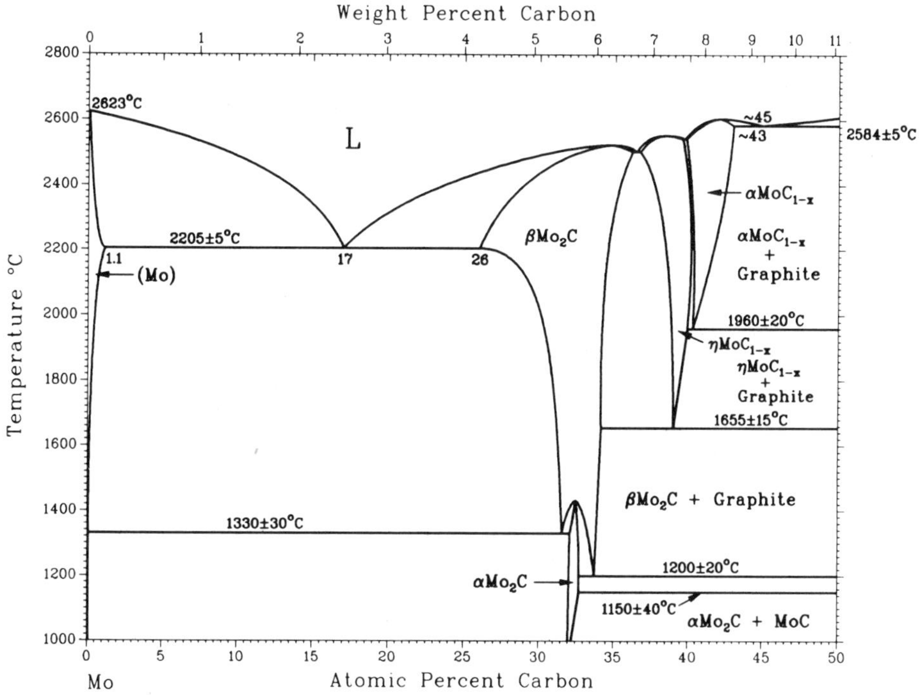

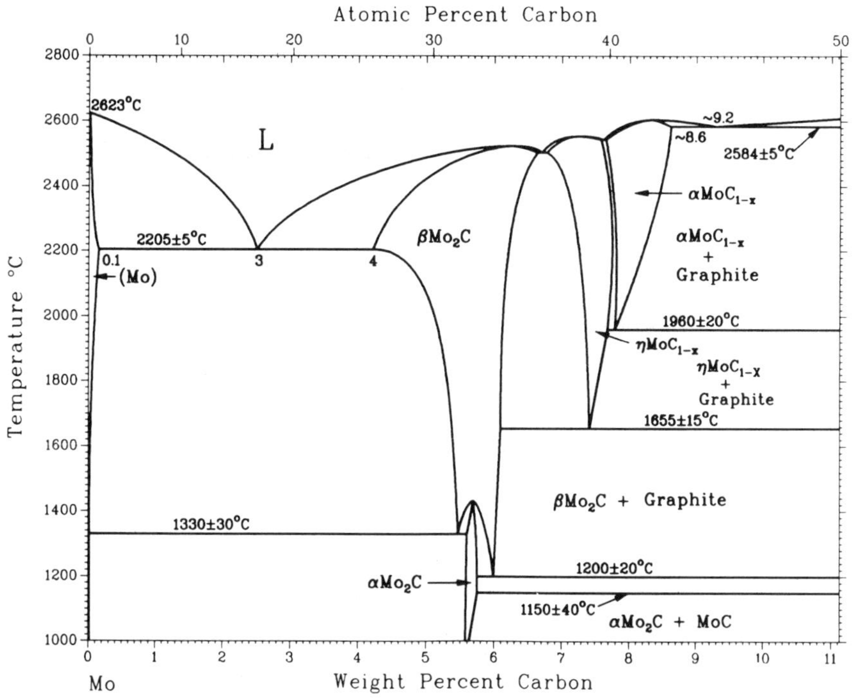

Nb-C Phase Diagram

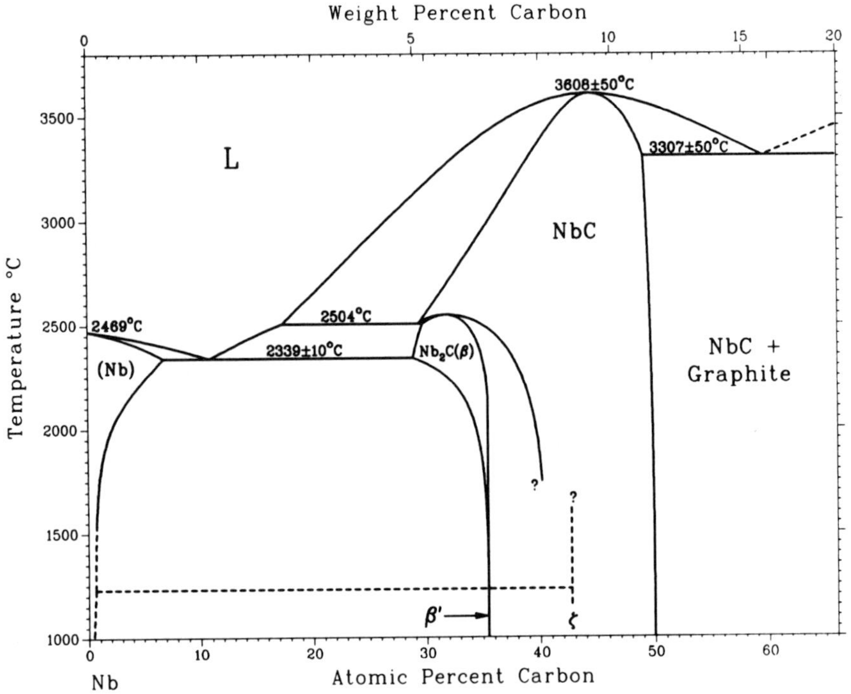

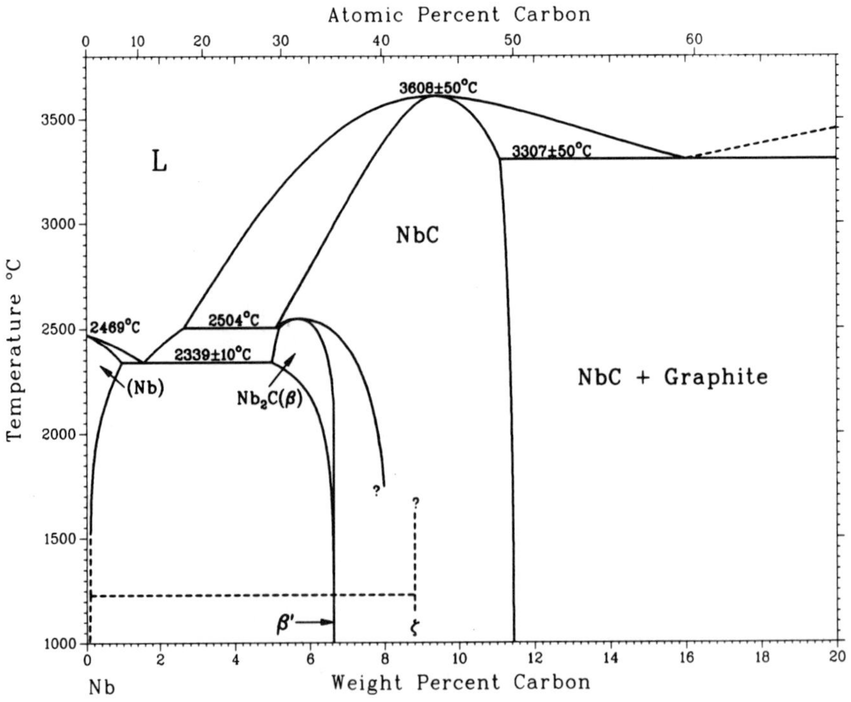

Pu-C Phase Diagram

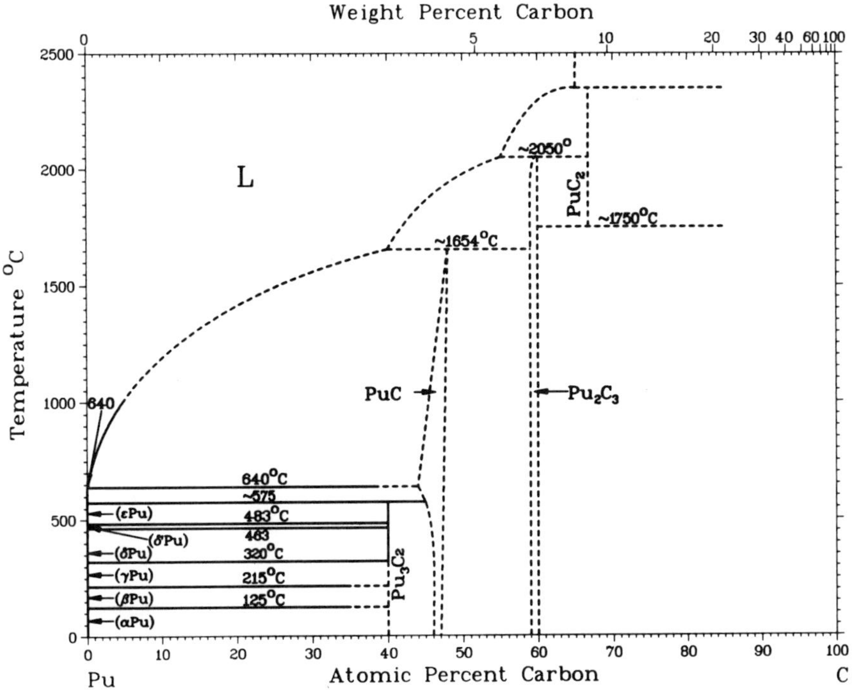

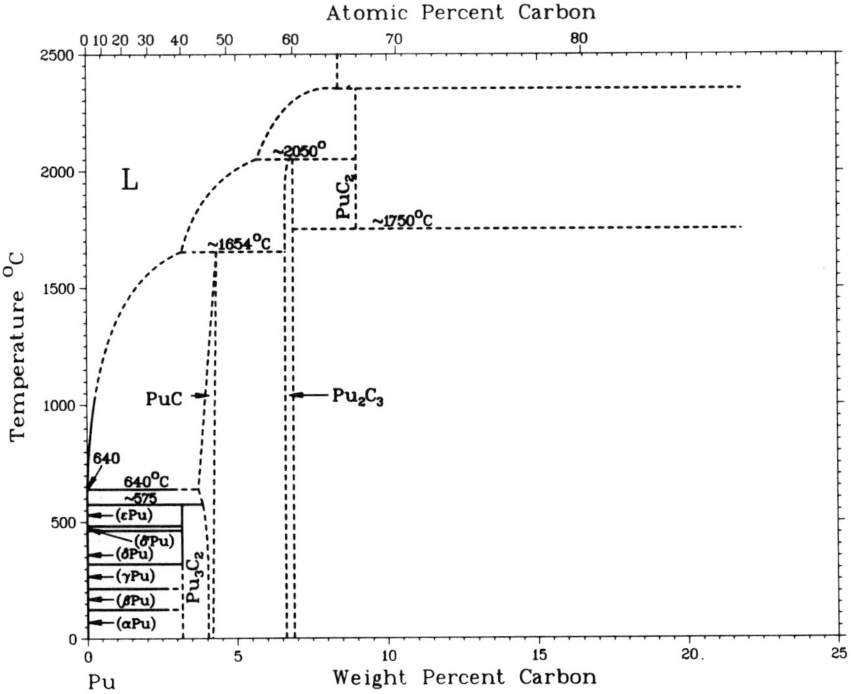

Assessed Si-C Phase Diagram

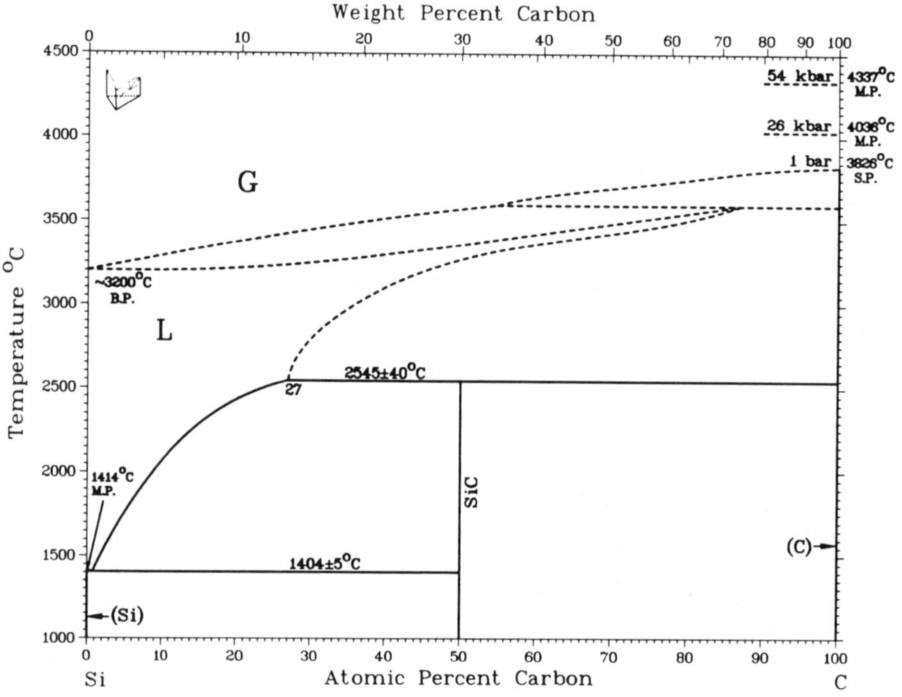

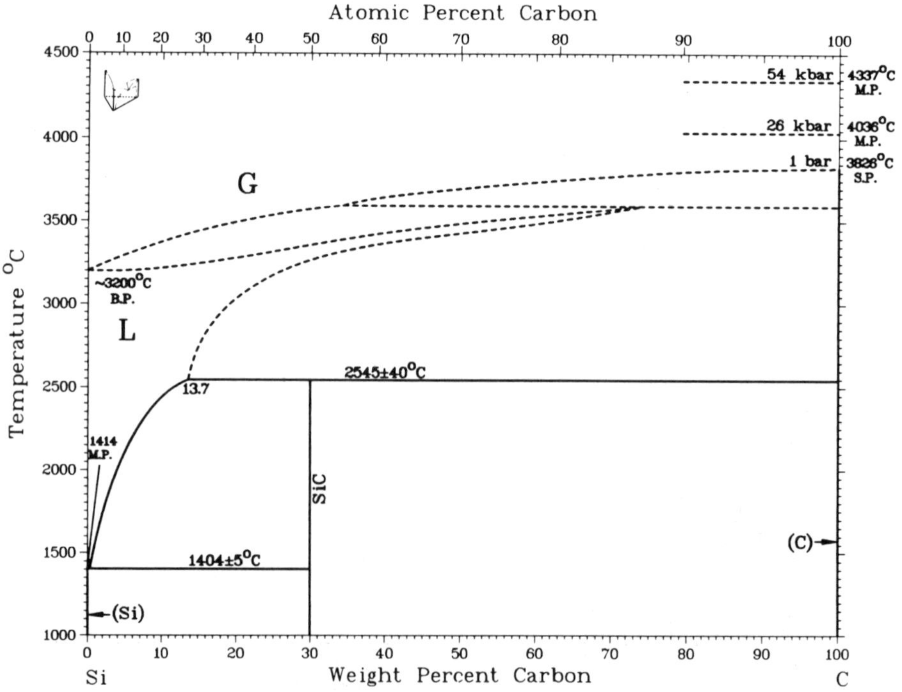

Ta-C Phase Diagram

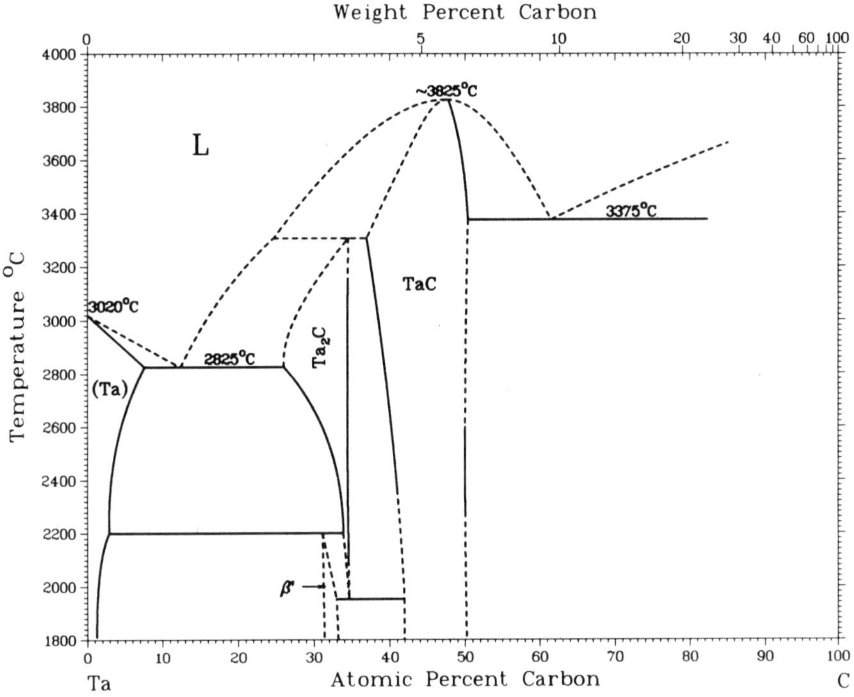

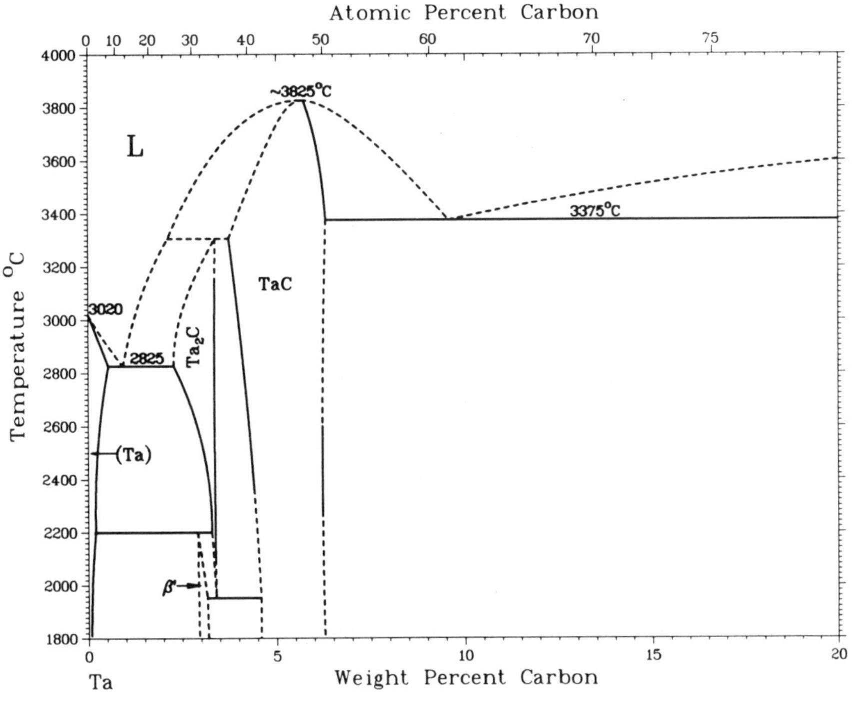

Th-C Phase Diagram

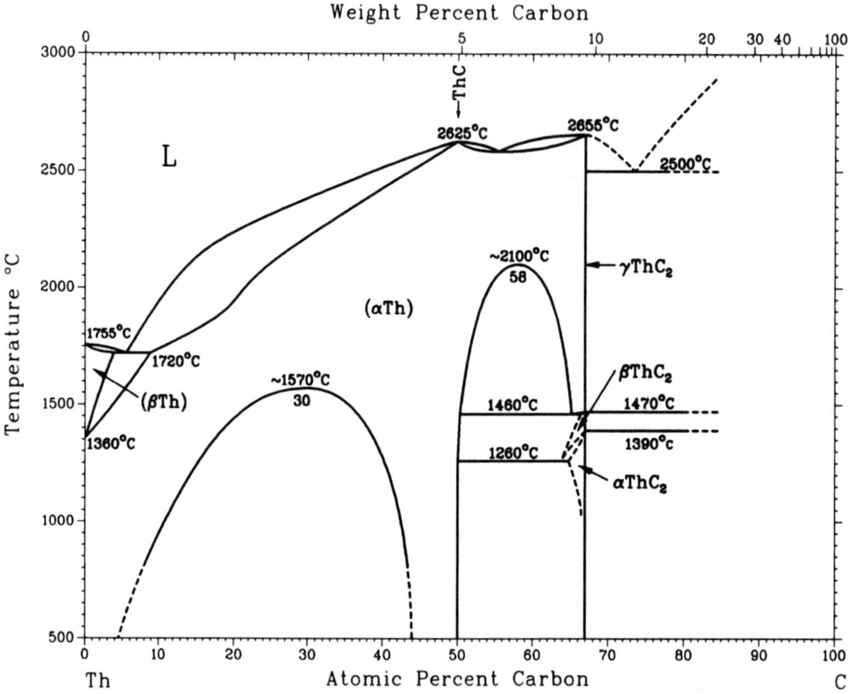

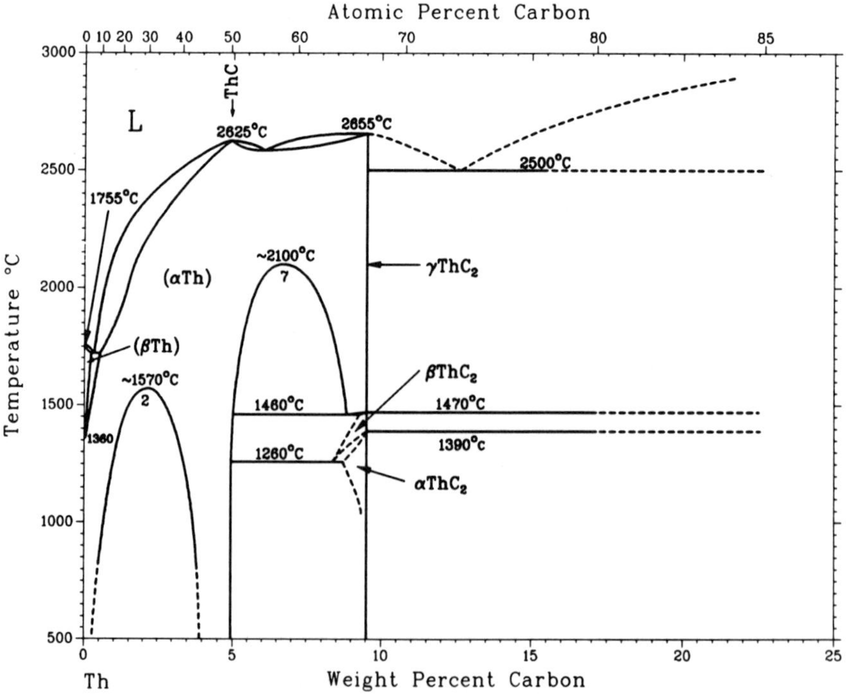

Assessed Ti-C Phase Diagram

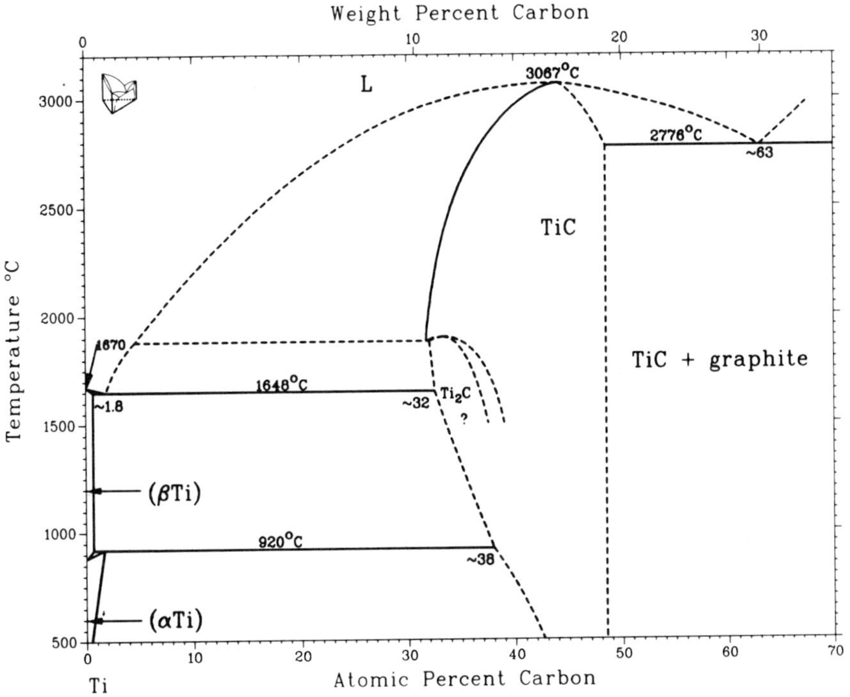

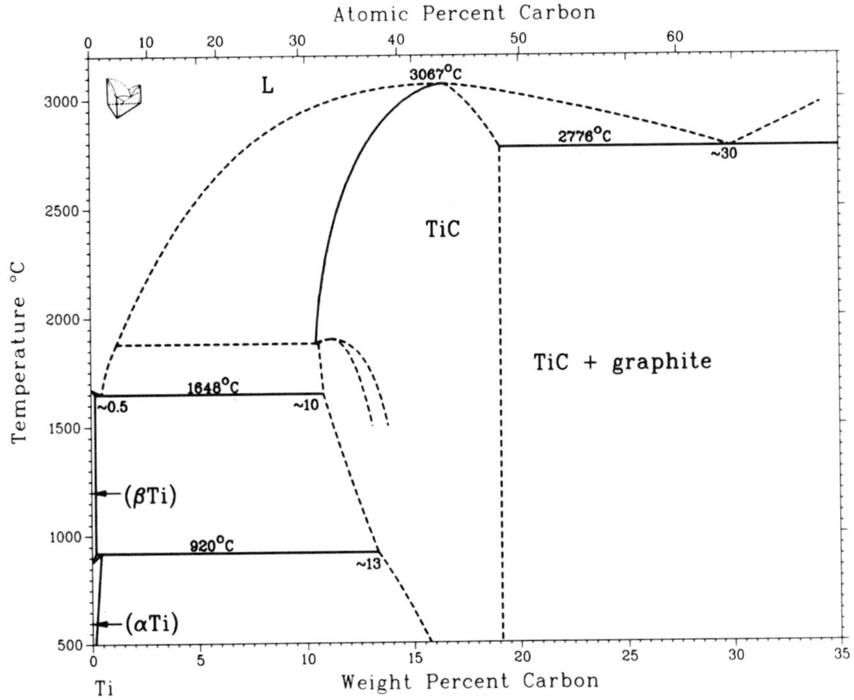

U-C Phase Diagram

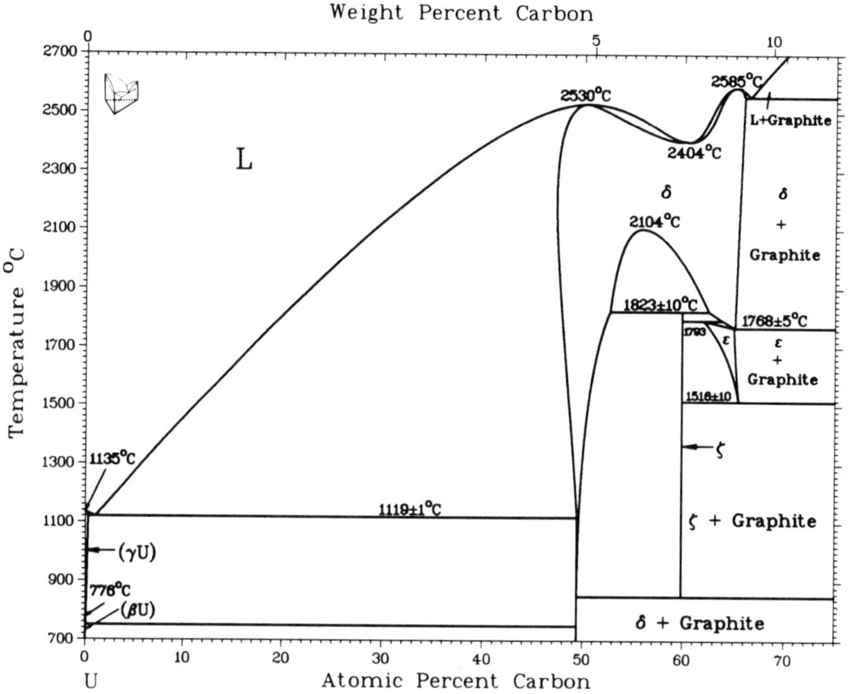

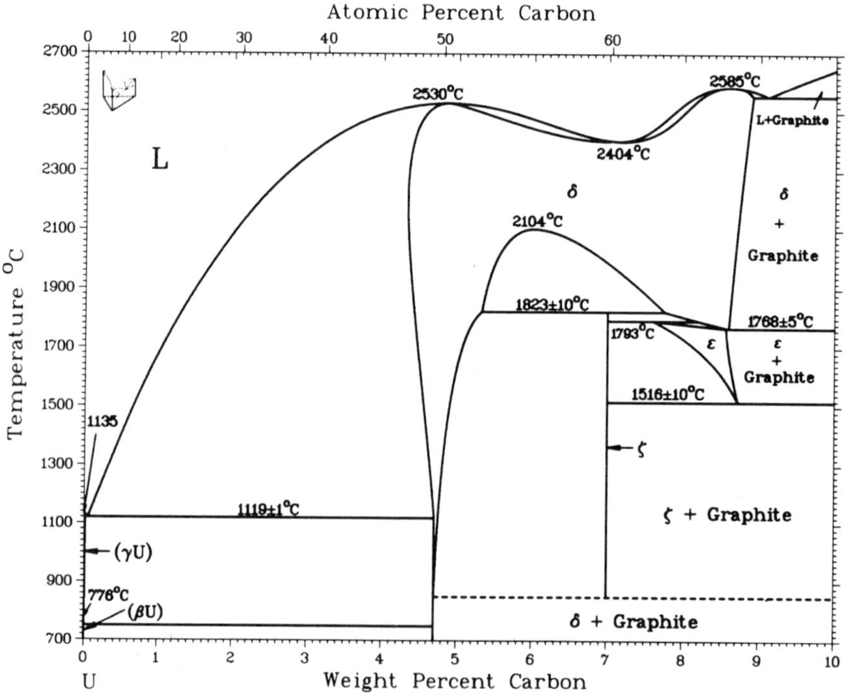

Assessed V-C Phase Diagram

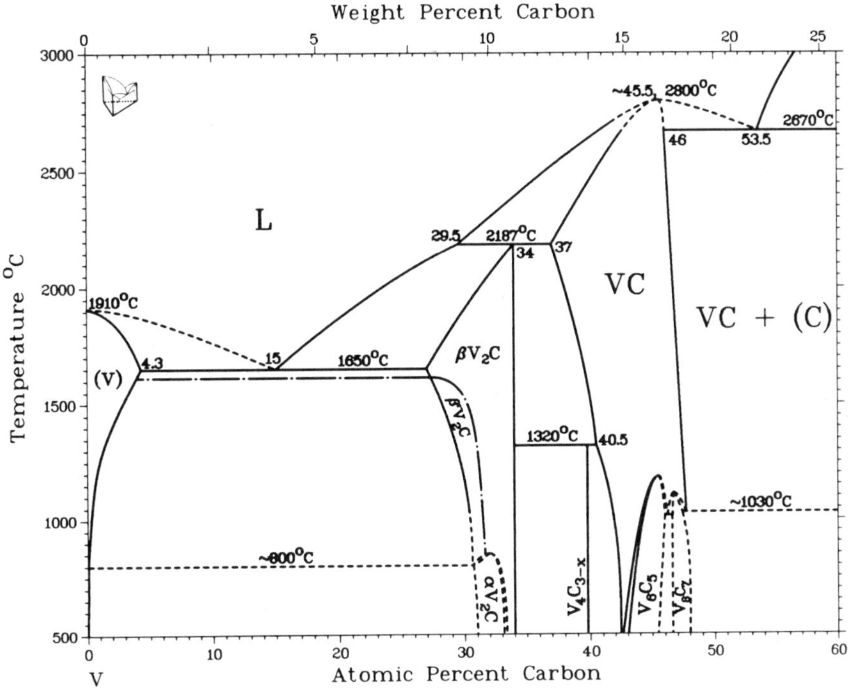

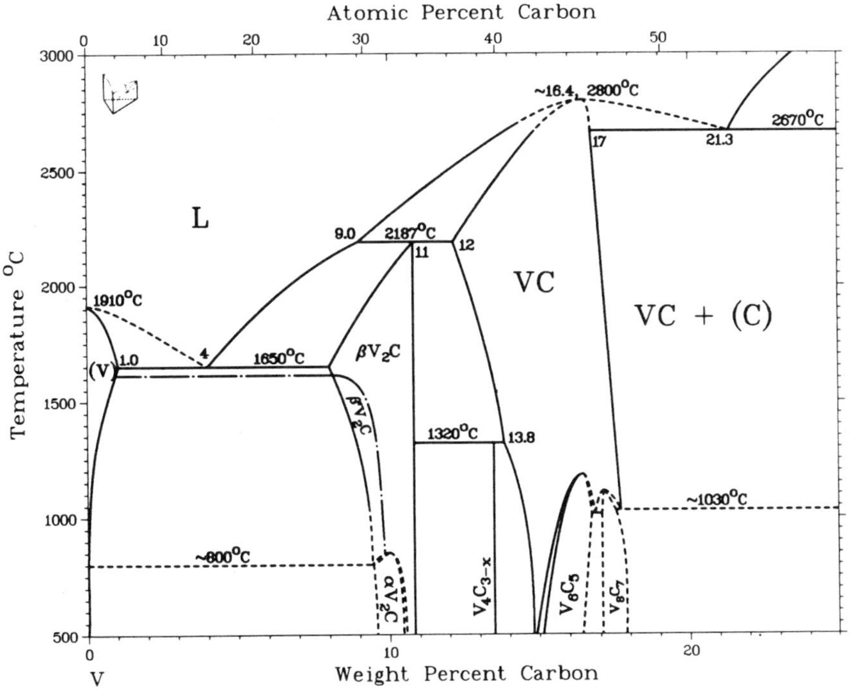

W-C Phase Diagram

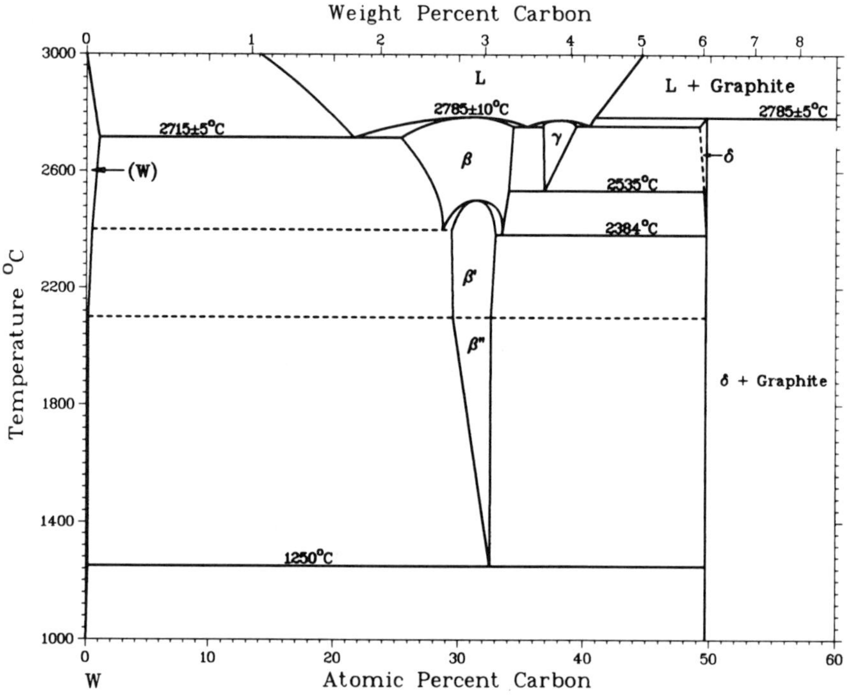

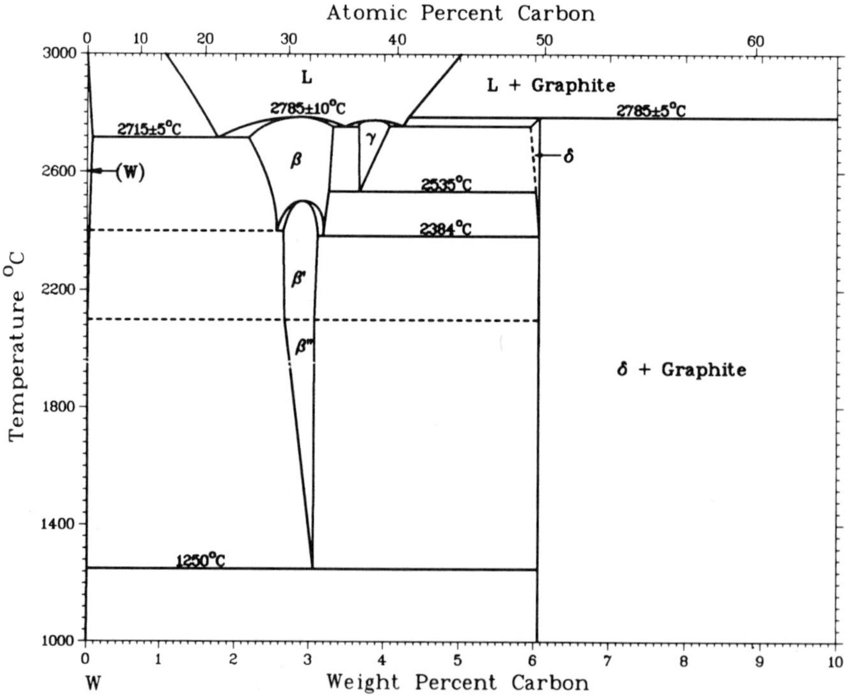

Zr-C Phase Diagram

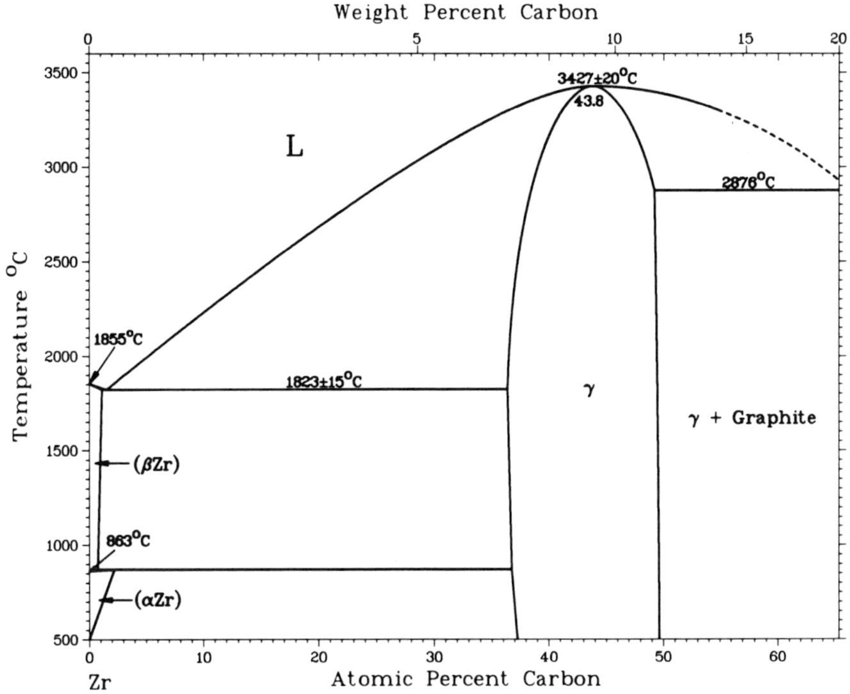

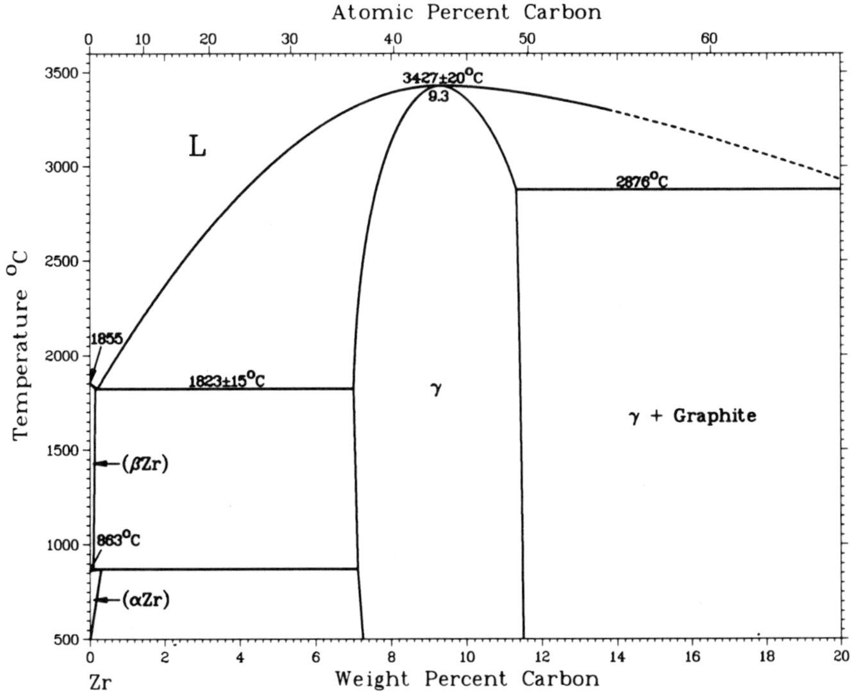

Ca-N Phase Diagram

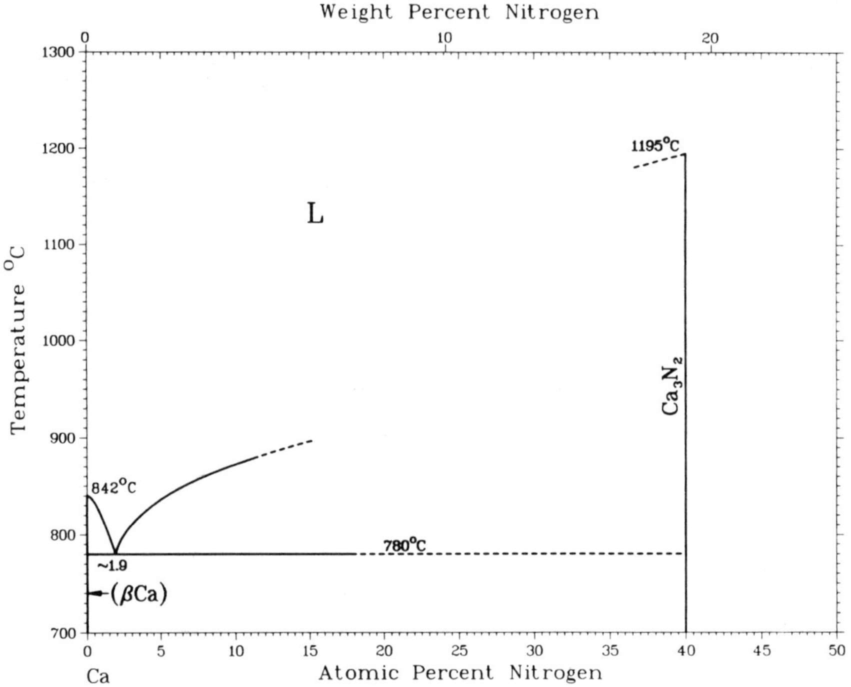

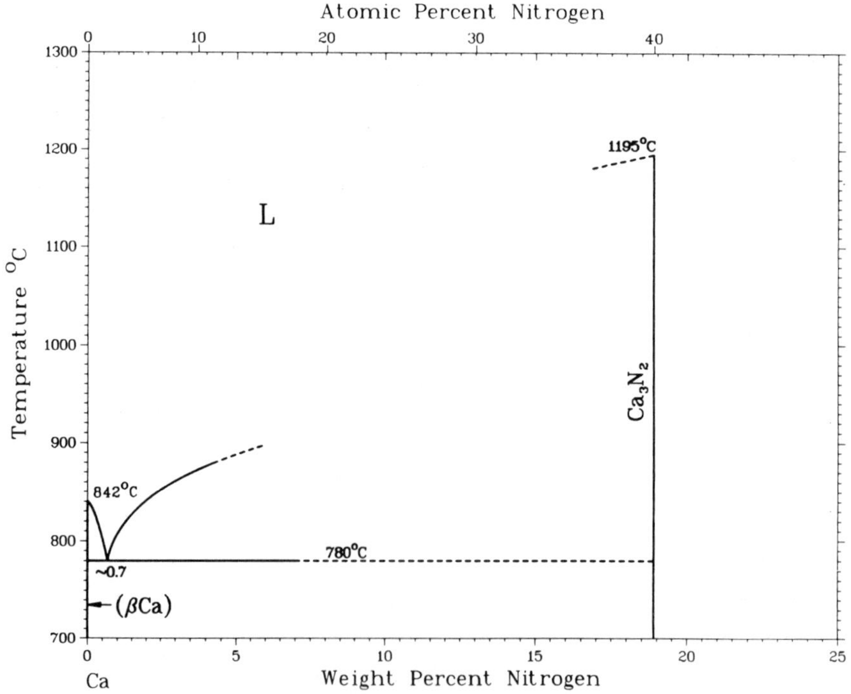

Assessed Ca-O Phase Diagram (Condensed System)

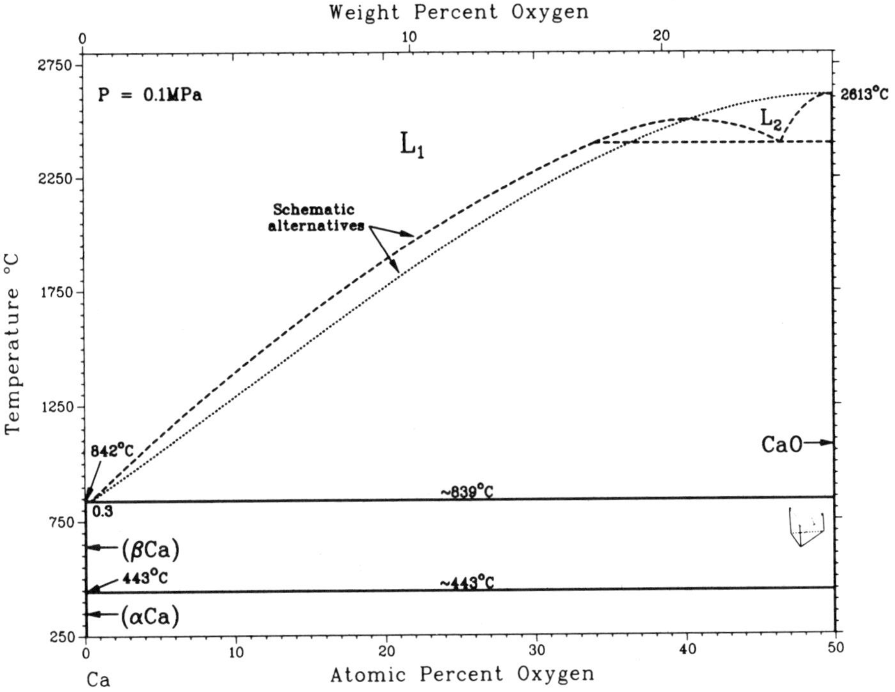

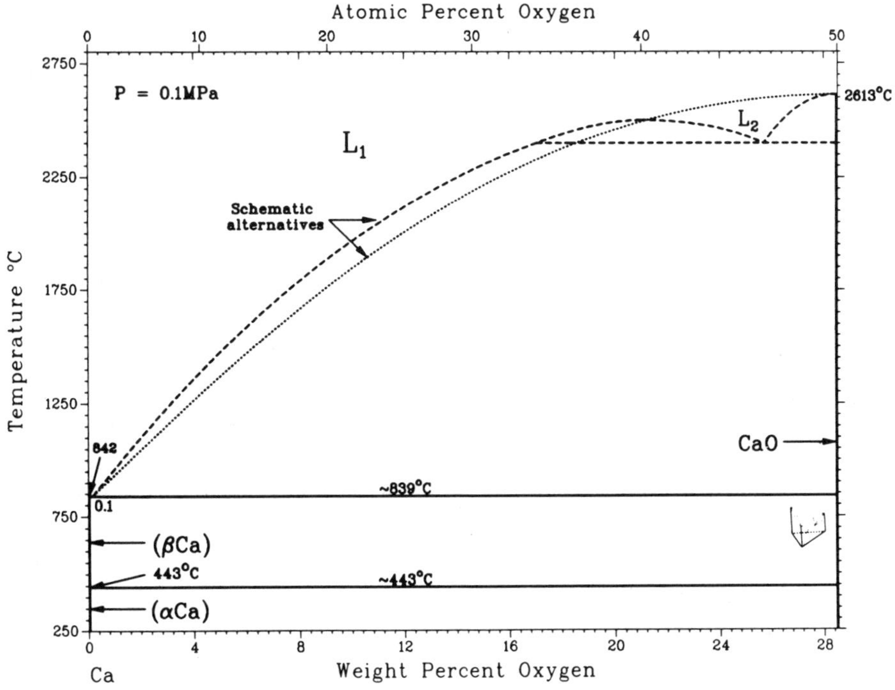

Cr-O Phase Diagram

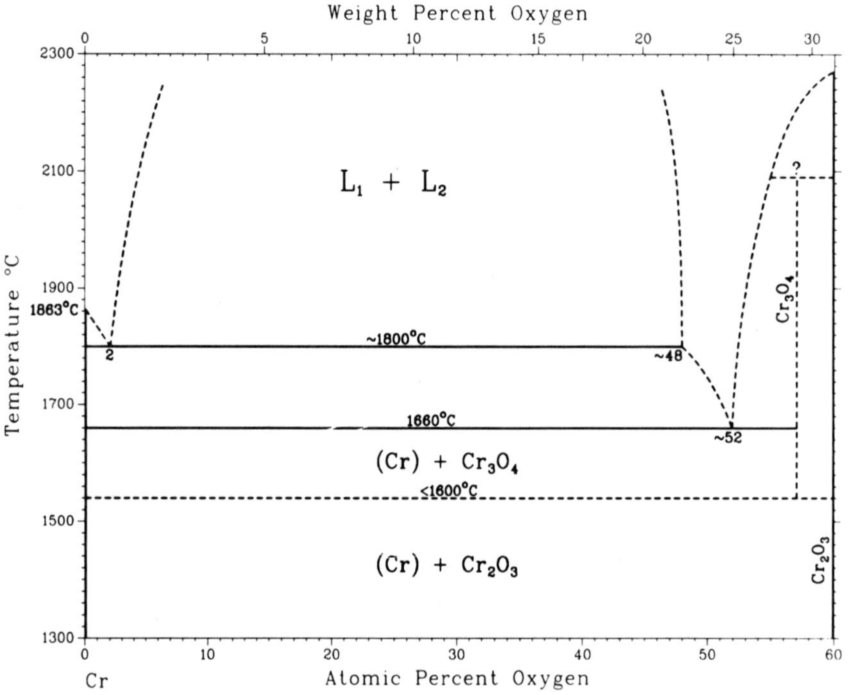

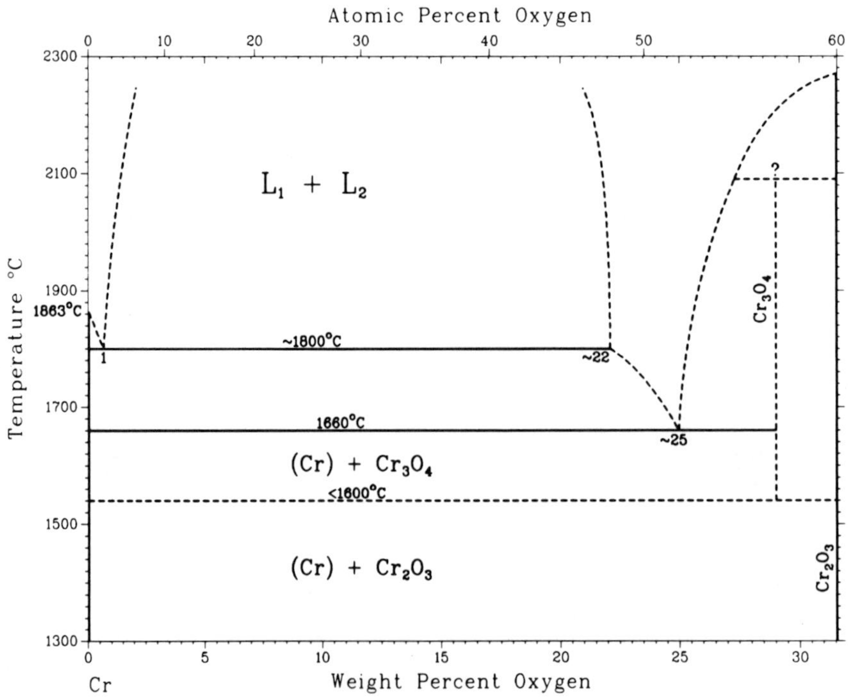

Hf-O Phase Diagram

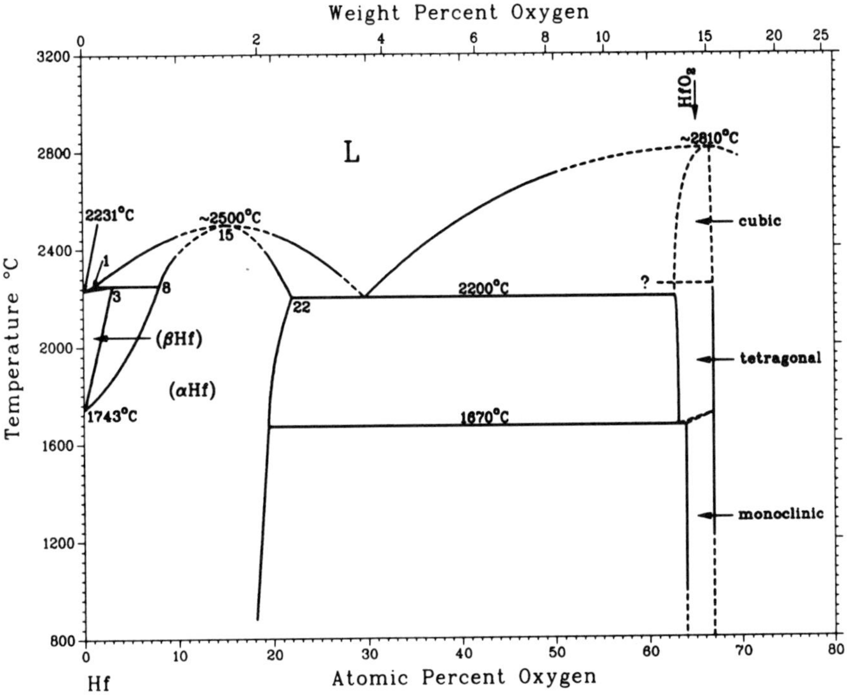

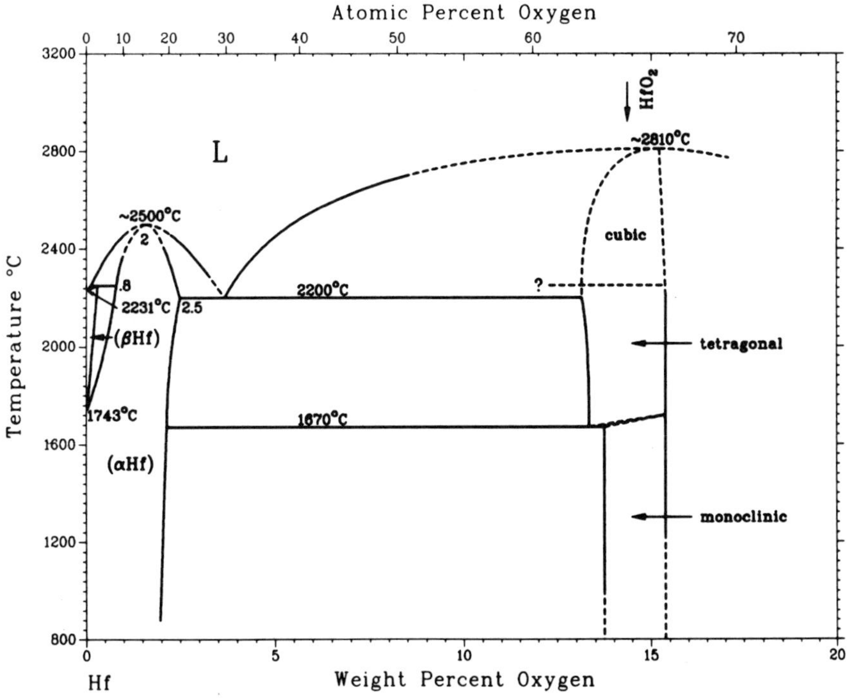

Assessed Mg-O Phase Diagram (Condensed System)

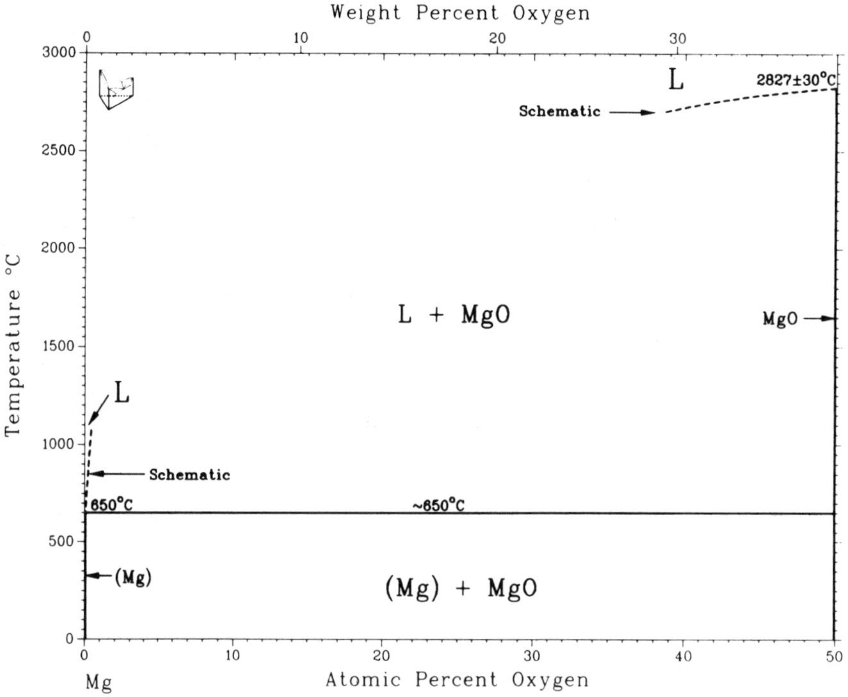

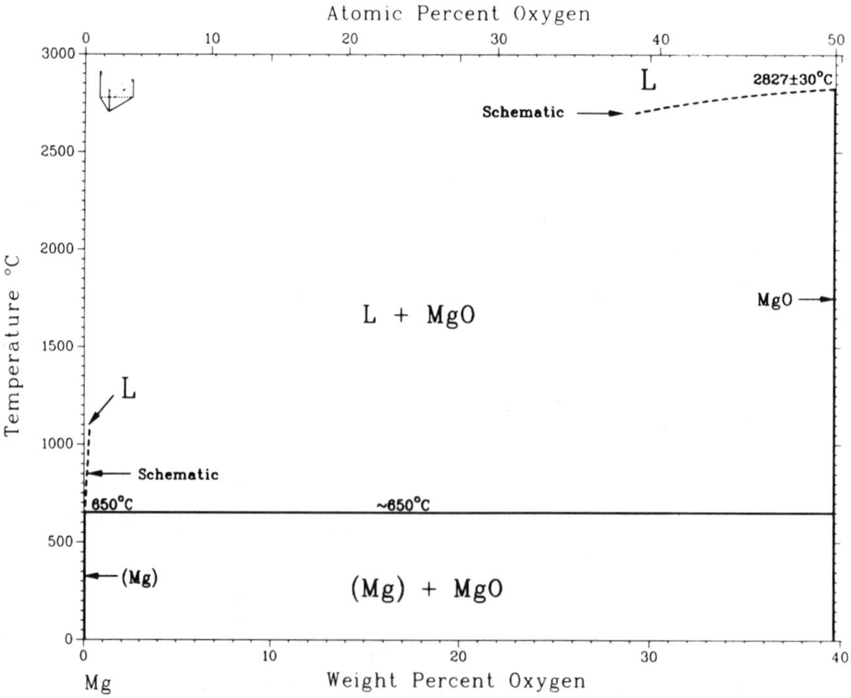

Assessed Mo-N Phase Diagram

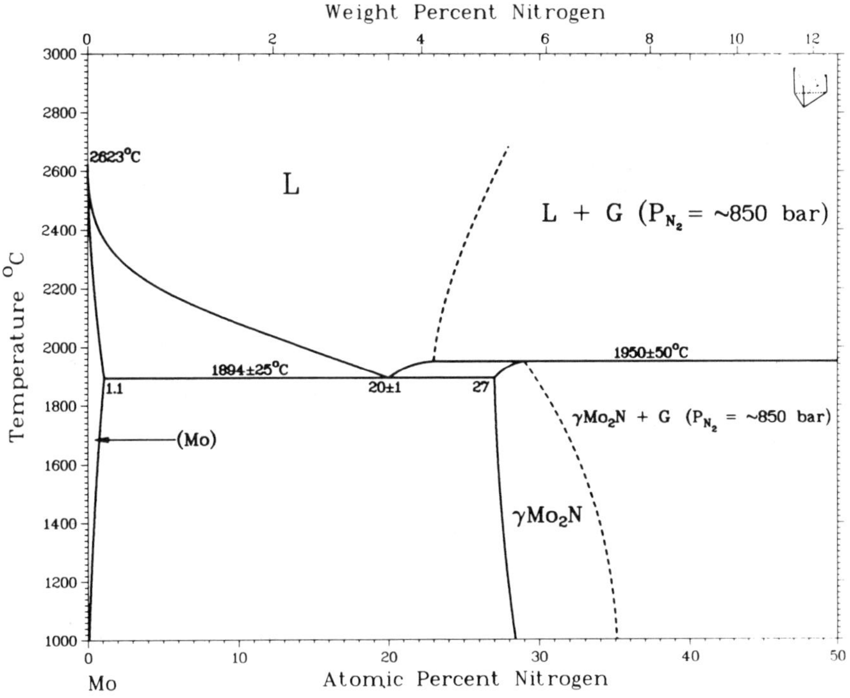

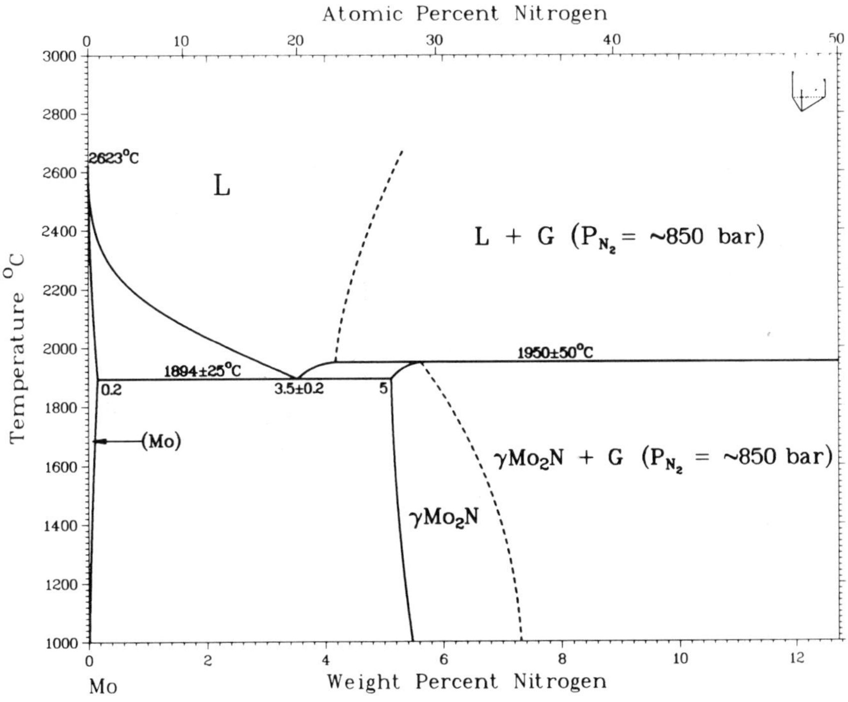

Nb-N Phase Diagram

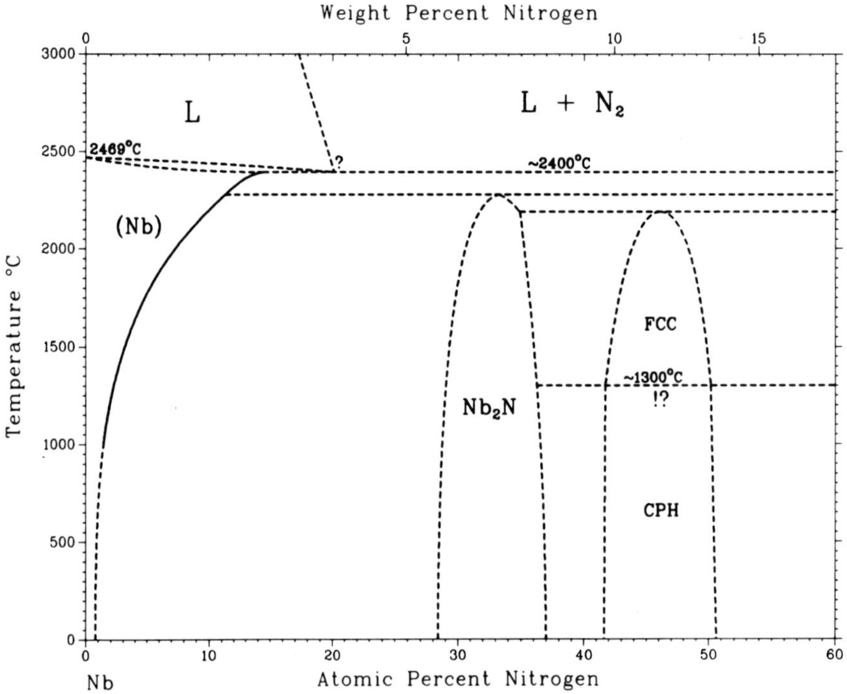

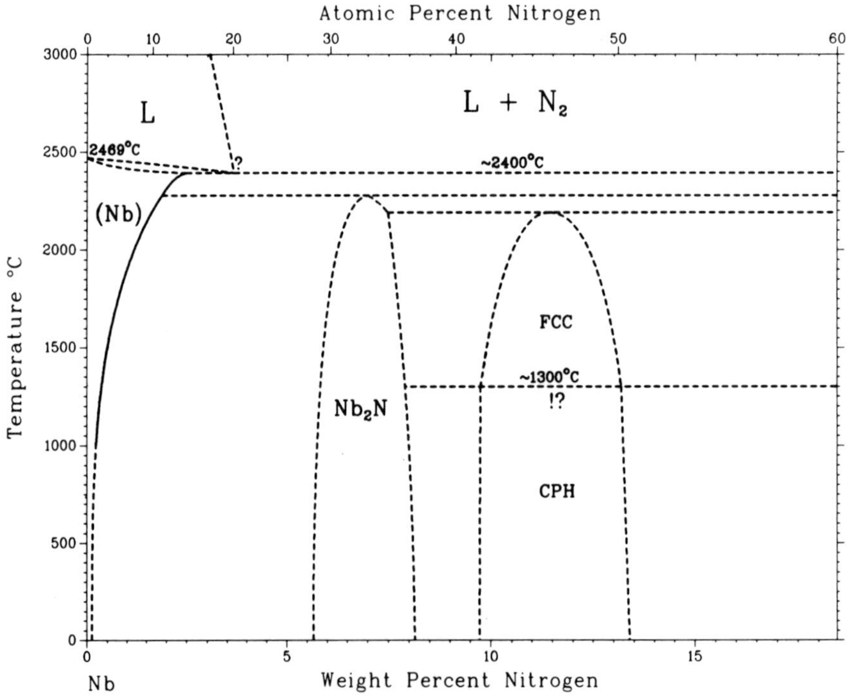

Ta-N Phase Diagram

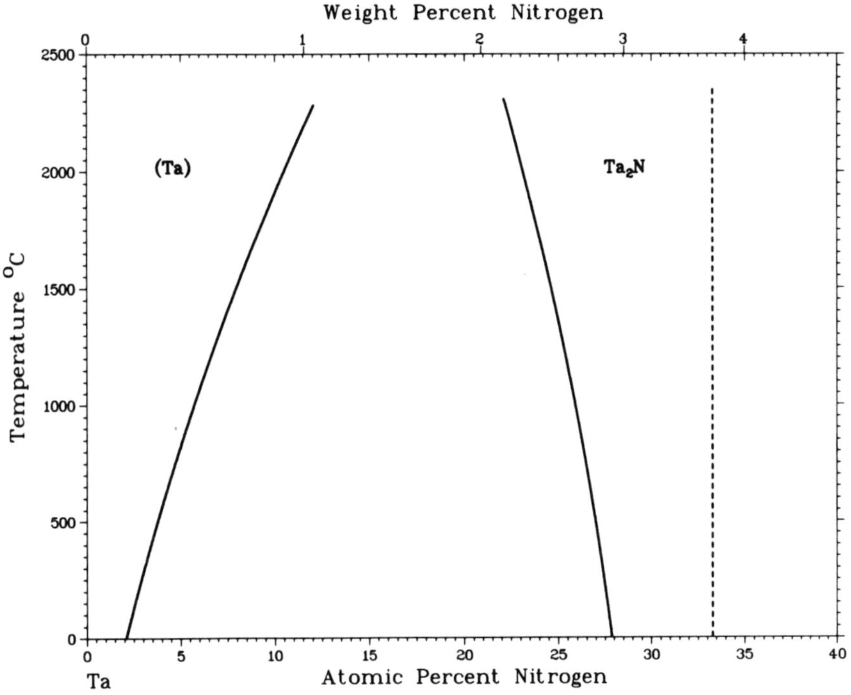

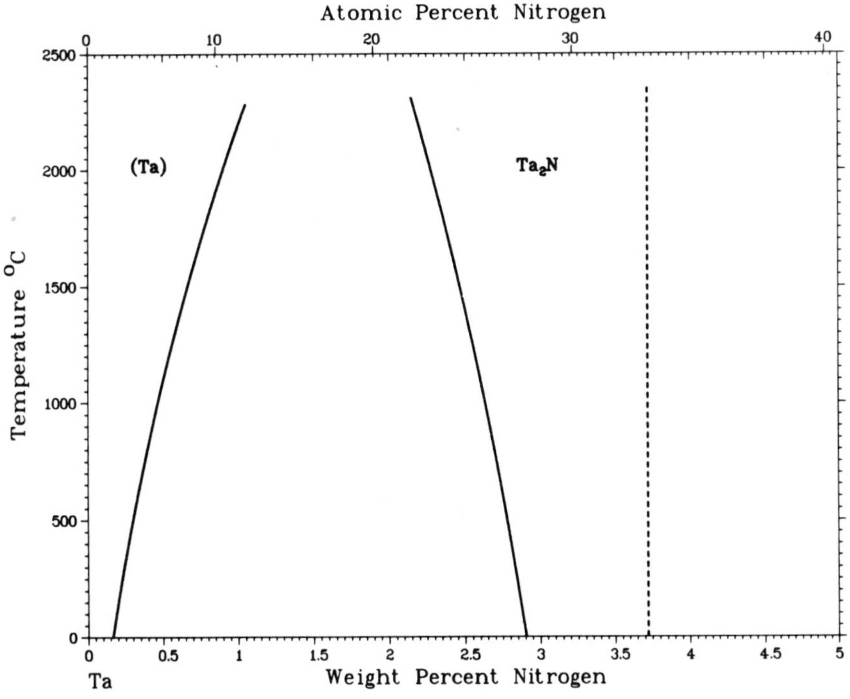

Th-N Phase Diagram

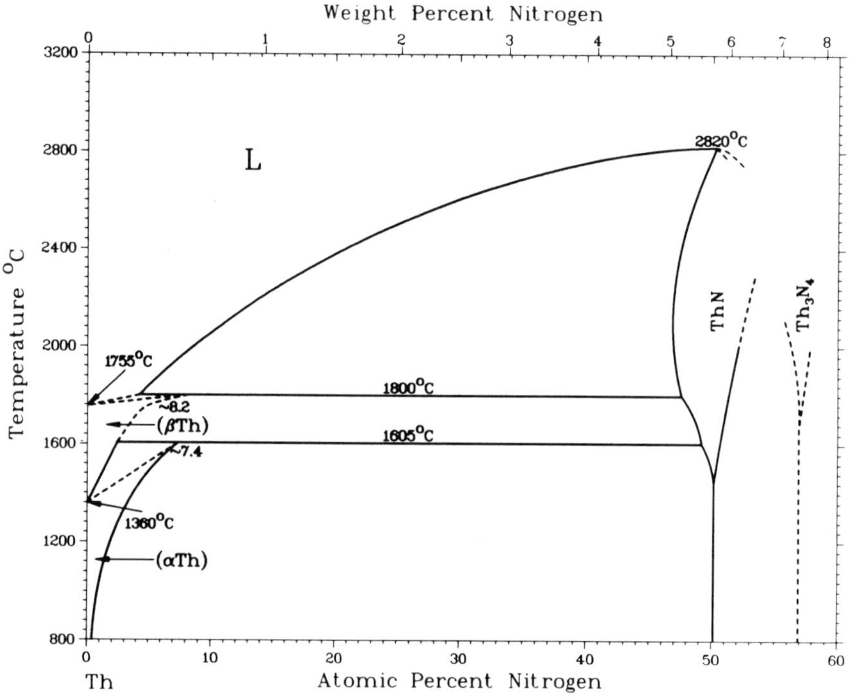

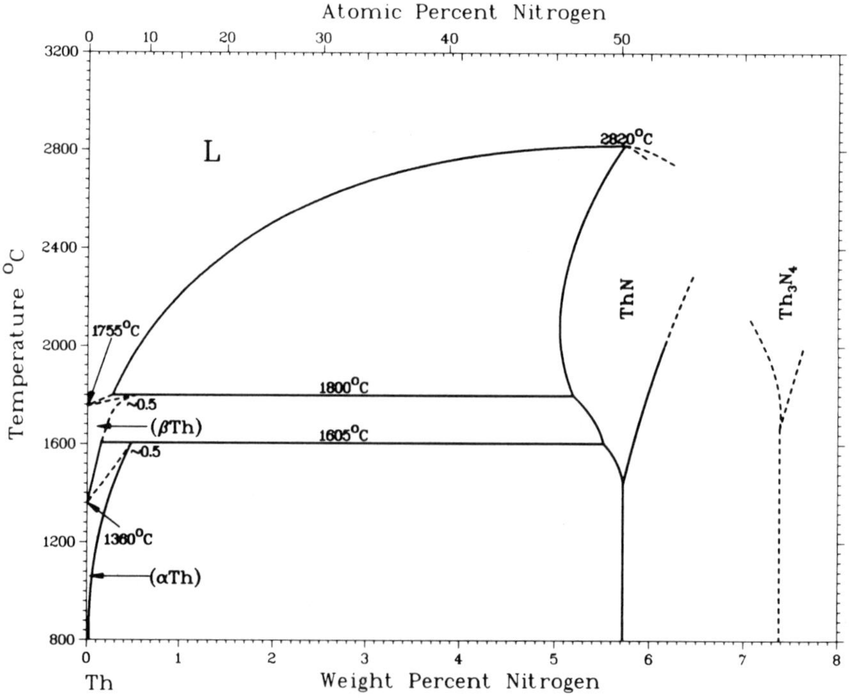

Assessed Ti-N Phase Diagram (Condensed System)

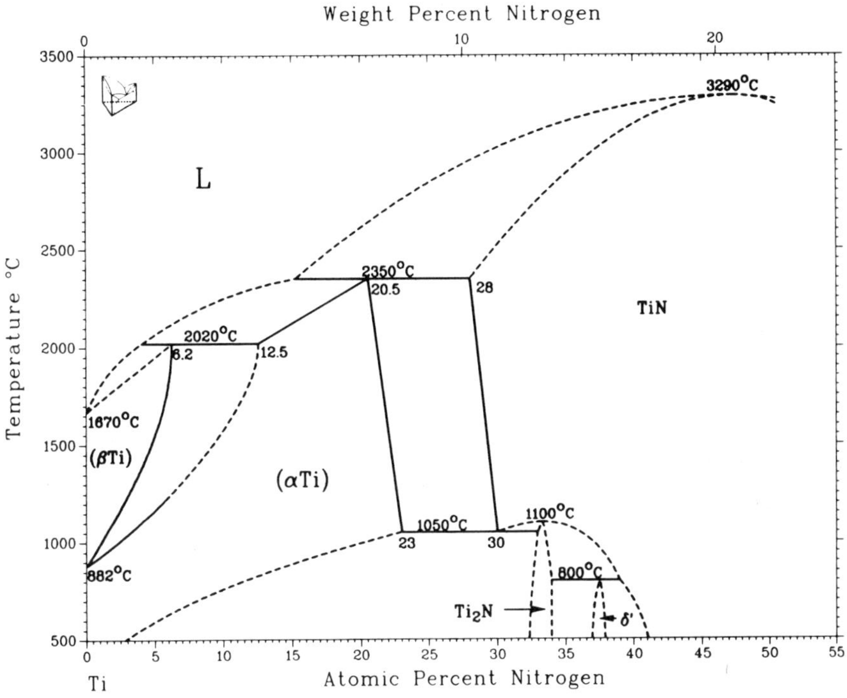

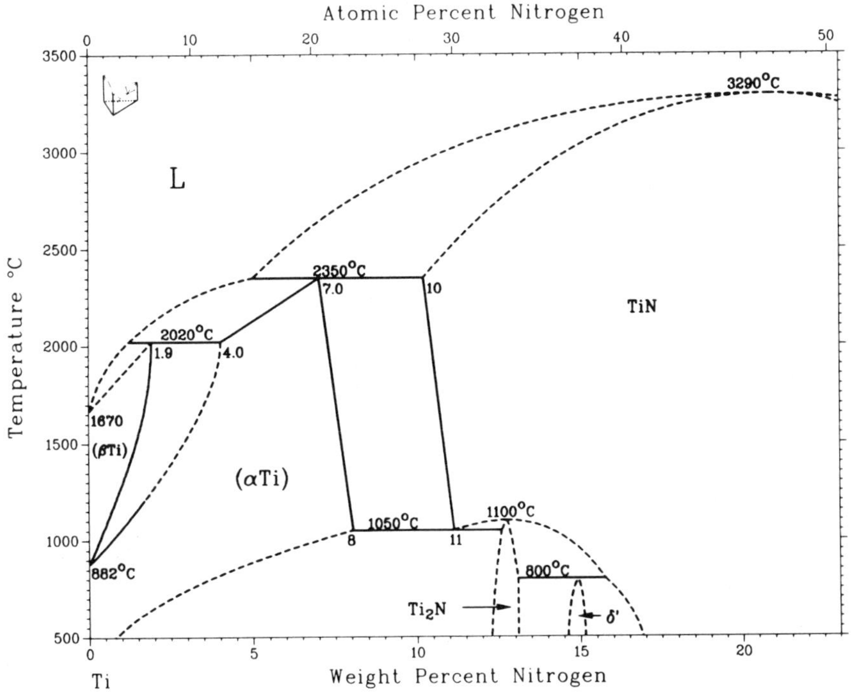

U-N Phase Diagram

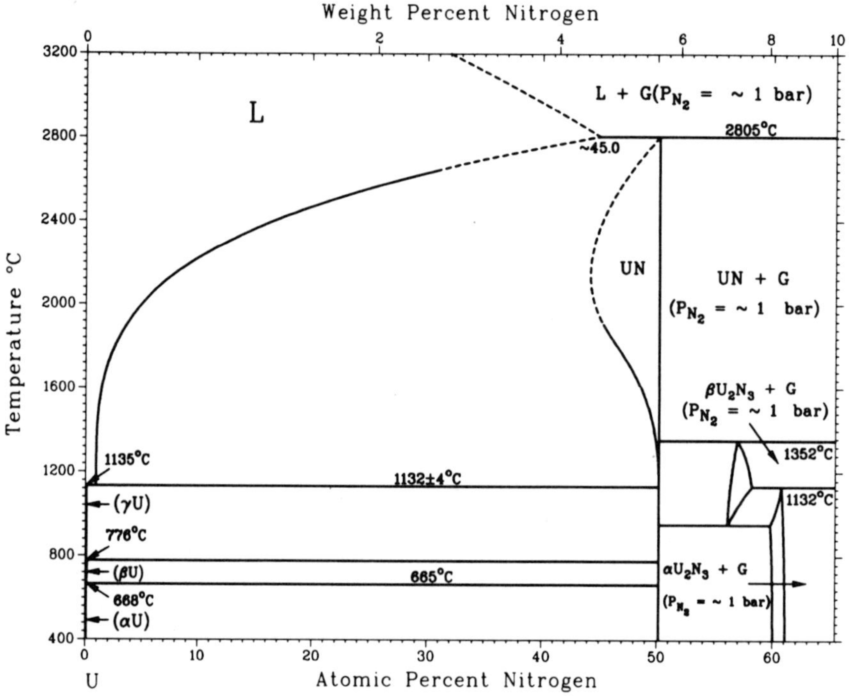

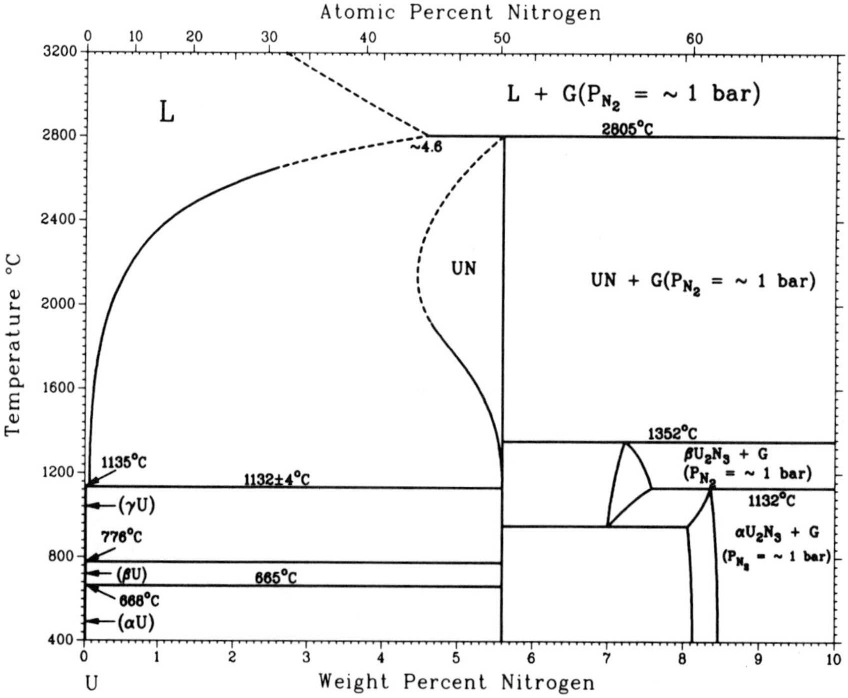

Assessed V-N Phase Diagram

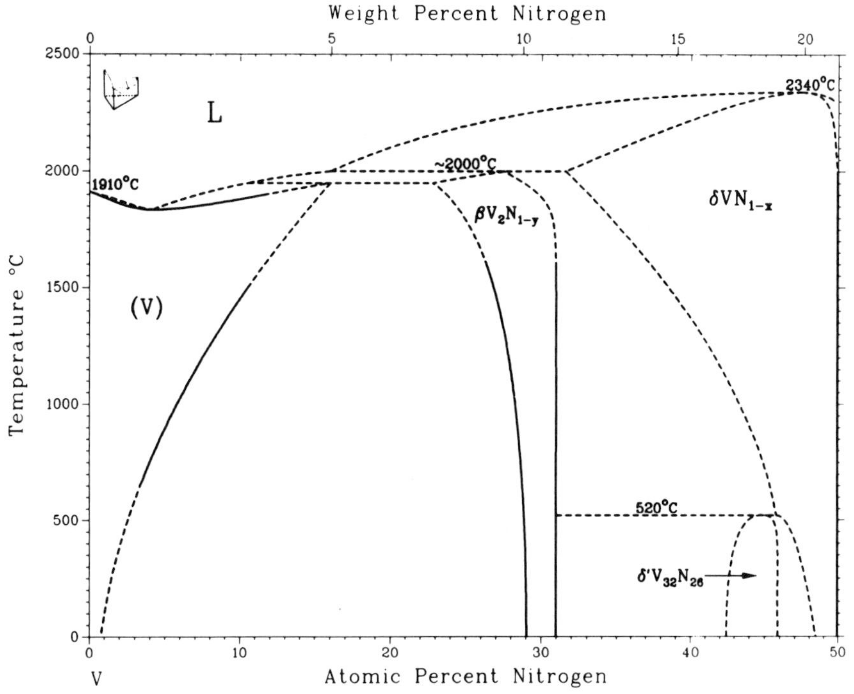

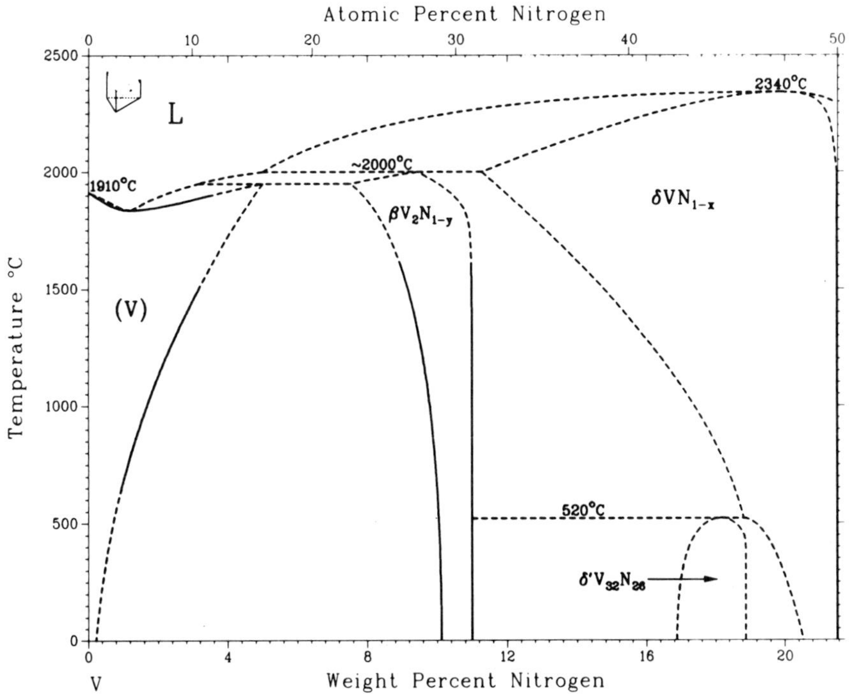

Zr-N Phase Diagram

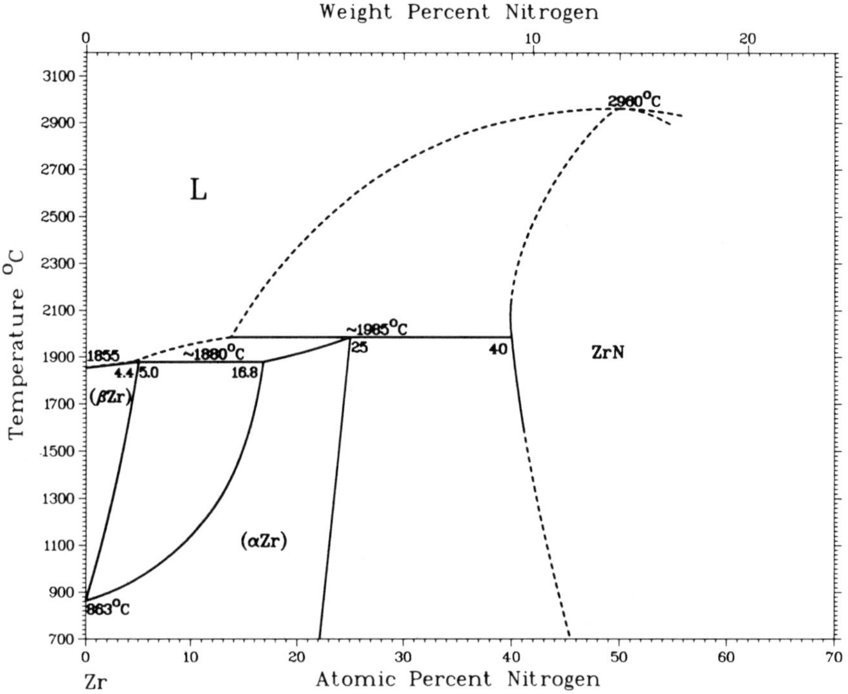

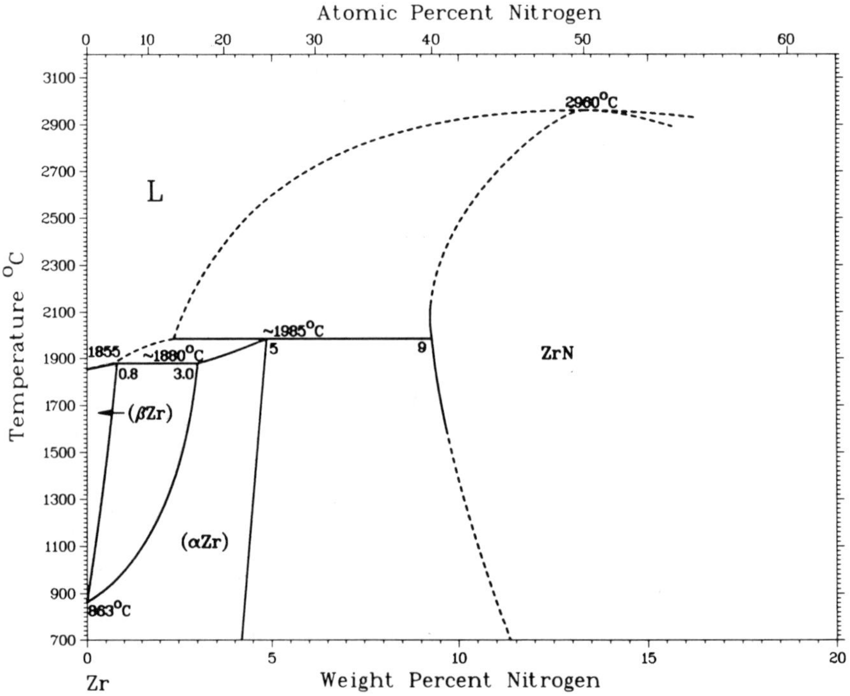

Nb-O Phase Diagram

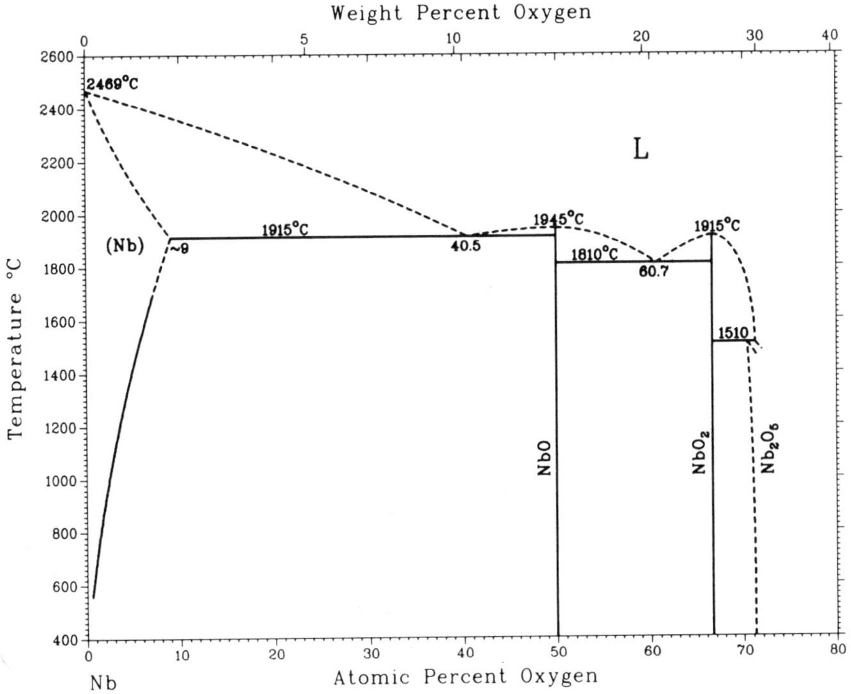

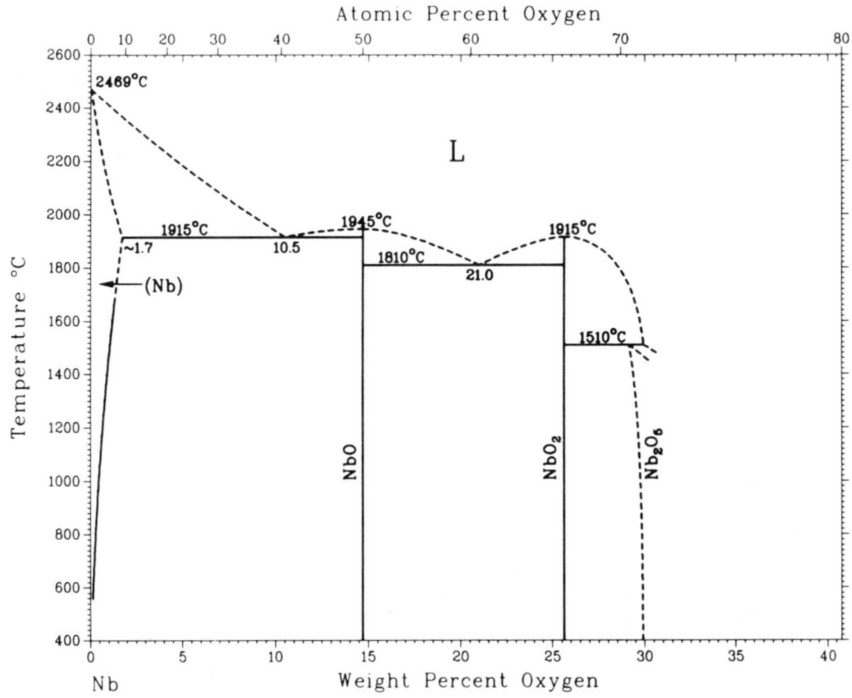

Pu-O Phase Diagram

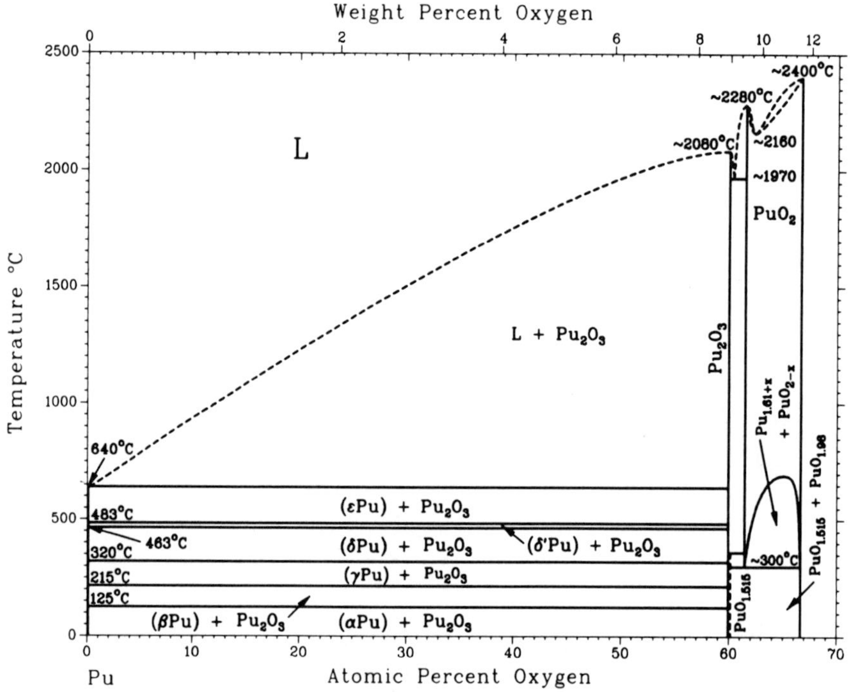

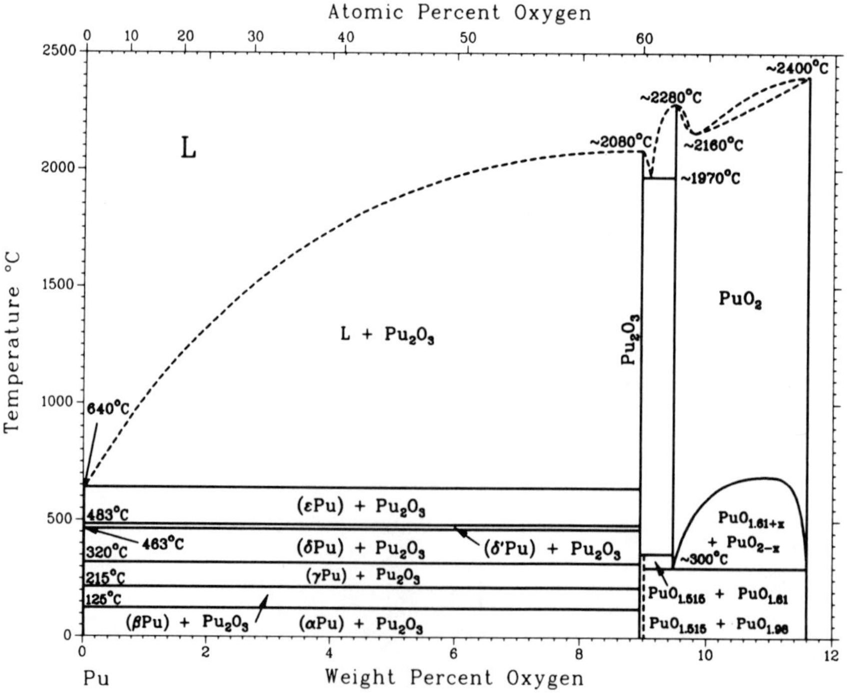

Si-O Phase Diagram

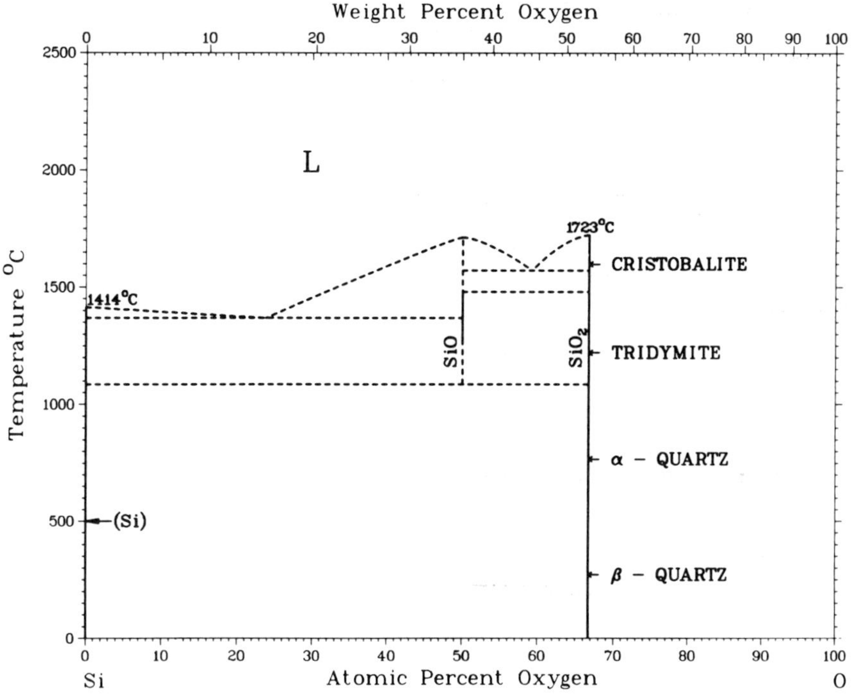

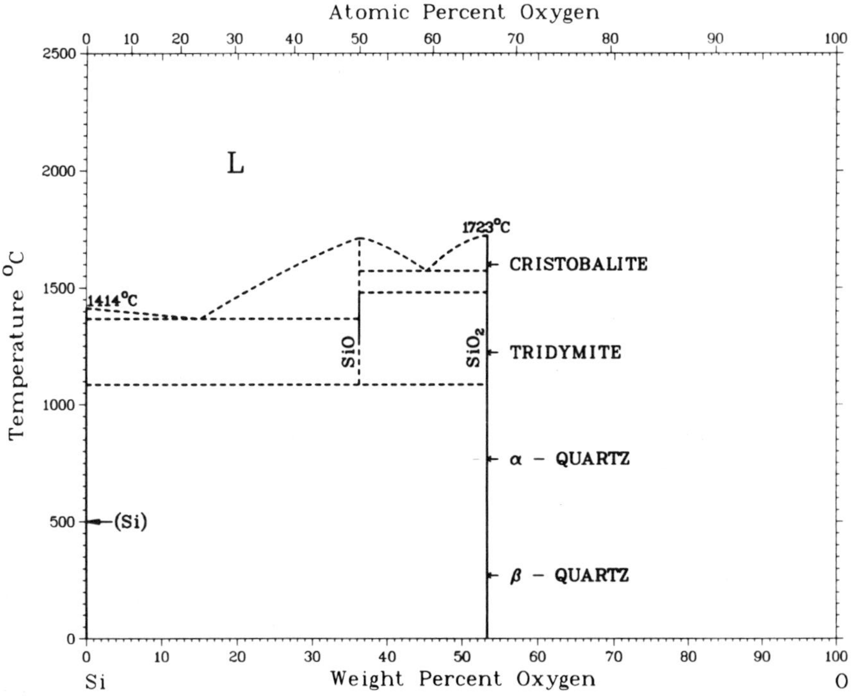

Ta-O Phase Diagram

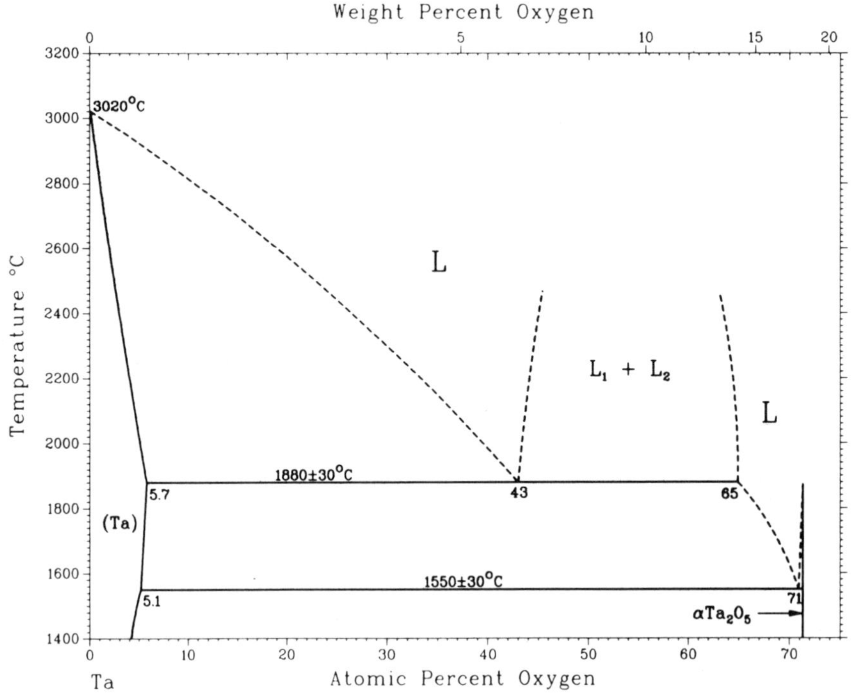

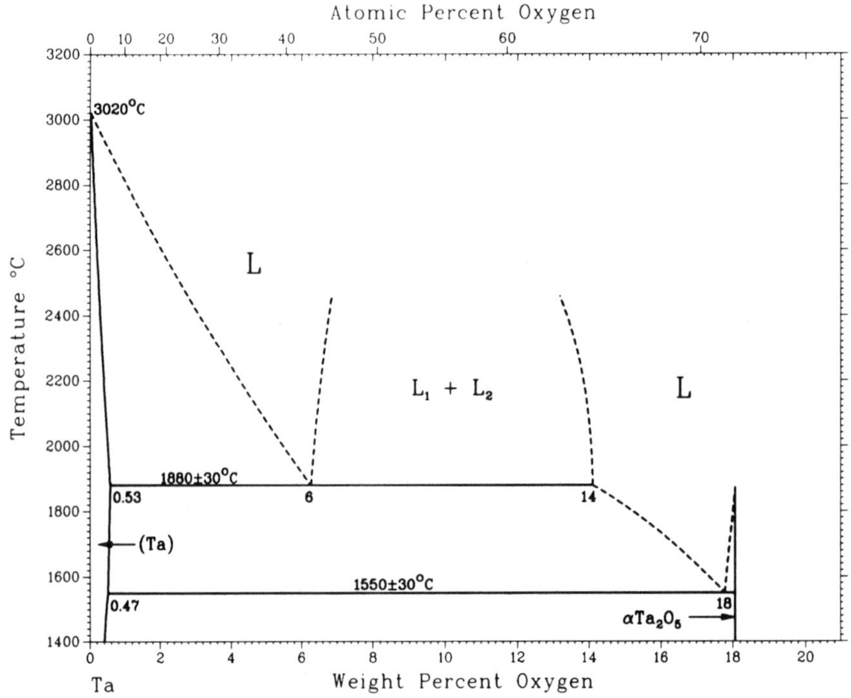

Th-O Phase Diagram

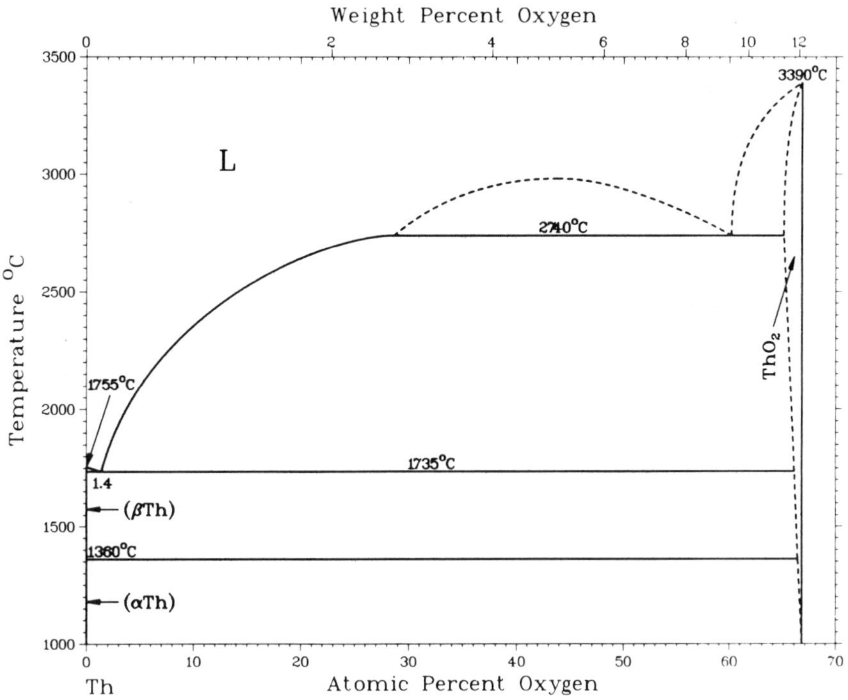

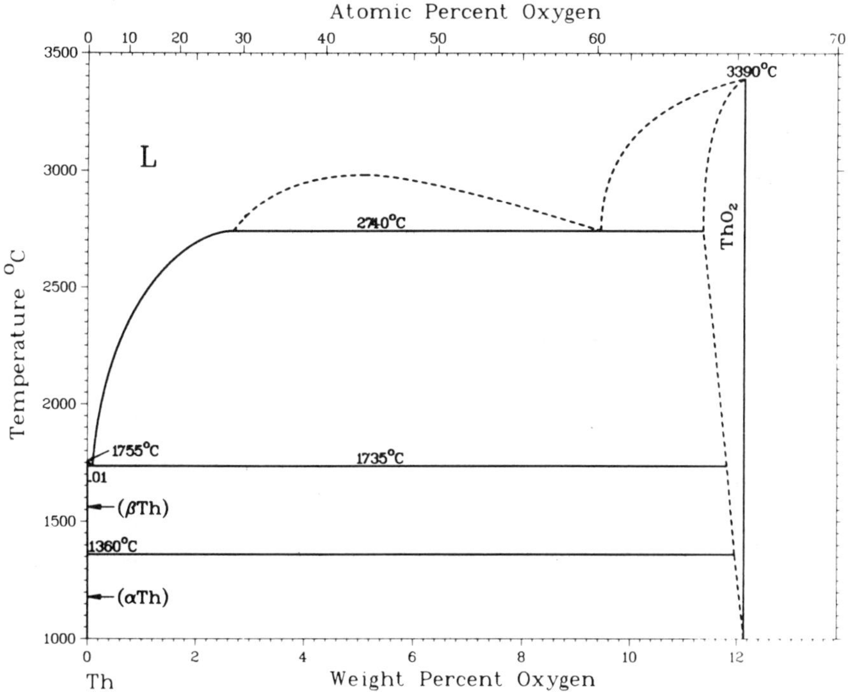

Assessed Ti-O Phase Diagram

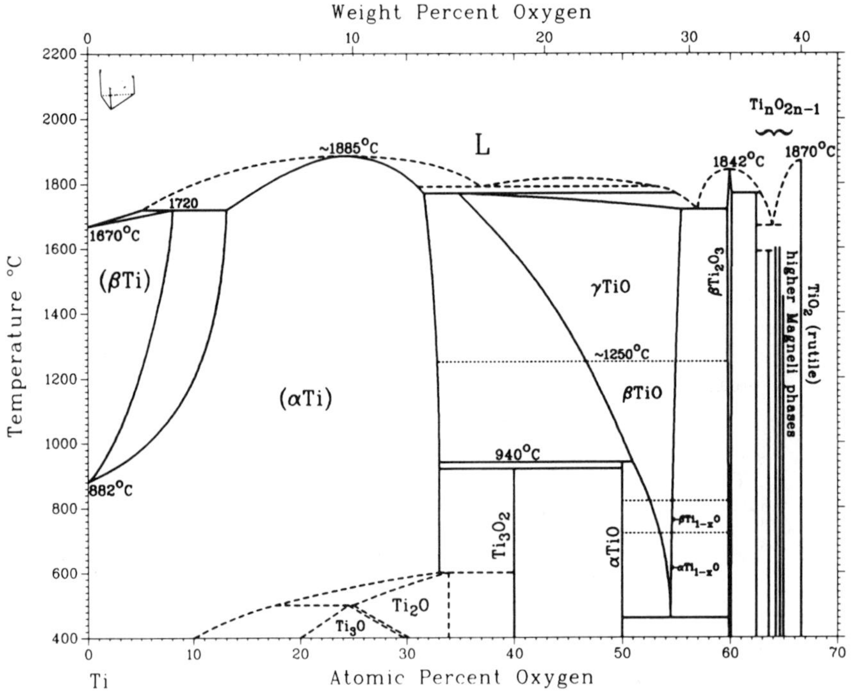

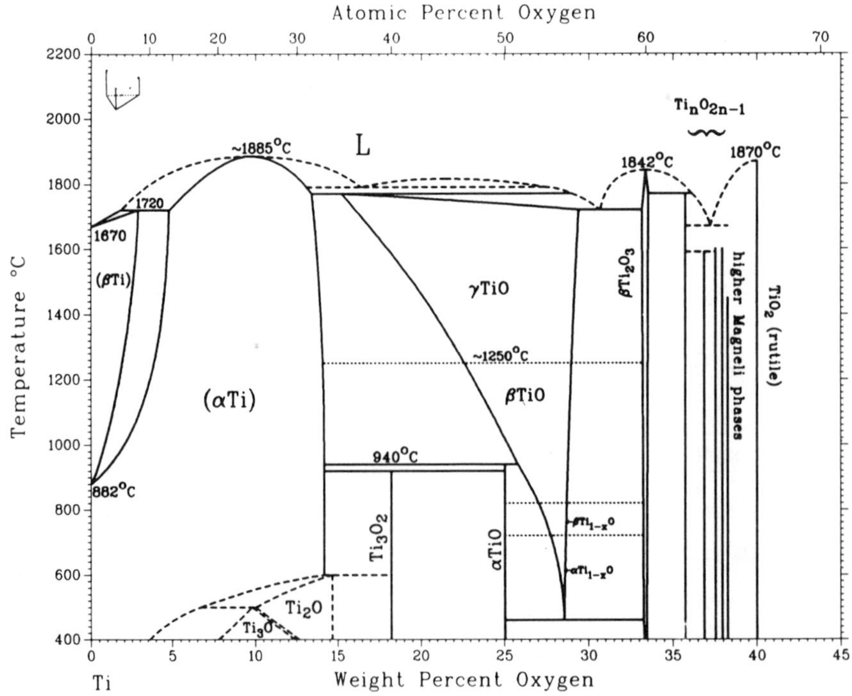

U-O Phase Diagram

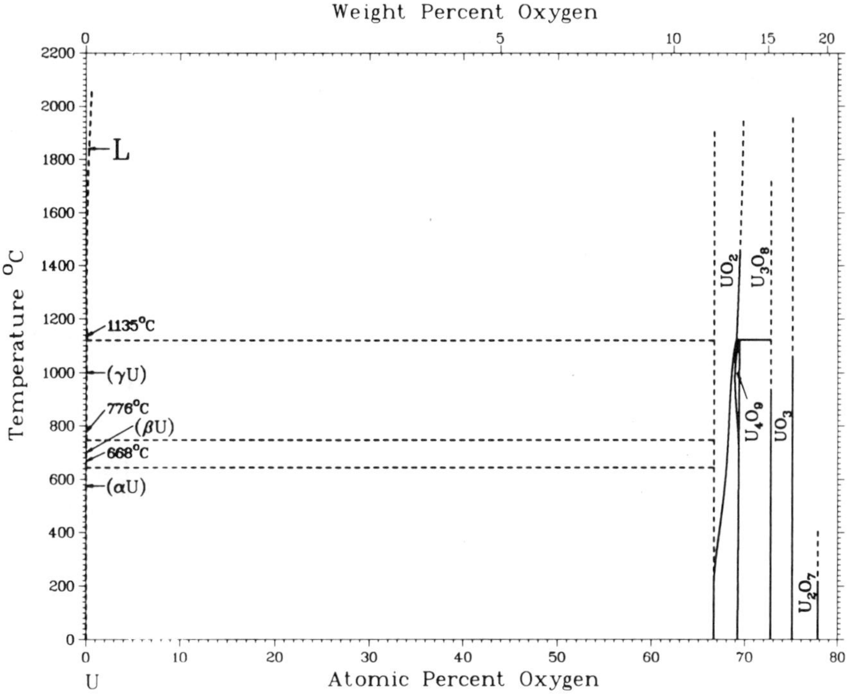

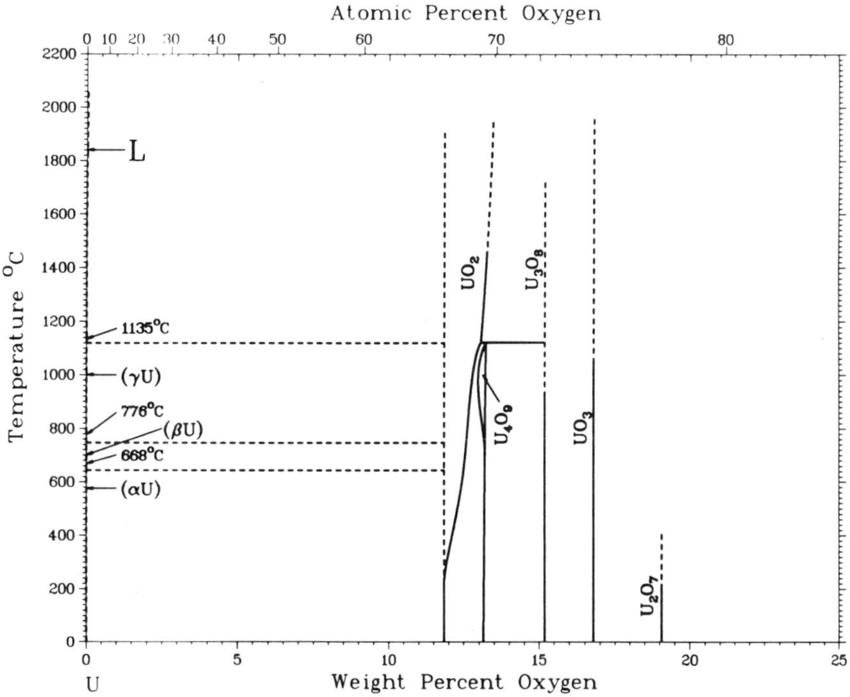

V-O Phase Diagram

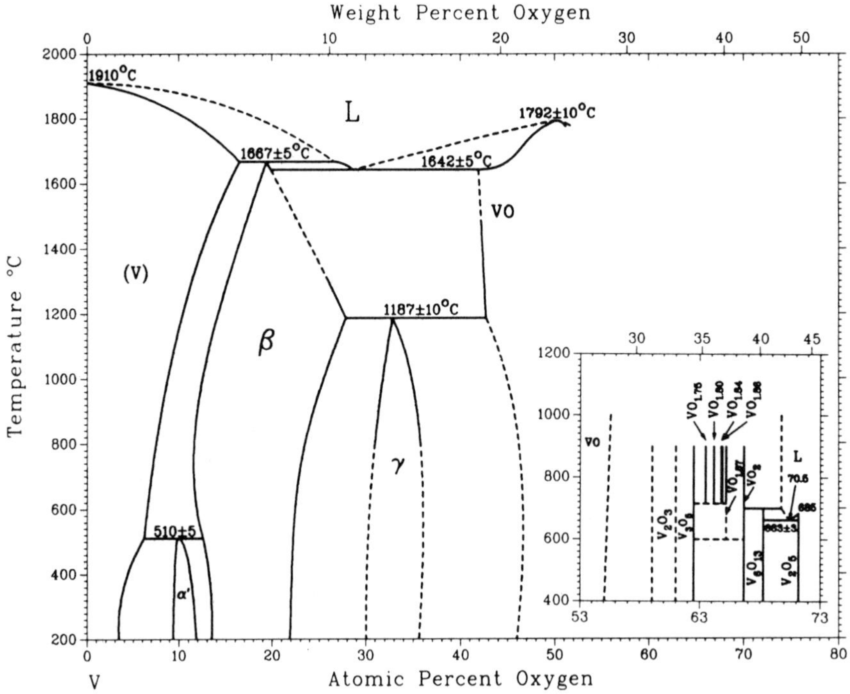

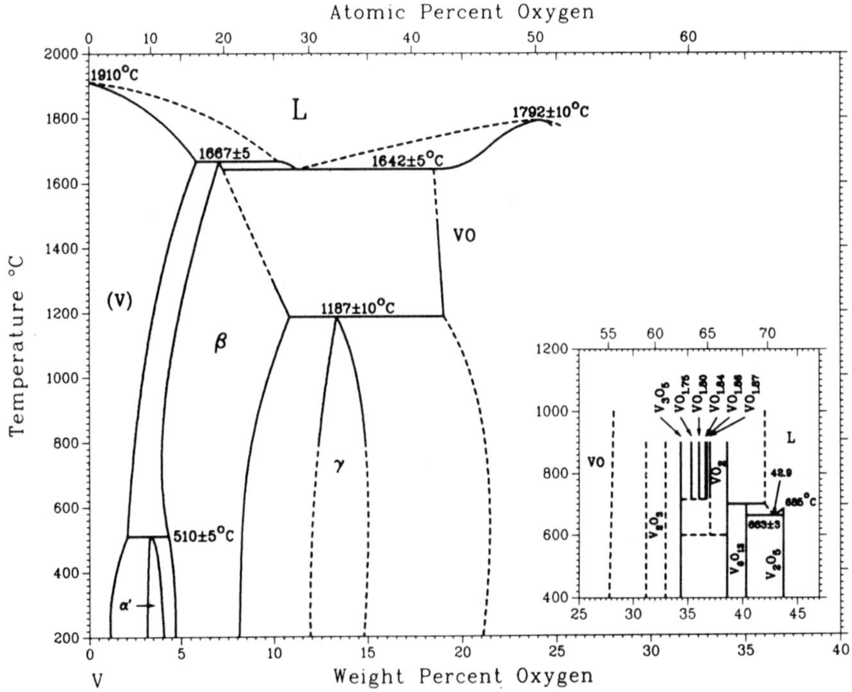

Y-O Phase Diagram

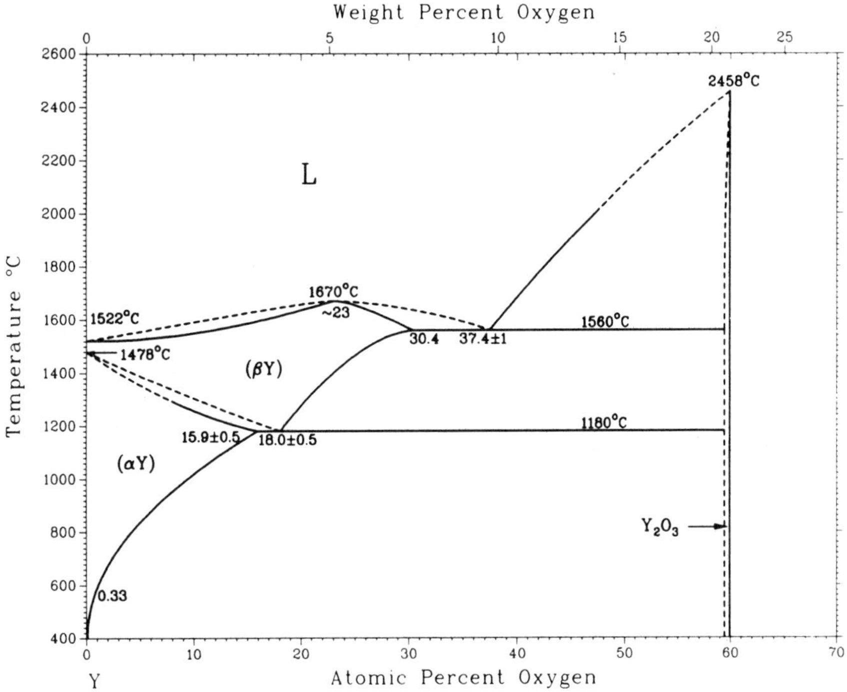

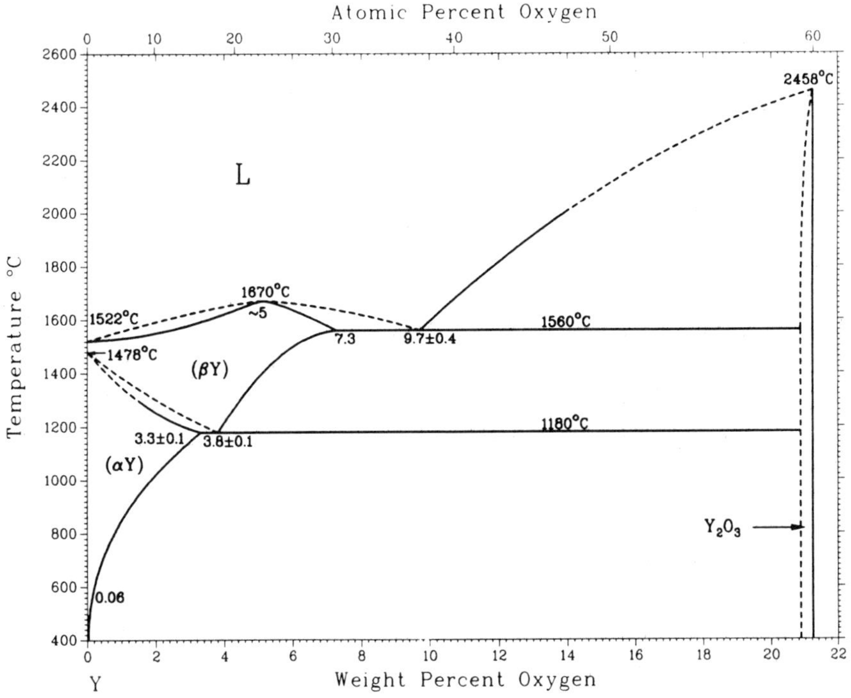

Assessed Zr-O Phase Diagram at 1 atm

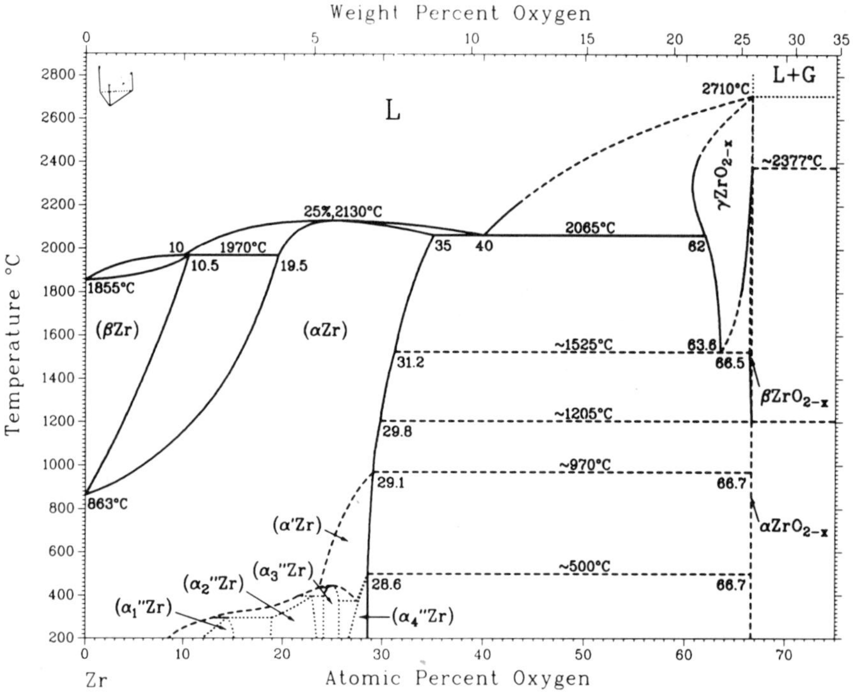

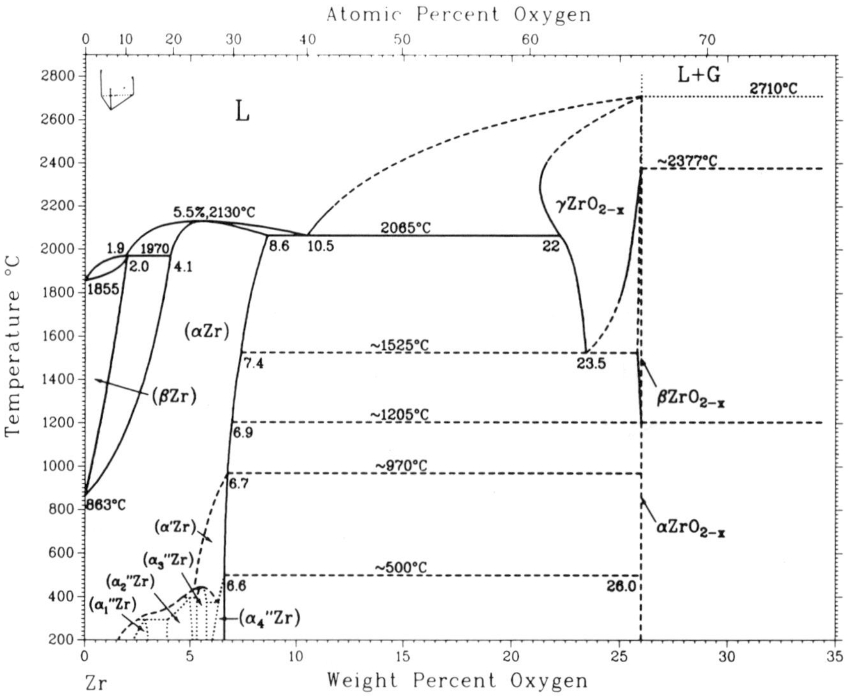

Plastics

PLASTICS

Plastics

General

Application areas generally associated with generic resin types, based on performance characteristics

Resin type	Applications	Typical neat resin properties
Polyester	Consumer products, tanks, pipes, pressure vessels, automotive structures	Tensile strength of 3.4 to 90 MPa (0.5 to 13 ksi); compressive strength of 90 to 210 MPa (13 to 30 ksi); up to 120 °C (250 °F) continuous use; low viscosity; fast reaction; can be catalyzed; high shrinkage
Vinyl ester	Consumer products, pipes, ducts, stacks, automotive structures, flooring, linings	Tensile strength of 60 to 90 MPa (9 to 13 ksi); elongation of 2-6%; up to 120 °C (250 °F) continuous use; low viscosity; fast reaction; can be catalyzed; intermediate shrinkage
Polybutadiene	Resin modifiers, coatings, adhesives, potting compounds	Good chemical resistance; up to 120 °C (250 °F) continuous use; high viscosity; fast reaction; can be catalyzed; low moisture pick-up
Epoxy	Adhesives, tooling, electronics, aerospace and automotive structures	Tensile strength of 55 to 130 MPa (8 to 19 ksi); excellent chemical resistance; up to 175 °C (350 °F) continuous use; high viscosity; can be catalyzed; intermediate reaction; low shrinkage
Polyimide	Primary and secondary aerospace structures in high-temperature areas, electronics	Tensile strength of 55 to 120 MPa (8 to 17 ksi); up to 315 °C (600 °F) continuous use; high viscosity; can be catalyzed; slow reaction; reaction by-products; microcracking
Bismaleimide	Similar to polyimide	Similar to polyimide, except that continuous use only up to 230 °C (450 °F); no reaction by-product
Low-performance thermoplastic	Automotive panels, appliance housings, gears, bearings, fixtures, consumer products	Amorphous or semicrystalline; high toughness; up to 120 °C (250 °F) continuous use; high processing temperatures and pressures; high viscosity
Engineering-grade thermoplastic	Automotive and aerospace structures	High toughness; up to 230 °C (450 °F) continuous use; high processing temperatures and pressures; high viscosity; amorphous or semicrystalline

Source: Ref 6, p 169

Values of fracture toughnness and glass transition temperature for selected plastics

Plastic	Fracture energy, kJ/m^2	Fracture energy, $ft{\cdot}lb/in.^2$	T_g, °C	T_g, °F
Thermosets				
Polyimide-1	0.20	0.095	350	662
Polyimide-2	0.12	0.057	360	680
Tetrafunctional epoxy	0.076	0.036	260	500
Thermoplastics				
Polyarylsulfone	5.5	2.6	...	...
Poly(amide-imide)	3.4	1.6	275	527
Polysulfone	3.1	1.5	174	345
Polyethersulfone	2.6	1.2	230	446
Polyimide-4	2.1	1.0	365	689
Polyimide-3	0.81	0.38	326	619
Polyphenylene sulfide	0.21	0.10	...	...

Source: Ref 31, Table 4

Common thermoplastic and thermoset molding compounds

Base polymer	Principal applications	Potential subprocess problem areas(a) Feeding	Transporting	Injecting	Flowing
Thermoplastics					
ABS	Furniture, cabinets, containers, trim	B	H I	K L N O	Q
Acetal	Clock gears, miniature engineered parts	B	F H I J	O	P S
Acrylic	Automobile light lenses, plastic glazing	B	H I	K L N O	Q
Cellulose	Esters trim, moldings, screwdrivers	B	H I	K L N O	Q R
Polycarbonate	Auto bumpers, traffic lights, lenses	B C	F H I J	K L N O	Q
Polyester	Appliance parts, pump and electrical housings	B C	H I J	O	P S
Polyethylene	Houseware, food storage, dunnage	B	I	O	P
Fluoroplastics	Corrosion/solvent-resistant parts	B C	F I J	K L N O	Q
Polyimide	Aerospace items, electrical insulators	B C	H I J	K L N O	Q
Ionomer	Bumper rub strips, golf ball covers	B	H I	K L N O	P
Nylon	Auto parts, bearing retainers, appliances	B C	H I J	O	P S
Polyphenylene oxide and alloys	Auto instrument panels	B	H I (J)	K L N O	Q R
Polypropylene	Battery cases, auto parts, containers	B	H I	O	P S
Polystyrene	Toys, advertising displays, picture frames	B	H I	K L N O	Q
Polysulfone	Camera cases, aircraft parts, connectors	B C	H I J	K L N O	Q
Polyvinyl chloride	Soft steering wheels, trim items	B	F H I J	K L N O	Q R
Thermosets					
Alkyd	Switches, motor housings, pot/pan handles	A B	G H I J	K L M N O	Q R S T
Allyl	Electrical connectors, circuit boards	A B	G H I J	K L M N O	Q R S T
Epoxy	Electrical insulators, electronic cases	B E	G H I J	K L M N O	Q R S T
Polyester	Automotive structural parts	D E	G I J	K L M N O	P Q R S T
Polyimide	Aircraft components, aerospace parts	A D E	G H I J	K L M N O	P Q R S T
Melamine	Dinnerware, microwave cookware	B	G H I J	K L M N O	Q R S T
Phenolic	Distributor caps, plastic ash trays	A B	G H I J	K L M N O	Q R S T
Urethane	Automotive body panels, bumpers	A D E	G I	L M O	Q R T
Vinylester	Composite car/truck springs, wheels	B D E	G I J	K L M N O	P Q R S T

(a) Feeding
- A—Moisture may chemically react to degrade the polymer base
- B—Drying is recommended to avoid splay in molded product
- C—Drying is essential to prevent molecular weight attrition
- D—Drying may volatilize monomers essential to the curing reaction
- E—Fiber reinforcement breakage may occur during force feeding

Transporting
- F—Overheating may cause explosive depolymerization
- G—Overheating may cause premature curing of (thermoset) compound
- H—Venting is recommended to remove volatiles and reduce splay
- I—Fiber reinforcement lengths may be severely reduced
- J—Overheating may produce chemical changes in the base polymer

Injecting
- K—Fast injection of the molding compound can lead to serious overheating
- L—Filled or reinforced compounds will exhibit much higher viscosity
- M—Open runners required as material can cure in closed channels
- N—Melt fracture may occur with high injection speeds
- O—Fiber-filler orientation will occur if molding compound is reinforced

Flowing
- P—Major sink marks may develop if part sections are thick, not uniform
- Q—Weak knitlines may develop if compound packing pressure is low
- R—Large mold vents recommended to allow volatiles to escape
- S—Hot molds required to promote cure, crystalline growth
- T—Curing reaction may produce peak exotherm leading to degradation

Source: Ref 6, p 165

Advantages and disadvantages of thermoset and thermoplastic materials as housings for electronic assemblies

Housing or container material	Advantages	Disadvantages
Molded thermosets (epoxy, alkyd, phenolic, diallyl phthalate, etc.)	Many standard sizes available; good insulators; corrosion resistant; color or identification can be molded in; terminals can sometimes be molded in; cut off of resin-filled shell easier than for metal cans; same type of material can be used for shell and filling resin, resulting in good compatibility	Does not always adhere to resin, especially if silicone mold releases used to make shell; sealing of leakage joints can be difficult; physically weaker than steel, especially in thin sections; molding flash can cause fitting problems; cleaning of resin spillage can break shells
Molded thermoplastics (nylon, polyethylene, polystyrene, etc.)	Same as listed for thermosets except last two items; often less prone to cracking than thermosetting shells, although this depends on resiliency of material	Same as first three items listed for thermosets; adhesion can be poor due to excellent release characteristics of most thermoplastics; shell can distort from heat; cut off can be a problem, due to melting or softening of thermoplastics under mechanically generated heat

Source: Ref 5, p 405

Advantages and disadvantages of various plastics as mold materials

Mold material	Advantages	Disadvantages
Cast epoxy	Good dimensional control; surface can be polished; can be made for inserts and multiple-part molds: long life and low maintainance	Dimensional control not quite as good as for machined-metal molds; requires mold release and cleaning; low thermal conductivity compared with metals
Cast plastisols	Parts easily removed from molds: molds are easy to make	Short useful life; poor dimensional control
Cast RTV silicone rubber	Same as for plastisols; better life than plastisols	Poor dimensional control, though better than plastisols
Machined TFE fluorocarbon	No mold release required; convenient to make for short runs and simple shapes; withstands high temperature cures	Poor dimensional control
Machined polyethylene and polypropylene	Same as listed for TFE fluorocarbon except high-temperature capability and lower cost	Poor dimensional control
Molded polyethylene and polypropylene	Same as listed for TFE fluorocarbon except high-temperature capability and lower cost	Poor dimensional control

Source: Ref 5, p 404

Chemical and environmental resistance of various plastics(a)

Plastic	Hydrocarbons: Aliphatic	Hydrocarbons: Aromatic	Hydrocarbons: Fully halo genated	Hydrocarbons: Partially halo genated	Alcohols: Mono-hydric	Alcohols: Poly-hydric	Oils, fats, waxes	Phenols	Ketones	Esters	Ethers
ABS	3	4	4	4	3	2	2	4	4	4	4
Acrylic	2	4	4	4	4	2	2	4	4	4	4
Allyic esters	1	2	2	2	1	1	1	3-4	3	3	2
Cellulosic esters	3	4	4	4	4	1	3	4	4	4	3
Chlorinated poly-alkylene ether	2	2	3-4	4	2	1	2	2	3-4	3	2
Epoxy	1	1-2	1	2	1	1	1	3	3	2	3
FEP	1	1	1	2	1	1	1	1	1	1	2
Melamine formaldehyde	2	1-2	1	1	1	1	2	2	2	2	1
Phenolic	1	1	1	1	2	1	1	1	2	2	1
Polyacetal	2	2	2	2-3	2	2	2	4	2	2	2
Polyalkylene ether	3	4	4	4	4	4	3-4	4	4	4	4
Polyamide (nylon)	2	3	3	4	2	2	2	4	2	3	2
Polyaromatic ether	3	4	4	4	4	2	2	1	4	4	4
Polybutadiene (1, 2)	2	3	—	3	1	1	—	2	2	2	2
Polycarbonate	3-4	4	2-3	4	2	2	3-4	4	4	4	4
Polyesters (TP)	2	3-4	2-4	4	1-2	1	—	4	4	3-4	4
Polyesters (Uns)	2	4	4	4	2	2	2	4	4	4	2
Polyethylene	3	3	4	4	3	2	1	1	2	2	2
Polyimide	1	1	1	1	1	1	1	2	1	1	1
Polyphenylene	2	2	2	2	1	1	1	2	2	2	1
Polyphenylene sulfide	1	2	3	4	1	1	—	2	2	1	2
Polypropylene	3	3	4	3-4	2	1	2	2	2	2	3
Polystyrene	3-4	4	4	4	2	2	4	4	4	4	4
Polysulfone	4	4	4	4	3-4	2	3	4	4	4	4
Polyurethane	3	4	4	4	4	2-3	2	4	4	4	4
Polyvinyl acetate	3-4	4	4	4	4	3	3	4	4	4	4
Polyvinyl chloride	2	4	4	4	2	2	2	4	4	4	4
Polyvinyl chloride vinylidene chloride	2	2-3	4	4	1	2	2	3	4	4	4
Polyvinyl fluoride	1	2	1	1	1	1	1	2	4	3	2
Polyvinylidene fluoride	1	2	2	2-3	1	1	1	3	3	3	2
Silicone	3	3-4	4	4	2	1	1	4	2	3	3
Styrene acrylonitrile	2	4	4	4	3	2	2	4	4	4	4
TFE	1	1	1	2	1	1	1	1	1	1	2
Urea-formaldehyde	2	2	1	1	1-2	1	2	2	2	2-3	1
Polybutylene	2	2	4	3	2	1	2	2	2	2	2
Polyxylylene	1	1	1	2	1	1	1	2	3	3	2
Vinyl esters	2	2	1	2	1	1	1	3	3	3	3

(a) 1 - excellent; 2 - good; 3 - fair; 4 - poor.

Source: Ref 15, p A102-A107

General characteristics of amorphous and (partially) crystalline plastics

Material	Characteristics
Amorphous plastics	Transparent Low mold shrinkage Low and uniform coefficient of thermal expansion Low dependence of properties on temperature
Crystalline plastics	Resistance to organic solvents Good dynamic fatigue strength Load-bearing temperature range markedly increased by inorganic fiber reinforcement Enhancement of strength by orientation possible

Source: Ref 24, p 236

Chemical and environmental resistance of various plastics (continued)

Acids						Bases		Salts			
Inorganic		Organic		Oxidizing							
Concentrated	Dilute	Concentrated	Dilute	Concentrated	Dilute	Concentrated	Dilute	Acid	Neutral	Base	Sunlight and weathering
2	2	4	3-4	4	2	2	2	2	1	2	2-3
4	2	4	3	4	3	3-4	2	2	2	2	2
3	3	3	2	4	3	3	2	2	1	2	1
4	3	4	4	4	4	4	3	2	2	3	2-3
3	2	3	2	4	3	3	2	2	2	2	3
3	1	3	2	4	3	1	1	1	1	1	2
1	1	1-2	1	1	1	1	1	1	1	1	1
4	3	2	2	4	3	4	2	1	1	2	2
3-4	2-3	2	3	4	4	4	4	1	1	3	2
4	4	4	3	4	4	4	4	3	2	2	2-3
4	4	4	4	4	4	4	4	4	4	4	4
4	2	4	3	4	4	2	1	4	2	3	2-3
2	1	2	1	4	2	1	1	1	1	1	1
3	1	2-3	1	3	2	2	2	1	1	2	—
3	2	3	3	4	2	4	4	2	1	3	2
2-3	2	3-4	2	4	3-4	4	3-4	2	2	3-4	2
2	1	2	2	4	3	3	2	1	1	2	2
1-2	1	1	1	4	2	2	2	1	1	1	4
3	2	2	1	4	2	4	3	2	2	4	1
2	1	2	1	2	1	2	1	1	1	1	1
2	2	2	2	4	3	2	1	2	2	2	2
1	1	2	1	4	2	1	1	1	1	1	4
3	2	4	3	4	3	3	1	2	1	2	3
2	2	3	2	4	2	2	1	2	1	2	1-2
4	3	4	3	4	4	2	2	2	1	1	3
4	4	4	4	4	4	4	4	4	4	4	2-3
2	1	4	2	2	1	1	1	1	1	1	2
2	2	4	2	3	2	3	1	2	2	2	2
3	1	3	1	2	1	1	1	1	1	1	1
1	1	2	1	1-2	1	1	1	1	1	1	1
2	2	3	2	4	1	2	1	1	1	1	1
2	2	3	2	4	2	2	1	2	1	2	3
1	1	1-2	1	1	1	1	1	1	1	1	1
4	3	2	2	4	3	4	3	2	2	2	2
1	1	2	1	4	2	1	1	1	1	1	4
1	1	1	1	4	2	1	1	1	1	1	2
2	1	2	2	4	3	3	2	1	1	2	2

Thermoplastics

General

Melt viscosities of various thermoplastics

Material	Temperature °C	Temperature °F		Viscosity(a) Pa·s	Viscosity(a) P
Udel P-1700	240	460	....	10^5	10^6
	300	570	....	10^4	10^5
	400	750	....	10^3	10^4
Ultem	305	580	....	10^5	10^6
Ryton	313	595	....	10^4	10^5
	328	620	....	10^3	10^4
Torlon	350	660	....	10^6	10^7
PEEK	400	750	....	10^3	10^4
For reference:					
Molasses	25	80	....	10^2	10^3
5208 Al	100	212	....	10^{-1}	10^0

(a) At low shear rates.

Source: Ref 6, p 102

Characteristics of high-temperature thermoplastics

Product	Producer	Chemistry	Glass transition temperature °C	Glass transition temperature °F	Melting point °C	Melting point °F	Form
Avimid K	Du Pont	Polymide	210	410	...	None	Amorphous
Ryton PAS	Phillips	Polyarylene sulfide	145	293	170	345	Crystalline
Torlon	Amoco	Polyamideimide	335	638	...	None	Amorphous
Ultem	General Electric	Polyetherimide	215	422	...	None	Amorphous
PEEK	ICI	Polyether etherketone	180	360	370	700	Crystalline

Source: Ref 6, p 138

Glass transition temperatures and melting points of selected high-temperature thermoplastics

Polymer	Symbol	Glass transition temperature °C	Glass transition temperature °F	Melting point °C	Melting point °F
Polybutylene terephthalate	PBT	40	105	228	440
Polyethylene terephthalate	PET	80	175	265	510
Polysulfone	PS	190	375	(a)	
Polyphenylene sulfide	PPS	93	200	288	550
Polyether sulfone	PES	230	445	(a)	
Polyether etherketone	PEEK	143	290	340	645
Polyimides	PI	280	535	(a)	
Polyetherimide	PEI	210	410	(a)	

(a) Amorphous polymer.

Source: Ref 6, p 142

Factors affecting the conductivities of thermoplastic compounds

Aspect ratio: ....A continuous pathway of particles is essential to electrical conductivity. The greater the aspect ratio, the less additive needed to make the resin conductive.

Loading level: ...The amount of additive used in a compound. The higher the loading, the greater the conductivity. The lowest loading of an additive required to initiate conductivity is called "critical concentration."

Resin matrix:....Depending on the type of resin chosen, more or less additive may be required to meet conductivity requirements.

Processing:......Improper processing can cause the fiber additives to break into shorter lengths and decrease their effectiveness. Conductivity is enhanced by proper tooling, gating, and processing.

Conductivity: ...The more conductive the additive is, the more conductive the finished product will be.

*From Wilson-Fiberfil International.

Source: Ref 16, p 6.4

Coefficients of thermal expansion for selected thermoplastics

Resin	Glass content, %	Coefficient of thermal expansion(a) $10^{-6}/°C$	$10^{-6}/°F$
Polybutylene terephthalate (PBT)	30	25	14
Polyethylene terephthalate (PET)	30	29	16
	43	22	12
Polyphenylene sulfide (PPS)	40	20	11
Polyetherimide (PEI)	30	20	11
Polyamide (6/6 nylon)	25	23	13
Liquid crystal polymer (LCP)	30	5	3

(a) In direction of flow.

Source: Ref 14, p 99

Values of Poisson's ratio for selected unfilled thermoplastics

Polymer	Poisson's ratio (μ)
Acetal	0.35
Nylon 6/6	0.39
Modified PPO	0.38
Polycarbonate	0.36
Polystyrene	0.33
PVC	0.38
TFE (tetrafluorethylene)	0.46
FEP (fluorinated ethylene propylene)	0.48

Source: Ref 17, p 35

Interlaminar fracture toughness of thermoplastic resins

Material	J/m^2	$ft \cdot lbf/ft^2$
Polyphenylene sulfide	720	50
Polyetherimide	950	65
Polyamideimide	1050	70
Polysulfone	1175	80
Polyether etherketone	1600	110

Source: Ref 6, p 293

Properties of selected plastics

Plastic	Flammability(a)	Heat deflection (DTUL) (b) °C at 1.82 MPa	Heat deflection (DTUL) (b) °F at 264 psi	Thermal conductivity(c) W/m·K	Thermal conductivity(c) Btu·in./ft²·h·°F	Dielectric strength(d) V/μm	Dielectric strength(d) V/mil	Volume resistivity(e), Ω/cm	Relative permittivity(f), at 60 Hz	Arc resistance(g), s	Mold shrinkage(h), mm/mm (in./in.)
Acetal	HB	128	230	0.225	1.56	19.7	500	10^{15}	3.70	129	0.020
Nylon 6	HB	86-103	155-185	0.173	1.20	12.0	305	4.5×10^{13}	4.00	...	0.005
Nylon 6/6	V-2	93	167	0.245	1.70	15.2	385	10^{15}	4.00	120	0.008
Polycarbonate	V-2	150	270	0.195	1.35	15.0	380	$>10^{16}$	2.96	120	0.005-0.007
Polypropylene	HB	64-78	115-140	0.175	1.21	23.6	600	$>10^{17}$	2.20	125	0.020
Polyphenylene sulfide	V-0	153	275	0.241	1.67	15.0	380	10^{16}	...	...	0.007
Acrylonitrile butadiene styrene (ABS)	HB	111-122	200-220	0.138	0.96	13.8-19.7	350-500	2.7×10^{16}	2.80	...	0.004-0.009
Polyphenylene oxide (PPO)	V-1	118	212	0.133	0.92	15.7	400	$>10^{16}$	2.65	75	0.005-0.007
Polystyrene acrylonitrile (SAN)	HB	122	220	0.101	0.70	20.3	515	4.4×10^{16}	3.50	65	...
Polyester (PBT)	HB	68-103	122-185	0.147-0.241	1.02-1.67	16.5-21.7	420-550	4×10^{16}	3.10	184	0.015-0.020
Polyester (PET)	HB	56-59	100-106	0.125	0.87	...	...	$>10^{16}$	...	...	0.02-0.025

(a) ASTM test method UL-94. (b) ASTM test method D648. (c) ASTM test method C177. (d) ASTM test method D149. (e) ASTM test method D257. (f) ASTM test method D150. (g) ASTM test method D495. (h) ASTM test method D955.

Source: Ref 18, p 1-6

Properties of thermoplastics

Material	Supplier	Cost, $/lb	T_g °C	T_g °F	T_m °C	T_m °F	Tensile Strength MPa	Tensile Strength ksi	Tensile Modulus MPa	Tensile Modulus ksi	Fracture toughness, G_{Ic} J/m²	Fracture toughness, G_{Ic} ft·lbf/ft²
Udel P-1700	Union Carbide	4.50	190	375	None	...	76	11	2200	320	3200	220
Radel	Union Carbide	...	220	430	None	...	...	...	...	...	5500	380
Vitrx PES 200P	ICI	...	220	430	None	...	83	12	2410	350	2600	180
Ultem	General Electric	6.80	220	430	None	...	110	16	3300	480	3700	250
Torlon	Amoco	22.00	275	530	None	...	193	28	4800	700	3400	230
Ryton	Phillips	...	85	185	285	545	65	9.5(a)	3800	550(a)	210(b)	15
PEEK	ICI	28.00	143	290	343	650	...	...	...	...	...	...
Avimid K-II	Du Pont	...	277	530	None	...	110	16	2850	415	14000	960
CM-X	Ausimont U.S.A.	100.00	...	...	327	620	37	5.5(b)	3800	550	(b)	...

(a) 15% crystallinity. (b) High crystallinity

Source: Ref 6, p 101

Impact resistance of selected engineering thermoplastics

Thermoplastic	Notched Izod 23 °C (73 °F) J/cm	Notched Izod 23 °C (73 °F) ft·lb/in.	Notched Izod -29 °C (-20 °F) J/cm	Notched Izod -29 °C (-20 °F) ft·lb/in.	Notched Izod -40 °C (-40 °F) J/cm	Notched Izod -40 °C (-40 °F) ft·ln/in.	Falling dart(a) 23 °C (73 °F) J	Falling dart(a) 23 °C (73 °F) in.·lb	Falling dart(a) -29 °C (-20 °F) J	Falling dart(a) -29 °C (-20 °F) in.·lb	Falling dart(a) -40 °C (-40 °F) J	Falling dart(a) -40 °C (-40 °F) in.·lb
Acetal	0.53	1.0	...	...	...	...	...	...	...	...	...	...
Nylon amorphous	5.9-11	11-21	2.1-4.8	4-9	...	...	...	...	...	...	...	...
Nylon 6(b)	0.53-0.80	1.0-1.5	...	...	...	...	...	...	...	...	...	...
Impact-modified nylon 6(b)	9.6	18	2.9	5.5	...	...	65.1	576	57.0	504	...	...
Nylon 6/6(b)	0.3-0.6	0.5-1.2	...	...	...	...	...	...	...	...	...	...
Toughened nylon 6/6(b)	9.1-12	17-22	...	...	...	...	...	...	...	...	...	...
33% glass-toughened nylon 6/6(b)	2.2	4.1	...	...	...	...	...	...	...	...	...	...
Nylon/ABS	9.98	18.7	1.0	1.8	...	...	73.2	648	35.3	312	...	...
Nylon/PPO	2.1	4.0	1.3	2.5	...	...	50.8	450	39.6	350	...	...
Polyarylate	2.9	5.5	...	...	2.1	4.0	...	...	...	...	...	...
Polycarbonate	6.4-8.5	12-16	...	...	1.3-5.98	2.5-11.2	...	...	...	...	...	...
Polycarbonate/ABS	4.5-6.4	8.5-12	2.7-6.14	5-11.5	...	...	38.0-61.0	336-540	29.8-46.1	264-408	...	...
Polycarbonate/PBT	6.9-8.5	13-16	...	...	1.6-6.4	3-12	55.6	492	48.8	432	...	...
Polycarbonate/PET	9.1	17	6.9	13	...	...	57.0	504	50.2	444	...	...
Polyester, aromatic (LCP)	0.53-5.3	1-10	...	...	...	...	...	...	...	...	...	...
Thermoplastic polyester, PBT	0.4-0.53	0.8-1.0	...	...	...	...	...	...	...	...	...	...
Impact-modified PBT	8.0-8.5	15-16	7.5-8.0	14-15	...	...	43.4-54.2	384-480	43.4	384	...	...
Impact-modified PBT/PET	8.0-8.5	15-16	...	...	...	...	54.2	480	...	...	...	...
30% glass-reinforced PET, thermoplastic polyester	1.0	1.9	...	...	1.0	1.8	...	...	...	...	...	...
Toughened 30% glass PET	1.4-2.3	2.6-4.4	...	...	1.2-1.6	2.3-3.0	...	...	...	...	...	...
Polyether ether ketone	0.59-0.80	1.1-1.5	...	...	...	...	...	...	...	...	...	...
Polyetherimide	0.3-0.53	0.6-1.0	...	...	...	...	33.9-36.2	300-320	...	...	...	...
Polyether sulfone	0.53-0.80	1.0-1.5	...	...	...	...	...	...	...	...	...	...
Polyphenylene oxide, modified	2.7-5.3	5-10	...	...	1.3-1.9	2.5-3.5	14.7-33.9	130-300	...	...	3.4-11.3	30-100
Polyphenylene sulfide	0.3-0.75	0.5-1.4	...	...	...	...	...	...	...	...	...	...
Polyphthalate carbonate	5.3	10	...	...	...	...	...	...	...	...	...	...
Polysulfone	0.69	1.3	...	...	0.64	1.2	...	...	...	...	...	...
Polyurethane, engineering thermoplastic	0.5-2.1	1.4	...	...	...	...	...	...	...	...	...	...

(a) Gardner or falling dart. (b) Dry, as molded.

Source: Ref 4, p 293

Wear factors, coefficients of friction, and PV limits of reinforced and lubricated thermoplastics against steel

Material	PTFE content, wt %	Silicone content, wt %	Glass fiber content, wt %	Carbon fiber content, wt %	Aramid fiber content, wt %	Wear factor(a)	Coefficient of friction(b)	PV limit at 100 fpm psi·fpm
ABS(c)			Unmodified			3500	0.35	...
	15	...	...	...	...	300	0.16	4,000
	...	2	...	...	...	80	0.14	...
SAN(d)			Unmodified			3000	0.33	...
	15	...	...	...	...	200	0.14	5,000
	...	2	...	...	...	70	0.13	...
Polystyrene			Unmodified			3000	0.32	1,500
	...	2	...	...	...	37	0.08	9,000
	15	...	...	...	...	175	0.14	...
Polycarbonate			Unmodified			2500	0.38	500
	15	...	...	...	...	75	0.15	20,000
	13	2	...	...	...	40	0.09	23,000
	20	...	...	...	...	70	0.14	22,000
	13	2	30	...	...	27	0.19	30,000
	...	...	...	30	...	85	0.17	5,500
Polyetherimide			Unmodified			4000	0.17	...
	...	...	30	...	...	130	0.24	...
	...	...	...	30	...	75	0.23	...
	15	...	30	...	...	35	0.20	...
Polyethylene	20	...	...	...	...	45	0.13	...
Polysulfone			Unmodified			1500	0.37	5,000
	15	...	...	...	...	46	0.14	...
	...	...	30	...	...	160	0.22	...
	15	...	30	...	...	55	0.19	35,000
Polyether sulfone	15	...	30	...	...	60	0.20	30,000
	15	...	...	30	...	40	0.17	33,000
Polyimide			Unmodified			100	0.29	300,000
	10	...	...	15	...	28	0.12	1,000,000
Acetal			Unmodified			65	0.21	3,500
	15	...	...	...	...	20	0.16	...
	...	2	...	...	...	20	0.11	12,000
	18	2	...	...	...	7	0.10	18,000
	...	...	...	20	...	40	0.14	20,000
	20	...	...	...	...	13	0.12	16,000
	...	2	...	...	...	27	0.12	9,000
Polypropylene	20	...	...	...	...	33	0.11	5,000
	15	...	30	...	...	36	0.09	12,000
Polyphenylene sulfide			Unmodified			540	0.24	3,000
	20	...	...	...	...	55	0.10	...
	15	...	...	30	...	75	0.15	35,000
	...	...	...	30	...	160	0.20	20,000
	13	2	30	...	...	50	0.22	30,000
Nylon 6/6			Unmodified			200	0.28	2,500
	20	...	...	...	...	12	0.18	17,500
	...	2	...	...	...	40	0.09	6,000
	18	2	...	...	...	6	0.08	30,000
	...	...	...	30	...	20	0.20	27,000
	...	...	30	...	...	75	0.31	10,000
	13	2	...	30	...	6	0.11	43,000
	...	...	...	...	20	62	0.25	...
	10	...	...	...	15	23	0.25	...
Polyurethane			Unmodified			340	0.37	1,500
	...	...	...	60	0.32			
	...	2	...	...	...	55	0.31	...
	15	...	30	...	...	35	0.25	10,000

(continued)

Wear factors, coefficients of friction, and PV limits of reinforced and lubricated thermoplastics against steel (continued)

Material	PTFE content, wt %	Silicone content, wt %	Glass fiber content, wt %	Carbon fiber content, wt %	Aramid fiber content, wt %	Wear factor(a)	Coefficient of friction(b)	PV limit at 100 fpm psi·fpm
Polyether (PBT)	Unmodified					210	0.25	...
	20	...	...	...	...	15	0.17	15,500
	18	2	...	...	...	9	0.13	...
	...	...	...	30	...	24	0.15	22,000
	13	2	30	...	...	12	0.12	24,000
	...	...	...	...	20	95	0.21	...
Polyester elastomer	Unmodified					1000	0.59	...
	...	2	...	...	...	30	0.22	...
	15	...	...	...	...	40	0.25	...
	13	2	...	...	...	5	0.21	...
Modified PPO(e)	Unmodified					3000	0.39	500
	15	...	...	...	...	100	0.16	...
	15	...	30	...	...	45	0.22	22,000
Polyether ether keytone	...	...	...	30	...	60	0.13	...
	20	...	...	...	...	130	0.23	...
	15	...	...	15	...	60	0.20	40,000
Bronze (oil impregnated)	...	...	...	...	...	100	0.20	...
Phenolic:								
Wood flour and PTFE	...	...	...	...	...	30	0.26	...
Cellulose and PTFE	...	...	...	...	...	300	0.16	35,000
Teflon PTFE (milled glass)	...	...	...	...	...	8	0.16	...

(a) 10^{-10} in.^{3} · min/ft·lb·h. (b) Dynamic, at 40 psi and 50 fpm. (c) Acrylonitrile butadiene styrene. (d) Styrene acrylonitrile. (e) Polyphenylene oxide.

Source: Ref 19, p 231; Ref 4, p 90

Retention of tensile strength by selected thermoplastics after exposure to various chemical media

Chemical medium	Retention of tensile strength(a), %, by: PPS(b)	Nylon 6/6	PC(c)	PSO(d)	Modified PPO(e)
Acids:					
Hydrochloric, 37%	100	0	0	100	100
Sulfuric, 30%	100	0	100	100	100
Acetic, glacial	98	0	67	91	78
Bases:					
Ammonium hydroxide, 28%	100	85	0	100	100
Sodium hydroxide, 30%	100	89	7	100	100
N-butylamine	49	91	0	0	0
Aniline	96	85	0	0	0
Hydrocarbons:					
Cyclohexane	100	90	75	99	0
Toluene	98	76	0	0	0
Diesel fuel	100	87	100	100	36
Gasoline	100	80	99	100	0
Organic solvents:					
Chloroform	87	57	0	0	0
Chlorobenzene	100	73	0	0	0
Ethylene chloride	72	65	0	0	0
Butyl alcohol	100	87	94	100	84
Cyclohexanol	100	84	74	95	27
Phenol	100	0	0	0	0
Methyl ethyl ketone	100	87	0	0	0
Ethyl acetate	100	89	0	0	0

(a) After exposure for 24 h at 93 °C (200 °F). (b) Polyphenylene sulfide. (c) Polycarbonate. (d) Polysulfone. (e) Polyphenylene oxide.

Source: Ref 4, p 112

Fatigue endurance of reinforced engineering thermoplastics

Material	Glass fiber content, %	Carbon fiber content, %	Cyclic failure stress(a) At 10^4 cycles MPa	ksi	At 10^7 cycles MPa	ksi
Acetal copolymer	30	...	62	9.0	48	7.0
Nylon 6(b)	30	...	48	7.0	40	5.75
Nylon 6/6	...	...	45	6.5	36	5.2
	40	...	72	10.5	63	9.1
Nylon 6/6(b)	...	...	23	3.4	21	3.1
	30	...	55	8.0	41	5.9
	40	...	62	9.0	48	7.0
	...	30	90	13.0	55	8.0
	...	40	103	15.0	59	8.5
Nylon 6/10(b)	30	...	47	6.8	40	5.5
	40	...	55	8.0	48	7.0
Polycarbonate	20	...	62	9.0	34	5.0
	40	...	100	14.5	41	6.0
Polyester, PBT	30	...	76	11.0	35	5.1
	...	30	90	13.0	45	6.5
Polyether ether ketone	...	30	124	18.0	121	17.5
Polyether sulfone	30	...	110	16.0	34	5.0
	40	...	131	19.0	43	6.2
	...	30	152	22.0	46	6.7
Modified polyphenylene oxide	30	...	50	7.2	33	4.75
Polyphenylene sulfide	...	30	90	13.0	66	9.5
Polysulfone	30	...	97	14.0	31	4.5
	40	...	110	16.0	38	5.5

(a) Specimens tested in accordance with ASTM D671 at 1800 cycles per minute. (b) Moisture conditioned, 50% relative humidity.

Source: Ref 4, p 69

Melt viscosities of thermoplastic resins

Material	Temperature °C	°F	Viscosity Pa·s	P
Polyamideimide	350	662	10^6	10^7
Polysulfone	300	570	10^4	10^5
Polyetherimide	305	580	10^5	10^6
Polyphenylene sulfide	313	595	10^4	10^5
Polyether ether ketone	400	750	10^3	10^4

Source: Ref 6, p 294

Acetal

Molecular structure of polyacetal resin

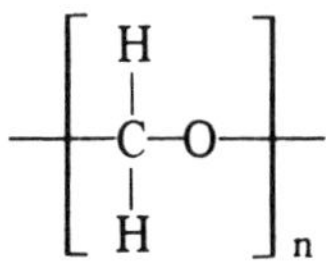

Source: Ref 20, p 50

Selected properties of acetal (injection molded or extruded)

Property	ASTM test	Reported value
Mechanical properties		
Yield strength, MPa (ksi)	D638	62-69 (9-10)
Tensile modulus, GPa (10^6 psi)	D638	2.8-3.6 (0.41-0.52)
Elongation (break), %	D638	25-40
Compressive strength, MPa (ksi)	D695	36 (5.2)
Flexural yield strength, MPa (ksi)	D790	90-97 (13-14)
Flexural modulus, GPa (10^6 psi)	D790	2.8 (0.40)
Impact strength, Izod(a), J/cm (ft·lb/in.)	D256	0.53-0.80 (1.0-1.5)
Hardness, Rockwell	...	M80-90
Thermal properties		
Coefficient of linear thermal expansion, 10^{-5}/°C (10^{-5}/°F)	D696	15.3-18.0 (8.5-10.0)
Specific heat, kJ/kg·K (Btu/lb·°F)	C351	1.7(0.4)
Continuous service temperature, °C (°F)	...	91 (195)
Electrical properties		
Volume resistivity, Ω·cm	D257	10^{15}
Dielectric strength, V/10^{-3} mm (V/10^{-3} in.)	D149	20 (500)
Dielectric constant, (50-100 Hz)	D150	3.7
Dissipation factor (50-100 Hz)	D150	0.005
Processing parameters		
Processing temperature, °C (°F)	...	170-230 (340-450)
Density, g/cm^3	D792	1.42
Specific volume, cm^3/kg (in.^{3}/lb)	D792	685 (19)
Linear mold shrinkage, mm/mm (in./in.)	D955	0.02-0.025
Water absorption, % in 24 h	D570	0.3

(a) Notched, at room temperature.

Source: Ref 15, p 1-8

Recommended processing factors for injection-moldable acetal resins

Grade	Melt-temperature range		Melt viscosity(a), Pa·s	Injection speed	Mold temperature		Maximum, regrind, %	Gating diameter(b)		Shrinkage range, mm/mm (in./in.)	Minimum cycle time(b), s
	°C	°F			°C	°F		mm	in.		
Homopolymers											
Unmodified base resin(c)	205-225	400-440	240-1300	Low to medium	80-105	180-220	30(d)	1.6-1.9	0.063-0.075	0.020	20
Impact modified	195-215	380-420	330-1300	Low to medium	10-70	50-160	35	1.6-1.9	0.063-0.075	0.020	20
Copolymers											
Unmodified base resin:											
Low viscosity(e)	175-200	350-390	180	Medium to high	65-95	150-200	25	2.03 × 3.81	0.080 × 0.150	0.020-0.022	20
Medium viscosity(f)	180-200	360-390	250	Medium	65-95	150-200	25	2.54 × 3.81	0.100 × 0.150	0.020-0.023	30
High viscosity(g)	190-205	370-400	400	Medium	65-95	150-200	25	3.18 × 5.08	0.125 × 0.200	0.018-0.022	35
Glass fiber reinforced:											
12.5%	190-205	370-400	325	Medium	65-95	150-200	25	2.54 × 3.81	0.100 × 0.150	0.006-0.020	30
25%	195-210	380-410	400	Medium	65-95	150-200	25	3.18 × 5.08	0.125 × 0.200	0.004-0.018	30
Mineral filled:											
Low filler, high flow	180-200	360-390	200	Low	65-95	150-200	25	2.54 × 3.81	0.100 × 0.150	0.016-0.020	25
High filler, medium flow	190-205	370-400	280	Low	65-95	150-200	25	3.18 × 5.08	0.125 × 0.200	0.014-0.018	30
Impact modified:											
Medium	180-200	360-390	...	Medium	65-95	150-200	25	2.54 × 3.81	0.100 × 0.150	0.018-0.020	30
High	190-205	370-400	...	Medium	65-95	150-200	25	3.18 × 5.08	0.125 × 0.200	0.017-0.020	30

(a) Melt viscosity measured at 1000/s at the median temperature of the recommended melt-temperature range. (b) Data based on a wall thickness of 3.18 mm (1.8 in.). (c) Includes glass filled, PTFE filled, all viscosities. (d) 100% regrind can be used if clean and free from deterioration. (e) 45 melt index. (f) 9.0 melt index. (g) 2.5 melt index.

Source: Ref 19, p 34-35

Thermal, physical, and chemical properties of acetal polymers (POM)

Property	Homopolymer	Copolymer	25% glass-reinforced copolymer
Heat-deflection temperature at 1.82 MPa, °C	125	110	160
Maximum resistance to continuous heat, °C	100	110	125
Coefficient of linear thermal expansion, 10^{-5}/°C	10.0	8.5	5.0
Compressive strength, MPa	106	110	117
Flexural strength, MPa	97	90	193
Impact strength, Izod, cm·N/cm of notch	80.1	69.4	96.1
Tensile strength, MPa	69	62	129
Elongation, %	30	50	3
Hardness, Rockwell	M94	M78	M79
Specific gravity	1.142	1.41	1.61
Dielectric constant	3.2	3.7	4.0
Water absorption, %	0.25	0.25	0.3
Resistance to chemicals at 25 °C(a):			
Nonoxidizing acids (20% H_2SO_4)	U	U	U
Oxidizing acids (10% HNO_3)	U	U	U
Aqueous salt solutions (NaCl)	S	S	S
Aqueous alkalis (C_2NOH)	S	S	S
Polar solvents (C_2H_5OH)	S	S	S
Nonpolar solvents (C_6H_6)	Q	Q	Q
Water	S	S	S

(a) S, satisfactory; Q, questionable; U, unsatisfactory.

Source: Ref 13, p 129

Acrylics

Molecular structure of polyacrylonitrile

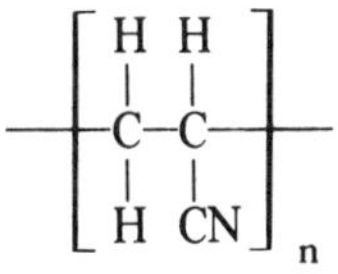

Source: Ref 20, p 48

Typical pultrusion formulation for an acrylic IPN

Material	Content, wt %
Interpol 047-1000	83.5
t-butylperbenzoate	0.5
t-butylperoctoate	1.5
Mold release	1.0
Stypol 040-7263	15.5

Source: Ref 22, p 248

Typical prepreg formulation for an acrylic IPN

Material	Content, wt %
Interpol 047-1050	82.0
USP-245 (peroxide catalyst)	1.0
Stypol (040-7263)	17.0

Viscosity build

Time, h	CP
0	140
0.5	175
1.0	250
2.0	500
20.0	48,000
72.0	900,000

Source: Ref 22, p 248

Thermal, physical, and chemical properties of typical acrylics

Property	Cast acrylics	Acrylic-PVC alloy
Heat-deflection temperature at 1.82 MPa, °C	95	70
Maximum resistance to continuous heat, °C	75	60
Coefficient of linear thermal expansion, 10^{-5}/°C	7.0	6.0
Compressive strength, MPa,	103	58
Flexural strength, MPa,	97	72
Impact strength, Izod, cm·N/cm of notch	21.4	8.0
Tensile strength, MPa	66	45
Elongation, %	4	100
Hardness, Rockwell	M80	R100
Specific gravity	1.18	1.25
Dielectric constant	3.0	3.5
Resistance to chemicals at 25 °C(a):		
Nonoxidizing acids (20% H_2SO_4)	S	S
Oxidizing acids (10% HNO_3)	U	Q
Aqueous salt solutions (NaCl)	S	S
Aqueous alkalis (NaOH)	S	S
Polar solvents (C_2H_5OH)	S	S
Nonpolar solvents (C_6H_6)	Q	Q
Water	S	S

(a) S, satisfactory; Q, questionable; U, unsatisfactory.

Source: Ref 23, p 62

Selected properties of acrylics

Property	ASTM test	Reported value for: Cast acrylic Molded acrylic	Molded high-sheet and rod	Molded high-impact acrylic
Mechanical properties				
Yield strength, MPa (ksi)	D638	48-69 (7.0-10.0)	55-76 (8.0-11.0)	38-62 (5.5-9.0)
Tensile modulus, GPa (10^6 psi)	D638	2.6-3.1 (0.38-0.45)	2.4-3.1 (0.35-0.45)	1.4-2.4 (0.20-0.35)
Elongation (break), %	D638	2-9	2-7	8-13
Compressive strength, MPa (ksi)	D695	83-125 (12-18)	76-130 (11-19)	41-97 (6-14)
Flexural yield strength, MPa (ksi)	D790	90-115 (13-17)	83-115 (12-17)	55-90 (8-13)
Flexural modulus, GPa (10^6 psi)	D790	2.8-3.4 (0.040-0.50)	2.8-3.4 (0.40-0.50)	2.1-2.8 (0.30-0.40)
Impact strength, Izod(a), J/cm (ft·lb/in.)	D256	0.2-0.3 (0.3-0.5)	0.2-0.3 (0.3-0.6)	0.4-1.3 (0.8-2.4)
Hardness, Rockwell	...	M85-105	M80-100	R105-120
Thermal properties				
Thermal conductivity, W/m·K (Btu·in./ft²·h·°F)	C177	0.19-0.29 (1.3-2.0)	0.19-0.29 (1.3-2.0)	0.17-0.26 (1.2-1.8)
Coefficient of linear thermal expansion, 10^{-5}/°C (10^5/°F)	D696	9.0-16 (5.0-9.0)	9.0-16 (5.0-9.0)	9.0-14 (5.0-8.0)
Continuous service temperature, °C (°F)	...	68-88 (155-190)	60-71 (140-160)	71-82 (160-180)
Electrical properties				
Volume resistivity, Ω·cm	D257	>10^{15}	>10^{15}	2.0 x 10^{16}
Dielectric strength, V/10^{-3} mm (V/10^{-3} in.)	D149	18-21 (450-530)	17 (430)	16-20 (400-500)
Dielectric constant (50-100 Hz)	D150	3.5-4.5	3.5-4.5	3.5-3.9
Dissipation factor (50-100 Hz)	D150	0.050-0.060	0.050-0.060)	0.030-0.040
Processing parameters				
Processing temperature, °C (° F)	...	165-230 (325-450)	...	150-205 (300-400)
Density, g/cm³	D792	1.17-1.19	1.17-1.20	1.11-1.18
Specific volume, cm³/kg (in.³/lb)	D792	865-830 (24-23)	865-795 (24-22)	940-865 (26-24)
Linear mold shrinkage, mm/mm (in./in.)	D955	0.001-0.004	...	0.004-0.008
Water absorption, % in 24 h	D570	0.3-0.4	0.3-0.4	0.2-0.3

(a) Notched, at room temperature.

Source: Ref 15, 1-9

ABS

Molecular structure of acrylonitrile-butadiene-styrene copolymer

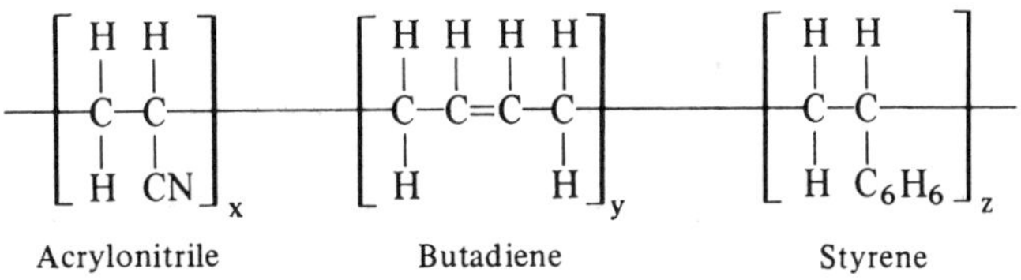

Source: Ref 20, p 49

Thermal, physical, and chemical properties of typical ABS plastics

Property	Extrusion-grade ABS	20% glass reinforced ABS
Heat deflection temperature at 1.82 MPa, °C	90	100
Maximum resistance to continuous heat, °C	80	90
Coefficient of linear thermal expansion, 10^{-5}/°C	9.5	2.0
Compressive strength, MPa	48	97
Flexural strength, MPa,	62	103
Impact strength, Izod, cm·N/cm of notch	320.3	53.4
Tensile strength, MPa	34	76
Elongation, %	60	5
Hardness, Rockwell	R60	M85
Specific gravity	1.03	1.2
Dielectric constant	0.25	0.4
Resistance to chemicals at 25 °C(a):		
Nonoxidizing acids (20% H_2SO_4)	S	S
Oxidizing acids (10% HNO_3)	U	U
Aqueous salt solutions (NaCl)	S	S
Aqueous alkalis (NaOH)	S	S
Polar solvents (C_2H_5OH)	Q	Q
Nonpolar solvents (C_6H_6)	Q	Q

(a) S, satisfactory; Q, questionable; U, unsatisfactory.

Source: Ref 13, p 123

Typical conditions for injection molding of ABS

Process parameter	Setting
Drying temperature, °C (°F)	82-88 (180-190)
Drying time, h	1-2
Mold temperature, °C (°F)	49-66 (120-150)
Melt temperature, °C (°F)	220-260 (425-500)
Barrel sets, °C (°F):	
Rear	195 (380)
Middle	210-215 (410-420)
Front	220-230 (430-450)
Fill speed	low-medium
Screw speed, rpm	50-60
Back pressure, kPa (psi)	690 (100)
Injection pressure, MPa (ksi)	69-83 (10-12)

Source: Ref 24, p 361

Selected properties of extruded and molded ABS thermoplastics

Property	ASTM test	Reported value for: Extruded ABS	Molded high-impact ABS	Molded medium-impact ABS	Molded heat-resistant ABS	Molded flame-resistant ABS	Molded transparent ABS
Mechanical properties							
Yield strength, MPa (ksi)	D638	22-54 (3.2-7.8)	32-43 (4.7-6.3)	41-55 (6.0-8.0)	48-55 (7.0-8.0)	35-48 (5.1-7.0)	39-43 (5.6-6.2)
Tensile modulus, GPa (10^6 psi)	D638	0.97-2.5 (0.14-0.36)	1.4-2.2 (0.20-0.32)	1.5-2.8 (0.22-0.40)	2.4-2.9 (0.35-0.42)	2.4-2.8 (0.35-0.40)	2.0-2.3 (0.29-0.34)
Elongation (break), %	D638	30-90	10-50	10-25	10-20	15-20	40-60
Compressive strength, MPa (ksi)	D695	6.9-28 (1-4)	6.9-21 (1-3)	55-83 (8-12)	62-76 (9-11)	48-55 (7-8)	48-62 (7-9)
Flexural yield strength, MPa (ksi)	D790	34-97 (5-14)	41-69 (6-10)	69-83 (10-12)	76-90 (11-13)	62-90 (9-13)	69-76 (10-11)
Flexural modulus, GPa (10^6 psi)	D790	0.69-2.8 (0.10-0.40)	1.4-2.1 (0.20-0.30)	2.1-2.8 (0.30-0.40)	2.1-2.8 (0.30-0.40)	2.1-2.8 (0.30-0.40)	2.1-2.8 (0.30-0.40)
Impact strength, Izod(a), J/cm (ft·lb/in.)	D256	1.4-6.09 (2.6-11.4)	3.2-3.7 (6.0-7.0)	1.7-2.4 (3.1-4.5)	1.3-2.1 (2.5-4.0)	1.9-2.9 (3.5-5.5)	2.1-2.7 (4.0-5.0)
Hardness, Rockwell	...	R75-110	R85-100	R108-115	R105-115	R100-120	R98-105
Thermal properties							
Thermal conductivity, W/m·K (Btu·in./ft^2·h·°F)	C177	0.19-0.33 (1.3-2.3)	0.14-0.33 (1.0-2.3)	0.14-0.30 (0.98-2.1)	0.22-0.36 (1.5-2.5)	...	0.19-0.33 (1.3-2.3)
Coefficient of thermal expansion 10^{-5}/°C (10^{-5}/°F)	D696	10.8-23.0 (6.0-12.8)	11.0-16 (6.1-9.1)	8.1-11.3 (4.5-6.3)	7.7-10.8 (4.3-6.0)	...	...
Specific heat, kJ/kg·K (Btu/lb·°F)	C351	1.3-1.7 (0.3-0.4)	1.3-1.7 (0.3-0.4)	1.3-1.7 (0.3-0.4)	1.3-2.1 (0.3-0.5)	...	...
Continuous service temperature, °C (°F)	...	85 (185)	85 (185)	85 (185)	...	...	...
Electrical properties							
Volume resistivity, Ω·cm	D257	1.0-5.0 x 10^{16}	1.0-5.0 x 10^{16}	1.0-5.0 x 10^{16}	...	1.5 x 10^{14}	2.5 x 10^{15}
Dielectric strength, V/10^{-3} mm (V/10^{-3} in.)	D149	14-18 (350-450)	14-16 (350-400)	15-18 (375-450)	14-16 (360-400)	15-16 (380-400)	16 (400)
Dielectric constant (50-100 Hz)	D150	2.7-3.1	2.7-3.1	2.7-3.1	2.7-3.1	...	3.7
Dissipation factor (50-100 Hz)	D150	...	0.005-0.010	0.003-0.005	0.030-0.040	...	0.015
Processing parameters							
Processing temperature, °C (°F)	...	165-230 (325-450)	165-260 (325-500)	165-260 (325-500)	175-290 (350-550)	160-260 (320-500)	160-245 (320-475)
Density, g/cm^3	D792	1.02-1.06	1.01-1.05	1.03-1.06	1.06-1.08	1.16-1.22	1.07
Specific volume, cm^3/kg (in.^{3}/lb)	D792	975-940 (27-26)	975-940 (27-26)	975-940 (27-26)	940-905 (26-25)	865-795 (24-22)	940 (26)
Linear mold shrinkage, mm/mm (in./in.)	D955	0.005-0.009	0.005-0.009	0.005-0.009	0.005-0.009	0.005-0.008	0.005-0.008
Water absorption, % in 24 h	D570	0.2-0.4	0.2-0.4	0.2-0.4	0.2-0.4	...	0.4

(a) Notched, at room temperature.

Source: Ref 15, p 1-7

Typical conditions for extrusion of ABS

Process parameter	Setting
Drying temperature, °C (°F)	82 (180)
Drying time, h	3
Melt temperature, °C (°F)	205-245 (400-475)
Barrel sets, °C (°F):	
Rear	190-220 (375-425)
Middle	205-230 (400-450)
Front	205-245 (400-475)
Die temperature, °C (°F)	220-250 (425-480)
Roll temperatures,(a) °C (°F):	
Top	79-105 (175-225)
Middle	66-79 (150-175)
Bottom	79-105 (175-225)

(a) Down stack.

Source: Ref 24, p 363

Typical conditions for thermoforming of ABS

Process parameter	Setting
Sheet temperature,(a) °C (°F):	
Minimum any layer	110 (230)
Maximum surface	220 (425)
Normal	165 (325)
Heater temperature range, °C (°F)	260-815 (500-1500)
Heater normal temperature, °C (°F)	540 (1000)
Plug temperature for female mold, °C (°F)	95-150 (200-300)
Mold temperature, °C (°F):	
Female mold	49-115 (120-240)
Male mold	49-115 (120-240)
Plug speed, mm/s (in./s)	125-255 (5-10)
Plug size for fremale mold, % of mold	80-90
Heater distance from sheet, mm (in.):	
Top of sheet	150-305 (6-12)
Bottom of sheet	305-455 (12-18)
Maximum sheet heating rate, s/μm (s/mil) of sheet thickness:	
One-side heaters	0.02 (0.5)
Two-side heaters	0.03 (0.75)
Billow height, % of draw depth:	
Female mold	50
Male mold	75
Maximum sheet cooling rate, s/μm (s/mil) of sheet thickness:	
Forced air	0.01 (0.25)
Water fog	0.005 (0.125)

(a) As determined by optical pyrometer measurements. For sheet thicknesses greater than 4.44 mm (0.175 in.), minimum temperature must be higher than that shown. ABS/PVC resins should be processed at temperatures 11 to 17 °C (20 to 30 °F) lower. ABS/SMA resins should be processed at temperatures 8 to 17 °C (15 to 30 °F) higher.

Source: Ref 24, p 364

Cellulosics

Molecular structure of cellulose

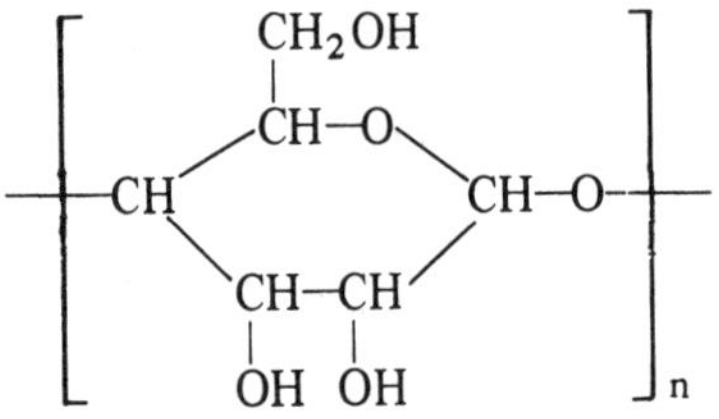

Source: Ref 20, p 50

Thermal, physical, and chemical properties of typical cellulosics

Property	Ethylcellulose	Cellulose triacetate	Cellulose acetate butyrate
Heat-deflection temperature at 1.82 MPa, °C	65	65	65
Maximum resistance to continuous heat, °C	60	60	60
Coefficient of linear thermal expansion, 10^{-5}/°C	15.0	12.5	14.0
Compressive strength, MPa	120	55	34.5
Flexural strength, MPa	41	55	41
Impact strength, Izod, cm·N/cm of notch	21.35	106.7	160
Tensile strength, MPa	34	42	34
Elongation, %	10	25	50
Hardness, Rockwell	R60	R80	R75
Specific gravity	1.1	1.3	1.2
Dielectric constant	3	4	4
Resistance to chemicals at 25 °C(a):			
Nonoxidizing acids (20% H_2SO_4)	Q	U	U
Oxidizing acids (10% HNO_3)	U	U	U
Aqueous salt solutions (NaCl)	S	S	S
Aqueous alkalis (NaOH)	S	S	S
Polar solvents (C_2H_5OH)	Q	Q	Q
Nonpolar solvents (C_6H_6)	U	U	U
Water	S	S	S

(a) S, satisfactory; Q, questionable; U, unsatisfactory.

Source: Ref 23, p 55

Selected properties of molded cellulosics

Property	ASTM test	Reported value for: Molded ethyl cellulose	Molded cellulose acetate	Molded cellulose acetate/butyrate	Molded cellulose acetate propionate	Molded cellulose nitrate
Mechanical properties						
Yield strength, MPa (ksi)	D638	14-55 (2.0-8.0)	21-48 (3.0-7.0)	17-55 (2.5-8.0)	28-45 (4.0-6.5)	48-55 (7.0-8.0)
Tensile modulus, GPa (10^6 psi)	D638	0.69-2.1 (0.10-0.30)	0.6-2.8 (0.08-0.40)	0.3-1.4 (0.05-0.20)	...	1.3-1.6 (0.19-0.23)
Elongation (break), %	D638	...	6-60	40-45	...	40-45
Compressive strength, MPa (ksi)	D695	69-240 (10-35)	14-62 (2-9)	14-140 (2-20)	...	150-240 (22-35)
Flexural yield strength, MPa (ksi)	D790	28-83 (4-12)	28-83 (4-12)	21-62 (3-9)	41-55 (6-8)	62-76 (9-11)
Flexural modulus, GPa (10^6 psi)	D790	...	1.4 (0.20)	0.69-1.4 (0.10-0.20)	0.69-1.4 (0.10-0.20)	...
Impact strength, Izod(a), J/cm (ft·lb/in.)	D256	1.1-4.5 (2.0-8.5)	1.6-3.7 (3.0-7.0)	0.5-2.7 (0.9-5.0)	...	2.7-3.7 (5.0-7.0)
Hardness, Rockwell	...	R50-115	R50-115	R30-100	...	R95-110
Thermal properties						
Thermal conductivity, W/m·K (Btu·in./ft^2·h·°F)	C177	0.16-0.29 (1.1-2.0)	0.17-0.32 (1.2-2.2)	0.16-0.32 (1.1-2.2)	0.17-0.33 (1.2-2.3)	0.19 (1.3)
Coefficient of linear thermal expansion 10^{-5}/°C (10^{-5}/°F)	D696	18.0-36.0 (10.0-20.0)	14.4-32.4 (8.0-18.0)	19.8-30.6 (11.0-17.0)	...	14.4-21.6 (8.0-12.0)
Specific heat, kJ/kg·K (Btu/lb·°F)	C351	...	1.3-1.7 (0.3-0.4)	1.3-1.7 (0.3-0.4)	1.3-1.7 (0.3-0.4)	1.3-1.7 (0.3-0.4)
Electrical properties						
Volume resistivity, Ω·cm	D257	10^{12}-10^{14}	10^{10}-10^{14}	10^{11}-10^{15}	10^{11}-10^{14}	10^{11}
Dielectric strength, V/10^{-3} mm (V/10^{-3} in.)	D149	14-20 (350-500)	8-20 (200-500)	10-16 (250-400)	10-16 (250-400)	12-22 (300-550)
Dielectric constant (50-100 Hz)	D150	3.0-4.2	3.5-7.5	3.5-6.2	3.5-6.2	7.0-7.5
Dissipation factor (50-100 Hz)	D150	0.005-0.020	0.010-1.00	0.020-0.050	0.020-0.050	0.090-0.120
Processing parameters						
Processing temperature, °C (°F)	...	...	125-215 (260-420)	140-200 (280-390)	...	85-120 (185-250)
Density, g/cm^3	D792	1.09-1.17	1.22-1.34	1.15-1.22	1.19-1.21	1.35-1.40
Specific volume, cm^3/kg (in.^{3}/lb)	D792	905-865 (25-24)	830-725 (23-20)	865-830 (24-23)	865-830 (24-23)	760-725 (21-20)
Linear mold shrinkage, mm/mm (in./in.)	D955	0.005-0.009	0.003-0.009	...	0.003-0.009	...
Water absorption, % in 24 h	D570					

(a) Notched, at room temperature.

Source: Ref 15, p 1-11

Fluoroplastics

Molecular structure of fluoroplastics

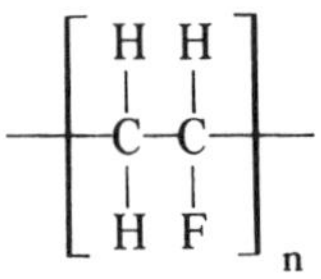

Polyvinyl fluoride

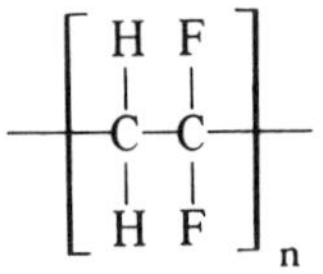

Polyvinylidene fluoride

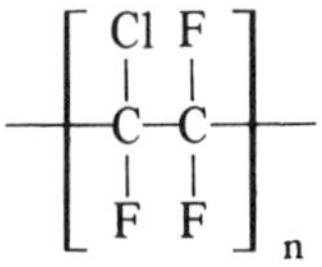

Chlorotrifluoroethylene

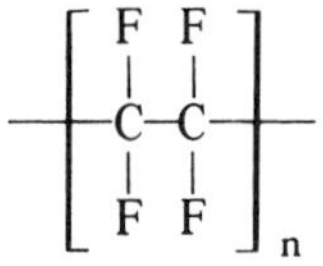

Polytetrafluoroethylene

Source: Ref 20, p 48

Molecular structure of TFE-HFP copolymer

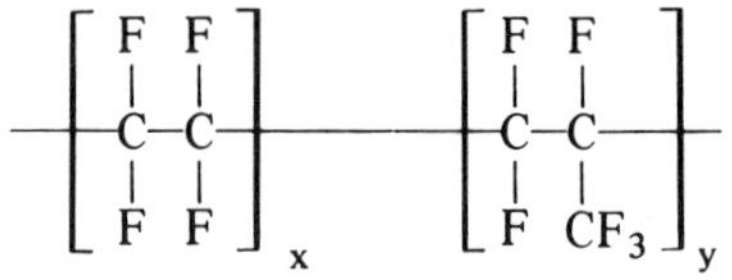

TFE HFP

Source: Ref 20, p 49

Thermal, physical and chemical properties of typical polyfluorocarbon plastics

Property	PTFE	PCTFE	PVDF	PVF	PE-CTFE	PE-TFE
Heat-deflection temperature at 1.82 MPa, °C	100	100	80	90	115	120
Maximum resistance to continuous heat, °C	250	200	150	125	100	160
Coefficient of linear thermal expansion, 10^{-5}/°C	10	14	8.5	10	8	7
Compressive strength, MPa	28	38	...	...	41	48
Flexural strength, MPa	...	60	...	...	48	38
Impact strength, Izod cm·N/cm of notch	160	133.5	...	...	...	...
Tensile strength, MPa	24	34	55	...	48	48
Elongation, %	200	100	200	...	200	250
Hardness, Rockwell	D52	R80	R110	D64	R95	R50
Specific gravity	2.16	2.1	1.76	1.4	1.7	1.7
Dielectric constant	2	2.5	8	8	5	2.7
Water absorption, %	0	0	0	0	0	0
Resistance to chemicals at 25 °C(a):						
Nonoxidizing acids (20% H_2SO_4)	S	S	S	S	S	S
Oxidizing acids (10% HNO_3)	S	S	S	Q	Q	Q
Aqueous salt solutions (NaCl)	S	S	S	S	S	S
Aqueous alkalis (NaOH)	S	S	S	S	S	S
Polar solvents (C_2H_5OH)	S	S	S	S	S	S
Nonpolar solvents (C_6H_6)	S	S	S	S	Q	Q
Water	S	S	S	S	S	S

(a) S, satisfactory; Q, questionable; U, unsatisfactory.

Source: Ref 13, p 125

Selected properties of fluoroplastics

Property	ASTM test	Reported value for: Molded fluorocarbon PCTFE	Molded fluorocarbon PTFE	Extruded or molded fluorocarbon FEP	Extruded or molded fluorocarbon PVF_2	Extruded or molded fluorocarbon ETFE/ECTFE	Extruded or molded fluorocarbon PFA
Mechanical properties							
Yield strength, MPa (ksi)	D638	32-39 (4.6-5.7)	17-45 (2.5-6.5)	19-21 (2.7-3.1)	50-59 (7.2-8.5)	...	29 (4.2)
Tensile modulus, GPa (10^6 psi)	D638	1.0-2.1 (0.15-0.30)	0.28-0.48 (0.04-0.07)	0.34 (0.05)	1.2-1.4 (0.17-0.20)	2.8-3.1 (0.40-0.45)	3.0 (0.43)
Elongation (break), %	D638	120-300	250-375	250-330	200-300	150-200	300
Compressive strength, MPa (ksi)	D695	7-14 (1-2)	7-14 (1-2)	14 (2)	62-97 (9-14)	14 (2)	...
Flexural yield strength, MPa (ksi)	D790	55-62 (8-9)	...	No break	14 (2)	17 (2.5)	...
Flexural modulus, GPa (10^6 psi)	D790	0.69 (0.10)	0.69 (0.10)	0.69 (0.10)	1.4 (0.20)	...	0.69 (0.10)
Impact strength, Izod(a), J/cm (ft·lb/in.)	D256	1.3-1.4 (2.4-2.7)	1.6 (3.0)	No break	1.9-2.1 (3.6-4.0)	No break	No break
Hardness, Rockwell	...	R90-110	D50-55	D58	...	...	...
Thermal properties							
Thermal conductivity, W/m·K (Btu·in./ft^2·h·°F)	C177	0.20-0.26 (1.4-1.8)	0.27 (1.9)	0.20 (1.4)	0.17 (1.2)	0.22 (1.5)	0.26 (1.8)
Coefficient of linear thermal expansion 10^{-5}/°C (10^{-5}/°F)	D696	8.1-9.4 (4.5-5.2)	14.4-16.2 (8.0-9.0)	10.8-18.9 (6.0-10.5)	15.3 (8.5)	9.0-13.5 (5.0-7.5)	14.4-21.6 (8.0-12.0)
Specific heat, kJ/kg·K (Btu/lb·°F)	C351	0.8 (0.2)	1.3 (0.3)	1.3 (0.3)	1.3 (0.3)	...	25 (6.0)
Continuous service temperature, °C (°F)	...	195 (380)	290 (550)	...	170 (340)	150-180 (300-355)	260 (500)
Electrical properties							
Volume resistivity, Ω·cm	D257	10^{18}	$>10^{18}$	10^{18}	5.0×10^{14}	10^{16}	10^{18}
Dielectric strength, V/10^{-3} mm (V/10^{-3} in.)	D149	18-22 (450-550)	16-20 (400-500)	20-24 (500-600)	10 (260)	19 (490)	79 (2000)
Dielectric constant (50-100 Hz)	D150	2.3-2.8	2.1	2.1	10	2.6	2.1
Dissipation factor (50-100 Hz)	D150	0.002	<0.0002	<0.0003	0.050	0.0007	0.0002
Processing parameters							
Processing temperature, °C (°F)	...	240-290 (460-550)	...	315-400 (600-750)	205-290 (400-550)	230-260 (450-500)	345-400 (650-750)
Density, g/cm^3	D792	2.10-2.20	2.14-2.30	2.13-2.17	1.77	1.70	2.10-2-20
Specific volume, cm^3/kg (in.^{3}/lb)	D792	470 (13)	470 (13)	470 (13)	540 (15)	580 (16)	470 (13)
Linear mold shrinkage, mm/mm (in./in.)	D955	0.010-0.015	0.010-0.015	0.030-0.060	0.030	0.020	0.040
Water absorption, % in 24 h	D570	0.0	0.0	...	0.1	...	0.1

(a) Notched, at room temperature.

Source: Ref 15, p 1-13

Polyamides (nylons)

Molecular structure of nylons

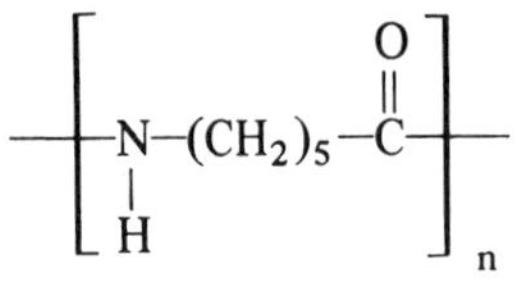

Nylon 6

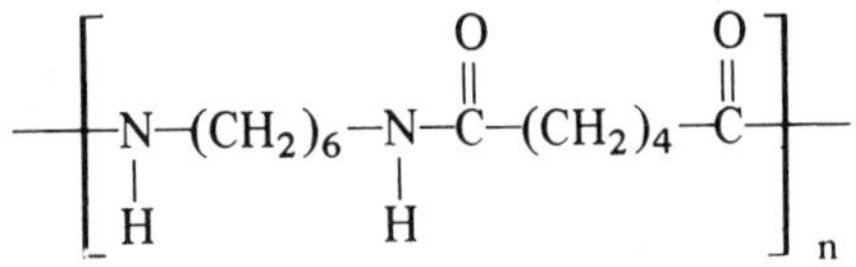

Nylon 6/6

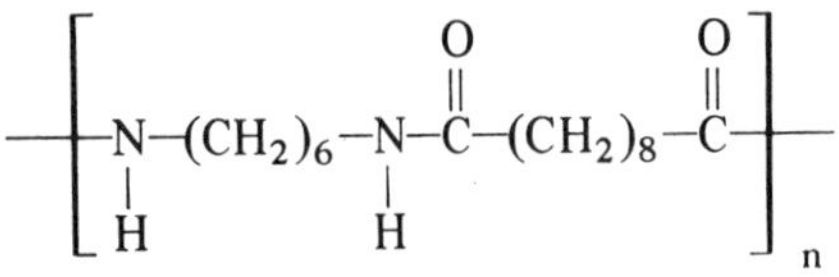

Nylon 6/10

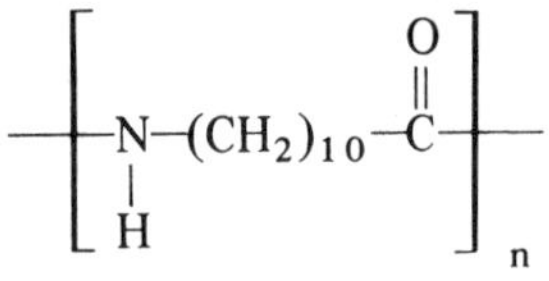

Nylon 11

Source: Ref 20, p 50

Thermal, physical, and chemical properties of typical polyamides

Property	Nylon-6/6	Nylon-6	Nylon-11	Nylon-6/10
Heat-deflection temperature at 1.82 MPa, °C	75	80	55	80
Maximum resistance to continuous heat, °C	65	70	60	75
Melting point, °C	265	215	185	220
Coefficient of linear thermal expansion, $10^{-5}/°C$	8.0	8.0	10	10
Tensile strength, MPa	83	62	50	55
Elongation, %	60	100	120	100
Specific gravity	1.14	1.13	1.05	1.09
Dielectric constant	4.0	4.0	3.5	4.5
Resistance to chemicals at 25 °C(a):				
Nonoxidizing acids (20% H_2SO_4)	U	U	U	U
Oxidizing acids (10% HNO_3)	U	U	U	U
Aqueous salt solutions (NaCl)	S	S	S	S
Aqueous alkalis (NaOH)	S	S	S	S
Polar solvents (C_2H_5OH)	Q	Q	Q	Q
Nonpolar solvents (C_6H_6)	S	S	S	S
Water	S	S	S	S

(a) S, satisfactory; Q, questionable; U, unsatisfactory.

Source: Ref 23, p 67

Recommended processing factors for injection-moldable nylons

Grade	Melt-temperature range °C	Melt-temperature range °F	Melt viscosity(a), Pa·s	Injection speed	Mold temperature °C	Mold temperature °F	Maximum regrind, %	Gating diameter(b) mm	Gating diameter(b) in.	Shrinkage range, mm/mm (in./in.)	Minimum cycle time(b), s
Nylon 6											
Unmodified base resin:											
Low viscosity	240-270	460-520	120	Low to high	10-95	50-200	50	1.0-1.8	0.040-0.070	0.009-0.015	12-15
Medium viscosity	250-280	480-540	...	Low to high	10-95	50-200	50	1.5-2.3	0.060-0.090	0.009-0.015	12-15
High viscosity	240-270	460-520	120	Low to high	80-95	180-200	50	1.0-1.8	0.040-0.070	0.007-0.013	9-12
Glass reinforced:											
15 %	245-275	470-530	...	High	80-95	180-200	25	1.5-2.3	0.060-0.090	0.004-0.008	12-18
33 %	260-295	500-560	200	High	80-95	180-200	25	1.5-3.18	0.060-0.125	0.003-0.006	12-18
44 %	265-300	510-570	...	High	80-95	180-200	25	1.5-3.18	0.060-0.125	0.002-0.005	12-18
Mineral reinforced, 35 to 40 %	270-300	520-570	185	High	80-105	180-220	25	1.5-3.18	0.060-0.125	0.008-0.012	12-18
Mineral/glass reinforced, 35 to 40 %	270-300	520-570	200	High	80-105	180-220	25	1.5-3.18	0.060-0.125	0.003-0.006	12-18
Impact modified:											
Low	250-280	480-540	160	Low to high	10-95	50-200	25	1.0-2.0	0.040-0.080	0.009-0.080	15-18
High	250-280	480-540	...	Low to high	10-95	50-200	25	1.5-2.3	0.060-0.090	0.010-0.016	15-18
Flame retarded:											
Unreinforced	240-260	460-500	...	High	80-95	180-200	25	1.5-2.3	0.060-0.090	0.009-0.015	12-15
25 % glass reinforced	260-295	500-560	...	High	80-95	180-200	25	1.5-3.18	0.060-0.125	0.003-0.006	12-15
Nylon 6/6											
Unmodified base resin, low viscosity	270-295	520-560	1000 poise	High	40-95	100-200	50	1.0-1.8	0.040-0.070	0.015-0.020	12-15
Glass reinforced, 13, 33, and 43 %	290-300	550-575	...	High	65-95	150-200	25	1.5-3.18	0.060-0.125	0.003-0.008	12-18
Mineral reinforced, 35-40 %	280-300	540-570	...	High	80-105	180-200	25	1.5-3.18	0.060-0.125	0.012-0.017	12-18
Mineral/glass reinforced, 35-40 %	280-300	540-570	...	High	80-105	180-220	25	1.5-3.18	0.060-0.125	0.010-0.015	12-18
Impact modified	290-295	550-560	...	High	40-95	100-200	25	1.5-3.18	0.060-0.125	0.015-0.020	12-15
Flame retarded:											
Unreinforced	250-270	480-520	...	High	40-95	100-200	25	1.5-2.3	0.060-0.090	0.015-0.020	12-15
25% glass reinforced	250-270	480-520	...	High	40-95	100-200	25	1.5-2.3	0.060-0.090	0.003-0.005	12-18
Supertough	290-295	550-560	...	High	40-95	100-200	25	1.5-2.3	0.060-0.090	0.015-0.020	12-15

(a) Melt viscosity measured at 1000/s at the median temperature of the recommended melt-temperature range. (b) Data based on a wall thickness of 3.18 mm (1/8 in.).

Source: Ref 19, p 36

Selected properties of nylons

Property	ASTM test	Reported value for: Extruded or molded, aromatic or transparent nylon	Extruded or molded nylon 6	Extruded or molded elastomer copolymer nylon 6	Cast nylon 6	Extruded or molded nylon 6/6	Extruded or molded high-impact modified nylon 6/6
Mechanical properties							
Yield strength, MPa (ksi)	D638	68-84 (9.8-12.2)	62-90 (9.0-13.0)	52-69 (7.5-10.0)	76-90(11.0-13.0)	76-83 (11.0-12.0)	14-55 (2.0-8.0)
Tensile modulus, GPa (10^6 psi)	D638	2.8 (0.40)	1.4-2.8 (0.20-0.40)	...	2.4-3.7 (0.35-0.54)	2.6-3.2 (0.38-0.47)	...
Elongation (break), %	D638	70-150	100-320	150-275	20-60	60-300	40-240
Compressive strength, MPa (ksi)	D695	97 (14)	69-90 (10-13)	...	62-105 (9-15)	105 (15)	34 (5)
Flexural yield strength, MPa (ksi)	D790	90 (13)	34-97 (5-14)	34-83 (5-12)	34-110 (5-16)	41-110 (6-16)	...
Flexural modulus, GPa (10^6 psi)	D790	2.8 (0.40)	2.8 (0.40)	0.69-2.1 (0.10-0.30)	1.4-3.4 (0.20-0.50)	1.4-4.1 (0.20-0.60)	1.4-2.1 (0.20-0.30)
Impact strength, Izod(a), J/cm (ft·lb/in.)	D256	...	0.53-1.6 (1.0-3.0)	No break	0.53-1.6 (1.0-3.0)	0.53-1.1 (1.0-2.0)	8.0-13.3 (15.0-25.0)
Hardness, Rockwell	...	M95	R90-120	R90-110	R95-120	R120	R112
Thermal properties							
Thermal conductivity, W/m·K (Btu·in./ft^2·h·°F)	C177	0.19 (1.3)	0.17 (1.2)	...	0.17 (1.2)	0.26 (1.8)	0.26 (1.8)
Coefficient of linear thermal expansion 10^{-5}/°C (10^{-5}/°F)	D696	5.0 (2.8)	9.0-14 (5.0-8.0)	...	9.0-16 (5.0-9.0)	14 (8.0)	...
Specific heat, kJ/kg·K (Btu/lb·°F)	C351	1.7 (0.4)	1.7 (0.4)	...	1.7 (0.4)	1.7 (0.4)	1.7 (0.4)
Continuous service temperature, °C (°F)	...	...	79-120 (175-250)	...	79-120 (175-250)	79-150 (175-300)	79-150 (175-300)
Electrical properties							
Volume resistivity, Ω·cm	D257	10^{15}	10^{11}-10^{14}	...	10^{14}	10^{11}-10^{14}	10^{13}-10^{14}
Dielectric strength, V/10^{-3} mm (V/10^{-3} in.)	D149	14 (350)	16 (400)	18 (450)	14 (350)	24 (600)	20 (500)
Dielectric constant (50-100 Hz)	D150	3.7	3.8-5.0	3.0-4.0	4.0	4.2	3.2
Dissipation factor (50-100 Hz)	D150	0.02	0.010-0.100	0.010-0.080	0.020	0.020	0.010-0.015
Processing parameters							
Processing temperature, °C (°F)	...	255-315 (490-600)	225-270 (440-520)	230-300 (450-575)	...	270-325 (520-620)	270-305 (520-580)
Density, g/cm^3	D792	1.12	1.12-1.14	1.08-1.14	1.15	1.13-1.15	1.08-1-11
Specific volume, cm^3/kg (in.^{3}/lb)	D792	865 (24)	905-865 (25-24)	905-865 (25-24)	865 (24)	865 (24)	905 (25)
Linear mold shrinkage, mm/mm (in./in.)	D955	0.005	0.002-0.005	0.008-0.020	...	0.005-0.015	0.012-0.019
Water absorption, % in 24 h	D570	0.4	1.7-1.8	0.8-1.4	0.6	1.5	1.5

(a) Notched, at room temperature.

(continued)

Selected properties of nylons (continued)

Property	ASTM test	Reported value for: Extruded or molded copolymer nylon 6/6-6	Extruded or molded nylon 6/9	Extruded or molded nylon 6/10	Extruded or molded nylon 6/12	Extruded or molded nylon 11	Extruded or molded nylon 12
Mechanical properties							
Yield strength, MPa (ksi)	D638	52-82.7 (7.5-12.0)	59 (8.5)	49-59 (7.1-8.5)	55-68.9 (8.0-10.0)	55 (8.0)	55-66 (8.0-9.5)
Tensile modulus, GPa (10^6 psi)	D638	1.0-2.8 (0.15-0.40)	1.9 (0.28)	1.2-1.9 (0.17-0.28)	1.2-2.1 (0.18-0.30)	1.0-1.4 (0.15-0.20)	1.2 (0.18)
Elongation (break), %	D638	100-285	1100	100-300	150-325	300	300
Compressive strength, MPa (ksi)	D695	...	...	...	...	...	...
Flexural yield strength, MPa (ksi)	D790	69-103 (10-15)	69-103 (10-15)	69-103 (10-15)	83-90 (12-13)	...	...
Flexural modulus, GPa (10^6 psi)	D790	1.4-2.8 (0.20-0.40)	1.4-2.1 (0.20-0.30)	1.4-2.1 (0.20-0.30)	1.4-2.8 (0.20-0.40)	2.1 (0.30)	1.4 (0.20)
Impact strength, Izod(a), J/cm (ft·lb/in.)	D256	0.53 (1.0)	0.59-1.1 (1.1-2.1)	0.5-1.2 (0.9-2.3)	0.53-0.80 (1.0-1.5)	0.96 (1.8)	1.1-2.9 (2.0-5.5)
Hardness, Rockwell	...	R90-115	R111	R111	R115	R90-110	R105-110
Thermal properties							
Thermal conductivity, W/m·K (Btu·in./ft^2·h·°F)	C177	0.25 (1.7)	...	0.20 (1.4)	0.20 (1.4)	0.23 (1.6)	0.20 (1.4)
Coefficient of linear thermal expansion 10^{-5}/°C (10^{-5}/°F)	D696	14 (8.0)	14-16 (8.0-9.0)	16 (9.0)	9.0 (5.0)	18.0 (10.0)	18.0 (10.0)
Specific heat, kJ/kg·K (Btu/lb·°F)	C351	1.7 (0.4)	1.7 (0.4)	1.7 (0.40)	1.3-1.7 (0.3-0.4)	2.1 (0.5)	1.3 (0.3)
Continuous service temperature, °C (°F)	...	79-135 (175-275)	...	79-120 (175-250)	79-150 (175-300)	79-150 (175-300)	...
Electrical properties							
Volume resistivity, Ω·cm	D257	10^{10}	10^{13}	10^{12}	10^{13}	10^{13}	10^{13}
Dielectric strength, V/10^{-3} mm (V/10^{-3} in.)	D149	16-24 (400-600)	24 (600)	16 (400)	16 (400)	17 (425)	18 (450)
Dielectric constant (50-100 Hz)	D150	...	3.7	3.9	3.9	3.7	4.2
Dissipation factor (50-100 Hz)	D150	...	0.020	0.040	0.020	...	0.04
Processing parameters							
Processing temperature, °C (°F)	...	175-205 (350-400)	230-290 (450-550)	230-290 (450-550)	230-290 (450-550)	200-260 (390-500)	180-260 (360-500)
Density, g/cm^3	D792	1.08-1.14	1.10	1.07-1.09	1.07	1.04	1.01-1.02
Specific volume, cm^3/kg (in.^{3}/lb)	D792	905-865 (25-24)	905 (25)	905 (25)	905 (25)	940 (26)	905 (25)
Linear mold shrinkage, mm/mm (in./in.)	D955	0.006-0.010	0.010-0.015	0.012	0.010	0.010	0.003-0.010
Water absorption, % in 24 h	D570	1.5-2.0	0.4	0.2	0.4	0.3	0.3

(a) Notched, at room temperature.

Source: Ref 15, p 1-14

Polyamideimides (PAI)

Molecular structures of polyamideimide

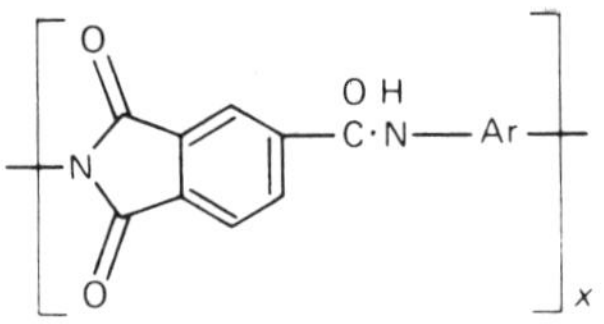

Source: Ref 6, p 101

Thermal, physical, and chemical properties of a typical polyamide-imide (PAI)

Heat-deflection temperature at 1.82 MPa, °C	275
Maximum resistance to continuous heat, °C	225
Coefficient of linear thermal expansion, $10^{-5}/°C$	3.6
Compressive strength, MPa	220
Flexural strength, MPa	210
Impact strength, Izod, cm·N/cm of notch	133.5
Tensile strength, MPa	186
Elongation, %	12
Hardness, Rockwell	M119
Specific gravity	1.4
Dielectric constant	4.0
Water absorption	0.2
Resistance to chemicals at 25% °C(a)	
Nonoxidizing acids (20% H_2SO_4)	S
Oxidizing acids (10% HNO_3)	U
Aqueous salt solutions (NaCl)	S
Aqueous alkalis (NaOH)	U
Polar solvents (C_2H_5OH)	S
Nonpolar solvents (C_6H_6)	S
Water	S

(a) S, satisfactory; Q, questionable; U, unsatisfactory.

Source: Ref 13, p 133

Selected properties of molded polyamideimide

Property	ASTM test	Reported value
Mechanical properties		
Yield strength, MPa (ksi)	D638	93.1 (13.5)
Elongation (break), %	D638	3
Compressive strength, MPa (ksi)	D695	240 (35)
Flexural yield strength, MPa (ksi)	D790	160-165 (23-24)
Flexural modulus, GPa (10^6 psi)	D790	4.8 (0.70)
Impact strength, Izod(a), J/cm (ft·lb/in.)	D256	0.53 (1.0)
Hardness, Rockwell	...	R104
Thermal properties		
Coefficient of linear thermal expansion, $10^{-5}/°C$ ($10^{-5}/°F$)	D696	6.3 (3.5)
Electrical properties		
Volume resistivity, Ω·cm	D257	10^{14}
Dissipation factor (50-100 Hz)	D150	0.005-0.007
Processing parameters		
Processing temperature, °C (°F)	...	345-385 (650-725)
Density, g/cm^3	D792	1.40
Specific volume, cm^3/kg (in.^{3}/lb)	D792	725 (20)
Linear mold shrinkage, mm/mm (in./in.)	D955	0.006
Water absorption, % in 24 h	D570	0.3

(a) Notched, at room temperature.

Source: Ref 15, p 1-18

Properties of polyamideimide engineering resins with various reinforcements and additives

Nominal composition	Properties/characteristics	Applications
High-strength composites		
3% TiO_2 ½%fluorocarbon	Best impact resistance, most elongation, and good mold-release and electrical properties	Connectors, switches, relays, thrust washers, spline liners, valve seats, poppets, mechanical linkages, bushings, wear rings, insulators, cams, picker fingers, ball bearings, rollers, thermal insulators
30% glass fiber 1% fluorocarbon	High stiffness, good retention of stiffness at elevated temperatures, very low creep, and high strength	Burn-in sockets, gears, valve plates, fairings, tube clamps, impellers, rotors, housings, back-up rings, terminal strips, insulators, brackets
30% glass fiber	Can be molded to greater thickness with some sacrifice in mechanical properties	Same as above, but for parts requiring thicker cross sections
30% glass fiber 4% TiO_2 1% fluorocarbon	High stiffness, good retention of stiffness at elevated temperatures, very low creep, and high strength	Structural, electrical, valve plates, metal replacement
30% graphite fiber 1% fluorocarbon	Best retention of stiffness at high temperatures, best fatigue resistance; electrically conductive	Metal replacement, housings, mechanical linkages, gears, fasteners, spline liners, cargo rollers, brackets, valves, labyrinth seals, fairings, tube clamps, standoffs, impellers, shrouds, potential use for EMI sheilding
33% carbon fiber 1% fluorocarbon	Can be molded to greater thickness with some sacrifice in mechanical properties; electrically conductive	Injection molding of normally troublesome thick cross sections
Proprietary blend of carbon fibers and fluorocarbons	High stiffness and lubricity	Service requiring high stiffness and some lubricity, especially sliding vanes; potential use for EMI shielding
Wear-resistant composites		
12% graphite powder 8% fluorocarbon	Good for reciprocating motion or bearings subject to high loads at low speed; best wear resistance	Bearings, thrust washers, wear pads, strips, piston rings, seals
12% graphite powder 8% fluorocarbon	Designed for bearing use; good wear resistance, low coefficient of friction, and high compressive strength	Bearings, thrust washers, wear pads, strips, piston rings, seals vanes, valve seats
20% graphite powder 3% fluorocarbon	Better wear resistance at high speeds	Bearings, thrust washers, wear pads, strips, piston rings, seals vanes, valve seats
High performance composite		
40% glass fiber 1% fluorocarbon	Best cost-to-performance ratio	Switiches, relays, terminal strips, wear bands, back-up rings, housings, impellers, brackets, thermal insulators

Source: Ref 16, p 6-25

Polyarylates

Thermal, physical, and chemical properties of typical crystalline polyarylates (Ekonol)

Heat-deflection temperature at 1.82 MPa, °C	300
Maximum resistance to continuous heat, °C	250
Coefficient of linear thermal expansion, 10^{-5}/°C	71
Compressive strength, MPa	265
Flexural strength, MPa	63
Impact strength, Izod, cm·N/cm of notch	25
Tensile strength, MPa	60
Elongation, %	15
Hardness, Rockwell	R130
Specific gravity	1.44
Dielectric constant	3
Water absorption	0.02
Resistance to chemicals at 25% °C(a):	
Nonoxidizing acids (20% H_2SO_4)	S
Oxidizing acids (10% HNO_3)	Q
Aqueous salt solutions (NaCl)	S
Aqueous alkalis (NaOH)	S
Polar solvents (C_2H_5OH)	S
Nonpolar solvents (C_6H_6)	S
Water	S

(a) S, satisfactory; Q, questionable; U, unsatisfactory.

Source: Ref 13, p 141

Thermal, physical, and chemical properties of typical polyarylates

Heat-deflection temperature at 1.82 MPa, °C	175
Maximum resistance to continuous heat, °C	150
Coefficient of linear thermal expansion, 10^{-5}/°C	6.5
Compressive strength, MPa	93
Flexural strength, MPa	80
Impact strength, Izod, cm·N/cm of notch	215
Tensile strength, MPa	68
Elongation, %	50
Hardness, Rockwell	R125
Specific gravity	1.2
Dielectric constant	0.7
Water absorption	0.26
Resistance to chemicals at 25% °C(a):	
Nonoxidizing acids (20% H_2SO_4)	S
Oxidizing acids (10% HNO_3)	Q
Aqueous salt solutions (NaCl)	S
Aqueous alkalis (NaOH)	S
Polar solvents (C_2H_5OH)	S
Nonpolar solvents (C_6H_6)	S
Water	S

(a) S, satisfactory; Q, questionable; U, unsatisfactory.

Source: Ref 13, p 140

Polyarylsulfones

Molecular structure of polyarylsulfone

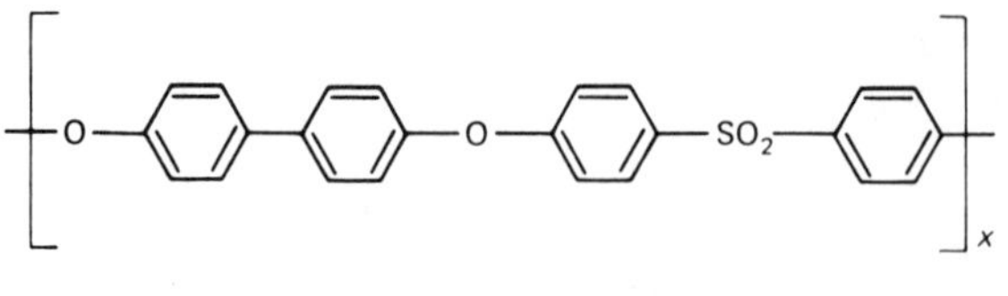

Source: Ref 6, p 101

Selected properties of polyarylsulfone (molded or extruded)

Property	ASTM test	Reported value
Mechanical properties		
Yield strength, MPa (ksi)	D638	89.6 (13.0)
Tensile modulus, GPa (10^6 psi)	D638	2.6 (0.37)
Elongation (break), %	D638	13-20
Compressive strength, MPa (ksi)	D695	125 (18)
Flexural yield strength, MPa (ksi)	D790	115 (17)
Flexural modulus, GPa (10^6 psi)	D790	2.8 (0.40)
Impact strength, Izod(a), J/cm (ft·lb/in.)	D256	0.53-1.1 (1.0-2.0)
Hardness, Rockwell	...	M110
Thermal properties		
Thermal conductivity, W/m·K (Btu·in./ft^{2}h·°F)	C177	0.19 (1.3)
Coefficient of linear thermal expansion, 10^{-5}/°C (10^{-5}/°F)	D696	8.5 (4.7)
Continuous service temperature, °C (°F)	...	190 (375)
Electrical properties		
Volume resistivity, Ω·cm	D257	10^{16}
Dielectric strength, V/10^{-3} mm (V/10^{-3} in.)	D149	14 (350)
Dielectric constant, (50-100 Hz)	D150	3.5
Dissipation factor (50-100 Hz)	D150	0.003
Processing parameters		
Processing temperature, °C (°F)	...	370-425 (700-800)
Density, g/cm^3	D792	1.36
Specific volume, cm^3/kg (in.^{3}/lb)	D792	725 (20)
Linear mold shrinkage, mm/mm (in./in.)	D955	0.007-0.009
Water absorption, % in 24 h	D570	1.1

(a) Notched, at room temperature.

Source: Ref 15, p 1-20

Polycarbonates

Molecular structure of polycarbonate

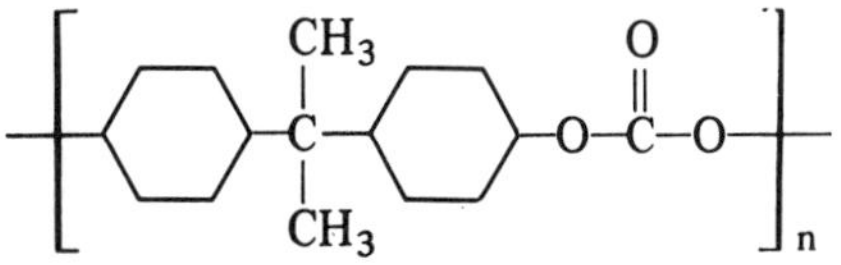

Source: Ref 20, p 50

Selected properties of polycarbonate (extruded or molded)

Property	ASTM test	Reported value
Mechanical properties		
Yield strength, MPa (ksi)	D638	55-66 (8.0-9.5)
Tensile modulus, GPa (10^6 psi)	D638	2.1-2.4 (0.30-0.35)
Elongation (break), %	D638	100-125
Compressive strength, MPa (ksi)	D695	83-90 (12-13)
Flexural yield strength, MPa (ksi)	D790	90-97 (13-14)
Flexural modulus, GPa (10^6 psi)	D790	2.1 (0.30)
Impact strength, Izod(a), J/cm (ft·lb/in.)	D256	6.41-9.61 (12.0-18.0)
Hardness, Rockwell	...	M70-82
Thermal properties		
Thermal conductivity, W/m·K (Btu·in./ft^2·h·°F)	C177	0.19 (1.3)
Coefficient of linear thermal expansion, 10^{-5}/°C (10^{-5}/°F)	D696	11-13 (6.0-7.0)
Specific heat, kJ/kg·K (Btu/lb·°F)	C351	1.3 (0.3)
Continuous service temperature, °C (°F)	...	120-135 (250-275)
Electrical properties		
Volume resistivity, Ω·cm	D257	2.0×10^{16}
Dielectric strength, V/10^{-3} mm (V/10^{-3} in.)	D149	15 (380)
Dielectric constant, (50-100 Hz)	D150	3.0
Dissipation factor (50-100 Hz)	D150	0.0007
Processing parameters		
Processing temperature, °C (°F)	...	250-345 (480-650)
Density, g/cm^3	D792	1.20
Specific volume, cm^3/kg (in.^{3}/lb)	D792	830 (23)
Linear mold shrinkage, mm/mm (in./in.)	D955	0.005-0.007
Water absorption, % in 24 h	D570	0.2

(a) Notched, at room temperature.

Source: Ref 15, p 1-25

Thermal, physical, and chemical properties of typical polycarbonates

Property	Unfilled polycarbonate	20% glass-filled polycarbonate
Heat-deflection temperature, °C	130	145
Maximum resistance to continuous heat, °C	115	130
Coefficient of linear thermal expansion, 10^{-5}/°C	6.8	2.2
Compressive strength, MPa	86	124
Flexural strength, MPa	93	158
Impact strength, Izod, cm·N/cm of notch	534	106
Tensile strength, MPa	72	131
Elongation, %	110	4
Hardness, Rockwell	M70	M92
Specific gravity	1.2	1.4
Dielectric constant	3	4
Water absorption, %	0.15	0.25
Resistance to chemicals at 25 °C(a):		
Nonoxidizing acids (20% H_2SO_4)	Q	Q
Oxidizing acids (10% HNO_3)	U	U
Aqueous salt solutions (NaCl)	S	S
Aqueous alkalis (NaOH)	U	U
Polar solvents (C_2H_5OH)	S	S
Nonpolar solvents (C_6H_6)	U	U
Water	S	S

(a) S, satisfactory; Q, questionable; U, unsatisfactory.

Source: Ref 13, p 136

Recommended processing factors for injection-moldable polycarbonate

Grade	Melt-temperature range °C	°F	Mold temperature °C	°F	Shrinkage range, mm/mm (in./in.)	Minimum cycle time(a), s
Unmodified base resin:						
Fast cycling	270-315	520-600	65-95	150-200	0.005-0.007	15-20
Low viscosity	275-315	530-600	65-95	150-200		15-20
Medium viscosity	290-315	550-600	65-95	150-200		15-20
High viscosity	315-345	600-650	65-95	150-200	0.005-0.007	15-20
High heat (polythalate carbonate)	330-360	630-680	65-95	150-200	0.007-0.010	20-25
Glass reinforced:						
10%	300-325	570-620	80-115	180-240	0.002-0.004	20-25
20%	315-340	600-640	80-115	180-240	0.003	20-25
30%	315-345	600-650	80-115	180-240	0.0025	20-25
40%	315-350	600-660	95-125	200-260	0.002	20-25
Impact modified	290-315	550-600	70-95	160-200	0.005-0.007	20-25
Flame retarded:						
Fast cycling	270-305	520-580	65-95	150-200	0.005-0.007	15-20
Low viscosity	275-295	530-560	70-95	160-200		20-25
Medium viscosity	290-315	550-600	70-95	160-200		20-25
High viscosity	315-345	600-650	80-115	180-240	0.005-0.007	20-25

(a) Data based on a wall thickness of 3.18 mm (1/8 in.). Low to medium injection speed. Maximum regrind, 20-25%.

Source: Ref 19, p 39

Polybutylene terephthalate

Selected properties of polybutylene (extruded or molded)

Property	ASTM test	Reported value
Mechanical properties		
Yield strength, MPa (ksi)	D638	21-28 (3.0-4.0)
Tensile modulus, GPa (10^6 psi)	D638	0.21-0.34 (0.03-0.05)
Elongation (break), %	D638	350-450
Flexural modulus, GPa (10^6 psi)	D790	0.69 (0.10)
Impact strength, Izod(a), J/cm (ft·lb/in.)	D256	No break
Hardness, Rockwell	...	D55-65
Thermal properties		
Thermal conductivity, W/m·K (Btu·in./ft^2·h·°F)	C177	0.22 (1.5)
Coefficient of linear thermal expansion, 10^{-5}/°C (10^{-5}/°F)	D696	16 (9.0)
Specific heat, kJ/kg·K (Btu/lb·°F)	C351	1.7-2.1 (0.4-0.5)
Continuous service temperature, °C (°F)	...	105 (225)
Electrical properties		
Dielectric constant, (50-100 Hz)	D150	2.2
Dissipation factor (50-100 Hz)	D150	0.005
Processing parameters		
Processing temperature, °C (°F)	...	93-195 (199-380)
Density, g/cm^3	D792	0.91
Specific volume, cm^3/kg (in.^{3}/lb)	D792	1080 (30)
Linear mold shrinkage, mm/mm (in./in.)	D955	0.003
Water absorption, % in 24 h	D570	0.01

(a) Notched, at room temperature.

Source: Ref 15, p 1-24

Thermal, physical, and chemical properties of typical polybutylene terephthalate (PBT)

Property	Unfilled PBT	PBT with 30% fibrous glass
Heat-deflection temperature at 1.82 MPa, °C	65	200
Maximum resistance to continuous heat, °C	60	150
Coefficient of linear thermal expansion, $10^{-5}/°C$	7.0	2.5
Compressive strength, MPa	75	120
Flexural strength, MPa,	96	110
Impact strength, Izod, cm·N/cm of notch	53.4	50
Tensile strength, MPa	55	117
Elongation, %	100	4
Hardness, Rockwell	M70	M90
Specific gravity	1.35	1.5
Dielectric constant	4.0	4.0
Water absorption, %	0.05	0.05
Resistance to chemicals at 25 °C(a):		
Nonoxidizing acids (20% H_2SO_4)	S	S
Oxidizing acids (10% HNO_3)	Q	Q
Aqueous salt solutions (NaCl)	S	S
Aqueous alkalis (NaOH)	S	S
Polar solvents (C_2H_5OH)	S	S
Nonpolar solvents (C_6H_6)	S	S
Water	S	S

(a) S, satisfactory; Q, questionable.

Source: Ref 13, p 138

Thermoplastic polyesters

Molecular structure of a typical vinyl ester

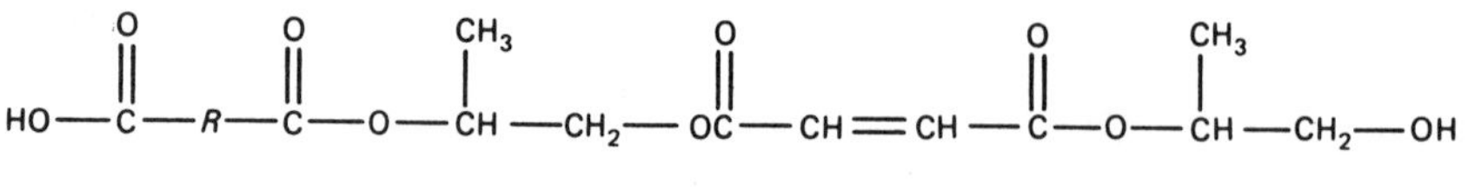

Source: Ref 6, p 137

Selected properties of thermoplastic polyester (extruded or molded)

Property	ASTM test	Reported value
Mechanical properties		
Yield strength, MPa (ksi)	D638	55-59 (8.0-8.5)
Tensile modulus, GPa (10^6 psi)	D638	1.9 (0.28)
Elongation (break), %	D638	100-300
Compressive strength, MPa (ksi)	D695	55-105 (8-15)
Flexural yield strength, MPa (ksi)	D790	83-110 (12-16)
Flexural modulus, GPa (10^6 psi)	D790	2.1-2.8 (0.30-0.40)
Impact strength, Izod(a), J/cm (ft·lb/in.)	D256	0.4-0.53 (0.8-1.0)
Hardness, Rockwell	...	M80-95
Thermal properties		
Thermal conductivity, W/m·K (Btu·in./ft^2·h·°F)	C177	0.19-0.29 (1.3-2.0)
Coefficient of linear thermal expansion, 10^{-5}/°C (10^{-5}/°F)	D696	11-16 (6.0-9.0)
Specific heat, kJ/kg·K (Btu/lb·°F)	C351	1.3-2.1 (0.3-0.5)
Continuous service temperature, °C (°F)	...	93-120 (200-250)
Electrical properties		
Volume resistivity, Ω·cm	D257	10^{15}
Dielectric strength, V/10^{-3} mm (V/10^{-3} in.)	D149	23 (580)
Dielectric constant, (50-100 Hz)	D150	3.2
Dissipation factor (50-100 Hz)	D150	0.002
Processing parameters		
Processing temperature, °C (°F)	...	205-275 (400-525)
Density, g/cm^3	D792	1.30-1.40
Specific volume, cm^3/kg ($in.^3$/lb)	D792	940-865 (26-24)
Linear mold shrinkage, mm/mm (in./in.)	D955	0.015-0.020
Water absorption, % in 24 h	D570	0.1

(a) Notched, at room temperature.

Source: Ref 15, p 1-26

Thermal, physical, and chemical properties of a typical polyester thermoplastic elastomer

Tensile strength, MPa	39
Elongation, %	350
Hardness, Shore D	72
Specific gravity	1.25
Izod notched impact strength, J/cm	2.1
Heat-deflection temperature at 500 kPa, °C	166
Softening point, °C	203
Coefficient of linear thermal expansion, 10^{-5}/°C	21
Water absorption, %	0.3
Resistance to chemicals at 25% °C(a):	
Nonoxidizing acids (20% H_2SO_4)	S
Oxidizing acids (10% HNO_3)	U
Aqueous salt solutions (NaCl)	S
Aqueous alkalis (NaOH)	Q
Polar solvents (C_2H_5OH)	S
Nonpolar solvents (C_6H_6)	Q

(a) S, satisfactory; Q, questionable; U, unsatisfactory.

Source: Ref 23, p 88

Thermal, physical, and chemical properties of a typical polyester

Heat-deflection temperature at 1.82 MPa, °C	65
Maximum resistance to continuous heat, °C	60
Melting point °C	240
Coefficient of linear thermal expansion, 10^{-5}/°C	7
Tensile strength, MPa	60
Elongation, %	50
Specific gravity	1.35
Dielectric constant	3.0
Resistance to chemicals at 25% °C(a):	
Nonoxidizing acids (20% H_2SO_4)	Q
Oxidizing acids (10% HNO_3)	Q
Aqueous salt solutions (NaCl)	S
Aqueous alkalis (NaOH)	Q
Polar solvents (C_2H_5OH)	Q
Nonpolar solvents (C_6H_6)	U
Water	S

(a) S, satisfactory; Q, questionable; U, unsatisfactory.

Source: Ref 23, p 70

Recommended processing factors for injection-moldable thermoplastic polyesters

Grade	Melt-temperature range °C	Melt-temperature range °F	Melt viscosity(a) Pa·s	Injection speed	Mold temperature °C	Mold temperature °F	Maximum regrind, %	Gating diameter(b) mm	Gating diameter(b) in.	Shrinkage range(c), mm/mm (in./in.)	Minimum cycle time(b), s
Polybutylene terephthalate (PBT)											
Unmodified base resin	245-255	470-490	...	Medium to high	91-121	195-250	25	1.3	0.050	0.018-0.022	35-50
Glass reinforced, 20-30%	250-265	480-510	...	High	79-91	175-195	25	1.5	0.060	0.004-0.007(F); 0.007-0.012(T)	30-40
Mineral/glass reinforced	250-260	480-500	...	High	79-91	175-195	25	1.5	0.060	0.007-0.010(F); 0.017-0.022(T)	30-45
Impact modified, unfilled	230-255	450-490	...	High	88-91	190-195	0	2.4	0.095	0.017-0.022	50-100
Flame retarded, 20-30%	250-265	480-510	...	High	79-91	175-195	25	1.5	0.060	0.004-0.007(F); 0.007-0.012(T)	30-45
Polyethylene terephthalate (PET)											
Glass reinforced:											
15%	270-295	520-560	...	Medium to high	82-121	180-250	25	0.51-2.54	0.020-0.100	0.004-0.008	15-18
30%	280-310	540-590	120-350	Medium to high	82-121	180-250	25	0.51-2.54	0.020-0.100	0.003-0.006	15-18
45%	280-310	540-590	110-350	Medium to high	82-121	180-250	25	0.51-2.54	0.020-0.100	0.002-0.004	15-18
55%	280-310	540-590	110-400	Medium to high	82-121	180-250	25	0.51-2.54	0.020-0.100	0.002-0.003	15-18
Mineral/glass reinforced, 35-40%	280-310	540-590	105-500	Medium to high	82-121	180-250	25	0.51-2.54	0.020-0.100	0.003-0.006	15-18
Impact modified:											
30%	275-300	530-575	600	High	93-121	200-250	25	0.51-2.54	0.020-0.100	0.002-0.009	15-18
35%	270-300	520-570	450	High	85-121	185-250	25	0.51-2.54	0.020-0.100	0.002-0.009	15-18
Flame retarded:											
15% glass reinforced	270-295	520-560	...	High	82-121	180-250	25	0.51-2.54	0.020-0.100	0.004-0.008	15-18
30% glass reinforced	280-300	540-570	120-300	High	82-121	180-250	25	0.51-2.54	0.020-0.100	0.015-0.009	15-18
35% mineral filled	280-300	540-570	110-300	High	82-121	180-250	25	0.51-2.54	0.020-0.100	0.015-0.009	15-18

(a) Melt viscosity measured at 1000/s at the median temperature of the recommended melt-temperature range. (b) Data based on a wall thickness of 3.18 mm (1/8 in.). (c) F, flow direction; T, transverse direction.

Source: Ref 19, p 46-47

Polyetheretherketones (PEEK)

Molecular structure of polyetheretherketone (PEEK)

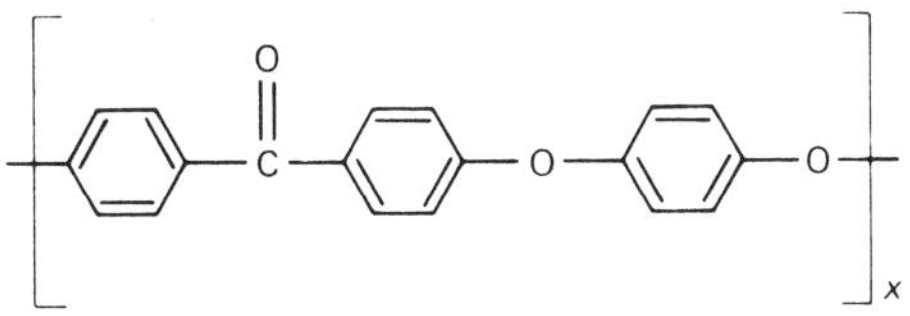

Source: Ref 6, p 101

Thermal, physical, and chemical properties of a typical polyetheretherketone (PEEK)

Property	Unfilled PEEK	40% glass-filled PEEK
Heat-deflection temperature at 1.82 MPa, °C	150	300
Maximum resistance to continuous heat, °C	125	225
Coefficient of linear thermal expansion, 10^{-5}/°C	5.5	2.2
Compressive strength, MPa	90	125
Flexural strength, MPa	110	250
Impact strength, Izod, cm·N/cm of notch	50	75
Tensile strength, MPa	70	107
Elongation, %	50	2
Hardness, Rockwell	R123	R123
Specific gravity	1.3	1.5
Dielectric constant	3.2	3.5
Water absorption, %	0.15	0.12
Resistance to chemicals at 25 °C(a):		
Nonoxidizing acids (20% H_2SO_4)	S	S
Oxidizing acids (10% HNO_3)	S	S
Aqueous salt solutions (NaCl)	S	S
Aqueous alkalis (NaOH)	S	S
Polar solvents (C_2H_4OH)	S	S
Nonpolar solvents (C_6H_6)	S	S
Water	S	S

(a) S, satisfactory.

Source: Ref 13, p 146

Recommended processing factors for injection-moldable polyetherether-ketone (PEEK) and polyethersulfone (PES) (a)

Material	Melt-temperature range °C	Melt-temperature range °F	Melt viscosity(a) Pa·s	Shrinkage range(c), mm/mm (in./in.)	Minimum cycle time(b), s
PEEK					
Unmodified base resin:					
General purpose	350-380	660-720	330	0.010	10-15
General purpose	350-380	660-720	250	0.010	10-15
Glass reinforced:					
20%	360-385	680-725	450	0.007	10-15
30%	360-385	680-725	480	0.005	10-15
Carbon fiber reinforced	370-395	700-740	550	0.005	10-15
PES					
Unmodified base resin:					
Low viscosity	340-360	645-680	180	0.006	10-15
General purpose	340-360	645-680	220	0.006	10-15
Medium viscosity	350-370	660-700	400	0.006	10-15
High viscosity	350-375	660-710	550	0.006	10-15
Glass reinforced:					
Easy flow, 20% glass	335-355	635-670	200	0.003	10-15
Easy flow 30% glass	335-355	635-670	250	0.002	10-15
General purpose, 20% glass	350-370	660-700	250	0.003	10-15
General purpose, 30% glass	350-375	660-710	300	0.002	10-15
High viscosity, 20% glass	350-375	660-710	600	0.003	10-15

(a) Medium to high injection speed. Mold temperature, 160-170 °C (320-340 °F). Maximum regrind, 30%. Gating diameter, 1.0-2.0 mm (0.004-0.008 in.) (b) Melt viscosity measured at 1000/s at the median temperature of the recommended melt-temperature range. (c) Data based on a wall thickness of 3.18 mm (1/8 in.).

Source: Ref 19, p 40-41

Polyetherimides (PEI)

Molecular structure of polyetherimide (PEI)

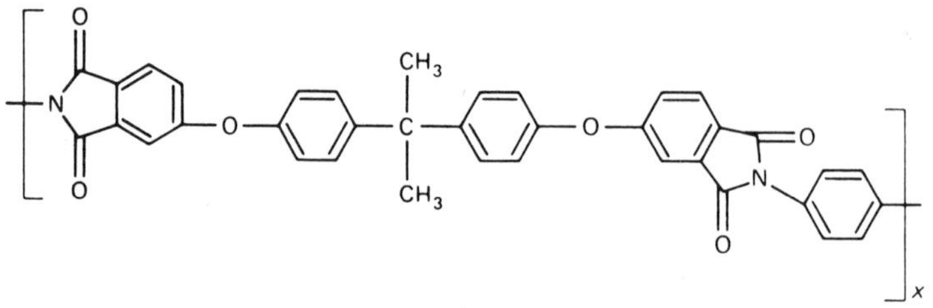

Source: Ref 6, p 101

Recommended processing factors for injection-moldable polyetherimide(a)

Material	Melt-temperature range °C	Melt-temperature range °F	Melt viscosity(b), Pa·s	Mold temperature °C	Mold temperature °F	Maximum regrind, %
Unmodified base resin	345-425	650-800	450	65-150	150-300	50
Glass reinforced:						
10% glass	355-425	675-800	400	65-195	150-380	40
20% glass	355-425	675-800	450	65-195	150-380	40
30% glass	355-425	675-800	500	65-195	150-380	40
Mineral reinforced	355-425	675-800	...	65-195	150-380	40
Mineral/glass reinforced	355-425	675-800	...	65-195	150-380	40
Impact modified	290-310	550-590	...	65-120	150-250	...
Flame retarded	290-310	550-590	...	-9-175	15-350	...

(a) Data based on a wall thickness of 3.18 mm (0.125 in.). Maximum injection speed, 262 cm^3/s (16 $in.^3/s$). (b) Melt viscosity measured at 1000/s at the median temperature of the recommended melt-temperature range.

Source: Ref 19, p 41

Thermal, physical, and chemical properties of a typical polyetherimide (PEI)

Property	Unmodified PEI	10% glass-reinforced PEI	20% glass-reinforced PEI	30% glass-reinforced PEI
Heat-deflection temperature at 1.82 MPa, °C	190	200	205	210
Maximum resistance to continuous heat, °C	170	175	180	185
Coefficient of linear thermal expansion, 10^{-5}/°C	5.6	4.4	3.2	2.0
Compressive strength, MPa	140	155	169	176
Flexural strength, MPa	145	195	205	225
Impact strength, Izod, cm·N/cm of notch	133.5	146	213	267
Tensile strength, MPa	104	114	138	169
Elongation, %	6.0	6.0	3.0	3.0
Hardness, Rockwell	M109	M115	M120	M125
Specific gravity	1.27	1.35	1.45	1.6
Dielectric constant	3.1	3.3	3.5	3.7
Water absorption, %	0.06	0.1	0.15	0.2
Resistance to chemicals at 25 °C(a):				
Nonoxidizing acids (20% H_2SO_4)	S	S	S	S
Oxidizing acids (10% HNO_3)	U	U	U	U
Aqueous salt solutions (NaCl)	S	S	S	S
Aqueous alkalis (NaOH)	Q	Q	Q	Q
Polar solvents (C_2H_5OH)	S	S	S	S
Nonpolar solvents (C_6H_6)	S	S	S	S
Water	S	S	S	S

(a) S, satisfactory; Q, questionable; U, unsatisfactory.

Source: Ref 13, p 135

Polyether sulfones

Molecular structure of polyether sulfone (PES)

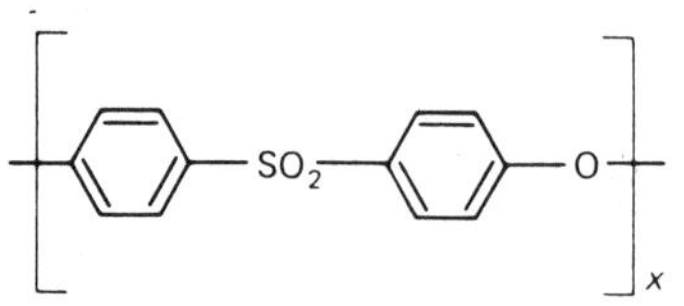

Source: Ref 6, p 101

Selected properties of polyether sulfone (extruded or molded)

Property	ASTM test	Reported value
Mechanical properties		
Yield strength, MPa (ksi)	D638	85.5 (12.4)
Tensile modulus, GPa (10^6 psi)	D638	2.4 (0.35)
Elongation (break), %	D638	50-100
Compressive strength, MPa (ksi)	D695	97 (14)
Flexural yield strength, MPa (ksi)	D790	125-130 (18-19)
Flexural modulus, GPa (10^6 psi)	D790	2.8 (0.40)
Impact strength, Izod(a), J/cm (ft·lb/in.)	D256	0.85 (1.6)
Hardness, Rockwell	...	M88
Thermal properties		
Thermal conductivity, W/m·K (Btu·/ft^2·h·°F)	C177	0.14-0.22 (1.0-1.5)
Coefficient of linear thermal expansion, 10^{-5}/°C (10^{-5}/°F)	D696	7.2 (4.0)
Specific heat, kJ/kg·K (Btu/lb·°F)	C351	1.3 (0.3)
Continuous service temperature, °C (°F)	...	170 (340)
Electrical properties		
Volume resistivity, Ω·cm	D257	10^{17}
Dielectric strength, V/10^{-3} mm (V/10^{-3} in.)	D149	16 (400)
Dielectric constant, (50-100 Hz)	D150	3.1
Dissipation factor (50-100 Hz)	D150	0.001
Processing parameters		
Processing temperature, °C (°F)	...	300-355 (575-675)
Density, g/cm^3	D792	1.40
Specific volume, cm^3/kg (in.^{3}/lb)	D792	725 (20)
Linear mold shrinkage, mm/mm (in./in.)	D955	0.007
Water absorption, % in 24 h	D570	0.4

(a) Notched, at room temperature.

Source: Ref 15, p 1-21

Polyethylenes

Molecular structure of polyethylene

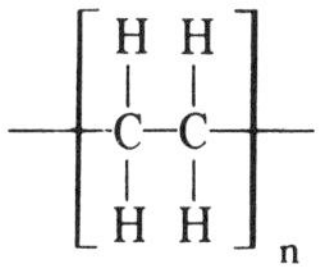

Source: Ref 20, p 48

Thermal, physical, and chemical properties of polyethylene

Property	LDPE	HDPE	UHMWPE
Heat-deflection temperature at 1.82 MPa, °C	40	85	85
Maximum resistance to continuous heat, °C	40	80	80
Coefficient of linear thermal expansion, 10^{-5}/°C	10.0	12.0	12.0
Compressive strength, MPa	...	21	...
Impact strength, Izod, cm·N/cm of notch	No break	106.7	No break
Tensile strength, MPa	6	28	38
Elongation, %	100	30	400
Hardness, Rockwell	D40	D40	R50
Specific gravity	0.91	0.95	0.94
Dielectric constant	2.3	2.3	2.3
Resistance to chemicals at 25 °C(a):			
Nonoxidizing acids (20% H_2SO_4)	S	S	S
Oxidizing acids (10% HNO_3)	Q	Q	Q
Aqueous salt solutions (NaCl)	S	S	S
Polar solvents (C_2H_5OH)	S	S	S
Nonpolar solvents (C_6H_6)	Q	Q	Q
Water	S	S	S

(a) S, satisfactory; Q, questionable.

Source: Ref 23, p 57

Selected properties of molded polyethylenes

Property	ASTM test	Reported value for: Low-density polyethylene	Medium density polyethylene	High-density polyethylene	Polyallomer polyethylene	Cross-linkable polyethylene	UHMW polyethylene
Mechanical properties							
Yield strength, MPa (ksi)	D638	4.1-16 (0.6-2.3)	8.3-24 (1.2-3.5)	21-38 (3.1-5.5)	21-28 (3.0-4.1)	11-32 (1.6-4.6)	17-24 (2.5-3.5)
Tensile modulus, GPa (10^6 psi)	D638	0.07-0.28 (0.01-0.04)	0.14-0.41 (0.02-0.06)	0.41-1.2 (0.06-0.018)	...	0.34-3.4 (0.05-0.50)	0.14-0.76 (0.02-0.11)
Elongation (break), %	D638	100-800	50-600	30-1300	400-500	100-450	300-500
Compressive strength, MPa (ksi)	D695	...	...	21-28 (3-4)	...	18 (2.6)	...
Flexural yield strength, MPa (ksi)	D790	...	34-48 (5-7)	...	...	14-48 (2-7)	...
Flexural modulus, GPa (10^6 psi)	D790	0.07 (0.01)	0.69 (0.10)	0.69-2.1 (0.10-0.30)	0.69 (0.10)	0.69-2.8 (0.10-0.40)	0.69 (0.10)
Impact strength, Izod(a), J/cm (ft·lb/in.)	D256	No break	0.3-8.0 (0.5-15.0)	0.3-10.7 (0.5-20.0)	No break	No break	No break
Hardness, Rockwell	...	D40-50	D50-60	D60-70	R50-85	D55-80	D60-70
Thermal properties							
Thermal conductivity, W/m·K (Btu·in./ft^2·h·°F)	C177	0.30 (2.1)	0.30-0.42 (2.1-2.9)	0.43-0.52 (3.0-3.6)	0.087-0.17 (0.6-1.2)	...	...
Coefficient of thermal expansion 10^{-5}/°C (10^{-5}/°F)	D696	18.0-36.0 (10.0-20.0)	25.2-28.8 (14.0-16.0)	21.6-23.4 (12.0-13.0)	14.4-21.6 (8.0-12.0)	18.0-63.0 (10.0-35.0)	12.6 (7.0)
Specific heat, kJ/kg·K (Btu/lb·°F)	C351	2.1-2.5 (0.5-0.6)	2.1-2.5 (0.5-0.6)	2.1-2.5 (0.5-0.6)	2.1 (0.5)	...	...
Continuous service temperature, °C (°F)	...	60-90 (140-190)	60-90 (140-190)	60-90 (140-190)	...	...	...
Electrical properties							
Volume resistivity, Ω·cm	D257	10^{16}	10^{16}	10^{16}	10^{15}	...	10^{16}
Dielectric strength, V/10^{-3} mm (V/10^{-3} in.)	D149	18-20 (450-500)	18-20 (450-500)	18-20 (450-500)	0.08 (2)	...	...
Dielectric constant (50-100 Hz)	D150	2.2-2.4	2.2-2.4	2.2-2.4	...	...	...
Dissipation factor (50-100 Hz)	D150	<0.0005	<0.0005	<0.0005	0.005	...	0.0002
Processing parameters							
Processing temperature, °C (°F)	...	150-315 (300-600)	150-315 (300-600)	150-315 (300-600)	...	120-400 (250-750)	260-650 (500-1200)
Density, g/cm^3	D792	0.900-0.925	0.926-0.940	0.941-0.965	0.89	0.95-1.45	0.94
Specific volume, cm^3/kg (in.^{3}/lb)	D792	1080 (30)	1050 (29)	1010 (28)	...	...	1080 (30)
Linear mold shrinkage, mm/mm (in./in.)	D955	0.015-0.050	0.015-0.050	0.020-0.050	0.010-0.020	0.007	...
Water absorption, % in 24 h	D570	1.5	1.5	1.5	<0.1	<0.1	<0.1

(a) Notched, at room temperature.

Source: Ref 15, p 1-20

Polyethylene terephthalate

Thermal, physical, and chemical properties of typical polyethylene terephthalate (PET)

Property	PET	PET with 30% fibrous glass
Heat-deflection temperature at 1.82 MPa, °C	100	226
Maximum resistance to continuous heat, °C	100	160
Coefficient of linear thermal expansion, 10^{-5}/°C	6.5	2.9
Compressive strength, MPa	86	172
Flexural strength, MPa,	112	234
Impact strength, Izod, cm·N/cm of notch	26.7	50
Tensile strength, MPa	62	158
Elongation, %	100	2.5
Hardness, Rockwell	M96	M100
Specific gravity	1.35	1.56
Dielectric constant	3.6	4.0
Water absorption, %	0.2	0.05
Resistance to chemicals at 25 °C(a):		
Nonoxidizing acids (20% H_2SO_4)	S	S
Oxidizing acids (10% HNO_3)	Q	Q
Aqueous salt solutions (NaCl)	S	S
Aqueous alkalis (NaOH)	S	S
Polar solvents (C_2H_5OH)	S	S
Nonpolar solvents (C_6H_6)	S	S
Water	S	S

(a) S, satisfactory; Q, questionable.

Source: Ref 13, p 137

Thermoplastic polyimides

Thermal, physical, and chemical properties of typical polyimides (PI)

Property	Thermo-plastic	Glass-filled thermoset (50%)
Heat-deflection temperature at 1.82 MPa, °C	315	350
Maximum resistance to continuous heat, °C	300	325
Coefficient of linear thermal expansion, 10^{-5}/°C	5.0	1.3
Compressive strength, MPa	241	234
Flexural strength, MPa,	172	145
Impact strength, Izod, cm·N/cm of notch	80	294
Tensile strength, MPa	96.5	44
Elongation, %	8	0.5
Hardness, Rockwell	E60	M118
Specific gravity	1.4	1.6
Dielectric constant	3.4	3.5
Water absorption, %	0	0.2
Resistance to chemicals at 25 °C(a):		
Nonoxidizing acids (20% H_2SO_4)	Q	Q
Oxidizing acids (10% HNO_3)	Q	Q
Aqueous salt solutions (NaCl)	S	S
Aqueous alkalis (NaOH)	U	U
Polar solvents (C_2H_5OH)	S	S
Nonpolar solvents (C_6H_6)	S	S
Water	S	S

(a) S, satisfactory; Q, questionable.

Source: Ref 13, p 132

Liquid crystal polymers

Thermal, physical, and chemical properties of typical liquid crystal polymers (LCP)

Property	Unfilled LCP	50% talc-filled LCP
Heat-deflection temperature at 1.82 MPa, °C	350	325
Maximum resistance to continuous heat, °C	250	250
Compressive strength, MPa	42	42
Flexural strength, MPa	125	110
Impact strength, Izod, cm·N/cm of notch	135	70
Tensile strength, MPa	135	70
Elongation, %	4.0	3.0
Hardness, Rockwell	R60	R76
Specific gravity	1.35	1.84
Dielectric constant	3	3.5
Water absorption	0	0
Resistance to chemicals at 25 °C(a):		
Nonoxidizing acids (20% H_2SO_4)	S	S
Oxidizing acids (10% HNO_3)	S	S
Aqueous salt solutions (NaCl)	S	S
Aqueous alkalis (NaOH)	S	S
Polar solvents (C_2H_5OH)	S	S
Nonpolar solvents (C_6H_6)	S	S
Water	S	S

(a) S, satisfactory.

Source: Ref 13, p 148

Polyphenylene oxide

Thermal, physical, and chemical properties of a polyphenylene oxide (PPO)

Property	PPO	Glass-filled PPO
Heat-deflection temperature at 1.82 MPa, °C	100	145
Maximum resistance to continuous heat, °C	80	130
Coefficient of linear thermal expansion, 10^{-5}/°C	5.0	2.0
Compressive strength, MPa	96	123
Flexural strength, MPa	89	144
Impact strength, Izod, cm·N/cm of notch	270	107
Tensile strength, MPa	55	120
Elongation, %	50	4
Hardness, Rockwell	R115	R115
Specific gravity	1.1	1.1
Dielectric constant	2.8	3.0
Resistance to chemicals at 25 °C(a):		
Nonoxidizing acids (20% H_2SO_4)	S	S
Oxidizing acids (10% HNO_3)	Q	Q
Aqueous salt solutions (NaCl)	S	S
Aqueous alkalis (NaOH)	S	S
Polar solvents (C_2H_5OH)	S	S
Nonpolar solvents (C_6H_6)	U	U
Water	S	S

(a) S, satisfactory; Q, questionable; U, unsatisfactory.

Source: Ref 13, p 142

Molecular structure of polyphenylene oxide

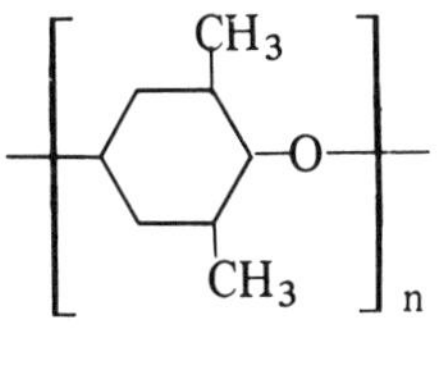

Source: Ref 20, p 50

Selected properties of phenylene oxides

Property	ASTM test	Reported value for: Extruded phenylene oxide	Molded phenylene oxide
Mechanical properties			
Yield strength, MPa (ksi)	D638	46-66 (6.6-9.5)	54-79.3 (7.8-11.5)
Tensile modulus, GPa (10^6 psi)	D638	2.5-2.6 (0.36-0.38)	2.5-2.6 (0.36-0.38)
Elongation (break), %	D638	50-60	50-60
Compressive strength, MPa (ksi)	D695	110-115 (16-17)	110-115 (16-17)
Flexural yield strength, MPa (ksi)	D790	90-97 (13-14)	90-97 (13-14)
Flexural modulus, GPa (10^6 psi)	D790	2.8 (0.40)	2.8 (0.40)
Impact strength, Izod(a) J/cm (ft·lb/in.)	D256	2.7-3.7 (5.0-7.0)	2.7 (5.0)
Hardness, Rockwell	...	R114	R115-119
Thermal properties			
Thermal conductivity, W/m·K (Btu·in./ft^2·h·°F)	C177	0.22 (1.5)	0.22 (1.5)
Coefficient of linear thermal 10^{-5}/°C (10^{-5}/°F)	D696	5.4-7.2 (3.0-4.0)	9.4 (5.2)
Specific heat, kJ/kg·K (Btu/lb·°F)	C351	1.3 (0.3)	1.3 (0.3)
Continuous service temperature, °C (°F)	...	80-105 (175-220)	80-105 (175-220)
Electrical properties			
Volume resistivity, Ω·cm	D257	10^{16}-10^{17}	10^{16}-10^{17}
Dielectric strength, V/10^{-3} mm (V/10^{-3} in.)	D149	16-24 (400-600)	16-22 (400-500)
Dielectric constant (50-100 Hz)	D150	2.6	2.6
Dissipation factor (50-100 Hz)	D150	0.0004	0.004
Processing parameters			
Processing temperature, °C (°F)	...	205-240 (400-460)	205-315 (400-600)
Density, g/cm^3	D792	1.06-1.10	1.06-1.10
Specific volume, cm^3/kg (in.^{3}/lb)	D792	940-905 (26-25)	940-905 (26-25)
Linear mold shrinkage, (in./in.)	D955	0.005-0.007	...
Water absorption, % in 24 h	D570	0.1	0.6

(a) Notched, at room temperature.

Source: Ref 15, p 1-16

Recommended processing factors for injection-moldable polyphenylene oxide (PPO)

Grade	Melt-temperature range °C	Melt-temperature range °F	Injection speed	Mold temperature °C	Mold temperature °F	Maximum regrind, %	Gating diameter(a) mm	Gating diameter(a) in.	Shrinkage range, mm/mm (in./in.)	Minimum cycle time(b), s
Unmodified base resin(c) ...	260-315	500-600	Medium to high	65-120	150-250	25	1.52-3.18 (T); 3.18-6.35 (W)	0.060-0.125 (T); 0.125-0.250 (W)	0.005-0.007	30-60
Glass reinforced	290-325	550-620	Low to high	90-105	190-220	25	1.52-3.18 (T); 3.18-6.35 (W)	0.060-0.125 (T); 0.125-0.250 (W)	0.002-0.005	30-60
Flame retarded	230-315	450-600	Medium to high	65-120	150-250	25	1.52-3.18 (T); 3.18-6.35 (W)	0.060-0.125 (T); 0.125-0.250 (W)	0.005-0.007	30-60
Automotive	275-310	530-590	Medium to high	75-105	170-220	25	Gates vary	Gates vary	...	50-90

(a) T, thickness; W, width. (b) Data based on a wall thickness of 3.18 mm (1/8 in.). (c) Multiple grades.

Source: Ref 19, p 42

Polyphenylene sulfide

Molecular structure of polyphenylene sulfide

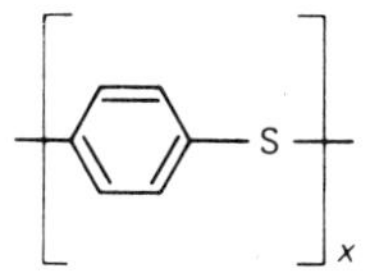

Source: Ref 6, p 101

Selected properties of molded phenylene sulfide

Property	ASTM test	Reported value
Mechanical properties		
Yield strength, MPa (ksi)	D638	68.9 (10.0)
Tensile modulus, GPa (10^6 psi)	D638	3.4 (0.50)
Elongation (break), %	D638	3
Flexural yield strength, MPa (ksi)	D790	140 (20)
Flexural modulus, GPa (10^6 psi)	D790	4.1 (0.60)
Impact strength, Izod(a), J/cm (ft·lb/in.)	D256	0.16 (0.30)
Hardness, Rockwell	...	R124
Thermal properties		
Thermal conductivity, W/m·K (Btu·in./ft^2·h·°F)	C177	0.29 (2.0)
Coefficient of linear thermal expansion, 10^{-5}/°C (10^{-5}/°F)	D696	9.9 (5.5)
Continuous service temperature, °C (°F)	...	260 (500)
Electrical properties		
Volume resistivity, Ω·cm	D257	10^{16}
Dielectric strength, V/10^{-3} mm (V/10^{-3} in.)	D149	24 (600)
Dielectric constant, (50-100 Hz)	D150	3.1
Dissipation factor (50-100 Hz)	D150	0.0004
Processing parameters		
Processing temperature, °C (°F)	...	330-370 (625-700)
Density, g/cm^3	D792	1.34
Specific volume, cm^3/kg (in.^{3}/lb)	D792	760 (21)
Linear mold shrinkage, mm/mm (in./in.)	D955	0.010
Water absorption, % in 24 h	D570	0.2

(a) Notched, at room temperature.

Source: Ref 15, p 1-17

Thermal, physical, and chemical properties of a typical polyphenylene sulfide (PPS)

Property	Unfilled PPS	40% glass-filled PPS
Heat-deflection temperature at 1.82 MPa, °C	135	250
Maximum resistance to continuous heat, °C	110	200
Coefficient of linear thermal expansion, 10^{-5}/°C	5.0	2.2
Compressive strength, MPa	110	145
Flexural strength, MPa	96	207
Impact strength, Izod, cm·N/cm of notch	21	75
Tensile strength, MPa	74	141
Elongation, %	1.1	1
Hardness, Rockwell	R123	R123
Specific gravity	1.3	1.6
Dielectric constant	3.8	4.6
Water absorption, %	0.02	0.03
Resistance to chemicals at 25 °C(a):		
Nonoxidizing acids (20% H_2SO_4)	S	S
Oxidizing acids (10% HNO_3)	S	S
Aqueous salt solutions (NaCl)	S	S
Aqueous alkalis (NaOH)	S	S
Polar solvents (C_2H_5OH)	S	S
Nonpolar solvents (C_6H_6)	S	S
Water	S	S

(a) S, satisfactory.

Source: Ref 13, p 143

Typical conditions for injection molding of polyphenylene sulfide molding compounds

Machine	Reciprocating screw machine preferred
Predrying of resin	Drying for 2 h at 150 to 165 °C (300 to 325 °F) recommended
Clamp tonnage	0.35 to 0.56 t/cm^2 (2.5 to 4 tons/in.^{2}) of projected area
Cylinder temperature	315 to 345 °C (600 to 650 °F) preferred
Injection pressure	Maximum allowable without flashing
Injection rate	High for good surface; moderate for minimum warpage
Cycle time	15 to 50 s
Screw speed	Medium range
Back pressure	345 kPa (50 psi) or less - just enough to yield consistent shot-to-shot weight
Mold temperature	95 to 150 °C (200 to 300 °F) preferred
Purge	Purge before and after with a low-melt-flow polyethylene
Mold release	Coat mold cavities with a high-temperature mold release, particularly in ribs and bosses. High-temperature silicones, fluorocarbon release sprays, and stearate dusts have been found to be effective. Use of release should be continued after each shot until the press is on cycle. Use of release may then be discontinued or, for difficult-to-eject parts, only periodic application may be required.

Source: Ref 24, p 222

Recommended processsing factors for injection-moldable polyphenylene sulfide(a)

Grade	Melt temperature range °C	°F	Melt viscosity(b), Pa·s	Gating diameter(c) mm	in.
Glass reinforced:					
30%	300-355	575-675	180	0.76	0.030
40-45%	300-355	575-675	200	0.76	0.030
40% glass filled, modified	300-355	575-675	470	0.76	0.030
Glass/mineral reinforced:					
30% glass/30% mineral	310-355	590-675	210	1.3	0.050
30% glass/35% mineral	310-355	590-675	240	1.3	0.050

(a) High injection speed, mold temperature, 135-150 °C (275-300 °F); maximum regrind, 30-35%. (b) Melt viscosity measured at 1000/s; data taken at 300 °C (572 °F). (c) Data based on a wall thickness of 3.18 mm (0.125 in.).

Source: Ref 19, p 43

Polypropylene

Molecular structure of polypropylene

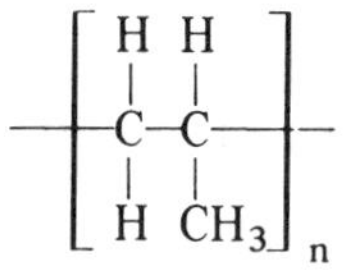

Source: Ref 20, p 48

Thermal, physical, and chemical properties of typical polypropylene and polymethyl pentene

Property	PP	TPX
Heat-deflection temperature at 1.82 MPa, °C	80	55
Maximum resistance to continuous heat, °C	70	50
Coefficient of linear thermal expansion, 10^{-5}/°C	9.0	11.7
Compressive strength, MPa	45	38
Flexural strength, MPa	48	34.5
Impact strength, Izod, cm·N/cm of notch	27	27
Tensile strength, MPa	34.5	24
Elongation, %	100	15
Hardness, Rockwell	R80	L70
Specific gravity	0.90	0.83
Dielectric constant	2.3	2.1
Resistance to chemicals at 25 °C(a):		
Nonoxidizing acids (20% H_2SO_4)	S	S
Oxidizing acids (10% HNO_3)	Q	Q
Aqueous salt solutions (NaCl)	S	S
Aqueous alkalis (NaOH)	S	S
Polar solvents (C_2H_5OH)	S	S
Nonpolar solvents (C_6H_6)	Q	Q
Water	S	S

(a) S, satisfactory; Q, questionable.

Source: Ref 23, p 60

Selected properties of stamped polypropylene

Property	Directionalized Primary direction	Directionalized Transverse direction	Isotropic
Flexural strength, MPa (ksi)	240 (35)	160 (23)	165 (24)
Flexural modulus, GPa (10^6 psi)	8.62 (1.25)	4.5 (0.65)	5.5 (0.80)
Tensile strength, MPa (ksi)	160 (23)	97 (14)	97 (14)
Tensile modulus, GPa (10^6 psi)	8.62 (1.25)	4.5 (0.65)	5.2 (0.75)
Deflection temperature at 1.82 MPa (264 psi), °C (°F)	155 (310)	155 (310)	155 (310)
Specific gravity	1.19	1.19	1.19
Glass content, %	40	40	40

Source: Ref 22, p 251

Selected properties of polypropylenes

Property	ASTM test	Reported value for: Extruded or molded polypropylene	Extruded or molded high-impact polypropylene
Mechanical properties			
Yield strength, MPa (ksi)	D638	30-38 (4.3-5.5)	19-31 (2.8-4.5)
Tensile modulus, GPa (10^6 psi)	D638	1.1-1.6 (0.16-0.23)	0.69-1.2 (0.10-0.18)
Elongation (break), %	D638	200-700	350-500
Compressive strength, MPa (ksi)	D695	41-55 (6-8)	28-48 (4-7)
Flexural yield strength, MPa (ksi)	D790	41-55 (6-8)	...
Flexural modulus, GPa (10^6 psi)	D790	1.4-2.1 (0.20-0.30)	0.69-1.4 (0.10-0.20)
Impact strength, Izod(a) J/cm (ft·lb/in.)	D256	0.27-1.2 (0.5-2.2)	0.53-6.41 (1.0-12.0)
Hardness, Rockwell	...	R80-110	R50-85
Thermal properties			
Thermal conductivity, W/m·K (Btu·in./ft^2·h·°F)	C177	0.10 (0.7)	0.12-0.17 (0.8-1.2)
Coefficient of linear thermal 10^{-5}/°C (10^{-5}/°F)	D696	11-18.0 (6.0-10.0)	11-16.2 (6.0-9.0)
Specific heat, kJ/kg·K (Btu/lb·°F)	C351	2.1 (0.5)	2.1 (0.5)
Continuous service temperature, °C (°F)	...	110 (230)	...
Electrical properties			
Volume resistivity, Ω·cm	D257	10^{16}	10^{16}
Dielectric strength, V/10^{-3} mm (V/10^{-3} in.)	D149	20-24 (500-600)	20-26 (500-650)
Dielectric constant (50-100 Hz)	D150	2.2-2.6	2.3
Dissipation factor (50-100 Hz)	D150	>0.0005	>0.0003
Processing parameters			
Processing temperature, °C (°F)	...	175-290 (350-550)	175-290 (350-550)
Density, g/cm^3	D792	0.90-0.91	0.89-0.90
Specific volume, cm^3/kg (in.^{3}/lb)	D792	1120 (31)	1120 (31)
Linear mold shrinkage, (in./in.)	D955	0.010-0.020	0.010-0.020
Water absorption, % in 24 h	D570	0.01	0.01

(a) Notched, at room temperature.

Source: Ref 15, p 1-31

Polystyrene

Molecular structure of polystyrene

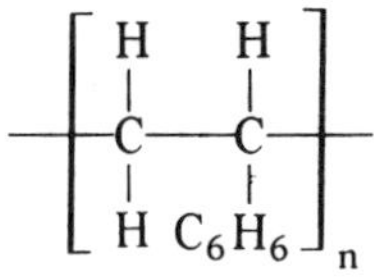

Source: Ref 20, p 48

Selected properties of molded polystyrenes

Property	ASTM test	Reported value for: Molded polystyrene	Heat- and impact-resistant molded polystyrene	Impact- and flame resistant molded polystyrene
Mechanical properties				
Yield strength, MPa (ksi)	D638	34-82.7 (5.0-12.0)	10-48 (1.5-7.0)	26-34 (3.8-5.0)
Tensile modulus, GPa (10^6 psi)	D638	2.8-3.4 (0.40-0.50)	1.0-3.4 (0.15-0.50)	1.9-2.2 (0.28-0.32)
Elongation (break), %	D638	1.0-2.5	10-90	13-25
Compressive strength, MPa (ksi)	D695	76-110 (11-16)	28-62 (4-9)	...
Flexural yield strength, MPa (ksi)	D790	55-97 (8-14)	21-83 (3-12)	34-48 (5-7)
Flexural modulus, GPa (10^6 psi)	D790	2.8-3.4 (0.40-0.50)	0.69-3.4 (0.10-0.50)	2.1 (0.30)
Impact strength, Izod(a) J/cm (ft·lb/in.)	D256	0.1-0.2 (0.2-0.4)	0.3-0.4 (0.5-0.8)	0.53-0.75 (1.0-1.4)
Hardness, Rockwell	...	M65-80	M10-50	M10-15
Thermal properties				
Thermal conductivity, W/m·K (Btu·in./ft^2·h·°F)	C177	0.10-0.14 (0.7-1.0)	0.04-0.12 (0.3-0.8)	...
Coefficient of linear thermal expansion, 10^{-5}/°C (10^{-5}/°F)	D696	11-14 (6.0-8.0)	9.0-37.8 (5.0-21.0)	...
Specific heat, kJ/kg·K (Btu/lb·°F)	C351	1.3 (0.3)	1.3 (0.3)	...
Continuous service temperature, °C (°F)	...	70-95 (160-205)	60-70 (140-160)	...
Electrical properties				
Volume resistivity, Ω·cm	D257	10^{16}	10^{16}	10^{15}
Dielectric strength, V/10^{-3} mm (V/10^{-3} in.)	D149	20-28 (500-700)	12-24 (300-600)	20 (500)
Dielectric constant (50-100 Hz)	D150	2.5-3.1	2.5-4.8	3.2
Dissipation factor	D150	0.0001-0.0006	0.0004-0.0020	0.0006
Processing parameters				
Processing temperature, °C (°F)	...	150-260 (300-500)	175-315 (350-600)	150-260 (300-500)
Density, g/cm^3	D792	1.04-1.09	1.04-1.10	1.10-1.20
Specific volume, cm^3/kg ($in.^3$/lb)	D792	940 (26)	940 (26)	905 (25)
Linear mold shrinkage, mm/mm (in./in.)	D955	0.001-0.006	0.002-0.006	0.002-0.006
Water absorption, 5 in 24 h	D570	0.0-0.1	0.1-0.6	...

(a) Notched, at room temperature.

Source: Ref 15, p 1-32

Thermal, physical, and chemical properties of typical polystyrenes

Property	Unfilled PS	Impact PS	30% glass-filled PS	SAN
Heat-deflection temperature at 1.82 MPa (ksi)	90	90	105	100
Maximum resistance to continuous heat, °C	75	70	95	85
Coefficient of linear thermal expansion, $10^{-5}/°C$	7.5	8.0	4.0	6.0
Compressive strength, MPa	90	45	103	90
Flexural strength, MPa	83	50	117	100
Impact strength, Izod cm·N/cm of notch	21	80	80	30
Tensile strength, MPa	41	41	83	60
Elongation, %	1.5	3	1	1.5
Hardness, Rockwell	M60	M35	M60	M80
Specific gravity	1.04	1.04	1.2	1.07
Dielectric constant	2.5	3.0	3.0	3.5
Resistance to chemicals at 25 °C(a):				
Nonoxidizing acids (20% H_2SO_4)	S	S	S	S
Oxidizing acids (10% HNO_3)	Q	Q	Q	Q
Aqueous salt solutions (NaCl)	S	S	S	S
Aqueous alkalis (NaOH)	S	S	Q	S
Polar solvents (C_2H_5OH)	S	S	S	S
Nonpolar solvents (C_6H_6)	U	U	U	U
Water	S	S	S	S

(a) S, satisfactory; Q, questionable; U, unsatisfactory.

Source: Ref 23, p 61

Polysulfones

Molecular structure of polysulfone

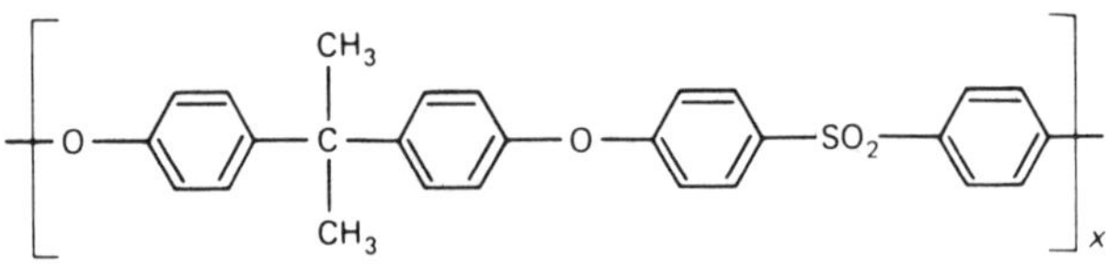

Source: Ref 6, p 101

Working stress levels for polysulfone in water at various temperatures

Water temperature °C	°F	Steady load MPa	ksi	Intermittent load MPa	ksi
22	72	20.7	3.0	24.1	3.5
60	140	10.3	1.5	13.8	2.0
82	180	3.4	0.5	6.9	1.0
99	210	0.34	0.05	...	...

Source: Ref 24, p 193

Selected properties of polysulfone (extruded or molded)

Property	ASTM test	Reported value
Mechanical properties		
Yield strength, MPa (ksi)	D638	68.9-75.8 (10.0-11.0)
Tensile modulus, GPa (10^6 psi)	D638	2.5 (0.36)
Elongation (break), %	D638	50-100
Compressive strength, MPa (ksi)	D695	90-97 (13-14)
Flexural yield strength, MPa (ksi)	D790	97-105 (14-15)
Flexural modulus, GPa (10^6 psi)	D790	2.8 (0.40)
Impact strength, Izod(a), J/cm (ft·lb/in.)	D256	0.7 (1.3)
Hardness, Rockwell	...	M69
Thermal properties		
Thermal conductivity, W/m·K (Btu·in./ft^2·h·°F)	C177	01. (0.8)
Coefficient of linear thermal expansion, 10^{-5}/°C (10^{-5}/°F)	D696	9.4-10 (5.2-5.6)
Specific heat, kJ/kg·K (Btu/lb·°F)	C351	1.3 (0.3)
Continuous service temperature, °C (°F)	...	175-190 (350-375)
Electrical properties		
Volume resistivity, Ω·cm	D257	5.0×10^{16}
Dielectric strength, V/10^{-3} mm (V/10^{-3} in.)	D149	17 (425)
Dielectric constant, (50-100 Hz)	D150	3.1
Dissipation factor (50-100 Hz)	D150	0.0008
Processing parameters		
Processing temperature, °C (°F)	...	290-400 (550-750)
Density, g/cm^3	D792	1.24
Specific volume, cm^3/kg (in.^{3}/lb)	D792	795 (22)
Linear mold shrinkage, mm/mm (in./in.)	D955	0.007
Water absorption, % in 24 h	D570	0.2

(a) Notched, at room temperature.

Source: Ref 15, p 1-33

Permeability of polysulfone to various gases (ASTM D1434)

Gas	Permeability 10^{-13}cm/s(a) at 25 °C
Ammonia (NH_3)	6,400
Carbon dioxide (CO_2)	5,700
Helium (He)	11,700
Hydrogen (H_2)	10,800
Methane (CH_4)	220
Nitrogen (N_2)	240
Oxygen (O_2)	1,380
Sulfur hexafluoride (SF_6)	10.8
Dicholorofluoromethane (CCl_2F_2)	3.5
Dicholorotetrafluoroethane ($C_2Cl_2F_4$)	1.5

(a) $\frac{\text{cm}^3 \text{ at STP·cm}}{\text{cm}^2 \cdot \text{cm Hg·s}}$

Source: Ref 24, p 193

Processing factors for injection-moldable polysulfone

Grade	Melt temperature range °C	°F	Melt viscosity(a), Pa·s	Injection speed g/s	oz/s	Mold temperature °C	°F
Unmodified base resin	330-400	625-750	800	15-20	0.5-0.7	95-150	200-300
Glass reinforced, 10 to 30%	355-400	670-750	850-950	18-24	0.6-0.8	120-150	250-300
Mineral reinforced	370-400	700-750	900-1000	10-15	0.4-0.5	120-150	250-300
Impact modified	260-315	500-600	600-750	15-25	0.5-0.9	70-120	160-250
Flame retarded	330-400	625-750	800	15-20	0.5-0.7	95-150	200-300
Other	270-295	520-560	700	18-24	0.6-0.8	120-250	250-300

(a) Melt viscosity measured at 1000/s at the median temperature of the recommended melt-temperature range. Maximum regrind, 25%.

Source: Ref 19, p 43

Thermal, physical, and chemical properties of typical polysulfones

Property	Polysulfone	Polyether sulfone
Heat deflection temperature at 1.82 MPa, °C	175	205
Maximum resistance to continuous heat, °C	150	165
Coefficient of linear thermal expansion, 10^{-5}/°C	5.4	5.5
Compressive strength, MPa	96	96
Flexural strength, MPa,	107	127
Impact strength, Izod, cm·N/cm of notch	80	80
Tensile strength, MPa	82	82
Elongation, %	25	25
Hardness, Rockwell	M69	M88
Specific gravity	1.24	1.37
Dielectric constant	3.1	3.1
Water absorption, %	0.3	0.4
Resistance to chemicals at 25 °C(a):		
Nonoxidizing acids (20% H_2SO_4)	S	S
Oxidizing acids (10% HNO_3)	U	U
Aqueous salt solutions (NaCl)	S	S
Aqueous alkalis (NaOH)	S	S
Polar solvents (C_2H_5OH)	S	S
Nonpolar solvents (C_6H_6)	Q	Q
Water	S	S

(a) S, satisfactory; Q, questionable; U, unsatisfactory.

Source: Ref 13, p 145

Polyurethanes

Molecular structure of polyurethanes

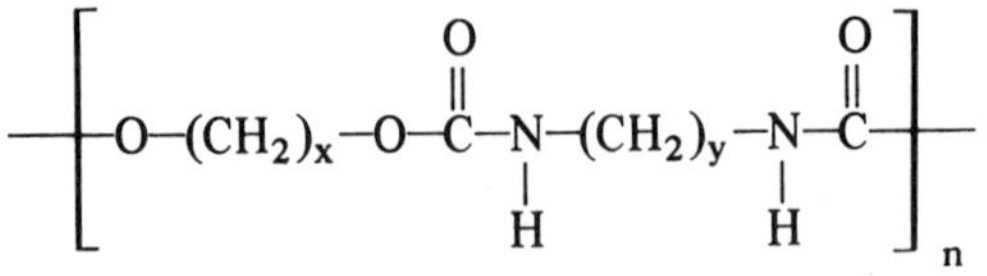

Source: Ref 20, p 50

Thermal, physical, and chemical properties of a typical thermoplastic polyurethane elastomer (TPU)

Heat-deflection temperature at 1.82 MPa, °C	70
Coefficient of linear thermal expansion, 10^{-5}/°C	15
Tensile strength, MPa	20
Elongation, %	600
Specific gravity	1.25
Hardness, Shore A	80
Resistance to chemicals at 25 °C(a):	
Nonoxidizing acids (20% H_2SO_4)	Q
Oxidizing acids (10% HNO_3)	U
Aqueous salt solutions (NaCl)	S
Aqueous alkalis (NaOH)	Q
Polar solvents (C_2H_5OH)	U
Nonpolar solvents (C_6H_6)	Q
Water	S

(a) S, satisfactory; Q, questionable; U, unsatisfactory.

Source: Ref 23, p 88

Selected properties of polyurethanes

Property	ASTM test	Reported value for: Cast liquid polyurethane	Thermoplastic elastomer polyurethane
Mechanical properties			
Yield strength, MPa (ksi)	D638	14-69 (2.0-10.0)	14-55 (2.0-8.0)
Tensile modulus, GPa (10^6 psi)	D638	0.69-6.9 (0.10-1.0)	0.069-2.1 (0.01-0.30)
Elongation (break), %	D638	200-1000	100-500
Compressive strength, MPa (ksi)	D695	140 (20)	140 (20)
Flexural yield strength, MPa (ksi)	D790	6.9-34 (1-5)	6.9-62 (1-9)
Flexural modulus, GPa (10^6 psi)	D790	0.69 (0.10)	2.1 (0.30)
Impact strength, Izod(a), J/cm (ft·lb/in.)	D256	>13.3 (>25.0)	No break
Hardness, Rockwell	...	D90	M29
Thermal properties			
Thermal conductivity, W/m·K (Btu·in./ft^2·h·°F)	C177	0.20 (1.4)	0.07-0.29 (0.5-2.0)
Coefficient of linear thermal expansion 10^{-5}/°C (10^{-5}/°F)	D696	18.0-36.0 (10.0-20.0)	18.0-36.0 (10.0-20.0)
Specific heat, kJ/kg·K (Btu/lb·°F)	C351	1.7 (0.4)	1.7-19 (0.4-4.5)
Electrical properties			
Volume resistivity, Ω·cm	D257	10^{12}-10^{15}	10^{11}-10^{13}
Dielectric strength, V/10^{-3} mm (V/10^{-3} in.)	D149	16-20 (400-500)	21 (525)
Dissipation factor (50-100 Hz)	D150	0.015-0.017	0.015-0.048)
Processing parameters			
Processing temperature, °C (° F)	...	85-120 (185-250)	150-230 (300-450)
Density, g/cm^3	D792	1.10-1.50	1.05-1.25
Specific volume, cm^3/kg ($in.^3$/lb)	D792	975-795 (27-22)	940-795 (26-22)
Linear mold shrinkage, mm/mm (in./in.)	D955	0.020	0.001-0.030
Water absorption, % in 24 h	D570	0.0-1.5	0.7-0.9

(a) Notched, at room temperature.

Source: Ref 15, p 1-34

Styrene Acrylonitrile

Molecular structure of styrene acrylonitrile copolymer

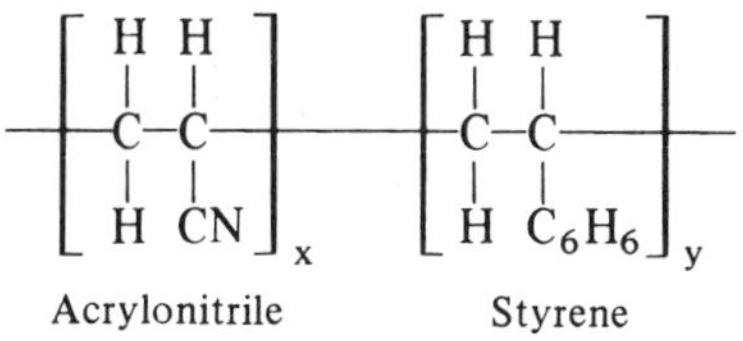

Acrylonitrile Styrene

Source: Ref 20, p 49

Selected properties of molded styrene acrylonitrile

Property	ASTM test	Reported value
Mechanical properties		
Yield strength, MPa (ksi)	D638	62-83 (9.0-12.0)
Tensile modulus, GPa (10^6 psi)	D638	2.8-3.9 (0.40-0.56)
Elongation (break), %	D638	1.5-4.0
Compressive strength, MPa (ksi)	D695	97-115 (14-17)
Flexural yield strength, MPa (ksi)	D790	97-130 (14-19)
Flexural modulus, GPa (10^6 psi)	D790	3.4-4.1 (0.50-0.60)
Impact strength, Izod(a), J/cm (ft·lb/in.)	D256	0.2-0.3 (0.3-0.5)
Hardness, Rockwell	...	M80-90
Thermal properties		
Thermal conductivity, W/m·K (Btu·in./ft^2·°F)	C177	0.1 (0.8)
Coefficient of linear thermal expansion, 10^{-5}/°C (10^{-5}/°F)	D696	6.5-6.8 (3.6-3.8)
Specific heat, kJ/kg·K (Btu/lb·°F)	C351	1.3 (0.3)
Continuous service temperature, °C (°F)	...	80-90 (175-190)
Electrical properties		
Volume resistivity, Ω·cm	D257	10^{16}
Dielectric strength, V/10^{-3} mm (V/10^{-3} in.)	D149	16-20 (400-500)
Dielectric constant, (50-100 Hz)	D150	2.6-3.4
Dissipation factor (50-100 Hz)	D150	0.006-0.008
Processing parameters		
Processing temperature, °C (°F)	...	175-300 (350-575)
Density, g/cm^3	D792	1.20-1.33
Specific volume, cm^3/kg (in.^{3}/lb)	D792	940-905 (26-25)
Linear mold shrinkage, mm/mm (in./in.)	D955	0.002-0.007
Water absorption, % in 24 h	D570	0.2-0.3

(a) Notched, at room temperature.

Source: Ref 15, p 1-36

Thermosets

General

Recommended wall thicknesses for molded thermoset plastic parts

Material	Small parts mm	Small parts in.	General parts mm	General parts in.	Large parts mm	Large parts in.
Phenolic	1.575-3.175	0.062-0.125	2.36-4.75	0.093-0.187	4.75-25.4	0.187-1.00
Urea	1.575-3.175	0.062-0.125	2.36-4.75	0.093-0.187	4.75-9.52	0.187-0.375
Melamine	1.575-3.175	0.062-0.125	2.36-4.75	0.093-0.187	4.75-9.52	0.187-0.375
DAP	1.143-2.362	0.045-0.093	1.98-3.96	0.078-0.156	3.175-9.52	0.125-0.375
Alkyd	1.98-3.175	0.078-0.125	2.54-4.75	0.100-0.187	4.75-12.7	0.187-0.500
Polyester:						
Granular	1.98-3.175	0.078-0.125	2.54-4.75	0.100-0.187	4.75-12.7	0.187-0.500
Bulk molding compound	1.143-2.632	0.045-0.093	1.98-3.96	0.100-0.187	3.175-9.52	0.125-0.375
Sheet molding compound	1.575-2.632	0.062-0.093	2.36-4.75	0.093-0.187	4.75-9.52	0.187-0.375

Source: Ref 25, p 240

Typical characteristics of thermoset adhesives

Type	Form	Cure temperature, °C (°F)	Maximum use temperature, °C (°F)	Advantages	Disadvantages
Epoxy	Two-part paste	Room or accelerated at 93-178 (200-350)	Generally below 82 (180)	Ease of storage at room temperature; ease of mixing and use; long shelf life; gap filling when filled	Not generally as strong or environmentally resistant as typical heat-cured epoxies
	One-part film	121 (250)	To 82 (180)	Covers large areas; bondline thickness control; wide variety of formulas; higher-temperature curing materials; better environmental properties	Store at 18 °C (0 °F); short shelf life; high-temperature cure; brittle and low peel strength
		149 (300) 178 (350)	149-177 (300-350)		
Acrylic	Two-part liquid or pastes	Room to 100 (212)	105 (221)	Fast setting; easy to mix and use; good moisture resistance; tolerant of surface contamination	Strong, objectionable odor; limited pot life
Polyurethane	One or two parts	Room or heat cure	...	Good peel; good for cryogenic use	Moisture sensitive before and after cure
Silicone	One- and two-part pastes	Room to 260 (500)	To 260 (500)	High peel and impact resistance; easy to use; good heat and moisture resistance	High cost; low strength
Hot melt	One-part	Melt at 190-232 (375-450)	18-171 (120-340)	Rapid application; fast setting; low cost; indefinite shelf life; nontoxic; no mixing	Poor heat resistance; special equipment required; poor creep resistance; low strength; high melt temperature
Bismaleimide (BMI)	One-part paste or film	>178 (350) and 246 (475) postcure	232 (450)	Structural bonds with bismaleimide composites; higher temperature than epoxies; no volatiles; good shelf life	Brittle and low peel; limited formulas available
Polyimide	Thermoplastic liquids; one- and two-part pastes	260 (500) and postcure	204-260 (400-500)	High-temperature resistance; structural strength	High cost; low peel strength; high cure and postcure temperatures; volatiles for some forms
Phenolic-based	One-part films	163-177 (325-350)	To 177 (350)	High-temperature use	Low peel strength

Source: Ref 6, p 684

Clear casting mechanical properties

Material	Barcol hardness	Tensile strength, MPa	Tensile strength, ksi	Tensile modulus, 10^{-2} Pa	Tensile modulus, 10^{-5} psi	Elongation, %	Flexural strength, MPa	Flexural strength, ksi	Flexural modulus, 10^{-2} Pa	Flexural modulus, 10^{-5} psi	Compressive strength, MPa	Compressive strength, ksi	Heat deflection temperature, °C	Heat deflection temperature, °F
Orthophthalic	...	55	8	34.5	5.0	2.1	80	12	34.5	5.0	...	...	80	175
Isophthalic	40	75	11	33.8	4.9	3.3	130	19	35.9	5.2	120	17	90	195
BPA fumarate	34	40	6	28.3	4.1	1.4	110	16	33.8	4.9	100	15	130	265
Chlorendic	40	20	3	33.8	4.9	...	120	17	39.3	5.7	100	15	140	285
Vinyl ester	35	80	12	35.9	5.2	4.0	140	20	37.2	5.4	...	...	100	212

Source: Ref 6, p 91

Use-temperature guide to structural adhesives

Peel: L, low; M, medium; H, high. Lap shear: P, poor; Mod, moderate; G, good; V, very good; E, excellent. Peel is indicated first, followed by lap shear: peel/lap shear.

	Use temperature, °C (°F)								
Adhesive	−253 (−423)	−196 (−320)	−73 (−100)	−54 (−65)	Room	82 (180)	149 (300)	216 (420)	260 (500)
Epoxy-nitrile modified	L/V	L/V	L/E	L/E	H/E	M/V	L/Mod	...	...
Epoxy-nylon	L/E	L/E	L/E	L/E	H/V	L/G	L/Mod	...	...
Epoxy-phenolic	...	...	L/V	L/V	L/V	L/V	L/G	G	G
Vinyl-phenolic	...	...	...	L/V	M/E	H/E	L/Mod	...	...
Nitrile-phenolic	Mod	E	E	L/E	H/V	M/G	L/Mod	...	...
Bismaleimides	...	...	...	Mod	L/G	L/G	L/G	L/V	...
Polyimides	...	...	...	L/V	L/G	L/G	L/G	L/G	L/G
Polyurethanes	H/V	H/V	H/V	H/G	H/G	H/Mod	H/P	...	...
Acrylics	...	...	...	L/P	H/E	M/G	L/P	...	...

Source: Ref 6, p 684

Comparative characteristics of thermoset adhesives

Resin base	Disadvantages	Advantages	Major uses
Polyesters	Considerable shrinkage, brittle on impact, poor hot strength	Fair strength, low viscosity, low-temperature cure, good electrically	Repairs, compatible with explosives and radomes
Epoxy	Generally rigid, poor hot strength, somewhat toxic	High strength, low shrinkage low-temperature cure, fair electrically	FRP-to-metal joints radomes, and aircraft structural parts
Phenolics	Requires solvents and high-temperature cures, bad electrically, may be corrosive	Good hot strength nontoxic, inexpensive	High-temperature ceramic-to-FRP joints
Rubber phenolics	Require solvents and high-temperature cures, bad electrically	Moderate strength and peel resistance	Metallic joints, shock-absorbing parts
Epoxy phenolics	Rigid; requires heat cures, bad electrically	High heat resistance, fair strength, good cryogenic strength	Aircraft structural parts where high temperature or extremely low temperature is required
Silicones	Low strength, requires solvents	Extremely high heat resistance, good arc resistance	High temperature, silicone adherends and in copolymer adhesives
Polyimides	Rigid, requires heat cure, may be corrosive	Extremely high heat resistance good electrically	High-temperature, long-age metal-to-metal aircraft parts

Source: Ref 26, Table 22.7

DAP

Thermal, physical, and chemical properties of a typical fibrous-glass-filled allylic plastic (DAP)

Heat-deflection temperature at 1.82 MPa, °C	200
Coefficient of linear thermal expansion, $10^{-5}/°C$	150
Compressive strength, MPa	186
Flexural strength, MPa	131
Impact strength, Izod, cm·N/cm of notch	106
Tensile strength, MPa	58
Elongation, %	4
Hardness, Rockwell	E80
Specific gravity	1.7
Water absorption	0.14
Dielectric constant	4
Resistance to chemicals at 25 °C(a):	
Nonoxidizing acids (20% H_2SO_4)	S
Oxidizing acids (10% HNO_3)	U
Aqueous salt solutions (NaCl)	S
Aqueous alkalis (NaOH)	Q
Polar solvents (C_2H_5OH)	S
Nonpolar solvents (C_6H_6)	U
Water	S

(a) S, satisfactory; Q, questionable; U, unsatisfactory.

Source: Ref 23, p 118

Amino

Thermal, physical, and chemical properties of typical amino plastics

Property	Cellulose filled MF(a)	Cellulose-filled UF
Heat-deflection temperature at 1.82 MPa, °C	150	130
Maximum resistance to continuous heat, °C	100	75
Coefficient of linear thermal expansion, $10^{-5}/°C$	4	3
Compressive strength, MPa	276	221
Flexural strength, MPa,	86	96.5
Impact strength, Izod, cm·N/cm of notch	16	16
Tensile strength, MPa	69	55
Elongation, %	0.7	0.7
Hardness, Rockwell	M115	M110
Specific gravity	1.5	1.5
Resistance to chemicals at 25 °C(a):		
Nonoxidizing acids (20% H_2SO_4)	S	S
Oxidizing acids (10% HNO_3)	U	U
Aqueous salt solutions (NaCl)	S	S
Aqueous alkalis (NaOH)	S	S
Polar solvents (C_2H_5OH)	S	S
Nonpolar solvents (C_6H_6)	Q	Q
Water	S	S

(a) S, satisfactory; Q, questionable; U, unsatisfactory.

Source: Ref 23, p 53

Epoxies

Epoxy resin synthesis

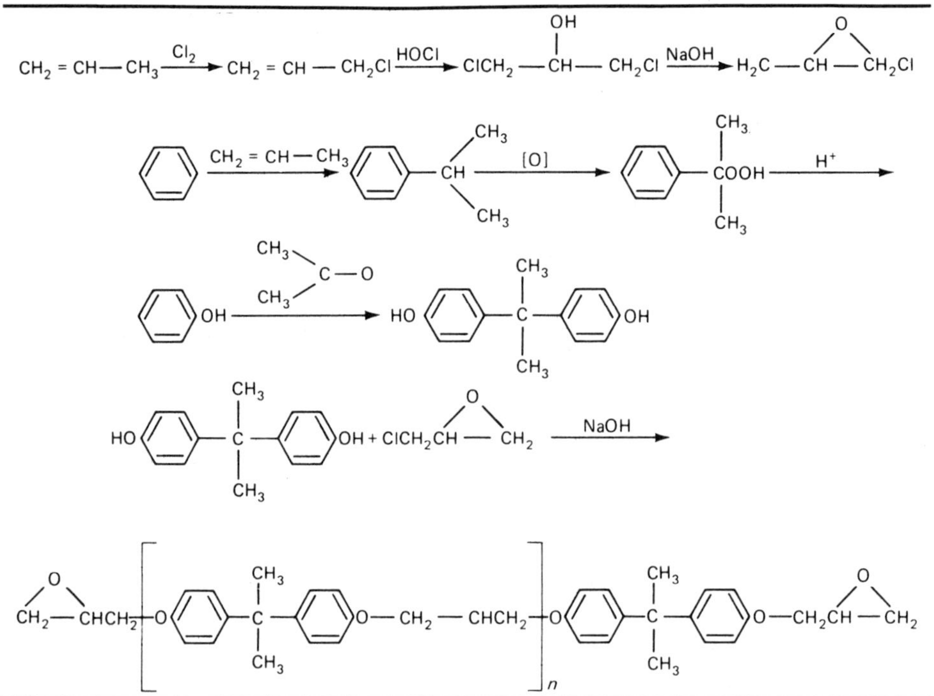

Source: Ref 6, p 67

Epoxy properties and tests

Ingredient	Property	Test method
Epoxy resins	Epoxide content	Titration
	Viscosity/softening point	Viscometer/Duran or rheometer
	Residual chlorides	Titration
	Moisture content	Titration/Karl Fisher
	Molecular weight distribution	GPC
	Characterization	HPLC/infrared spectroscopy
Hardener (amine)	Amine content	Titration
	Purity	Melting point refractive index, HPLC
Catalyst	Purity	Melting point
	Cation	Atomic absorption
Modifier (inorganic)	Particle size	Sedigraph/particle sizer
	Moisture	Moisture analyzer/Karl Fisher
Modifier (organic)	Viscosity	Rheometer
	Reactivity	Titration

GPC, gel permeation chromatography; HPLC, High-performance liquid chromatography

Source: Ref 6, p 289

Selected epoxy resins

Resin	Supplier	Formula
DER 332	Dow Chemical Co.	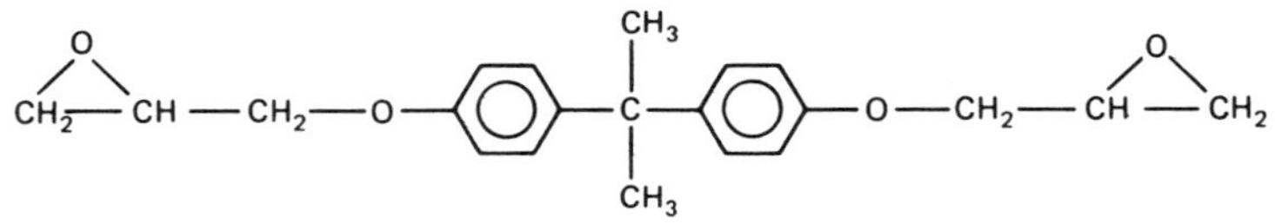
EPON 826	Shell Chemical Co.	
EPI-REZ 509	Interez Inc.	
Araldite GY 6008 (Diglycidyl ether of bisphenol A)	Ciba-Geigy Corp.	
EPN 1139	Ciba-Geigy Corp.	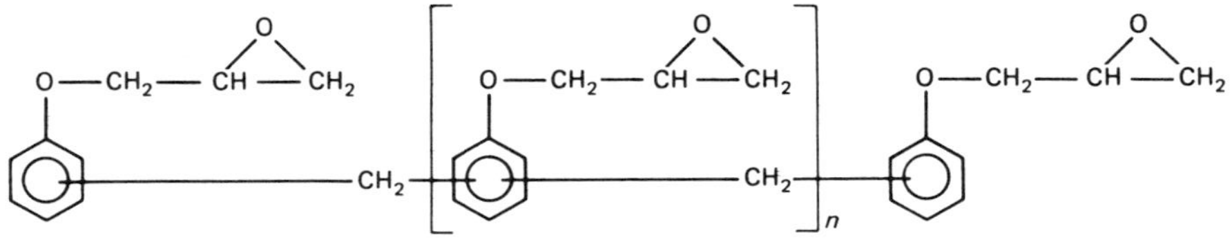
DEN 431 (Polyglycidyl ether of phenol-formaldehyde novolac)	Dow Chemical Co.	
EPI-REZ 5022	Interez Inc.	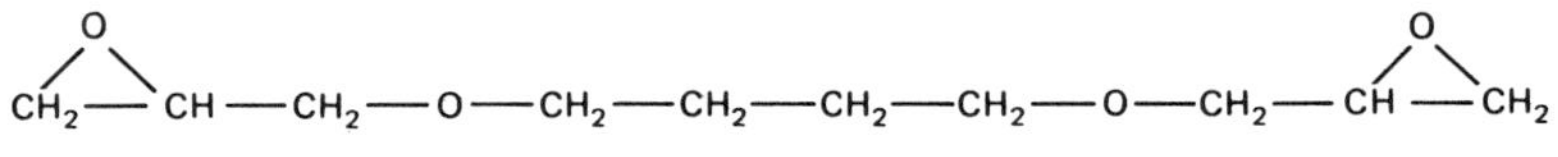
RD-2 (Diglycidyl ether of butanediol)	Ciba-Geigy Corp.	
Tonox 60/40		
40% *m*-phenylenediamine		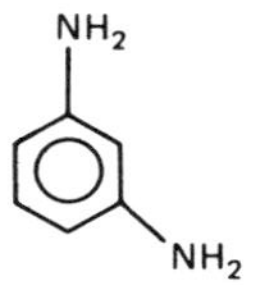
60% 4,4′-methylenedianiline	UniRoyal	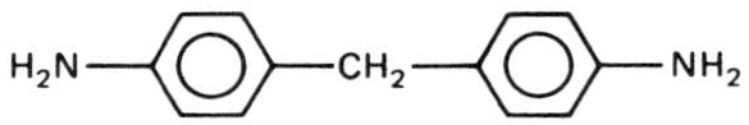

Source: Ref 6, p 136

Molecular structure of brominated epoxy resins

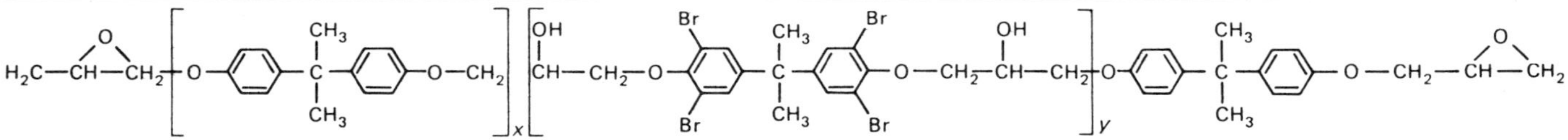

Source: Ref 6, p 67

Molecular structure of some commercial epoxy resins

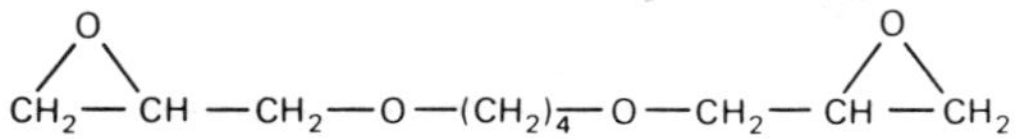

Butylene glycol diglycidyl ether (BGDGE)

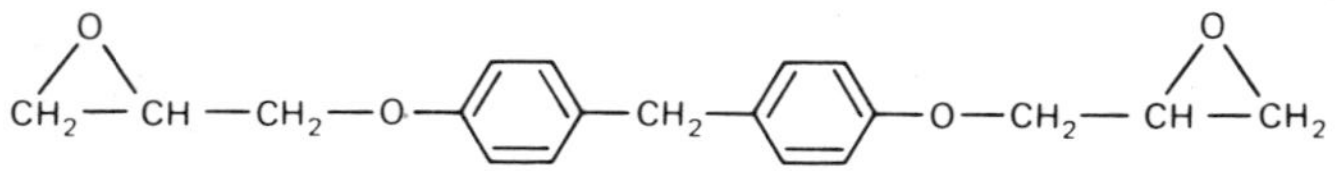

Diglycidyl ether of bisphenol F (DGEBPF)

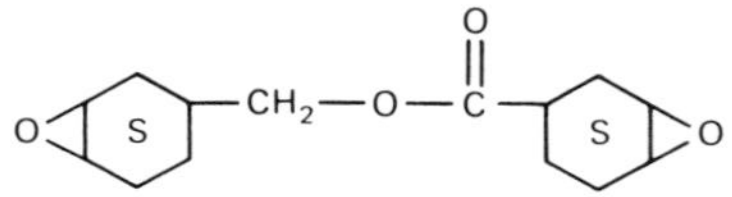

3, 4-epoxycyclohexyl methyl-
3′, 4′-epoxycyclohexane carboxylate

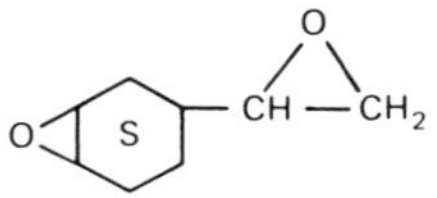

Vinyl cyclohexene diepoxide (VCDO)

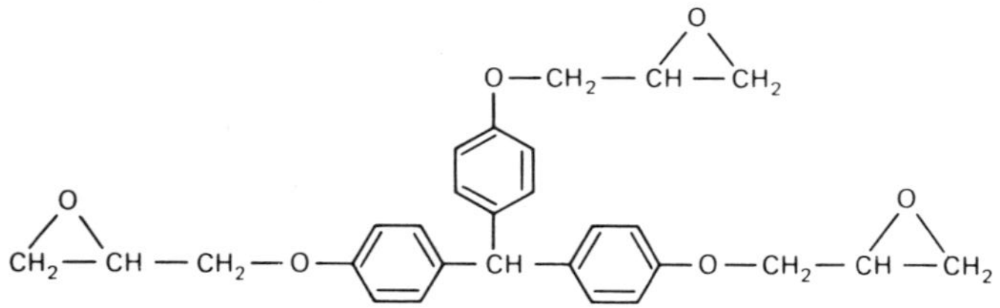

Triglycidyl ether of triphenyl methane (TGETPM)

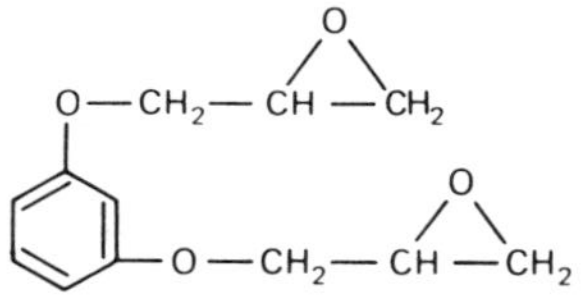

Resorcinol diglycidyl ether (RDE)

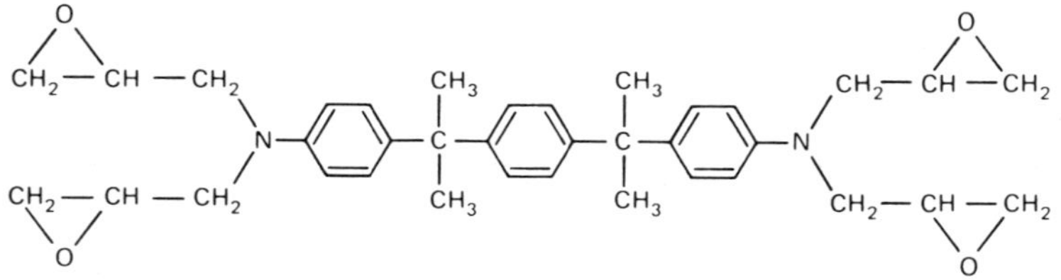

Tetraglycidyl-4, 4′ (4-aminophenyl)-*p*-diisopropyl benzene

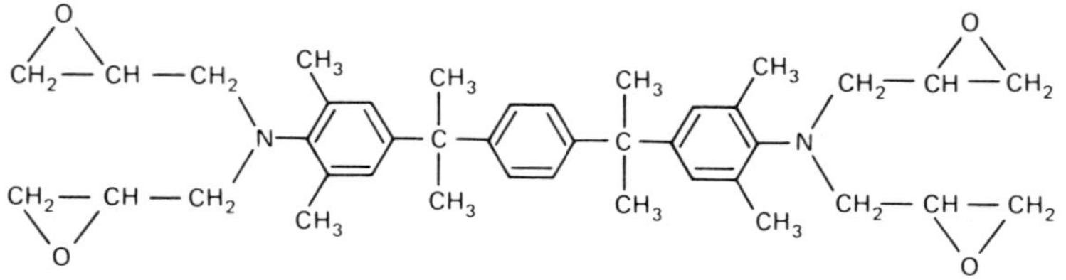

Tetraglycidyl-4, 4′ (4-amino-3, 5-dimethylphenyl)-*p*-diisopropylbenzene

Source: Ref 6, p 59

Epoxy resins used in aerospace prepregs

Name	Supplier	Formula
Araldite MY 0510 (Triglycidyl *p*-aminophenol	Ciba-Geigy Corp.	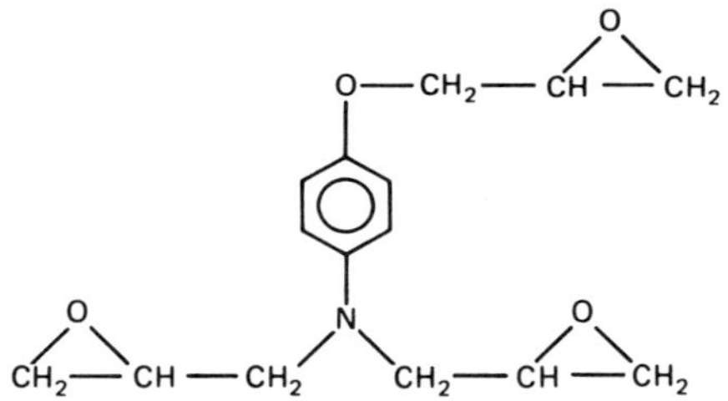
Araldite MY 720 (*N,N,N',N'*-tetraglycidyl-4',4'-methylenebisbenzenamine)	Ciba-Geigy Corp.	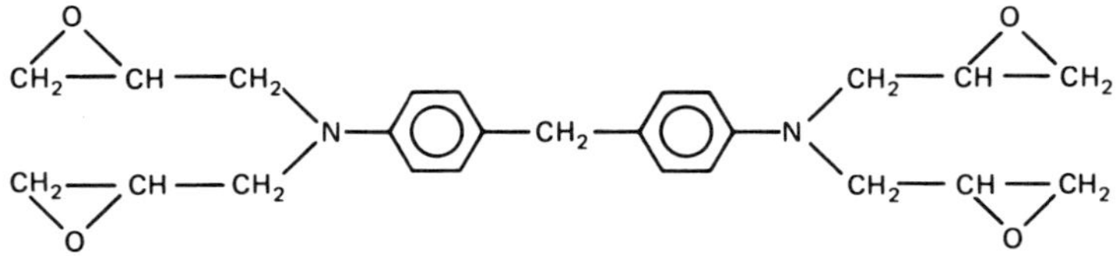
EPON 826 (Diglycidyl ether of bisphenol A)	Shell Chemical Co.	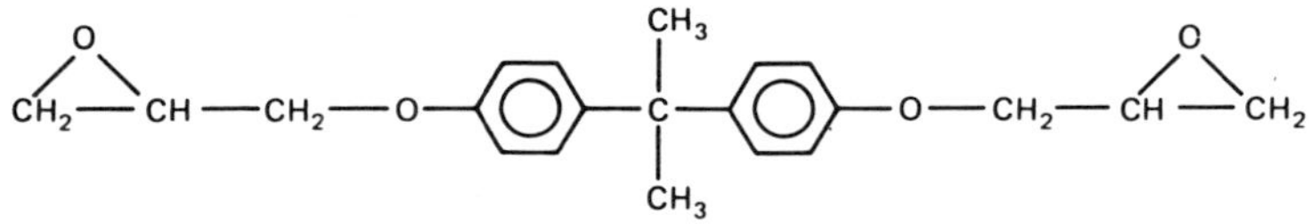
DER 330 (Diglycidyl ether of bisphenol A)	Dow Chemical Co.	
Epiclon 830 (Diglycidyl ether of bisphenol F)	Dinippon	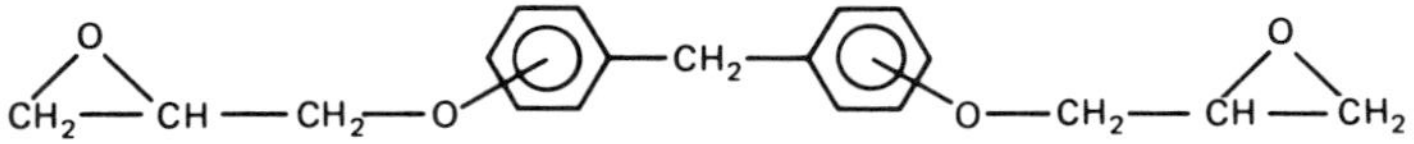
Araldite ECN 1235 (Epoxy novolac)	Ciba-Geigy Corp.	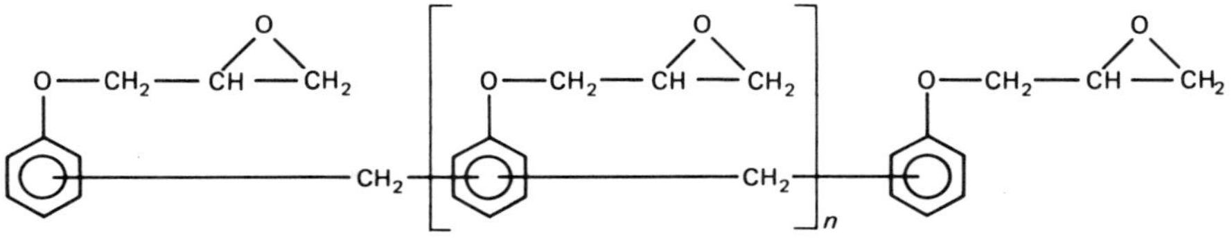

Source: Ref 6, p 140

Selected properties of epoxies

Property	ASTM test	Reported value for: Cast epoxy	Cast flexible epoxy	Molded epoxy novalac	Molded epoxy cycloaliphatic
Mechanical properties					
Yield strength, MPa (ksi)	D638	34-82.7 (5.6-12.0)	17-68.9 (2.5-10.0)	48-82.7 (7.0-12.0)	62-82.7 (9.0-12.0)
Tensile modulus, GPa (10^6 psi)	D638	2.4-2.8 (0.35-0.40)	...	2.8-3.4 (0.40-0.50)	3.3-3.4 (0.48-0.50)
Elongation (break), %	D638	3-6	25.0-70.0	2-6	2-8
Compressive strength, MPa (ksi)	D695	105-170 (15-25)	14-90 (2-13)	140-170 (20-25)	205-345 (30-50)
Flexural yield strength, MPa (ksi)	D790	90-145 (13-21)	14-83 (2-12)	69-83 (10-12)	83-90 (12-13)
Flexural modulus, GPa (10^6 psi)	D790	1.4-3.4 (0.20-0.50)	...	...	2.8-3.4 (0.40-0.50)
Impact strength, Izod(a), J/cm (ft·lb/in.)	D256	0.1-0.53 (0.2-1.0)	1.9-2.7 (3.5-5.0)	...	...
Hardness, Rockwell	...	M80-110	...	...	...
Thermal properties					
Thermal conductivity, W/m·K (Btu·in./ft²·h·°F)	C177	0.17-0.23 (1.2-1.6)	...	...	...
Coefficient of thermal expansion 10^{-5}/°C (10^{-5}/°F)	D696	8.1-11 (4.5-6.2)	3.6-14 (2.0-8.0)	3.1-4.0 (1.7-2.2)	2.9-5.4 (1.6-3.0)
Specific heat, kJ/kg·K (Btu/lb·°F)	C351	0.8-1.3 (0.2-0.3)	...	...	...
Continuous service temperature, °C (°F)	...	80-90 (175-190)	40-50 (100-125)	230-260 (450-500)	230-260 (450-500)
Electrical properties					
Volume resistivity, Ω·cm	D257	10^{12}-10^{17}	10^{14}	2.0-5.0 × 10^{14}	>10^{16}
Dielectric strength, V/10^{-3} mm (V/10^{-3} in.)	D149	12-20 (300-500)	10-16 (250-400)	11-16 (280-400)	16 (400)
Dielectric constant (50-100 Hz)	D150	3.5-5.0	3.0-5.0	4.5-5.5	3.4
Dissipation factor (50-100 Hz)	D150	0.002-0.010	0.010-0.040	0.003	0.006
Processing parameters					
Processing temperature, °C (°F)	...	...	...	120-165 (250-330)	...
Density, g/cm³	D792	1.10-1.40	1.05-1.30	1.20-1.70	1.22
Specific volume, cm³/kg (in.³/lb)	D792	760-725 (21-20)	760-725 (21-20)	795-540 (22-15)	795 (22)
Linear mold shrinkage, mm/mm (in./in.)	D955	0.001-0.010	0.001-0.010	...	...
Water absorption, % in 24 h	D570	0.1	...	0.1-0.2	0.1-0.7

(a) Notched, at room temperature.

Source: Ref 15, p 1-12

Properties of commercial grades of BPA epoxy resins

Average mol wt	Average wpe(a)	Approximate value of *n*	Viscosity at 25 °C (80 °F)	Softening point(b) °C	Softening point(b) °F
350	182	0	80	...	...
380	188	0.15	140	...	...
600	310	0.9	Semisolid	40	105
900	475	2.0	Solid	70	160
1400	900	3.7	Solid	100	212
2900	1750	9.0	Solid	130	265
3750	3200	11.9	Solid	150	300

(a) Weight per epoxide, that is, grams of resins needed to provide 1 molar equivalent of epoxide. Also referred to as EEW (epoxide equivalent weight) and EMM (epoxy molar mass). All three items are interchangeable. (b) Softening point by Durran's mercury method

Source: Ref 6, p 67

Thermal, physical, and chemical properties of a typical epoxy resin coating

Heat-deflection temperature at 1.82 MPa. °C	140
Maximum resistance to continuous heat, °C	130
Coefficient of linear thermal expansion, 10^{-5}/°C	5
Tensile strength, MPa	50
Elongation, %	5
Hardness, Shore D	85
Specific gravity	112
Water absorption, %	1
Dielectric constant	4
Resistance to chemicals at 25 °C(a):	
Nonoxidizing acids (20% H_2SO_4)	S
Oxidizing acids (10% HNO_3)	U
Aqueous salt solutions (NaCl)	S
Aqueous alkalis (NaOH)	S
Polar solvents (C_2H_5OH)	S
Nonpolar solvents (C_6H_6)	S
Water	S

(a) S, satisfactory; U, unsatisfactory.

Source: Ref 23, p 98

Thermal, physical, and chemical properties of typical molded epoxy plastics (EP)

Property	Epoxy plastic	Glass-filled EP	Glass-sphere-filled EP
Heat-deflection temperature at 1.82 MPa, °C	140	150	115
Maximum resistance to continuous heat, °C	120	135	110
Coefficient of linear thermal expansion, 10^{-5}/°C	2.5	2.0	2.5
Compressive strength, MPa	120	207	83
Flexural strength, MPa	124	103	41
Impact strength, Izod, cm·N/cm of notch	53.4	53.4	10.6
Tensile strength, MPa	52	83	41
Elongation, %	5	4	1
Hardness, Rockwell	M90	M105	...
Specific gravity	1.2	1.8	0.8
Dielectric constant	4	4	4
Water absorption	0.2	0.1	0.1
Resistance to chemicals at 25 °C(a):			
Nonoxidizing acids (20% H_2SO_4)	S	S	S
Oxidizing acids (10% HNO_3)	U	U	U
Aqueous salt solutions (NaCl)	S	S	S
Aqueous alkalis (NaOH)	S	S	S
Polar solvents (C_2H_5OH)	S	S	S
Nonpolar solvents (C_6H_6)	S	S	S
Water	S	S	S

(a) S, satisfactory; U, unsatisfactory.

Source: Ref 13, p 121

Phenolics

Phenolic properties and tests

Ingredient	Property	Test method
Phenolic resin	Phenol	Titration
	Molecular weight	GPC
	Characterization	HPLC/infrared spectroscopy
	Solids	Evaporation
Modifier (organic)	Viscosity	Rheometer
	Molecular weight	GPC

Source: Ref 6, p 290

Typical properties of phenolic resin

Specific gravity	1.08-1.09
Solids content, %	60-62
Viscosity at 25 °C (77 °F), Pa · s (cP)	0.12-0.20 (120-200)
Refractive index	1.518-1.525
Cure time at 165 °C, (329 °F), s	85-105
Free formaldehyde, %	0-0.5
Free phenol, %	11.5-13.5
Trace elements and sodium	< 5 ppm each
Potassium, lithium, iron	< 10 ppm total

Ref 6, p 916

Phenolic matrix properties and test methods

Material	Property	Test method(s)
Uncured resin	Composition	HPLC; infrared spectroscopy; GPC
	Processibility	Solids; gel; volatile content
	Chemical activity	DSC
Cured neat resin	Completeness of cure	Glass transition; solvent extraction
Uncured impregnated system	Characterization	HPLC; infrared spectroscopy; DSC
	Processibility	Resin content; volatile content; flow; gel; tack/drape; fiber weight
Cured impregnated system	Completeness of cure	DSC
	Thermal properties	TGA; flammability
	Electrical properties	Dielectric

Source: Ref 6, p 292

Selected properties of molded phenolics

Property	ASTM test	Reported value
Mechanical properties		
Yield strength, MPa (ksi)	D638	48-55 (7.0-8.0)
Tensile modulus, GPa (10^6 psi)	D638	5.2-6.9 (0.75-1.0)
Elongation (break), %	D638	1-2
Compressive strength, MPa (ksi)	D695	83-195 (12-28)
Flexural yield strength, MPa (ksi)	D790	83-105 (12-15)
Impact strength, Izod(a), J/cm (ft·lb/in.)	D256	0.1-0.2 (0.2-0.4)
Hardness, Rockwell	...	M124-128
Thermal properties		
Thermal conductivity, W/m·K (Btu·in./ft^2·°F)	C177	0.13-0.22 (0.9-1.5)
Coefficient of linear thermal expansion, 10^{-5}/°C (10^{-5}/°F)	D696	4.5-11 (2.5-6.0)
Specific heat, kJ/kg·K (Btu/lb·°F)	C351	1.7 (0.4)
Continuous service temperature, °C (°F)	...	150-175 (300-350)
Electrical properties		
Volume resistivity, Ω·cm	D257	10^{11}-10^{12}
Dielectric strength, V/10^{-3} mm (V/10^{-3} in.)	D149	12-16 (300-400)
Dielectric constant, (50-100 Hz)	D150	5.0-5.6
Dissipation factor (50-100 Hz)	D150	0.06-0.10
Processing parameters		
Processing temperature, °C (°F)	...	130-160 (270-320)
Density, g/cm^3	D792	1.25-1.30
Specific volume, cm^3/kg ($in.^3$/lb)	D792	795-760 (22-21)
Linear mold shrinkage, mm/mm (in./in.)	D955	0.010-0.012
Water absorption, % in 24 h	D570	0.1-0.2

(a) Notched, at room temperature.

Source: Ref 15, p 1-15

Thermal, physical, and chemical properties of typical phenolic plastics

Property	Wood flour-filled	Mineral-filled	Glass-reinforced
Heat deflection temperature at 1.82 MPa, °C	165	200	250
Maximum resistance to continuous heat, °C	160	175	175
Coefficient of linear thermal expansion, 10^{-5}/°C	3.0	2.0	1.5
Compressive strength, MPa	172	172	120
Impact strength, Izod, cm·N/cm of notch	21.5	21.5	75
Tensile strength, MPa	48	41	60
Elongation, %	0.5	0.5	0.2
Hardness, Rockwell	M100	M110	E70
Specific gravity	1.4	1.5	1.85
Water absorption, %	0.4	0.03	0,5
Dielectric constant	6	8	5
Resistance to chemicals at 25 °C(a):			
Nonoxidizing acids (20% H_2SO_4)	S	S	S
Oxidizing acids (10% HNO_3)	Q	Q	Q
Aqueous salt solutions (NaCl)	S	S	S
Aqueous alkalis (NaOH)	Q	Q	Q
Polar solvents (C_2H_5OH)	S	S	S
Nonpolar solvents (C_6H_6)	S	S	S
Water	S	S	S

(a) S, satisfactory; Q, questionable.

Source: Ref 23, p 116

Thermoset polyesters

Selected properties of cast thermoset polyesters

Property	ASTM test	Reported value for: Rigid TS polyester	Flexible TS polyester
Mechanical properties			
Yield strength, MPa (ksi)	D638	41-89.6 (6.0-13.0)	3.4-21 (0.5-3.0)
Tensile modulus, GPa (10^6 psi)	D638	2.1-4.1 (0.30-0.60)	...
Elongation, (break), %	D638	5	40-300
Compressive strength, MPa (ksi)	D695	90-205 (13-30)	14-105 (2-15)
Flexural yield strength, MPa (ksi)	D790	105-140 (15-20)	34-105 (5-15)
Flexural modulus, GPa (10^6 psi)	D790	1.4-5.5 (0.20-0.80)	0.69 (0.10)
Impact strength, Izod(a) J/cm (ft·lb/in.)	D256	0.1-0.2 (0.2-0.4)	3.7 (7.0)
Hardness	...	Rockwell M70-115	Shore D85-95
Thermal properties			
Thermal conductivity, W/m·K (Btu·in./ft^2·h·°F)	C177	0.17 (1.2)	...
Coefficient of linear thermal 10^{-5}/°C (10^{-5}/°F)	D696	9.0-18.0 (5.0-10.0)	...
Specific heat, kJ/kg·K (Btu/lb·°F)	C351	1.3-2.1 (0.3-0.5)	...
Continuous service temperature, °C (°F)	...	120-150 (250-300)	65-120 (150-250)
Electrical properties			
Volume resistivity, Ω·cm	D257	10^{13}	10^{12}
Dielectric strength, V/10^{-3} mm (V/10^{-3} in.)	D149	16-20 (400-500)	9.8-16 (250-400)
Dielectric constant (50-100 Hz)	D150	3.0-4.4	4.5-8.0
Dissipation factor (50-100 Hz)	D150	0.003-0.030	0.030-0.300
Processing parameters			
Density, g/cm^3	D792	1.10-1.50	1.00-1.20
Specific volume, cm^3/kg (in.^{3}/lb)	D792	905-685 (25-19)	975-830 (27-23)
Water absorption, % in 24 h	D570	0.1-0.6	0.5-2.5

(a) Notched, at room temperature.

Source: Ref 15, p 1-27

Polyester matrix properties and test methods

Material	Property	Test method
Uncured resin	Composition	Infrared spectroscopy
		HPLC
	Processibility	RDS viscosity
		Gel
		Volatile content
Cured neat resin	Completeness of cure	DSC
Uncured impregnated system	Characterization	HPLC
	Processibility	Resin content
		Flow
		Gel
		Tack/drape
		Fiber weight
Cured impregnated system	Completeness of cure	DSC
	Laminate properties	Ply thickness
		Fiber volume
		Hardness

Source: Ref 6, p 292

Types of polyester resin

Type	Anhydride	Glycol
General purpose	Orthophthalic, maleic	Ethylene, diethylene, or propylene
Corrosion resistant	Isophthalic, maleic, bisphenol A, chlorendic	Propylene
Flame resistant	Brominated tetra hydropththalic, tetrabromo-pththalic, chlorendic anhydride	Dibromoneopentyl glycol

Source: Ref 8

Ref 6, p 138

Components of polyester resins

Anhydride(a)	Diluent	Use
Orthophthalic	Styrene	General low cost
Isophthalic	Styrene	Better mechanical
Isophthalic	Vinyl toluene	Better mechanical, less volatile
Orthophthalic	Diallylphthalate	Improved electrical, less volatile
Isophthalic	Methyl methacrylate	Outdoor exposure
Tetrabromophthalic	Styrene	Fire retardant
Isophthalic & bisphenol A	Styrene	Corrosion resistance

(a) In addition to maleic or fumaric anhydride

Source: Ref 6, p 132

Polyester properties and tests

Ingredient	Property	Test method
Polyester	Reactivity	Titration of peroxide
	Molecular weight	GPC
	Purity	HPLC
		H_2O determinations

Source: Ref 6, p 290

Typical polyester formulations

% MEKP (0.5% cobalt naphthenate)	Gel time at 30 °C (86 °F), min
2.0	9.0
1.0	18.5
0.75	25.0
0.5	38.5

% BPO (0.2% dimethyl aniline)	Gel time at 30 °C (86 °F), min
2.0	4.5
1.0	7.5
0.5	12.0
0.25	21.0

Source: Ref 6, p 133

Thermal, physical, and chemical properties of glass-reinforced unsaturated polyesters

Heat-deflection temperature at 1.82 MPa, °C	200
Maximum resistance to continuous heat, °C	160
Coefficient of linear thermal expansion, 10^{-5}/°C	2.5
Compressive strength, MPa	172
Flexural strength, MPa	83
Impact strength, Izod cm·N/cm of notch	160
Tensile strength, MPa	69
Elongation, %	1.5
Hardness, Rockwell	M50
Specific gravity	2
Dielectric constant	5
Resistance to chemicals at 25 °C(a):	
Nonoxidizing acids (20% H_2SO_4)	S
Oxidizing acids (10% HNO_3)	Q
Aqueous salt solutions (NaCl)	S
Aqueous alkalis (NaOH)	Q
Polar solvents (C_2H_5OH)	S
Nonpolar solvents (C_6H_6)	S
Water	S

(a) S, satisfactory; Q, questionable.

Source: Ref 23, p 54

Catalyst-promotor-inhibitor systems for room-temperature-cure polyester resins

Application or end use	System, %	Gel time starting at room temperature, min	Approximate time at 21-24 °C (70-75 °F) for development of Barcol Hardness = 35
Gel coats	MEKP-1.5(a) Cobalt naphthenate-0.4(b) (assessory promoters usually omitted because of tendency to discolor)	30 (high filler content)	6-8 (can proceed with lay-up over gel coat in 30-45 min)
For normal lay-up resins	MEKP-1.0 Cobalt naphthenate-0.4	32	6-8
For fast-cure resins	MEKP-1.0 Cobalt naphthenate-0.4 Dimethyl aniline-0.1	16	2-2.5
For fast-cure resins	MEKP-1.0 Cobalt naphthenate-0.4 Quaternary ammonium salt-0.1	15	2-2.5
Alternate room-temperature cure	Cyclohexanone peroxide(c)-1.0 Cobalt naphthenate-0.4	30	~6-8
Alternate room-temperature cure	Bis-I-hydroxy cyclohexyl peroxide(c)-1.0	30	~6-8
Alternate room-temperature cure	Benzoyl peroxide-1.0 Dimethyl aniline-0.1	20	2
Effect of inhibitor	MEKP-1.0 Cobalt naphthanate-0.4 Hydroquinone-0.1	∞	∞

(a) Percentages based on 100 parts polyester resin. (b) Concentration of cobalt metal, 6%. (c) Peroxides costlier than MEKP.

Source: Ref 6, p 133

Bismaleimides

Selected properties of 5245C modified bismaleimide resin

Property	Reported value	Property	Reported value
Density, g/cm^3	1.25	Flexural modulus (dry), GPa (10^6 psi)	
T_g, °C (°F)	220 (428)	At RT	3.3 (0.49)
Tensile strength (dry) at RT, MPa (ksi)	84 (12.2)	At 93 °C (270 °F)	3.1 (0.46)
		At 130 °C (270 °F)	2.7 (0.40)
Tensile modulus (dry) at RT, GPa (10^6 psi)	33 (0.48)	Flexural modulus (wet)(a), GPa (10^6 psi):	
		At 93 °C (200 °F)	2.9 (0.43)
Tensile strain-to-failure (dry) at RT, %	2.9	At 130 °C (270 °F)	2.7 (0.39)
Flexural strength (dry), at MPa (ksi):		Izod impact strength, unnotched, J/m (ft·lbf/in.)	410 (7.7)
At RT	145 (21.0)	Fracture toughness, G_{Ic},	
At 93 °C (200 °F)	115 (16.7)	J/m^2 (in.·lbf/in.^{2})	67 (0.38)
At 132 °C (270 °F)	107 (15.5)	Moisture absorption (72-h water boil), %	1.7
Flexural strength (wet)(a), MPa (ksi)			
At 93 °C (200 °F)	96 (14.0)	CTE, 10^{-6}/K	72
At 130 °C (200 °F)	83 (12.1)		

(a) Wet condition, 40-h water boil, specimen held 5 min at test temperature before loading.

Source: Ref 6, p 88

Commercial bismaleimides

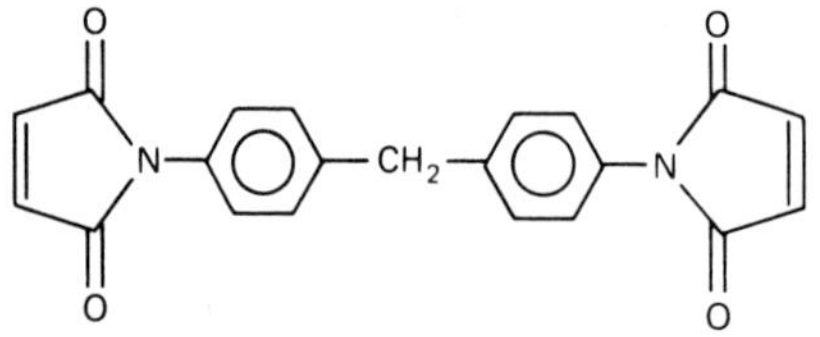

Kerimid 601

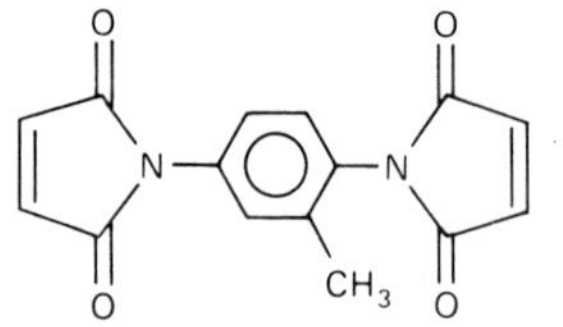

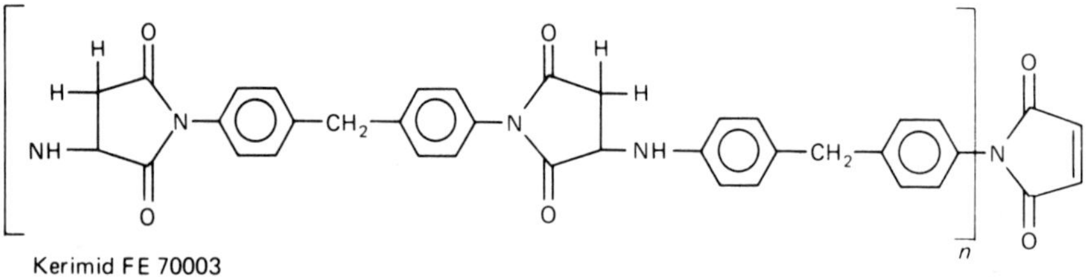

Kerimid FE 70003

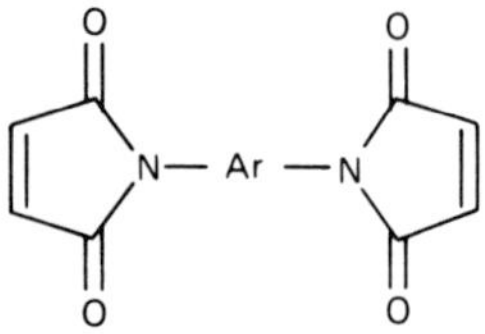

Completely aromatic

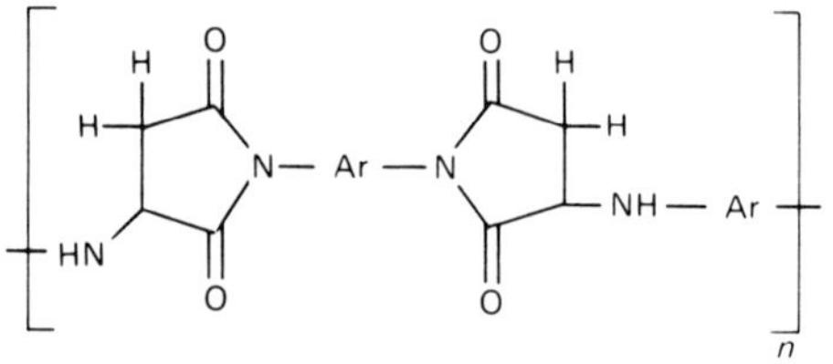

Rhone-Poulenc
Kerimid 353

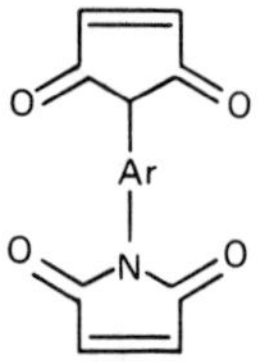

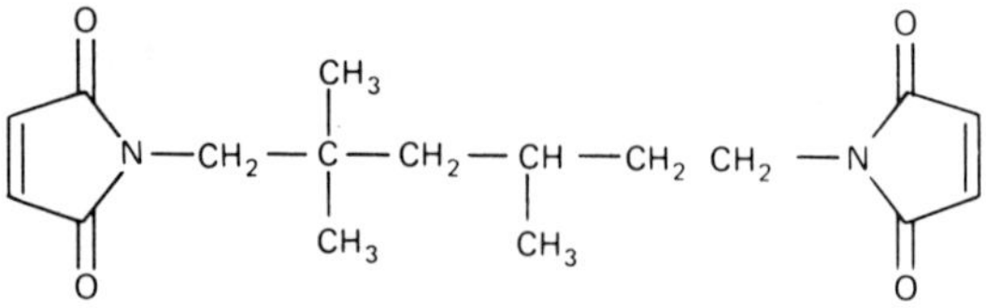

Ar =

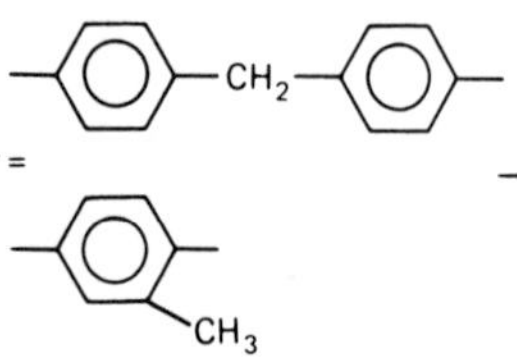

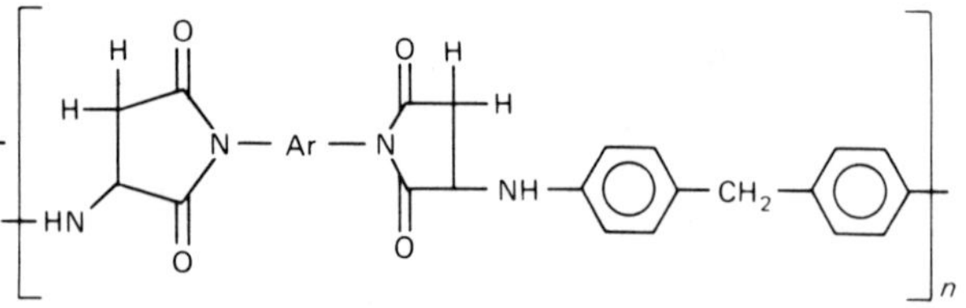

Source: Ref 6, p 88

Selected properties of compimide bismaleimide resins

Property	C795	C800	C183	C796	C353 /TM-122(a)	C796 /TM-122(b)
Density, g/cm^3	1.32	...	1.31	...	...	...
T_g (dry), °C (°F)	290 (554)	290 (554)	250 (482)	...	...	...
Morphology	Amorphous	Amorphous	Amorphous	Amorphous	Amorphous	Amorphous
Flexural strength (dry), MPa (ksi):						
At RT	110 (15.9)	92 (13.2)	106 (15.4)	76 (11.0)	110 (15.9)	114 (16.5)
At RT after isothermal aging for 500 h at 200 °C (392 °F) in circulating air	100 (14.5)	...	...	...	...	...
At RT after isothermal aging for for 500 h at 250 °C (478 °F) in circulating air	108 (15.6)	...	...	...	...	...
At RT after isothermal aging for 500 h at 250 °C (478 °F) in circulating air and then exposed to 200 °C (392 °F)	65 (9.4)	...	58 (8.4)	...	...	...
At 200 °C (392 °F) after isothermal aging for 500 h at 200 °C (392 °F) in circulating air	71 (10.2)	...	...	...	...	...
At 200 °C (392 °F) after isothermal aging for 500 h at 250 °C (478 °F) in circulating air	62 (8.9)	46 (6.6)	...	31 (4.5)	64 (9.3)	73 (10.5)
At 250 °C (478 °F) after isothermal aging for 500 h at 250 ° C (°F) in circulating air	41 (5.9)	...	...	...	...	...
Flexural modulus (dry), GPa (10^6 psi):						
At RT	5.5 (0.79)	3.9 (0.56)	4.1 (0.59)	4.6 (0.66)	3.7 (0.53)	3.9 (0.56)
At RT after isothermal aging for 500 h at 200 °C (392 °F) in circulating air	5.5 (0.79)	...	...	...	...	...
At RT after isothermal aging for 500 h at 250 °C (478 °F) in circulating air	5.3 (0.77)	...	...	...	...	...
At 200° C (392 °F) after isothermal aging for 500 h at 200 °C (392 °F) in circulating air	4.7 (0.68)	...	3.2 (0.47)	...	...	...
At 200 °C (392 °F) after isothermal aging for 500 h at 250 °C (478 °F)	3.4 (0.49)	2.1 (0.30)	..	3.0 (0.43)	2.5 (0.36)	2.62 (0.38)
At 250 °C (478 °F) after isothermal aging for 500 h at 250 °C (478 °F) in circulating air	4.5 (0.65)	...	...	...	...	...
Flexural elongation (dry), %:						
At RT	2.4	...	2.6	1.7	2.7	3.0
At 200 °C (392 °F)	...	...	1.8	...	...	...
At 250 °C (478 °F)	2.2	...	...	1.0	2.6	3.0
Tensile strength (dry) at RT, MPa (ksi)	89 (13)	...	...	...	...	...
Tensile strain-to-failure, %:						
Dry, at RT	2.2	...	...	...	...	...
Wet, at 250 °C ° (478 °F)	2.0	...	...	...	...	...
Fracture toughness, G_{Ic}, J/m^2 (in.·lbf/in.^{2})	40 (0.23)	160 (0.80)	180 (1.0)	63 (0.36)	389 (2.2)	230 (1.27)
Gel time at 170 °C (338 °F), min.	25	50	45	>30	...	...
Complex viscosity at 110 °C ° (230 °F), Pa·s (cP)	0.4-2.8 (400-2800)	0.2-2.5 (220 -2500)	2.0-8.0 (2000-8000)	0.4-3.5 (400-3500)	...	...
Heat of polymerization, J/g (Btu/lb)	265 (0.10)	260 (0.10)	260 (0.10)	>200 (>0.85)	...	...
Melting range, °C (°F)	110-120 (230-248)	...	100-140 (212-284)	...	...	...
CTE to 250 °C (478 °F), 10^{-6}/K	73.4	...	66	...	...	...
Moisture absorption, wt %	4.85	...	...	...	...	...

(a) Weight ratio for C353/TM-122, 76/24; TM-122 is 4,4′-bis (2-propenylphenoxy) diphenylsulfone. (b) Weight ratio for C796/TM-122, 70/30; TM-122 is 4,4′-bis (2-propenylphenoxy) diphenylsulfone.

Source: Ref 6, p 87

Bismaleimide properties and tests

Ingredient	Property	Test method
Bismaleimide resin	Viscosity	Rheometer
	Composition	HPLC/infrared spectroscopy
Modifier (organic)	Viscosity	Rheometer
	Molecular weight	GPC

Source: Ref 6, p 289

Bismaleimide matrix properties and test methods

Material	Property	Test method
Uncured resin	Composition	HPLC
		Infrared spectroscopy
		GPC
		DSC
	Processibility	RDS viscosity
		Gel time
		Volatile content
Cured neat resin	Completeness of cure	Glass transition
	H_2O/solvent resistance	Glass transition, wet
		Moisture weight gain
		Solvent weight gain
	Resin toughness	Cleavage (G_{Ic})
Uncured impregnated system	Characterization	HPLC
		Infrared spectroscopy
		DSC
	Processibility	Resin content
		Fiber content
		Flow
		Gel
		RDS viscosity
		Volatiles
		Tack/drape
Cured impregnated system	Completeness of cure	Glass transition
		DSC
	Moisture resistance	Weight gain
	Thermal properties	Thermal conductivity
	Laminate properties	Ply thickness
		Fiber volume

Source: Ref 6, p 291

Selected properties of bismaleimide resins

Material	Density, g/cm^3	Uncured melting or softening temperature °C	°F	Tensile strength (dry) at RT MPa	ksi	Tensile strength (dry) at 232 °C (450 °F) MPa	ksi
Hexcel F178	...	...	...	56	8.14	...	...
Narmco 5250-2	1.24	...	...	62	9.0	...	...
U.S. Polymeric V378A (Ref 50)	1.26	...	...	78	11.3	44	6.4
Univ. Dayton BPA-BMI (Ref 51)	1.26	70	155	48	7.0	...	...
Kerimid 70003 (Ref 52)	1.25	...	...	...	...	...	...

Material	Flexural modulus (dry) at RT GPa	10^6 psi	Fracture toughness at RT J/m^2	in. · lbf/in.^{2}	T_g, (dry) °C	°F
Hexcel F178	...	...	29.4	0.17	260	500
Narmco 5250-2	2.9	0.43	100	0.56	321	610
U.S. Polymeric V378A	...	...	...	...	...	...
Univ. Dayton BPA-BMI	...	...	...	...	280	536
Kerimid 70003	...	...	82	0.46	320 (Ref 19)	620

Selected properties of bismaleimide resins (continued)

Material	Tensile strain-to-failure (dry) at RT, %	Tensile modulus (dry) at RT GPa	10^6 psi	Flexural strength (dry) at RT MPa	ksi
Hexcel F178	1.3	...	...	...	...
Narmco 5250-2	1.7	2.7	0.39	138	20
U.S. Polymeric V378A (Ref 50)	6.6	...	...	...	...
Univ. Dayton BPA-BMI (Ref 51)	1.5	3.4	0.50	...	...
Kerimid 70003 (Ref 52)	2.2	...	...	...	...

Material	T_g, (wet) °C	°F	CTE, 10^{-6}/K	Moisture absorbtion, wt%
Hexcel F178	140	284	...	3.7
Narmco 5250-2	...	...	...	3.3 (steam 96 h)
U.S. Polymeric V378A	...	...	...	...
Univ. Dayton BPA-BMI	...	...	...	1.7
Kerimid 70003	...	...	21	1.7 (100 h BW)

Source: Ref 6, p 88

Selected properties of Matrimid 5292 bismaleimide

Property	Reported value(a)
Density, g/cm^3	1.23
Melting point	
Viscosity Component A	150-160 °C (302-320 °F)
Viscosity Component B, amber liquid, at 25 °C (77 °F), mPa·s	12000-20000
T_g, (by TMA), °C (°F)	273 (523)
T_g, (by DMA), °C (°F):	
Dry	295 (563)
Wet	305 (581)
Morphology	Amorphous, cross-linked
Tensile strength (dry), MPa (ksi):	
At RT	82 (11.9)
At 150 °C (300 °F)	51 (7.4)
At 204 °C (400 °F)	40 (5.8)
Tensile modulus (dry), GPa (10^6 psi):	
At RT	4.3 (0.62)
At 150 °C (300 °F)	2.4 (0.35)
At 204 °C (400 °)	2.0 (0.29)
Tensile strain-to-failure, %:	
At RT	2.3
At 150 °C (300 °F)	2.6
At 204 °C (400 °F)	2.3
Tensile strength (wet), MPa (ksi):	
At RT	66 (9.6)
At 150 °C (300 °F)	30 (4.3)
Tensile modulus (wet), GPa (10^6 psi);	
At RT	3.8 (0.55)
At 150 °C (300 °F)	1.9 (0.27)
Flexural strength (dry) at RT, MPa (ksi)	167 (24.2)
Flexural modulus (dry) at RT GPa (10^6 psi)	4.0 (0.59)
Moisture absorption, wt %	1.40
Fracture toughness (compact tension), J/m^2 (in.·lbf/in.^{2})	170 (0.97); 216 (1.22)

(a) Properties taken from commercial data sheet and technical literature.

Source: Ref 6, p 86

Selected properties of Keramid 601 and 353 bismaleimides

Property	601	353
Density, g/cm^3	1.30	...
Melting point, °C (°F)	80 (177)	70-125 (158-298)
T_g, °C (°F)	290 (554)	285 (545)
Morphology	Amorphous	Amorphous
Gel time at 170 °C (338 °F), min	...	30
Melt viscosity, Pa·s (cP)	...	0.115-0.130 (115-130)
Tensile strength (dry), MPa (ksi):		
At RT	63.4 (9.2)	...
At 200 °C (392 °F)	42.0 (6.1)	...
Tensile strain-to-failure, %:		
At RT	3.1	...
At 200 °C (392 °F)	4.9	...
Flexural strength (dry), MPa (ksi):		
At RT	150 (21.7)	60 (8.7)
At 250 °C (482 °F)	...	50 (7.3)
After 210 h at 250 °C (482 °F)	103 (15.0)	...
After 1650 h at 250 °C (482 °F)	82 (11.9)	...
Flexural modulus (dry), GPa (10^6 psi):		
At RT	...	5.6 (0.81)
At 250 °C (478 °F)	...	3.4 (0.49)
Flexural elongation (dry), %:		
At RT	...	1.8
At 250 °C (478 °F)	...	1.2
Fracture toughness at RT:		
K_{Ic}, $MPa\sqrt{m}$ ($ksi\sqrt{in.}$)	382 (348)	...
G_{Ic}, J/m^2 (in.·lbf/$in.^2$)	34 (0.19)	30 (0.17)
Thermogravimetric weight loss under nitrogen, wt %	1 up to 400 °C (750 °F)	...
Moisture absorption, wt %:		
At RT in 24 h	0.3	...
At 100 °C in 2 h	1.0	...
Coefficient of thermal expansion (CTE), $10^{-6}/K$	61	...

Source: Ref 6, p 86

Constituent properties of bismaleimide resins

Property	Fiberite X-86	Cycom 3100	Hysol EA9102	Hysol EA9655	Hexcel F650	Fiberite 987A
Density, g/cm^3	1.22	...	...	...	...	1.27
Uncured melting or softening temperature, °C (°F)	70-125 (159-257)	...	...	...	...	...
Tensile strength (dry), MPa (ksi):						
At RT	...	...	58.5 (8.5)	52 (7.5)	49.3 (7.15)	...
At 177 °C (350 °F)	...	...	43 (6.3)	27 (3.9)	...	...
Tensile strength (wet), at RT, MPa (ksi)	...	...	...	39 (5.7)	...	...
Tensile strain-to-failure (dry), %:						
At RT	...	2.1	2.2	1.6	...	1.2
At 177 °C (350 °F)	...	2.2	3.3	2.2	...	...
Tensile strain-to-failure (wet), at 177 °C (350 °F), %	...	3.0	...	...	...	...
Tensile modulus (dry), at RT, GPa (10^6 psi)	...	...	...	...	...	4.2 (0.61)
Flexural strength (dry), MPa (ksi):						
At RT	130 (18.9)	117 (17)	...	...	...	121 (17.6)
At 177 °C (350 °F)	85 (12.3)	93 (13.5)	...	...	...	53.7 (7.8)
At 232 °C (450 °F)	72 (10.5)	...	...	...	...	41 (6.0)
Flexural strength (wet), at 177 °C (350 °F), MPa (ksi)	46 (6.7)	58 (8.5)	...	...	...	38 (5.5)
Flexural modulus (dry), GPa (10^6 psi):						
At RT	4.6 (0.66)	...	...	...	...	4.3 (0.62)
At 177 °C (350 °F)	3.0 (0.44)	...	...	...	...	2.2 (0.32)
At 232 °C (450 °F)	2.6 (0.38)	...	...	...	...	1.8 (0.26)
Flexural modulus (wet), at 177 °C (350 °F), GPa (10^6 psi)	1.7 (0.25)	...	...	...	...	1.6 (0.23)
Fracture toughness at RT, G_{Ic}, J/m^2 (in.·lbf/in.^{2})	...	...	...	...	67 (0.38)	...
T_g (dry), °C (°F)	290 (554)	300 (527)	298 (568)	253 (489)	>316 (>600)	320 (608)
T_g (wet), °C (°F)	...	...	210 (410)	118 (244)	...	...
Moisture absorption, wt%	4.4(a)	...	...	...	4.3	2.97(a)

(a) Equilibrium water boil.

Source: Ref 6, p 88

Polyimides

Polyimide properties and tests

Ingredient	Property	Test method
Polyimide		
Resin	Ingredient ratio	HPLC/infrared titration
	Purity	HPLC
	Functional groups	Titration

Source: Ref 6, p 84

Condensation polyimides

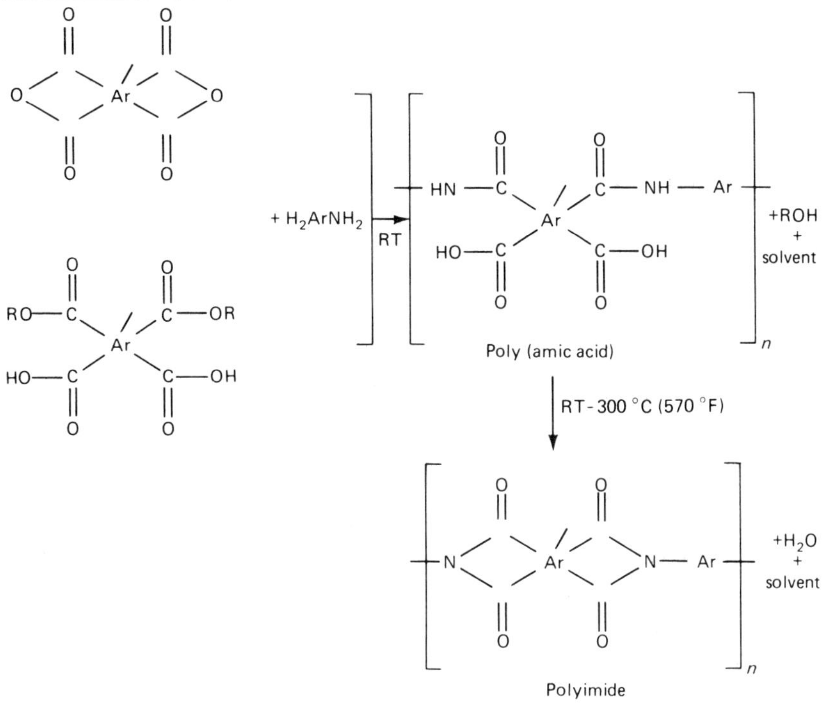

Ref 6, p 79

Condensation polyimide resins

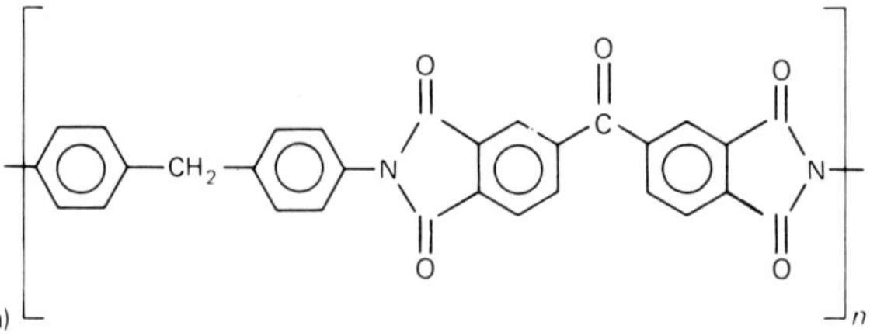
(a)

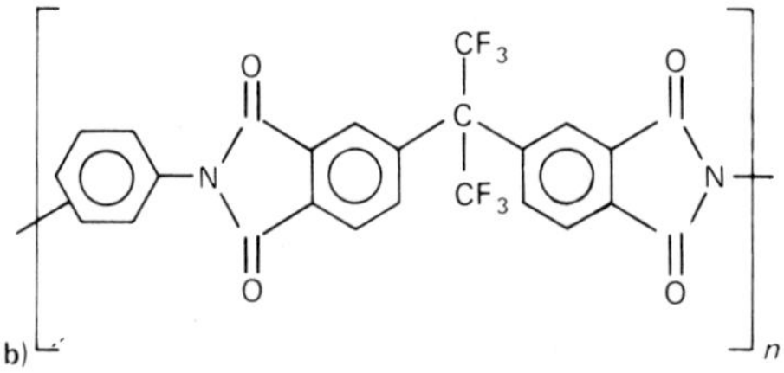
(b)

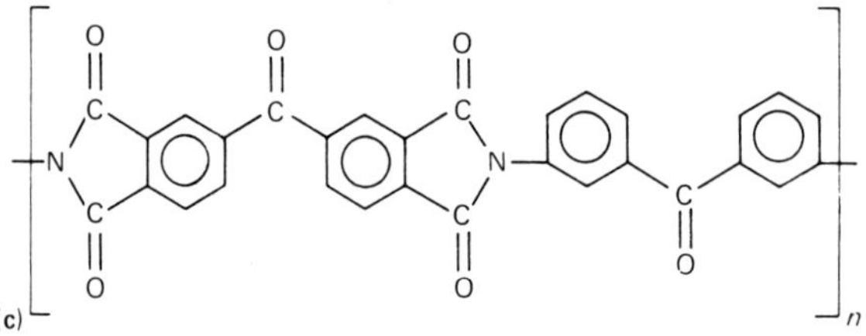
(c)

Source: Ref 6, p 79

Molecular structure of PMR acetylene end-capped Thermid AL-600

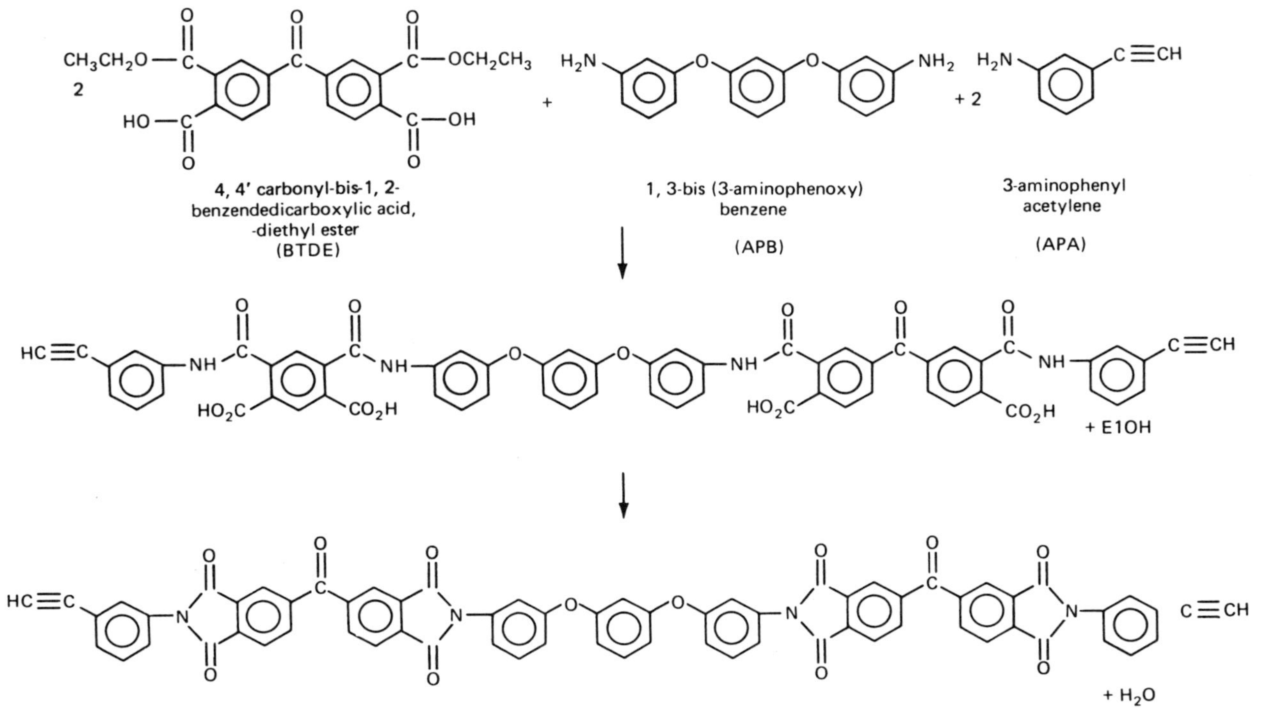

Ref 6, p 84

Chemistry of Compimide resins

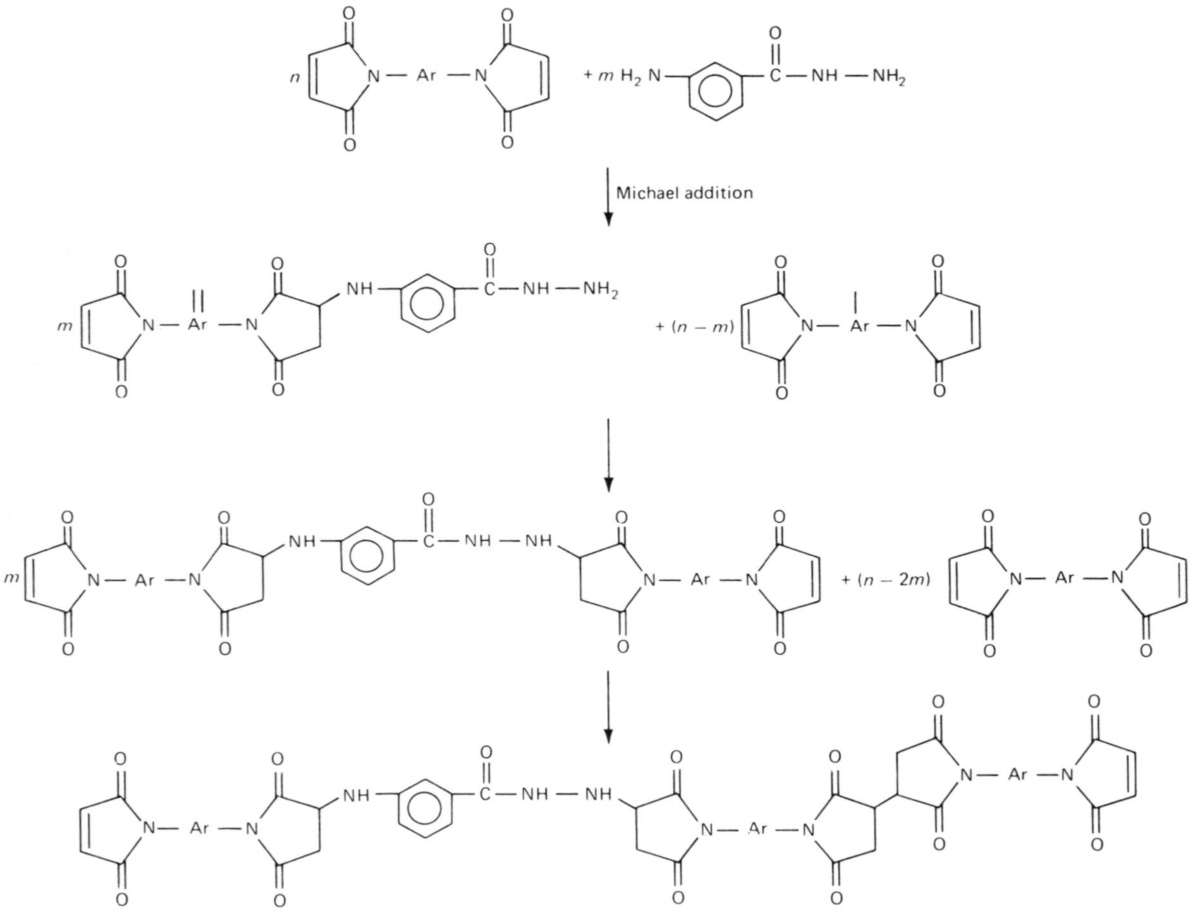

Source: Ref 6, p 81

Physical and mechanical properties of polyimides

Material	Density, g/cm³	Tensile strength MPa	Tensile strength ksi	Tensile modulus GPa	Tensile modulus 10^6 psi
Avimid K-III	1.31	102	15.0	3.8	0.55
Skybond 701	1.35	69	10.0	4.1	0.60
PMR-15	1.32	38.6	5.6	3.9	0.57
NR-150B2	1.40	110	16.0	4.1	0.60
Thermid MC-600	1.34	83	12.0	4.1	0.60
UpJohn 2080	1.40	120	17.1	1.3	0.19
BMIs	1.22-1.30	41-82	6-12	4.1-4.8	0.60-0.70
Ultem 1000	1.27	104	15.2	3.0	0.43
Torlon 4203	1.38	186	27.0	4.4	0.64

Material	Izod impact strength, notched J/m	Izod impact strength, notched ft · lbf/in.	Strain-to-failure, %
Avimid K-III	...	...	14
Skybond 701	53.4	1.0	1.00
PMR-15	53.4	1.0	1.5
NR-150B2	42.7	0.8	6.0
Thermid MC-600	48	0.9	1.5
UpJohn 2080	37.4	0.7	10.0
BMIs	...	...	1.3-2.3
Ultem 1000	53.4	1.0	60
Torlon 4203	133.5	2.5	20

Physical and mechanical properties of polyimides (continued)

Material	Flexural strength MPa	Flexural strength ksi	Flexural modulus GPa	Flexural modulus 10^6 psi
Avimid K-III	...	...	...	...
Skybond 701	...	...	...	...
PMR-15	176	25.5	4.0	0.58
NR-150B2	...	...	...	...
Thermid MC-600	145	21.0	4.5	0.66
UpJohn 2080	117	17.0	3.4	0.48
BMIs	76-145	11-21	3.4-4.8	0.50-0.70
Ultem 1000	145	21	3.4	0.48
Torlon 4203	211	30.7	4.5	0.66

Material	Glass transition temperature, T_g °C	Glass transition temperature, T_g °F	Fracture toughness, G_{Ic} J/m²	Fracture toughness, G_{Ic} in. · lb/in.²
Avimid K-III	250	482	1900	11.0
Skybond 701	330	626	...	...
PMR-15	340	644	280	1.57
NR-150B2	340	644	2400	13.4
Thermid MC-600	320	608	...	...
UpJohn 2080	280	536	...	...
BMIs	230-290	446	34-260	0.19-1.45
Ultem 1000	210	426	...	...
Torlon 4203	267	512	3900	21.9

Source: Ref 6, p 76

Selected properties of polyimides

Property	ASTM test	Reported value for: Molded polyimide	Encapsulated polyimide
Mechanical properties			
Yield strength, MPa (ksi)	D638	118 (17.1)	18 (2.6)
Tensile modulus, GPa (10^6 psi)	D638	1.3 (0.19)	...
Elongation, (break), %	D638	10	1
Compressive strength, MPa (ksi)	D695	205 (30)	69 (10)
Flexural yield strength, MPa (ksi)	D790	200 (29)	69 (10)
Flexural modulus, GPa (10^6 psi)	D790	3.4 (0.50)	4.1 (0.60)
Impact strength, Izod(a) J/cm (ft·lb/in.)	D256	0.37 (0.7)	0.37-0.43 (0.7-0.8)
Hardness, Rockwell	...	E99	D50
Thermal properties			
Thermal conductivity, W/m·K (Btu·in./ft^2·h·°F)	C177	0.1 (0.7)	0.3 (1.9)
Coefficient of linear thermal expansion 10^{-5}/°C (10^{-5}/°F)	D696	9.0 (5.0)	8.1 (4.5)
Specific heat, kJ/kg·K (Btu/lb·°F)	C351	1.3 (0.3)	...
Continuous service temperature, °C (°F)	...	260-425 (500-800)	...
Electrical properties			
Volume resistivity, Ω·cm	D257	10^{16}	10^{15}
Dielectric strength, V/10^{-3} mm (V/10^{-3} in.)	D149	22 (560)	29 (725)
Dielectric constant (50-100 Hz)	D150	3.4	...
Dissipation factor (50-100 Hz)	D150	0.0005	...
Processing parameters			
Processing temperature, °C (°F)	...	315 (600)	...
Density, g/cm^3	D792	1.43	1.55
Specific volume, cm^3/kg (in.^{3}/lb)	D792	685 (19)	650 (18)
Linear mold shrinkage, mm/mm (in./in.)	D955	...	0.003
Water absorption, % in 24 h	D570	0.3	0.1

(a) Notched, at room temperature.

Source: Ref 15, p 1-29

Selected properties of PMR-15 polyimide

Property	PMR-15 polimide
Density, g/cm^3	1.32
T_g after 316 °C (600 °F), 16-h postcure, °C (°F)	340 (662)
Morphology	Amorphous, cross-linked
Tensile strength (dry) at RT, MPa (ksi):	38.6 (5.6)
Tensile modulus (dry), at RT, GPa (10^6 psi):	39 (0.57)
Tensile strain-to-failure, %:	1.1
Flexural strength (dry), MPa (ksi):	
At RT	176 (25.5)
At 288 °C (550 °F)	73 (10.7)
At 316 °C (600 °F)	72 (10.4)
At 343 °C (650 °F)	52 (7.6)
Flexural modulus (dry), GPa (10^6 psi):	
At RT	4.0 (0.58)
At 288 °C (550 °F)	2.3 (0.34)
At 316 °C (600 °F)	1.9 (0.27)
At 343 °C (650 °F)	1.8 (0.26)
Fracture toughness:	
K_{Ic}, MPa $\sqrt{m}$ (ksi $\sqrt{in.}$)	1100 (1010); 648 (590)
G^3_{Ic}, J/m^2 (ft·lbf/in.^{2})	280 (1.6); 94 (0.52)
Izod impact strength, notched, J/m (ft·lbf/in.)	53.37 (1.0)
Moisture absorption, equilibrium moisture absorption (95% RH, 71 °C, or 160 °F), wt %	4.2
T_g, °C (°F):	
After 316 °C (1-h) cure	320 (608)
After 316 °C (16-h) cure	340 (662)
Weight loss after 1000 h at 288 °C (550 °F), %:	
In flowing air (100 cc/min)	0.3
After 2000 h	0.8
After 3000 h	2.0
CTE, 10^{-6}/K	14

Source: Ref 6, p 89

Polyimide matrix properties and test methods

Material	Property	Test method
Uncured resin	Composition	HPLC
		Infrared spectroscopy
		GPC
	Processibility	RDS viscosity
		Gel time
		Volatile content
Cured neat resin	Completeness of cure	Glass transition
		DSC
	H_2O/solvent resistance	Glass transition, wet
		Solvent weight gain
Uncured impregnated system	Characterization	HPLC
		Infrared spectroscopy
		DSC
	Processibility	Resin content
		Fiber weight
		Flow
		Gel
		RDS viscosity
		Volatiles
		Tack/drape
Cured impregnated system	Completeness of cure	Glass transition
		DSC
	Moisture resistance	Weight gain
	Thermal properties	TGA
		Thermal conductivity
	Laminate properties	Ply thickness
		Fiber volume

Source: Ref 6, p 292

Selected properties of Thermid resins

Property	Thermid MC-600	Thermid IP-600
Density, g/cm^3	1.37	1.34
Moisture absorption (24-h water boil), wt %	1.24	1.18
Uncured melting range, °C (°F)	190-210 (374-410)	149-171 (300-340)
T_g °C (°F),		
After 371 °C (700 °F), 8-h postcure	320 (608)	300 (572)
After 371 °C (700 °F), 16-h postcure	350 (662)	350 (662)
Morphology	Amorphous, cross-linked	Amorphous, cross-linked
Heat of reaction, J/g (Btu/lb)	335 (0.15)	335 (0.15)
Tensile strength (dry) MPa (ksi):		
At RT	82.7 (12.0)	58 (8.5)
Tensile modulus (dry), GPa (10^6 psi):		
At RT	4.1 (0.60)	5.0 (0.73)
At 316 °C (600 °F)	...	28 (4.01)
Tensile strain-to-failure, %:		
At RT	1.5	1.2
At 316 °C (600 °F)	...	4.2
Compressive strength (dry):		
At RT, MPa (ksi)	172 (25)	...
Flexural strength (dry), MPa (ksi):		
At RT	146.6 (21)	106 (15.3)
After 1000 h	92 (13.4)	...
Flexural strength (dry), MPa (ksi):		
At 316 °C (600 °F)	29 (4.2)	44 (6.4)
After 1000 h	18 (2.6)	...
Flexural modulus (dry), GPa (10^6 psi):	4.6 (0.66)	...
Izod impact, notched, J/m (ft·lbf/in.)	80 (1.5)	...
CTE, 10^{-6}/K	35-50	...
Weight loss, %		
After 500 h at 316 °C (600 °F)	2.89	...
After 1000 h at 316 °C (600 °F)	4.04	...
Dielectric constant (1 MHz)	3.496	...
Dissipation factor (1MHz)	0.0096	...

Source: Ref 6, p 89

Polyurethanes

Thermal, physical, and chemical properties of a typical PUR coating deposited from a two-package system

Heat-deflection temperature at 1.82 MPa, °C	75
Maximum resistance to continuous heat, °C	70
Coefficient of linear thermal expansion, 10^{-5}/°C	15.0
Tensile strength, MPa	6.89
Elongation, %	200
Hardness, Shore	D50
Specific gravity	1.2
Dielectric constant	6
Resistance to chemicals at 25 °C(a):	
Nonoxidizing acids (20% H_2SO_4)	Q
Oxidizing acids (10% HNO_3)	U
Aqueous salt solutions (NaCl)	S
Aqueous alkalis (NaOH)	Q
Polar solvents (C_2H_5OH)	U
Nonpolar solvents (C_6H_6)	Q
Water	S

(a) S, satisfactory; Q, questionable; U, unsatisfactory.

Source: Ref 23, p 101

Selected properties of liquid cast polyurethane

Property	ASTM test	Reported value
Mechanical properties		
Yield strength, MPa (ksi)	D638	14-69 (2.0-10.0)
Tensile modulus, GPa (10^6 psi)	D638	0.69-6.9 (0.10-1.0)
Elongation (break), %	D638	200-1000
Compressive strength MPa (ksi)	D695	140 (20)
Flexural yield strength, MPa (ksi)	D790	6.9-34 (1-5)
Flexural modulus, GPa (10^6 psi)	D790	0.69 (0.10)
Impact strength, Izod(a), J/cm (ft·lb/in.)	D256	>13.3 (>25.0)
Hardness, Rockwell	...	D90
Thermal properties		
Thermal conductivity, W/M·K (Btu·in./ft^2·h·°F)	C177	0.20 (1.4)
Coefficient of linear thermal expansion, 10^{-5}/°C (10^{-5}/°F)	D696	18.0-36.0 (10.0-20.0)
Specific heat, kJ/kg·K (Btu/lb·°F)	C351	1.7 (0.4)
Electrical properties		
Volume resistivity, Ω·cm	D257	10^{12}-10^{15}
Dielectric strength, V/10^{-3} mm (V/10^{-3} in.)	D149	16-20 (400-500)
Dissipation factor (50-100 Hz)	D150	0.015-0.017
Processing parameters		
Processing temperature, °C (°F)	...	85-120 (185-250)
Density, g/cm^3	D792	1.10-1.50
Specific volume, cm^3/kg (in.^{3}/lb)	D792	975-795 (27-22)
Linear mold shrinkage, mm/mm (in./in.)	D955	0.020
Water absorption, % in 24 h	D570	0.0-1.5

(a) Notched, at room temperature.

Source: Ref 15, p 1-34

Silicones

Thermal, physical, and chemical properties of typical silicone coatings

Maximum resistance to continuous heat, °C	250
Coefficient of linear thermal expansion, 10^{-5}/°C	40
Tensile strength, MPa	5
Elongation, %	200
Hardness, Shore A	65
Specific gravity	1.2
Dielectric constant	3
Water absorption, %	0.2
Resistance to chemicals at 25 °C(a):	
Nonoxidizing acids (20% H_2SO_4)	Q
Oxidizing acids (10% HNO_3)	U
Aqueous salt solutions (NaCl)	S
Aqueous alkalis (NaOH)	S
Polar solvents (C_2H_5OH)	S
Nonpolar solvents (C_6H_6)	Q
Water	S

(a) S, satisfactory; Q, questionable; U, unsatisfactory.

Source: Ref 23, p 102

Selected properties of flexible cast silicone

Property	ASTM test	Reported value
Mechanical properties		
Yield strength, MPa (ksi)	D638	2.1-6.9 (0.3-1.0)
Tensile modulus, GPa (10^6 psi)	D638	620 (90)
Elongation (break), %	D638	100-700
Compressive strength, MPa (ksi)	D695	0-6.9 (0-1)
Hardness, Rockwell	...	A15-60
Thermal properties		
Thermal conductivity, W/m·K (Btu·in./ft^2·h·°F)	C177	0.14-0.29 (1.0-2.0)
Coefficient of linear thermal expansion, 10^{-5}/°C (10^{-5}/°F)	D696	14-54 (8.0-30.0)
Density, g/cm^3	D792	0.99-1.50
Water absorption, % in 24 h	D570	0.1
Electrical properties		
Volume resistivity, Ω·cm	D257	10^{14}-10^{15}
Dielectric strength, V/10^{-3} mm (V/10^{-3} in.)	D149	22 (550)
Dissipation factor (50-100 Hz)	D150	0.001-0.025

Source: Ref 15, p 1-35

Vinyl esters

Typical vinyl ester

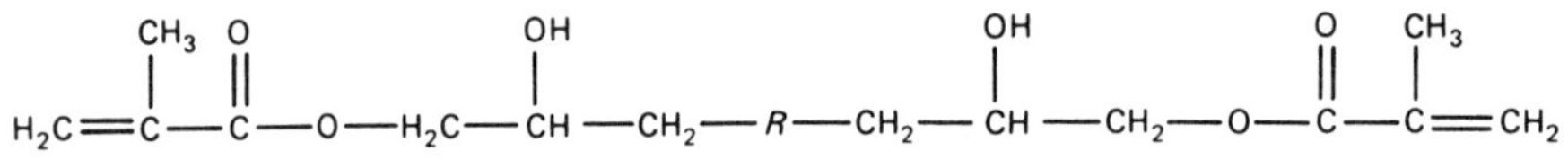

Source: Ref 6, p 137

Gel times of a vinyl ester resin with various initiators

Initiator	Gel time, min(a) at 25 °C (80 °F)	80 °C (175 °F)	120 °C (250 °F)
2% benzoylperoxide (BPO) (50%) + 0.1% dimethylaniline (DMA)	10	...	...
1.5% methy ethyl ketone peroxide (MEKP) (50%) + 3% Co-octoate (1%) + 0.1% DMA	14	...	...
1.5% MEKP + 3% Co-octoate (1%) + 0.015% DMA	34	...	...
2% BPO (50%) + 0.01% DMA	117	...	...
2% MEKP (50%)	700	6	2
2% MEKP (50%)	...	25	15
1% BPO (50%)	...	25	15
1% cumylhydroperoxide	...	32	10
1% *t*-butylperbenzoate	...	120	6
1% *t*-butylcumylperoxide	...	360	9

(a) 5g (140 oz.), isothermally.

Source: Ref 6, p 142

Typical vinyl ester formulation, by parts

(Percentages based on 100 parts vinyl ester resin)

Vinyl ester	100%
Promoter (6% cobalt naphthenate)	0.2-0.5%
Activator (100% dimethyl aniline)	0.0-0.2%
Catalyst (9% MEKP)	0.9-2.5%

Source: Ref 6, p 133

Gel time variation for Derakane 411-45 resin

Material: MEKP, wt%	Cobalt naphthanate, wt%	2,4 pentandione, wt%	Gel time, min
1.0	0.25	0.0	21
1.0	0.25	0.05	23
1.0	0.25	0.1	60
1.0	0.25	0.2	180
1.0	0.25	0.3	265

Source: Ref 6, p 137

Typical vinyl ester formulation, by weight

Component	Parts by weight
Vinyl ester resin	40
Styrene	30
Polyvinyl acetate	10
Calcium carbonate	150
AS P400 clay	17
Zinc stearate	4
1-butyl perbenzoate	0.5

Source: Ref 6, p 141

Delayed gel times for Derakane 411-45 resin

Material: MEKP, wt%	Material: Cobalt naphthenate, wt%	Material: 2,4 pentanedione, wt%	Typical gel time, min	Typical peak time, min	Typical exotherm °C	Typical exotherm °F
1.0	0.25	0.0	21	37	40	108
1.0	0.25	0.05	23	39	60	138
1.0	0.25	0.1	60	74	55	132
1.0	0.25	0.2	180	191	70	153
1.0	0.25	0.3	265	280	60	147

Source: Ref 6, p 133

Derakane vinyl ester resins

Product name	Type	Heat distortion temperature °C	Heat distortion temperature °F	Resin/styrene ratio	Applications
Derakane 411-45	Bisphenol A epoxy	100	215	55/45	General
Derakane 510N	Brominated bisphenol A epoxy	120	250	64/36	Flame retardant
Derakane 8084	Flexibilized bisphenol A epoxy	80	180	60/40	Toughened
Derakane 470-36	Epoxy novolac	140-150	290-300	64/36	High temperatures

Source: Ref 6, p 138

Selected properties of molded vinyls

Property	ASTM test	Reported value for: Vinyl PVA-PVC(a)	Vinyl PVC-PVC₂	Vinyl CI-PVC	Vinyl PVC-PVB
Mechanical properties					
Yield strength, MPa (ksi)	D638	41-52 (6.0-7.5)	21-34 (3.0-5.0)	52-62 (7.5-9.0)	3.4-21 (0.5-3.0)
Tensile modulus, GPa (10^6 psi)	D638	...	0.34-0.55 (0.05-0.08)	2.5-3.3 (0.36-0.48)	...
Elongation (break), %	D638	40-80	100-250	5-65	150-450
Compressive strength, MPa (ksi)	D695	55-90 (8-13)	14-21 (2-3)	62-150 (9-22)	...
Flexural yield strength, MPa (ksi)	D790	69-110 (10-16)	28-41 (4-6)	97-115 (14-17)	...
Flexural modulus, GPa (10^6 psi)	D790	...	...	2.8-31 (0.4-4.5)	...
Impact strength, Izod(a), J/cm (ft·lb.in.)	D256	0.21-11 (0.4-20)	0.16-0.53 (0.3-1.0)	0.53-3.0 (1.0-5.6)	...
Hardness	...	Rockwell D65-85	Rockwell M50-65	Rockwell 118-122	Shore A10-100
Thermal properties					
Thermal conductivity, W/m·K (Btu·in./ft^3·h·°F)	C177	0.14-0.22 (1.0-1.5)	0.07 (0.5)	0.13 (0.9)	...
Coefficient of linear thermal expansion 10^{-5}/°C (10^{-5}/°F)	D696	9.0-18.0 (5.0-10.0)	15 (8.5)	6.8 (3.8)	...
Specific heat, kJ/kg·K (Btu/lb·°F)	C351	0.8-1.7 (0.2-0.4)	1.3 (0.3)	1.3 (0.3)	...
Continuous service temperature, °C (°F)	...	65-75 (150-165)	75-100 (170-212)	110 (230)	...
Electrical properties					
Volume resistivity, Ω·cm	D257	10^{16}	10^{14}-10^{16}	10^{15}	5.0 x 10^{10}
Dielectric strength, V/10^{-3} mm (V/10^{-3} in.)	D149	14-20 (350-500)	16-24 (400-600)	47-59 (1200-1500)	14 (350)
Dielectric constant (50-100 Hz)	D150	3.2-4.0	4.5-6.0	3.1	5.6
Dissipation factor (50-100 Hz)	D150	0.007-0.020	0.030-0.045	0.020	0.115
Processing parameters					
Processing temperature, °C (°F)	...	150-215 (300-415)	120-205 (250-400)	175-225 (350-440)	140-170 (280-340)
Density, g/cm^3	D792	1.30-1.58	1.65-1.72	1.50-1.58	1.05
Specific volume, cm^3/kg (in.^{3}/lb)	D792	760-685 (21-19)	615-580 (17-16)	650 (18)	940 (26)
Linear mold shrinkage, mm/mm (in./in,)	D955	0.001-0.005	0.005-0.025	0.003-0.007	...
Water absorption, % in 24 h	D570	0.0-0.4	0.1	0.5	1.0-2.0

(a) Rigid material.
(b) Notched, at room temperature.

Source: Ref 15, p 1-37

Production and Machining Methods

Typical parts currently manufactured using RTM

Use	Part
Industrial	Solar collectors, electrostatic precipitator plates, fan blades, business-machine cabinetry, water tanks
Recreational	Canoe paddles, television antennae, snowmobiles
Construction	Seating, bathtubs, roof sections
Aerospace	Airplane wing ribs, cockpit hatch covers, airplane escape doors
Automobile	Crash members, leaf springs, car bodies, bus shelters

Source: Ref 6, p 168

Advantages and disadvantages of various RTM reinforcement forms

Reinforcement form	Advantages/disadvantages
Continuous strand mat	Good formability, wash resistance, high bulk factor, high part fill-out, uses glass fibers
Woven roving/fabric	High strength (biaxial), good formability
Unidirectional roving/fabric	High strength (unidirectional), high stiffness, good formability
Chopped-strand mat	Low formability, low wash resistance, low cost, high bulk factor, uses glass fibers
Preforms	Highly complex forms possible, little forming/handling necessary, high initial cost
Veils/surfacing mats	Good surface quality, wear resistance

Source: Ref 6, p 169

Breakdown of costs of RTM versus hand lay-up

Item	Resin transfer molding	Hand lay-up
Product weight, kg (lb)	30 (65)	33 (75)
Production rate, pieces/month	1000	1000
Direct laborers	14	30
Materials cost:		
Resin	28.4%	27.4%
Glass fibers	27.7%(a)	26.4%
Others	0.3%	2.7%
Subtotal	56.4%	56.5%
Depreciation cost:		
Mold	9.0%(b)	3.0%(c)
Equipment	1.8%	0.7%
Subtotal	10.8%	3.7%
Scrap	1.6%	0.0%
Manufacturing cost	14.4%	39.6%
Fuel cost	0.0%	0.2%
Subtotal	16.0%	39.8%
Total	83.2%	100.0%

(a) Including the costs of auxiliary materials, fuel, manufacturing, depreciation of preformer, and preforming screen. (b) Life of the mold is assumed to be 2000 pieces. (c) Life of the mold is assumed to be 200 pieces.

Source: Ref 6, p 170

Subjective assessment of thermoset (TS) and thermoplastic (TP) processing

Aspect	Advantage	Reasons
Prepreg formulaton:		
Viscosity	TS	Lower
Solvents	TS	Greater choice
Hand	TS	More flexible
Tack	TS	Prepolymer variable
Storage	TP	Not reactive
Quality control	TP	Fewer variables
Composite fabrication:		
Lay-up	TS	Ease of handling
Degassing	TP	Fewer volatiles
Temperature changes	TP	Fewer
Maximum temperature	TS	Lower
Pressure changes	TP	Fewer
Maximum pressure	TS	Lower
Cycle time	TP	Lower
Postcure cycle	TP	Not required
Repair	TP	Remelt
Post forming	TP	Remelt

Source: Ref 6, p 142

Pultrusion product characteristics

Size	Forming guide system and equipment pulling capacity influence size limitations.
Shape	Straight, constant cross sections; some curved sections possible
Length	No limit
Reinforcements	Fiberglass, aramid fiber, carbon fiber, and thermoplastic
Resin systems	Polyester, vinyl ester, and epoxy
Reinforcement, wt %	All roving, 40-80%; mat and roving, 25-50%; 55% woven roving or biaxial materials and mat, 40-70%
Mechanical strength	Medium to high, primarily unidirectional, approaching isotropic
Labor intensity	Low to medium
Mold cost	Low to medium
Production rate	Shape and thickness related

Source: Ref 6, p 542

Production methods for cellular polymers

Type of polymer	Extrusion	Expandable formulation	Spraying	Froth foaming	Compression molding	Injection molding	Sintering	Leaching
Cellulose acetate	X							
Epoxy resin		X	X	X				X
Phenolic resin		X						
Polyethylene	X	X			X	X	X	X
Polystyrene	X	X				X	X	
Silicones		X						
Urea-formaldehyde resin				X				
Urethane polymers		X	X	X		X		
Latex foam rubber				X				
Natural rubber	X	X			X			
Synthetic elastomers	X	X			X			
Polyvinyl chloride	X	X		X	X	X		X
Ebonite					X			
Polytetrafluroethylene							X	

Source: Ref 5, p 544

Design guidelines for use in RTM materials selection

General question	Specific areas of concern
Mold:	
How many parts are required from the process in a given time period?	Primarily, tooling and pumping/dispensing; secondarily, release, cleaners, resin
Is design life a consideration?	Tooling, pumping/dispensing, cleaners, release agents
What are the dimensional requirements?	Tooling, cleaners, release agents
What are the strength requirements?	Tooling, cleaners, release agents
What are the surface finish requirements?	Tooling, cleaning, release agents, resin system, reinforcement
Part:	
What are the performance requirements of the part?	Reinforcement, resin system
Production:	
What are the shop environmental requirements?	Resin system, cleaners, release agents, tooling, pumping/dispensing
What are the capabilities of personnel?	All
What are the cost objectives?	All

Source: Ref 6, p 168

Breakdown of costs of RTM versus other molding techiniques

Item	Resin transfer molding	Sheet molding compounds	Injection molding
Process operation:			
Production volume	5000-10,000/press	25,000/press	30,000/press
Fixed assets	Moderate	High	High
Labor	High	Moderate	Moderate
Skill dependency	Considerable	Very low	Lowest
Operation	Movements/intersections	Flowing, neat	Flowing, neat
Inspection/control:			
Raw materials	Yes	Compounds for degradation	Compounds for degradation
Products	Visual with attention	Visual, easy	Visual, easy
Finishing	Trim flash, and so on	Very little	Very little
Products:			
Complexity	Preform limit	Yes	Best
Size	Big parts for low investment	Big parts if flat	Not very big parts
Tolerence	Good	Very good	Very good
Surface appearance	Gel-coated	Very good	Very good
Voids/wrinkles	Occasional	Extremely rare	Least
Reproducibility	Skill dependent	Yes	Yes
Cores/inserts	Possible	Not easy	Possible
Strength	Moderate	Best	Very good
Material usage:			
Raw-material cost	Lowest	Highest	High
Handling/applying	Skill dependent	Easy	Automatic feed
Inventory	Raw materials	Dependent on number of types	Dependent on number of types
Precision	Skill dependent	Very good	Automatic feed
Waste	<3%	Very low	Attention runner
Scrap	Skill dependent	Cuts reusable	Low
Reinforcement flexibility	Yes	No	No
Mold:			
Initial cost	Moderate	Very high	Very high
Cycle life	3000-4000 parts	Years	Years
Handling	With care	Metal	Metal
Preparation	In-factory	Special shop	Special shop
Maintenance	In-factory	Special machine shop/equipment	Special machine shop/equipment

Source: Ref 6, p 170

Thread milling of plastics

Materials	Hardness, Rockwell	Condition(a)	Speed m/min	Speed fpm	Feed mm/tooth/rev	Feed in./tooth/rev	HSS tool material ISO	HSS tool material AISI
Thermoplastics								
Acrylic, acetal, polycarbonate, polysulfone, polystyrene	M60-120	C, M, or E	90	300	0.050	0.002	S4, S2	M2, M7
Acrylonitrile-butadiene-styrene (ABS), polyarylether, polypropylene, polyethylene, cellulose acetate	R50-120	C, M, or E	105	340	0.050	0.002	S4, S2	M2, M7
Fluorocarbons: tetrafluoroethylene (TFE), chlorotrifluoroethylene (CTFE)	R74-95	M or E	90	300	0.050	0.002	S4, S2	M2, M7
Polyamides (nylons): unfilled—types 6, 6/6, 6/12, 11, 12	R78-120	M or E	115	375	0.050	0.002	S4, S2	M2, M7
Thermosetting plastics								
Epoxy, melamine, phenolic	M100-128	C or M	115	375	0.050	0.002	S4, S2	M2, M7
Furan, polybutadiene	R40-100	C	58	190	0.038	0.0015	S9, S11	T15, M42
Silicone	Shore A15-65	C or M	46	150	0.038	0.0015	S9, S11	T15, M42
Polyimide	E40-50	M, or E	115	375	0.050	0.002	S4, S2	M2, M7
Polyurethane	Shore A65-95	C	58	190	0.038	0.0015	S9, S11	T15, M42
	Shore D55-75	C	69	225	0.050	0.002	S4, S2	M2, M7
Allyl (DAP)	M95-100	C	90	300	0.050	0.002	S4, S2	M2, M7
Allyl, fiber filled	M108-115	F and M	58	190	0.038	0.0015	S9, S11	T15, M42

(a) C, cast; M, molded; E, extruded; F, filled.

Source: Ref 21, p 2-267 to 2-269

Face milling of plastics

Materials	Hardness, Rockwell	Condition(a)	Depth of cut mm	Depth of cut in.	Speed m/min	Speed fpm	High speed steel tool: Feed tooth mm	Feed tooth in.	Tool material ISO	Tool material AISI
Thermoplastics										
Acrylic, acetal, polycarbonate, polysulfone, polystyrene	M60-120	C, M, or E	1	0.040	120	400	0.13	0.005	S4, S2	M2, M3
			4	0.150	105	350	0.20	0.008	S4, S2	M2, M3
			8	0.300	90	300	0.25	0.010	S4, S2	M2, M3
Acrylonitrile-butadiene-styrene (ABS), polyarylether, polypropylene, polyethylene, cellulose acetate	R50-120	C, M, or E	1	0.040	135	450	0.13	0.005	S4, S2	M2, M3
			4	0.150	120	400	0.20	0.008	S4, S2	M2, M3
			8	0.300	105	350	0.25	0.010	S4, S2	M2, M3
Fluorocarbons: tetrafluoroethylene (TFE), chlorotrifluoroethylene (CTFE)	R74-95	M or E	1	0.040	120	400	0.13	0.005	S4, S2	M2, M3
			4	0.150	105	350	0.20	0.008	S4, S2	M2, M3
			8	0.300	90	300	0.25	0.010	S4, S2	M2, M3
Polyamides (nylons): unfilled—types 6, 6/6, 6/12, 11, 12	R78-120	M or E	1	0.040	150	500	0.15	0.006	S4, S2	M2, M3
			4	0.150	135	450	0.25	0.010	S4, S2	M2, M3
			8	0.300	120	400	0.36	0.014	S4, S2	M2, M3
Polyamides (nylons): 35% glass reinforced—types 6, 6/6	R78-120	F and M	1	0.040	...	...	...	...	...	...
			4	0.150	...	...	...	...	...	...
			8	0.300	...	...	...	...	...	...
Polyamides (nylons): 35% glass reinforced—types 6/10, 6/12	E40-50	F and M	1	0.040	...	...	...	...	...	...
			4	0.150	...	...	...	...	...	...
			8	0.300	...	...	...	...	...	...
Thermosetting plastics										
Epoxy, melamine, phenolic	M100-128	C or M	1	0.040	150	500	0.13	0.005	S4, S2	M2, M7
			4	0.150	135	450	0.20	0.008	S4, S2	M2, M7
			8	0.300	105	350	0.25	0.010	S4, S2	M2, M7
Furan, polybutadiene	R40-100	C	1	0.040	76	250	0.13	0.005	S9, S11	T15, M42
			4	0.150	60	200	0.20	0.008	S9, S11	T15, M42
			8	0.300	46	150	0.25	0.010	S9, S11	T15, M42
Silcone	Shore A15-65	C or M	1	0.040	60	200	0.13	0.005	S9, S11	T15, M42
			4	0.150	53	175	0.20	0.008	S9, S11	T15, M42
			8	0.300	38	125	0.25	0.010	S9, S11	T15, M42
Silicone, glass filled	M80-90	F and M	1	0.040	...	...	...	...	...	...
			4	0.150	...	...	...	...	...	...
			8	0.300	...	...	...	...	...	...
Polyimide	E40-50	M or E	1	0.040	150	500	0.13	0.005	S4, S2	M2, M7
			4	0.150	135	450	0.20	0.008	S4, S2	M2, M7
			8	0.300	105	350	0.25	0.010	S4, S2	M2, M7

(continued)

Face milling of plastics (continued)

Materials	Hardness, Rockwell	Condition(a)	Depth of cut mm	Depth of cut in.	Speed m/min	Speed fpm	Feed tooth mm	Feed tooth in.	Tool material ISO	Tool material AISI
					High speed steel tool					
Polyimide, glass filled	M109-115	F and M	1	0.040	...	...	...	...	...	...
			4	0.150	...	...	...	...	...	...
			8	0.300	...	...	...	...	...	...
Polyurethane	Shore A65-95	C	1	0.040	76	250	0.13	0.005	S9, S11	T15, M42
			4	0.150	60	200	0.20	0.008	S9, S11	T15, M42
			8	0.300	46	150	0.25	0.010	S9, S11	T15, M42
	Shore D55-75	C	1	0.040	90	300	0.13	0.005	S4, S2	M2, M7
			4	0.150	76	250	0.20	0.008	S4, S2	M2, M7
			8	0.300	60	200	0.25	0.010	S4, S2	M2, M7
Allyl (DAP)	M95-100	C	1	0.040	120	400	0.13	0.005	S4, S2	M2, M3
			4	0.150	105	350	0.20	0.008	S4, S2	M2, M3
			8	0.300	90	300	0.25	0.010	S4, S2	M2, M3
Allyl, glass filled	E80-87	F and M	1	0.040	...	...	...	...	...	...
			4	0.150	...	...	...	...	...	...
			8	0.300	...	...	...	...	...	...
Allyl, fiber filled	M108-115	F and M	1	0.040	76	250	0.13	0.005	S9, S11	T15, M42
			4	0.150	60	200	0.20	0.008	S9, S11	T15, M42
			8	0.300	53	175	0.25	0.010	S9, S11	T15, M42

Materials	Hardness, Rockwell	Condition(a)	Depth of cut mm	Depth of cut in.	Speed, Brazed m/min	Speed, Brazed fpm	Speed, Indexable m/min	Speed, Indexable fpm	Feed per tooth mm	Feed per tooth in.	Tool material grade ISO	Tool material grade AISI
					Carbide tool (uncoated)							
Thermoplastics												
Acrylic, acetal, polycarbonate, polysulfone, polystyrene	M60-120	C, M, or E	1	0.040	200	650	200	650	0.13	0.005	K20, M20	C-2
			4	0.150	185	600	185	600	0.18	0.007	K20, M20	C-2
			8	0.300	170	550	170	550	0.23	0.009	K20, M20	C-2
Acrylonitrile-butadiene-styrene (ABS), polyarylether, polypropylene, polyethylene, cellulose acetate	R50-120	C, M, or E	1	0.040	230	750	230	750	0.102	0.004	K20, M20	C-2
			4	0.150	215	700	215	700	0.18	0.007	K20, M20	C-2
			8	0.300	200	650	200	650	0.23	0.009	K20, M20	C-2
Fluorocarbons: tetrafluoroethylene (TFE), chlorotrifluoroethylene (CTFE)	R74-95	M or E	1	0.040	200	650	200	650	0.102	0.004	K20, M20	C-2
			4	0.150	185	600	185	600	0.18	0.007	K20, M20	C-2
			8	0.300	170	550	170	550	0.23	0.009	K20, M20	C-2
Polyamides (nylons): unfilled—types 6, 6/6, 6/12, 11, 12	R78-120	M or E	1	0.040	260	850	260	850	0.15	0.006	K20, M20	C-2
			4	0.150	230	750	230	750	0.20	0.008	K20, M20	C-2
			8	0.300	215	700	215	700	0.25	0.010	K20, M20	C-2
Polyamides (nylons): 35% glass reinforced—types 6, 6/6	R78-120	F and M	1	0.040	200	650	200	650	0.102	0.004	K20, M20	C-2
			4	0.150	185	600	185	600	0.15	0.006	K20, M20	C-2
			8	0.300	170	550	170	550	0.20	0.008	K20, M20	C-2
Polyamides (nylons): 35% glass reinforced—types 6/10, 6/12	E40-50	F and M	1	0.040	170	550	170	550	0.102	0.004	K20, M20	C-2
			4	0.150	150	500	150	500	0.15	0.006	K20, M20	C-2
			8	0.300	135	450	135	450	0.20	0.008	K20, M20	C-2

(continued)

Face milling of plastics (continued)

Materials	Hardness, Rockwell	Condition(a)	Depth of cut mm	Depth of cut in.	Speed, Brazed m/min	Speed, Brazed fpm	Speed, Indexable m/min	Speed, Indexable fpm	Carbide tool (uncoated): Feed per tooth mm	Feed per tooth in.	Tool material grade ISO	Tool material grade AISI
Thermosetting plastics												
Epoxy, melamine, phenolic	M100-128	C or M	1	0.040	260	850	260	850	0.102	0.004	K20, M20	C-2
			4	0.150	230	750	230	750	0.15	0.006	K20, M20	C-2
			8	0.300	215	700	215	700	0.20	0.008	K20, M20	C-2
Furan, polybutadiene	R40-100	C	1	0.040	145	475	145	475	0.102	0.004	K20, M20	C-2
			4	0.150	130	420	130	425	0.15	0.006	K20, M20	C-2
			8	0.300	115	375	115	375	0.20	0.008	K20, M20	C-2
Silcone	Shore A15-65	C or M	1	0.040	145	475	145	475	0.102	0.004	K20, M20	C-2
			4	0.150	130	425	130	425	0.15	0.006	K20, M20	C-2
			8	0.300	115	375	115	375	0.20	0.008	K20, M20	C-2
Silicone, glass filled	M80-90	F and M	1	0.040	130	425	130	425	0.102	0.004	K20, M20	C-2
			4	0.150	115	375	115	375	0.15	0.006	K20, M20	C-2
			8	0.300	100	325	100	325	0.20	0.008	K20, M20	C-2
Polyimide	E40-50	M or E	1	0.040	260	850	260	850	0.13	0.005	K20, M20	C-2
			4	0.150	230	750	230	750	0.20	0.008	K20, M20	C-2
			8	0.300	200	650	200	650	0.25	0.010	K20, M20	C-2
Polyimide, glass filled	M109-115	F and M	1	0.040	160	525	160	525	0.13	0.005	K20, M20	C-2
			4	0.150	145	475	145	475	0.20	0.008	K20, M20	C-2
			8	0.300	130	425	130	425	0.25	0.010	K20, M20	C-2
Polyurethane	Shore A65-95	C	1	0.040	145	475	145	475	0.102	0.004	K20, M20	C-2
			4	0.150	130	425	130	425	0.15	0.006	K20, M20	C-2
			8	0.300	115	375	115	375	0.20	0.008	K20, M20	C-2
	Shore D55-75	C	1	0.040	160	525	160	525	0.102	0.004	K20, M20	C-2
			4	0.150	145	475	145	475	0.15	0.006	K20, M20	C-2
			8	0.300	130	425	130	425	0.20	0.008	K20, M20	C-2
Allyl (DAP)	M95-100	C	1	0.040	200	650	200	650	0.15	0.006	K20, M20	C-2
			4	0.150	185	600	185	600	0.20	0.008	K20, M20	C-2
			8	0.300	170	550	170	550	0.25	0.010	K20, M20	C-2
Allyl, glass filled	E80-87	F and M	1	0.040	130	425	130	425	0.15	0.006	K20, M20	C-2
			4	0.150	115	375	115	375	0.20	0.008	K20, M20	C-2
			8	0.300	100	325	100	325	0.25	0.010	K20, M20	C-2
Allyl, fiber filled	M108-115	F and M	1	0.040	160	525	160	525	0.15	0.006	K20, M20	C-2
			4	0.150	145	475	145	475	0.20	0.008	K20, M20	C-2
			8	0.300	130	425	130	425	0.25	0.010	K20, M20	C-2

(a) C, cast; M, molded; E, extruded; F, filled.

Source: Ref 21, p 2-46 to 2-48

Side and slot milling of plastics (with arbor-mounted cutters)

Materials	Hardness, Rockwell	Condition(a)	Depth of cut mm	Depth of cut in.	Speed m/min	Speed fpm	High speed steel tool: Feed tooth mm	Feed tooth in.	Tool material ISO	Tool material AISI
Thermoplastics										
Acrylic, acetal, polycarbonate, polysulfone, polystyrene	M60-120	C, M, or E	1	0.040	90	300	0.102	0.004	S4, S2	M2, M3
			4	0.150	79	260	0.13	0.005	S2, S5	M2, M3
			8	0.300	67	220	0.18	0.007	S4, S2	M2, M3
Acrylonitrile-butadiene-styrene (ABS), polyarylether, polypropylene, polyethylene, cellulose acetate	R50-120	C, M, or E	1	0.040	105	340	0.102	0.004	S4, S2	M2, M3
			4	0.150	90	300	0.13	0.005	S4, S2	M2, M3
			8	0.300	79	260	0.18	0.007	S4, S2	M2, M3
Fluorocarbons: tetrafluoroethylene (TFE), chlorotrifluoroethylene (CTFE)	R74-95	M or E	1	0.040	90	300	0.102	0.004	S4, S2	M2, M3
			4	0.150	79	260	0.13	0.005	S4, S2	M2, M3
			8	0.300	67	220	0.18	0.007	S4, S2	M2, M3
Polyamides (nylons): unfilled—types 6, 6/6, 6/12, 11, 12	R78-120	M or E	1	0.040	115	380	0.102	0.004	S4, S2	M2, M3
			4	0.130	105	340	0.15	0.006	S4, S2	M2, M3
			8	0.300	90	300	0.20	0.008	S4, S2	M2, M3
Polyamides (nylons): 35% glass reinforced—types 6, 6/6	R78-120	F and M	1	0.040	...	...	...	...	...	...
			4	0.150	...	...	...	...	...	...
			8	0.300	...	...	...	...	...	...
Polyamides (nylons): 35% glass reinforced—types 6/10, 6/12	E40-50	F and M	1	0.040	...	...	...	...	...	...
			4	0.150	...	...	...	...	...	...
			8	0.300	...	...	...	...	...	...
Thermosetting plastics										
Epoxy, melamine, phenolic	M100-128	C or M	1	0.040	115	380	0.102	0.004	S4, S2	M2, M7
			4	0.150	105	340	0.13	0.005	S4, S2	M2, M7
			8	0.300	79	260	0.18	0.007	S4, S2	M2, M7
Furan, polybutadiene	R40-100	C	1	0.040	60	200	0.102	0.004	S9, S11	T15, M42
			4	0.150	46	150	0.13	0.005	S9, S11	T15, M42
			8	0.300	35	115	0.18	0.007	S9, S11	T15, M42
Silcone	Shore A15-65	C or M	1	0.040	46	150	0.102	0.004	S9, S11	T15, M42
			4	0.150	41	135	0.13	0.005	S9, S11	T15, M42
			8	0.300	29	95	0.18	0.007	S9, S11	T15, M42
Silicone, glass filled	M80-90	F and M	1	0.040	...	...	...	...	...	...
			4	0.150	...	...	...	...	...	...
			8	0.300	...	...	...	...	...	...
Polyimide	E40-50	M or E	1	0.040	115	380	0.102	0.004	S4, S2	M2, M7
			4	0.150	105	340	0.13	0.005	S4, S2	M2, M7
			8	0.300	79	260	0.18	0.007	S4, S2	M2, M7

(continued)

Side and slot milling of plastics (with arbor-mounted cutters) (continued)

Materials	Hardness, Rockwell	Condition(a)	Depth of cut mm	Depth of cut in.	Speed m/min	Speed fpm	High speed steel tool: Feed tooth mm	Feed tooth in.	Tool material ISO	Tool material AISI
Polyimide, glass filled	M109-115	F and M	1	0.040	...	...	...	...	...	...
			4	0.150	...	...	...	...	...	...
			8	0.300	...	...	...	...	...	...
Polyurethane	Shore A65-95	C	1	0.040	60	200	0.102	0.004	S9, S11	T15, M42
			4	0.150	46	150	0.13	0.004	S9, S11	T15, M42
			8	0.300	35	115	0.18	0.007	S9, S11	T15, M42
	Shore D55-75	C	1	0.040	67	220	0.102	0.004	S4, S2	M2, M7
			4	0.150	60	200	0.13	0.005	S4, S2	M2, M7
			8	0.300	46	150	0.18	0.007	S4, S2	M2, M7
Allyl (DAP)	M95-100	C	1	0.040	90	300	0.102	0.004	S4, S2	M2, M3
			4	0.150	79	260	0.13	0.005	S4, S2	M2, M3
			8	0.300	67	220	0.18	0.007	S4, S3	M2, M3
Allyl, glass filled	E80-87	F and M	1	0.040	...	...	...	...	...	...
			4	0.150	...	...	...	...	...	...
			8	0.300	...	...	...	...	...	...
Allyl, fiber filled	M108-115	F and M	1	0.040	60	200	0.102	0.004	S9, S11	T15, M42
			4	0.150	46	150	0.13	0.005	S9, S11	T15, M42
			8	0.300	41	135	0.18	0.007	S9, S11	T15, M42

Side and slot milling of plastics (with arbor-mounted cutters) (continued)

Materials	Hardness, Rockwell	Condition(a)	Depth of cut mm	Depth of cut in.	Carbide tool (uncoated): Speed, Brazed m/min	Brazed fpm	Indexable m/min	Indexable fpm	Feed per tooth mm	Feed per tooth in.	Tool material grade ISO	Tool material grade AISI
Thermoplastics												
Acrylic, acetal, polycarbonate, polysulfone, polystyrene	M60-120	C, M, or E	1	0.040	145	475	145	475	0.102	0.004	K20, M20	C-2
			4	0.150	135	450	135	450	0.13	0.005	K20, M20	C-2
			8	0.300	120	400	120	400	0.15	0.006	K20, M20	C-2
Acrylonitrile-butadiene-styrene (ABS), polyarylether, polypropylene, polyethylene, cellulose acetate	R50-120	C, M, or E	1	0.040	170	550	170	550	0.102	0.004	K20, M20	C-2
			4	0.150	160	525	160	525	0.13	0.005	K20, M20	C-2
			8	0.300	145	475	145	475	0.15	0.006	K20, M20	C-2
Fluorocarbons: tetrafluoroethylene (TFE), chlorotrifluoroethylene (CTFE)	R74-95	M or E	1	0.040	145	475	145	475	0.075	0.003	K20, M20	C-2
			4	0.150	135	450	135	450	0.102	0.004	K20, M20	C-2
			8	0.300	120	400	120	400	0.13	0.005	K20, M20	C-2
Polyamides (nylons): unfilled—types 6, 6/6, 6/12, 11, 12	R78-120	M or E	1	0.040	200	650	200	650	0.102	0.004	K20, M20	C-2
			4	0.150	170	550	170	550	0.13	0.005	K20, M20	C-2
			8	0.300	160	525	160	525	0.18	0.007	K20, M20	C-2
Polyamides (nylons): 35% glass reinforced—types 6, 6/6	R78-120	F and M	1	0.040	145	475	145	475	0.075	0.003	K20, M20	C-2
			4	0.150	135	450	135	450	0.102	0.004	K20, M20	C-2
			8	0.300	120	400	120	400	0.13	0.005	K20, M20	C-2
Polyamides (nylons): 35% glass reinforced—types 6/10, 6/12	E40-50	F and M	1	0.040	120	400	120	400	0.075	0.003	K20, M20	C-2
			4	0.150	115	380	115	380	0.102	0.004	K20, M20	C-2
			8	0.300	105	340	105	340	0.13	0.005	K20, M20	C-2

Side and slot milling of plastics (with arbor-mounted cutters) (continued)

Materials	Hardness, Rockwell	Condition(a)	Depth of cut mm	Depth of cut in.	Speed, Brazed m/min	Speed, Brazed fpm	Speed, Indexable m/min	Speed, Indexable fpm	Feed per tooth mm	Feed per tooth in.	Tool material grade ISO	Tool material grade AISI
									Carbide tool (uncoated)			
Thermosetting plastics												
Epoxy, melamine, phenolic	M100-128	C or M	1	0.040	200	650	200	650	0.075	0.003	K20, M20	C-2
			4	0.150	170	550	170	550	0.102	0.004	K20, M20	C-2
			8	0.300	160	525	160	525	0.13	0.005	K20, M20	C-2
Furan, polybutadiene	R40-100	C	1	0.040	110	360	110	360	0.075	0.003	K20, M20	C-2
			4	0.150	100	320	100	320	0.102	0.004	K20, M20	C-2
			8	0.300	85	280	85	280	0.13	0.005	K20, M20	C-2
Silcone	Shore A15-65	C or M	1	0.040	110	360	110	360	0.075	0.003	K20, M20	C-2
			4	0.150	100	320	100	320	0.102	0.004	K20, M20	C-2
			8	0.300	85	280	85	280	0.13	0.005	K20, M20	C-2
Silicone, glass filled	M80-90	F and M	1	0.040	100	320	100	320	0.075	0.003	K20, M20	C-2
			4	0.150	85	280	85	280	0.102	0.004	K20, M20	C-2
			8	0.300	73	240	73	240	0.13	0.005	K20, M20	C-2
Polyimide	E40-50	M or E	1	0.040	200	650	200	650	0.102	0.004	K20, M20	C-2
			4	0.150	170	550	170	550	0.13	0.005	K20, M20	C-2
			8	0.300	145	475	145	475	0.18	0.007	K20, M20	C-2
Polyimide, glass filled	M109-115	F and M	1	0.040	120	390	120	390	0.102	0.004	K20, M20	C-2
			4	0.150	110	360	110	360	0.13	0.005	K20, M20	C-2
			8	0.300	100	320	100	320	0.18	0.007	K20, M20	C-2
Polyurethane	Shore A65-95	C	1	0.040	110	360	110	360	0.075	0.003	K20, M20	C-2
			4	0.150	100	320	100	320	0.102	0.004	K20, M20	C-2
			8	0.300	85	280	85	280	0.13	0.005	K20, M20	C-2
	Shore D55-75	C	1	0.040	120	390	120	390	0.075	0.003	K20, M20	C-2
			4	0.150	110	360	110	360	0.102	0.004	K20, M20	C-2
			8	0.300	100	320	100	320	0.13	0.005	K20, M20	C-2
Allyl (DAP)	M95-100	C	1	0.040	145	475	145	475	0.102	0.004	K20, M20	C-2
			4	0.150	135	450	135	450	0.13	0.005	K20, M20	C-2
			8	0.300	120	400	120	400	0.18	0.007	K20, M20	C-2
Allyl, glass filled	E80-87	F and M	1	0.040	100	320	100	320	0.102	0.004	K20, M20	C-2
			4	0.150	85	280	85	280	0.13	0.005	K20, M20	C-2
			8	0.300	73	240	73	240	0.18	0.007	K20, M20	C-2
Allyl, fiber filled	M108-115	F and M	1	0.040	120	390	120	390	0.102	0.004	K20, M20	C-2
			4	0.150	110	360	110	360	0.13	0.005	K20, M20	C-2
			8	0.300	100	320	100	320	0.18	0.007	K20, M20	C-2

(a) C, cast; M, molded; E, extruded; F, filled.

Source: Ref 21, p 2-123 to 2-125

Drilling of plastics

Materials	Hardness, Rockwell	Condition(a)	Speed m/min	Speed fpm	Feed, for reamer diameter of: 1.5 mm (1/16) in.) mm/rev	1.5 mm ipr	3 mm (1/8 in.) mm/rev	3 mm ipr	6 mm (1/4 in.) mm/rev	6 mm ipr	12 mm (1/2 in.) mm/rev	12 mm ipr
Thermoplastics												
Acrylic, acetal, polycarbonate, polysulfone, polystyrene	M60-120	C, M, or E	30	100	0.025	0.001						
			60	200	...	...	0.050	0.002	0.102	0.004	0.13	0.005
Acrylonitrile-butadiene-styrene (ABS), polyarylether, polypropylene, polyethylene, cellulose acetate	R50-120	C, M, or E	46	150	0.025	0.001						
			76	250	...	...	0.050	0.002	0.102	0.004	0.13	0.005
Fluorocarbons: tetrafluoroethylene (TFE), chlorotrifluoroethylene (CTFE)	R74-95	M or E	30	100	0.025	0.001						
			60	200	...	...	0.050	0.002	0.102	0.004	0.13	0.005
Polyamides (nylons): unfilled—types 6, 6/6, 6/12, 11, 12	R78-120	M or E	46	100	0.025	0.001						
			76	250	...	...	0.050	0.002	0.102	0.004	0.13	0.005
Polyamides (nylons): 35% glass reinforced—types 6, 6/6	R78-120	F and M	23	75	0.025	0.001						
			46	100	...	...	0.050	0.002	0.102	0.004	0.13	0.005
Polyamides (nylons): 35% glass reinforced—types 6/10, 6/12	E40-50	F and M	20	65	0.025	0.001						
			38	125	...	...	0.050	0.002	0.102	0.004	0.13	0.005
Thermosetting plastics												
Epoxy, melamine, phenolic	M100-128	C or M	30	100	0.025	0.001						
			60	200	...	...	0.050	0.002	0.075	0.003	0.102	0.004
Furan, polybutadiene	R40-100	C	30	100	0.025	0.001						
			60	200	...	...	0.050	0.002	0.075	0.003	0.102	0.004
Silicone	Shore A15-65	C or M	20	65	0.025	0.001						
			38	125	...	...	0.050	0.002	0.075	0.003	0.102	0.004
Silicone, glass filled	M80-90	F and M	20	65	0.025	0.001						
			38	125	...	..	0.025	0.001	0.050	0.002	0.075	0.003
Polyimide	E40-50	M or E	30	100	0.025	0.001						
			60	200	...	...	0.050	0.002	0.102	0.004	0.13	0.005
Polymide, glass filled	M109-115	F and M	20	65	0.025	0.001						
			38	125	...	...	0.025	0.001	0.102	0.002	0.075	0.003
Polyurethane	Shore A65-95	C	30	100	0.025	0.001						
			60	200	...	...	0.050	0.002	0.075	0.003	0.102	0.004
	Shore D55-75	C	46	150	0.025	0.001						
			76	250	...	...	0.050	0.002	0.102	0.004	0.13	0.005
Allyl (DAP)	M95-100	C	30	100	0.025	0.001						
			60	200	...	...	0.050	0.002	0.102	0.004	0.13	0.005
Allyl, glass filled	E80-87	F and M	20	65	0.025	0.001						
			30	100	...	...	0.050	0.002	0.102	0.004	0.13	0.005
Allyl, fiber filled	M108-115	F and M	23	75	0.025	0.001						
			46	150	...	...	0.050	0.002	0.102	0.004	0.15	0.006

(continued)

Drilling of plastics (continued)

Materials	Hardness, Rockwell	Condition(a)	Feed, for reamer diameter of: 18 mm (3/4 in.) m/rev	ipr	25 mm (1 in.) mm/rev	ipr	35 mm (1-1/2 in.) mm/rev	ipr	50 mm (2 in.) mm/rev	ipr	Tool material grade ISO	AISI or C
Thermoplastics												
Acrylic, acetal, polycarbonate, polysulfone, polystyrene	M60-120	C, M, or E	0.15	0.006	0.20	0.008	0.25	0.010	0.30	0.012	S2, S3	M10, M7, M1
Acrylonitrile-butadiene-styrene (ABS), polyarylether, polypropylene, polyethylene, cellulose acetate	R50-120	C, M, or E	0.15	0.006	0.20	0.008	0.25	0.010	0.30	0.012	S2, S3	M10, M7, M1
Fluorocarbons: tetrafluoroethylene (TFE), chlorotrifluoroethylene (CTFE)	R74-95	M or E	0.15	0.006	0.20	0.008	0.25	0.010	0.30	0.012	S2, S3	M10, M7, M1
Polyamides (nylons): unfilled—types 6, 6/6, 6/12, 11, 12	R78-120	M or E	0.15	0.006	0.20	0.008	0.25	0.010	0.30	0.012	S2, S3	M10, M7, M1
Polyamides (nylons): 35% glass reinforced-types 6, 6/6	R78-120	F and M	0.15	0.006	0.20	0.008	0.25	0.010	0.30	0.012	S9, S11	T15, M42
Polyamides (nylons): 35% glass reinforced-types 6/10, 6/12	E40-50	F and M	0.15	0.006	0.20	0.008	0.25	0.010	0.30	0.012	S9, S11	T15, M42
Thermosetting plastics												
Epoxy, melamine, phenolic	M100-128	C or M	0.13	0.005	0.15	0.006	0.20	0.008	0.25	0.010	S2, S3	M10, M7, M1
Furan, polybutadiene	R40-100	C	0.13	0.005	0.15	0.006	0.20	0.008	0.25	0.010	S2, S3	M10, M7, M1
Silicone	Shore A15-65	C or M	0.15	0.006	0.20	0.008	0.20	0.008	0.25	0.010	S2, S3	M10, M7, M1
Silicone, glass filled	M80-90	F and M	0.102	0.004	0.15	0.006	0.20	0.008	0.25	0.010	S9, S11	T15, M42
Polyimide	E40-50	M or E	0.15	0.006	0.20	0.008	0.25	0.010	0.30	0.012	S2, S3	M10, M7, M1
Polymide, glass filled	M109-115	F and M	0.102	0.004	0.15	0.006	0.20	0.008	0.25	0.010	S9, S11	T15, M42
Polyurethane	Shore A65-95	C	0.13	0.005	0.15	0.006	0.20	0.008	0.25	0.010	S2, S3	M10, M7, M1
	Shore D55-75	C	0.15	0.006	0.20	0.008	0.25	0.010	0.30	0.012	S2, S3	M10, M7, M1
Allyl (DAP)	M95-100	C	0.15	0.006	0.20	0.008	0.25	0.010	0.30	0.012	S2, S3	M10, M7, M1
Allyl, glass filled	E80-87	F and M	0.15	0.006	0.20	0.008	0.25	0.010	0.30	0.012	S9, S11	T15, M42
Allyl, fiber filled	M108-115	F and M	0.20	0.008	0.25	0.010	0.30	0.012	0.40	0.015	S9, S11	T15, M42

(a) C, cast; M, molded; E, extended; F, filled.

Source: Ref 21, p 3-48 to 3-50

Reaming of plastics (roughing or finishing)

Materials	Hardness, Rockwell	Condition(a)	Speed m/min	Speed fpm	Feed, 3 mm (1/8 in.) mm/rev	Feed, 3 mm (1/8 in.) ipr	Feed, 6 mm (1/4 in.) mm/rev	Feed, 6 mm (1/4 in.) ipr	Feed, 12 mm (1/2 in.) mm/rev	Feed, 12 mm (1/2 in.) ipr
Thermoplastics										
Acrylic, acetal, polycarbonate, polysulfone, polystyrene	M60-120	C, M, or E	20 40	65 130	0.050	0.002	0.050	0.002	0.102	0.004
Acrylonitrile-butadiene-styrene (ABS), polyarylether, polypropylene, polyethylene, cellulose acetate	R50-120	C, M, or E	30 50	100 165	0.050	0.002	0.050	0.002	0.102	0.004
Fluorocarbons: tetrafluoroethylene (TFE), chlorotrifluoroethylene (CTFE)	R74-95	M or E	20 40	65 130	0.050	0.002	0.050	0.002	0.102	0.004
Polyamides (nylons): unfilled—types 6, 6/6, 6/12, 11, 12	R78-120	M or E	20 50	65 165	0.050	0.002	0.050	0.002	0.102	0.004
Polyamides (nylons): 35% glass reinforced—types 6, 6/6	R78-120	F and M	15 30	50 100	0.050	0.002	0.050	0.002	0.075	0.003
Polyamides (nylons): 35% glass reinforced—types 6/10, 6/12	E40-50	F and M	14 26	45 85	0.050	0.002	0.050	0.002	0.075	0.003
Thermosetting plastics										
Epoxy, melamine, phenolic	M100-128	C or M	20 40	65 130	0.050	0.002	0.050	0.002	0.102	0.004
Furan, polybutadiene	R40-100	C	20 40	65 130	0.050	0.002	0.050	0.002	0.102	0.004
Silicone	Shore A15-65	C or M	14 26	45 85	0.050	0.002	0.050	0.002	0.102	0.004
Silicone, glass filled	M80-90	F and M	14 26	45 85	0.050	0.002	0.050	0.002	0.075	0.003
Polyimide	E40-50	M or E	20 40	65 130	0.050	0.002	0.050	0.002	0.102	0.004
Polymide, glass filled	M109-115	F and M	14 26	45 85	0.050	0.002	0.050	0.002	0.075	0.003
Polyurethane	Shore A65-95	C	20 40	65 130	0.050	0.002	0.050	0.002	0.075	0.003
	Shore D55-75	C	30 50	100 165	0.050	0.002	0.050	0.002	0.102	0.004
Allyl (DAP)	M95-100	C	20 40	65 130	0.050	0.002	0.050	0.002	0.102	0.004
Allyl, glass filled	E80-87	F and M	14 26	45 85	0.050	0.002	0.050	0.002	0.075	0.003
Allyl, fiber filled	M108-115	F and M	14 26	45 85	0.050	0.002	0.050	0.002	0.075	0.003

(continued)

Reaming of plastics (roughing or finishing) (continued)

Materials	Hardness, Rockwell	Condition(a)	Feed, for reamer diameter of: 25 mm (1 in.) mm/rev	ipr	35 mm (1-1/2 in.) mm/rev	ipr	50 mm (2 in.) mm/rev	ipr	Tool material grade ISO	AISI or C
Thermoplastics										
Acrylic, acetal, polycarbonate, polysulfone, polystyrene	M60-120	C, M, or E	0.15	0.006	0.20	0.008	0.25	0.010	S3, S4, S2 K10	M1, M2, M7 C-2
Acrylonitrile-butadiene-styrene (ABS), polyarylether, polypropylene, polyethylene, cellulose acetate	R50-120	C, M, or E	0.15	0.006	0.20	0.008	0.25	0.010	S3, S4, S2 K10	M1, M2, M7 C-2
Fluorocarbons: tetrafluoroethylene (TFE), chlorotrifluoroethylene (CTFE)	R74-95	M or E	0.15	0.006	0.20	0.008	0.25	0.010	S3, S4, S2 K10	M1, M2, M7 C-2
Polyamides (nylons): unfilled—types 6, 6/6, 6/12, 11, 12	R78-120	M or E	0.15	0.006	0.20	0.008	0.25	0.010	S3, S4, S2 K10	M1, M2, M7 C-2
Polyamides (nylons): 35% glass reinforced-types 6, 6/6	R78-120	F and M	0.075	0.003	0.102	0.004	0.13	0.005	S9, S11 K10	T15, M42 C-2
Polyamides (nylons): 35% glass reinforced-types 6/10, 6/12	E40-50	F and M	0.075	0.003	0.102	0.004	0.13	0.005	S9, S11 K10	T15, M42 C-2
Thermosetting plastics										
Epoxy, melamine, phenolic	M100-128	C or M	0.15	0.006	0.20	0.008	0.25	0.010	S3, S4, S2 K10	M1, M2, M7 C-2
Furan, polybutadiene	R40-100	C	0.15	0.006	0.20	0.008	0.25	0.010	S3, S4, S2 K10	M1, M2, M7 C-2
Silicone	Shore A15-65	C or M	0.15	0.006	0.20	0.008	0.25	0.010	S9, S11 K10	T15, M42 C-2
Silicone, glass filled	M80-90	F and M	0.075	0.003	0.102	0.004	0.13	0.005	S9, S11 K10	T15, M42 C-2
Polyimide	E40-50	M or E	0.15	0.006	0.20	0.008	0.25	0.010	S9, S11 K10	M1, M2, M7 C-2
Polymide, glass filled	M109-115	F and M	0.075	0.003	0.102	0.004	0.13	0.005	S9, S11 K10	T15, M42 C-2
Polyurethane	Shore A65-95	C	0.075	0.003	0.102	0.004	0.13	0.005	S3, S4, S2 K10	M1, M2, M7 C-2
	Shore D55-75	C	0.15	0.006	0.20	0.008	0.25	0.010	S3, S4, S2 K10	M1, M2, M7 C-2
Allyl (DAP)	M95-100	C	0.15	0.006	0.20	0.008	0.25	0.010	S3, S4, S2 K10	M1, M2, M7 C-2
Allyl, glass filled	E80-87	F and M	0.075	0.003	0.102	0.004	0.13	0.005	S9, S11 K10	T15, M42 C-2
Allyl, fiber filled	M108-115	F and M	0.075	0.003	0.102	0.004	0.13	0.005	S9, S11 K10	T15, M42 C-2

(a) C, cast; M, molded; E, extended; F, filled.

Source: Ref 21, p 3-218 to 3-219

Boring of plastics

Materials	Hardness, Rockwell	Condition(a)	Depth of cut mm	Depth of cut in.	Speed m/min	Speed fpm	High speed steel tool: Feed tooth mm	Feed tooth in.	Tool material ISO	Tool material AISI
Thermoplastics										
Acrylic, acetal, polycarbonate, polysulfone, polystyrene	M60-120	C, M, or E	0.25	0.010	120	400	0.050	0.002	S4, S5	M2, M3
			1.25	0.050	100	325	0.102	0.004	S4, S5	M2, M3
			2.5	0.100	84	275	0.15	0.006	S4, S5	M2, M3
Acrylonitrile-butadiene-styrene (ABS), polyarylether, polypropylene, polyethylene, cellulose acetate	R50-120	C, M, or E	0.25	0.010	135	435	0.050	0.002	S4, S5	M2, M3
			1.25	0.050	105	350	0.102	0.004	S4, S5	M2, M3
			2.5	0.100	100	325	0.15	0.006	S4, S5	M2, M3
Fluorocarbons: tetrafluoroethylene (TFE), chlorotrifluoroethylene (CTFE)	R74-95	M or E	0.25	0.010	125	405	0.050	0.002	S4, S5	M2, M3
			1.25	0.050	100	325	0.102	0.004	S4, S5	M2, M3
			2.5	0.100	84	275	0.15	0.006	S4, S5	M2, M3
Polyamides (nylons): unfilled—types 6, 6/6, 6/12, 11, 12	R78-120	M or E	0.25	0.010	150	500	0.050	0.002	S4, S5	M2, M3
			1.25	0.050	120	400	0.102	0.004	S4, S5	M2, M3
			2.5	0.100	105	350	0.20	0.008	S4, S5	M2, M3
Polyamides (nylons): 35% glass reinforced—types 6, 6/6	R78-120	F and M	0.25	0.010	...	...	...	...	...	...
			1.25	0.050	...	...	...	...	...	...
			2.5	0.100	...	...	...	...	...	...
Polyamides (nylons): 35% glass reinforced—types 6/10, 6/12	E40-50	F and M	0.25	0.010	...	...	...	...	...	...
			1.25	0.050	...	...	...	...	...	...
			2.5	0.100	...	...	...	...	...	...
Thermosetting plastics										
Epoxy, melamine, phenolic	M100-128	C or M	0.25	0.010	150	500	0.050	0.002	S4, S5	M2, M3
			1.25	0.050	120	400	0.102	0.004	S4, S5	M2, M3
			2.5	0.100	110	360	0.20	0.008	S4, S5	M2, M3
Furan, polybutadiene	R40-100	C	0.25	0.010	76	250	0.050	0.002	S9, S11	T15, M42
			1.25	0.050	60	200	0.102	0.004	S9, S11	T15, M42
			2.5	0.100	49	160	0.20	0.008	S9, S11	T15, M42
Silcone	Shore A15-65	C or M	0.25	0.010	60	200	0.050	0.002	S9, S11	T15, M42
			1.25	0.050	49	160	0.102	0.004	S9, S11	T15, M42
			2.5	0.100	43	140	0.20	0.008	S9, S11	T15, M42
Silicone, glass filled	M80-90	F and M	0.25	0.010	...	...	...	...	...	...
			1.25	0.050	...	...	...	...	...	...
			2.5	0.100	...	...	...	...	...	...
Polyimide	E40-50	M or E	0.25	0.010	150	500	0.050	0.002	S4, S5	M2, M3
			1.25	0.050	120	400	0.102	0.004	S4, S5	M2, M3
			2.5	0.100	110	360	0.20	0.008	S4, S5	M2, M3

(continued)

Boring of plastics (continued)

Materials	Hardness, Rockwell	Condition(a)	Depth of cut mm	Depth of cut in.	High speed steel tool: Speed m/min	Speed fpm	Feed tooth mm	Feed tooth in.	Tool material ISO	Tool material AISI
Polyimide, glass filled	M109-115	F and M	0.25	0.010	...	...	...	...	...	...
			1.25	0.050	...	...	...	...	...	...
			2.5	0.100	...	...	...	...	...	...
Polyurethane	Shore A65-95	C	0.25	0.020	76	250	0.050	0.002	S9, S11	T15, M42
			1.25	0.050	60	200	0.102	0.004	S9, S11	T15, M42
			2.5	0.100	49	160	0.20	0.008	S9, S11	T15, M42
	Shore D55-75	C	0.25	0.010	90	300	0.050	0.002	S4, S5	M2, M3
			1.25	0.050	73	240	0.102	0.004	S4, S5	M2, M3
			2.5	0.100	60	200	0.20	0.008	S4, S5	M2, M3
Allyl (DAP)	M95-100	C	0.25	0.010	120	400	0.050	0.002	S4, S5	M2, M3
			1.25	0.050	100	325	0.102	0.004	S4, S5	M2, M3
			2.5	0.300	84	275	0.15	0.006	S4, S5	M2, M3
Allyl, glass filled	E80-87	F and M	0.25	0.010	...	...	...	...	...	...
			1.25	0.050	...	...	...	...	...	...
			2.5	0.100	...	...	...	...	...	...
Allyl, fiber filled	M108-115	F and M	0.25	0.010	76	250	0.050	0.002	S9, S11	T15, M42
			1.25	0.050	60	200	0.102	0.004	S9, S11	T15, M42
			2.5	0.100	49	160	0.20	0.008	S9, S11	T15, M42

Boring of plastics (continued)

Materials	Hardness, Rockwell	Condition(a)	Depth of cut mm	Depth of cut in.	Carbide tool (uncoated): Speed, Brazed m/min	Brazed fpm	Indexable m/min	Indexable fpm	Feed per tooth mm	Feed per tooth in.	Tool material grade ISO	Tool material grade AISI
Thermoplastics												
Acrylic, acetal, polycarbonate, polysulfone, polystyrene	M60-120	C, M, or E	0.25	0.010	135	450	160	530	0.050	0.002	K20, M20	C-2
			1.25	0.050	110	360	130	425	0.102	0.004	K20, M20	C-2
			2.5	0.100	100	325	115	385	0.20	0.008	K20, M20	C-2
Acrylonitrile-butadiene-styrene (ABS), polyarylether, polypropylene, polyethylene, cellulose acetate	R50-120	C, M, or E	0.25	0.010	160	530	190	625	0.050	0.002	K20, M20	C-2
			1.25	0.050	130	425	150	500	0.102	0.004	K20, M20	C-2
			2.5	0.100	115	385	140	455	0.20	0.008	K20, M20	C-2
Fluorocarbons: tetrafluoroethylene (TFE), chlorotrifluoroethylene (CTFE)	R74-95	M or E	0.25	0.010	135	450	160	530	0.050	0.002	K20, M20	C-2
			1.25	0.050	110	360	130	425	0.102	0.004	K20, M20	C-2
			2.5	0.100	100	325	115	385	0.20	0.008	K20, M20	C-2
Polyamides (nylons): unfilled—types 6, 6/6, 6/12, 11, 12	R78-120	M or E	0.25	0.010	180	595	215	700	0.050	0.002	K20, M20	C-2
			1.25	0.050	145	475	170	560	0.102	0.004	K20, M20	C-2
			2.5	0.100	125	415	150	490	0.20	0.008	K20, M20	C-2
Polyamides (nylons): 35% glass reinforced—types 6, 6/6	R78-120	F and M	0.25	0.010	135	450	160	530	0.050	0.002	K20, M20	C-2
			1.25	0.050	110	360	130	425	0.102	0.004	K20, M20	C-2
			2.5	0.100	100	325	115	385	0.15	0.006	K20, M20	C-2
Polyamides (nylons): 35% glass reinforced—types 6/10, 6/12	E40-50	F and M	0.25	0.010	115	370	135	435	0.050	0.002	K20, M20	C-2
			1.25	0.050	90	300	105	350	0.102	0.004	K20, M20	C-2
			2.5	0.100	84	275	95	315	0.15	0.006	K20, M20	C-2

(continued)

Boring of plastics (continued)

Materials	Hardness, Rockwell	Condition(a)	Depth of cut mm	Depth of cut in.	Speed Brazed m/min	Speed Brazed fpm	Speed Indexable m/min	Speed Indexable fpm	Carbide tool (uncoated) Feed per tooth mm	Feed per tooth in.	Tool material grade ISO	Tool material grade AISI
Thermosetting plastics												
Epoxy, melamine, phenolic	M100-128	C or M	0.25	0.010	185	600	215	700	0.050	0.002	K20, M20	C-2
			1.25	0.050	145	475	170	560	0.102	0.004	K20, M20	C-2
			2.5	0.100	125	415	150	490	0.20	0.008	K20, M20	C-2
Furan, polybutadiene	R40-100	C	0.25	0.010	105	350	120	400	0.050	0.002	K20, M20	C-2
			1.25	0.050	84	275	95	315	0.102	0.004	K20, M20	C-2
			2.5	0.100	72	235	84	275	0.20	0.006	K20, M20	C-2
Silcone	Shore A15-65	C or M	0.25	0.010	105	340	120	400	0.050	0.002	K20, M20	C-2
			1.25	0.050	84	275	95	315	0.102	0.004	K20, M20	C-2
			2.5	0.100	72	235	84	275	0.15	0.006	K20, M20	C-2
Silicone, glass filled	M80-90	F and M	0.25	0.010	90	300	105	345	0.050	0.002	K20, M20	C-2
			1.25	0.050	72	235	84	275	0.102	0.004	K20, M20	C-2
			2.5	0.100	64	210	75	245	0.15	0.006	K20, M20	C-2
Polyimide	E40-50	M or E	0.25	0.010	180	585	210	690	0.050	0.002	K20, M20	C-2
			1.25	0.050	140	465	170	550	0.102	0.004	K20, M20	C-2
			2.5	0.100	130	425	150	500	0.20	0.008	K20, M20	C-2
Polyimide, glass filled	M109-115	F and M	0.25	0.010	115	375	135	440	0.050	0.002	K20, M20	C-2
			1.25	0.050	90	300	105	350	0.102	0.004	K20, M20	C-2
			2.5	0.100	84	275	95	315	0.20	0.008	K20, M20	C-2
Polyurethane	Shore A65-95	C	0.25	0.020	105	350	120	400	0.050	0.002	K20, M20	C-2
			1.25	0.050	84	275	95	315	0.102	0.004	K20, M20	C-2
			2.5	0.100	72	235	84	275	0.15	0.006	K20, M20	C-2
	Shore D55-75	C	0.25	0.010	115	375	135	440	0.050	0.002	K20, M20	C-2
			1.25	0.050	90	300	105	350	0.102	0.004	K20, M20	C-2
			2.5	0.100	84	275	95	315	0.20	0.008	K20, M20	C-2
Allyl (DAP)	M95-100	C	0.25	0.010	135	450	160	530	0.050	0.002	K20, M20	C-2
			1.25	0.050	110	360	130	425	0.102	0.004	K20, M20	C-2
			2.5	0.300	100	320	115	375	0.20	0.008	K20, M20	C-2
Allyl, glass filled	E80-87	F and M	0.25	0.010	90	295	105	345	0.050	0.002	K20, M20	C-2
			1.25	0.050	72	235	84	275	0.102	0.004	K20, M20	C-2
			2.5	0.100	64	210	75	245	0.20	0.008	K20, M20	C-2
Allyl, fiber filled	M108-115	F and M	0.25	0.010	115	370	135	435	0.050	0.002	K20, M20	C-2
			1.25	0.050	90	295	105	350	0.102	0.004	K20, M20	C-2
			2.5	0.100	81	265	95	315	0.20	0.008	K20, M20	C-2

(a) C, cast; M, molded; E, extruded; F, filled.

Source: Ref 21, p 3-266 to 3-267

End milling-slotting of plastics

Materials	Hardness, Rockwell	Condition(a)	Axial depth of cut mm	Axial depth of cut in.	Speed m/min	Speed fpm	Feed per tooth, 10 mm (3/8 in.) slot, mm	Feed per tooth, 10 mm (3/8 in.) slot, in.	Feed per tooth, 12 mm (1/2 in.) slot, mm	Feed per tooth, 12 mm (1/2 in.) slot, in.
Thermoplastics										
Acrylic, acetal, polycarbonate, polysulfone, polystyrene	M60-120	C, M, or E	0.75	0.030	120	400	0.075	0.003	0.102	0.004
			3	0.125	105	350	0.075	0.003	0.075	0.003
			Diam/2	Diam/2	90	300	0.050	0.002	0.050	0.002
			Diam/1	Diam/1	76	250	0.025	0.001	0.025	0.001
Acrylonitrile-butadiene-styrene (ABS), polyarylether, polypropylene, polyethylene, cellulose acetate	R50-120	C, M, or E	0.75	0.030	150	500	0.075	0.003	0.102	0.004
			3	0.125	135	450	0.075	0.003	0.075	0.003
			Diam/2	Diam/2	120	400	0.050	0.002	0.050	0.002
			Diam/1	Diam/1	105	350	0.025	0.001	0.025	0.001
Fluorocarbons: tetrafluoroethylene (TFE), chlorotrifluoroethylene (CTFE)	R74-95	M or E	0.75	0.030	120	400	0.075	0.003	0.102	0.004
			3	0.125	105	350	0.075	0.003	0.075	0.003
			Diam/2	Diam/2	90	300	0.050	0.002	0.050	0.002
			Diam/1	Diam/1	76	250	0.025	0.001	0.025	0.001
Polyamides (nylons): unfilled—types 6, 6/6, 6/12, 11, 12	R78-120	M or E	0.75	0.030	150	500	0.075	0.003	0.102	0.004
			3	0.125	135	450	0.075	0.003	0.075	0.003
			Diam/2	Diam/2	120	400	0.050	0.002	0.050	0.002
			Diam/1	Diam/1	105	350	0.025	0.001	0.025	0.001
Polyamides (nylons): 35% glass reinforced—types 6, 6/6	R78-120	F and M	0.75	0.030	120	400	0.075	0.003	0.075	0.003
			3	0.125	105	350	0.075	0.003	0.075	0.003
			Diam/2	Diam/2	90	300	0.050	0.002	0.050	0.002
			Diam/1	Diam/1	76	250	0.025	0.001	0.025	0.001
Polyamides (nylons): 35% glass reinforced—types 6/10, 6/12	E40-50	F and M	0.75	0.030	105	350	0.075	0.003	0.075	0.003
			3	0.125	90	300	0.075	0.003	0.075	0.003
			Diam/2	Diam/2	76	250	0.050	0.002	0.050	0.002
			Diam/1	Diam/1	60	200	0.025	0.001	0.025	0.001

(continued)

End milling-slotting of plastics (continued)

Materials	Hardness, Rockwell	Condition(a)	Axial depth of cut mm	Axial depth of cut in.	Speed m/min	Speed fpm	Feed per tooth, 10 mm (3/8 in.) slot, mm	10 mm (3/8 in.) in.	12 mm (1/2 in.) mm	12 mm (1/2 in.) in.
Thermosetting plastics										
Epoxy, melamine, phenolic	M100-128	C or M	0.75	0.030	150	500	0.075	0.003	0.102	0.004
			3	0.125	135	450	0.075	0.003	0.102	0.004
			Diam/2	Diam/2	120	400	0.050	0.002	0.075	0.003
			Diam/1	Diam/1	105	350	0.025	0.001	0.050	0.002
Furan, polybutadiene	R40-100	C	0.75	0.030	90	300	0.075	0.003	0.102	0.004
			3	0.125	84	275	0.075	0.003	0.102	0.004
			Diam/2	Diam/2	76	250	0.050	0.002	0.075	0.003
			Diam/1	Diam/1	60	300	0.025	0.001	0.050	0.002
Silcone	Shore A15-65	C or M	0.75	0.030	90	300	0.075	0.003	0.102	0.004
			3	0.125	84	275	0.075	0.003	0.102	0.004
			Diam/2	Diam/2	76	250	0.050	0.002	0.075	0.003
			Diam/1	Diam/1	60	200	0.025	0.001	0.050	0.002
Silicone, glass filled	M80-90	F and M	0.75	0.030	105	350	0.075	0.003	0.102	0.004
			3	0.125	90	300	0.075	0.003	0.102	0.004
			Diam/2	Diam/2	84	275	0.050	0.002	0.075	0.003
			Diam/1	Diam/1	76	250	0.025	0.001	0.050	0.002
Polyimide	E40-50	M or E	0.75	0.030	150	500	0.075	0.003	0.102	0.004
			3	0.125	135	450	0.075	0.003	0.102	0.004
			Diam/2	Diam/2	120	400	0.050	0.002	0.075	0.003
			Diam/1	Diam/1	105	350	0.025	0.001	0.050	0.002
Polyimide, glass filled	M109-115	F and M	0.75	0.030	135	450	0.075	0.003	0.102	0.004
			3	0.125	120	600	0.075	0.003	0.102	0.004
			Diam/2	Diam/2	103	350	0.050	0.002	0.075	0.003
			Diam/1	Diam/1	90	300	0.025	0.001	0.050	0.002
Polyurethane	Shore A65-95	C	0.75	0.030	90	300	0.075	0.003	0.102	0.004
			3	0.125	84	275	0.075	0.003	0.102	0.004
			Diam/2	Diam/2	76	250	0.050	0.002	0.075	0.003
			Diam/1	Diam/1	60	200	0.025	0.001	0.050	0.002
	Shore D55-75	C	0.75	0.030	105	350	0.075	0.003	0.102	0.004
			3	0.125	90	300	0.075	0.003	0.102	0.004
			Diam/2	Diam/2	84	275	0.050	0.002	0.075	0.003
			Diam/1	Diam/1	76	250	0.025	0.001	0.050	0.002
Allyl (DAP)	M95-100	C	0.75	0.030	105	350	0.075	0.003	0.102	0.004
			3	0.125	90	300	0.075	0.003	0.102	0.004
			Diam/2	Diam/2	84	275	0.050	0.002	0.075	0.003
			Diam/1	Diam/1	76	250	0.025	0.001	0.050	0.002
Allyl, glass filled	E80-87	F and M	0.75	0.030	120	400	0.075	0.003	0.102	0.004
			3	0.125	105	350	0.075	0.003	0.102	0.004
			Diam/2	Diam/2	90	300	0.050	0.002	0.075	0.003
			Diam/1	Diam/1	76	250	0.025	0.001	0.050	0.002
Allyl, fiber filled	M108-115	F and M	0.75	0.030	135	450	0.075	0.003	0.102	0.004
			3	0.125	130	425	0.075	0.003	0.102	0.004
			Diam/2	Diam/2	120	400	0.050	0.002	0.075	0.003
			Diam/1	Diam/1	105	350	0.025	0.001	0.050	0.002

(continued)

End milling-slotting of plastics (continued)

Materials	Hardness, Rockwell	Condition(a)	Axial depth of cut mm	Axial depth of cut in.	Feed per tooth, 18 mm (3/4 in.) slot, m/min	Feed per tooth, 18 mm (3/4 in.) slot, fpm	Feed per tooth, 25-50 mm (1-2 in.) slot, mm	Feed per tooth, 25-50 mm (1-2 in.) slot, in.	HSS tool material (except as noted) ISO	HSS tool material (except as noted) AISI
Thermoplastics										
Acrylic, acetal, polycarbonate, polysulfone, polystyrene	M60-120	C, M, or E	0.75	0.030	0.13	0.005	0.15	0.006	S4, S5, S2	M2, M3, M7
			3	0.125	0.102	0.004	0.13	0.005		
			Diam/2	Diam/2	0.075	0.003	0.102	0.004		
			Diam/1	Diam/1	0.050	0.002	0.075	0.003		
Acrylonitrile-butadiene-styrene (ABS), polyarylether, polypropylene, polyethylene, cellulose acetate	R50-120	C, M, or E	0.75	0.030	0.13	0.005	0.15	0.006	S4, S5, S2	M2, M3, M7
			3	0.125	0.102	0.004	0.13	0.005		
			Diam/2	Diam/2	0.075	0.003	0.102	0.004		
			Diam/1	Diam/1	0.050	0.002	0.075	0.003		
Fluorocarbons: tetrafluoroethylene (TFE), chlorotrifluoroethylene (CTFE)	R74-95	M or E	0.75	0.030	0.13	0.005	0.15	0.006	S4, S5, S2	M2, M3, M7
			3	0.125	0.102	0.004	0.13	0.005		
			Diam/2	Diam/2	0.075	0.003	0.102	0.004		
			Diam/1	Diam/1	0.050	0.002	0.075	0.003		
Polyamides (nylons): unfilled—types 6, 6/6, 6/12, 11, 12	R78-120	M or E	0.75	0.030	0.13	0.005	0.15	0.006	S4, S5, S2	M2, M3, M7
			3	0.125	0.102	0.004	0.13	0.005		
			Diam/2	Diam/2	0.075	0.003	0.102	0.004		
			Diam/1	Diam/1	0.050	0.002	0.075	0.003		
Polyamides (nylons): 35% glass reinforced—types 6, 6/6	R78-120	F and M	0.75	0.030	0.102	0.004	0.13	0.005	K10 carbide	C-2 carbide
			3	0.125	0.075	0.003	0.13	0.005		
			Diam/2	Diam/2	0.075	0.003	0.102	0.004		
			Diam/1	Diam/1	0.050	0.002	0.075	0.003		
Polyamides (nylons): 35% glass reinforced—types 6/10, 6/12	E40-50	F and M	0.75	0.030	0.102	0.004	0.13	0.005	K10 carbide	C-2 carbide
			3	0.125	0.075	0.003	0.13	0.005		
			Diam/2	Diam/2	0.075	0.003	0.102	0.004		
			Diam/1	Diam/1	0.050	0.002	0.075	0.003		
Thermosetting plastics										
Epoxy, melamine, phenolic	M100-128	C or M	0.75	0.030	0.20	0.008	0.25	0.010	S4, S5, S2	M2, M3, M7
			3	0.125	0.20	0.008	0.25	0.010		
			Diam/2	Diam/2	0.15	0.006	0.20	0.008		
			Diam/1	Diam/1	0.102	0.004	0.15	0.006		
Furan, polybutadiene	R40-100	C	0.75	0.030	0.13	0.005	0.15	0.006	S9, S11	T15, M42
			3	0.125	0.13	0.005	0.15	0.006		
			Diam/2	Diam/2	0.102	0.004	0.13	0.005		
			Diam/1	Diam/1	0.075	0.003	0.102	0.004		
Silcone	Shore A15-65	C or M	0.75	0.030	0.13	0.005	0.15	0.006	S9, S11	T15, M42
			3	0.125	0.13	0.005	0.15	0.006		
			Diam/2	Diam/2	0.102	0.004	0.13	0.005		
			Diam/1	Diam/1	0.075	0.003	0.102	0.004		
Silicone, glass filled	M80-90	F and M	0.75	0.030	0.13	0.005	0.15	0.006	K10 carbide	C-2 carbide
			3	0.125	0.13	0.005	0.15	0.006		
			Diam/2	Diam/2	0.102	0.004	0.13	0.005		
			Diam/1	Diam/1	0.075	0.003	0.102	0.004		

(continued)

End milling-slotting of plastics (continued)

Materials	Hardness, Rockwell	Condition(a)	Axial depth of cut mm	Axial depth of cut in.	Feed per tooth, 18 mm (3/4 in.) m/min	Feed per tooth, 18 mm (3/4 in.) fpm	Feed per tooth, 25-50 mm (1-2 in.) mm	Feed per tooth, 25-50 mm (1-2 in.) in.	HSS tool material (except as noted) ISO	HSS tool material (except as noted) AISI
Polyimide	E40-50	M or E	0.75	0.030	0.20	0.008	0.25	0.010	S4, S5, S2	M2, M3, M7
			3	0.125	0.20	0.008	0.25	0.010		
			Diam/2	Diam/2	0.15	0.006	0.20	0.008		
			Diam/1	Diam/1	0.102	0.004	0.15	0.006		
Polyimide, glass filled	M109-115	F and M	0.75	0.030	0.13	0.005	0.15	0.006	K10 carbide	C-2 carbide
			3	0.125	0.13	0.005	0.15	0.008		
			Diam/2	Diam/2	0.102	0.004	0.13	0.005		
			Diam/1	Diam/1	0.075	0.003	0.102	0.004		
Polyurethane	Shore A65-95	C	0.75	0.030	0.13	0.005	0.15	0.006	S9, S11	T15, M42
			3	0.125	0.13	0.005	0.15	0.006		
			Diam/2	Diam/2	0.102	0.004	0.13	0.005		
			Diam/1	Diam/1	0.075	0.003	0.102	0.004		
	Shore D55-75	C	0.75	0.030	0.15	0.006	0.20	0.008	S4, S5, S2	M2, M3, M7
			3	0.125	0.15	0.006	0.20	0.008		
			Diam/2	Diam/2	0.13	0.005	0.18	0.007		
			Diam/1	Diam/1	0.102	0.004	0.15	0.006		
Allyl (DAP)	M95-100	C	0.75	0.030	0.15	0.006	0.20	0.008	S4, S5, S2	M2, M3, M7
			3	0.125	0.15	0.006	0.20	0.008		
			Diam/2	Diam/2	0.13	0.005	0.15	0.006		
			Diam/1	Diam/1	0.102	0.004	0.13	0.005		
Allyl, glass filled	E80-87	F and M	0.75	0.030	0.13	0.005	0.15	0.006	K10 carbide	C-2 carbide
			3	0.125	0.13	0.005	0.15	0.006		
			Diam/2	Diam/2	0.102	0.004	0.13	0.005		
			Diam/1	Diam/1	0.075	0.003	0.13	0.005		
Allyl, fiber filled	M108-115	F and M	0.75	0.030	0.13	0.005	0.15	0.006	K10 carbide	C-2 carbide
			3	0.125	0.13	0.005	0.15	0.006		
			Diam/2	Diam/2	0.102	0.004	0.13	0.005		
			Diam/1	Diam/1	0.075	0.003	0.13	0.005		

(a) C, cast; M, molded; F, filled.

Source: Ref 21, p 2-229 to 2-231

Circular sawing of plastics (with HSS blade)

Materials	Hardness, Rockwell	Condition(a)	Solid stock diameter or thickness mm	in.	Pitch mm/tooth	in./tooth	Cutting speed m/min	fpm	Feed mm/tooth	in./tooth	HSS tool material ISO	AISI
Thermoplastics												
Acrylic, acetal, polycarbonate, polysulfone, polystyrene	M60-120	C, M, or E	6-80	¼-3	3-12	0.12-0.50	90	300	0.102	0.004		
			80-160	3-6	10-18	0.40-0.70	76	250	0.102	0.004	S4, S2	M2, M7
			160-250	6-9	15-20	0.60-0.80	60	200	0.15	0.006		
			250-400	9-15	18-25	0.70-1.00	46	150	0.15	0.006		
Acrylonitrile-butadiene-styrene (ABS), polyarylether, polypropylene, polyethylene cellulose acetate	R50-120	C, M, or E	6-80	¼-3	3-12	0.12-0.50	105	350	0.102	0.004		
			80-160	3-6	10-18	0.40-0.70	90	300	0.102	0.004	S4, S2	M2, M7
			160-250	6-9	15-20	0.60-0.80	67	220	0.15	0.006		
			250-400	9-15	18-25	0.70-1.00	52	170	0.15	0.006		
Fluorocarbons: tetrafluoroethylene (TFE), chlorotrifluoroethylene (CTFE)	R74-95	M or E	6-80	¼-3	3-12	0.12-0.50	90	300	0.102	0.004		
			80-160	3-6	10-18	0.40-0.70	76	250	0.102	0.004	S4, S2	M2, M7
			160-250	6-9	15-20	0.60-0.80	60	200	0.15	0.006		
			250-400	9-15	18-25	0.70-1.00	46	150	0.15	0.006		
Polyamide (nylons): unfilled—types 6, 6/6, polystyrene 6/12, 11, 12	R78-120	M or E	6-80	¼-3	3-12	0.12-0.50	105	350	0.102	0.004		
			80-160	3-6	10-18	0.40-0.70	90	300	0.102	0.004	S4, S2	M2, M7
			160-250	6-9	15-20	0.60-0.80	67	220	0.15	0.006		
			250-400	9-15	18-25	0.70-1.00	52	170	0.15	0.006		

(continued)

Circular sawing of plastics (with HSS blade) (continued)

Materials	Hardness, Rockwell	Condition(a)	Solid stock diameter or thickness mm	in.	Pitch mm/tooth	in./tooth	Cutting speed m/min	fpm	Feed mm/tooth	in./tooth	HSS tool material ISO	AISI
Thermosetting plastics												
Epoxy, melamine, phenolic	M100-128	C or M	6-80	¼-3	3-12	0.12-0.50	105	350	0.102	0.004		
			80-160	3-6	10-18	0.40-0.70	90	300	0.102	0.004	S4, S2	M2, M7
			160-250	6-9	15-20	0.60-0.80	67	220	0.15	0.006		
			250-400	9-15	18-25	0.70-1.00	52	170	0.15	0.006		
Furan, polybutadiene	R40-100	C	6-80	¼-3	3-12	0.12-0.50	46	150	0.075	0.003		
			80-160	3-6	10-18	0.40-0.70	37	120	0.075	0.003	S4, S2	M2, M7
			160-250	6-9	15-20	0.60-0.80	30	100	0.13	0.005		
			250-400	9-15	18-25	0.70-1.00	24	80	0.13	0.005		
Silicone	Shore A15-65	C or M	6-80	¼-3	3-12	0.12-0.50	30	100	0.075	0.003		
			80-160	3-6	10-18	0.40-0.70	24	80	0.075	0.003	S4, S2	M2, M7
			160-250	6-9	15-20	0.60-0.80	15	50	0.13	0.005		
			250-400	9-15	18-25	0.70-1.00	15	50	0.13	0.005		
Polyimide	E40-50	M or E	6-80	¼-3	3-12	0.12-0.50	105	350	0.102	0.004		
			80-160	3-6	10-18	0.40-0.70	90	300	0.102	0.004	S4, S2	M2, M7
			160-250	6-9	15-20	0.60-0.80	67	220	0.15	0.006		
			250-400	9-15	18-25	0.70-1.00	52	170	0.15	0.006		
Polyurethane	Shore A65-95	C	6-80	¼-3	3-12	0.12-0.50	46	150	0.075	0.003		
			80-160	3-6	10-18	0.40-0.70	37	120	0.075	0.003	S4, S2	M2, M7
			160-250	6-9	15-20	0.60-0.80	30	100	0.102	0.004		
			250-400	9-15	18-25	0.70-1.00	24	80	0.102	0.004		
	Shore D55-75	C	6-80	¼-3	3-12	0.12-0.50	60	200	0.102	0.004		
			80-160	3-6	10-18	0.40-0.70	46	150	0.102	0.004	S4, S2	M2, M7
			160-250	6-9	15-20	0.60-0.80	37	120	0.15	0.006		
			250-400	9-15	18-25	0.70-1.00	30	100	0.15	0.006		
Allyl (DAP)	M95-100	C	6-80	¼-3	3-12	0.12-0.50	76	250	0.102	0.004		
			80-160	3-6	10-18	0.40-0.70	60	200	0.102	0.004	S4, S2	M2, M7
			160-250	6-9	15-20	0.60-0.80	46	150	0.15	0.006		
			250-400	9-15	18-25	0.70-1.00	37	120	0.15	0.006		
Allyl, fiber filled	M108-115	F and M	6-80	¼-3	3-12	0.12-0.50	46	150	0.075	0.003		
			80-160	3-6	10-18	0.40-0.70	37	120	0.075	0.003	S4, S2	M2, M7
			160-250	6-9	15-20	0.60-0.80	30	100	0.13	0.005		
			250-400	9-15	18-25	0.70-1.00	24	80	0.13	0.005		

(a) C, cast; M, molded; E, extruded; F, filled.

Source: Ref 21, p 6-118 to 6-119

Electronic Materials

ELECTRONIC MATERIALS

Electronic Materials

General Information

Selected constants

Constant	Symbol	cgs unit	SI units
Gases			
Avogadro's number	N_A	6.02×10^{23} mole $^{-1}$	6.02×10^{23} mole $^{-1}$
Boltzmann constant	k_B	1.38×10^{-16} erg/K	1.38×10^{-23} J/K
Gas constant	R	1.987 cal(mole·K)$^{-1}$	8.31 J(mole·K)$^{-1}$
Atomic			
Planck's constant	h	6.63×10^{-27} erg·s	6.63×10^{-34} J·s
Electron volt	eV	1.60×10^{-12} erg 23.05 kcal·mole^{-1}	1.60×10^{-19} J 5.50 kJ·mole^{-1}
Rydberg	Ry	13.60 eV	1.097×10^{7} m^{-1}
Bohr radius	r_0	0.529×10^{-8} cm	0.529×10^{-10} m
Particles			
Electron rest mass	m	9.11×10^{-28} g	9.11×10^{-31} kg
Electron charge	e	4.80×10^{-10} esu	1.60×10^{-19} C
Temperature			
International Practical Scale	t	0 °C	+ 273.15 K
Thermodynamic scale	T	-273.15 °C	0 K
Electrical			
Resistivity	ρ	Ω-cm	Ω-meter
Conductivity	σ	(Ω-cm)$^{-1}$	(Ω-meter)$^{-1}$
Other constants			
Angstrom	Å	10^{-8} cm	10^{-10} m
Micron	μ	10^{-4} cm	10^{-6} m
Speed of light in vacuum	c	2.998×10^{10} cm·s^{-1}	2.998×10^{8} m·s^{-1}
Dielectric constant of vacuum	ϵ_0	1 (dimensionless)	8.85×10^{-12} Fm^{-1}

Source: Ref 12, p 275

Designation of electron states

n	Maximum ℓ	Maximum number of states	Designation of states
1	0	2	*s*
2	1	6	*p*
3	2	10	*d*
4	3	14	*f*

Source: Ref 12, p 62

Calculated Fermi energies and temperatures for free electrons(a)

Element	E_F, (eV)(a)	T_F, (K·10^{-4})
Li	4.7	5.5
Na	3.1	3.7
K	2.1	2.4
Rb	1.8	2.1
Cs	1.5	1.8
Cu	7.0	8.2
Ag	5.5	6.4
Au	5.5	6.4

(a) 1 eV = 23,050 cal/mole.

Source: Ref 12, p 78

Allowed combinations of quantum numbers, up to n = 4

n	ℓ	Designation	m_ℓ	m_s	Number	Total
1	0	1s	0	±½	2	2
2	0	2s	0	±½	2	8
2	1	2p	−1, 0, 1	±½	6	
3	0	3s	0	±½	2	
3	1	3p	−1, 0, 1	±½	6	18
3	2	3d	−2, −1, 0, 1, 2	±½	10	
4	0	4s	0	±½	2	
4	1	4p	−1, 0, 1	±½	6	32
4	2	4d	−2, −1, 0, 1, 2	±½	10	
4	3	4f	−3, −2, −1, 0, 1, 2, 3	±½	14	

Source: Ref 12, p 66

Comparison of enthalpies of atomization, fusion, and formation of ionic, metallic, and covalent solids(a)

Crystal type	Example	ΔH^{at}	ΔH^F	$\Delta H^F/\Delta H^{at}$	ΔH_f^0	$\Delta H^F/\Delta H_f^0$
Ionic	NaCl	77	3.4	0.044	49	0.070
	KF	87	3.4	0.044	68	0.050
Metallic	Sn	71	1.72	0.024	...	...
	Ga	62	1.34	0.022	...	...
	Al	73	2.55	0.035	...	...
Covalent	Si	106	7.2	0.068	...	...
	Ge	80	6.7	0.084	...	...
III-V	AlSb	83	9.77	0.117	11.5	0.848
	GaAs	73	12.6	0.173	10	0.795
	GaSb	69	7.78	0.113	5.52	0.675
	InAs	65	8.79	0.135	6.5	0.739
	InSb	64	5.70	0.089	4.0	0.703

(a) All energies are in Kcal (g-atom)$^{-1}$.

Source: Ref 13, p 18

The equivalent conductances of the separate ions

Ion	0°	18°	25°	50°	75°	100°	128°	156°	Ion	0°	18°	25°	50°	75°	100°	128°	156°
K	40.4	64.6	74.5	115	159	206	263	317									
Na	26	43.5	50.9	82	116	155	203	249	½SO_4	41	68	79	125	177	234	303	370
NH_4	40.2	64.5	74.5	115	159	207	264	319									
Ag	32.9	54.3	63.5	101	143	188	245	299	½C_2O_4	39	63	73	115	163	213	27.5	336
½Ba	33	55	65	104	149	200	262	322									
½Ca	30	51	60	98	142	191	252	312	½$C_4H_5O_7$	36	60	70	113	161	214	...	...
½La	35	61	72	119	173	235	312	388	¼Fe(CN)$_4$	58	95	111	173	244	321	...	...
Cl	41.1	65.5	75.5	116	160	207	264	318	H	240	314	350	465	565	644	722	777
NO_2	40.4	61.7	70.6	104	140	178	222	263	OH	105	172	192	284	360	439	525	592
$C_2H_2O_2$	20.3	34.6	40.8	67	96	130	171	211									

Source: Ref 14, p 138

Ground state electron configuration of atoms

$Z^{(a)}$	Element	Outer configuration
1	H	$1s$
2	He	$1s^2$
3	Li	$2s$
4	Be	$2s^2$
5	B	$2s^2\ 2p$
6	C	$2s^2\ 2p^2$
7	N	$2s^2\ 2p^3$
8	O	$2s^2\ 2p^4$
9	F	$2s^2\ 2p^5$
10	Ne	$2s^2\ 2p^6$
11	Na	$3s$
12	Mg	$3s^2$
13	Al	$3s^2\ 3p$
14	Si	$3s^2\ 3p^2$
15	P	$3s^2\ 3p^3$
16	S	$3s^2\ 3p^4$
17	Cl	$3s^2\ 3p^5$
18	Ar	$3s^2\ 3p^6$
19	K	$4s$
20	Ca	$4s^2$
21	Sc	$3d\ 4s^2$
22	Ti	$3d^2\ 4s^2$
23	V	$3d^3\ 4s^2$
24	Cr	$3d^5\ 4s$
25	Mn	$3d^5\ 4s^2$
26	Fe	$3d^6\ 4s^2$
27	Co	$3d^7\ 4s^2$
28	Ni	$3d^8\ 4s^2$
29	Cu	$3d^{10}\ 4s$
30	Zn	$4s^2$
31	Ga	$4s^2\ 4p$
32	Ge	$4s^2\ 4p^2$
33	As	$4s^2\ 4p^3$
34	Se	$4s^2\ 4p^4$
35	Br	$4s^2\ 4p^5$
36	Kr	$4s^2\ 4p^6$
37	Rb	$5s$
38	Sr	$5s^2$
39	Y	$4d\ 5s^2$
40	Zr	$4d^2\ 5s^2$
41	Nb	$4d^4\ 5s$
42	Mo	$4d^5\ 5s$
43	Tc	$4d^6\ 5s$
44	Ru	$4d^7\ 5s$
45	Rh	$4d^8\ 5s$
46	Pd	$4d^{10}$
47	Ag	$4d^{10}\ 5s$
48	Cd	$5s^2$
49	In	$5s^2\ 5p$
50	Sn	$5s^2\ 5p^2$
51	Sb	$5s^2\ 5p^3$
52	Te	$5s^2\ 5p^4$
53	I	$5s^2\ 5p^5$
54	Xe	$5s^2\ 5p^6$
55	Cs	$6s$
56	Ba	$6s^2$
57	La	$5d\ 6s^2$
58	Ce	$4f^2\ 6s^2$
59	Pr	$4f^3\ 6s^2$
60	Nd	$4f^4\ 6s^2$
61	Pm	$4f^5\ 6s^2$
62	Sm	$4f^6\ 6s^2$
63	Eu	$4f^7\ 6s^2$
64	Gd	$4f^7\ 5d\ 6s^2$
65	Tb	$4f^8\ 5d\ 6s^2$
66	Dy	$4f^9\ 5d\ 6s^2$
67	Ho	$4f^{10}\ 5d\ 6s^2$
68	Er	$4f^{11}\ 5d\ 6s^2$
69	Tm	$4f^{12}\ 5d\ 6s^2$
70	Yb	$4f^{13}\ 5d\ 6s^2$
71	Lu	$4f^{14}\ 5d\ 6s^2$
72	Hf	$5d^2\ 6s^2$
73	Ta	$5d^3\ 6s^2$
74	W	$5d^4\ 6s^2$
75	Re	$5d^5\ 6s^2$
76	Os	$5d^6\ 6s^2$
77	Ir	$5d^9$
78	Pt	$5d^9\ 6s$
79	Au	$5d^{10}\ 6s$
80	Hg	$6s^2$
81	Tl	$6s^2\ 6p$
82	Pb	$6s^2\ 6p^2$
83	Bi	$6s^2\ 6p^3$
84	Po	$6s^2\ 6p^4$
85	At	$6s^2\ 6p^5$
86	Rn	$6s^2\ 6p^6$
87	Fr	$7s$
88	Ra	$7s^2$
89	Ac	$6d\ 7s^2$
90	Th	$6d^2\ 7s^2$
91	Pa	$6d^3\ 7s^2$
92	U	$6d^4\ 7s^2$
93	Np	$5f^5\ 7s^2$
94	Pu	$5f^5\ 6d\ 7s^2$
95	Am	$5f^6\ 6d\ 7s^2$
96	Cm	$5f^7\ 6d\ 7s^2$
97	Bk	$5f^8\ 6d\ 7s^2$
98	Cf	$5f^9\ 6d\ 7s^2$
99	Es	$5f^{10}\ 6d\ 7s^2$
100	Fm	$5f^{11}\ 6d\ 7s^2$

(a) Atomic number.

Source: Ref 12, p 68

Electrical resistivities and temperature coefficients of some elements near room temperature (20 °C)

Element	ρ (Ω-cm × 10^6)	α (Ω/Ω-deg × 10^3)
Aluminum	2.6548	4.29
Antimony	39.0 (0 °C)	3.6
Beryllium	4	25
Bismuth	106.5 (0 °C)	5.6
Cadmium	6.83 (0 °C)	4.2
Calcium	3.91 (0 °C)	4.02 (0 °C)
Carbon (graphite)	13.75 (0 °C)	–
Chromium	12.9 (0 °C)	3
Cobalt	6.24	5.3
Copper	1.6730	4.3
Germanium (impure)	46	–
Gold	2.35	3.5
Indium	8.0 (0 °C)	5
Iodine	1.3 × 10^{15}	–
Iridium	5.3	3.93
Iron	9.71	6.51
Lead	20.648	3.68
Lithium	9.35	5
Magnesium	4.45	3.7
Mercury	98.4 (50 °C)	0.97
Molybdenum	5.2 (0 °C)	5.3
Nickel	6.84	6.92
Palladium	10.8	3.77
Platinum	9.85	3.927
Plutonium	141.4 (107 °C)	−2.08 (107 °C)
Rhenium	19.3	3.95
Rhodium	4.51	4.3
Silicon (impure)	10 (0 °C)	–
Silver	1.59	4.1
Sodium	4.69	–
Sulfur (yellow)	2 × 10^{23}	–
Tantalum	13.5	3.83
Tellurium	4.35 × 10^5 (23 °C)	–
Thallium	15 (0 °C)	–
Thorium	15.7 (25 °C)	3.8
Tin	11 (0 °C)	3.64
Titanium	42	3.5
Tungsten	5.3 (27 °C)	4.5
Uranium	11 (0 °C)	2.1 (27 °C)
Zinc	5.916	4.19

Source: Ref 12, p 123

Changes in resistivity per atomic percent of alloying element in Au, Ag, and Cu

Alloying element	Change in resistivity in: Au	Ag	Cu
Cu	0.485(a)	0.068(a)	...
Ni	1.00	...	1.25(a)
Co	6.1	...	6.4
Fe	7.66	...	9.3
Mn	2.41	...	2.83
Cr	4.25	...	...
Ti	14.4	...	...
Ag	0.38	...	0.14
Pd	0.407	0.436	0.89
Rh	4.2	...	4.40
Au	...	0.38	0.55
Pt	1.02	1.59	2.51
Ir	...	...	6.1
Zn	0.96	0.62	0.335
Ga	2.2	2.28	1.40
Ge	5.2	5.52	3.75
As	...	8.46	6.8
Cd	0.64	0.382	0.21
In	1.41	1.78	1.10
Sn	3.63	4.32	2.85
Sb	...	7.26	5.45
Hg	0.41	0.79	1.00
Tl	...	2.27	...
Pb	...	4.64	...
Bi	...	7.3	...

(a) In $\mu\Omega$-cm per atomic percent at 18 °C.

Source: Ref 12, p 136

Corrosion of electronic materials

Material	Device or component	Corrosion
Gold	Integrated circuits	Uniform films
Ni-plated Al	Electrical connectors	Galvanic
Aluminum	Bond pads	Galvanic
Aluminum	Metallized components	Pitting
Al-Mg alloy	Wire	Intergranular
Palladium	Electrical connectors	Fretting
Tin and tin/lead	Electrical connectors	Uniform
Stainless steel	Electrical connectors	Passive films
Gallium arsenide	Electrical connectors	Complex
Ni undercoat Au	Electrical connectors	Pitting
Tin/tin-lead alloy	Connector	Uniform

Source: Ref 14, p 17

Solubility parameters, molar volumes, and electronegatives of selected elements

Element	δ	V	χ
Al	86	10.0	1.5
Ga	74	11.8	1.6
In	67	15.7	1.6
P	75	17.8	2.0
As	66	13.1	2.0
Sb	59	18.2	1.9
Si	88	11.7	1.8
Ge	76	13.7	1.8
Sn	65	16.3	1.8
Pb	51	18.3	1.8
S	60.6	15.5	2.5
Se	54.8	16.5	2.4
Te	47.4	20.4	2.1
Zn	58	9.2	1.5
Cd	45	13	1.5
Fe	117	7.1	1.8
Ni	124	6.6	1.7
Co	126	6.6	1.8

Source: Ref 13, p 19

Electronic Materials Property Data

Superconductivies of selected compounds

Compound	T_C, K	Compound	T_C, K
Nb_3Sn	18.05	V_3Ga	16.5
Nb_3Ge	23.2	V_3Si	17.1
Nb_3Al	17.5	$Pb_1Mo_{5.1}S_6$	14.4
NbN	16.0	Ti_2Co	3.44
$(SN)_x$ polymer	0.26	La_3In	10.4

Source: Ref 12, p 175

Lattice parameters, temperatures, and entropies of fusion of III-V compounds

Compound	a_0, Å	Experimental T^F, K	Experimental ΔS^F, eu	Calculated ΔS^F, eu	T^F, K	Values used in calculation, ΔS^F, eu
AlP	5.451	2823	...	17.2	1853	17.2
AlAs	5.662	2013	...	16.7	1719	16.7
AlSb	6.136	1330	14.7	15.2	1344	14.7
GaP	5.451	1740	...	16.7	1623	16.7
GaAs	5.653	1513	16.64	16.2	1340	16.64
GaSb	6.096	985	15.8	14.7	983	15.8
InP	5.869	1333	...	15.2	1357	14.0
InAs	6.058	1210	14.52	14.7	993	14.7
InSb	6.479	797	14.3	13.2	745	13.2

Source: Ref 13, p 14

Superconducting properties of some selected superconducting elements

Element	ρ(a)	T_C, K	H_0, Oe	Element	ρ(a)	T_C, K	H_0, Oe
W	5.3	0.012	1070	Th (α)	15.7	1.37	162
Be	4	0.026	...	Pa	10.8	1.4	...
Ir	5.3	0.14	19	Re	19.3	1.7	193
Hf (α)	35.1	0.165	...	Tl	15	2.4	171
Ti (α)	42	0.39	56	In	8	3.4	293
Ru	7.6	0.49	66	Sn (β)	11	3.7	309
Cd	6.8	0.52	30	Hg	98	4.15	412
Os	9.5	0.65	65	Ta	13.5	4.48	830
U (α)	30	0.68(?)	...	V	24.8	5.3	1020
Zr (α)	45	0.55	47	La (β)	61.5	5.9	1600
Zn	5.9	0.85	52	Pb	20.6	7.2	803
Mo	5.2	0.92	98	Tc	18.5	8.2	1410
Ga	15	1.09	59	Nb	15	9.2-9.4	1950
Al	2.6	1.19	99				

(a) In units of μΩ-cm near room temperature.

Source: Ref 12, p 175

Characteristics of conductor materials

Material	Resistance, Ω·cm	Coefficient of thermal expansion, 10^{-6}/°C	Thermal conductivity, cal/cm·s·°C	Maximum temperature to resist, °C	Elastic constant, 10^4 kg/mm^2	Bending strength, kg/mm^2
Copper	1.67×10^{-6}	17	0.94	1084	1.1	21
Gold	2.4×10^{-6}	14	0.71	1064	0.78	12
Silver	1.6×10^{-6}	19	1.00	962	0.83	12-16
Aluminum	2.69×10^{-6}	23.5	0.57	660	0.63	48
Molybdenum	4.8×10^{-6}	5.9	0.34	2610	3.5	52
Tungsten	5.5×10^{-6}	4.4	0.38	3387	3.5	40
Silicon	2.3×10^{4}	3.5	0.2-0.35	1414	...	...

Source: Ref 13, p 273

Calculated values of critical temperature without, T_c, and with, T_s, the inclusion of coherency strain for quaternary III-V alloys

System	Highest T_c for bounding ternary, K	T_c, K
$Ga_xIn_{1-x}As_yP_{1-y}$	908	1081
$Ga_xIn_{1-x}As_ySb_{1-y}$	856	1428
$Al_xIn_{1-x}As_yP_{1-y}$	973	1019
$Al_xIn_{1-x}As_ySb_{1-y}$	735	1599
$Al_xIn_{1-x}P_ySb_{1-y}$	2105	2619
$Ga_xIn_{1-x}P_ySb_{1-y}$	1965	2470
$Al_xGa_{1-x}As_ySb_{1-y}$	973	973
$Al_xGa_{1-x}P_ySb_{1-y}$	1965	2045
$InP_xAs_ySb_{1-x-y}$	1319	1319
$GaP_xAs_ySb_{1-y-x}$	1996	1996
$AlP_xAs_ySb_{1-x-y}$	2105	2105
$Al_xGa_yIn_{1-x-y}P$	973	973
$Al_xGa_yIn_{1-x-y}As$	735	735
$Al_xGa_yIn_{1-x-y}Sb$	462	462

Source: Ref 13, p 32

Properties of some lightly doped semiconductors at 300 K

Type	Semiconductor	E_g, eV	μ_e, $cm^2/V \cdot s$	μ_h, $cm^2/V \cdot s$	m_e^*/m_0	m_h^*/m_0
Element (IV)	C	5.3	1800	1600	...	...
	Si	1.1	1350	475	0.23	0.12
	Ge	0.7	3900	1900	0.02	0.08
	SiC	2.8	400	50	0.60	1.20
III-V	AlS	2.2	180	...	...	...
	AlP	3.0	80	...	...	...
	AlSb	1.6	200	420	0.30	0.40
	BN	4.6	...	...	...	...
	BP	6.0	...	300	...	...
	GaAs	1.4	8500	400	0.07	0.09
	GaP	2.3	110	75	0.12	0.50
	GaSb	0.7	4000	1400	0.20	0.39
	InAs	0.4	33000	460	0.03	0.02
	InP	1.3	4600	150	0.07	0.69
	InSb	0.2	80000	750	0.01	0.18
II-VI	CdS	2.6	340	18	0.21	0.80
	CdSe	1.7	600	...	0.13	0.45
	CdTe	1.5	300	65	0.14	0.37
	ZnS	3.6	120	5	0.40	...
	ZnSe	2.7	530	16	0.10	0.60
	ZnTe	2.3	530	900	0.10	0.60
IV-VI	PbS	0.4	600	200	0.25	0.25
	PbSe	0.3	1400	1400	0.33	0.34
	PbTe	0.3	6000	4000	0.22	0.29
II-IV	Mg_2Ge	0.7	530	110	...	...
	Mg_2Si	0.8	370	65	...	0.46
	Mg_2Sn	0.4	210	150	...	...
II-V	Cd_3As_2	0.1	...	15000	0.05	...
	CdSb	0.5	300	1000	0.16	0.10
	Zn_3As_2	0.9	...	10	...	...
	ZnSb	0.5	10	350	0.15	...

Source: Ref 12, p 214

Typical properties of dielectric materials

Material	Electrical resistivity, Ω-cm	Dielectric strength, 10^5 V/cm
Ceramics:		
Alumina	10^{11} - 10^{14}	0.16-0.64
Corderite	10^{12} - 10^{14}	0.16-0.99
Forsterite	10^{14}	0.94
Porcelains:		
Dry process	10^{12} - 10^{14}	0.16-0.95
Wet process	10^{12} - 10^{14}	0.35-1.6
Zirconia	10^{13} - 10^{15}	0.99-1.6
Steatite	10^{13} - 10^{15}	0.79-1.6
Titanates (Ba, Sr, Ca, Mg, Pb)	10^{8} - 10^{15}	0.02-1.2
Titanium dioxide	10^{12} - 10^{18}	0.39-0.83
Plastics, resins:		
Allyl, cast	10^{13} - 10^{14}	1.5
Aniline formaldehyde	10^{16} - 10^{17}	2.5
Epoxy, cast	10^{16} - 10^{17}	1.6
Melamine formaldehyde	10^{11} - 10^{14}	1.5
Methyl methacrylate	10^{14} - 10^{15}	1.8
Nylons	10^{13} - 10^{15}	1.2-2.4
Phenol formaldehyde	10^{12} - 10^{13}	1.4
Polyethylene	10^{12} - 10^{13}	1.8
Polystyrene	10^{17} - 10^{19}	2.0
Rubber, hard	10^{15}	1.9
Shellac	10^{9}	0.8
Vinyl chloride	10^{14} - 10^{16}	2.4
Fiber-reinforced resins:		
Polyester	10^{13} - 10^{14}	1.4-2.0
Phenolic	10^{11} - 10^{12}	0.6-1.5
Epoxy	10^{14} - 10^{15}	1.4
Melamine	10^{10} - 10^{11}	0.7-1.2
Polyurethane	10^{11} - 10^{14}	1.3-3.5
Rubbers and elastomers:		
Polyisoprene	10^{15}	2.4
Butadiene	10^{15}	...
Styrene-butadiene	10^{14}	2.4
Acrylonite butadiene	10^{10}	1.9
Polychloroprene	10^{11}	2.0
Isobutylene-isoprene	10^{17}	3.0
Polysufide	10^{8}	1.3
Polymethane	10^{11}	2.0

Source: Ref 12, p 269

Characteristics of insulator materials

Material	Composition	Coefficient of thermal expansion, 10^{-6}/°C	Thermal conductivity, cal/cm·s·°C	Dielectric constant, MHz	Maximum temperature to resist, °C	Elastic constant, 10^4 kg/mm^2	Bending strength, kg/mm^2
Alumina	Al_2O_3	6.0-6.5	0.03-0.04	8.5-9.5	Over 1000	3.7-4.1	30-35
Silicon carbide	SiC	3.5	0.64	40	Over 1000	4.2	45
Silicon nitride	Si_3N_4	2.8-3.2	0.03-0.05	7.5	Over 1000	3.2	28-95
Aluminitride	AlN	5.7	0.14	8.9	Over 1000	3.5	50
Quartz glass	SiO_2	0.6	0.004	3.9	Over 800	0.7	6-10
Epoxy glass	...	10-30	0.001	5.0	150	...	45-52
Polyimide	...	66	0.0005	3.5	400	...	13-17

Source: Ref 13, p 271

Characteristics of solder/braze materials

Material	Composition	Coefficient of thermal expansion, $10^{-6}/°C$	Thermal conductivity, cal/cm·s·°C	Maximum temperature to resist, °C	Elastic constant, 10^4 kg/mm^2	Bending strength, kg/mm^2
Au-Sn eutectic	0.80 Au, 0.20 Sn	16	0.6	232	...	...
Au-Si eutectic	0.97 Au, 0.02-0.03 Si	10-12.9	0.68	370	0.71-0.77	26-31
Low-melting point solder	0.63 Sn, 0.37 Pb	24.7	0.121	183	...	4.9
High-melting point solder	0.95 Pb, 0.05 Sn	28.7	0.085	300	0.185	9.5
Silver wax	0.72 Ag	20.4	0.5	779	...	37

Source: Ref 13, p 273

Semiconductor Devices

Refractory metals and silicides deposited by PECVD

Material	Gas ambient	Deposition temperature, °C	As-deposited structure(a)	Minimum resistivity (after annealing), μΩ-cm
Tungsten	WF_6/H_2	350	c	7
Molybdenum	$MoCl_5/H_2$	400	a	10
W-silicide	$WF_6/SiH_4/He$	230	a	40
Mo-silicide	$MoCl_5/SiH_4/Ar/H_2$	400	a	...
Ta-silicide	$TaCl_5/SiH_2Cl_2/H_2$	<540	a	55
		>580	c	
Ti-silicide	$TiCl_4/SiH_4/Ar$	300-350	a	20
Ti-silicide (nonplasma)	$TiCl_4/SiH_4$	650-700	c	22(b)

(a) a = amorphous; c = crystalline. (b) As deposited.

Source: Ref 13, p 112

Characteristics of single-chip carriers

Carrier type	Greatest no. of I/O	Occupying space for pins, cm^2	Materials
Pin-through-hole (PTH).			
DIP(a)	64	7.74	Plastics and ceramics
PGA(b)	300	2.8	Plastics and ceramics
Surface-mount techonology (SMT)			
SOP(c)	40	3.9	Plastics
FPP(d)	120	2.4	Plastics
LCC(e)	132	1.9	Ceramics
PLCC(f)	68	2.0	Plastics

(a) Dual-in-line packages. (b) Pin grid array. (c) Small outline package. (d) Flat plastic package. (e) Leadless chip carrier. (f) Plastic leaded chip carrier.

Source: Ref 13, p 268

Problem areas in silicon device production

Oxidation	
The SiO_2/Si interface morphology and its movement as a function of processing conditions: correlation with electrical parameters	
Stress distribution at the oxide edges	
Integrity of the oxide after implantation doping	
Availability of silicon interstitials, interaction with impurities, OSF, and OED, gettering, precipitation, etc.	
Ion implantation	
Wafer-heating effect on amorphization	
Control of amorphous/crystalline interface	
Oxide-edge effect: curving of defect layer toward the junction	
Oxygen recoil: effect on defect stabilization	
Shallow junction degradation channeling enhanced, diffused, etc.	
Reliable preamorphization procedures	
Dopant precipitation, reduced electrical activation	
Gettering	
Control of oxygen precipitation	
Control of intrinsic gettering	
Understanding relationship between intrinsic and extrinsic gettering	
Annealing out of secondary defects in extrinsic gettering process	
Silicides	
Right choice of metal	
Stability at processing temperatures	
Degradation by oxidation	
Surface preparation	
Basic understanding of metal/Si → silicide/Si phase transformations	

Source: Ref 13, p 227

Typical NMOS process flowchart

- Si wafer
- Oxidation — pad oxide
- Backside damage/gettering
- Isolation:
  - Photolithography
  - Patterning
  - Nitridation
  - Field oxide
- Gate electrodes:
  - Gate oxide
  - LPCVD Si (poly)
  - Silidation
  - LPCVD oxidation
  - I.I (a) As junction
- Contacts:
  - Al-Si-Cu
  - Metallization
  - Silicides, etc.

(a) Ion implantation

Source: Ref 13, p 202

Electronic materials produced by chemical vapor deposition

Electronic Materials	
Superconductors	Nb_3Ga
Semiconductors	Si, GaAs, ZnS
Insulators	SiO_2, Si_3N_4
Metals	W, Ta
Magnetic garnets	$Y_3GaFe_4O_{12}$
High-Temperature Materials and High-Hardness Materials	
Metal films	W, Mo, Ta, B, V
Carbon films	Pyrolitic carbon
Carbides	SiC, B_4C, ZrC, TaC
Oxides	Al_2O_3, Y_2O_3
Nitrides	TaN_x
Silicides	MoSi, TaSi
Alloys	Rh-W, W-Mo

Source: Ref 13, p 65

Dimensionless numbers essential to chemical vapor deposition

Dimensionless number	Symbol	Physical meaning	Formula(a)	Range epi reactors
Reynolds	*Re*	Inertia force / Viscous force	$\frac{\rho VD}{\nu}$	25-200
Grashof	*Gr*	Inertia × bouyancy / Viscous force	$\frac{\beta g \Delta TL^3}{\nu^2}$	100-1000
Prandtl	*Pr*	Momemtum diffusivity / Thermal diffusivity	$\frac{\nu}{\alpha}$	0.4-0.8
Rayleigh	*Ra*	Gravity / Thermal diffusivity	$\frac{\beta g \Delta TL^3}{\alpha\nu}$	40-800

(a) ρ = density; V = velocity; D = diameter; μ = viscosity; β = temperature coefficient of expansion; T = temperature difference; ν = kinematic viscosity; α = thermal conductivity; g = gravitational constant; L = linear dimension.

Source: Ref 13, p 78

Deposition parameters for PECVD of silicon epitaxy

Deposition parameter	Value reported by: Townsend and Uddin	Susuki and Itoh	Donahue, Burger, and Reif
Deposition temperature, °C	800	750	775
Gas ambient	SiH_4/H_2	SiH_4	SiH_4
Operating pressure, torr	0.2-0.6	1×10^{-2}	1.5×10^{-2}
SiH_4 partial pressure, torr	...	1×10^{-2}	1.5×10^{-2}
Discharge frequency, MHz	27	13.56	13.56
RF power, W	350	200	20 0
Deposition rate, Å/min	200	1980	450 340

W.G. Townsend and M.E. Uddin, *Solid State Electronics*, Vol 16, 1973, p 39; S. Suzuki and T. Itoh, *J. Appl. Phys.*, Vol 54, 1983, p 1466; T.J. Donahue, W.R. Burger and R. Reif, *Appl. Phys. Lett.*, Vol 44, 1984, p 346; T.J. Donahue and R. Reif, *J. Appl. Phys.*, Vol 57, 1985, p 2757.

Source: Ref 13, p 109

Resistivity change of LPCVD and PECVD undoped polysilicon films upon annealing(a)

Deposition process	As-deposited resistivity, 10^6 Ω-cm	Annealed resistivity, 10^6 Ω-cm
LPCVD	13.8	2.1
PECVD	0.78	0.96

(a) The films were deposited at 650 °C and annealed for 1 h at 1100 °C in nitrogen.

Source: Ref 13, p 106

Diffusion and Electromigration

Characteristics of boron diffusion

Property	Result
B^-V^+ migration energy is ~0.6 eV	Relatively fast diffuser dominated by $[V^+]$
Small tetrahedral covalent radius	1. Small diffusion entropy 2. Misfit strain a. Dislocations b. Good gettering c. Reduced diffusivity
2.26 eV required to make B interstitial	Oxidation-enhanced diffusion
Forms stable oxides B_2O_3 and HBO_2	Segregation into growing SiO_2 from Si

Source: Ref 13, p 143

Electromigration of selected materials

Material	Behavior
Teflon	No migration
Polystyrene	No migration
Polyethylene	No migration
Polyvinyl chloride	No migration
Hard rubber (sand blasted)	No migration
Bakelite	No migration
Glass-filled silicone	No migration
Glass filled epoxy	Slight surface migration
Glass microscope slide	Slight surface migration
Glazed steatite	Slight surface migration
Laminated phenolic paper	Heavy migration in fibers
Cellophane	Heavy migration
Glass-filled melamine	Heavy migration
Glass-bonded mica	Heavy surface migration
Nylon 6	Extensive surface migration

Source: Ref 14, p 44

Estimated interstitial formation energies in silicon

Element	Interstitial formation energy, eV
Si	2.2
Al^{2+}	2.21
B	2.26
P	2.4
As	2.5

Source: Ref 13, p 131

Interdiffusion in solids

Distance of diffusion	Sample	Relationship between flux and driving force	Compound formation
~1 Å	Monolayer of metallic atoms of silicon	...	...
~10 Å	Man-made superlattices	Nonlinear	Single
~1000 Å	Bilayer thin films, film/substrate	Linear	Single
~1 mm	Bulk samples	Linear	Multiple

Source: Ref 13, p 149

Fractional interstitialcy components of diffusion via self-interstitials in silicon at 1000 to 1100 °C

Element	Fair	Antoniadis	$f_i = D_i^I/D_i^Y$ Matsumoto	Gosele	Mathiot
B	0.17	0.32	0.41	0.8-1.0	0.18
Al	0.2	...	...	0.6-0.7	...
P	0.12	0.40	0.35-0.5	0.5-1.0	0.19
As	0.09	0.43	0.45-0.75	0.2-0.5	0.16
Sb	0.13	0.15	...	0.02	...

R.B. Fair, *J. Appl. Phys.*, Vol 51, 1980, p 5828; D.A. Antoniadis and I. Moskowitz, *J. Appl. Phys.*, Vol 53, 1982, p 9214; S. Matsumoto, Y. Ishikawa and T. Niimi, Extended Abstracts of 15th Conf. on Solid State Devices and Materials, in *Jap. J. Appl. Phys.*, 1983, p 19; U. Gosele and T.Y. Tan, *Defects in Semiconductors II,* edited by S. Mahajan and J.W. Corbett, North Holland Press, Amsterdam, 1983; D. Mathiot and J.C. Pfister, *J. Appl. Phys.*, Vol 55, 1984, p 3518.

Source: Ref 13, p 131

Normalized average antimony diffusivity and calculated interstitial concentration at 1100 ° C

Oxidation time, min	D/D_i, measured	D/D_i^0, calculated	$(C_I)/C_I^0$	$(C_V)/C_V^0$
5	1.11	0.91	5.1	0.27
10	0.92	0.76	4.4	0.19
20	0.66	0.63	3.9	0.13
30	0.52	0.57	3.6	0.1
60	0.38	0.38	3.2	0.05

Source: Ref 13, p 134

Characteristics of arsenic diffusion

Property	Result
$\Delta S_D^- \gg \Delta S_D^x$	As^+V^- pair diffusion dominates at $T > 1050$ °C
$As^+V^=$ binding energy is ~0.25 eV less than $P^+V^=$	Fewer $As^+V^=$ pairs than $P^+V^=$ a. Less gettering of metal donors b. No emitter push above 700 °C c. No effect on $[O_i]$ precipitation
VAs_2 pair binding energy = 1.6 eV	1. Significant $[VAs_2]$ form at $n = 2 \times 10^{20}$ cm^{-3} at 1000 °C 2. Reduces n solubility 3. Causes retarded base diffusion in some cases
2.5 eV required to make As interstitial	Little oxidation-enhanced diffusion

Source: Ref 13, p 139

Characteristics of phosphorous diffusion

Property	Result
Large $\Delta S_D^x(11.6\,k)$	Diffusion via V^x is dominant at low concentrations
Large $P^+V^=$ pair binding energy (1.57 eV)	$P^+V^=$ pair dominates high concentration diffusion a. $P^+V^=$ pairs compensate monatomic P^+ b. Gettering of donor metal ions $\rightarrow P^+M^-$
$P^+V^=$ dissociation at $E_c - E_f = 0.11$ eV	1. Emitter push effect 2. Defect shrinkage near junction 3. Reduced $[O_i]$ precipitation
Small covalent radius	1. Misfit strain a. Gettering by misfit dislocations b. Reduced diffusivity through band gap narrowing 2. $\Delta S_{P-V}^- < \Delta S_{As-V}^-$

Source: Ref 13, p 140

Binary Phase Diagrams of Electronic Materials

All the following material has been extracted from Ref 11, Vol. 1 and Vol. 2

Al-As (Aluminum-Arsenic)

26.98154 74.9216

Few data are available for the Al-As system. However, the phase diagram is topologically equivalent to those of other III-V systems (for example, In-As and Ga-As). The equilibrium condensed phases consist of: (1) the liquid; (2) (Al), based on fcc Al; (3) (As), based on rhombohedral αAs (the high-temperature solid structure, with several other low-temperature forms occurring); and (4) the cubic compound AlAs, isotypic with ZnS, having a narrow or nil solubility range, and melting congruently at about 1760 °C.

The assessed phase diagram has been constructed from the liquid and AlAs model free energies of [81Kau], with the triple point used as the melting point for As. The studies of [72Fos] and [64Kis] indicated good agreement.

Al-As Crystal Structure Data

Phase	Composition, at.% As	Pearson symbol	Space group	Strukturbericht designation	Prototype
(Al)	0	*cF*4	*Fm*3*m*	*A*1	Cu
AlAs	50	*cF*8	*F*43*m*	*B*3	ZnS
(As)	100	*hR*2	*R*3*m*	*A*7	αAs

64Kis: W. Kischio, *Z. Anorg. Chem.*, *328*, 187-193 (1964) in German.

72Fos: L.M. Foster, J.E. Scardefield, and J.F. Woods, *J. Electrochem. Soc., Solid State Sci. Technol.*, *119*, 765-766 (1972).

81Kau: L. Kaufman, J. Nell, K. Taylor, and F. Hayes, *Calphad*, *5*, 185–215 (1981).

Complete evaluation contains 1 figure, 1 table, and 16 references.

Al-Nb (Aluminum-Niobium)

26.98154 92.9064

The Al-Nb system is characterized by three intermediate phases—Al_3Nb, $AlNb_2$, and $AlNb_3$. Al_3Nb melts congruently, and $AlNb_3$ is formed peritectically. Al_3Nb, the Al-rich compound in the system, has a very narrow homogeneity range. The assessed phase diagram is based primarily on the work of [66Lun]. A calculated diagram produced from thermodynamic data by [78Kau] is in qualitative agreement with the experimentally determined diagram.

Al-Nb Crystal Structure Data

Phase	Composition, at.% Nb	Pearson symbol	Space group	Strukturbericht designation	Prototype
(Al)	<0.3	*cF*4	*Fm*3*m*	*A*1	Cu
Al_3Nb	25	*tI*8	*I*4/*mmm*	$D0_{22}$	Al_3Ti
$AlNb_2$	58.7 to 68	*tP*30	$P4_2/mnm$	$D8_b$	σCrFe
$AlNb_3$	68 to 82	*cP*8	*Pm*3*n*	*A*15	Cr_3Si
(Nb)	77 to 100	*cI*2	*Im*3*m*	*A*2	W

66Lun: C.E. Lundin and A.S. Yamamoto, *Trans. AIME*, *236*, 863-872 (1966).

78Kau: L. Kaufman and H. Nesor, *Calphad*, *2*(4), 325-348 (1978).

Complete evaluation contains 9 figures, 2 tables, and 31 references.

Al-P (Aluminum-Phosphorus)

26.98154 30.97376

The phase diagram of the Al-P system in equilibrium with its vapor is assumed to be topologically similar to other III-V systems, such as In-Sb and As-Ga. Therefore, in addition to the vapor, four phases should appear:

(1) the liquid; (2) the fcc (Al) solid solution; (3) the (P) solid solution, based on the complex cubic white αP allotrope, which melts metastably with respect to the red and black forms, whose thermodynamic properties are not as well characterized; and (4) the intermediate phase, AlP, which has the cubic ZnS structure.

The assessed phase diagram has been constructed from model Gibbs energies for AlP and the liquid and is based primarily on the work of [44Whi], with review of the work of [72Pan], [73Ile], and [74Ile] on ternary and quaternary diagrams. Good agreement was found between the calculated boundaries and the few available experimental data.

Al-P Crystal Structure Data

Phase	Composition, at.% P	Pearson symbol	Space group	Struktur-bericht designation	Prototype
(Al)	0	*cF*4	*Fm*3*m*	*A*1	Cu
AlP	50	*cF*8	*F*43*m*	*B*3	ZnS (sphalerite)
(P)	100	(a)	...	...	αP

(a) Complex cubic.

44Whi: W.E. White and A.H. Bushey, *J. Am. Chem. Soc., 66,* 1666-1672 (1944).
72Pan: M.B. Panish and M. Ilegems, *Prog. Solid State Chem., 7,* 39-83 (1972).
73Ile: M. Ilegems and M.B. Panish, *J. Cryst. Growth, 20,* 77-81 (1973).
74Ile: M. Ilegems and M.B. Panish, *J. Phys. Chem. Solids, 35,* 409-420 (1974).

Complete evaluation contains 1 figure, 3 tables, and 15 references.

Al-S (Aluminum-Sulfur)

26.98154 32.06

αAl_2S_3: *hP**. γAl_2S_3: *D*5_1. βAl_2S_3: *B*4, stable only in the presence of carbon.

The Al-S phase diagram is redrawn from [Hansen].

Al-Sb (Aluminum-Antimony)

26.98154 121.76

Four equilibrium phases have been observed in the Al-Sb system: (1) the liquid; (2) the fcc (Al) solid solution; (3) the rhombohedral (Sb) solid solution; and (4) the intermetallic compound AlSb with the cubic ZnS structure at low pressure, which transforms to the bct βSn structure at about 120 kbar. Only negligible amounts of Sb dissolve in (Al), and the solubility of Al in (Sb) appears to be nil.

The assessed phase diagram has been obtained by optimizing model Gibbs energy functions and is based primarily on evaluation of the work of [06Tam], [19Ura], [33Gue], and [69Lic]. The modeling in the present evaluation, assuming nil solubility at both ends of the diagram, has placed the L ⇄ (Al) + AlSb eutectic at 657.5 °C and 0.43 at.% Sb, with the L ⇄ AlSb + (Sb) eutectic at 625.0 °C and 98.3 at.% Sb. The present parameterization gives good fit from 0 to 50 at.% Sb, as well as for the Sb-rich liquidus and the eutectic.

Al-Sb Invariant Points

Reaction	Composition, at.% Sb			Temperature, °C	Reaction type
L ⇄ (Al) + AlSb	0.40 ± 0.20	0	0.50	657.0 ± 2.5	Eutectic
L ⇄ AlSb + (Sb)	98.5 ± 0.5	0.50	100	627.0 ± 2.5	Eutectic
L ⇄ AlSb		...		1058 ± 10	Congruent melting
L ⇄ Al		...		660.452	Melting
L ⇄ Sb		...		630.755	Melting

Al-Sb Crystal Structure Data

Phase	Composition, at.% Sb	Pearson symbol	Space group	Strukturbericht designation	Prototype
(Al)	0	*cF*4	*Fm*3*m*	*A*1	Cu
AlSb	50	*cF*8	*F*43*m*	*B*3	ZnS
(Sb)	100	*hR*2	*R*3*m*	*A*7	αAs
High pressure					
AlSb(a)	50	*tI*4	*I*4_1/*amd*	*A*5	βSn

(a) At 120 kbar.

06Tam: Tamman, *Z. Anorg. Chem., 48,* 53-60 (1906) in German.
19Ura: G.G. Urasow, *Zh. Fiz. Khim., 51,* 461-471 (1919); also *Inst. Fiz. Khim. Anal., 1,* 416-417 (1921).
33Gue: W. Guertler and A. Bergmann, *Z. Metallkd., 25,* 81-84 (1933).
69Lic: B.D. Lichter and P. Sommelet, *Trans. Met. Soc. AIME, 245,* 99-105 (1969).

Complete evaluation contains 1 figure, 4 tables, and 25 references.

As-Cd (Arsenic-Cadmium)

74.9216 112.41

αCd_3As_2: *D*5_9. βCd_3As_2: *D*5_5?
The Cd-As phase diagram is redrawn from [Hansen].

As-Ga (Arsenic-Gallium)

74.9216 69.72

GaAs: B_3.
The Ga-As phase diagram is redrawn from [Hansen].

As-In (Arsenic-Indium)

74.9216 114.82

β: *B*3.
The In-As phase diagram is redrawn from [Metals].

As-Zn (Arsenic-Zinc)

74.9216 65.38

αZn_3As_2: *D*5_3. βZn_3As_2: *D*5_9. $ZnAs_2$: monoclinic.

The Zn-As phase diagram is redrawn from [Hansen].

Cd-S (Cadmium-Sulfur)

112.41 32.06

CdS: *B*4 is stable at ambient pressure; *B*3 can also be synthesized, e.g., from $CdSO_4 + H_2S$; *B*1 is stable at pressures above ~20 kbar.

The Cd-S phase diagram is redrawn from [Shunk].

Cd-Se (Cadmium-Selenium)

112.41 78.96

CdSe: *B*3, both *B*3 and *B*4 in thin films, *B*1 at ~30 kbar.

The Cd-Se phase diagram is redrawn from [Shunk].

Cd-Te (Cadmium-Tellurium)

112.41 127.60

β: *B*3, *B*1 above 36 kbar, Sn above 200 kbar, *B*4 in thin films.

The Cd-Te phase diagram is redrawn from [Metals].

Co-Ti (Cobalt-Titanium)

58.9332 47.88

The equilibrium solid phases of the Ti-Co system are: (1) the cph solid solutions, (αTi) and (ϵCo), where (αTi) is stable below 882 °C and (αCo) is stable below approximately 422 °C; (2) the bcc solid solution (βTi), stable in pure Ti above 882 °C, with a maximum solubility of Co in (βTi) of 14.5 at.% at 1020 °C; (3) the fcc solid solution (αCo), stable in pure Co above 422 °C, with a maximum solubility of Ti in (αCo) of 14.1 at.% at 1190 °C; (4) Ti_2Co, an ordered fcc structure with a homogeneity range of not more than about 0.3 at.% near stoichiometry; (5) TiCo, with the CsCl structure, which melts congruently at 1325 °C and has a homogeneity range of 49 ± 1 to 55 ± 0.5 at.% Co at 1200 °C; (6) cubic (*C*15) and hexagonal (*C*36) Laves phases of approximate stoichiometry $TiCo_2$, designated in the present evaluation as $TiCo_2$(c) and $TiCo_2$(h), respectively, where $TiCo_2$(h) is slightly richer in Co, because the homogeneity ranges of $TiCo_2$(c) and $TiCo_2$(h) are approximately 66.5 to 67.0 at.% Co and 68.75 to 72 at.% Co, respectively; and (7) $TiCo_3$, with the ordered fcc $AuCu_3$ structure and a maximum homogeneity range of 75.5 to 80.7 at.% Co.

The assessed phase diagram contains two "deep eutectics," in which amorphous alloys can be formed, and the higher order ferromagnetic transition has a pronounced effect on the solvus. However, some features of the diagram have received only cursory examination, or are the subject of controversy. Notably, the liquidus has not been determined from 0 to 20 at.% Co, and there are discrepancies between 20 and 80 at.% Co. There are conflicting reports about which of the $TiCo_2$ Laves phases are stable phases, and it has been suggested (based on observation of polytypism) that this part of the diagram may be more complex than had been imagined.

Conflicting reports suggest that only a hexagonal Laves phase $TiCo_2$(h) [50Duw] or only a cubic Laves phase $TiCo_2$(c) [83Uhr] is stable. By X-ray diffraction, [50Duw] found only $TiCo_2$(h) in alloys containing 63 to 68 at.% Co. [83Uhr] found only $TiCo_2$(c) formed from the liquid at 71 at.% Co. [59Fou] found only $TiCo_2$(h) in an as-cast 66.1 at.% Co sample.

Special Points of the Assessed Ti-Co Phase Diagram

Reaction	Composition, at.% Co			Temperature, °C	Reaction type
L ⇄ (βTi) + Ti_2Co	23.2	14.5	32.9	1020	Eutectic
(βTi) ⇄ (αTi) + Ti_2Co	7.0	0.86	33.2	685	Eutectoid
L + TiCo ⇄ Ti_2Co	27.1	49.0	33.1	1058	Peritectic
L ⇄ TiCo		50		1325	Congruent
TiCo + L ⇄ $TiCo_2$(c)	55.2	67.2	66.5	1235	Peritectic
$TiCo_2$(c) + L ⇄ $TiCo_2$(h)	67.0	71.0	68.75	1210	Peritectic
L ⇄ $TiCo_2$(h) + $TiCo_3$	75.8	72.0	77.2	1170	Eutectic
L + (αCo) ⇄ $TiCo_3$	79.3	85.9	80.7	1210	Peritectic
L ⇄ (βTi)		0		1670	Melting point
(βTi) ⇄ (αTi)		0		882	Allotropic transformation
L ⇄ (αCo)		100		1495	Melting point
(αCo) ⇄ (ϵCo)		100		421	Allotropic transformation

Ti-Co Crystal Structure Data

Phase	Composition, at.% Co	Pearson symbol	Space group	Strukturbericht designation	Prototype
(αTi)	0 to 0.8	*hP*2	*P*6$_3$/*mmc*	*A*3	Mg
(βTi)	0 to 14.5	*cI*2	*Im*3*m*	*A*2	W
Ti_2Co	32.9 to 33.3	*cF*96	*Fd*3*m*	*E*9$_3$	Fe_3W_3C
TiCo	49 to 55	*cP*2	*Pm*3*m*	*B*2	CsCl
$TiCo_2$(c)	66.5 to 67	*cF*24	*Fd*3*m*	*C*15	$MgCu_2$
$TiCo_2$(h)	68.75 to 72	*hP*24	*P*6$_3$/*mmc*	*C*36	$MgNi_2$
$TiCo_3$	75.5 to 80.7	*cP*4	*Pm*3*m*	*L*1$_2$	$CuAu_3$
(ϵCo)	~99 to 100	*hP*2	*P*6$_3$/*mmc*	*A*3	Mg
(αCo)	85.6 to 100	*cF*4	*Fm*3*m*	*A*1	Cu
ω(a)	(b)	*hP*3	*P*6/*mmm*	...	ωMnTi
(α''Co)	(b)	(c)	...	...	...

(**a**) The "ideal" ω structure is hexagonal, but a distorted trigonal form has also been observed in some Ti systems. The structure of ω in Ti-Co has not been definitively established. (**b**) Metastable. (**c**) Rhombohedral.

Using electron microscopy, [72All] found that Ti-Co exhibits polytypism: the stable hexagonal form can be described as a 4*H* stacking and 6*H* and 12-layer stackings were also found. These contradictory observations are at present unresolvable.

The minimum Co content for which (βTi) can be retained metastably during quenching ranges from 4.5 [59Yak] to 10 at.% Co [58Swa]. The composition reported depends on the investigator's judgment of whether (βTi) was fully retained, as well as on the quenching rate. [77Nik] found that whereas cph Co martensite is formed from 100 to 98 at.% Co, a faulted structure appeared between 98 and 96 at.% Co and that from 96 to 93 at.% Co the martensite has a 126-layer rhombohedral structure.

[80Ino] produced mixed crystalline and amorphous phases near 23 and 88 at.% Co by rapid solidification. Between 87 and 89 at.% Co, only the amorphous phase was present without any trace of the crystalline phase.

50Duw: P. Duwez and J.L. Taylor, *Trans. AIME, 188*, 1173-1176 (1950).

58Swa: P.R. Swann and J.G. Parr, *Trans. AIME, 212*, 276-279 (1958).

59Fou: R.W. Fountain and W.D. Forgeng, *Trans. AIME, 215*, 998-1008 (1959).

59Yak: F.W. Yakymyshyn, G.R. Prudy, R. Taggart, and J.G. Parr, *Trans. ASM, 53*, 283-294 (1959).

72All: C.W. Allen, P. Delavignette, and Amelinckx, *Phys. Status Solidi A, 9*, 237-246 (1972).

77Nik: B.I. Nikolin, *Dokl. Akad. Nauk SSSR, 233*, 587-590 (1977) in Russian; TR: *Sov. Phys. Dokl., 22*(4), 226-228 (1977).

80Ino: A. Inoue, K. Kobayashi, C. Suryanarayana, and T. Masumoto, *Scr. Metall., 14,* 119-123 (1980).

83Uhr: B. Uhrenius and K. Forsen, *Z. Metallkd., 74*(9), 610-615 (1983).

Complete evaluation contains 4 figures, 8 tables, and 60 references.

Ga-Nb (Gallium-Niobium)

69.72 92.9064

Nb_3Ga: *A*15. Nb_5Ga_3: $D8_m$; $D8_8$ with $O^=$ contamination. Nb_3Ga_2: $D5_a$. $NbGa_3$: $D0_{22}$.

The Nb-Ga phase diagram is redrawn from [Niobium].

Ga-Ni (Gallium-Nickel)

69.72 58.69

Ni_3Ga: $L1_2$. $\delta(Ni_{16}Ga_9)$: Ga_3Pt_5. $(\gamma,\gamma'?)Ni_{59}Ga_{41}$: In_3Pt_4. Ni_2Ga_3: $D5_{13}$. $NiGa_4$: $D8_2$.

The Ni-Ga phase diagram is redrawn from [Hansen] and [Elliott].

Ga-P (Gallium-Phosphorus)

69.72 30.97376

GaP: *B*3, *C*19 in thin film annealed at 773 K.

The Ga-P phase diagram is redrawn from [Shunk].

Ga-Pu (Gallium-Plutonium)

69.72 (244)

The assessed phase diagram for the Pu-Ga system is based primarily on the work of [64Ell], [67Hoc], and [76Che]. Elemental Pu metal exhibits six allotropes between ambient temperature and the melting point. The phase stability of these modifications is often a function of pressure and/or relatively small amounts of another element.

[64Ell] identified six intermediate phases—$Pu_3Ga(\zeta')$, Pu_5Ga_3, $PuGa(\iota)$, Pu_2Ga_3, $PuGa_2$, and $PuGa_3(\mu')$—that are stable at room temperature, and three phases—$Pu_3Ga(\zeta)$, $PuGa_3(\mu)$, and η—that are stable at elevated temperatures. Two additional phases, $PuGa_4$ and $PuGa_6(\xi)$, were also observed by [65Lan]. [67Hoc] confirmed the existence of these phases (except Pu_2Ga_3), as well as that of a $PuGa(\iota')$ polymorph existing at high temperatures. [76Che] confirmed all of these phases and also reported several new Ga-rich compounds at approximate compositions of Pu_2Ga_7, (Pu_3Ga_{11}-Pu_4Ga_{15}) and Pu_2Ga_{15}, as well as three $PuGa_3$ and two $PuGa_6$ modifications.

[64Ell] reported no significant solubility of Ga in αPu, βPu, γPu, or δ'Pu, with measured Ga solubilities

Special Points of the Pu-Ga Phase Diagram

Reaction	Composition, at.% Ga			Temperature, °C	Reaction type
$L \rightleftarrows \epsilon Pu$		0		640 ± 2	Fusion
$\epsilon Pu \rightleftarrows \delta' Pu$		0		483 ± 4	Allotropic
$\delta' Pu \rightleftarrows \delta Pu$		0		463 ± 2	Allotropic
$\delta Pu \rightleftarrows \gamma Pu$		0		320 ± 1	Allotropic
$\gamma Pu \rightleftarrows \beta Pu$		0		215 ± 1	Allotropic
$\beta Pu \rightleftarrows \alpha Pu$		0		125 ± 1	Allotropic
$(\epsilon Pu) + Pu_3Ga \rightleftarrows (\delta Pu)$	7.5	25	12.5	645 ± 5	Peritectoid
$L + \eta \rightleftarrows (\epsilon Pu)$	10	21	13	715 ± 5	Peritectic
$\eta \rightleftarrows (\epsilon Pu) + Pu_3Ga(\zeta)$	18.5	8.5	25	655 ± 4	Eutectoid
$\eta + (Pu_5Ga_3) \rightleftarrows Pu_3Ga(\zeta)$	22	35.5	25	677 ± 2	Peritectoid
$Pu_3Ga(\zeta) \rightleftarrows Pu_3Ga(\zeta')$		25		363 ± 10	Polymorphic
$\eta \rightleftarrows Pu_5Ga_3 + PuGa(\iota')$	39.5	37.5	50	767 ± 5	Peritectoid
$\eta \rightleftarrows Pu_5Ga_3$		37.5		790 ± 10	Congruent
$L + PuGa(\iota') \rightleftarrows \eta$	40	50	41.6	928 ± 7	Peritectic
$PuGa(\iota') \rightleftarrows PuGa(\iota)$		50		570 ± 5	Polymorphic
$L + PuGa_2 \rightleftarrows PuGa(\iota')$	49.5	66.7	50	979 ± 7	Peritectic
$PuGa(\iota') + PuGa_2 \rightleftarrows Pu_2Ga_3$	50	66.7	60	720	Peritectoid
$L \rightleftarrows PuGa_2$		66.7		1264 ± 10	Congruent
$PuGa_2 + L \rightleftarrows PuGa_3(\mu)$	66.7	78	75	1105 ± 10	Peritectic
$PuGa_3(\mu) \rightleftarrows PuGa_3(\mu')$		75		922 ± 1	Polymorphic
$PuGa_3(\mu') \rightleftarrows PuGa_3(\mu'')$		75		400	Polymorphic
$PuGa_3(\mu'') + PuGa_{3.7} \rightleftarrows Pu_2Ga_7$	75	78.7	77.8	200	Peritectoid
$PuGa_3(\mu') + L \rightleftarrows PuGa_{3.7}$	75	99.6	78.7	475 ± 5	Peritectic
$PuGa_{3.7} + L \rightleftarrows PuGa_4$	78.7	99.8	80	420 ± 2	Peritectic
$PuGa_4 + L \rightleftarrows PuGa_6(\xi)$	80	99.9	85.7	315 ± 5	Peritectic
$PuGa_4 \rightleftarrows PuGa_{3.7} + PuGa_6(\xi)$	78.7	80	85.7	150	Eutectoid
$PuGa_6(\xi) \rightleftarrows PuGa_6(\xi')$		85.7		75	Polymorphic
$PuGa_6(\xi') + L \rightleftarrows Pu_2Ga_{15}$	85.7	99.9	88.2	50	Peritectic
$L \rightleftarrows Ga$		100		29.7741	Fusion

Solubility of Pu in Ga

Temperature, °C	Solubility, at.% Pu
496 ± 5	0.24
500 ± 5	0.17
597 ± 5	1.01
706 ± 5	3.2
811 ± 5	6.4
907 ± 5	10.6

Based on [64Ell].

Pu-Ga Crystal Structure Data

Phase	Composition, at.% Ga	Pearson symbol	Space group	Strukturbericht designation	Prototype
(εPu)	0 to 13	*cI*2	*Im*3*m*	*A*2	W
(δ′Pu)	0.25	*tI*2	*I*4/*mmm*	*A*6	In
(δPu)	0 to 12.5	*cF*4	*Fm*3*m*	*A*1	Cu
(γPu)	0	*oF*8	*Fddd*	···	γPu
(βPu)	0	*mC*34	*I*2/*m*	···	βPu
(αPu)	0	*mP*16	$P2_1/m$	···	αPu
η	18.5 to 42	···	$I2_13$(a)	···	···
Pu_3Ga(ζ)	25	*cP*4	*Pm*3*m*	$L1_2$	$AuCu_3$
Pu_3Ga(ζ′)	25	*tP*4	*P*4/*mmm*		Pb_3Sr
Pu_5Ga_3	36 to 37.5	*tI*38	*I*4/*mcm*	$D8_m$	Si_3W_5
		(b)	···	···	···
PuGa(ι′)	50	*cI*2	*Im*3*m*	*A*2	W
		(c)	*I*4/*mmm*	···	···
PuGa(ι)	50	(d)	*I*4*mm*	···	···
Pu_2Ga_3	60	(e)	···	···	···
$PuGa_2$	66.7	*hP*3	*P*6/*mmm*	*C*32	AlB_2
$PuGa_3$(μ)	75	Unknown	···	···	···
$PuGa_3$(μ′)	75	*hP*8	$P6_3/mmc$	$D0_{19}$	Ni_3Sn
$PuGa_3$(μ″)	75	···	*R*3*m*	···	···
Pu_2Ga_7	77.8	(c)	···	···	···
$PuGa_{3.7}$	78.7	Unknown	···	···	···
$PuGa_4$	80	*oI*20	*Imma*	$D1_b$	Al_4U
$PuGa_6$(ξ)	85.7	···	*P*4/*nbm*	···	$PuGa_6$(ξ)
			*P*4/*mbm*		
$PuGa_6$(ξ′)	85.7	Unknown	···	···	···
Pu_2Ga_{15}	88.2	(c)	···	···	···
(Ga)	100	*oC*8	*Cmca*	*A*11	Ga

(**a**) Partially ordered. (**b**) Face-centered cubic. (**c**) Tetragonal. (**d**) Body-centered tetragonal. (**e**) Hexagonal.

in γPu and δ′Pu of <0.25 at.%. They found that the maximum Ga solubilities in (δPu) and (εPu) were 12.5 and 20 at.%, respectively.

[64Ell] and [67Hoc] determined that (δPu) alloys containing between 2 and 8 at.% Ga are retained to room temperature. Conversely, from examinations of samples annealed under high pressure, [76Che] concluded that retained (δPu) is metastable at atmospheric pressure and indicated that (δPu) undergoes a eutectoid decomposition at about 100 °C to form αPu and a Pu_3Ga compound. Phase transformations under high pressure are difficult to interpret. [65Ell], [67Gar], and [67Rou] concluded that (δPu) transforms to (αPu) under high pressures, but that the transformation is reversible at Ga concentrations above 3.3 at.% Ga. Retained (δPu) is assumed to be stable down to room temperature.

64Ell: F.H. Ellinger, C.C. Land, and V.O. Struebing, *J. Nucl. Mater., 12,* 226-236 (1964).

65Ell: R.O. Elliott and K.A. Gschneidner, Jr., Los Alamos Sci. Lab Rep. LA-2312 (1965).

65Lan: C.C. Land, F.H. Ellinger, and K.A. Johnson, *J. Nucl. Mater., 16,* 87 (1965).

67Gar: H.R. Gardner, *Plutonium 1965,* Proc. 3rd Int. Conf. Plutonium, London, 1965, A.E. Kay and M.B. Waldron, Ed., Chapman and Hall, London, 118-132 (1967).

67Hoc: B. Hocheid, A. Tanon, S. Bedere, J. Despres, S. Hay, and F. Miard, *Plutonium 1965,* Proc. 3rd Int. Conf. Plutonium, London, 1965, A.E. Kay and M.B. Waldron, Ed., Chapman and Hall, London, 321-340 (1967).

67Rou: C. Roux, P. LeRoux, and M. Rapin, *Mem. Sci., R. de Met., 62,* 691-692, in French; *Plutonium 1965,* Proc. 3rd Int. Conf. Plutonium, London, 1965, A.E. Kay and M.B. Waldron, Ed., Chapman and Hall, London, 133-136 (1967).

76Che: N.T. Chebotarev, E.S. Smotriskaya, M.A. Andrianov, and O.E. Kostyuk, *Plutonium 1975 and Other Actinides,* H. Blank and R. Lindner, Ed., North-Holland, Amsterdam, 37-45 (1976).

Complete evaluation contains 1 figure, 7 tables, and 27 references.

Ga-Sb (Gallium-Antimony)

69.72 121.75

β: *B*3.

The Ga-Sb phase diagram is redrawn from [Metals].

Ga-V (Gallium-Vanadium)

69.72 50.9415

The equilibria and crystallography of the V-Ga system are affected by relatively minor amounts of interstitial contaminants, particularly oxygen and nitrogen. Consequently, significant differences exist among published diagrams. As such, the assessed V-Ga phase diagram represents a critical selection from the composite of available data. It is based primarily on a review of the work of [63Sch], [64Hau], [64Sav], [65Mei], [66Mai], [67Sav], and [68Saz]. In the composition range 0 to 40 at.% Ga, the essential features of the equilibria are well established; the uncertainties are in the precision associated with the temperature and composition coordinates of the phase boundaries.

There is evidence for the existence of five intermediate phases, and peritectic melting occurs for both V_2Ga_5 and V_8Ga_{41}. However, in the composition range 40 to 95 at.% Ga in the temperature interval 900 to 1300 °C, a diversity of equilibria have been reported, and the details, particularly the monotectic reaction, are more uncertain.

V_3Ga has the cubic Cr_3Si-type structure with a superconducting transition temperature near 16.8 K. Below room temperature, the Cr_3Si-type structure of V_3Ga persists down to about 20 K, where the phase undergoes a martensitic transformation analogous with the transformations in other V_3X and Nb_3X superconductors.

There is general agreement that eutectoidal equilibrium exists between V_3Ga, the bcc terminal solution, and V_6Ga_5 at some temperature in the range 1000 to 1100 °C. V_6Ga_5 is hexagonal and isostructural with αTi_6Sn_5. A congruent solid-state transformation to the bcc terminal solution of Ga in (V) has been reported for the phase, but a peritectoidal decomposition of V_6Ga_5 to form the bcc terminal solution plus a more Ga-rich phase near V_6Ga_7 also has been reported and is more likely. Because of a high probability of oxygen contamination, a phase reported at the stoichiometry of V_5Ga_3 [67Sav] is unlikely to be a true equilibrium phase in the binary system.

The decomposition of V_6Ga_7 at 1015 °C was reported to be by eutectoidal reaction and at 1150 °C by peritectic reaction. The temperatures for these reactions in the assessed diagram represent reasonable compromises among the composite data; most of the investigations show the stoichiometry of the phase at both the eutectoidal and peritectic reactions as being nearer to V_6Ga_7 than to VGa. The monotectic reaction in association with V_6Ga_7 is the least certain feature of the diagram.

V_2Ga_5 is tetragonal with the Mn_2Hg_5-type structure. This phase undergoes peritectic melting between 1015 and 1085 °C, and an eclectic temperature of 1050 ± 20 °C was selected for the assessed phase diagram.

Solubility of V in solid (Ga) is virtually nil, so the equilibrium between solid (Ga), V_8Ga_{41}, and the liquid occurs, for practical purposes, at the melting point of Ga.

V-Ga Crystal Structure Data

Phase	Composition, at.% Ga	Pearson symbol	Space group	Strukturbericht designation	Prototype
(V)	0 to 41	*cI*2	*Im*3*m*	*A*2	W
V_3Ga	20 to 32	*cP*8	*Pm*3*n*	*A*15	Cr_3Si
V_6Ga_5	45.5	*hP*22	...	...	αTi_6Sn_5
V_6Ga_7	51 to 55	*cI*52	$I\bar{4}3m$	$D8_2$	Cu_5Zn_8
V_2Ga_5	71.4	*tP*14	...	...	Mn_2Hg_5
V_8Ga_{41}	83.7	*hR*49	...	...	V_8Ga_{41}
(Ga)	100	*oC*8	*Cmca*	*A*11	αGa

63Sch: K. Schubert, K. Frank, R. Gohle, A. Maldonado, A. Raman, and W. Rossteutscher, *Naturwissenschaften, 50,* 41 (1963).

64Hau: J.J. Hauser, *Phys. Rev. Lett., 13,* 470-471 (1964).

64Sav: E.M. Savitskii, P.I. Kripyakevich, V.V. Baron, and Yu. V. Efimov, *Zh. Neorg. Khim., 9*(5), 1155 (1964) in Russian; TR: *Russ. J. Inorg. Chem., 9,* 631-633 (1964).

65Mei: H.G. Meissner and K. Schubert, *Z. Metallkd., 56,* 475-484 (1965).

66Mai: R.G. Maier, Y. Uzel, and H. Kandler, *Z. Naturforsch, 21,* 531-540 (1966).

67Sav: E.M. Savitskii, P.I. Kripyakevich, V.V. Baron, and Yu. V. Efimov, *Izv. Akad. Nauk SSSR, Neorg. Mater., 3,* 35-42 (1967) in Russian; TR: *Inorg. Mater. (USSR), 3*(1), 45 (1967).

68Saz: N.P. Sazhin, N.S. Vorobeva, Y.N. Kunakov, and G.N. Ronami, *Dokl. Akad. Nauk SSSR, 178,* 341-344 (1968) in Russian; TR: *Sov. Phys.-Dokl., 13*(1), 66-70 (1968).

Complete evaluation contains 1 figure, 1 table, and 29 references.

Ge-Mg (Germanium-Magnesium)

72.59 24.305

The equilibrium phases of the Mg-Ge system are: (1) the liquid; (2) the terminal solid solution, (Mg), with the *hP*2-type structure and with negligible solid solubility of Ge in Mg; (3) the terminal solid solution, (Ge), with the *cF*8-type structure (the solid solubility of Mg in Ge is very low); and (4) the Mg_2Ge compound, a semiconductor, of essentially stoichiometric composition with the antifluoride *cF*12-type structure. The assessed phase diagram is based primarily on the thermal analyses of [68Gef] and [40Ray], with evaluation of the work of [41Kle], [66Bea], [66Eld], and [71Rao].

The eutectic composition and temperature on the Mg-rich side, L $\rightleftarrows$ (Mg) + Mg_2Ge, was given by [68Gef] as 1.15 at.% Ge and 635.6 °C, and on the Ge-rich side, L $\rightleftarrows$ Mg_2Ge + (Ge), as 64.3 at.% Ge and 696.7 °C. The melting point of Mg_2Ge was placed at 1117.4 °C by [68Gef]. Between the two eutectic points, the thermodynamic analysis of [66Eld], using the isopiestic technique for alloys, placed the liquidus curves at temperatures close to those of [68Gef].

At pressures higher than 2.06 MPa and at temperatures up to ~1500 °C, Mg_2X (where X = Si, Ge, and Sn) exhibited polymorphic transformations. For magnesium germanide (Mg_2Ge), its cubic *Fm3m* antifluoride structure transformed to the hexagonal crystal structure, after exposure to temperatures in the range of 600 to 1200 °C and pressures of 2.53 to 5.57 MPa. The new structure acted as a semiconductor as well, but with a different resistivity than the cubic Mg_2Ge.

The Mg_2Ge compound is a semiconductor with an energy gap of 0.74 eV. The high-temperature and pressure structure was noted to be indefinitely metastable at ambient conditions after the release of pressure. At high pressures, the cubic structure of Ge transforms to a tetragonal structure.

Mg-Ge Crystal Structure Data

Phase	Composition, at.% Ge	Pearson symbol	Space group	Strukturbericht designation	Prototype
(Mg)	~0	*hP*2	$P6_3/mmc$	*A*3	Mg
Mg_2Ge	33.33	*cF*12	*Fm*3*m*	*C*1	CaF_2
(Ge)	~100	*cF*8	*Fd*3*m*	*A*4	C (diamond)

The calculated and the assessed Mg-Ge phase diagrams are in good agreement, with deviations of less than 10 °C, except in a few areas.

40Ray: G.V. Raynor, *J. Inst. Met., 66,* 403-426 (1940).
41Kle: W. Klemm and H. Westlinning, *Z. Anorg. Chem., 245,* 365-380 (1941).
66Bea: P. Beardmore, B.W. Howlett, B.D. Lichter, and M.B. Bever, *Trans. AIME, 236,* 102-108 (1966).
66Eld: J.M. Eldridge, E. Miller, and K.L. Komarek, *Trans. AIME, 236,* 1094-1098 (1966).
68Gef: R. Geffken and E. Miller, *Trans. AIME, 242,* 2323-2328 (1968).
71Rao: Y.K. Rao and G.R. Belton, *Metall. Trans., 2,* 2215-2219 (1971).

Complete evaluation contains 3 figures, 7 tables, and 28 references.

Ge-Nb (Germanium-Niobium)

72.59 92.9064

Nb_6Ge? or β: *h*∗∗. Nb_5Ge_3: $D8_l$ or $D8_m$? $NbGe_2$: *C*40.

The Nb-Ge phase diagram is redrawn from [Moffatt].

Ge-Si (Germanium-Silicon)

72.59 28.0855

The equilibrium phases of the Ge-Si system are: (1) the liquid; and (2) the cubic, diamond-type, substitutional solid solution. The assessed phase diagram is based primarily on the X-ray and thermal data of [39Sto].

The solid solution may transform to a two-phase mixture at low temperatures and may develop a bct structure at higher pressures. Anomalous thermal arrests, at approximately 10 °C below the melting point of Ge, were also reported by [39Sto] during normal cooling of various alloys from the molten state, but not during heating.

The Ge-Si system has also been studied theoretically, using an electronic theory based on pseudopotentials [74Bub, 75Alt]. Under high pressure, the Ge-Si solid solution was found to transform to a two-phase mixture at low temperatures (below 300 K). A bct structure of the Ge-Si solid solution, at high pressures (above 300 kbar), was found to be more stable than the diamond-type structure at lower pressures [82Som].

Amorphous Ge-Si films, a few micrometers thick, of various compositions have been prepared on fused silica substrates by RF sputtering in an argon atmosphere.

Phase equilibria based on strictly regular solid and liquid solutions agree well with the measured diagram, with enthalpy of mixing coefficients in the solid and liquid of 6.50 and 3.50 kJ/mol, respectively.

Ge-Si Crystal Structure Data

Phase	Composition, at.% Si	Pearson symbol	Space group	Strukturbericht designation	Prototype
(Ge,Si)	0 to 100	*cF*8	*Fd*3*m*	*A*4	C (diamond)
High-pressure phases					
Ge	...	*tI*4	*I*4$_1$/*amd*	*A*5	βSn
Si	...	*tI*4	*I*4$_1$/*amd*	*A*5	βSn

Note: At 25 °C.

39Sto: H. Stohr and W. Klemm, *Z. Anorg. Chem.*, *241*, 305-323 (1939) in German.
74Bub: V.T. Bublik, S.S. Gorelik, A.A. Zaitsev, and A.Y. Polyakov, *Phys. Status Solidi (b)*, *65*, K79-K84 (1974).
75Alt: A.M. Altshuler, Yu.Kh. Vekilov, and G.R. Umarov, *Phys. Status Solidi (b)*, *69*, 661-671 (1975).
82Som: T. Soma, H. Iwanami, and H. Matsuo, *Solid State Commun.*, *42*, 469-471 (1982).

Complete evaluation contains 2 figures, 3 tables, and 27 references.

In-La (Indium-Lanthanum)

114.82 138.9055

In_3La: *L*1$_2$. In_2La: Cu_2Ce. In_5La_3: Pd_5Pu_3. InLa: *B*2. $InLa_2$: *B*8$_2$. $InLa_3$: *L*1$_2$.

The In-La phase diagram is redrawn from [Moffatt].

In-P (Indium-Phosphorus)

114.82 30.97376

β: *B*3.

The In-P phase diagram is redrawn from [Metals].

In-Sb (Indium-Antimony)

114.82 121.75

β: *B*3.

The In-Sb phase diagram is redrawn from [Metals].

Mg-Si (Magnesium-Silicon)

24.305 28.0855

The equilibrium phases of the Mg-Si system are: (1) the liquid, L; (2) the terminal Mg solid solution, (Mg), with a maximum solid solubility of 0.003 at.% Si; (3) the stoichiometric intermetallic compound, Mg_2Si; and (4) the terminal Si solid solution, (Si), with negligible solid solubility of Mg.

[09Vog] determined the liquidus across the phase diagram by thermal analysis of 18 Mg-Si alloys made of low-grade Si. [40Ray] accurately determined the (Mg) liquidus. Recently, [68Gef] and [77Sch] redetermined the liquidus by thermal analysis of high-purity alloys in the composition ranges of 10 to 63 and 0 to 100 at.% Si, respectively. The liquidus curves of these investigators, in the common range of composition, are only a few degrees apart. Consequently, in the assessed phase diagram, the (Mg) liquidus has been taken from [40Ray], and the liquidus for the remainder of the diagram has been based on select data of [68Gef] and [77Sch]. The work of [26Woh] and [81Rao] was also evaluated in construction of the assessed diagram.

The general characteristics of the Mg-Si equilibrium phase diagram, first reported by [09Vog], are well established and generally accepted. The controversy of two different liquidus curves 30 °C apart, as reported in [Hansen], has been settled.

[35Elc] noted the existence of a metastable Mg-$MgZn_2$-Mg_4Si system and concluded that the Mg-Mg_4Si system is a metastable system with a eutectic at 2.34 at.% Si and 575 °C. [77Sch] reported the possible existence of the MgSi compound that would decompose on heating by a peritectoid reaction ($Mg_2Si + Si \rightleftarrows 2MgSi$) at ≈805 °C.

At high pressure, metallic Si has a tetragonal structure. After being subjected to 16.0 or 20.0 GPa pressure at room temperature, Si had a bcc structure. When heated in air at 200 to 600 °C for up to 3 days, bcc Si formed a new hexagonal structure. AB_2 substances, crystallizing with the fluorite (*Fm3m*) structure, may exhibit any or all of the following types of transformations: (1) a martensitic transformation to a hexagonal phase; (2) a polymorphic transformation from the *Fm3m* structure to the structure of either *A* or *B* in the AB_2 compound; and (3) another transformation based on the other of the two components.

At pressures above ≈2.5 GPa and temperatures above 900 °C, Mg_2Si transforms from a cubic fluorite structure to a hexagonal structure. The high-pressure hexagonal phase was noted as being indefinitely metastable. Following the release of pressure, the hexagonal structure did not change after more than a year at ambient conditions. An alloy of Mg−50 at.% Si, subjected to 3.0 GPa pressure at 900 °C, exhibited a texture similar to that of a martensitic steel.

At ordinary temperatures (25 to 400 °C) and high pressures (up to 10.0 GPa), the Mg_2Si compound undergoes polymorphic transformations. The phase change (from cubic to hexagonal) was accompanied by a large hysteresis. The region where the two phases (cubic and hexagonal Mg_2Si) coexisted extended up to 10.0 GPa at 400 °C. Similar transformations were found in intermetallic compounds of Mg with other elements of group IVB (Mg_2Sn and Mg_2Ge).

Invariant Reactions in the Mg-Si System

Reaction	Composition of L, at.% Si	Temperature, °C	Reaction type
$L \rightleftarrows (Mg) + Mg_2Si$	1.16	637.6	Eutectic
$L \rightleftarrows Mg_2Si + (Si)$	53.00	945.6	Eutectic
$L \rightleftarrows Mg_2Si$	33.3	1085	Congruent

Mg-Si Crystal Structure Data

Phase	Composition, at.% Si	Pearson symbol	Space group	Strukturbericht designation	Prototype
(Mg) or α	~0	*hP*2	*P*6_3/*mmc*	*A*3	Mg
Mg_2Si	33.33	*cF*12	*Fm*3*m*	*C*1	CaF_2
(Si) or β	~100	*cF*8	*Fd*3*m*	*A*4	C (diamond)
High-pressure phases					
Si	100	...	...	...	βSn
	100(a)	...	...	...	...
	100(b)	...	...	...	...
Mg_2Si(c)	33.33	...	...	...	...

(**a**) Subjected to 16 or 20 GPa at room temperature. (**b**) bcc Si, heated in air, 200 to 600 °C for 3 days, to form hexagonal structure. (**c**) Above ~2.5 GPa and 900 °C, it forms hexagonal structure.

09Vog: R. Vogel, *Z. Anorg. Chem., 61,* 46-53 (1909) in German.
26Woh: L. Wohler and O. Schliephake, *Z. Anorg. Chem., 151,* 1-20 (1926) in German.
35Elc: E. Elchardus and P. Laffitte, *C.R. Hebd. Seances Acad. Sci., 200,* 1938-1940 (1935).
40Ray: G.V. Raynor, *J. Inst. Met., 66,* 403-426 (1940).
68Gef: R. Gefficen and E. Miller, *Trans. Met. Soc. AIME, 242,* 2323-2328 (1968).
77Sch: E. Schurmann and A. Fischer, *Giessereiforschung, 29,* 111-113 (1977) in German.
81Rao: Y.K. Rao and G.R. Belton, *Chemical Metallurgy—A Tribute to Carl Wagner,* N.A. Gokcen, Ed., The Metallurgical Society of AIME, 75-96 (1981).

Complete evaluation contains 2 figures, 8 tables, and 49 references.

Mg-Sn (Magnesium-Tin)

24.305 118.69

Two eutectic reactions and a congruently melting line compound, Mg_2Sn, characterize the Mg-Sn system. The phase diagram originally was determined by [05Gru] and [05Kur], and the assessed diagram is based primarily on their work, with modifications based on the work of [40Ray], [41Vos], [64Ste], [68Nay], [69Nay], and [73Ell].

Calculation of the Mg-Sn phase diagram by the present authors is in good agreement with thermodynamic modeling of the Mg-Sn liquidus. Modeling of the Mg-Sn system indicated that liquid Mg-Sn alloys can be modeled satisfactorily by assuming an ideal mixture of four species (Mg, Sn, MgMg-Sn, and MgSn-Mg associates). Modeling based on an ideal mixture of three species (Mg, Sn, and Mg_2Sn associates) was inadequate.

The present authors have calculated the liquidus, assuming (Mg) to be a Raoultian solid solution, the heat of mixing to be independent of temperature, and nil solid solubility of Mg in Sn.

Despite a favorable size factor, the solubility of Sn in Mg is restricted by the formation of the stable Mg_2Sn compound. The maximum solid solubility of Sn in Mg is 3.35 at.% Sn at 561.2 °C. The maximum solid solubility of Mg in βSn has not been determined, but appears to be infinitesimally small.

A metastable phase has been reported in Sn-rich Mg-Sn alloys that were rapidly quenched from the melt to low temperatures. This metastable phase was unstable at room temperature and decomposed on heating at −100 to −40 °C. The structure of the metastable phase at −190 °C was found to be hexagonal.

Formation of a metastable phase with an fcc structure was reported in splat-cooled Mg-rich Mg-Sn alloys. In an Mg–15 at.% Sn alloy, the lattice parameter of the metastable fcc phase was determined to be 0.4516 ± 9 nm at room temperature. This metastable phase decomposed at room temperature to the equilibrium phases.

Another metastable phase was found on the Sn-rich side of the Mg_2Sn compound. X-ray analysis of this phase produced diffraction pattern lines that were in fair agreement with those reported for a phase obtained in vapor-deposited films of a Mg-Sn alloy. This phase appeared to be metastable from room temperature to ~150 °C for Mg-rich specimens and from room temperature to ~325 °C for Sn-rich specimens.

The cubic Mg_2Sn, when exposed to temperatures of 600 to 1200 °C and pressures of 2.5 to 5.5 GPa, transforms to a hexagonal structure [64Can]. This transformation occurs readily and at a faster rate above 770.5 °C, the melting point of Mg_2Sn at atmospheric pressure. Large grains of the new phase (hexagonal Mg_2Sn) were formed on cooling. The temperature-pressure relationship of this transformation has been determined [72Dyu, 76Dyu]. The high-pressure structure can be preserved by quenching from a high temperature, followed by the release of pressure.

Invariant Transformations in the Mg-Sn System

Reaction	Composition, at.% Sn			Temperature, °C	Reaction type
L ⇄ (Mg) + Mg_2Sn	10.7	3.35	...	561.2 ± 0.3	Eutectic
L ⇄ Mg_2Sn		33.33		770.5	Congruent
L ⇄ Mg_2Sn + (βSn)	90.4	...	100	203.5 ± 0.3	Eutectic
Pure components					
L ⇄ (Mg)		~0		650	Congruent
L ⇄ (βSn)		~100		231.9681(a)	Congruent
(βSn) ⇄ (αSn)		~100		13.05	Allotropic

(a) Defined fixed point on IPTS-68.

Mg-Sn Crystal Structure Data

Phase	Composition, at.% Sn	Pearson symbol	Space group	Strukturbericht designation	Prototype
(Mg)	0 to 3.35	*hP*2	$P6_3/mmc$	*A*3	Mg
Mg_2Sn	33.33	*cF*12	*Fm*3*m*	*C*1	CaF_2
(βSn), white Sn	100	*tI*4	$I4_1/amd$	*A*5	Sn
(αSn), gray Sn	100	*cF*8	*Fd*3*m*	*A*4	C (diamond)
Pressure stabilized					
SnII	...	*tI*2	...	...	...
SnIII	...	*cI*2	...	...	...
Mg_2Sn	...	(a)	...	...	...

(a) Hexagonal.

05Gru: G. Grube, *Z. Anorg. Chem., 46,* 76-84 (1905) in German.

05Kur: N.S. Kurnakow and N.J. Stepanow, *Z. Anorg. Chem., 46,* 177-192 (1905) in German.

40Ray: G.V. Raynor, *J. Inst. Met., 6,* 403-426 (1940).

41Vos: H. Vosskuhler, *Metallwirtschaft, 20,* 805-808 (1941) in German.

64Can: P. Cannon and E.T. Conlin, *Science, 145,* 487-489 (1964).

64Ste: A. Steiner, E. Miller, and K.L. Komarek, *Trans. Met. Soc. AIME, 230,* 1361-1367 (1964).

68Nay: A.K. Nayak and W. Oelsen, *Trans. Indian Inst. Met.,* 53-58 (1969).

72Dyu: T.I. Dyuzheva, S.S. Kabalkina, and L.F. Vereshchagin, *Kristallografiya, 17*(4), 804-811 (1972) in Russian; TR: *Sov. Phys. Crystallogr., 17*(4), 705-710 (1973).

76Dyu: T.I. Dyuzheva, S.S. Kabalkina, and L.F. Vereshchagin, *Proc. Akad. Nauk SSSR, 228*(5), 1073-1075 (1976) in Russian.

73Ell: J. Ellmer, K.E. Hall, R.W. Kamphefner, J.T. Pfeifer, V. Stamboni, and C.D. Graham, *Metall. Trans., 4,* 889-891 (1973).

Complete evaluation contains 2 figures, 15 tables, and 75 references.

Mo-Si (Molybdenum-Silicon)

95.94 28.0855

The equilibrium phases of the Mo-Si system are: (1) the liquid, L; (2) the terminal solid solution, (Mo), with a maximum solubility of approximately 4 at.% at the peritectic temperature of 2025 °C; (3) the terminal solid solution, (Si), with negligible solubility; (4) the cubic intermediate phase, Mo_3Si, which decomposes peritectically at 2025 °C; (5) the tetragonal phase, Mo_5Si_3, which melts congruently at 2180 °C; and (6) the tetragonal phase, $\alpha MoSi_2$, which transforms to hexagonal $\beta MoSi_2$ at 1900 °C and melts congruently at 2020 °C. The assessed phase diagram is based primarily on the data of [71Sve] and [74Cha].

The present diagram has been modeled using select thermodynamic data from [74Cha]. The (Mo) phase was assumed to obey Henry's law, and the solid solubility of Mo in (Si) was assumed to be nil. The liquidus is partially speculative due to the lack of experimental data. However, the present thermodynamic calculations indicate the enthalpy of mixing in the liquid to be strongly negative.

[71Sve] determined that the eutectic between Mo_3Si and Mo_5Si_3 occurs at 2020 °C and 26.4 at.% Si and also determined that the intermediate $MoSi_2$ phase undergoes a polymorphic transformation at 1900 °C. [71Sve] further indicated the presence of a reaction involving $\beta MoSi_2$, Mo_5Si_3, and $\alpha MoSi_2$ at 1850 °C, which the present evaluators view as speculative, requiring further investigation.

Monovariant and Invariant Equilibria in the Mo-Si System

Reaction	Composition, at.% Si			Temperature, °C	Reaction type
Mo ⇄ L	0			2623	Melting point
(Mo) + L ⇄ Mo_3Si	4	25.72	25	2025	Peritectic
L ⇄ Mo_3Si + Mo_5Si_3	26.4	25	37	2020	Eutectic
Mo_5Si_3 ⇄ L		37.5		2180	Congruent
L ⇄ Mo_5Si_3 + $\beta MoSi_2$	54	40	66.7	1900	Eutectic
$\beta MoSi_2$ ⇄ L		66.7		2020	Congruent
$\beta MoSi_2$ ⇄ $\alpha MoSi_2$		66.7		1900	Polymorphic
L ⇄ $\alpha MoSi_2$ + (Si)	98.3	66.7	100	1400	Eutectic
Si ⇄ L		100		1414	Melting point

Mo-Si Crystal Structure Data

Phase	Composition, at.% Si	Pearson symbol	Space group	Strukturbericht designation	Prototype
(Mo)	0	*cI*2	*Im*3*m*	*A*2	W
Mo_3Si	25	*cP*8	*Pm*3*n*	*A*15	Cr_3Si
Mo_5Si_3	37.5	*tI*38	*I*4/*mcm*	$D8_m$	W_5Si_3
$\alpha MoSi_2$	66.7	*tI*6	*I*4/*mmm*	$C11_b$	$MoSi_2$
$\beta MoSi_2$	66.7	···	$C6_22$	···	···
(Si)	100	*cF*8	*Fd*3*m*	*A*4	C (diamond)

71Sve: V.N. Svechnikov, Yu.A. Kocherzhinskii, and L.M. Yupko, *Diagrammy Sostoyaniya Metal Sist. Nauka*, N.V. Ageev, Ed., Moscow, 116-119 (1971) in Russian.

74Cha: T.G. Chart, *Met. Sci.*, (8), 344-348 (1974).

Complete evaluation contains 1 figure, 4 tables, and 37 references.

Nb-Sn (Niobium-Tin)

92.9064 118.69

$NbSn_2$: C_b. Nb_6Sn_5: complex body-centered orthorhombic. Nb_3Sn: *A*15.

The Nb-Sn phase diagram is redrawn from [Shunk].

Ni-Si (Nickel-Silicon)

58.69 28.0855

The assessed phase diagram for the Ni-Si system is based mainly on the work of [36Iwa], [40For], [62Aal], [63Yam], and [83Oya] for the liquidus and solidus, and on the data of [36Iwa], [62Aal], [63Yam], [64Gra], [74Bad], and [83Oya] for the solid state equilibria. The liquidus and solidus in the region of 30 to 40 at.% Si is tentative. The liquidus from 40 to 100 at.% Si is based on the data of [36Iwa]. The polymorphic transformation in Ni_3Si_2 at 820 °C on the Ni-rich side and at 800 °C on the Si-rich side [74Bad] has been incorporated in the assessed diagram, although the phase boundaries other than the invariants have not been established.

The (Ni) solvus in the range from the eutectic to the $\beta_3 \rightleftarrows \beta_2$ transition is given by [38Oka] as increasing in solute content with decreasing temperature from 16.4 to 17.6 at.% Si. [83Oya] shows a slight decrease in solubility with decreasing temperature. No systematic investigation of the incoherent (Ni) solvus has been undertaken over a wide range of temperatures, and it cannot be considered well established.

The composition limits for the β_1 phase were established as 22.8 to 24.5 at.% Si [83Oya]. The β_2 and β_3 phases have narrow composition ranges: ~1 at.% Si centered close to the stoichiometric composition of 25 at.% Si. The γ phase has a narrow composition range covering the stoichiometric composition of 27.9 at.% Si. The composition ranges of the phases $Ni_{31}Si_{12}$ (γ), Ni_2Si, Ni_3Si_2 (ϵ), and NiSi are taken from the work of [74Bad], but cannot be considered well established. All of the remaining intermediate phases are considered line compounds. The existence of $Ni_{31}Si_{12}$ (designated Ni_5Si_2), Ni_2Si, Ni_3Si_2, and NiSi was reported in bulk diffusion couples.

Rapid solidification of Ni-rich alloys extends the solid solubility of Si in (Ni) to about 18 at.%, although some faint diffraction lines of a second unidentified phase were noted. An increase in short-range order was reported on annealing deformed Ni–7 at.% Si alloys; however, on longer annealing times, the short-range order diminished. It is therefore a metastable transition state.

Ni-Si Crystal Structure Data

Phase	Composition, at.% Si	Pearson symbol	Space group	Strukturbericht designation	Prototype
(Ni)	0 to 15.8	*cF*4	*Fm*3*m*	*A*1	Cu
$\beta_1(Ni_3Si)$	22.8 to 24.5	*cP*4	*Pm*3*m*	$L1_2$	$AuCu_3$
$\beta_2(Ni_3Si)$	~24.5 to 25.5	*mC*16	...	...	$GePt_3$
$\beta_3(Ni_3Si)$	~24.5 to 25.5	*mC*16	...	...	$GePt_3$
γ ($Ni_{31}Si_{12}$)	~26.5 to 29.5	*hP*14	...	...	...
ϑ (Ni_2Si)	...	*hP*6	...	...	...
δ (Ni_2Si)	32.5 to 34.5	*oP*12	...	...	...
$Ni_3Si_2(\epsilon)$	39 to 42	*oP*80	...	...	...
NiSi	49.5 to 51	*oP*8	*Pnma*	*B*31	MnP
$\beta NiSi_2$	66.6	?	...	...	?
$\alpha NiSi_2$	66.6	*cF*12	*Fm*3*m*	*C*1	CaF_2
(Si)	~100	*cF*8			...

Ni-Si Invariant Reactions

Reaction	Composition, at.% Si			Temperature, °C	Reaction type
$L \rightarrow (Ni) + \beta_3$	21.4	15.8	25	1143	Eutectic
$\beta_2 \rightleftarrows \beta_3$		...		~1115	Polymorphic
$(Ni) + \beta_2 \rightleftarrows \beta_1$	14.7	25.1	23.7	1035	Peritectoid
$\beta_2 \rightleftarrows \beta_1 + \gamma$	25.2	24.5	27.9	990	Eutectoid
$L + \gamma \rightleftarrows \beta_3$	22	27.9	25.2	1170	Peritectic
$L \rightleftarrows \gamma$		27.9		1242	Congruent
$L \rightleftarrows \gamma + \delta$	29.8	27.9	33.3	1215	Eutectic
$L + \vartheta \rightleftarrows \delta$	30.8	33.4	33.3	1255	Peritectic
$L \rightleftarrows \vartheta$		~33.5		1306	Congruent
$\vartheta \rightleftarrows \delta + \epsilon'$	37.8	33.3	39.2	825	Eutectoid
$\epsilon' \rightleftarrows \delta + \epsilon$	39.2	33.3	39.5	820	Eutectoid
$\epsilon' \rightleftarrows \epsilon + NiSi$	41	40.7	50	800	Eutectoid
$\vartheta + NiSi \rightleftarrows \epsilon'$	38.5	50	40	845	Peritectoid
$\epsilon \rightleftarrows \epsilon'$		40		~830	Polymorphic
$L \rightleftarrows \vartheta + NiSi$	46	41	50	964	Eutectic
$L \rightleftarrows NiSi$		50		992	Congruent
$L \rightleftarrows NiSi + \alpha NiSi_2$	56.2	50	66.7	966	Eutectic
$L + (Si) \rightleftarrows \beta NiSi_2$	59	~100	66.7	993	Peritectic
$\beta NiSi_2 \rightleftarrows \alpha NiSi_2$		66.7		981	Polymorphic

36Iwa: K. Iwase and M. Okamoto, *Sci. Rep. Tohoku Imp. Univ., K. Honda Anniv.*, 777-792 (1936).

38Oka: M. Okamoto, *Nippon Kinzoku Gakkai-Shi, 2*, 544-551 (1938) in Japanese.

40For: A.C. Forsyth and R.L. Dowdell, *Trans. AIME, 137*, 373-387 (1940).

62Aal: J.H. Aalberts and M.L. Verheijke, *Appl. Phys. Lett., 1*(1), 19-20 (1962).

63Yam: Y. Yamaguchi, M. Yoshida, and H. Aoki, *Jpn. J. Appl. Phys., 2*(11), 714-718 (1963).

64Gra: I. Gray and G.P. Miller, *J. Inst. Met., 93*, 315-316 (1964).

74Bad: E.B. Badtiev, O.S. Petrushkova, and L.A. Panteleimonov, *Vestn. Mosk. Univ. Khim., 15*(3), 367-368 (1974) in Russian.

83Oya: Y. Oya and T. Suzuki, *Z. Metallkd.*, 21-24 (1983).

Complete evaluation contains 5 figures, 8 tables, and 61 references.

Pb-S (Lead-Sulfur)

207.2 32.06

The Pb-S system is characterized by an intermediate phase (PbS) with a narrow homogeneity range, which is formed at 50 at.% S and melts congruently at 1115 °C. A eutectic close to pure Pb exists in the Pb-PbS region. A liquid miscibility gap, resulting in the formation of a monotectic reaction, exists in the PbS-S region with a eutectic close to pure S.

The assessed phase diagram is based on thermodynamic modeling and calculations with review of the work of [05Fri], [61Bla], [66Mil], and [69Kul]. The calculated liquidus is in good agreement with the experimental data. The liquidus in the Pb-PbS region has an inflection point, which is indicative of a metastable liquid miscibility gap at lower temperatures. This miscibility gap has been calculated, based on the thermodynamic model, and is shown by dotted lines. It has not been confirmed experimentally.

The Pb-S phase diagram has been calculated by assuming that the PbS phase is a line compound. Agreement between the calculated and the experimental phase diagram is good.

The eutectic in the Pb-PbS region is very close to pure Pb. The eutectic in the PbS-S region is very close to pure S and has not been experimentally determined.

S has negligible solubility in solid Pb. The solubility of S in Pb at 300 °C has been determined to be less than 0.0006 at.% S [39Gre]. The solubility of Pb in solid S is also negligible. S in the solid state exists in two allotropic forms: αS, stable up to 95.5 °C, and βS, stable from 95.5 °C to its melting point (115.22 °C).

The homogeneity range of PbS is very small. [74Leu] determined the deviation from stoichiometry to be ~1.7×10^{-5} mole of Pb per mole of PbS at 380 °C. PbS is a semiconducting compound, which transforms from the cubic (NaCl-type) to the orthorhombic (GeS-type) structure at high pressures by a first-order transformation at ~25 kbar.

Pb-S Crystal Structure Data

Phase	Composition, at.% S	Pearson symbol	Space group	Strukturbericht designation	Prototype
(Pb)	~0	*cF*4	*Fm*3*m*	*A*1	Cu
PbS	50	*cF*8	*Fm*3*m*	*B*1	NaCl
PbS	50	*oP*8	*Pnma*	*B*16	GeS
High-pressure phases					
(βS)	100	*mP*48	*P*21/*a*	...	...
(αS)	100	*oP*128	*Fddd*	...	...

Pb-S Invariant Points

Reaction	Composition, at.% S			Temperature, °C	Reaction type
$L_1 \rightleftarrows$ (Pb) + PbS	0.006	~0	50	327	Eutectic
$L_1 \rightleftarrows$ PbS + L_2	70.7	50	98.1	800	Monotectic
$L_2 \rightleftarrows$ PbS + (βS)	≈100	50	~100	~115	Eutectic

05Fri: K. Friedrich and A. Leroux, *Metallurgie, 2,* 536-539 (1905) in German.

39Gre: J.N. Greenwood and H.W. Worner, *J. Inst. Met., 115*(2), 435-445 (1939).

61Bla: R.F. Blanks and G.M. Wills, *Physical Chemistry of Process Metallurgy,* Part 2, Interscience Publishers, New York, 991-1028 (1961).

66Mil: E. Miller and K.L. Komarek, *Trans. TMS-AIME, 236,* 832-840 (1966).

69Kul: G. Kullerud, *Am. J. Sci., 267*(A), 233-256 (1969).

74Leu: A.P. Leushina and M.V. Simonova, *Z. Fiz. Khim., 48*(5), 1187-1189 (1974) in Russian; TR: *Russ. J. Phys. Chem., 48*(5), 687-688 (1974).

Complete evaluation contains 7 figures, 9 tables, and 41 references.

Pb-Se (Lead-Selenium)

207.2 78.96

The Pb-Se system contains an intermediate phase, PbSe, with a narrow composition range of stability at 50 at.% Se. In the Pb-PbSe region, a eutectic reaction occurs close to pure Pb. A liquid miscibility gap, a monotectic reaction, and a eutectic reaction occur in the PbSe-Se region. The solubility of Se in (Pb) is 0.004 at.% Se at 300 °C, but there are no reported measurements on the solubility of Pb in Se.

The assessed phase diagram is based on thermodynamic modeling and calculations, with review of the work of [55Noz], [61Sei], [66Sei], [66Mil], and [74Sch]. It has been calculated assuming that PbSe is a line compound. There is fairly good agreement between the calculated liquidus and the available experimental data. The liquidus in the Pb-PbSe region has an inflection point, which is indicative of a metastable liquid miscibility gap at lower temperatures.

The eutectic in the Pb-PbSe region is very close to pure Pb (0.013 at.% Se and 0.2 °C below the melting point

Pb-Se Crystal Structure Data

Phase	Composition, at.% Se	Pearson symbol	Space group	Strukturbericht designation	Prototype
(Pb)	~0	*cF*4	*Fm*3*m*	*A*1	Cu
PbSe	50	*cF*8	*Fm*3*m*	*B*1	NaCl
PbSe(HP)	50	*oP*8	*Pnma*	*B*16	GeS
(Se)	~100	*hP*3	$P3_121$	*A*8	γSe

of Pb). The eutectic in the PbSe-Se region is very close to pure Se and has not been experimentally determined.

The homogeneity range of PbSe is vary narrow, and there are no reported direct measurements of the stability range of PbSe. PbSe, a semiconductor compound, has a cubic (NaCl-type) structure at atmospheric pressure and transforms to an orthorhombic (GeS-type) structure at ~43 kbar.

55Noz: R. Nozato and K. Isaki, *Bull. Naniwa Univ. (Jpn.), A3,* 125-141 (1955).

Pb-Se Invariant Points

Reaction	Composition, at.% Se			Temperature, °C	Reaction type
$L_1 \rightleftarrows$ (Pb) + PbSe	0.002	~0	50	327	Eutectic
$L_1 \rightleftarrows$ PbSe + L_2	75.9	50	98.8	678	Monotectic
$L_2 \rightleftarrows$ PbSe + (Se)	99.992	50	~100	220	Eutectic

61Sei: D.N. Seidman, I. Cadoff, K.L. Komarek, and E. Miller, *Trans. TMS-AIME, 221,* 1269-1270 (1961).
66Mil: E. Miller and K.L. Komarek, *Trans. TMS-AIME, 236,* 832-840 (1966).
66Sei: D.N. Seidman, *Trans. TMS-AIME, 236,* 1361-1362 (1966).
74Sch: M. Schneider and J.C. Guillaume, *J. Phys. Chem. Solids, 35,* 471-478 (1974) in French.

Complete evaluation contains 6 figures, 8 tables, and 36 references.

Pb-Te (Lead-Tellurium)

207.2 127.60

The Pb-Te system is characterized by the existence of an intermediate phase, PbTe, with a very narrow range of homogeneity, which is formed at 50 at.% Te, and two eutectic reactions, one each in the Pb-PbTe and PbTe-Te portions of the phase diagram. The eutectic in the Pb-PbTe region is very close to pure Pb. The assessed phase diagram is based primarily on the work of [02Fay], [15Kim], [39Gre], [65Lug], and [66Mil] and has been obtained by thermodynamic modeling. It was calculated by assuming the PbTe phase to be a line compound. Agreement between the calculated and the experimental phase diagram is good.

The mutual solid solubilities of Pb and Te are limited. [39Gre] determined the solubility of Te in (Pb) to be 0.0065 at.% Te at 300 °C. [65Lug] reported the solubility of Pb in (Te) to be 2.5 at.% Pb at 330 °C, 1.5 at.% Pb at 260 °C, and 0.85 at.% Pb at room temperature. The measurements of [65Lug] were not conducted under equilibrium conditions. Consequently, the present evaluators have assessed the solubility of Pb in Te to be negligible.

PbTe is a semiconductor. It transforms from a cubic (NaCl-type) to an orthorhombic (GeS-type) structure at high pressures (41 kbar at room temperature). A two-phase region (cubic plus orthorhombic) has been reported between 50 to 80 kbar, and the volume change for the cubic to orthorhombic transformation was reported to be 5.2%.

During crystallization studies of an amorphous 80 at.% Te alloy obtained by rapid cooling from the liquid phase, crystallization was reported to begin at ~64 °C. The first phase to appear during crystallization had a hexagonal structure and may be considered a supersaturated solution of Pb in Te.

The liquidus in the Pb-PbTe region has an inflection point, which is indicative of a metastable liquid miscibility gap at lower temperature This miscibility gap calculation based on a thermodynamic model ar has not been confirmed experime tally.

Invariant Equilibria in the Pb-Te System

Reaction	Composition of L, at.% Te	Temperature, °C	Reactic type
L $\rightleftarrows$ (Pb) + PbTe	0.04	326.95	Eutecti
L $\rightleftarrows$ PbTe + (Te)	89.13	410.9	Eutect

Pb-Te Crystal Structure Data

Phase	Composition, at.% Te	Pearson symbol	Space group	Strukturbericht designation	Prototyp
(Pb)	0	*cF*4	*Fm*3*m*	*A*1	Cu
PbTe	50	*cF*8	*Fm*3*m*	*B*1	NaCl
PbTe(HP)	50	*oP*8	*Pnma*	*B*16	GeS
(Te)	100	*hP*3	$P3_121$	*A*8	Se

02Fay: H. Fay and C.B. Gillson, *A Chem. J., 27,* 81-95 (1902).
15Kim: M. Kimura, *Mem. Coll. Sci. Kyo Univ., 1,* 149-152 (1915) in German.
39Gre: J.N. Greenwood and H.W. W ner, *J. Inst. Met., 115,* 435-445 (1939
66Mil: E. Miller and K.L. Komarek, *Tra AIME, 236,* 832-840 (1966).

Complete evaluation contains 5 figures, tables, and 49 references.

Sb-Zn (Antimony-Zinc)

121.75 65.38

β: B_e. Sb_3Zn_4 (δ,ϵ?): *hR*22 or *oP*28 or *mC*∗? Sb_2Zn_3(ϵ,ζ,η?): *oI*∗ (HT), *oP*30?

The Sb-Zn phase diagram is redrawn from [Metals].

Se-Zn (Selenium-Zinc)

78.96 65.38

ZnSe: *B*3.

The Zn-Se phase diagram is redrawn from [Hansen].

Si-Ta (Silicon-Tantalum)

28.0855 180.9479

Ta_2Si: *C*16. αTa_5Si_3: $D8_l$. βTa_5Si_3: $D8_m$. $TaSi_2$: *C*40.

The Ta-Si phase diagram is redrawn from [Hansen].

Si-V (Silicon-Vanadium)

28.0855 50.9415

The V-Si system is characterized by the presence of four intermediate phases—V_3Si, V_5Si_3, VSi_2, and V_6Si_5. The assessed phase diagram is based on review of the experimental data [56Kie, 63Efi, 67Bru, 73Koc, 74Koc, 78Sav, 81Smi1, 81Smi2], with major revisions based on the recent work of [82Jor] and [85Sto].

Early work showed V_3Si as decomposing by a peritectic reaction near 2000 °C, and only the diagram of [73Koc] and [74Koc] showed congruent melting. The recent study of [82Jor], however, strongly supports congruent melting and other details of their diagram. In addition, thermodynamic measurements of [85Sto] show that V_6Si_5 is indeed a stable phase, but its stability is restricted in temperature range. Although limited to the composition range 15 to 31 at.% Si, [82Jor] found that V_3Si melts congruently at 1925 ± 10 °C at a composition of 24.5 at.% Si.

[82Jor] also reported the eutectic equilibrium on the V-rich side of V_3Si to occur at 1870 ± 10 °C, with the V-rich Si content of V_3Si being ~19 at.%

Special Points of the Assessed V-Si Phase Diagram

Reaction	Composition, at.% Si			Temperature, °C	Reaction type
L ⇄ (V) + V_3Si	~13	~7	~19	1870 ± 10	Eutectic
L ⇄ V_3Si + V_5Si_3	~29	~25.5	~37.5	1895 ± 10	Eutectic
L + V_5Si_3 ⇄ V_6Si_5	~57	~37.5	~45	1670	Peritectic
V_6Si_5 ⇄ V_5Si_3 + VSi_2	~45	~37.5	66.7	1160 ± 100	Eutectoid
L ⇄ V_6Si_5 + VSi_2	~59	~45	66.7	1640	Eutectic
L ⇄ VSi_2 + Si	~97	66.7	100	1400	Eutectic
L ⇄ V_3Si		24.5		1925	Congruent
L ⇄ V_5Si_3		37.5		2010	Congruent
L ⇄ VSi_2		66.7		1677	Congruent
L ⇄ (V)		0		1910 ± 10	Melting point
L ⇄ Si		100		1414	Melting point

V-Si Crystal Structure Data

Phase	Composition, at.% Si	Pearson symbol	Space group	Strukturbericht designation	Prototype
(V)	0 to ~7	*cI*2	*Im*3*m*	*A*2	W
V_3Si	~19 to ~25.5	*cP*8	*Pm*3*n*	*A*15	Cr_3Si
V_5Si_3	37.5	*tI*32	*I*4/*mcm*	$D8_m$	Si_3W_5
V_5Si_3	(a)	*hP*16	$P6_3/mcm$	$D8_8$	Mn_5Si_3
V_6Si_5	~45	*oI*44	*Immm*	...	Nb_6Sn_5
VSi_2	66.7	*hP*9	$P6_222$	*C*40	$CrSi_2$
Si	100	*cF*8	*Fm*3*m*	*A*4	C (diamond)

(a) Carbon-stabilized polymorph.

Si. There is general agreement that the liquid composition in this equilibrium reaction is near 13 at.% Si, whereas in contrast the eutectic temperature has been reported over the range 1780 to 1870 °C. The selection of ~7 at.% Si for the composition of the terminal solid solution, (V), at the eutectic equilibrium is a compromise among the available data, but favors the work of [73Koc] and [74Koc]. The data of [82Jor] and [85Sto] indicate that the Si composition along the V-rich boundary of the V_3Si phase increases from ~19 at.% Si at the 1870 °C eutectic temperature to between 21.5 and 22 at.% Si in the range 1400 through 1700 °C, where the change with temperature becomes slight.

On the Si-rich side of V_3Si, [82Jor] found eutectic equilibrium occurring at 1895 ± 10 °C, with liquid of ~29 at.% Si decomposing to V_5Si_3 and V_3Si, with a slight excess Si content of ~25.5 at.% Si in V_3Si. [82Jor] and [85Sto] reported that the Si content along the Si-rich boundary of V_3Si decreases from 25.5 at.% Si at the 1895 °C eutectic temperature to near stoichiometry at 1800 °C. Below 1800 °C, the boundary remains near stoichiometry, possibly becoming very slightly Si-deficient at lower temperatures.

With the exception of the temperature associated with the V-rich eutectic reaction that [73Koc] and [74Koc] placed near 1890 °C, there is good agreement between their diagram and that of [82Jor]. [73Koc] and [74Koc] placed the V-rich eutectic composition at 13 at.% Si, the congruent melting temperature of V_3Si at 1935 °C, and the eutectic composition on the Si-rich side of V_3Si at 28 at.% Si with an associated temperature of 1890 °C. Accordingly, the assessed diagram shows a melting temperature for V_5Si_3 of 2010 °C.

Measurements of thermodynamic activities by [85Sto] in the temperature range 1300 to 1725 °C indicate that V_6Si_5 is stable in that temperature region, but extrapolation of the activities to lower temperatures indicates instability with respect to VSi_2 and V_5Si_3 at temperatures below 1160 ± 100 °C. There is a large uncertainty in the decomposition temperature, because it was determined from the intersection of semilogarithmic plots of thermodynamic activities vs reciprocal temperatures for the two-phase regions VSi_2-V_6Si_5 and V_6Si_5-V_5Si_3. [85Sto] also confirmed that the homogeneity range of VSi_2 is negligible, and it is likely that V_6Si_5 and V_5Si_3 also have negligible homogeneity ranges.

There is appreciable interest in the low-temperature superconductivity of V_3Si. For stoichiometric material, the martensitic and superconducting transitions are most often reported in the ranges 21.7 to 21.9 K and 16.7 to 16.9 K, respectively. Both transitions are correlated with soft phonon modes. Not all V_3Si samples under these conditions become superconducting, but the superconducting transition temperature is sample dependent.

56Kie: R. Kieffer, F. Benesovsky, and H. Schmid, *Z. Metallkd., 47,* 247-253 (1956) in German.

63Efi: Yu.V. Efimov, *Zh. Neorg. Khim., 8,* 1522-1524 (1963) in Russian; TR: *Russ. J. Inorg. Chem., 8,* 790-792 (1963).

67Bru: H.A.C.M. Bruning, *Philips Res. Rep., 22,* 349-354 (1967).

73Koc: Yu.A. Kocherzhinskii, O.G. Kulik, and E. Shiskin, *Dokl. Akad. Nauk SSSR, 209,* 1347-1349 (1973) in Russian.

74Koc: Yu.A. Kocherzhinskii, O.G. Kulik, and E. Shiskin, *Strukt. Faz., Fazovye Prevrasch. Diagr. Sostoyaniyz Met. Sist.,* O.S. Ivanov, Ed., Izd. Nauka, Moscow, 136-139 (1974) in Russian.

78Sav: E.M. Savitskii, Yu.V. Efimov, K. Eichler, and P. Paufler, *Wissenschaft, Z. Tech. Univ. (Dresden), 27,* 675-676 (1978) in German.

81Smi1: J.F. Smith, *Bull. Alloy Phase Diagrams, 2*(1), 42-48 (1981).

81Smi2: J.F. Smith, *Bull. Alloy Phase Diagrams, 2*(1), 40-41 and *2*(2), 172 (1981).

82Jor: J.L. Jorda and J. Muller, *J. Less-Common Met., 84,* 39-48 (1982).

85Sto: E.K. Storms and C.E. Myers, *High-Temp. Sci.,* in press (1985).

Complete evaluation contains 1 figure, 3 tables, and 48 references.

Te-Zn (Tellurium-Zinc)

127.60 65.38

TeZn: *B*1, *hP*∗ in thin films.

The Te-Zn phase diagram is redrawn from [Shunk].

Assessed Al-As Phase Diagram (Constrained Vapor Pressure)

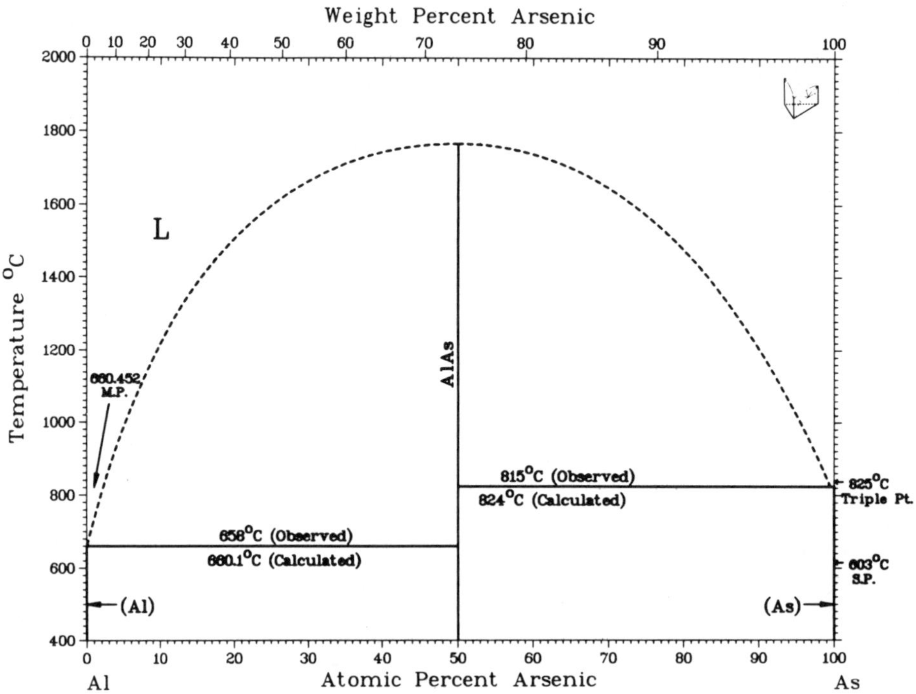

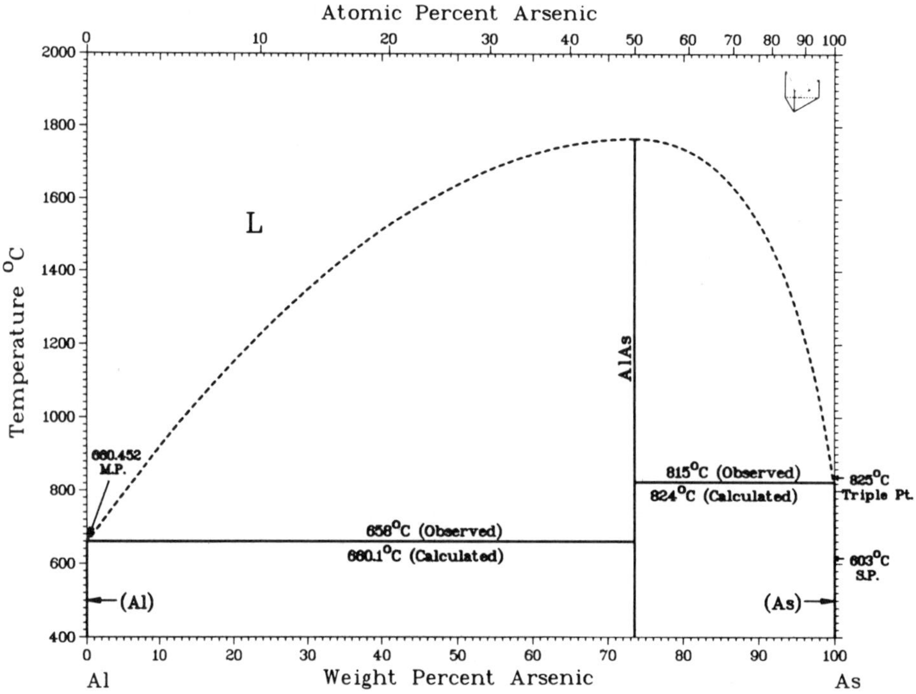

Speculative.

Assessed Al-Nb Phase Diagram

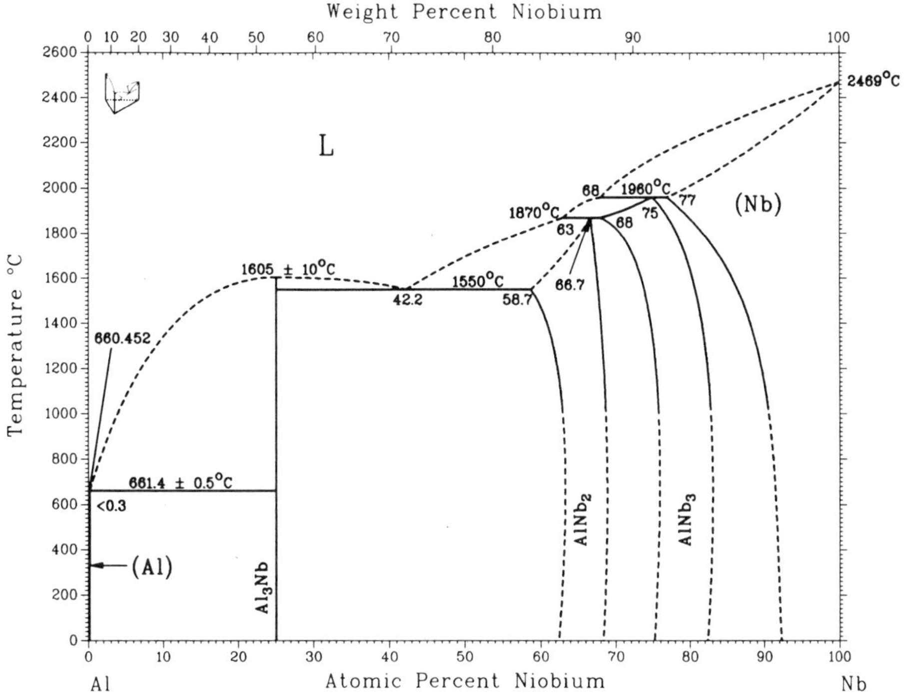

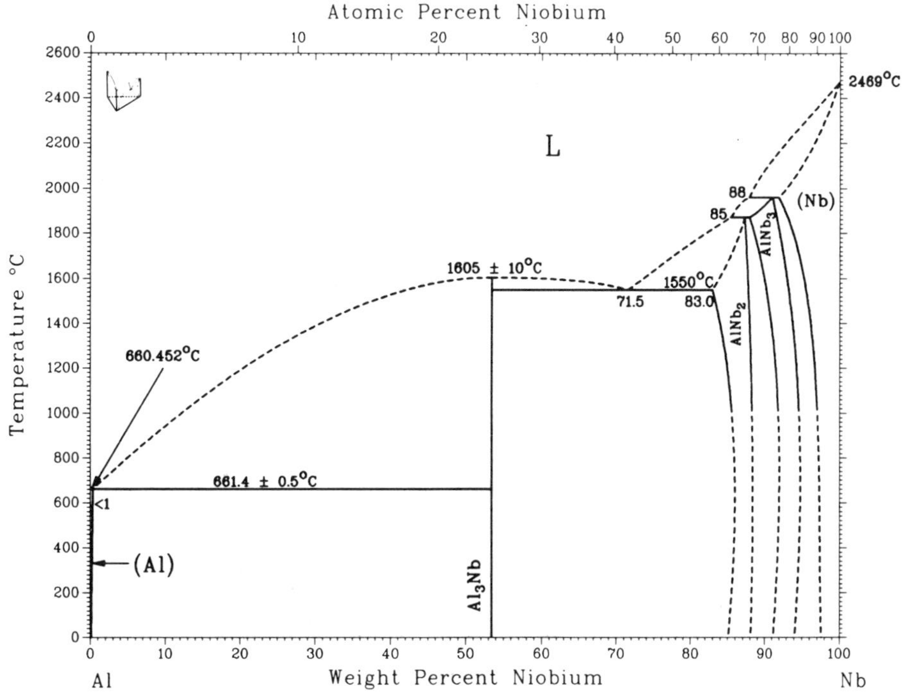

Assessed Al-P Phase Diagram (Constrained Vapor Pressure)

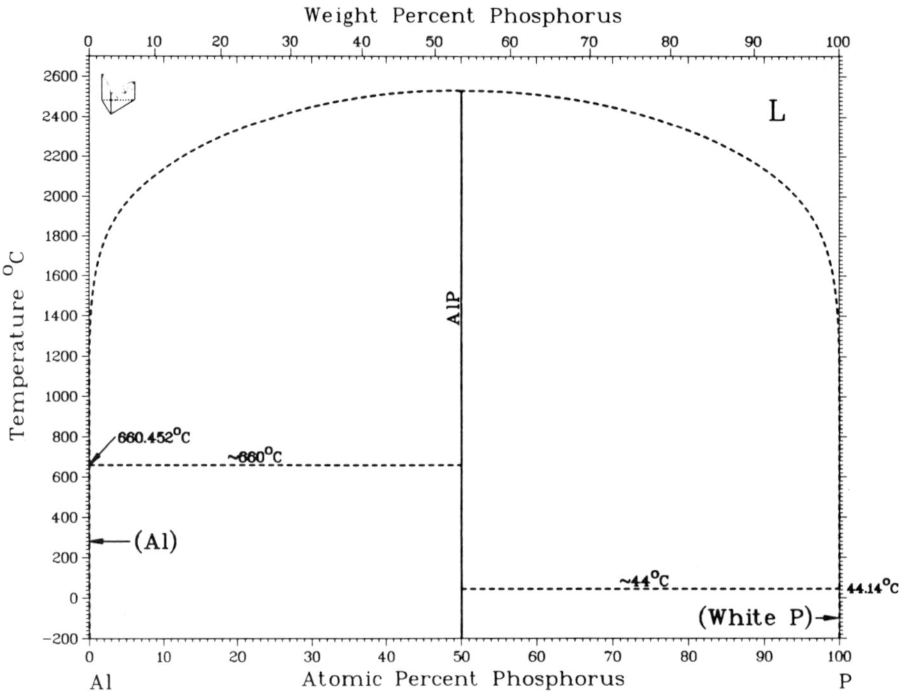

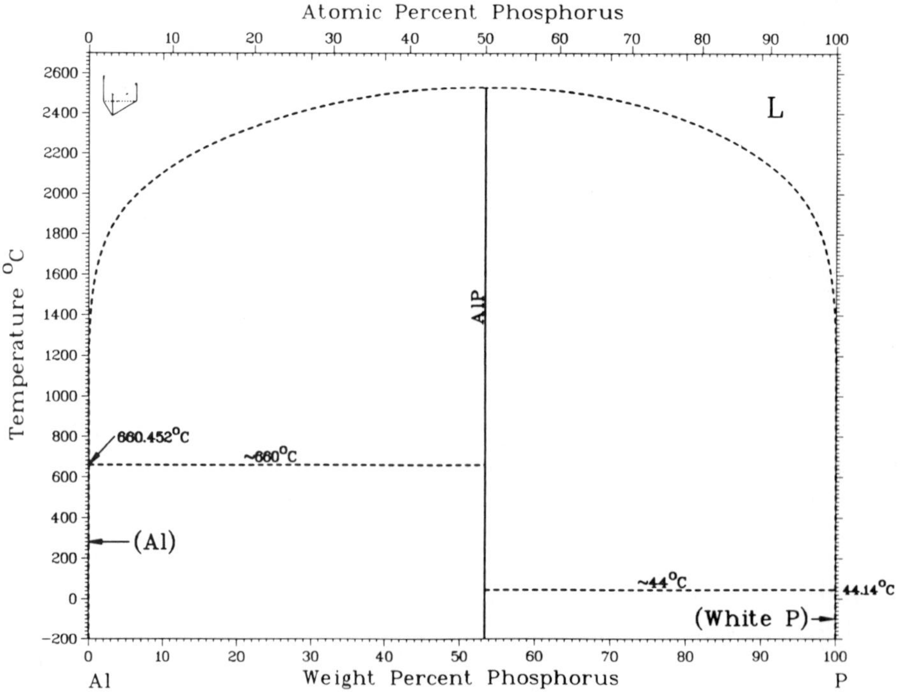

Speculative.

Al-S Phase Diagram

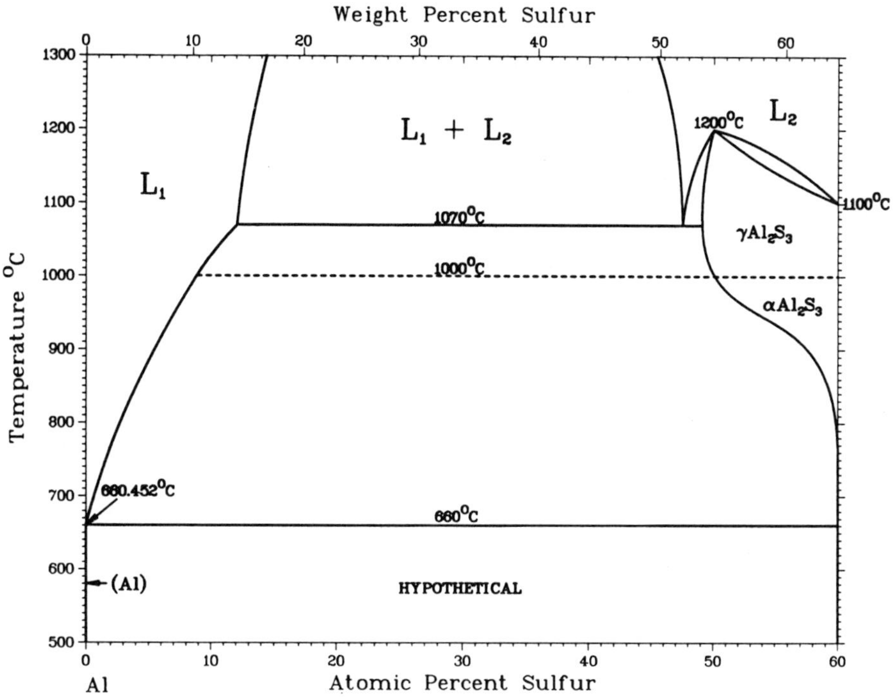

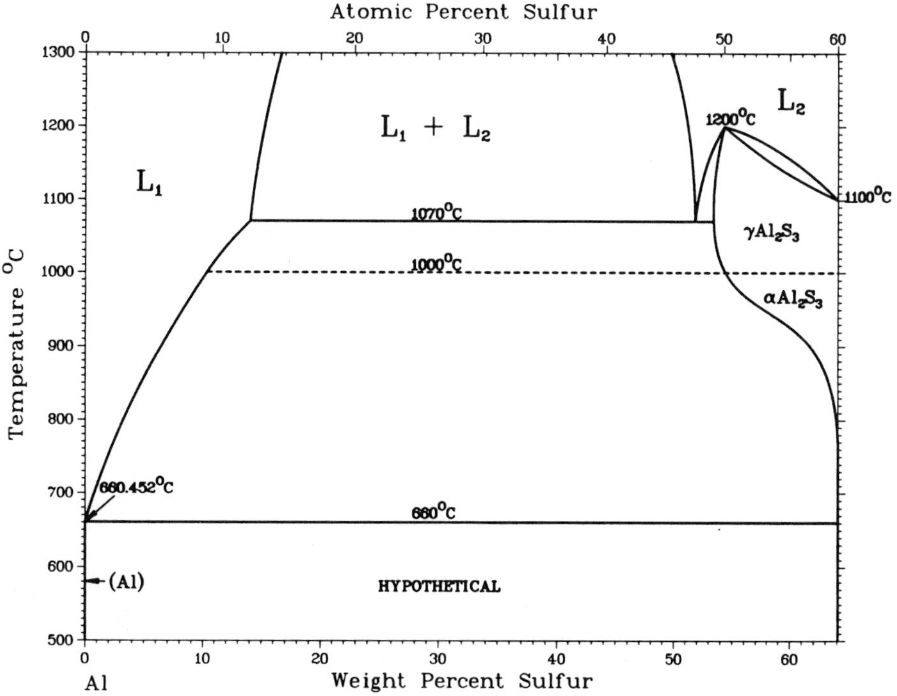

Assessed Al-Sb Phase Diagram

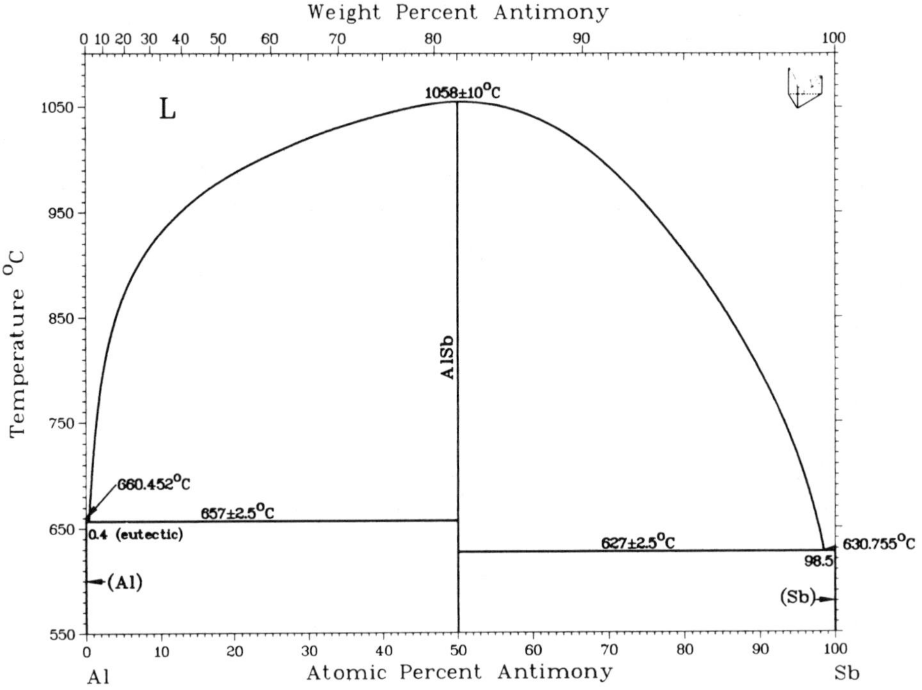

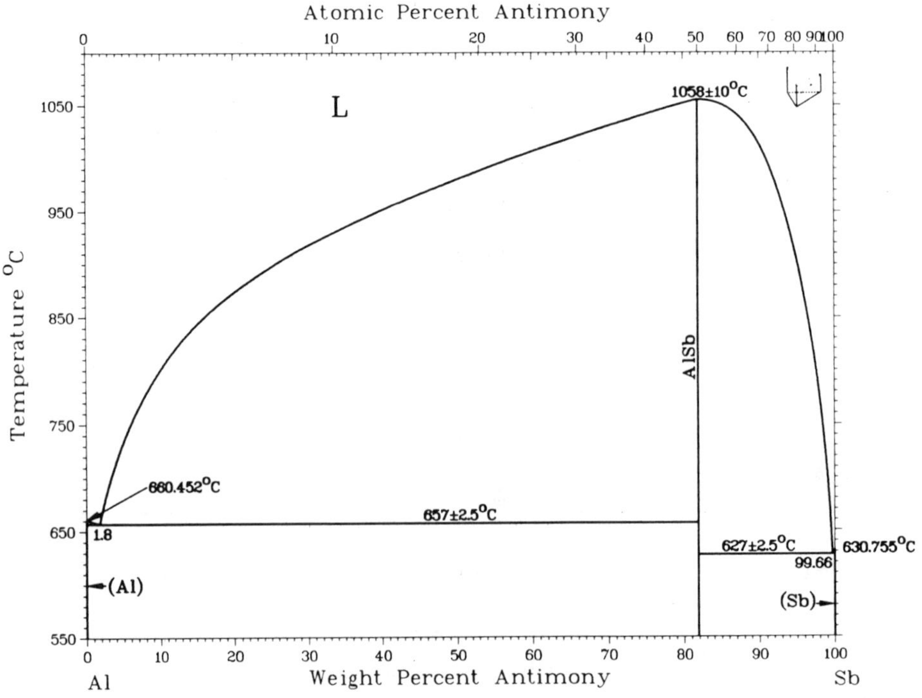

Cd-As Phase Diagram

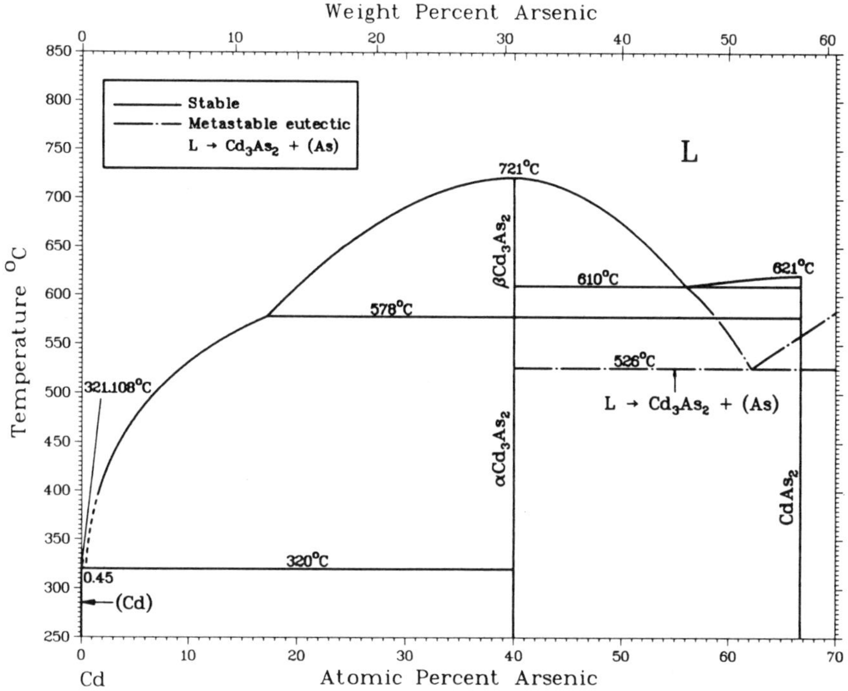

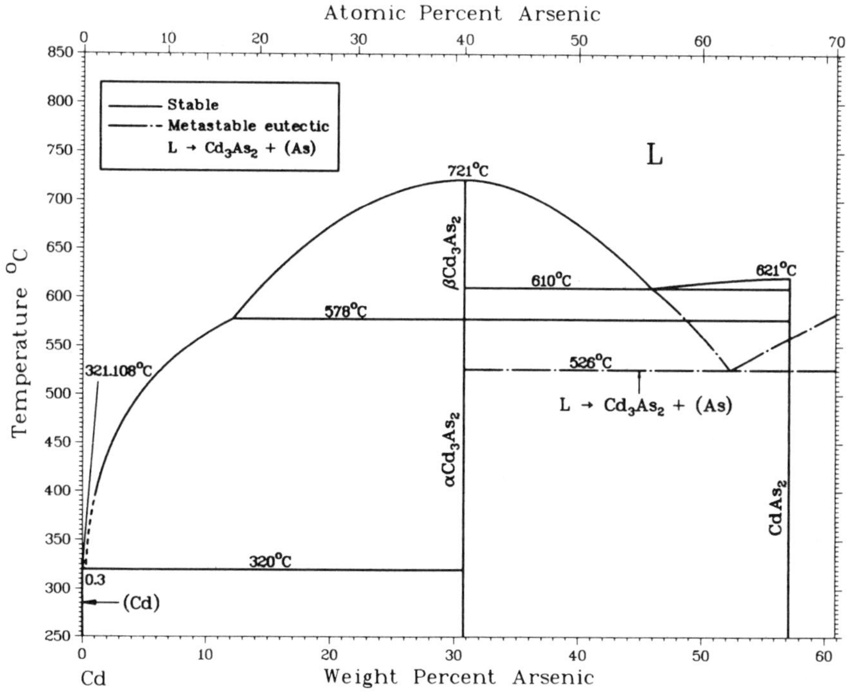

Ga-As Phase Diagram

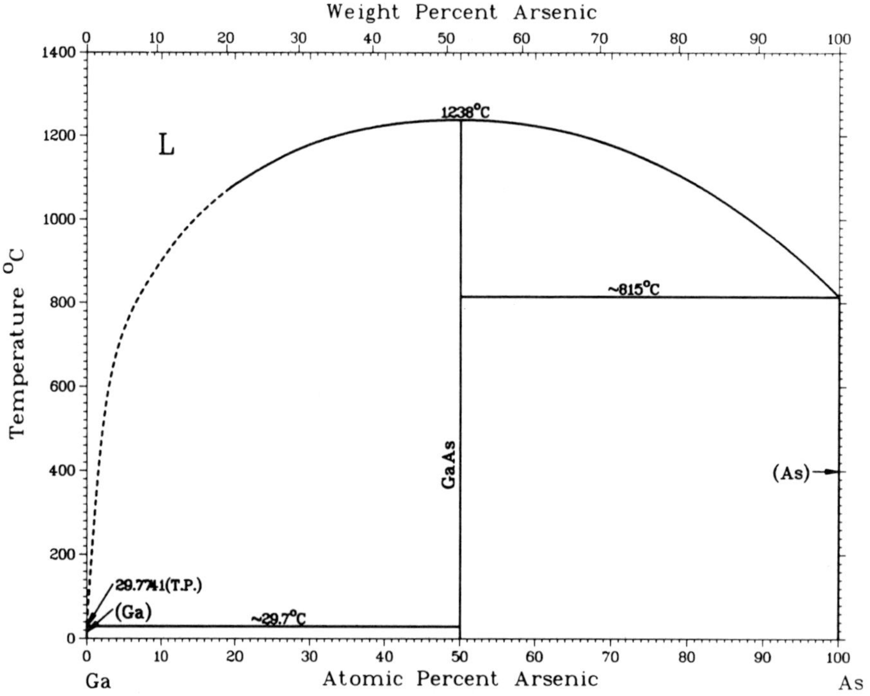

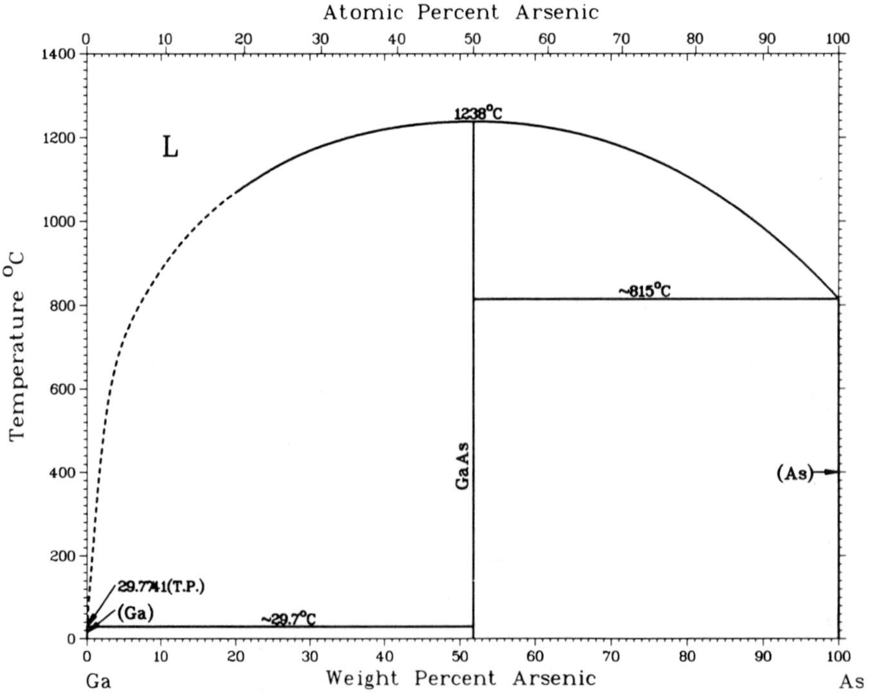

In-As Phase Diagram

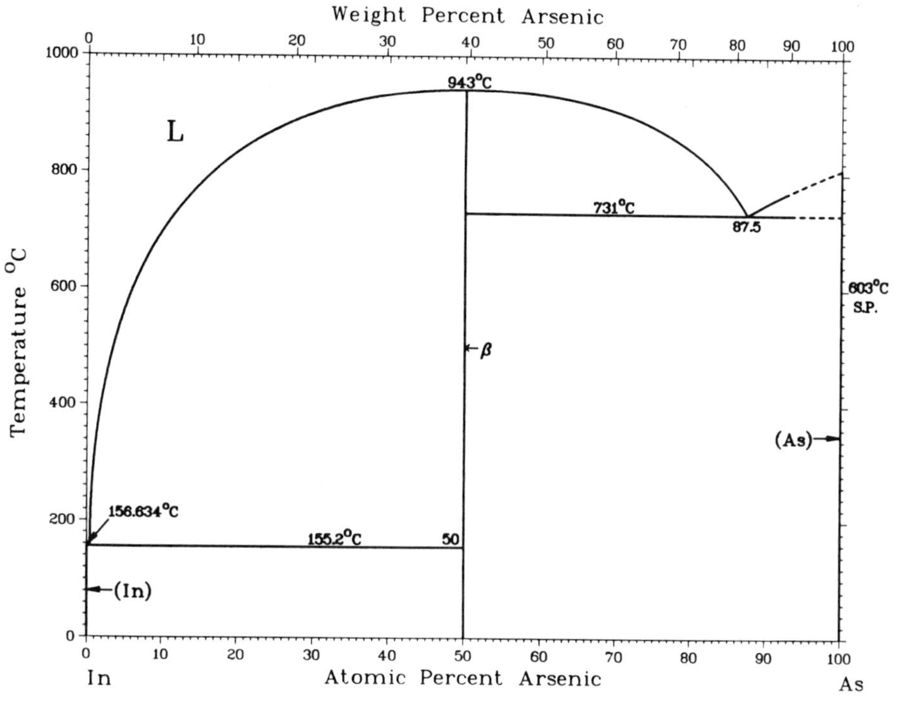

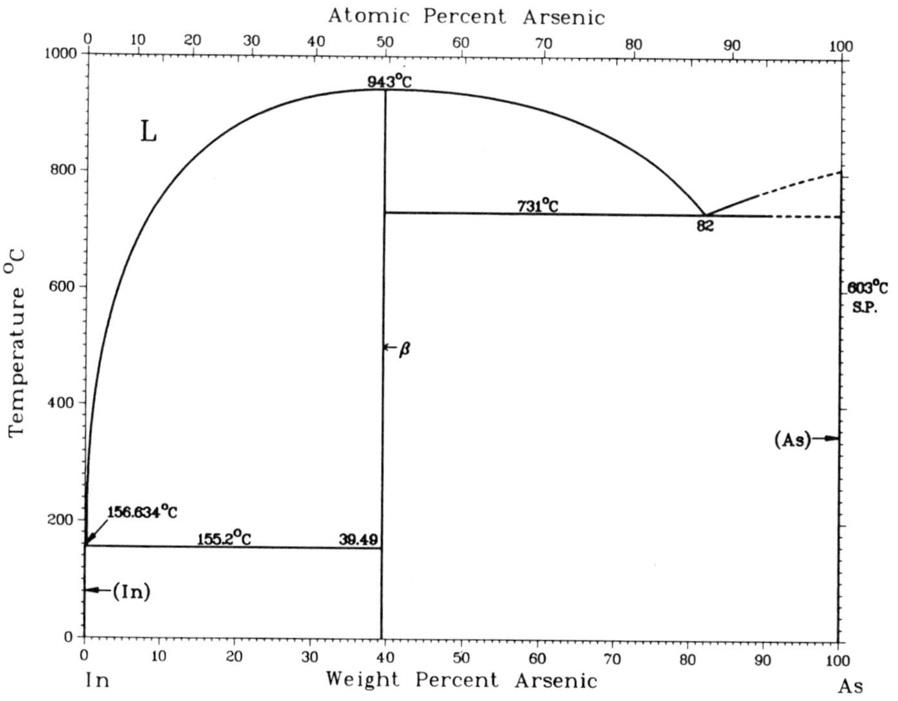

Zn-As Phase Diagram

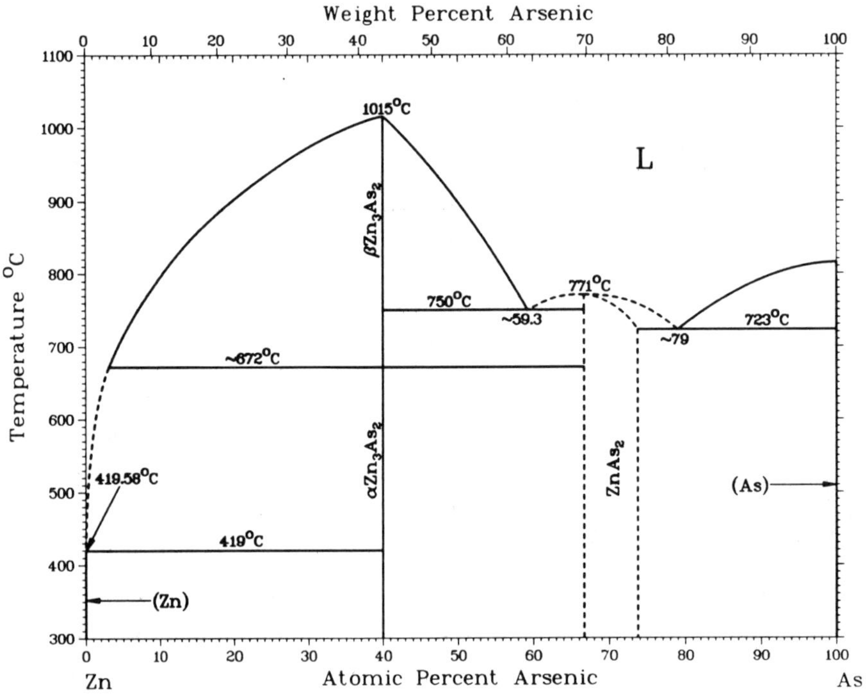

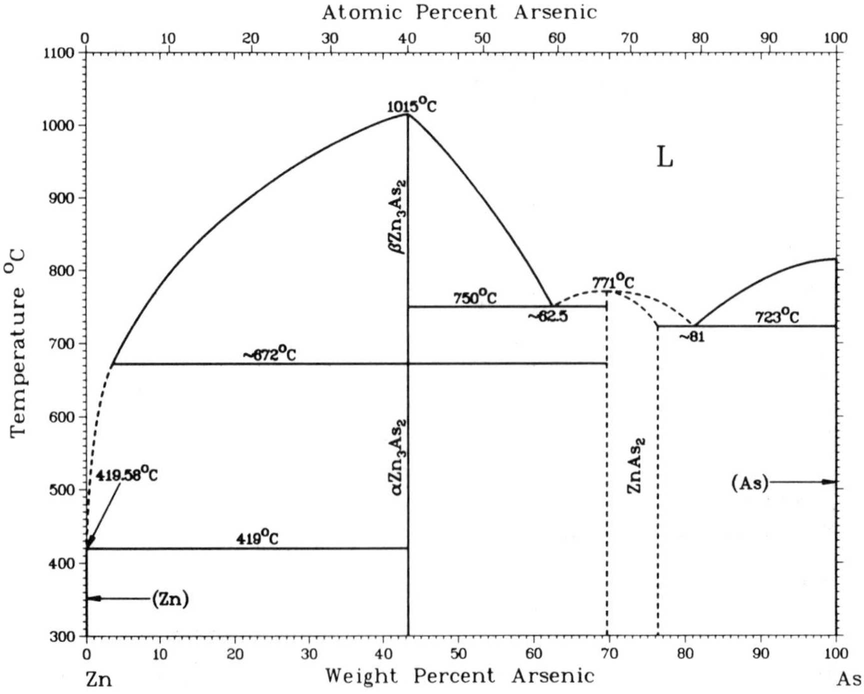

Cd-S Phase Diagram

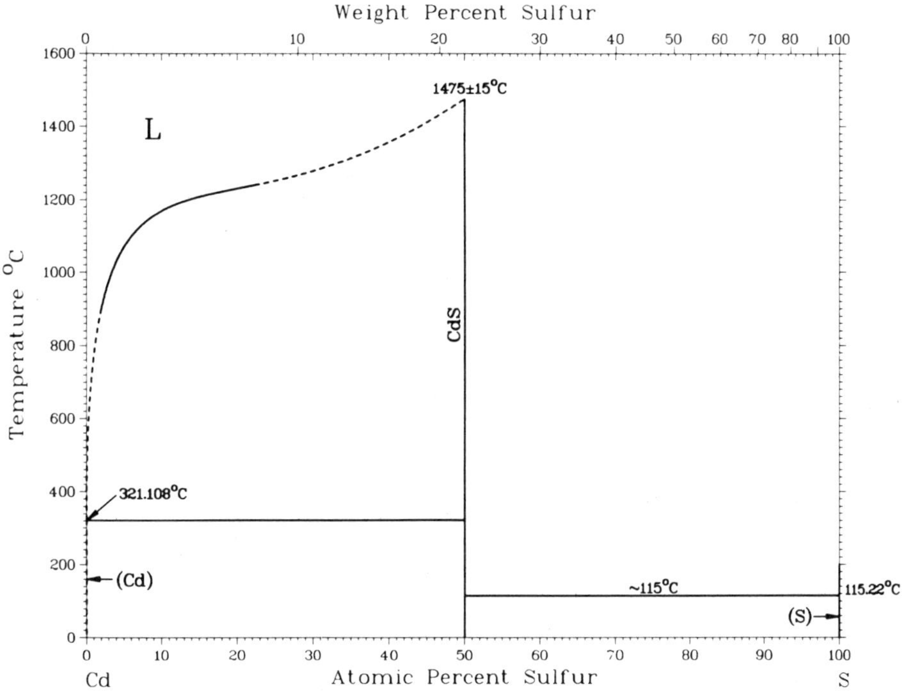

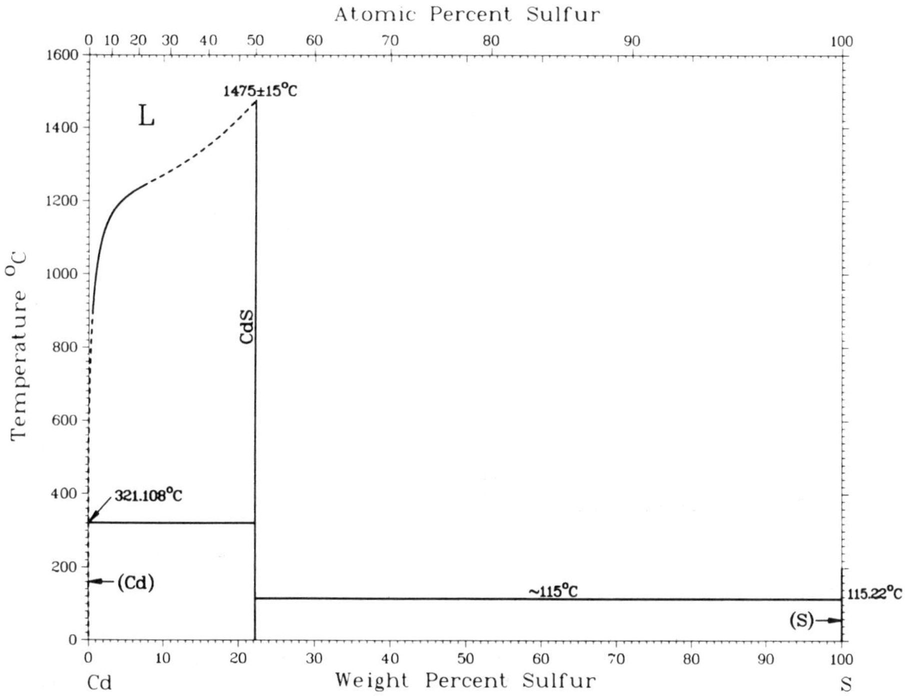

Cd-Se Phase Diagram

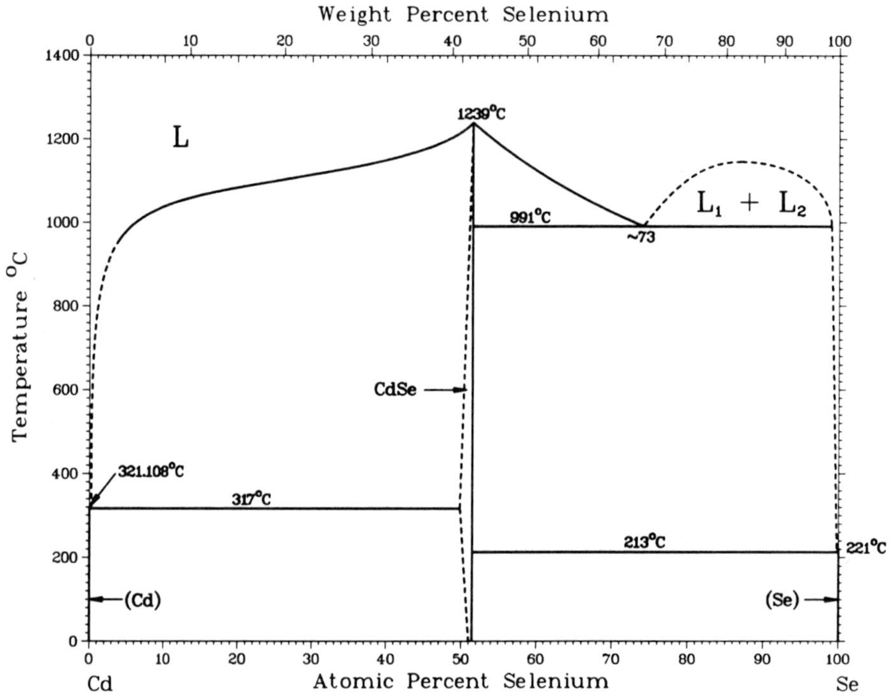

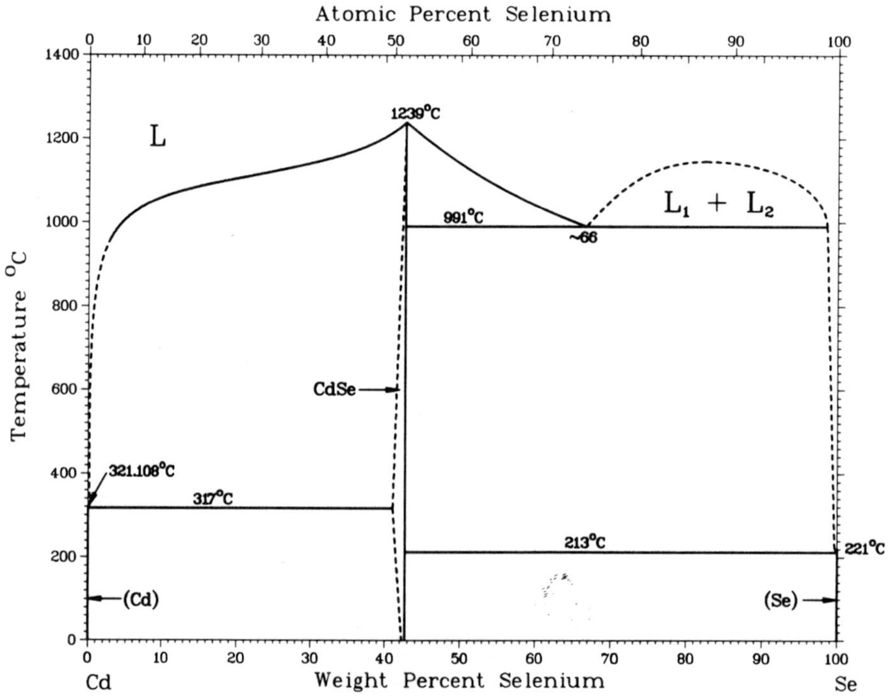

Cd-Te Phase Diagram

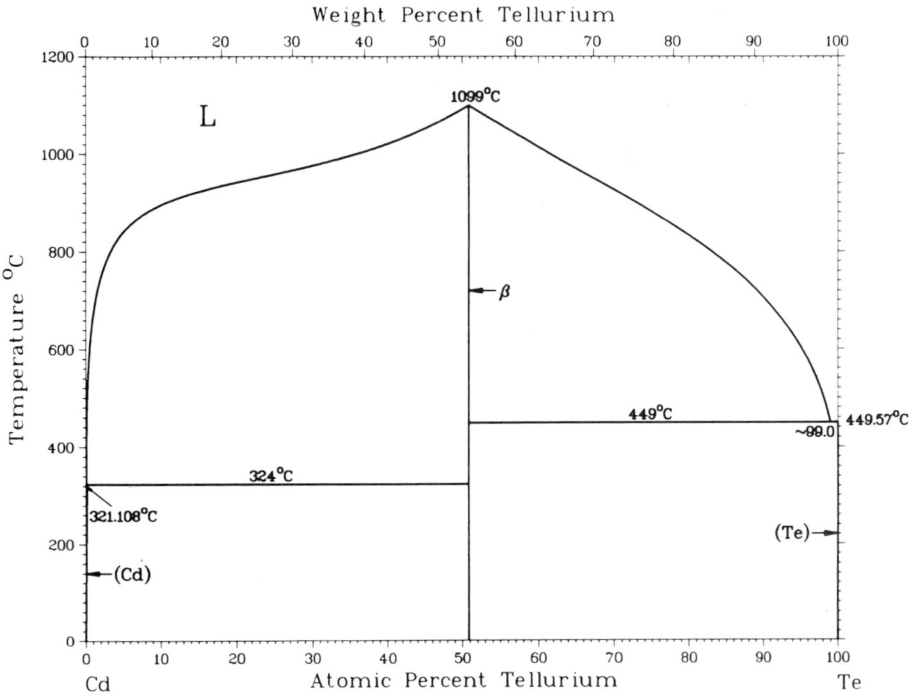

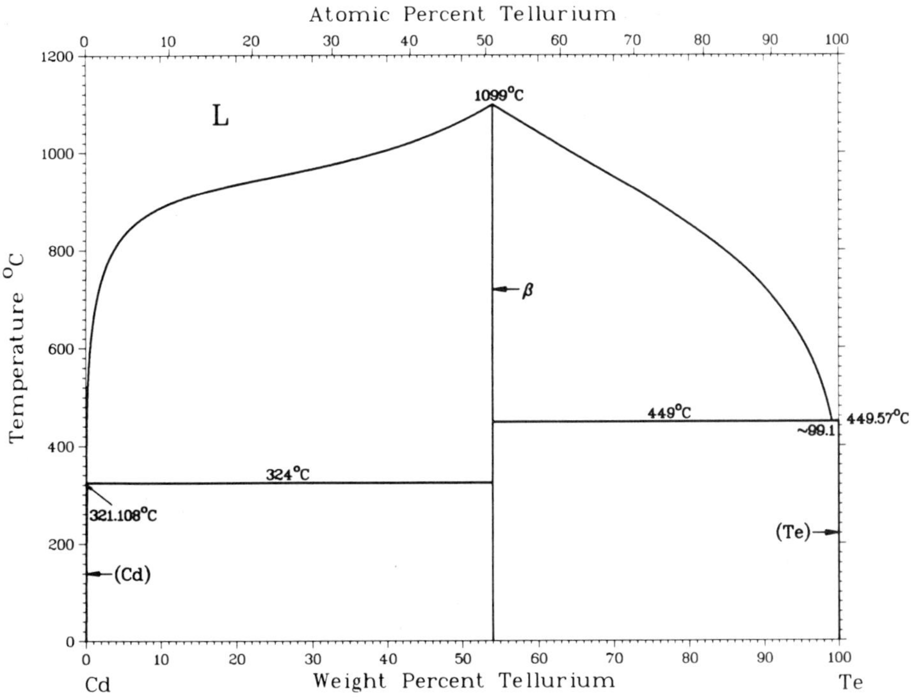

Assessed Ti-Co Phase Diagram

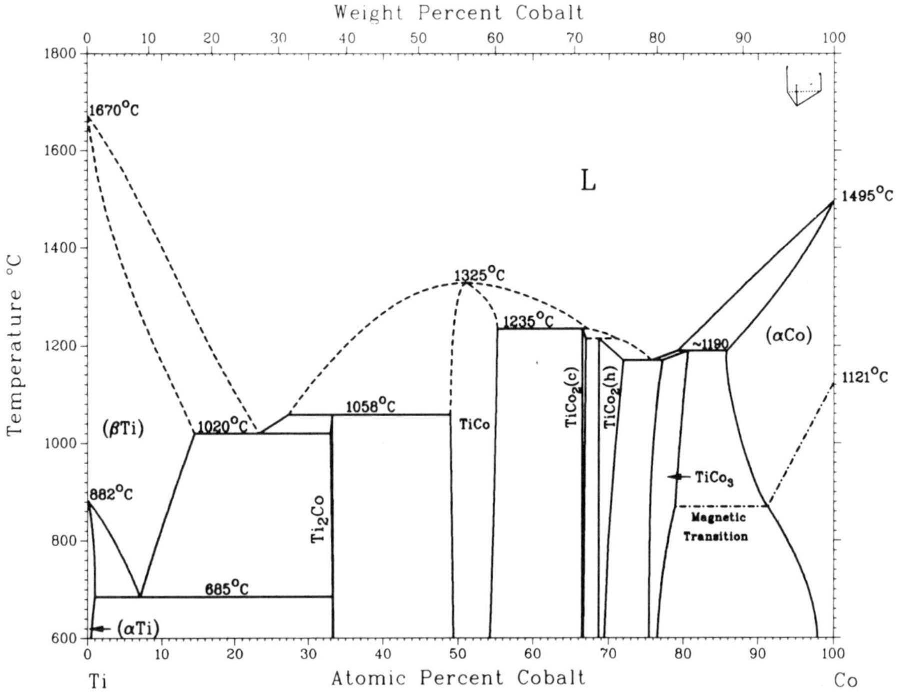

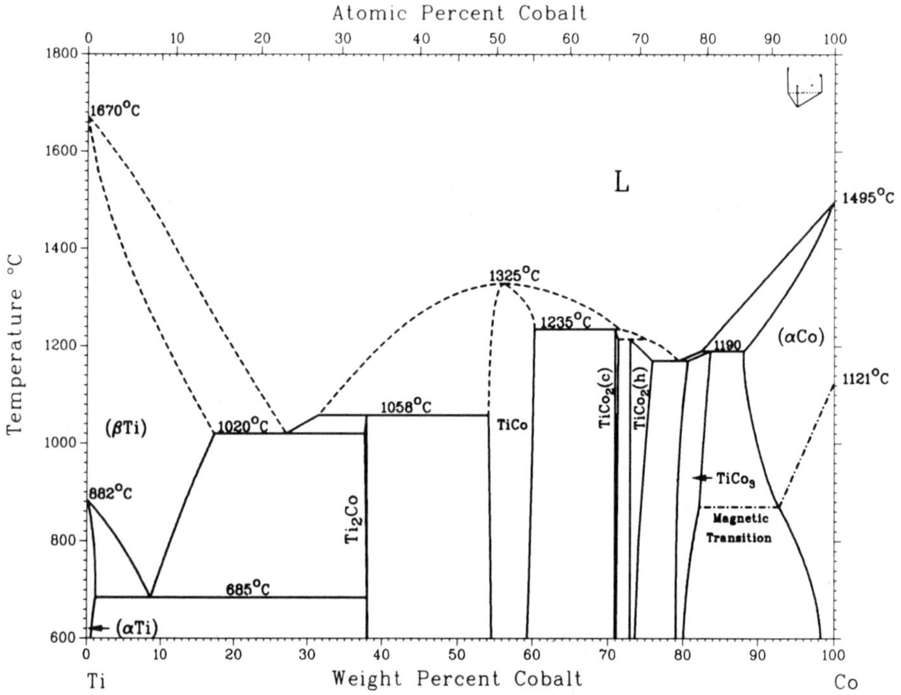

Nb-Ga Phase Diagram

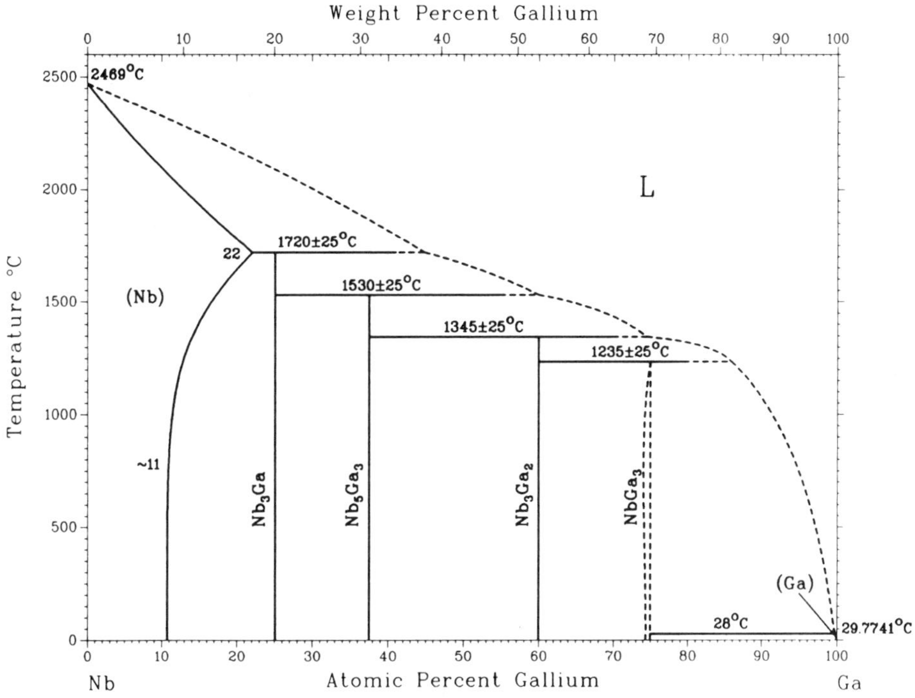

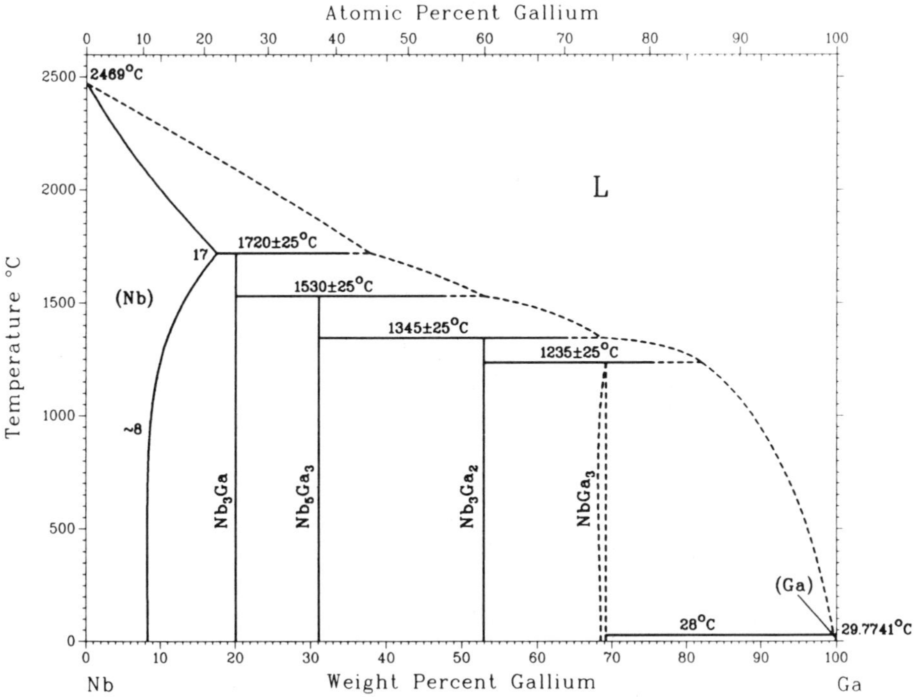

Ni-Ga Phase Diagram

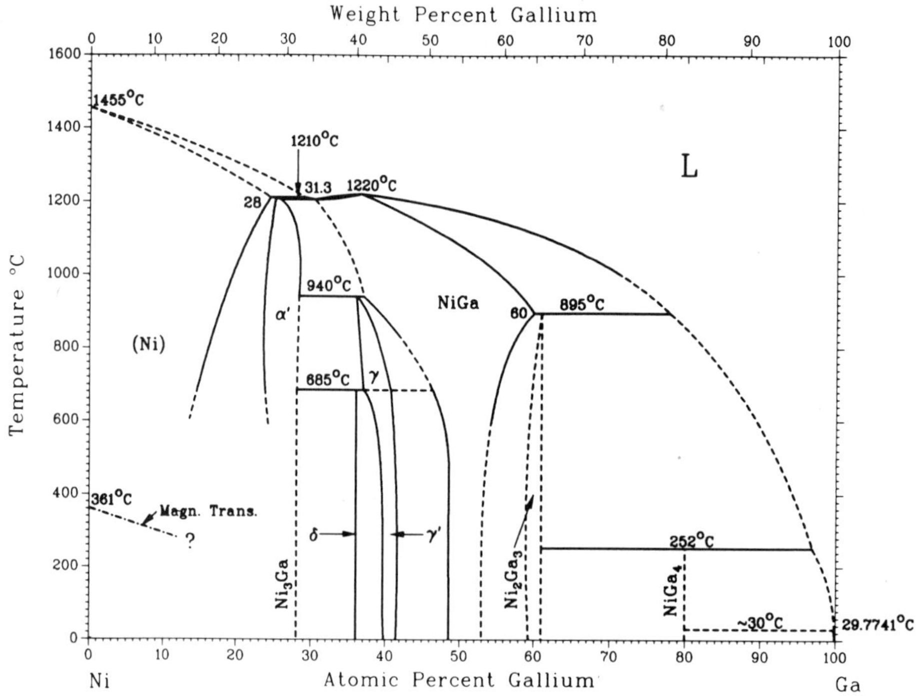

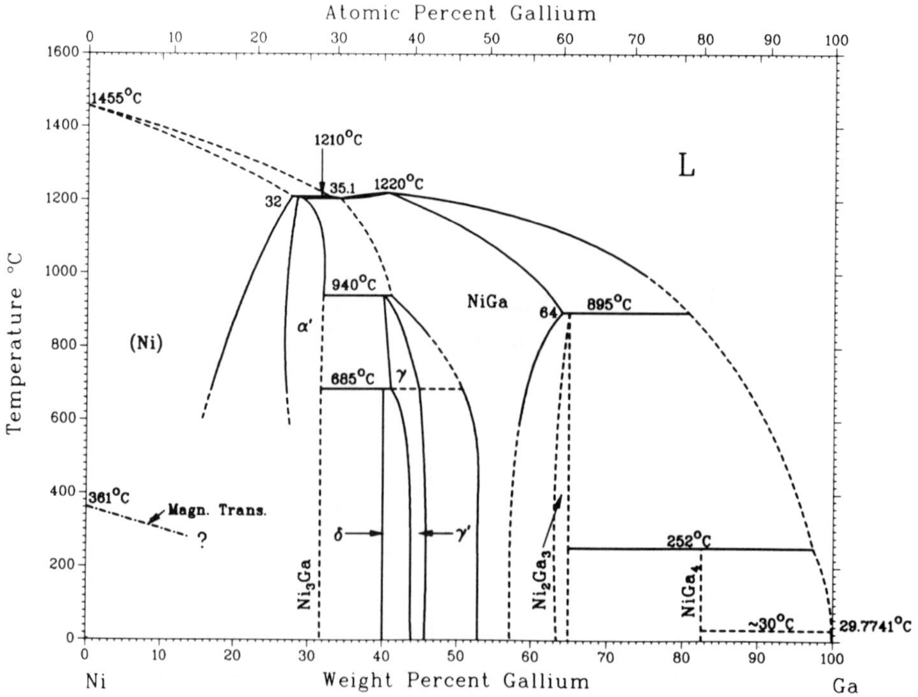

Ga-P Phase Diagram

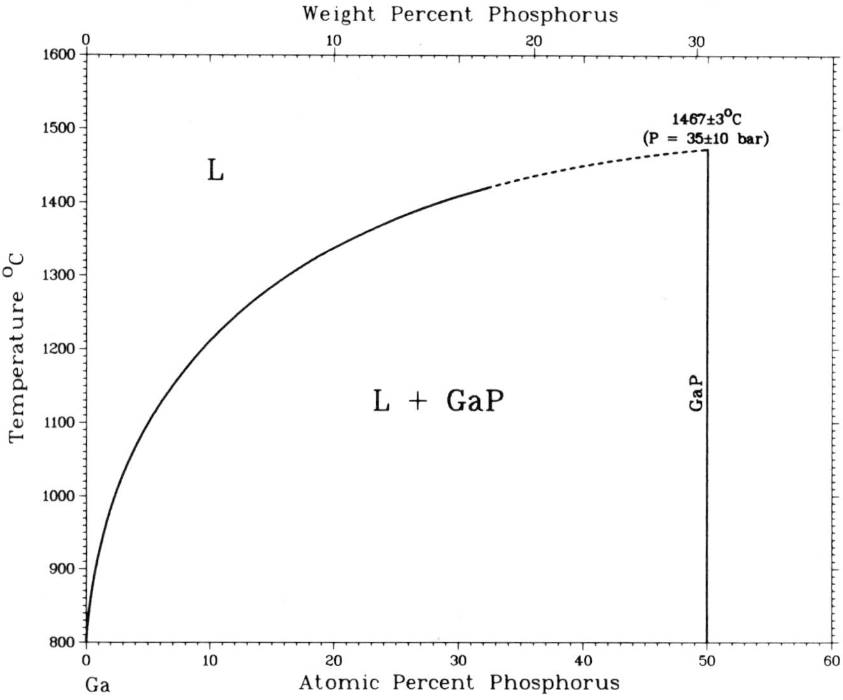

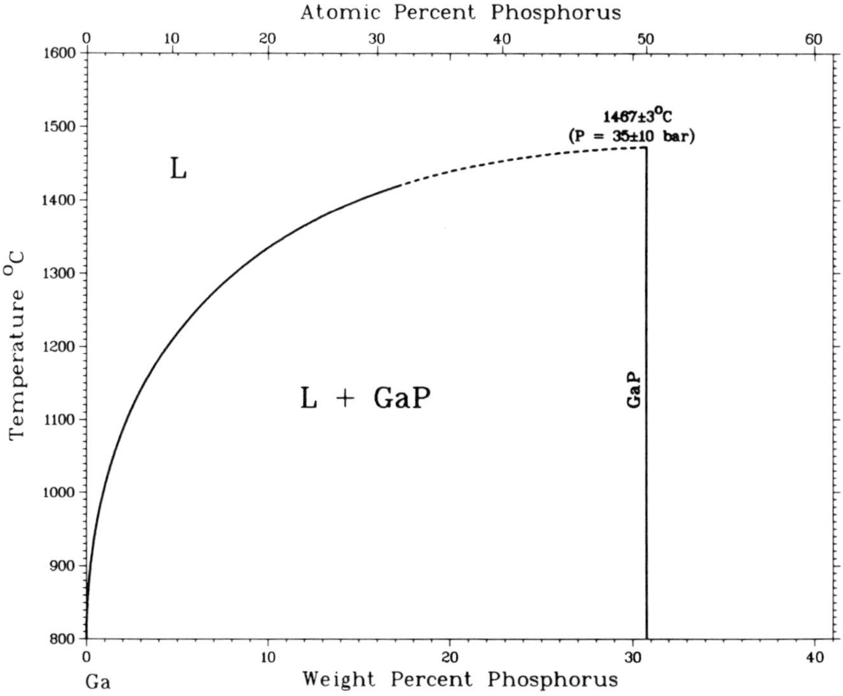

Assessed Pu-Ga Phase Diagram **Provisional**

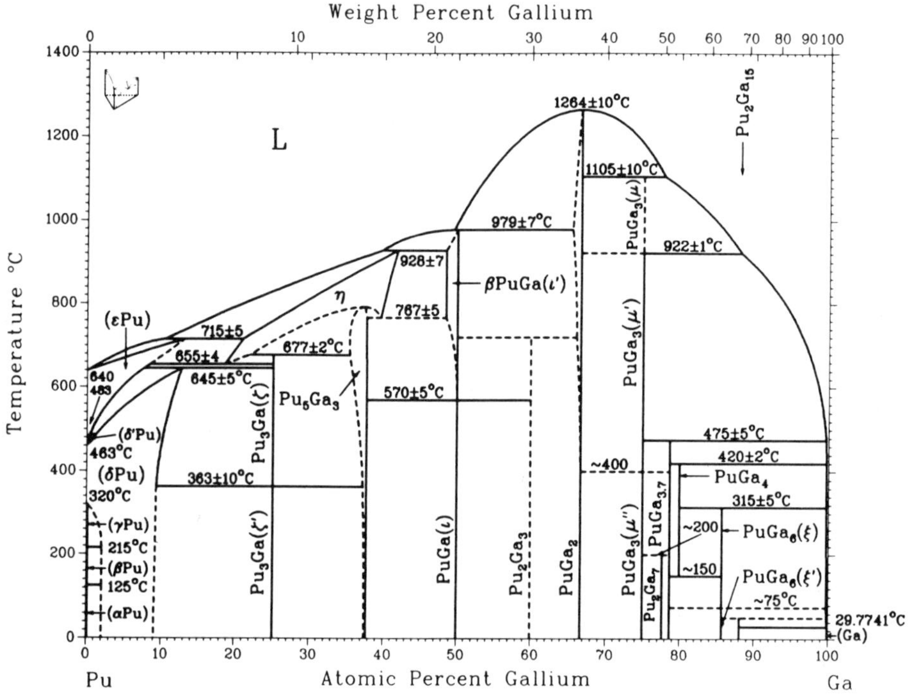

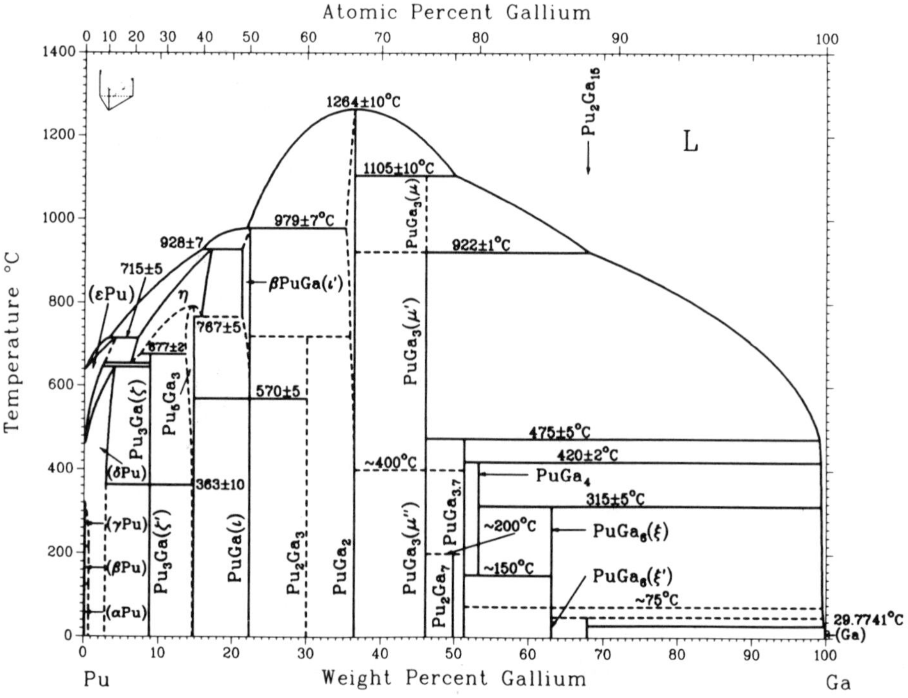

Ga-Sb Phase Diagram

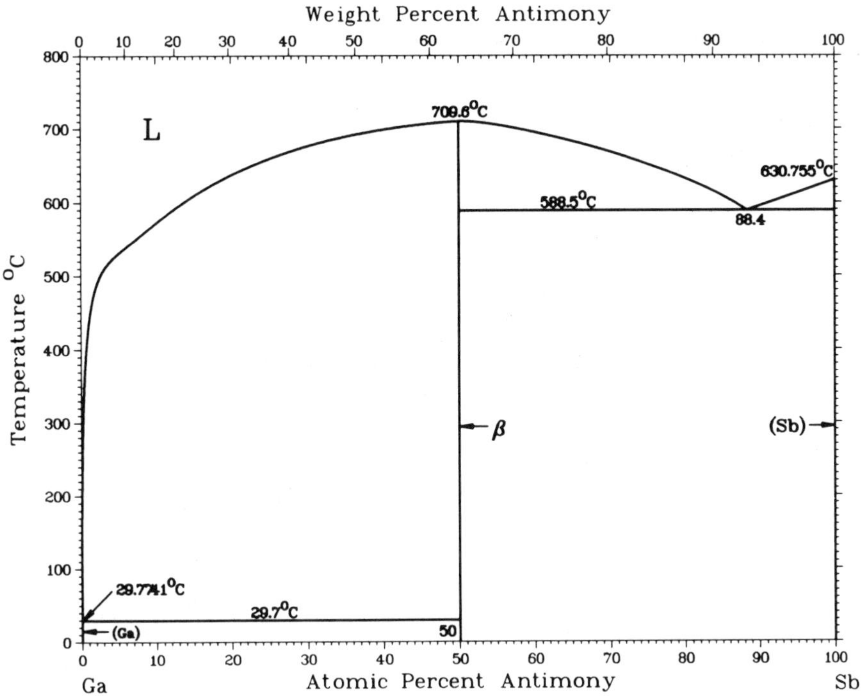

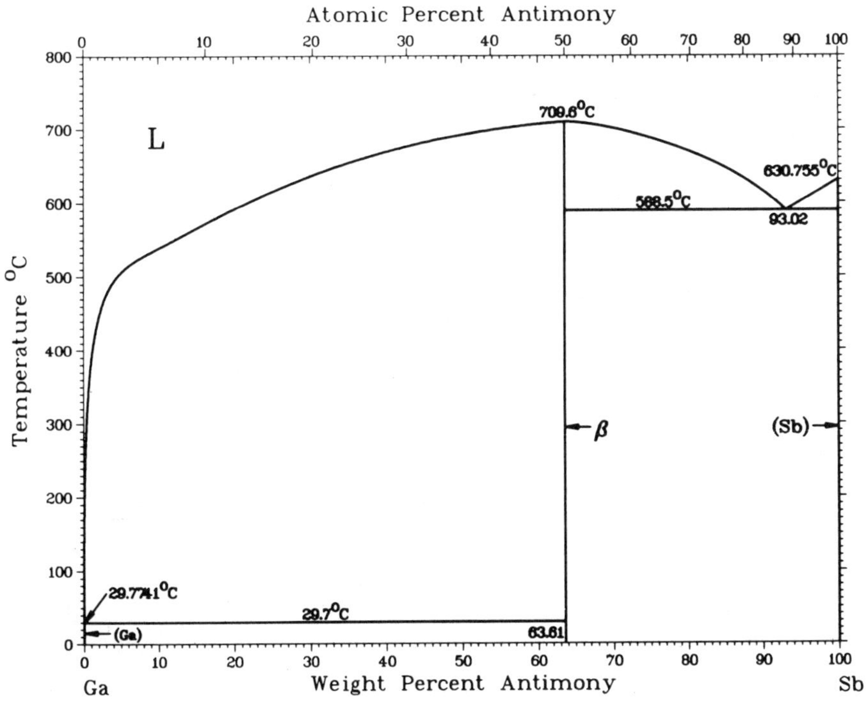

Assessed V-Ga Phase Diagram

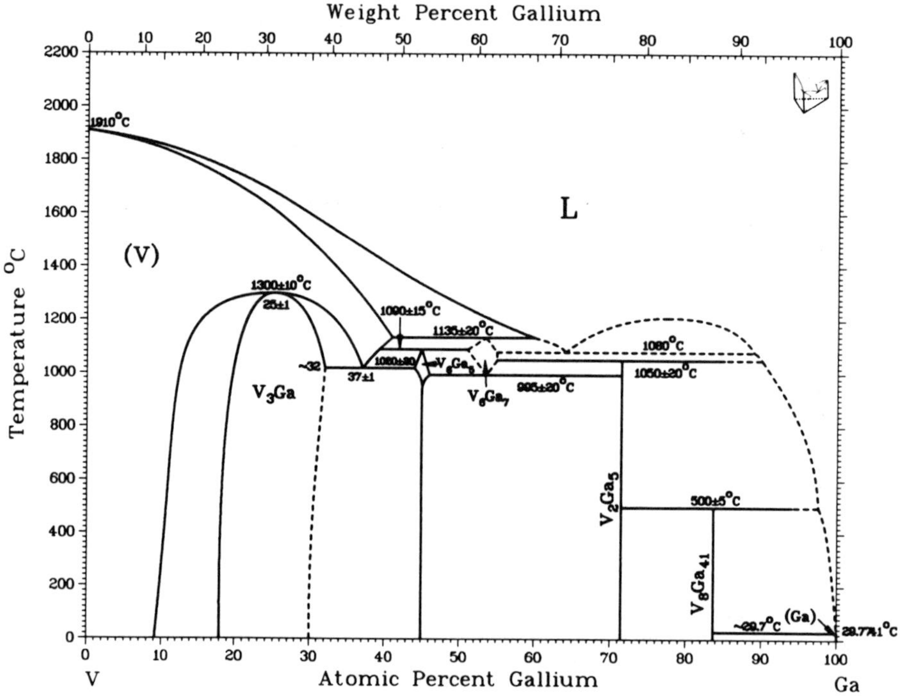

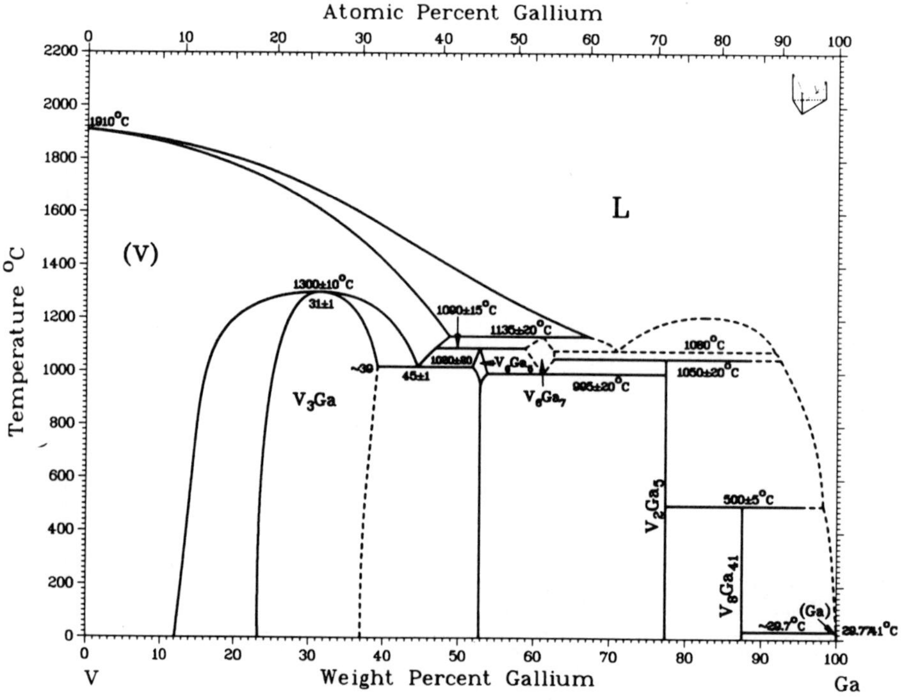

Assessed Mg-Ge Phase Diagram

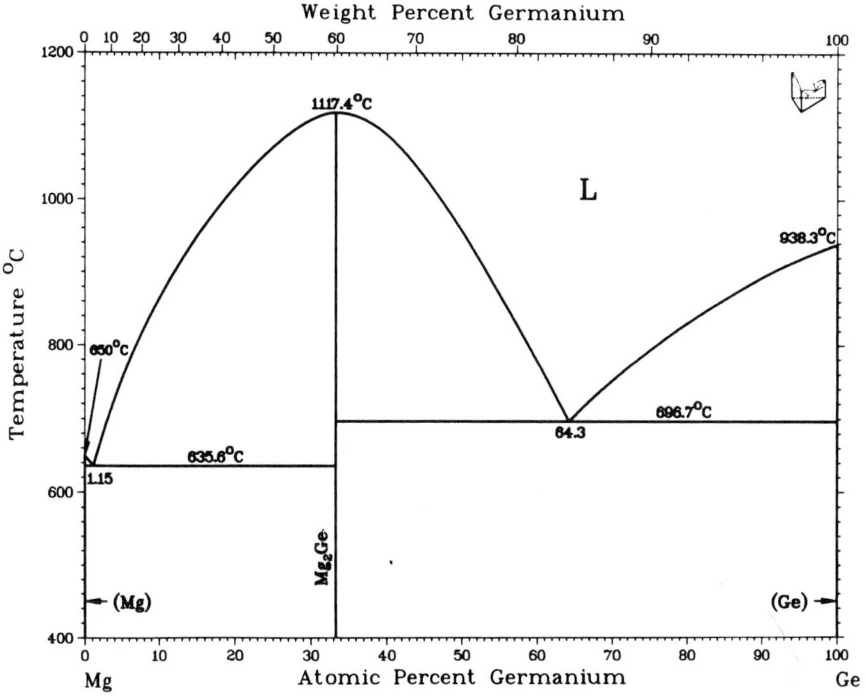

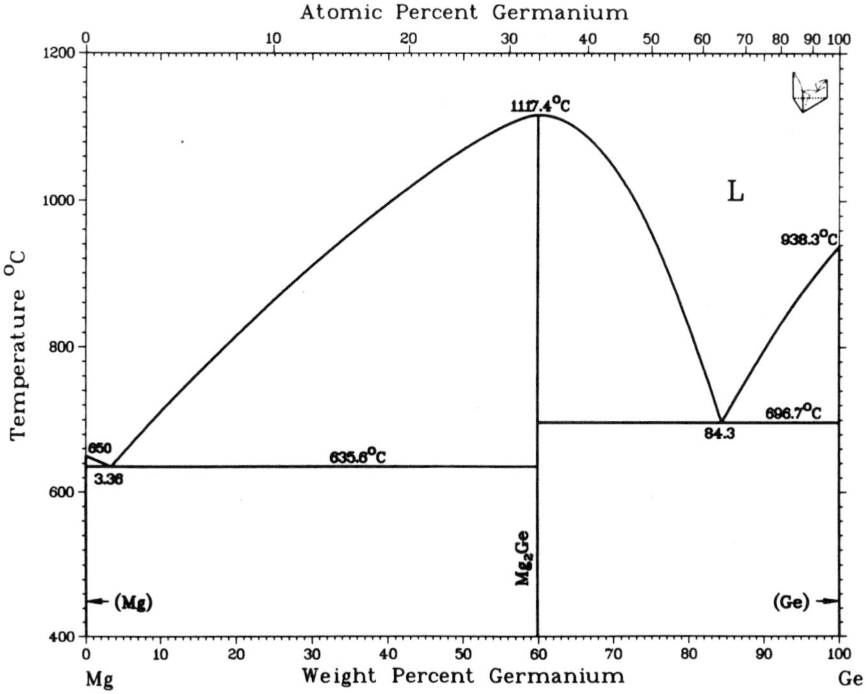

Nb-Ge Phase Diagram

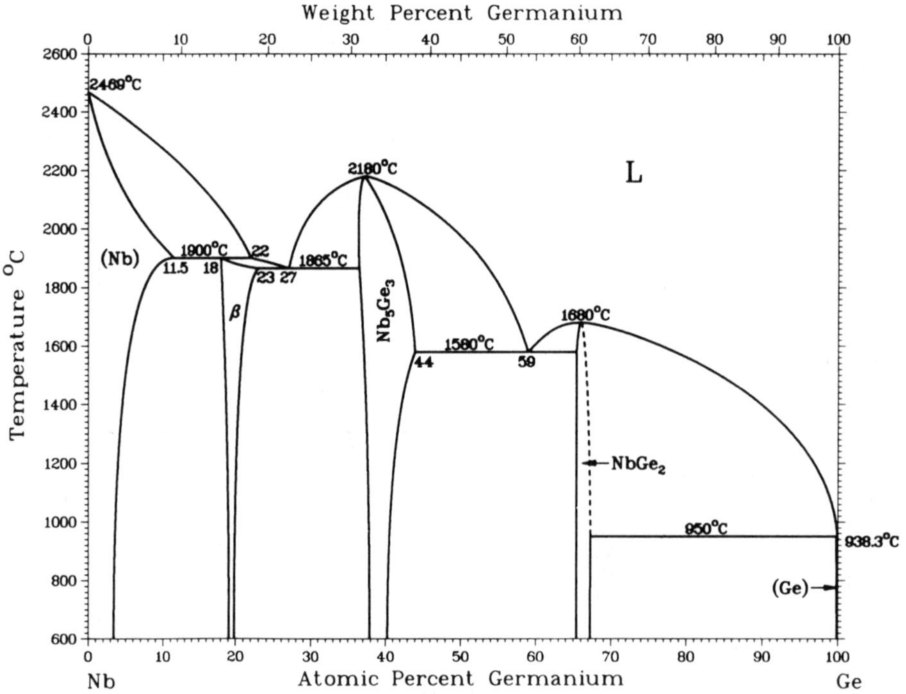

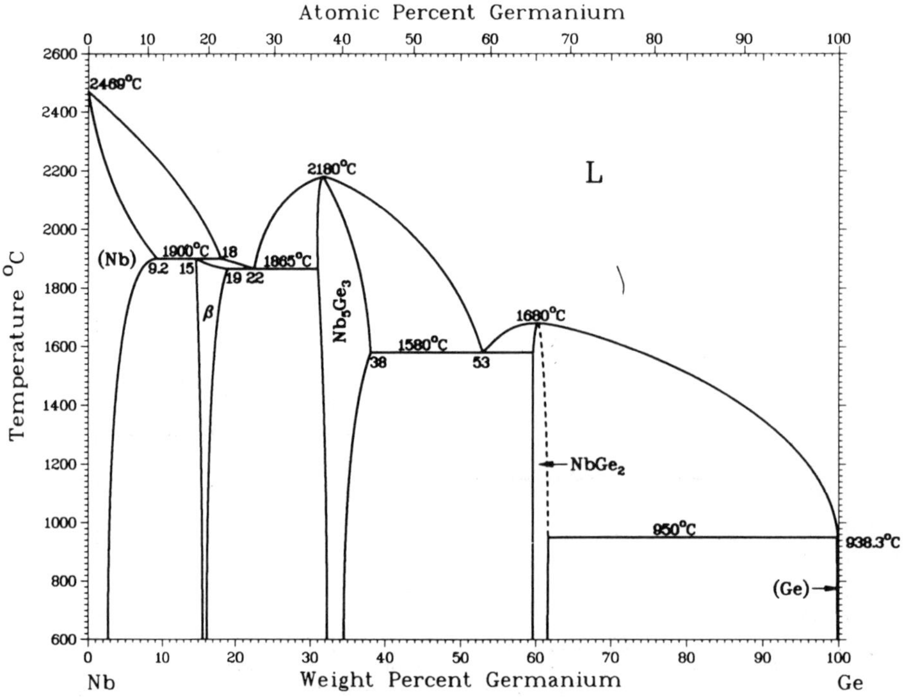

Assessed Ge-Si Phase Diagram

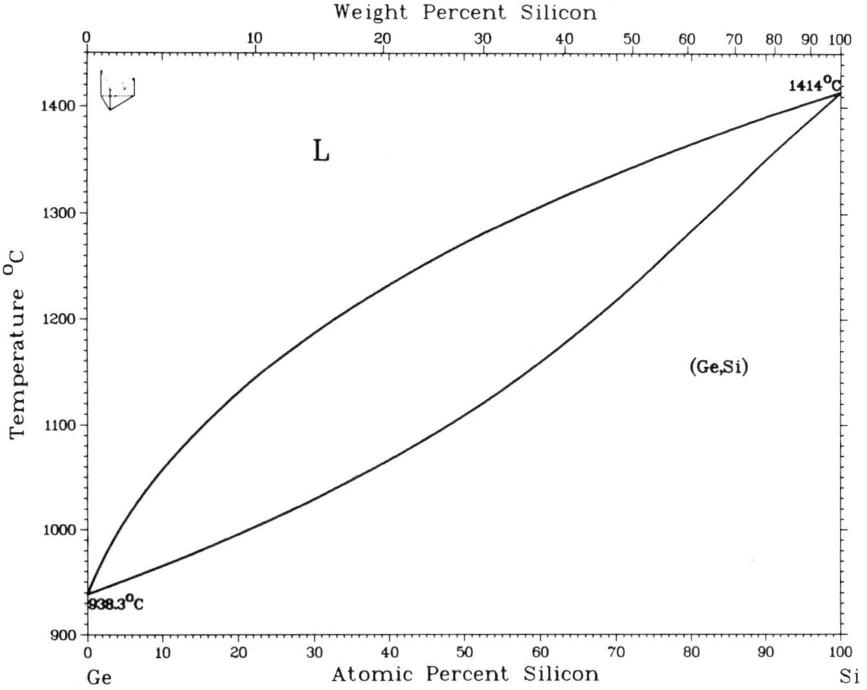

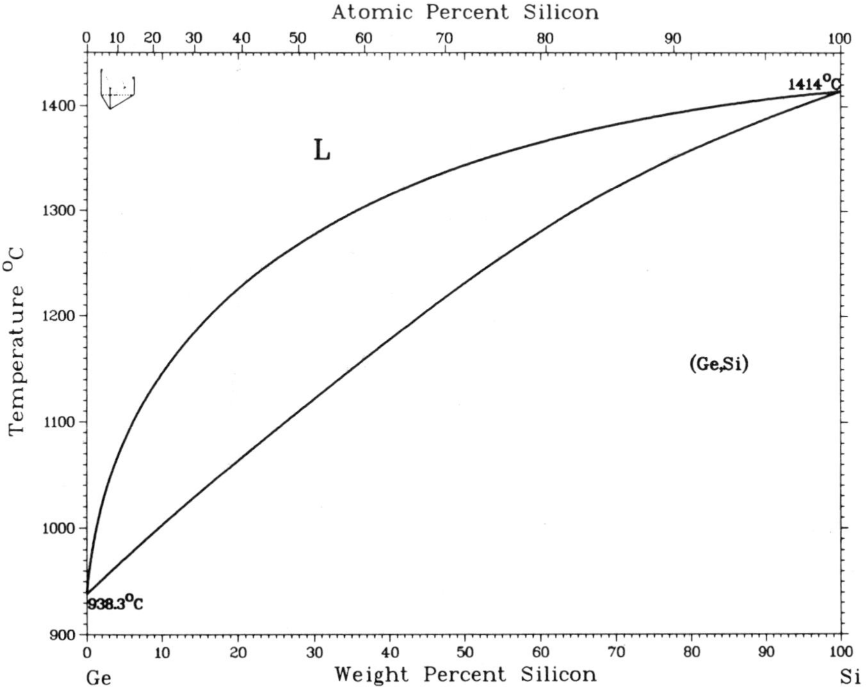

In-La Phase Diagram

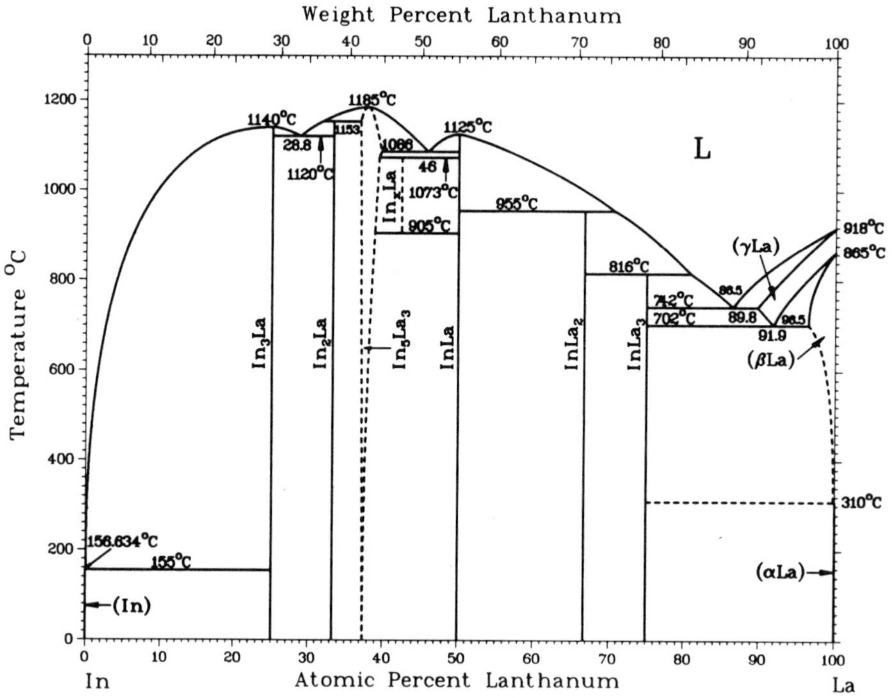

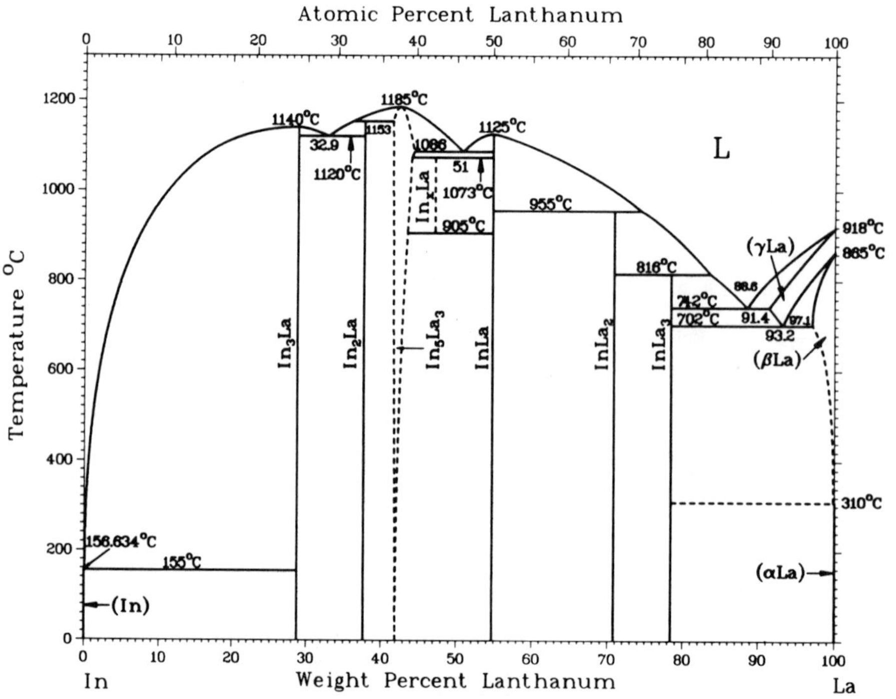

In-P Phase Diagram

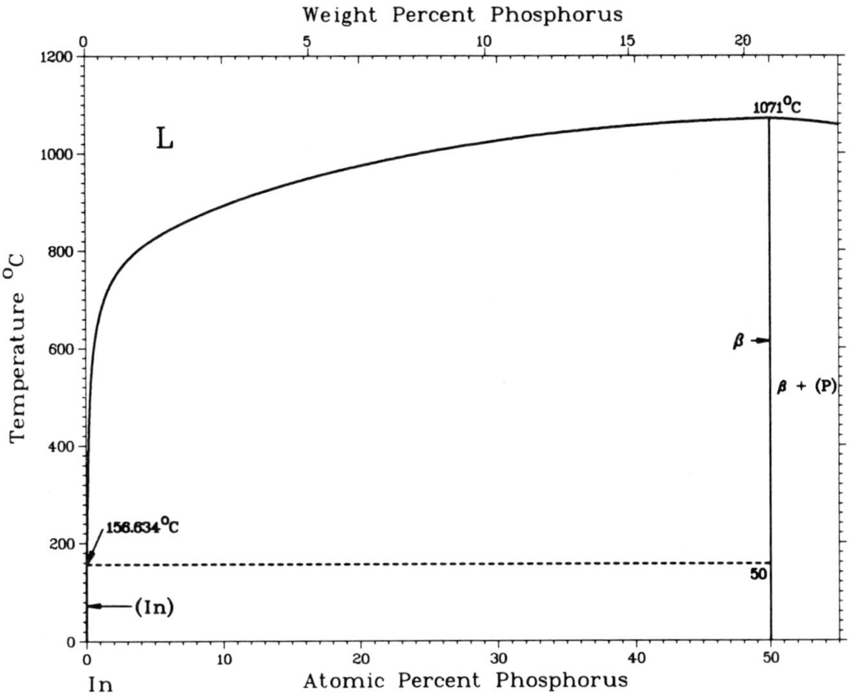

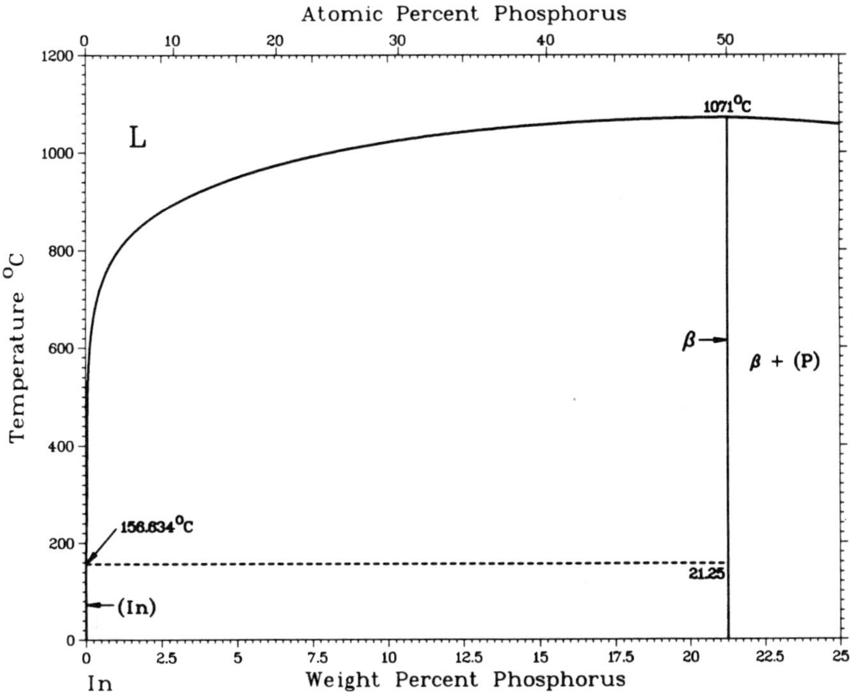

In-Sb Phase Diagram

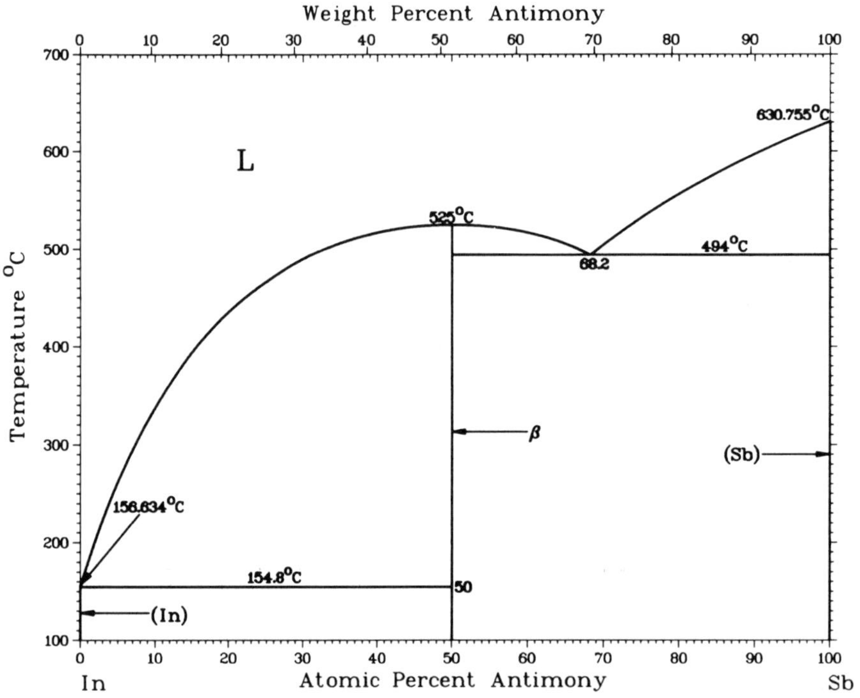

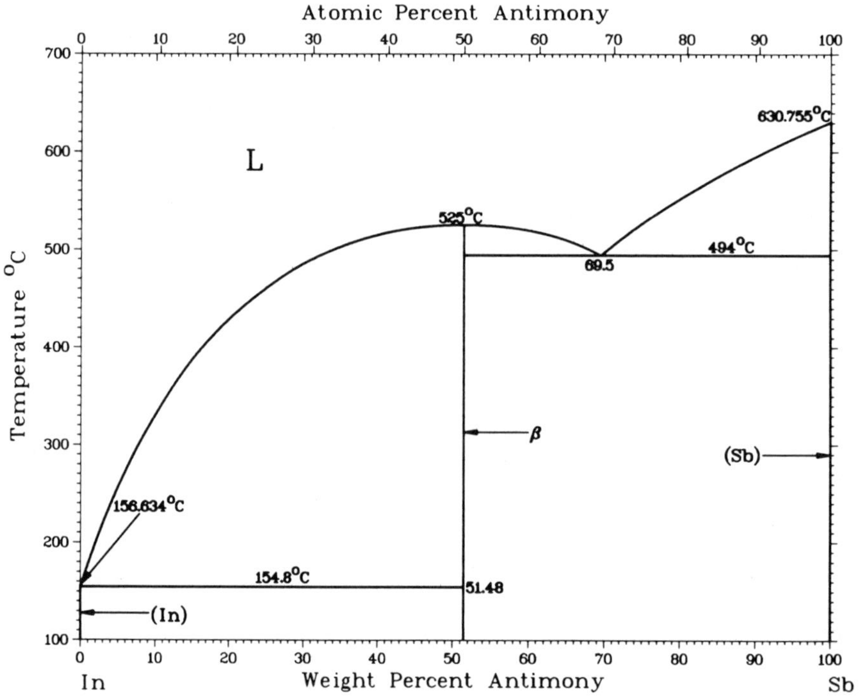

Assessed Mg-Si Phase Diagram

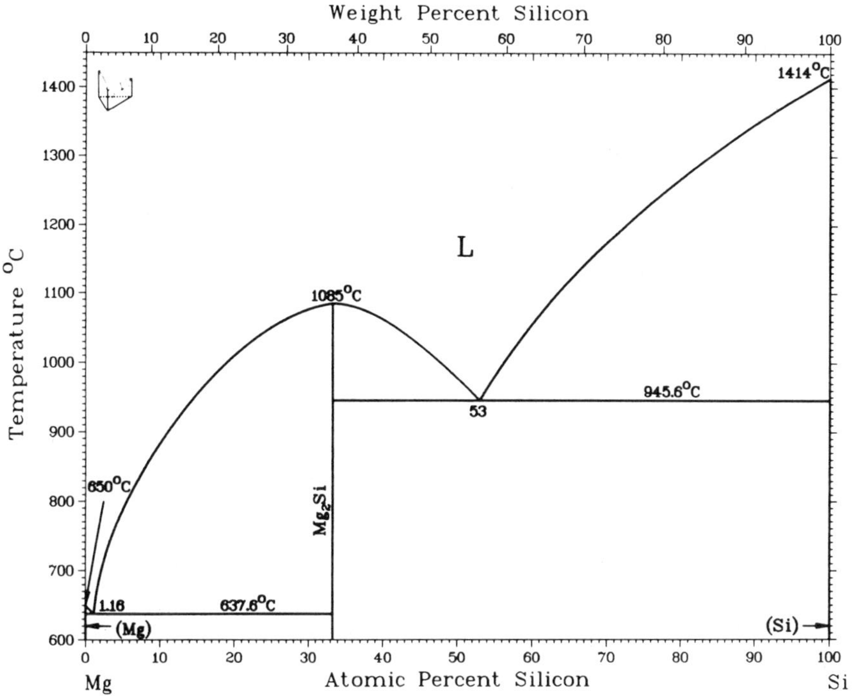

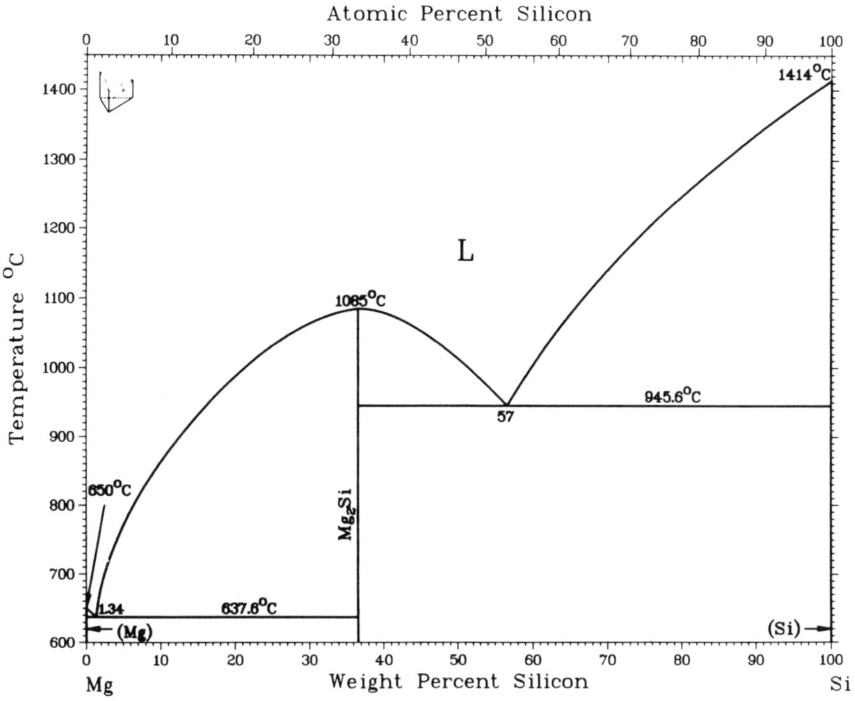

Assessed Mg-Sn Phase Diagram

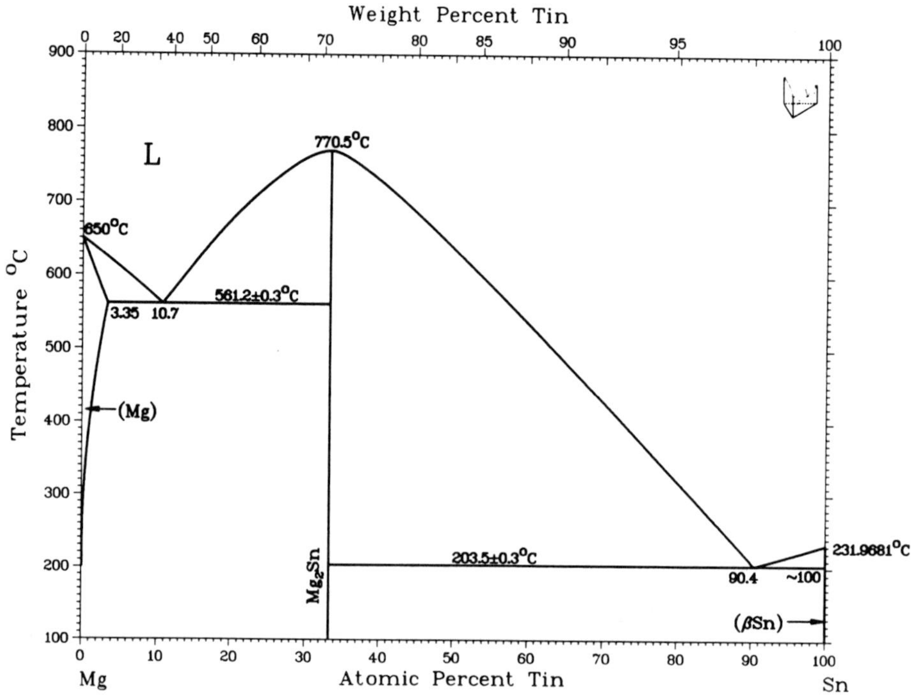

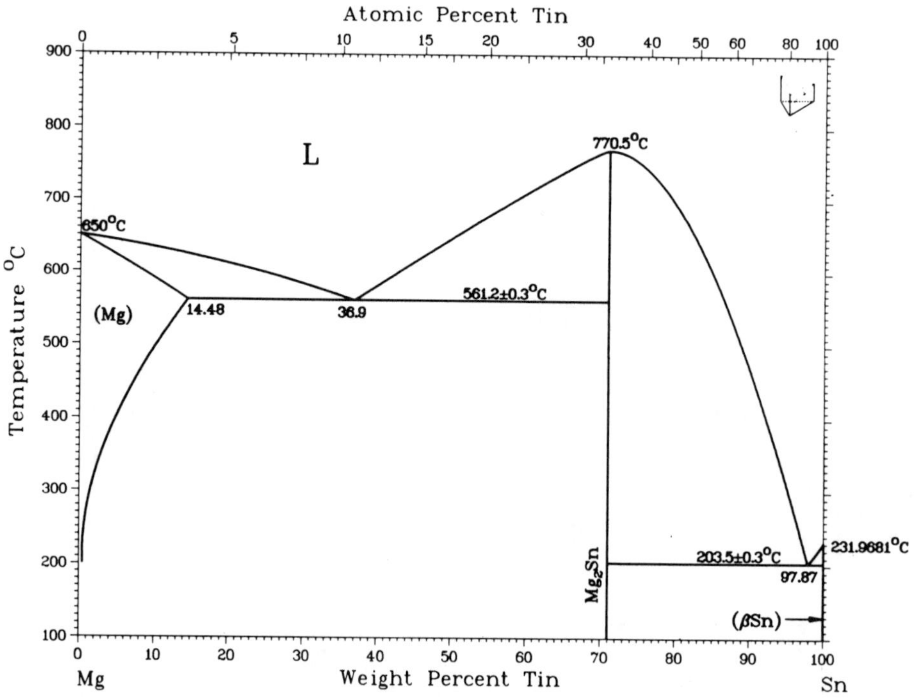

Assessed Mo-Si Phase Diagram

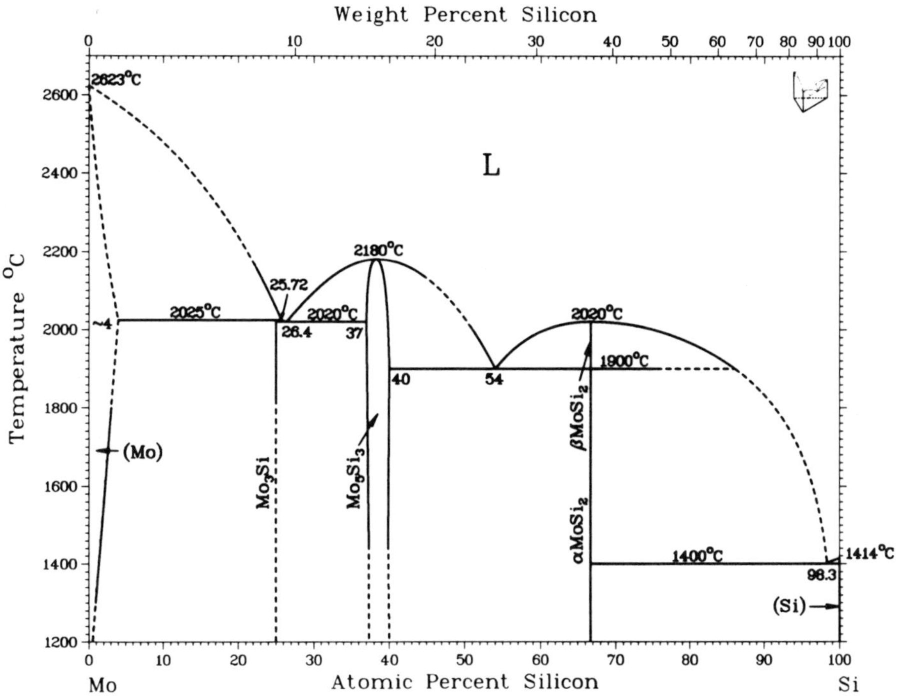

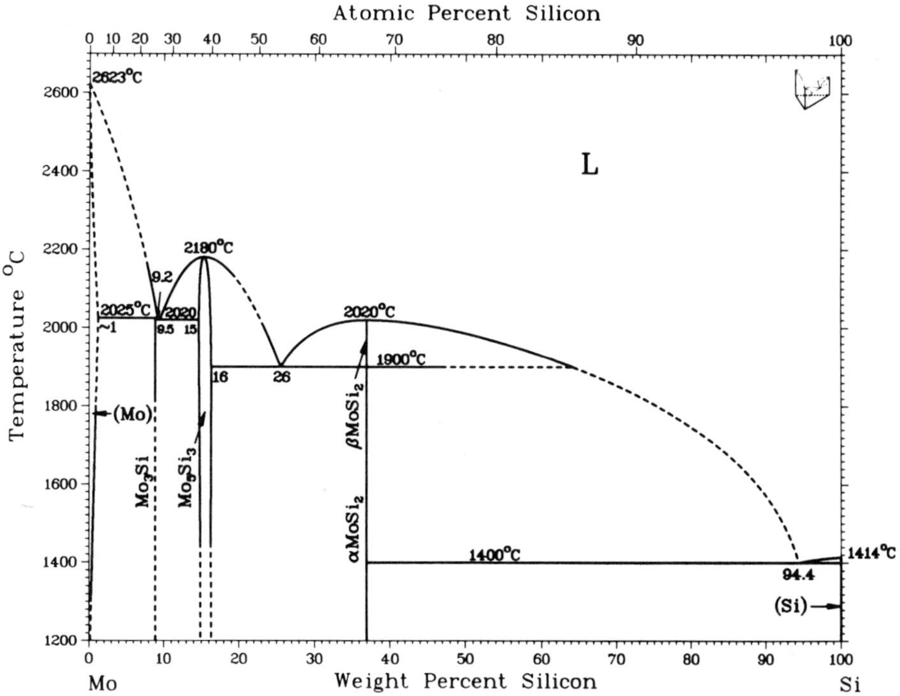

Nb-Sn Phase Diagram

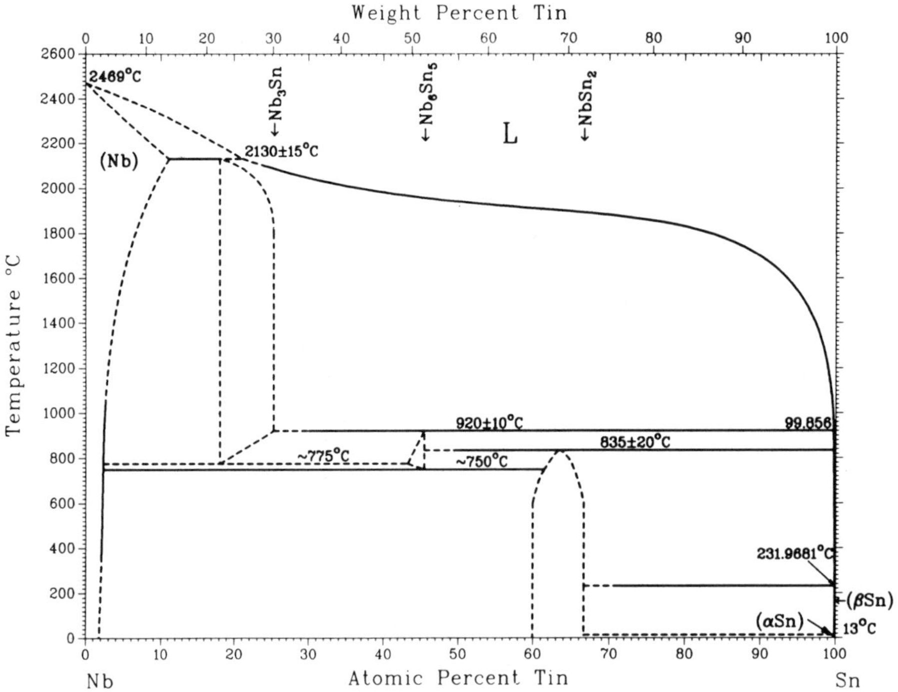

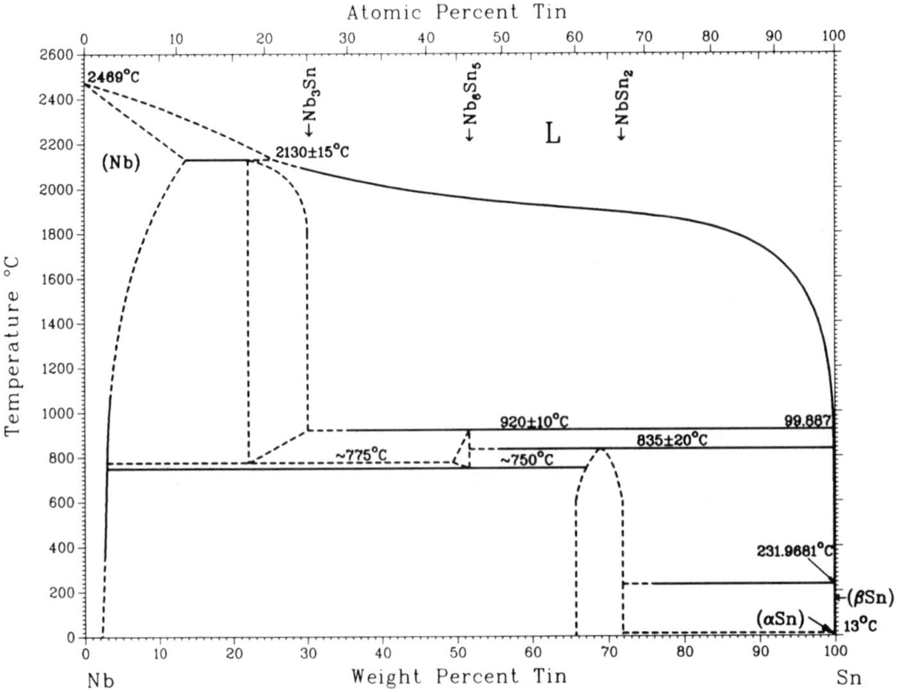

Assessed Ni-Si Phase Diagram

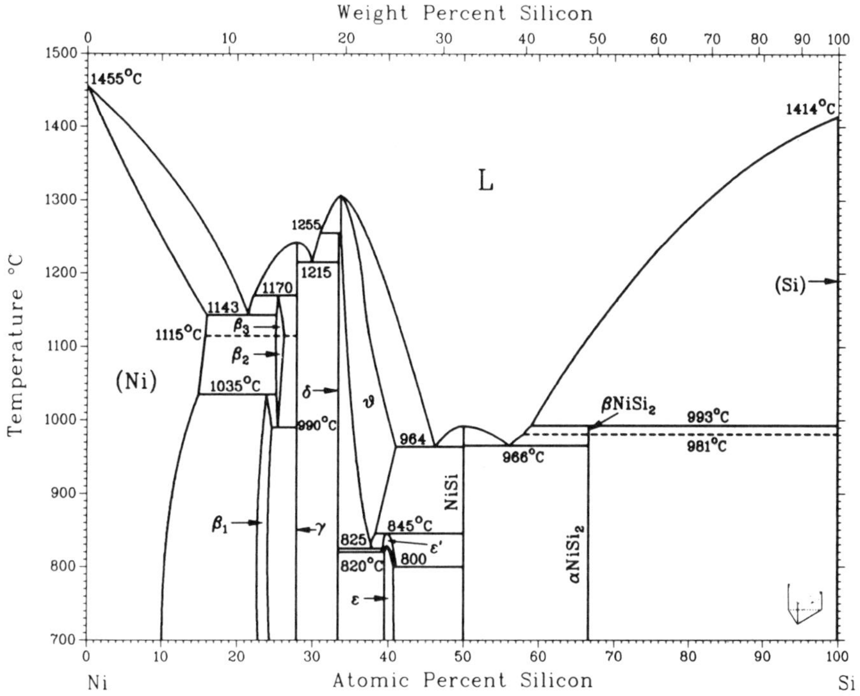

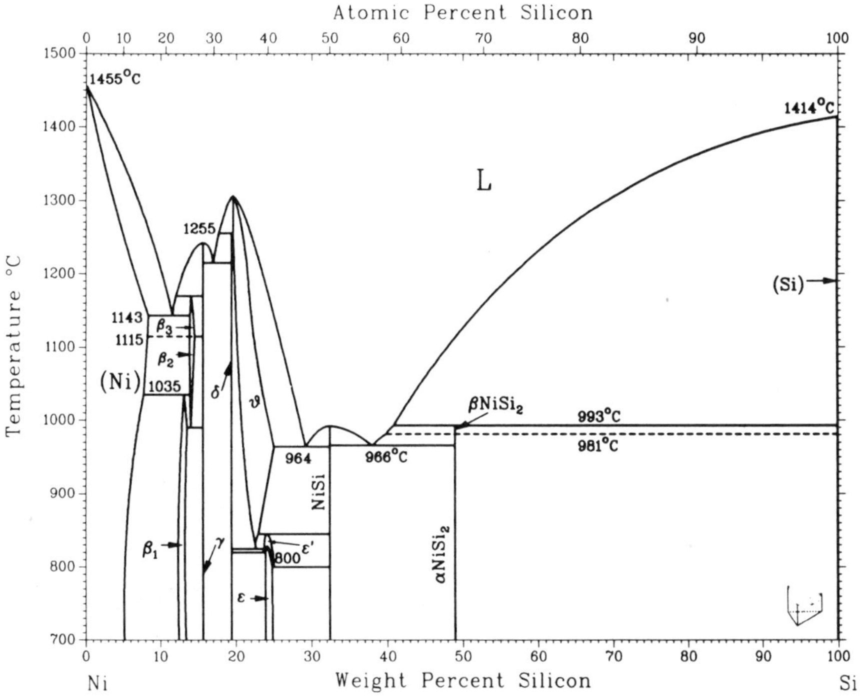

Assessed Pb-S Phase Diagram

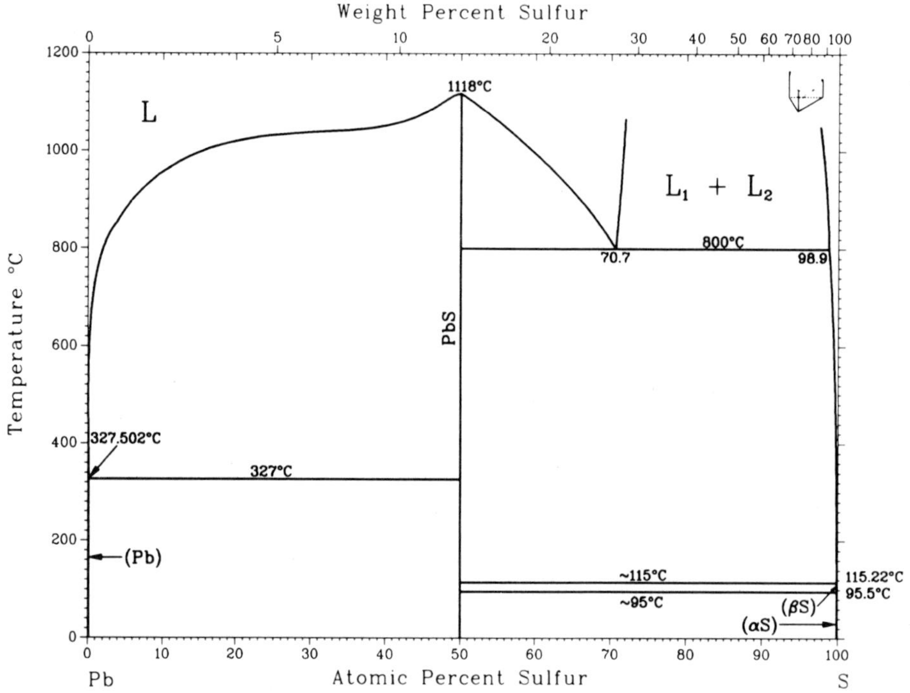

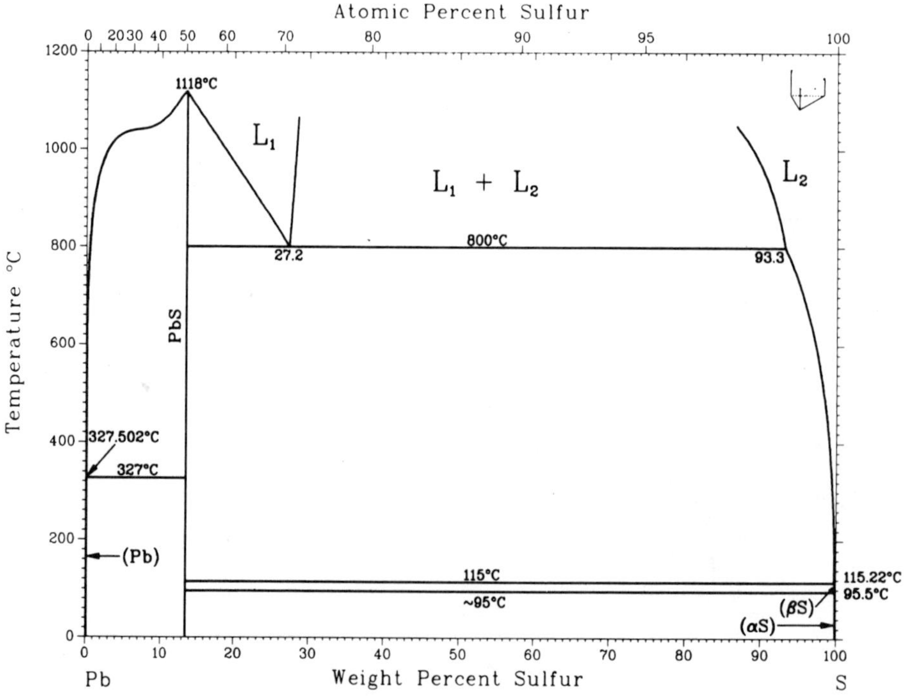

Assessed Pb-Se Phase Diagram

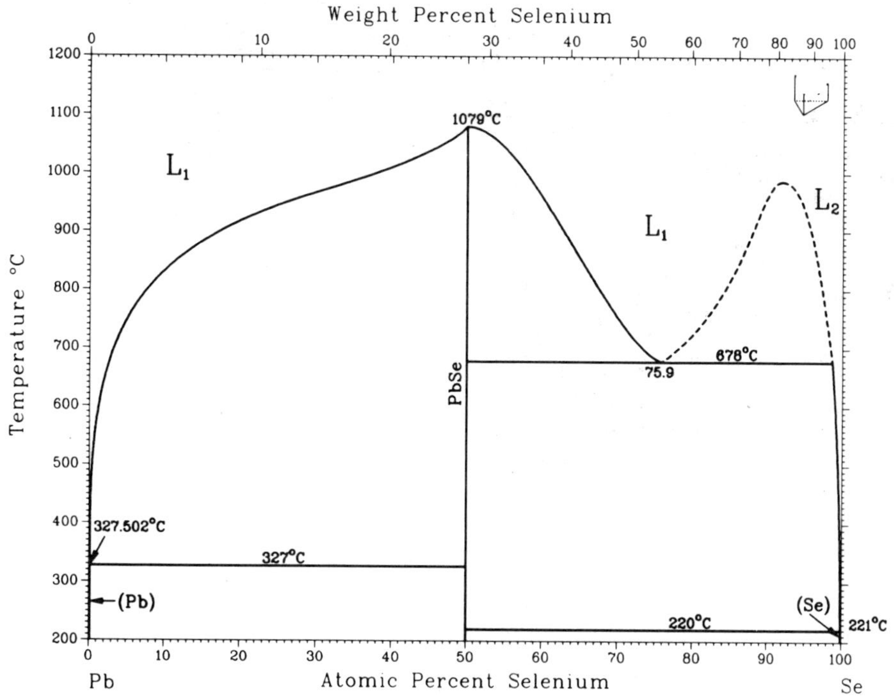

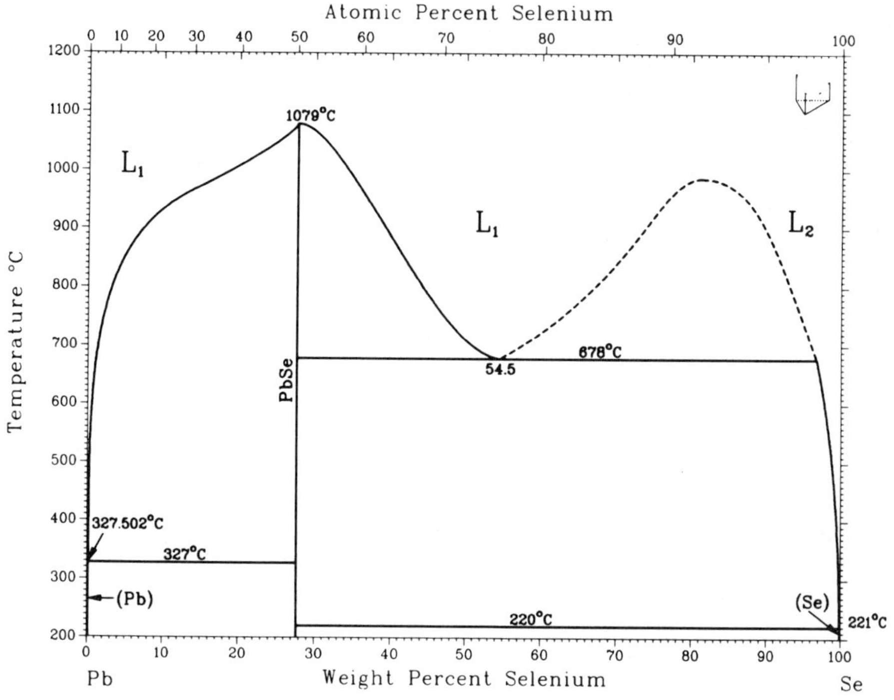

Assessed Pb-Te Phase Diagram

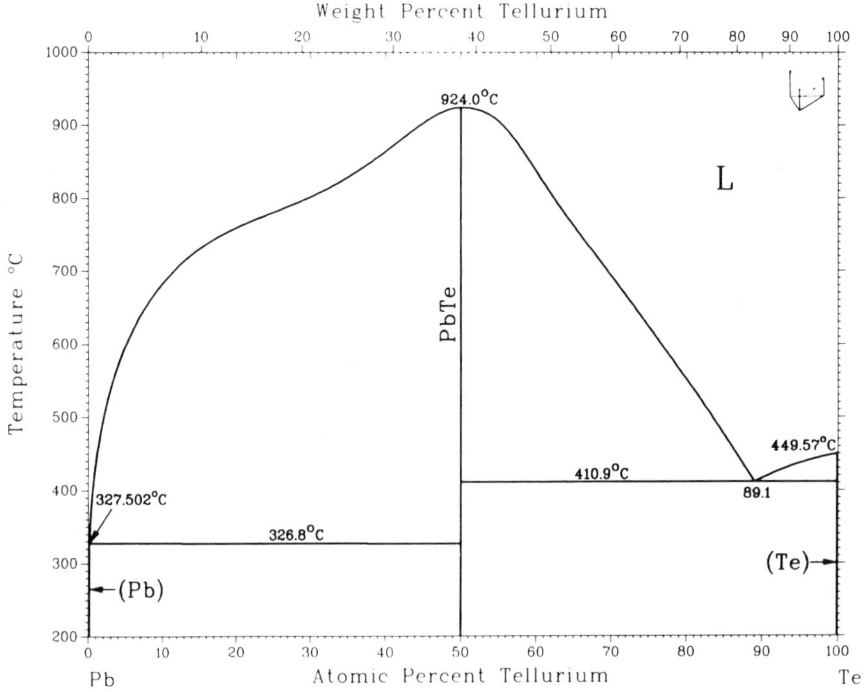

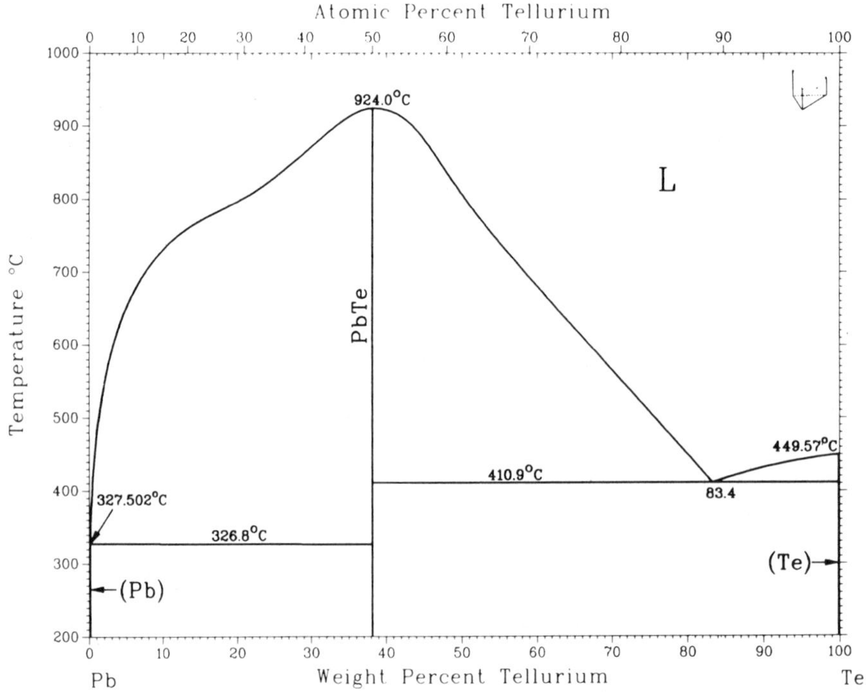

Sb-Zn Phase Diagram

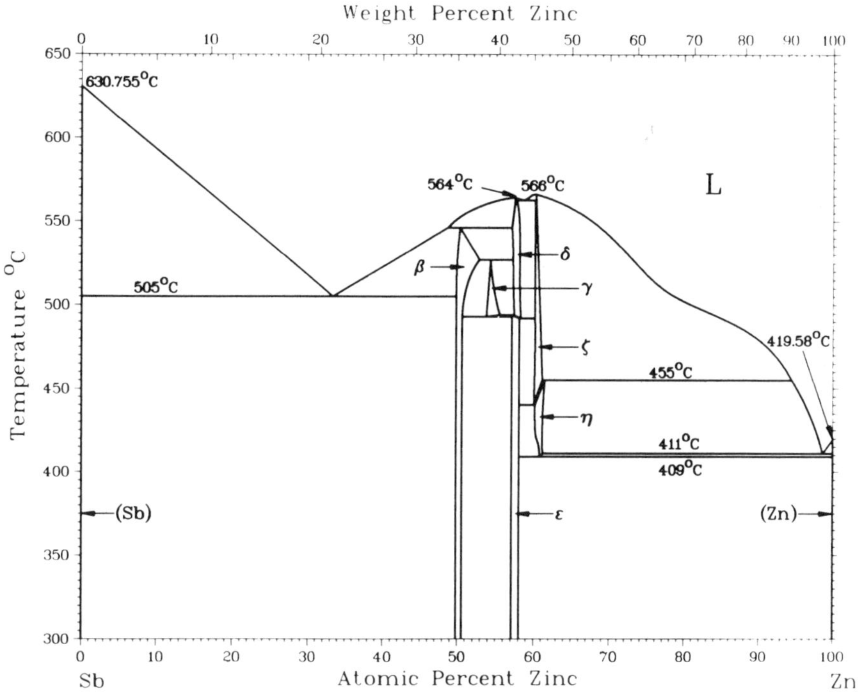

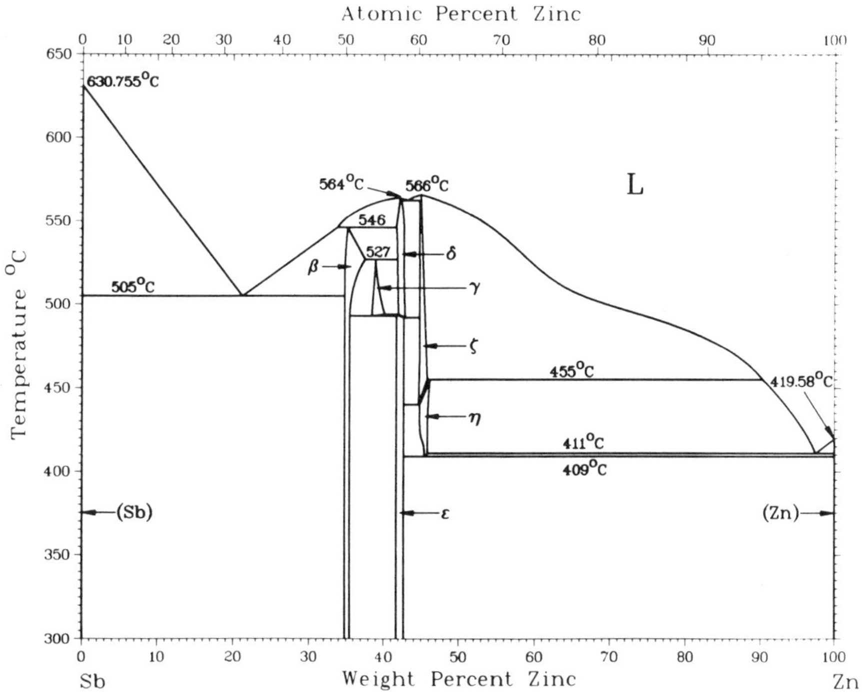

Zn-Se Phase Diagram

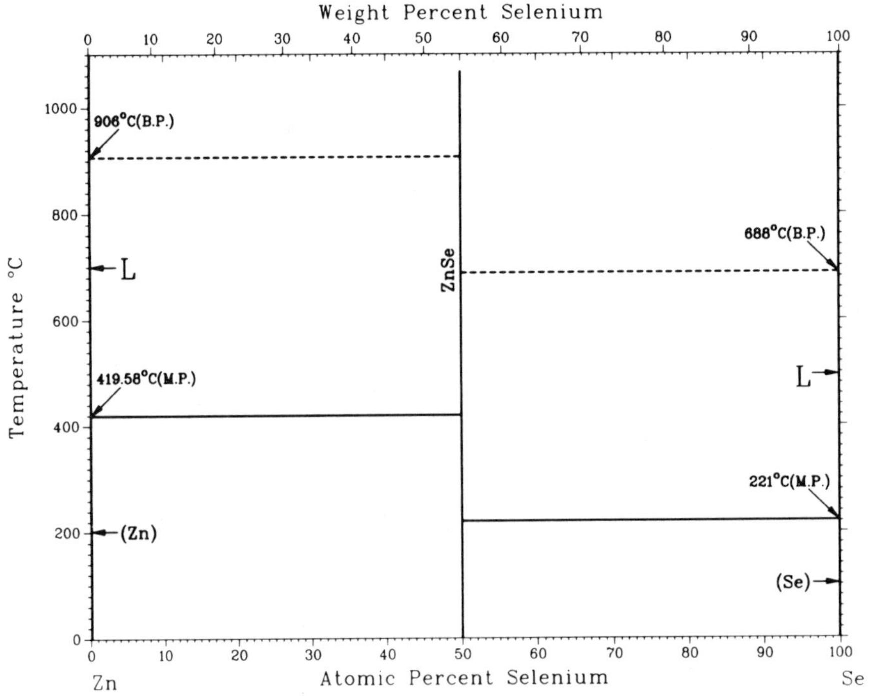

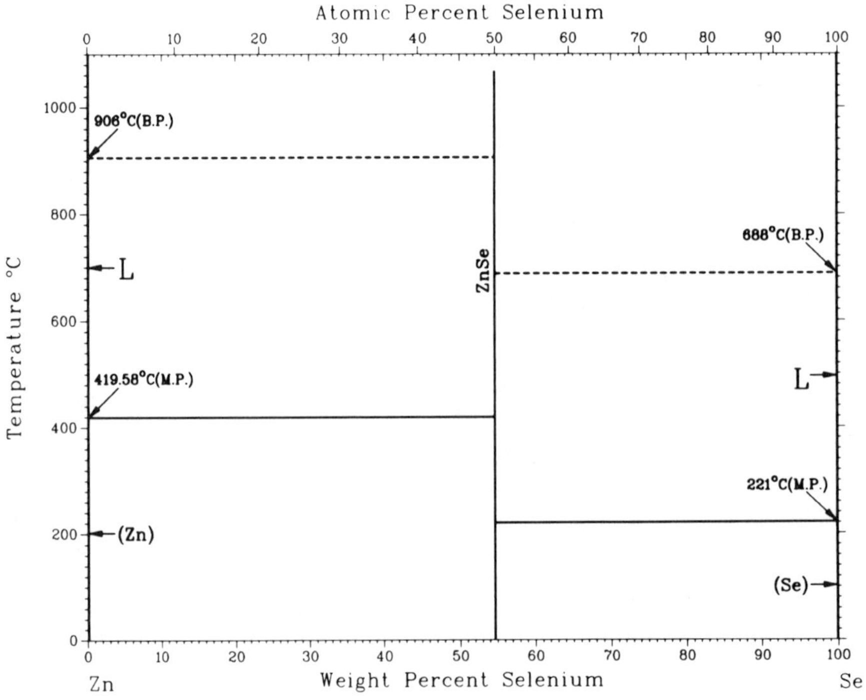

Ta-Si Phase Diagram

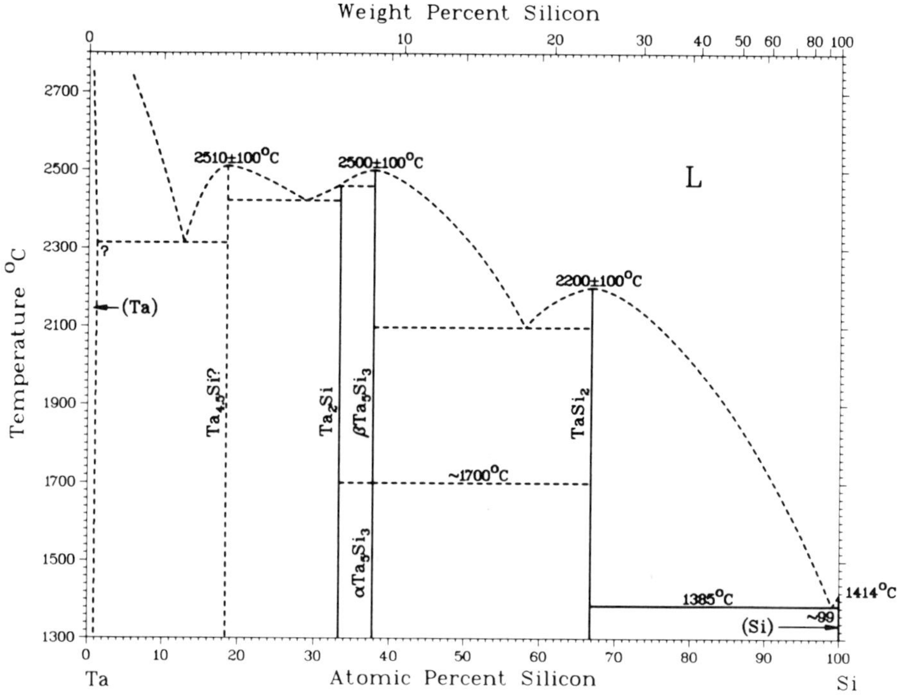

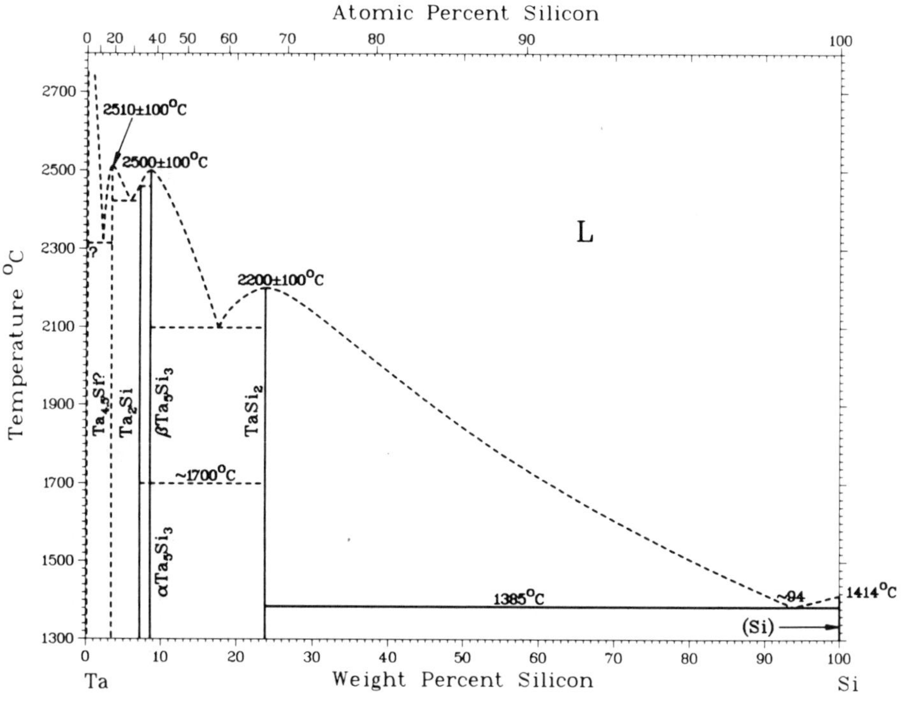

Assessed V-Si Phase Diagram

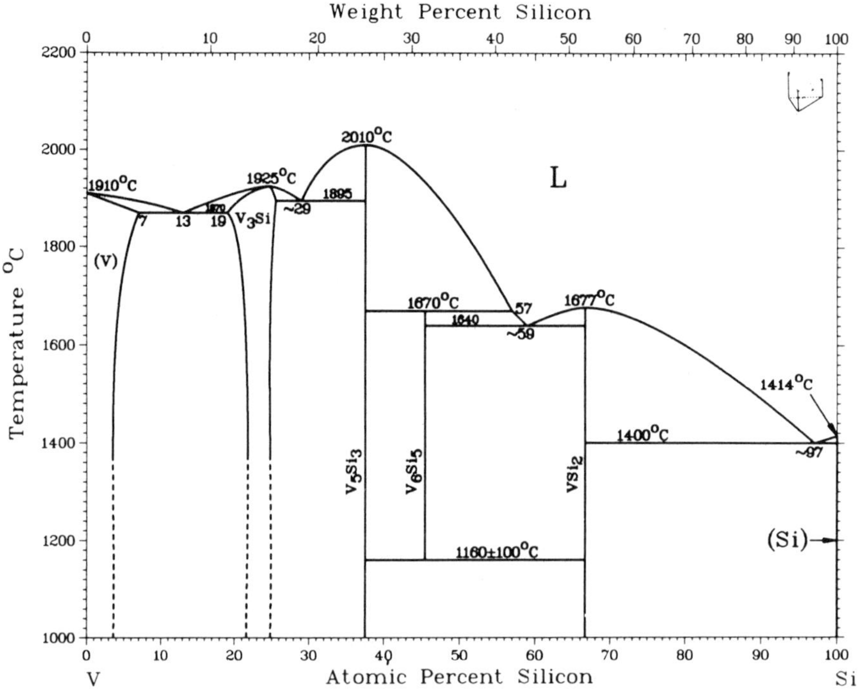

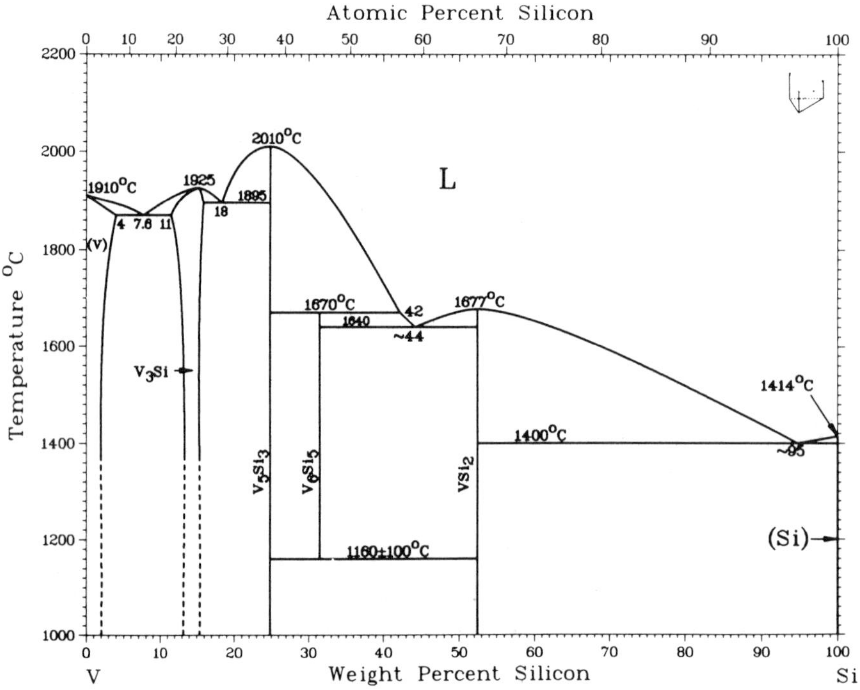

Te-Zn Phase Diagram

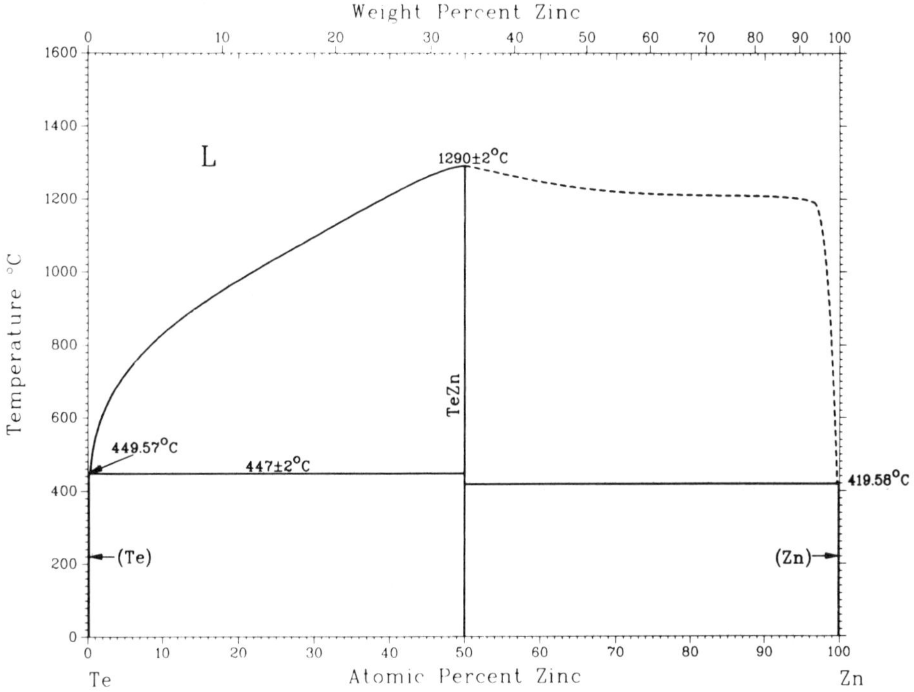

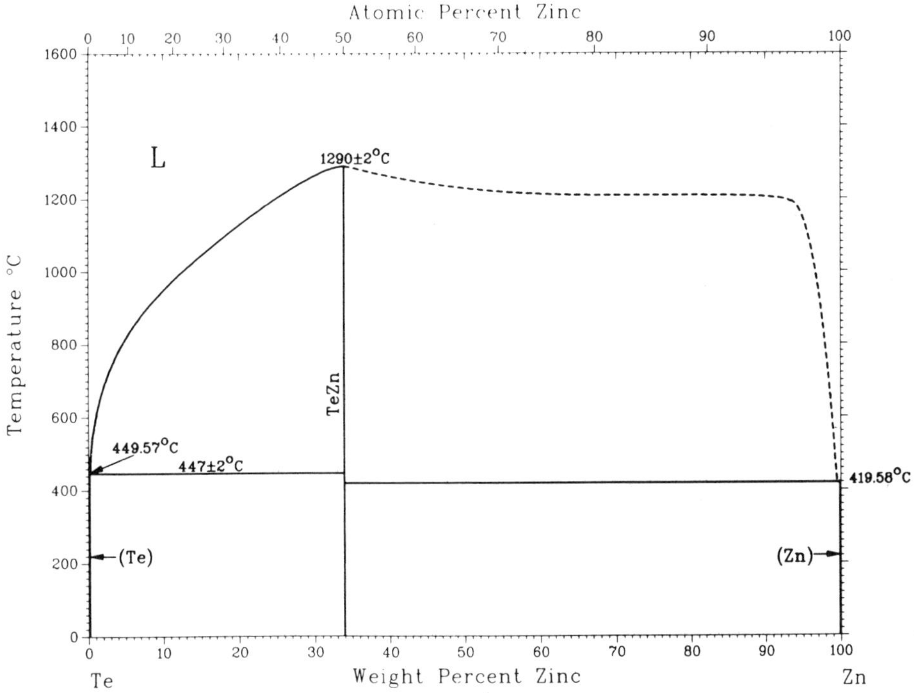

Appendices

APPENDICES

References

1. Engineering Applications of Ceramic Materials Source Book, Mel Schwartz, ed., American Society for Metals, Metals Park, OH, 1985

2. Product literature bulletin no. 502-PD1, Duramic Abrasive Products, Inc., 24135-T Gibson, Warren, MI 48089

3. Introduction to Ceramics, second edition, W.D. Kingery, H.K. Bowen and D.R. Uhlmann, John Wiley and Sons, New York, 1976

4. Guide to Selected Engineering Materials, A Special Issue of Advanced Materials and Processes, June 1987, ASM International, Metals Park, OH

5. Encyclopedia of Composite Materials and Components, Martin Grayson, ed., John Wiley and Sons, New York, 1983

6. ASM Engineered Materials Handbook, Vol. 1: Composites, ASM International, Metals Park, OH, 1987.

7. Advanced Materials and Processes, March 1976, ASM International, Metals Park, OH

8. Calcined Aluminas Product Data, Alcoa, Industrial Chemical Division, P.O. Box 300, Bauxite, Saline County, AR 72011

9. Engineering Property Data on Selected Ceramics Vol. 1, Vol. 2 and Vol. 3, Metals and Ceramics Information Center, Battelle Columbus Laboratories, Columbus, OH, August 1979

10. Ceramic Industry Data Book Issue, Vol. 125, No. 4, September 15, 1985, Cahners Publishing Co., Des Plaines, IL

11. Binary Alloy Phase Diagrams, Vol. 1 and Vol. 2, T.B. Massalski, ed., American Society for Metals, Metals Park, OH, 1986

12. Electrical Conduction in Solids: An Introduction, Daniel D. Pollock, American Society for Metals, Metals Park, OH, 1985

13. Advances in Electronic Materials, B.W. Wessels and G.Y. Chin, eds., American Society for Metals, Metals Park, OH, 1986

14. Electronic Packaging and Corrosion in Microelectronics, Morris E. Nicholson, ed., ASM International, Metals Park, OH, 1987

15. Plastics, Eighth Edition, Thermoplastics and Thermosets, D.A.T.A., Inc., a Cordura Company, San Diego, CA, 1986

16. Engineers' Guide to Composite Materials, American Society for Metals, Metals Park, OH, 1985

17. Plastics Gearing, Selection and Application, Clifford E. Adams, Marcel Dekker, Inc., New York, 1986

18. Hydrated Alumina Product Data, Alcoa, Industrial Chemicals Division, P.O. Box 300, Bauxite, Saline County, AR 72011

19. Reinforced Plastics for Commercial Composites Source Book, Gerry Shook, ed., American Society for Metals, Metals Park, OH, 1986

20. Plastics Engineering Handbook of the Society of the Plastics Industry, Inc., Fourth Edition, Joel Frados, ed., Van Nostrand Reinhold Co., New York, 1976

21. Machining Data Handbook, Third Edition, Vol. 1, Machinability Data Center, Metcut Research Associates, Inc., Cincinnati, OH, 1980

22. Advanced Composites: The Latest Developments, Conference Proceedings, ASM International, Metals Park, OH, 1986

23. Polymers for Engineering Applications, Raymond B. Seymour, ASM International, Metals Park, OH, 1987

24. Engineering Thermoplastics, Properties and Applications, James M. Margolis, ed., Marcel Dekker, Inc., New York, 1985

25. Plastics Products Design Handbook Part B, Processes and Design for Processes, Edward Miller, ed., Marcel Dekker, Inc., New York, 1983

26. Handbook of Composites, George Lubin, ed., Van Nostrand Reinhold Co., New York, 1982

27. Role of Interfaces on Materials Damping, Conference Proceedings, B.B. Rath, M.S. Misra, eds., American Society for Metals, Metals Park, OH, 1985

28. Advanced Composites III: Expanding the Technology, Conference Proceedings, ASM International, Metals Park, OH, 1987

29. Fabrication of Composite Materials Source Book, Mel Schwartz, ed., American Society for Metals, Metals Park, OH, 1985

30. Product literature, Fillite USA, Inc., Huntington, WV

31. Torlon Engineering Polymers Design Manual, AMOCO Chemicals Corporation, 1007 E. Randolph Dr., Chicago, IL, 60601

Abbreviations

a crack length
A ampere
A area; ratio of the alternating stress amplitude to the mean stress
Å angstrom
ABS acrylonitrile-butadiene-styrene
ac alternating current
ACM advanced cure monitor
APD avalanche photodiode
ARC accelerated rate calorimeter
at.% atomic percent
AS designation for surface-treated fiber
ATL automated tape layer
atm atmosphere (pressure)
AU designation for untreated fiber

BF_3MEA borontrifluoro-monoethylamine
BGDGE butylene glycol diglycidyl ether
BMC bulk molding compound
BMI bismaleimide (resin)
BPSG borophosphosilicate glasses
BT bismaleimide-triazine (resin)
Btu British thermal unit

c composite specific heat (C_p = constant pressure, C_v = constant volume)
C-C carbon-carbon
CAD/CAM computer-aided design/computer-aided manfacturing
CCA composite cylinder assemblage
CAT computer-aided tomography
CFRP carbon fiber reinforced plastic
cm centimeter
cpm cycles per minute
cps cycles per second
CTE coefficient of thermal expansion
CVD chemical vapor deposition
CVN Charpy V-notch (impact test or specimen)
CZ Czochralski (grown crystals)

d an operator used in mathematical expressions involving a derivative (denotes rate of change)
d depth; diameter
DADPS diaminodiphenylsufone
DAIP diallyl isophthalate
DAP diallyl phthalate
DBTT ductile-brittle transition temperature
dc direct current
DDA dynamic dielectric analysis
DGA diglycidyl aniline
DGEBA diglycidyl ether of bisphenol A
DGEBF diglycidyl ether of bisphenol F
DGT dynamic gel temperature
DH double heterostructure (lasers or LEDs)
diam diameter
DIB diiodobutane
DIP dual-in-line package
DLD dark line defects
DLP delta lattice parameter (phase diagram models)
DMA dynamic mechanical analysis
DP dissolution pit
DSC differential scanning calorimetry
DSD dark spot defect
DTA differential thermal analysis

e natural log base, 2.71828
E modulus of elasticity; Young's modulus
EL2 electrical compensation (doping)
epi epilayers, epitaxy or epitaxial
Eq equation
et al. and others
ESCA electron spectroscopy for chemical analysis

f fiber
f frequency
F force
FET field effect transistor
FMW formulated molecular weight
FP polycrystalline alumina fiber
FPP flat plastic package
FRP fiber-reinforced plastic
FRS fiber-reinforced superalloys
ft foot
FTIR Fourier transform infrared
FZ floatzone (-grown crystals)

g gram
G shear modulus
G' storage modulus
G" loss modulus
gal gallon
GPa gigapascal
GPC gel permeation chromatography
gpd grams per denier
gr grain
G_{XY} in-plane shear modulus of laminate
G_{IC} interlaminar fracture toughness (mode I, peel; mode II, shear; mode III, scissor shear)

h hour
h plate thickness of composite
H height
HERF high-energy-rate forging
HIP hot isostatic press
HM high modulus
HPLC high-performance liquid chromatography
H_r heat of reaction
HRTEM high-resolution transmission electron microscopy
HT high tensile
Hz hertz

IC integrated circuit
ID inside diameter
IM intermediate modulus
IR infrared (radiation)

J joule

k notch sensitivity factor
K Kelvin
K coefficient of thermal conductivity; modulus of elasticity
K_I stress-intensity factor
K_c plane-stress fracture toughness
K_{Ic} plane-strain fracture toughness; mode I critical stress-intensity factor
K_{Id} dynamic fracture toughness
K_{Iscc} threshold stress intensity for stress-corrosion cracking
K_t stress-concentration factor
K_t^{∞} stress-concentration factor for infinate plate
K_{th} threshold crack tip stress-intensity factor
kg kilogram
km kilometer
kPa kilopascal
ksi kips (1000 lb) per square inch
kV kilovolt

L liter; longitudinal direction
L length
lb pound
LCC leadless chip carrier
LEC liquid-encapsulated Czochralski (crystal growth)
LED light-emitting diode
LEFM linear-elastic fracture mechanics
ln natural logarithm (base e)
LPCVD low-pressure chemical vapor deposition
LPE liquid phase epitaxy

m matrix
MBE molecular beam epitaxy
MCM multichip module
MCO melt carryover
MDA methylenedianiline
MEKP methyl ethyl ketone peroxide
Mg megagram
min minute; minimum
MJ megajoule
mL milliliter
MLC multilayer ceramic (substrate or module)
mm millimeter
MMA methyl methacrylate
MMC metal matrix composite
MODFET modulation doped field effect transistor
mol % mole percent
MOS metal-oxide-semiconductor
MPa megapascal
mph miles per hour
MVT moisture vapor transmission

N Newton
N fatigue life (number of cycles)
NDE nondestructive evaluation
NDI nondestructive inspection
NDT nondestructive testing
nm nanometer
No. number

OD outside diameter
OED oxidation-enhanced diffusion
OMVPE organometallic vapor phase epitaxy
ORD oxidation-retarded diffusion
OSF oxidation-induced stacking faults
oz ounce

P applied load; pressure
Pa pascal
PAI polyamideimide
PAN polyacrylonitrile
PAS polyarylsulfone
PBI polybenzimidazole
PBT polybutylene terephthalate
PCB printed circuit board
PDCP polydicyclopentadiene
PECVD plasma-enhanced chemical vapor deposition
PEEK polyether etherketone
PEI polyetherimide
PES polyether sulfone
PET polyethylene terephthalate
PGA pin grid array
PI polyimide
PIC pressure-impregnation-carbonization
PLCC plastic-leaded chip carrier
P/M powder metallurgy

PMR *in-situ* polymerization of monomer reactants
ppb parts per billion
ppm parts per million
PPS polyphenylene sulfide
PS polysulfone
PSG phosphosilicate glass
psi pounds per square inch
psia pounds per square inch absolute
psid pounds per square inch differential
psig pounds per square inch gage
PTFE polytetrafluoroethylene
PTH pin-through-hole technology
PVA polyvinyl alcohol
PVC polyvinyl chloride

R radius; ratio of the minimum stress to the maximum stress
r rate of reaction
RA reduction of area
RDS rheometric dynamic scanning
Ref reference
RGA residual gas analysis
RH relative humidity
RIM reaction injection molding
RMS root mean square
ROM rule of mixtures; rough order of magnitude
rpm revolutions per minute
RRIM reinforced reaction injection molding
RTA rapid thermal annealing
RTD room temperature, dry
RTM resin transfer molding
RTW room temperature, wet
RTV room-temperature vulcanizing
RVE representative volume element
RDGE resorcinol diglycidyl ether

s second
SBS short beam shear
SCM single-chip module
SEM scanning electron microscope or microscopy
SF stacking fault
SMC sheet molding compound
SMT surface mount technology
S-N stress-number of cycles
SOP small outline package
SPF superplastic forming
sp gr specific gravity

t thickness; time
T transverse direction
T temperature; tenacity
TAB tape automated bonding
TCE thermal coefficient of expansion
TCL transmission cathodolumines cence (image)
TCM thermal conduction module
TEM transmission electron microscope or microscopy
TFT thin film transistor
T_g glass transition temperature
TGA thermogravimetric analysis
TGAP triglycidyl p-laminophenol
TGETPM triglycidyl ether of triphenyl methane
TLC thin-layer chromatography
T_m melting temperature
TMA thermomechanical analysis
TPI turns per inch
tan equal to ratio of the loss modulus to the storage modulus
TTU through-transmission ultrasonics

UDC unidirectional composite
UTS ultimate tensile strength
UV ultraviolet

V_f volume fraction of fiber
V_m volume fraction of matrix
V_v volume fraction of void content
VCDO vinyl cyclohexene diepoxide
vol volume
vol% volume percent
VLS vapor feed gases-liquid catalyst-solid crystalline whisker growth
VLSI very large scale integration

w whisker
W watt
W width
WPE weight per epoxide
wt% weight percent

XPS x-ray photoelectron spectroscopy

yr year

Metric Conversion Guide

This Section is intended as a guide for expressing weights and measures in the Système International d'Unités (SI). The purpose of SI units, developed and maintained by the General Conference of Weights and Measures, is to provide a basis for world-wide standardization of units and measure. For more information on metric conversions, the reader should consult the following references:

- "Standard for Metric Practice," E 380, *Annual Book of ASTM Standards,* Vol 14.02, 1987, American Society for Testing and Materials, 1916 Race Street, Philadelphia, PA 19103
- "Metric Practice," ANSI/IEEE 268–1982, American National Standards Institute, 1430 Broadway, New York, NY 10018
- *Metric Practice Guide—Units and Conversion Factors for the Steel Industry,* 1978, American Iron and Steel Institute, 1000 16th Street NW, Washington, DC 20036
- *The International System of Units,* SP 330, 1986, National Bureau of Standards. Order from Superintendent of Documents, U.S. Government Printing Office, Washington, DC 20402-9325
- *Metric Editorial Guide,* 4th ed. (revised), 1985, American National Metric Council, 1010 Vermont Avenue NW, Suite 320, Washington, DC 20005-4960
- *ASME Orientation and Guide for Use of SI (Metric) Units,* ASME Guide SI 1, 9th ed., 1982, The American Society of Mechanical Engineers, 345 East 47th Street, New York, NY 10017

Base, Supplementary, and Derived SI Units

Measure	Unit	Symbol
Base units		
Amount of substance	mole	mol
Electric current	ampere	A
Length	meter	m
Luminous intensity	candela	cd
Mass	kilogram	kg
Thermodynamic temperature	kelvin	K
Time	second	s
Supplementary units		
Plane angle	radian	rad
Solid angle	steradian	sr
Derived units		
Absorbed dose	gray	Gy
Acceleration	meter per second squared	m/s^2
Activity (of radionuclides)	becquerel	Bq
Angular acceleration	radian per second squared	rad/s^2
Angular velocity	radian per second	rad/s
Area	square meter	m^2
Capacitance	farad	F
Concentration (of amount of substance)	mole per cubic meter	mol/m^3
Conductance	siemens	S
Current density	ampere per square meter	A/m^2
Density, mass	kilogram per cubic meter	kg/m^3
Electric charge density	coulomb per cubic meter	C/m^3
Electric field strength	volt per meter	V/m
Electric flux density	coulomb per square meter	C/m^2
Electric potential, potential difference, electromotive force	volt	V
Electric resistance	ohm	Ω
Energy, work, quantity of heat	joule	J
Energy density	joule per cubic meter	J/m^3
Entropy	joule per kelvin	J/K
Force	newton	N
Frequency	hertz	Hz
Heat capacity	joule per kelvin	J/K
Heat flux density	watt per square meter	W/m^2
Illuminance	lux	lx
Inductance	henry	H
Irradiance	watt per square meter	W/m^2
Luminance	candela per square meter	cd/m^2
Luminous flux	lumen	lm
Magnetic field strength	ampere per meter	A/m
Magnetic flux	weber	Wb
Magnetic flux density	tesla	T
Molar energy	joule per mole	J/mol
Molar entropy	joule per mole kelvin	J/mol · K
Molar heat capacity	joule per mole kelvin	J/mol · K
Moment of force	newton meter	N · m
Permeability	henry per meter	H/m
Permittivity	farad per meter	F/m
Power, radiant flux	watt	W
Pressure, stress	pascal	Pa
Quantity of electricity, electric charge	coulomb	C
Radiance	watt per square meter steradian	$W/m^2 \cdot sr$
Radiant intensity	watt per steradian	W/sr
Specific heat capacity	joule per kilogram kelvin	J/kg · K
Specific energy	joule per kilogram	J/kg
Specific entropy	joule per kilogram kelvin	J/kg · K
Specific volume	cubic meter per kilogram	m^3/kg
Surface tension	newton per meter	N/m
Thermal conductivity	watt per meter kelvin	W/m · K
Velocity	meter per second	m/s
Viscosity, dynamic	pascal second	Pa · s
Viscosity, kinematic	square meter per second	m^2/s
Volume	cubic meter	m^3
Wavenumber	1 per meter	1/m

Conversion Factors

To convert from	to	multiply by
Area		
in.^{2}	mm^2	6.451 600 E + 02
in.^{2}	cm^2	6.451 600 E + 00
in.^{2}	m^2	6.451 600 E − 04
ft^2	m^2	9.290 304 E − 02
Bending moment or torque		
lbf · in.	N · m	1.129 848 E − 01
lbf · ft	N · m	1.355 818 E + 00
kgf · m	N · m	9.806 650 E + 00
ozf · in.	N · m	7.061 552 E − 03
Bending moment or torque per unit length		
lbf · in./in.	N · m/m	4.448 222 E + 00
lbf · ft/in.	N · m/m	5.337 866 E + 01
Current density		
A/in.^{2}	A/cm^2	1.550 003 E − 01
A/in.^{2}	A/mm^2	1.550 003 E − 03
A/ft^2	A/m^2	1.076 400 E + 01
Electric field strength		
V/mil	kV/m	3.937 008 E + 01
Electricity and magnetism		
gauss	T	1.000 000 E − 04
maxwell	μWb	1.000 000 E − 02
mho	S	1.000 000 E + 00
Oersted	A/m	7.957 700 E + 01
Ω · cm	Ω · m	1.000 000 E − 02
Ω circular-mil/ft	μΩ · m	1.662 426 E − 03
Energy (impact, other)		
ft · lbf	J	1.355 818 E + 00
Btu (thermochemical)	J	1.054 350 E + 03
cal (thermochemical)	J	4.184 000 E + 00
kW · h	J	3.600 000 E + 06
W · h	J	3.600 000 E + 03
Flow rate		
ft^3/h	L/min	4.719 475 E − 01
ft^3/min	L/min	2.831 000 E + 01
gal/h	L/min	6.309 020 E − 02
gal/min	L/min	3.785 412 E + 00
Force		
lbf	N	4.448 222 E + 00
kip (1000 lbf)	N	4.448 222 E + 03
tonf	kN	8.896 443 E + 00
kgf	N	9.806 650 E + 00
Force per unit length		
lbf/ft	N/m	1.459 390 E + 01
lbf/in.	N/m	1.751 268 E + 02
kip/in.	N/m	1.751 268 E + 05
Fracture toughness		
ksi $\sqrt{\text{in.}}$	MPa$\sqrt{\text{m}}$	1.098 800 E + 00
Heat content		
Btu/lb	kJ/kg	2.326 000 E + 00
cal/g	kJ/kg	4.186 800 E + 00
Heat input		
J/in.	J/m	3.937 008 E + 01
kJ/in.	kJ/m	3.937 008 E + 01
Impact energy per unit area		
ft · lbf/ft^2	J/m^2	1.459 002 E + 01
Length		
Å	nm	1.000 000 E − 01
μin.	μm	2.540 000 E − 02
mil	μm	2.540 000 E + 01
in.	mm	2.540 000 E + 01
in.	cm	2.540 000 E + 00
ft	m	3.048 000 E − 01
yd	m	9.144 000 E − 01
mile	km	1.609 300 E + 00
Length per unit mass		
in./lb	m/kg	5.599 740 E − 02
yd/lb	m/kg	2.015 907 E + 00
Mass		
oz	kg	2.834 952 E − 02
lb	kg	4.535 924 E − 01
ton (short, 2000 lb)	kg	9.071 847 E + 02
ton (short, 2000 lb)	kg × 10^3(a)	9.071 847 E − 01
ton (long, 2240 lb)	kg	1.016 047 E + 03
Mass per unit area		
oz/in.^{2}	kg/m^2	4.395 000 E + 01
oz/ft^2	kg/m^2	3.051 517 E − 01
oz/yd^2	kg/m^2	3.390 575 E − 02
lb/ft^2	kg/m^2	4.882 428 E + 00
Mass per unit length		
lb/ft	kg/m	1.488 164 E + 00
lb/in.	kg/m	1.785 797 E + 01
denier	kg/m	1.111 111 E − 07
tex	kg/m	1.000 000 E − 06
Mass per unit time		
lb/h	kg/s	1.259 979 E − 04
lb/min	kg/s	7.559 873 E − 03
lb/s	kg/s	4.535 924 E − 01
Mass per unit volume (includes density)		
g/cm^3	kg/m^3	1.000 000 E + 03
lb/ft^3	g/cm^3	1.601 846 E − 02
lb/ft^3	kg/m^3	1.601 846 E + 01
lb/in.^{3}	g/cm^3	2.767 990 E + 01
lb/in.^{3}	kg/m^3	2.767 990 E + 04
oz/in.^{3}	kg/m^3	1.729 994 E + 03
Power		
Btu/s	kW	1.055 056 E + 00
Btu/min	kW	1.758 426 E − 02
Btu/h	W	2.928 751 E − 01
erg/s	W	1.000 000 E − 07
ft · lbf/s	W	1.355 818 E + 00
ft · lbf/min	W	2.259 697 E − 02
ft · lbf/h	W	3.766 161 E − 04
hp (550 ft · lbf/s)	kW	7.456 999 E − 01
hp (electric)	kW	7.460 000 E − 01
Power density		
W/in.^{2}	W/m^2	1.550 003 E + 03
Pressure (fluid)		
atm (standard)	Pa	1.013 250 E + 05
bar	Pa	1.000 000 E + 05
in. Hg (32 °F)	Pa	3.386 380 E + 03
in. Hg (60 °F)	Pa	3.376 850 E + 03
lbf/in.^{2} (psi)	Pa	6.894 757 E + 03
torr (mm Hg, 0 °C)	Pa	1.333 220 E + 02
Specific area		
ft^2/lb	m^2/kg	2.048 161 E − 01
Specific energy		
cal/g	J/g	4.186 800 E + 00
Btu/lb	kJ/kg	2.326 000 E + 00
Specific heat capacity		
Btu/lb · °F	J/kg · K	4.186 800 E + 03
cal/g · °C	J/kg · K	4.186 800 E + 03
Stress (force per unit area)		
tonf/in.^{2} (tsi)	MPa	1.378 951 E + 01
kgf/mm^2	MPa	9.806 650 E + 00
ksi	MPa	6.894 757 E + 00
lbf/in.^{2} (psi)	MPa	6.894 757 E − 03
MN/m^2	MPa	1.000 000 E + 00
Temperature		
°F	°C	5/9 · (°F − 32)
°R	K	5/9
°F	K	5/9 · (°F + 459.67)
°C	K	°C + 273.15
Temperature interval		
°F	°C	5/9
Thermal conductivity		
Btu · in./s · ft^2 · °F	W/m · K	5.192 204 E + 02
Btu/ft · h · °F	W/m · K	1.730 735 E + 00
Btu · in./h · ft^2 · °F	W/m · K	1.442 279 E − 01
cal/cm · s · °C	W/m · K	4.184 000 E + 02
Thermal expansion		
μin./in. · °C	10^{-6}/K	1.000 000 E + 00
μin./in. · °F	10^{-6}/K	1.800 000 E + 00
Velocity		
ft/h	m/s	8.466 667 E − 05
ft/min	m/s	5.080 000 E − 03
ft/s	m/s	3.048 000 E − 01
in./s	m/s	2.540 000 E − 02
km/h	m/s	2.777 778 E − 01
mph	km/h	1.609 344 E + 00
Viscosity (dynamic and kinematic)		
poise (P)	Pa · s	1.000 000 E − 01
cP	Pa · s	1.000 000 E − 03
lbf · s/in.^{2}	Pa · s	6.894 757 E + 03
ft^2/s	m^2/s	9.290 304 E − 02
in.^{2}/s	mm^2/s	6.451 600 E + 02
Volume		
in.^{3}	m^3	1.638 706 E − 05
ft^3	m^3	2.831 685 E − 02
fluid oz	m^3	2.957 353 E − 05
gal (U.S. liquid)	m^3	3.785 412 E − 03
Volume per unit time		
ft^3/min	m^3/s	4.719 474 E − 04
ft^3/s	m^3/s	2.831 685 E − 02
in.^{3}/min	m^3/s	2.731 177 E − 07
Wavelength		
Å	nm	1.000 000 E − 01

(a) kg × 10^3 = 1 metric ton

Metric Stress or Pressure Conversions

The middle column of figures (in bold-faced type) contains the reading (in MPa or ksi) to be converted. If converting from ksi to MPa, read the MPa equivalent in the column headed "MPa". If converting from MPa to ksi, read the ksi equivalent in the column headed "ksi". 1 ksi = 6.894757 MPa. 1 psi = 6.894757 kPa.

ksi		MPa	ksi		MPa	ksi		MPa	ksi		MPa
0.14504	**1**	6.895	8.2672	**57**	393.00	33.359	**230**	1585.8	114.58	**790**	. . .
0.29008	**2**	13.790	8.4122	**58**	399.90	34.809	**240**	1654.7	116.03	**800**	. . .
0.43511	**3**	20.684	8.5572	**59**	406.79	36.259	**250**	1723.7	117.48	**810**	. . .
0.58015	**4**	27.579	8.7023	**60**	413.69	37.710	**260**	1792.6	118.93	**820**	. . .
0.72519	**5**	34.474	8.8473	**61**	420.58	39.160	**270**	1861.6	120.38	**830**	. . .
0.87023	**6**	41.369	8.9923	**62**	427.47	40.611	**280**	1930.5	121.83	**840**	. . .
1.0153	**7**	48.263	9.1374	**63**	434.37	42.061	**290**	1999.5	123.28	**850**	. . .
1.1603	**8**	55.158	9.2824	**64**	441.26	43.511	**300**	2068.4	124.73	**860**	. . .
1.3053	**9**	62.053	9.4275	**65**	448.16	44.962	**310**	2137.4	126.18	**870**	. . .
1.4504	**10**	68.948	9.5725	**66**	455.05	46.412	**320**	2206.3	127.63	**880**	. . .
1.5954	**11**	75.842	9.7175	**67**	461.95	47.862	**330**	2275.3	129.08	**890**	. . .
1.7405	**12**	82.737	9.8626	**68**	468.84	49.313	**340**	2344.2	130.53	**900**	. . .
1.8855	**13**	89.632	10.008	**69**	475.74	50.763	**350**	2413.2	131.98	**910**	. . .
2.0305	**14**	96.527	10.153	**70**	482.63	52.214	**360**	2482.1	133.43	**920**	. . .
2.1756	**15**	103.42	10.298	**71**	489.53	53.664	**370**	2551.1	134.89	**930**	. . .
2.3206	**16**	110.32	10.443	**72**	496.42	55.114	**380**	2620.0	136.34	**940**	. . .
2.4656	**17**	117.21	10.588	**73**	503.32	56.565	**390**	2689.0	137.79	**950**	. . .
2.6107	**18**	124.11	10.733	**74**	510.21	58.015	**400**	2757.9	139.24	**960**	. . .
2.7557	**19**	131.00	10.878	**75**	517.11	59.465	**410**	2826.9	140.69	**970**	. . .
2.9008	**20**	137.90	11.023	**76**	524.00	60.916	**420**	2895.8	142.14	**980**	. . .
3.0458	**21**	144.79	11.168	**77**	530.90	62.366	**430**	2964.7	143.59	**990**	. . .
3.1908	**22**	151.68	11.313	**78**	537.79	63.817	**440**	3033.7	145.04	**1000**	. . .
3.3359	**23**	158.58	11.458	**79**	544.69	65.267	**450**	3102.6	147.94	**1020**	. . .
3.4809	**24**	165.47	11.603	**80**	551.58	66.717	**460**	3171.6	150.84	**1040**	. . .
3.6259	**25**	172.37	11.748	**81**	558.48	66.168	**470**	3240.5	153.74	**1060**	. . .
3.7710	**26**	179.26	11.893	**82**	565.37	69.618	**480**	3309.5	156.64	**1080**	. . .
3.9160	**27**	186.16	12.038	**83**	572.26	71.068	**490**	3378.4	159.54	**1100**	. . .
4.0611	**28**	193.05	12.183	**84**	579.16	72.519	**500**	3447.4	162.44	**1120**	. . .
4.2061	**29**	199.95	12.328	**85**	586.05	73.969	**510**	. . .	165.34	**1140**	. . .
4.3511	**30**	206.84	12.473	**86**	592.95	75.420	**520**	. . .	168.24	**1160**	. . .
4.4962	**31**	213.74	12.618	**87**	599.84	76.870	**530**	. . .	171.14	**1180**	. . .
4.6412	**32**	220.63	12.763	**88**	606.74	78.320	**540**	. . .	174.05	**1200**	. . .
4.7862	**33**	227.53	12.909	**89**	613.63	79.771	**550**	. . .	176.95	**1220**	. . .
4.9313	**34**	234.42	13.053	**90**	620.53	81.221	**560**	. . .	179.85	**1240**	. . .
5.0763	**35**	241.32	13.198	**91**	627.42	82.672	**570**	. . .	182.75	**1260**	. . .
5.2214	**36**	248.21	13.343	**92**	634.32	84.122	**580**	. . .	185.65	**1280**	. . .
5.3664	**37**	255.11	13.489	**93**	641.21	85.572	**590**	. . .	188.55	**1300**	. . .
5.5114	**38**	262.00	13.634	**94**	648.11	87.023	**600**	. . .	191.45	**1320**	. . .
5.6565	**39**	268.90	13.779	**95**	655.00	88.473	**610**	. . .	194.35	**1340**	. . .
5.8015	**40**	275.79	13.924	**96**	661.90	89.923	**620**	. . .	197.25	**1360**	. . .
5.9465	**41**	282.69	14.069	**97**	668.79	91.374	**630**	. . .	200.15	**1380**	. . .
6.0916	**42**	289.58	14.214	**98**	675.69	92.824	**640**	. . .	203.05	**1400**	. . .
6.2366	**43**	296.47	14.359	**99**	682.58	94.275	**650**	. . .	205.95	**1420**	. . .
6.3817	**44**	303.37	14.504	**100**	689.48	95.725	**660**	. . .	208.85	**1440**	. . .
6.5267	**45**	310.26	15.954	**110**	758.42	97.175	**670**	. . .	211.76	**1460**	. . .
6.6717	**46**	317.16	17.405	**120**	827.37	98.626	**680**	. . .	214.66	**1480**	. . .
6.8168	**47**	324.05	18.855	**130**	896.32	100.08	**690**	. . .	217.56	**1500**	. . .
6.9618	**48**	330.95	20.305	**140**	965.27	101.53	**700**	. . .	220.46	**1520**	. . .
7.1068	**49**	337.84	21.756	**150**	1034.2	102.98	**710**	. . .	223.36	**1540**	. . .
7.2519	**50**	344.74	23.206	**160**	1103.2	104.43	**720**	. . .	226.26	**1560**	. . .
7.3969	**51**	351.63	24.656	**170**	1172.1	105.88	**730**	. . .	229.16	**1580**	. . .
7.5420	**52**	358.53	26.107	**180**	1241.1	107.33	**740**	. . .	232.06	**1600**	. . .
7.6870	**53**	365.42	27.557	**190**	1310.0	108.78	**750**	. . .	234.96	**1620**	. . .
7.8320	**54**	372.32	29.008	**200**	1379.0	110.23	**760**	. . .	237.86	**1640**	. . .
7.9771	**55**	379.21	30.458	**210**	1447.9	111.68	**770**	. . .	240.76	**1660**	. . .
8.1221	**56**	386.11	31.908	**220**	1516.8	113.13	**780**	. . .	243.66	**1680**	. . .

Metric Stress or Pressure Conversions (continued)

ksi		MPa	ksi		MPa	ksi		MPa	ksi		MPa
246.56	**1700**	· · ·	278.47	**1920**	· · ·	310.38	**2140**	· · ·	342.29	**2360**	· · ·
249.46	**1720**	· · ·	281.37	**1940**	· · ·	313.28	**2160**	· · ·	345.19	**2380**	· · ·
252.37	**1740**	· · ·	284.27	**1960**	· · ·	316.18	**2180**	· · ·	348.09	**2400**	· · ·
255.27	**1760**	· · ·	287.17	**1980**	· · ·	319.08	**2200**	· · ·	350.99	**2420**	· · ·
258.17	**1780**	· · ·	290.08	**2000**	· · ·	321.98	**2220**	· · ·	353.89	**2440**	· · ·
261.07	**1800**	· · ·	292.98	**2020**	· · ·	324.88	**2240**	· · ·	356.79	**2460**	· · ·
263.97	**1820**	· · ·	295.88	**2040**	· · ·	327.79	**2260**	· · ·	359.69	**2480**	· · ·
266.87	**1840**	· · ·	298.78	**2060**	· · ·	330.69	**2280**	· · ·	362.59	**2500**	· · ·
269.77	**1860**	· · ·	301.68	**2080**	· · ·	333.59	**2300**	· · ·			
272.67	**1880**	· · ·	304.58	**2100**	· · ·	336.49	**2320**	· · ·			
275.57	**1900**	· · ·	307.48	**2120**	· · ·	339.39	**2340**	· · ·			

Metric Stress-Intensity Conversions

The middle column of figures (in bold-faced type) contains the reading (in MPa$\sqrt{m}$ or ksi$\sqrt{in.}$) to be converted. If converting from ksi$\sqrt{in.}$ to MPa$\sqrt{m}$, read the MPa$\sqrt{m}$ equivalent in the column headed "MPa$\sqrt{m}$". If converting from MPa$\sqrt{m}$ to ksi$\sqrt{in.}$, read the ksi$\sqrt{in.}$ equivalent in the column headed "ksi$\sqrt{in.}$". 1 ksi$\sqrt{in.}$ = 1.098845 MPa$\sqrt{m}$.

ksi√in.		MPa√m	ksi√in.		MPa√m	ksi√in.		MPa√m	ksi√in.		MPa√m	ksi√in.		MPa√m
0.91005	**1**	1.0988	37.312	**41**	45.051	73.714	**81**	89.003	110.12	**121**	132.95	146.52	**161**	176.91
1.8201	**2**	2.1976	38.222	**42**	46.150	74.624	**82**	90.102	111.03	**122**	134.05	147.43	**162**	178.01
2.7301	**3**	3.2964	39.132	**43**	47.248	75.534	**83**	91.200	111.94	**123**	135.15	148.34	**163**	179.10
3.6402	**4**	4.3952	40.042	**44**	48.347	76.444	**84**	92.300	112.85	**124**	136.25	149.25	**164**	180.20
4.5502	**5**	5.4940	40.952	**45**	49.446	77.354	**85**	93.398	113.76	**125**	137.35	150.16	**165**	181.30
5.4603	**6**	6.5928	41.862	**46**	50.545	78.264	**86**	94.497	114.67	**126**	138.45	151.07	**166**	182.40
6.3703	**7**	7.6916	42.772	**47**	51.644	79.174	**87**	95.596	115.58	**127**	139.55	151.98	**167**	183.50
7.2804	**8**	8.7904	43.682	**48**	52.742	80.084	**88**	96.694	116.49	**128**	140.65	152.89	**168**	184.60
8.1904	**9**	9.8892	44.592	**49**	53.841	80.994	**89**	97.793	117.40	**129**	141.75	153.80	**169**	185.70
9.1005	**10**	10.988	45.502	**50**	54.940	81.904	**90**	98.892	118.31	**130**	142.84	154.71	**170**	186.80
10.011	**11**	12.087	46.412	**51**	56.039	82.814	**91**	99.991	119.22	**131**	143.94	155.62	**171**	187.90
10.921	**12**	13.186	47.322	**52**	57.138	83.724	**92**	101.09	120.13	**132**	145.04	156.53	**172**	189.00
11.831	**13**	14.284	48.232	**53**	58.236	84.634	**93**	102.19	121.04	**133**	146.14	157.44	**173**	190.10
12.741	**14**	15.383	49.143	**54**	59.335	85.544	**94**	103.29	121.95	**134**	147.24	158.35	**174**	191.19
13.651	**15**	16.482	50.053	**55**	60.434	86.454	**95**	104.39	122.86	**135**	148.34	159.26	**175**	192.29
14.561	**16**	17.581	50.963	**56**	61.533	87.364	**96**	105.48	123.77	**136**	149.44	160.17	**176**	193.39
15.471	**17**	18.680	51.873	**57**	62.632	88.275	**97**	106.58	124.68	**137**	150.54	161.08	**177**	194.49
16.381	**18**	19.778	52.783	**58**	63.730	89.185	**98**	107.68	125.59	**138**	151.63	161.99	**178**	195.59
17.291	**19**	20.877	53.693	**59**	64.829	90.095	**99**	108.78	126.50	**139**	152.73	162.90	**179**	196.69
18.201	**20**	21.976	54.603	**60**	65.928	91.005	**100**	109.88	127.41	**140**	153.83	163.81	**180**	197.78
19.111	**21**	23.075	55.513	**61**	67.027	91.915	**101**	110.98	128.32	**141**	154.93	164.72	**181**	198.88
20.021	**22**	24.174	56.423	**62**	68.126	92.825	**102**	112.08	129.23	**142**	156.03	165.63	**182**	199.98
20.931	**23**	25.272	57.333	**63**	69.224	93.735	**103**	113.18	130.14	**143**	157.13	166.54	**183**	201.08
21.841	**24**	26.371	58.243	**64**	70.323	94.645	**104**	114.28	131.05	**144**	158.23	167.45	**184**	202.18
22.751	**25**	27.470	59.153	**65**	71.422	95.555	**105**	115.37	131.96	**145**	159.33	168.36	**185**	203.28
23.661	**26**	28.569	60.063	**66**	72.521	96.465	**106**	116.47	132.87	**146**	160.42	169.27	**186**	204.38
24.571	**27**	29.668	60.973	**67**	73.620	97.375	**107**	117.57	133.78	**147**	161.52	170.18	**187**	205.48
25.481	**28**	30.766	61.883	**68**	74.718	98.285	**108**	118.67	134.69	**148**	162.62	171.09	**188**	206.57
26.391	**29**	31.865	62.793	**69**	75.817	99.195	**109**	119.77	135.60	**149**	163.72	172.00	**189**	207.67
27.301	**30**	32.964	63.703	**70**	76.916	100.11	**110**	120.87	136.51	**150**	164.82	172.91	**190**	208.77
28.211	**31**	34.063	64.613	**71**	78.015	101.02	**111**	121.97	137.42	**151**	165.92	173.82	**191**	209.87
29.121	**32**	35.162	65.523	**72**	79.114	101.93	**112**	123.07	138.33	**152**	167.02	174.73	**192**	210.97
30.032	**33**	36.260	66.433	**73**	80.212	102.84	**113**	124.16	139.24	**153**	168.12	175.64	**193**	212.07
30.942	**34**	37.359	67.343	**74**	81.311	103.75	**114**	125.26	140.15	**154**	169.22	176.55	**194**	213.17
31.852	**35**	38.458	68.253	**75**	82.410	104.66	**115**	126.36	141.06	**155**	170.31	177.46	**195**	214.27
32.762	**36**	39.557	69.164	**76**	83.509	105.57	**116**	127.46	141.97	**156**	171.41	178.37	**196**	215.36
33.672	**37**	40.656	70.074	**77**	84.608	106.48	**117**	128.56	142.88	**157**	172.51	179.28	**197**	216.46
34.582	**38**	41.754	70.984	**78**	85.706	107.39	**118**	129.66	143.79	**158**	173.61	180.19	**198**	217.56
35.492	**39**	42.853	71.893	**79**	86.805	108.30	**119**	130.76	144.70	**159**	174.71	181.10	**199**	218.66
36.402	**40**	43.952	72.804	**80**	87.904	109.21	**120**	131.86	145.61	**160**	175.81	182.01	**200**	219.76

Metric Energy Conversions

The middle column of figures (in bold-faced type) contains the reading (in J or ft·lb) to be converted. If converting from ft·lb to J, read the J equivalent in the column headed "J". If converting from J to ft·lb, read the equivalent in the column headed "ft·lb". 1 ft·lb = 1.355818 J.

ft·lb		J	ft·lb		J	ft·lb		J	ft·lb		J
0.7376	**1**	1.3558	28.7649	**39**	52.8769	56.7923	**77**	104.3980	129.0734	**175**	237.2681
1.4751	**2**	2.7116	29.5025	**40**	54.2327	57.5298	**78**	105.7538	132.7612	**180**	244.0472
2.2127	**3**	4.0675	30.2400	**41**	55.5885	58.2674	**79**	107.1096	136.4490	**185**	250.8263
2.9502	**4**	5.4233	30.9776	**42**	56.9444	59.0050	**80**	108.4654	140.1368	**190**	257.6054
3.6878	**5**	6.7791	31.7152	**43**	58.3002	59.7425	**81**	109.8212	143.8246	**195**	264.3845
4.4254	**6**	8.1349	32.4527	**44**	59.6560	60.4801	**82**	111.1771	147.5124	**200**	271.1636
5.1629	**7**	9.4907	33.1903	**45**	61.0118	61.2177	**83**	112.5329	154.8880	**210**	284.7218
5.9005	**8**	10.8465	33.9279	**46**	62.3676	61.9552	**84**	113.8887	162.2637	**220**	298.2799
6.6381	**9**	12.2024	34.6654	**47**	63.7234	62.6928	**85**	115.2445	169.6393	**230**	311.8381
7.3756	**10**	13.5582	35.4030	**48**	65.0793	63.4303	**86**	116.6003	177.0149	**240**	325.3963
8.1132	**11**	14.9140	36.1405	**49**	66.4351	64.1679	**87**	117.9562	184.3905	**250**	338.9545
8.8507	**12**	16.2698	36.8781	**50**	67.7909	64.9055	**88**	119.3120	191.7661	**260**	352.5126
9.5883	**13**	17.6256	37.6157	**51**	69.1467	65.6430	**89**	120.6678	199.1418	**270**	366.0708
10.3259	**14**	18.9815	38.3532	**52**	70.5025	66.3806	**90**	122.0236	206.5174	**280**	379.6290
11.0634	**15**	20.3373	39.0908	**53**	71.8583	67.1182	**91**	123.3794	213.8930	**290**	393.1872
11.8010	**16**	21.6931	39.8284	**54**	73.2142	67.8557	**92**	124.7452	221.2686	**300**	406.7454
12.5386	**17**	23.0489	40.5659	**55**	74.5700	68.5933	**93**	126.0911	228.6442	**310**	420.3036
13.2761	**18**	24.4047	41.3035	**56**	75.9258	69.3308	**94**	127.4469	236.0199	**320**	433.8617
14.0137	**19**	25.7605	42.0410	**57**	77.2816	70.0684	**95**	128.8027	243.3955	**330**	447.4199
14.7512	**20**	27.1164	42.7786	**58**	78.6374	70.8060	**96**	130.1585	250.7711	**340**	460.9781
15.4888	**21**	28.4722	43.5162	**59**	79.9933	71.5435	**97**	131.5143	258.1467	**350**	474.5363
16.2264	**22**	29.8280	44.2537	**60**	81.3491	72.2811	**98**	132.8702	265.5224	**360**	488.0944
16.9639	**23**	31.1838	44.9913	**61**	82.7049	73.0186	**99**	134.2260	272.8980	**370**	501.6526
17.7015	**24**	32.5396	45.7288	**62**	84.0607	73.7562	**100**	135.5818	280.2736	**380**	515.2108
18.4390	**25**	33.8954	46.4664	**63**	85.4165	77.4440	**105**	142.3609	287.6492	**390**	528.7690
19.1766	**26**	35.2513	47.2040	**64**	86.7723	81.1318	**110**	149.1400	295.0248	**400**	542.3272
19.9142	**27**	36.6071	47.9415	**65**	88.1282	84.8196	**115**	155.9191	302.4005	**410**	555.8854
20.6517	**28**	37.9629	48.6791	**66**	89.4840	88.5075	**120**	162.6982	309.7761	**420**	569.4435
21.3893	**29**	39.3187	49.4167	**67**	90.8398	92.1953	**125**	169.4772	317.1517	**430**	583.0017
22.1269	**30**	40.6745	50.1542	**68**	92.1956	95.8831	**130**	176.2563	324.5273	**440**	596.5599
22.8644	**31**	42.0304	50.8918	**69**	93.5514	99.5709	**135**	183.0354	331.9029	**450**	610.1181
23.6020	**32**	43.3862	51.6293	**70**	94.9073	103.2587	**140**	189.8145	339.2786	**460**	623.6762
24.3395	**33**	44.7420	52.3669	**71**	96.2631	106.9465	**145**	196.5936	346.6542	**470**	637.2344
25.0771	**34**	46.0978	53.1045	**72**	97.6189	110.6343	**150**	203.3727	354.0298	**480**	650.7926
25.8147	**35**	47.4536	53.8420	**73**	98.9747	114.3221	**155**	210.1518	361.4054	**490**	664.3508
26.5522	**36**	48.8094	54.5796	**74**	100.3305	118.0099	**160**	216.9308	368.7811	**500**	677.9090
27.2898	**37**	50.1653	55.3172	**75**	101.6863	121.6977	**165**	223.7099			
28.0274	**38**	51.5211	56.0547	**76**	103.0422	125.3856	**170**	230.4890			

Conversion of Inches to Millimeters

Inches	Millimeters	Inches	Millimeters	Inches	Millimeters
0.001	0.025	0.290	7.37	0.660	16.76
0.002	0.051	0.300	7.62	0.670	17.02
0.003	0.076	0.310	7.87	0.680	17.17
0.004	0.102	0.320	8.13	0.690	17.53
0.005	0.127	0.330	8.38	0.700	17.78
0.006	0.152	0.340	8.64	0.710	18.03
0.007	0.178	0.350	8.89	0.720	18.29
0.008	0.203	0.360	9.14	0.730	18.54
0.009	0.229	0.370	9.40	0.740	18.80
0.010	0.254	0.380	9.65	0.750	19.05
0.020	0.508	0.390	9.91	0.760	19.30
0.030	0.762	0.400	10.16	0.770	19.56
0.040	1.016	0.410	10.41	0.780	19.81
0.050	1.270	0.420	10.67	0.790	20.07
0.060	1.524	0.430	10.92	0.800	20.32
0.070	1.778	0.440	11.18	0.810	20.57
0.080	2.032	0.450	11.43	0.820	20.83
0.090	2.286	0.460	11.68	0.830	21.08
0.100	2.540	0.470	11.94	0.840	21.34
0.110	2.794	0.480	12.19	0.850	21.59
0.120	3.048	0.490	12.45	0.860	21.84
0.130	3.302	0.500	12.70	0.870	22.10
0.140	3.56	0.510	12.95	0.880	22.35
0.150	3.81	0.520	13.21	0.890	22.61
0.160	4.06	0.530	13.46	0.900	22.86
0.170	4.32	0.540	13.72	0.910	23.11
0.180	4.57	0.550	13.97	0.920	23.37
0.190	4.83	0.560	14.22	0.930	23.62
0.200	5.08	0.570	14.48	0.940	23.88
0.210	5.33	0.580	14.73	0.950	24.13
0.220	5.59	0.590	14.99	0.960	24.38
0.230	5.84	0.600	15.24	0.970	24.64
0.240	6.10	0.610	15.49	0.980	24.89
0.250	6.35	0.620	15.75	0.990	25.15
0.260	6.60	0.630	16.00	1.000	25.40
0.270	6.86	0.640	16.26	...	...
0.280	7.11	0.650	16.51	...	...

Conversion of Millimeters to Inches

Millimeters	Inches	Millimeters	Inches	Millimeters	Inches
0.01	0.0004	0.35	0.0138	0.68	0.0268
0.02	0.0008	0.36	0.0142	0.69	0.0272
0.03	0.0012	0.37	0.0146	0.70	0.0276
0.04	0.0016	0.38	0.0150	0.71	0.0280
0.05	0.0020	0.39	0.0154	0.72	0.0283
0.06	0.0024	0.40	0.0157	0.73	0.0287
0.07	0.0028	0.41	0.0161	0.74	0.0291
0.08	0.0031	0.42	0.0165	0.75	0.0295
0.09	0.0035	0.43	0.0169	0.76	0.0299
0.10	0.0039	0.44	0.0173	0.77	0.0303
0.11	0.0043	0.45	0.0177	0.78	0.0307
0.12	0.0047	0.46	0.0181	0.79	0.0311
0.13	0.0051	0.47	0.0185	0.80	0.0315
0.14	0.0055	0.48	0.0189	0.81	0.0319
0.15	0.0059	0.49	0.0193	0.82	0.0323
0.16	0.0063	0.50	0.0197	0.83	0.0327
0.17	0.0067	0.51	0.0201	0.84	0.0331
0.18	0.0071	0.52	0.0205	0.85	0.0335
0.19	0.0075	0.53	0.0209	0.86	0.0339
0.20	0.0079	0.54	0.0213	0.87	0.0343
0.21	0.0083	0.55	0.0217	0.88	0.0346
0.22	0.0087	0.56	0.0220	0.89	0.0350
0.23	0.0091	0.57	0.0224	0.90	0.0354
0.24	0.0094	0.58	0.0228	0.91	0.0358
0.25	0.0098	0.59	0.0232	0.92	0.0362
0.26	0.0102	0.60	0.0236	0.93	0.0366
0.27	0.0106	0.61	0.0240	0.94	0.0370
0.28	0.0110	0.62	0.0244	0.95	0.0374
0.29	0.0114	0.63	0.0248	0.96	0.0378
0.30	0.0118	0.64	0.0252	0.97	0.0382
0.31	0.0122	0.65	0.0256	0.98	0.0386
0.32	0.0126	0.66	0.0260	0.99	0.0390
0.33	0.0130	0.67	0.0264	1.00	0.0394
0.34	0.0134	...	...	...	...

Metric Length and Weight Conversion Factors

Unit	Inches to millimeters	Millimeters to inches	Pounds to kilograms	Kilograms to pounds
1	25.400 1	0.039 371	0.453 59	2.204 62
2	50.800 1	0.078 742	0.907 19	4.409 24
3	76.200 2	0.118 112	1.360 78	6.613 86
4	101.600 2	0.157 483	1.814 37	8.818 49
5	127.000 3	0.196 854	2.267 96	11.023 11
6	152.400 3	0.236 225	2.721 56	13.227 73
7	177.800 4	0.275 596	3.175 15	15.432 35
8	203.200 4	0.314 966	3.628 74	17.636 97
9	228.600 5	0.354 337	4.082 33	19.841 59
10	254.000 6	0.393 708	4.355 92	22.046 22

SI Prefixes-Names and Symbols

Exponential expression	Multiplication factor	Prefix	Symbol
10^{18}	1 000 000 000 000 000 000	exa	E
10^{15}	1 000 000 000 000 000	peta	P
10^{12}	1 000 000 000 000	tera	T
10^{9}	1 000 000 000	giga	G
10^{6}	1 000 000	mega	M
10^{3}	1 000	kilo	k
10^{2}	100	hecto(a)	h
10^{1}	10	deka(a)	da
10^{0}	1	BASE UNIT	
10^{-1}	0.1	deci(a)	d
10^{-2}	0.01	centi(a)	c
10^{-3}	0.001	milli	m
10^{-6}	0.000 001	micro	μ
10^{-9}	0.000 000 001	nano	n
10^{-12}	0.000 000 000 001	pico	p
10^{-15}	0.000 000 000 000 001	femto	f
10^{-18}	0.000 000 000 000 000 001	atto	a

(a) Nonpreferred. Prefixes should be selected in steps of 10^3 so that the resultant number before the prefix is between 0.1 and 1000. These prefixes should not be used for units of linear measurement, but may be used for higher order units. For example, the linear measurement, decimeter, is nonpreferred, but square decimeter is acceptable.

Symbols

- $\rightleftarrows$ direction of reaction
- $\div$ divided by
- $=$ equals
- ˆ circumflex
- $\approx$ approximately equals
- $\neq$ not equal to
- $\equiv$ identical with
- $>$ greater than
- $\gg$ much greater than
- $\geq$ greater than or equal to
- ∞ infinity
- $\propto$ is proportional to; varies as
- $\int$ integral of
- $<$ less than
- $\ll$ much less than
- $\leq$ less than or equal to
- $\pm$ maximum deviation
- $-$ minus; negative ion charge
- $\times$ diameters (magnification); multiplied by
- $\cdot$ multiplied by
- Ω ohm
- / per
- % percent
- + plus; positive ion charge
- $\sqrt{\ }$ square root of
- $\sim$ approximately; similar to
- ° angular measure; degree
- °**C** degree Celsius (centigrade)
- °**F** degree Fahrenheit
- **0°** fiber direction
- **90°** perpendicular to fiber direction
- α coefficient of thermal expansion
- Δ change in quantity; an increment; a range
- η viscosity
- ϵ strain
- γ shear strain
- μ**in.** microinch
- μ**m** micrometer (micron)
- υ Poisson's ratio
- π pi (3.141592)
- ψ damping
- ρ density
- σ tensile stress
- τ shear stress
- θ angle

Greek Alphabet

- **A**, α alpha
- **B**, β beta
- Γ, γ gamma
- Δ, δ delta
- **E**, ϵ epsilon
- **Z**, ζ zeta
- **H**, η eta
- Θ, θ theta
- **I**, ι iota
- **K**, κ kappa
- Λ, λ lambda
- **M**, μ mu
- **N**, ν nu
- Ξ, ξ xi
- **O**, o omicron
- Π, π pi
- **P**, ρ rho
- Σ, σ sigma
- **T**, τ tau
- **Y**, υ upsilon
- Φ, ϕ phi
- **X**, χ chi
- Ψ, ψ psi
- Ω, ω omega

Temperature Conversions

The general arrangement of this table was devised by Sauveur and Boylston more than 40 years ago. The middle column of figures (in bold-faced type) contains the reading (°F or °C) to be converted. If converting from degrees Fahrenheit to degrees Centigrade, read the Centigrade equivalent in the column headed "°C". If converting from Centigrade to Fahrenheit, read the Fahrenheit equivalent in the column headed "°F". °C = 5/9 (°F − 32)

°F		°C	°F		°C	°F		°C	°F		°C	°F		°C
···	**−458**	−272.22	···	**−358**	−216.67	−432.4	**−258**	−161.11	−252.4	**−158**	−105.56	−72.4	**−58**	−50.00
···	**−456**	−271.11	···	**−356**	−215.56	−428.8	**−256**	−160.00	−248.8	**−156**	−104.44	−68.8	**−56**	−48.89
···	**−454**	−270.00	···	**−354**	−214.44	−425.2	**−254**	−158.89	−245.2	**−154**	−103.33	−65.2	**−54**	−47.78
···	**−452**	−268.89	···	**−352**	−213.33	−421.6	**−252**	−157.78	−241.6	**−152**	−102.22	−61.6	**−52**	−46.67
···	**−450**	−267.78	···	**−350**	−212.22	−418.0	**−250**	−156.67	−238.0	**−150**	−101.11	−58.0	**−50**	−45.56
···	**−448**	−266.67	···	**−348**	−211.11	−414.4	**−248**	−155.56	−234.4	**−148**	−100.00	−54.4	**−48**	−44.44
···	**−446**	−265.56	···	**−346**	−210.00	−410.8	**−246**	−154.44	−230.8	**−146**	−98.89	−50.8	**−46**	−43.33
···	**−444**	−264.44	···	**−344**	−208.89	−407.2	**−244**	−153.33	−227.2	**−144**	−97.78	−47.2	**−44**	−42.22
···	**−442**	−263.33	···	**−342**	−207.78	−403.6	**−242**	−152.22	−223.6	**−142**	−96.67	−43.6	**−42**	−41.11
···	**−440**	−262.22	···	**−340**	−206.67	−400.0	**−240**	−151.11	−220.0	**−140**	−95.56	−40.0	**−40**	−40.00
···	**−438**	−261.11	···	**−338**	−205.56	−396.4	**−238**	−150.00	−216.4	**−138**	−94.44	−36.4	**−38**	−38.89
···	**−436**	−260.00	···	**−336**	−204.44	−392.8	**−236**	−148.89	−212.8	**−136**	−93.33	−32.8	**−36**	−37.78
···	**−434**	−258.89	···	**−334**	−203.33	−389.2	**−234**	−147.78	−209.2	**−134**	−92.22	−29.2	**−34**	−36.67
···	**−432**	−257.78	···	**−332**	−202.22	−385.6	**−232**	−146.67	−205.6	**−132**	−91.11	−25.6	**−32**	−35.56
···	**−430**	−256.67	···	**−330**	−201.11	−382.0	**−230**	−145.56	−202.0	**−130**	−90.00	−22.0	**−30**	−34.44
···	**−428**	−255.56	···	**−328**	−200.00	−378.4	**−228**	−144.44	−198.4	**−128**	−88.89	−18.4	**−28**	−33.33
···	**−426**	−254.44	···	**−326**	−198.89	−374.8	**−226**	−143.33	−194.8	**−126**	−87.78	−14.8	**−26**	−32.22
···	**−424**	−253.33	···	**−324**	−197.78	−371.2	**−224**	−142.22	−191.2	**−124**	−86.67	−11.2	**−24**	−31.11
···	**−422**	−252.22	···	**−322**	−196.67	−367.6	**−222**	−141.11	−187.6	**−122**	−85.56	−7.6	**−22**	−30.00
···	**−420**	−251.11	···	**−320**	−195.56	−364.0	**−220**	−140.00	−184.0	**−120**	−84.44	−4.0	**−20**	−28.89
···	**−418**	−250.00	···	**−318**	−194.44	−360.4	**−218**	−138.89	−180.4	**−118**	−83.33	−0.4	**−18**	−27.78
···	**−416**	−248.89	···	**−316**	−193.33	−356.8	**−216**	−137.78	−176.8	**−116**	−82.22	+3.2	**−16**	−26.67
···	**−414**	−247.78	···	**−314**	−192.22	−353.2	**−214**	−136.67	−173.2	**−114**	−81.11	+6.8	**−14**	−25.56
···	**−412**	−246.67	···	**−312**	−191.11	−349.6	**−212**	−135.56	−169.6	**−112**	−80.00	+10.4	**−12**	−24.44
···	**−410**	−245.56	···	**−310**	−190.00	−346.0	**−210**	−134.44	−166.0	**−110**	−78.89	+14.0	**−10**	−23.33
···	**−408**	−244.44	···	**−308**	−188.89	−342.4	**−208**	−133.33	−162.4	**−108**	−77.78	+17.6	**−8**	−22.22
···	**−406**	−243.33	···	**−306**	−187.78	−338.8	**−206**	−132.22	−158.8	**−106**	−76.67	+21.2	**−6**	−21.11
···	**−404**	−242.22	···	**−304**	−186.67	−335.2	**−204**	−131.11	−155.2	**−104**	−75.56	+24.8	**−4**	−20.00
···	**−402**	−241.11	···	**−302**	−185.56	−331.6	**−202**	−130.00	−151.6	**−102**	−74.44	+28.4	**−2**	−18.89
···	**−400**	−240.00	···	**−300**	−184.44	−328.0	**−200**	−128.89	−148.0	**−100**	−73.33	+32.0	**±0**	−17.78
···	**−398**	−238.89	···	**−298**	−183.33	−324.4	**−198**	−127.78	−144.4	**−98**	−72.22	+35.6	**+2**	−16.67
···	**−396**	−237.78	···	**−296**	−182.22	−320.8	**−196**	−126.67	−140.8	**−96**	−71.11	+39.2	**+4**	−15.56
···	**−394**	−236.67	···	**−294**	−181.11	−317.2	**−194**	−125.56	−137.2	**−94**	−70.00	+42.8	**+6**	−14.44
···	**−392**	−235.56	···	**−292**	−180.00	−313.6	**−192**	−124.44	−133.6	**−92**	−68.89	+46.4	**+8**	−13.33
···	**−390**	−234.44	···	**−290**	−178.89	−310.0	**−190**	−123.33	−130.0	**−90**	−67.78	+50.0	**+10**	−12.22
···	**−388**	−233.33	···	**−288**	−177.78	−306.4	**−188**	−122.22	−126.4	**−88**	−66.67	+53.6	**+12**	−11.11
···	**−386**	−232.22	···	**−286**	−176.67	−302.8	**−186**	−121.11	−122.8	**−86**	−65.56	+57.2	**+14**	−10.00
···	**−384**	−231.11	···	**−284**	−175.56	−299.2	**−184**	−120.00	−119.2	**−84**	−64.44	+60.8	**+16**	−8.89
···	**−382**	−230.00	···	**−282**	−174.44	−295.6	**−182**	−118.89	−115.6	**−82**	−63.33	+64.4	**+18**	−7.78
···	**−380**	−228.89	···	**−280**	−173.33	−292.0	**−180**	−117.78	−112.0	**−80**	−62.22	+68.0	**+20**	−6.67
···	**−378**	−227.78	···	**−278**	−172.22	−288.4	**−178**	−116.67	−108.4	**−78**	−61.11	+71.6	**+22**	−5.56
···	**−376**	−226.67	···	**−276**	−171.11	−284.8	**−176**	−115.56	−104.8	**−76**	−60.00	+75.2	**+24**	−4.44
···	**−374**	−225.56	···	**−274**	−170.00	−281.2	**−174**	−114.44	−101.2	**−74**	−58.89	+78.8	**+26**	−3.33
···	**−372**	−224.44	−457.6	**−272**	−168.89	−277.6	**−172**	−113.33	−97.6	**−72**	−57.78	+82.4	**+28**	−2.22
···	**−370**	−223.33	−454.0	**−270**	−167.78	−274.0	**−170**	−112.22	−94.0	**−70**	−56.67	+86.0	**+30**	−1.11
···	**−368**	−222.22	−450.4	**−268**	−166.67	−270.4	**−168**	−111.11	−90.4	**−68**	−55.56	+89.6	**+32**	±0.00
···	**−366**	−221.11	−446.8	**−266**	−165.56	−266.8	**−166**	−110.00	−86.8	**−66**	−54.44	+93.2	**+34**	+1.11
···	**−364**	−220.00	−443.2	**−264**	−164.44	−263.2	**−164**	−108.89	−83.2	**−64**	−53.33	+96.8	**+36**	+2.22
···	**−362**	−218.89	−439.6	**−262**	−163.33	−259.6	**−162**	−107.78	−79.6	**−62**	−52.22	+100.4	**+38**	+3.33
···	**−360**	−217.78	−436.0	**−260**	−162.22	−256.0	**−160**	−106.67	−76.0	**−60**	−51.11	+104.0	**+40**	+4.44

(continued)

Temperature Conversions (continued)

°F		°C	°F		°C	°F		°C	°F		°C	°F		°C
107.6	**42**	5.56	305.6	**152**	66.67	503.6	**262**	127.78	701.6	**372**	188.89	899.6	**482**	250.00
111.2	**44**	6.67	309.2	**154**	67.78	507.2	**264**	128.89	705.2	**374**	190.00	903.2	**484**	251.11
114.8	**46**	7.78	312.8	**156**	68.89	510.8	**266**	130.00	708.8	**376**	191.11	906.8	**486**	252.22
118.4	**48**	8.89	316.4	**158**	70.00	514.4	**268**	131.11	712.4	**378**	192.22	910.4	**488**	253.33
122.0	**50**	10.00	320.0	**160**	71.11	518.0	**270**	132.22	716.0	**380**	193.33	914.0	**490**	254.44
125.6	**52**	11.11	323.6	**162**	72.22	521.6	**272**	133.33	719.6	**382**	194.44	917.6	**492**	255.56
129.2	**54**	12.12	327.2	**164**	73.33	525.2	**274**	134.44	723.2	**384**	195.56	921.2	**494**	256.67
132.8	**56**	13.33	330.8	**166**	74.44	528.8	**276**	135.56	726.8	**386**	196.67	924.8	**496**	257.78
136.4	**58**	14.44	334.4	**168**	75.56	532.4	**278**	136.67	730.4	**388**	197.78	928.4	**498**	258.89
140.0	**60**	15.56	338.0	**170**	76.67	536.0	**280**	137.78	734.0	**390**	198.89	932.0	**500**	260.00
143.6	**62**	16.67	341.6	**172**	77.78	539.6	**282**	138.89	737.6	**392**	200.00	935.6	**502**	261.11
147.2	**64**	17.78	345.2	**174**	78.89	543.2	**284**	140.00	741.2	**394**	201.11	939.2	**504**	262.22
150.8	**66**	18.89	348.8	**176**	80.00	546.8	**286**	141.11	744.8	**396**	202.22	942.8	**506**	263.33
154.4	**68**	20.00	352.4	**178**	81.11	550.4	**288**	142.22	748.4	**398**	203.33	946.4	**508**	264.44
158.0	**70**	21.11	356.0	**180**	82.22	554.0	**290**	143.33	752.0	**400**	204.44	950.0	**510**	265.56
161.6	**72**	22.22	359.6	**182**	83.33	557.6	**292**	144.44	755.6	**402**	205.56	953.6	**512**	266.67
165.2	**74**	23.33	363.2	**184**	84.44	561.2	**294**	145.56	759.2	**404**	206.67	957.2	**514**	267.78
168.8	**76**	24.44	366.8	**186**	85.56	564.8	**296**	146.67	762.8	**406**	207.78	960.8	**516**	268.8[illegible]
172.4	**78**	25.56	370.4	**188**	86.67	568.4	**298**	147.78	766.4	**408**	208.89	964.4	**518**	270.00
176.0	**80**	26.67	374.0	**190**	87.78	572.0	**300**	148.89	770.0	**410**	210.00	968.0	**520**	271.11
179.6	**82**	27.78	377.6	**192**	88.89	575.6	**302**	150.00	773.6	**412**	211.11	971.6	**522**	272.22
183.2	**84**	28.89	381.2	**194**	90.00	579.2	**304**	151.11	777.2	**414**	212.22	975.2	**524**	273.33
186.8	**86**	30.00	384.8	**196**	91.11	582.8	**306**	152.22	780.8	**416**	213.33	978.8	**526**	274.44
190.4	**88**	31.11	388.4	**198**	92.22	586.4	**308**	153.33	784.4	**418**	214.44	982.4	**528**	275.56
194.0	**90**	32.22	392.0	**200**	93.33	590.0	**310**	154.44	788.0	**420**	215.56	986.0	**530**	276.67
197.6	**92**	33.33	395.6	**202**	94.44	593.6	**312**	155.56	791.6	**422**	216.67	989.6	**532**	277.78
201.2	**94**	34.44	399.2	**204**	95.56	597.2	**314**	156.67	795.2	**424**	217.78	993.2	**534**	278.89
204.8	**96**	35.56	402.8	**206**	96.67	600.8	**316**	157.78	798.8	**426**	218.89	996.8	**536**	280.00
208.4	**98**	36.67	406.4	**208**	97.78	604.4	**318**	158.89	802.4	**428**	220.00	1000.4	**538**	281.11
212.0	**100**	37.78	410.0	**210**	98.89	608.0	**320**	160.00	806.0	**430**	221.11	1004.0	**540**	282.22
215.6	**102**	38.89	413.6	**212**	100.00	611.6	**322**	161.11	809.6	**432**	222.22	1007.6	**542**	283.33
219.2	**104**	40.00	417.2	**214**	101.11	615.2	**324**	162.22	813.2	**434**	223.33	1011.2	**544**	284.44
222.8	**106**	41.11	420.8	**216**	102.22	618.8	**326**	163.33	816.8	**436**	224.44	1014.8	**546**	285.56
226.4	**108**	42.22	424.4	**218**	103.33	622.4	**328**	164.44	820.4	**438**	225.56	1018.4	**548**	286.67
230.0	**110**	43.33	428.0	**220**	104.44	626.0	**330**	165.56	824.0	**440**	226.67	1022.0	**550**	287.78
233.6	**112**	44.44	431.6	**222**	105.56	629.6	**332**	166.67	827.6	**442**	227.78	1040.0	**560**	293.33
237.2	**114**	45.56	435.2	**224**	106.67	633.2	**334**	167.78	831.2	**444**	228.89	1058.0	**570**	298.89
240.8	**116**	46.67	438.8	**226**	107.78	636.8	**336**	168.89	834.8	**446**	230.00	1076.0	**580**	304.44
244.4	**118**	47.78	442.4	**228**	108.89	640.4	**338**	170.00	838.4	**448**	231.11	1094.0	**590**	310.00
248.0	**120**	48.89	446.0	**230**	110.00	644.0	**340**	171.11	842.0	**450**	232.22	1112.0	**600**	315.56
251.6	**122**	50.00	449.6	**232**	111.11	647.6	**342**	172.22	845.6	**452**	233.33	1130.0	**610**	321.11
255.2	**124**	51.11	453.2	**234**	112.22	651.2	**344**	173.33	849.2	**454**	234.44	1148.0	**620**	326.67
258.8	**126**	52.22	456.8	**236**	113.33	654.8	**346**	174.44	852.8	**456**	235.56	1166.0	**630**	332.22
262.4	**128**	53.33	460.4	**238**	114.44	658.4	**348**	175.56	856.4	**458**	236.67	1184.0	**640**	337.78
266.0	**130**	54.44	464.0	**240**	115.56	662.0	**350**	176.67	860.0	**460**	237.78	1202.0	**650**	343.33
269.6	**132**	55.56	467.6	**242**	116.67	665.6	**352**	177.78	863.6	**462**	238.89	1220.0	**660**	348.89
273.2	**134**	56.67	471.2	**244**	117.78	669.2	**354**	178.89	867.2	**464**	240.00	1238.0	**670**	354.44
276.8	**136**	57.78	474.8	**246**	118.89	672.8	**356**	180.00	870.8	**466**	241.11	1256.0	**680**	360.00
280.4	**138**	58.89	478.4	**248**	120.00	676.4	**358**	181.11	874.4	**468**	242.22	1274.0	**690**	365.56
284.0	**140**	60.00	482.0	**250**	121.11	680.0	**360**	182.22	878.0	**470**	243.33	1292.0	**700**	371.11
287.6	**142**	61.11	485.6	**252**	122.22	683.6	**362**	183.33	881.6	**472**	244.44	1310.0	**710**	376.67
291.2	**144**	62.22	489.2	**254**	123.33	687.2	**364**	184.44	885.2	**474**	245.56	1328.0	**720**	382.22
294.8	**146**	63.33	492.8	**256**	124.44	690.8	**366**	185.56	888.8	**476**	246.67	1346.0	**730**	387.78
298.4	**148**	64.44	496.4	**258**	125.56	694.4	**368**	186.67	892.4	**478**	247.78	1364.0	**740**	393.33
302.0	**150**	65.56	500.0	**260**	126.67	698.0	**370**	187.78	896.0	**480**	248.89	1382.0	**750**	398.89

(continued)

Temperature Conversions (continued)

°F		°C	°F		°C	°F		°C	°F		°C	°F		°C
1400.0	**760**	404.44	2390.0	**1310**	710.00	3380.0	**1860**	1015.6	4370.0	**2410**	1321.1	5450.0	**3010**	1654.4
1418.0	**770**	410.00	2408.0	**1320**	715.56	3398.0	**1870**	1021.1	4388.0	**2420**	1326.7	5468.0	**3020**	1660.0
1436.0	**780**	415.56	2426.0	**1330**	721.11	3416.0	**1880**	1026.7	4406.0	**2430**	1332.2	5486.0	**3030**	1665.6
1454.0	**790**	421.11	2440.0	**1340**	726.67	3434.0	**1890**	1032.2	4424.0	**2440**	1337.8	5504.0	**3040**	1671.1
1472.0	**800**	426.67	2462.0	**1350**	732.22	3452.0	**1900**	1037.8	4442.0	**2450**	1343.3	5522.0	**3050**	1676.7
1490.0	**810**	432.22	2480.0	**1360**	737.78	3470.0	**1910**	1043.3	4460.0	**2460**	1348.9	5540.0	**3060**	1682.2
1508.0	**820**	437.76	2498.0	**1370**	743.33	3488.0	**1920**	1048.9	4478.0	**2470**	1354.4	5558.0	**3070**	1687.8
1526.0	**830**	443.33	2516.0	**1380**	748.89	3506.0	**1930**	1054.4	4496.0	**2480**	1360.0	5576.0	**3080**	1693.3
1544.0	**840**	448.89	2534.0	**1390**	754.44	3524.0	**1940**	1060.0	4514.0	**2490**	1365.6	5594.0	**3090**	1698.9
1562.0	**850**	454.44	2552.0	**1400**	760.00	3542.0	**1950**	1065.6	4532.0	**2500**	1371.1	5612.0	**3100**	1704.4
1580.0	**860**	460.00	2570.0	**1410**	765.56	3560.0	**1960**	1071.1	4550.0	**2510**	1376.7	5702.0	**3150**	1732.2
1598.0	**870**	465.56	2588.0	**1420**	771.11	3578.0	**1970**	1076.7	4568.0	**2520**	1382.2	5792.0	**3200**	1760.0
1616.0	**880**	471.11	2606.0	**1430**	776.67	3596.0	**1980**	1082.2	4586.0	**2530**	1387.8	5882.0	**3250**	1787.7
1634.0	**890**	476.67	2624.0	**1440**	782.22	3614.0	**1990**	1087.8	4604.0	**2540**	1393.3	5972.0	**3300**	1815.5
1652.0	**900**	482.22	2642.0	**1450**	787.78	3632.0	**2000**	1093.3	4622.0	**2550**	1398.9	6062.0	**3350**	1843.3
1670.0	**910**	487.78	2660.0	**1460**	793.33	3650.0	**2010**	1098.9	4640.0	**2560**	1404.4	6152.0	**3400**	1871.1
1688.0	**920**	493.33	2678.0	**1470**	798.89	3668.0	**2020**	1104.4	4658.0	**2570**	1410.0	6242.0	**3450**	1898.8
1706.0	**930**	498.89	2696.0	**1480**	804.44	3686.0	**2030**	1110.0	4676.0	**2580**	1415.6	6332.0	**3500**	1926.6
1724.0	**940**	504.44	2714.0	**1490**	810.00	3704.0	**2040**	1115.6	4694.0	**2590**	1421.1	6422.0	**3550**	1954.4
1742.0	**950**	510.00	2732.0	**1500**	815.56	3722.0	**2050**	1121.1	4712.0	**2600**	1426.7	6512.0	**3600**	1982.2
1760.0	**960**	515.56	2750.0	**1510**	821.11	3740.0	**2060**	1126.7	4730.0	**2610**	1432.2	6602.0	**3650**	2010.0
1778.0	**970**	521.11	2768.0	**1520**	826.67	3758.0	**2070**	1132.2	4748.0	**2620**	1437.8	6692.0	**3700**	2037.7
1796.0	**980**	526.67	2786.0	**1530**	832.22	3776.0	**2080**	1137.8	4766.0	**2630**	1443.3	6782.0	**3750**	2065.5
1814.0	**990**	532.22	2804.0	**1540**	837.78	3794.0	**2090**	1143.3	4784.0	**2640**	1448.9	6872.0	**3800**	2093.3
1832.0	**1000**	537.78	2822.0	**1550**	843.33	3812.0	**2100**	1148.9	4802.0	**2650**	1454.4	6962.0	**3850**	2121.1
1850.0	**1010**	543.33	2840.0	**1560**	848.89	3830.0	**2110**	1154.4	4820.0	**2660**	1460.0	7052.0	**3900**	2148.8
1868.0	**1020**	548.89	2858.0	**1570**	854.44	3848.0	**2120**	1160.0	4838.0	**2670**	1465.6	7142.0	**3950**	2176.6
1886.0	**1030**	554.44	2876.0	**1580**	860.00	3866.0	**2130**	1165.6	4856.0	**2680**	1471.1	7232.0	**4000**	2204.4
1904.0	**1040**	560.00	2894.0	**1590**	865.56	3884.0	**2140**	1171.1	4874.0	**2690**	1476.7	7322.0	**4050**	2232.2
1922.0	**1050**	565.56	2912.0	**1600**	871.11	3902.0	**2150**	1176.7	4892.0	**2700**	1482.2	7412.0	**4100**	2260.0
1940.0	**1060**	571.11	2930.0	**1610**	876.67	3920.0	**2160**	1182.2	4910.0	**2710**	1487.8	7502.0	**4150**	2287.7
1958.0	**1070**	576.67	2948.0	**1620**	882.22	3938.0	**2170**	1187.8	4928.0	**2720**	1493.3	7592.0	**4200**	2315.5
1976.0	**1080**	582.22	2966.0	**1630**	887.78	3956.0	**2180**	1193.3	4946.0	**2730**	1498.9	7682.0	**4250**	2343.3
1994.0	**1090**	587.78	2984.0	**1640**	893.33	3974.0	**2190**	1198.9	4964.0	**2740**	1504.4	7772.0	**4300**	2371.1
2012.0	**1100**	593.33	3002.0	**1650**	898.89	3992.0	**2200**	1204.4	4982.0	**2750**	1510.0	7862.0	**4350**	2398.8
2030.0	**1110**	598.89	3020.0	**1660**	904.44	4010.0	**2210**	1210.0	5000.0	**2760**	1515.6	7952.0	**4400**	2426.6
2048.0	**1120**	604.44	3038.0	**1670**	910.00	4028.0	**2220**	1215.6	5018.0	**2770**	1521.1	8042.0	**4450**	2454.4
2066.0	**1130**	610.00	3056.0	**1680**	915.56	4046.0	**2230**	1221.1	5036.0	**2780**	1526.7	8132.0	**4500**	2482.2
2084.0	**1140**	615.56	3074.0	**1690**	921.11	4064.0	**2240**	1226.7	5054.0	**2790**	1532.2	8222.0	**4550**	2510.0
2102.0	**1150**	621.11	3092.0	**1700**	926.67	4082.0	**2250**	1232.2	5072.0	**2800**	1537.8	8312.0	**4600**	2537.7
2120.0	**1160**	626.67	3110.0	**1710**	932.22	4100.0	**2260**	1237.8	5090.0	**2810**	1543.3	8402.0	**4650**	2565.5
2138.0	**1170**	632.22	3128.0	**1720**	937.78	4118.0	**2270**	1243.3	5108.0	**2820**	1548.9	8492.0	**4700**	2593.3
2156.0	**1180**	637.78	3146.0	**1730**	943.33	4136.0	**2280**	1248.9	5126.0	**2830**	1554.4	8582.0	**4750**	2621.1
2174.0	**1190**	643.33	3164.0	**1740**	948.89	4154.0	**2290**	1254.4	5144.0	**2840**	1560.0	8672.0	**4800**	2648.8
2192.0	**1200**	648.89	3182.0	**1750**	954.44	4172.0	**2300**	1260.0	5162.0	**2850**	1565.6	8762.0	**4850**	2676.6
2210.0	**1210**	654.44	3200.0	**1760**	960.00	4190.0	**2310**	1265.6	5180.0	**2860**	1571.1	8852.0	**4900**	2704.4
2228.0	**1220**	660.00	3218.0	**1770**	965.56	4208.0	**2320**	1271.1	5198.0	**2870**	1576.7	8942.0	**4950**	2732.2
2246.0	**1230**	665.56	3236.0	**1780**	971.11	4226.0	**2330**	1276.7	5216.0	**2880**	1582.2	9032.0	**5000**	2760.0
2264.0	**1240**	671.11	3254.0	**1790**	976.67	4244.0	**2340**	1282.2	5234.0	**2890**	1587.8	9122.0	**5050**	2787.7
2282.0	**1250**	676.67	3272.0	**1800**	982.22	4262.0	**2350**	1287.8	5252.0	**2900**	1593.3	9212.0	**5100**	2815.5
2300.0	**1260**	682.22	3290.0	**1810**	987.78	4280.0	**2360**	1293.3	5270.0	**2910**	1598.9	9302.0	**5150**	2843.3
2318.0	**1270**	687.78	3308.0	**1820**	993.33	4298.0	**2370**	1298.9	5288.0	**2920**	1604.4	9392.0	**5200**	2871.1
2336.0	**1280**	693.33	3326.0	**1830**	998.89	4316.0	**2380**	1304.4	5306.0	**2930**	1610.0	9482.0	**5250**	2898.8
2354.0	**1290**	698.89	3344.0	**1840**	1004.4	4334.0	**2390**	1310.0	5324.0	**2940**	1615.6	9572.0	**5300**	2926.6
2372.0	**1300**	704.44	3362.0	**1850**	1010.0	4352.0	**2400**	1315.6	5342.0	**2950**	1621.1	9662.0	**5350**	2954.4
									5360.0	**2960**	1626.7	9752.0	**5400**	2982.2
									5378.0	**2970**	1632.2	9842.0	**5450**	3010.0
									5396.0	**2980**	1637.8	9932.0	**5500**	3037.7
									5414.0	**2990**	1643.3	10 022.0	**5550**	3065.5
									5432.0	**3000**	1648.9	10 112.0	**5600**	3093.3

Periodic Table of the Elements

Metals ← → Nonmetals

Key to chart: Atomic Number → 50; Symbol → Sn; Atomic Weight → 118.69; Oxidation States → +2 +4; Electron Configuration → -18-18-4

Transition Elements (groups IIIb through VIII)

Iᵃ	IIᵃ	IIIᵇ	IVᵇ	Vᵇ	VIᵇ	VIIᵇ	VIII	VIII	VIII	Iᵇ	IIᵇ	IIIᵃ	IVᵃ	Vᵃ	VIᵃ	VIIᵃ	0	Orbit
1 +1 −1 H 1.0079 1																	2 0 He 4.00260 2	K
3 +1 Li 6.939 2 1	4 +2 Be 9.0122 2 2											5 +3 B 10.81 2-3	6 +2 +4 −4 C 12.011 2-4	7 +1 +2 +3 +4 +5 −1 −2 −3 N 14.0067 2-5	8 −2 O 15.9994 2-6	9 −1 F 18.998403 2-7	10 0 Ne 10.17₉ 2-8	K-L
11 +1 Na 22.9898 2 8 1	12 +2 Mg 24.312 2 8 2											13 +3 Al 26.98154 2-8-3	14 +2 +4 −4 Si 28.08 2-8-4	15 +3 +5 −3 P 30.97376 2-8-5	16 +4 +6 −2 S 32.06 2-8-6	17 +1 +5 +7 −1 Cl 35.453 2-8-7	18 0 Ar 39.948 2-8-8	K-L-M
19 +1 K 39.09 8 8 1	20 +2 Ca 40.08 8 8 2	21 +3 Sc 44.9559 8 9 2	22 +2 +3 +4 Ti 47.9 8 10 2	23 +2 +3 +4 +5 V 50.941 8 11 2	24 +2 +3 +6 Cr 51.996 8 13 1	25 +2 +3 +4 +7 Mn 54.9380 8 13 2	26 +2 +3 Fe 55.847 8 14 2	27 +2 +3 Co 58.9332 8 15 2	28 +2 +3 Ni 58.71 8 16 2	29 +1 +2 Cu 63.54 8 18 1	30 +2 Zn 65.38 8 18 2	31 +3 Ga 69.72 8 18 3	32 +2 +4 Ge 72.59 8 18 4	33 +3 +5 −3 As 74.9216 8 18 5	34 +4 +6 −2 Se 78.96 8 18 5	35 +1 +5 −1 Br 79.904 8 18 7	36 0 Kr 83.80 8 18 8	-L-M-N
37 +1 Rb 85.467 18 8 1	38 +2 Sr 87.62 18 8 2	39 +3 Y 88.9059 18 9 2	40 +4 Zr 91.22 -18-10-2	41 +3 +5 Nb 92.9064 -18-12-1	42 +6 Mo 95.94 -18-13-1	43 +4 +6 +7 Tc 98.9062 18-13-2	44 +3 Ru 101.07 -18-15-1	45 +3 Rh 102.905 18 16 1	46 +2 +4 Pd 106.4 -18-18-0	47 +1 Ag 107.868 18-18-1	48 +2 Cd 112.40 -18-18-2	49 +3 In 114.82 -18 18 3	50 +2 +4 Sn 118.69 -18-18-4	51 +3 +5 −1 Sb 121.75 18-18 5	52 +4 +6 −2 Te 127.60 18 18 6	53 +1 +5 +7 −3 I 126.9045 -18-18-7	54 0 Xe 131.30 -18-18 8	-M-N-O
55 +1 Cs 132.9054 -18-8-1	56 +2 Ba 137.3 -18-8-2	57* +3 La 138.9055 -18-9-2	72 +4 Hf 178.49 -32-10-2	73 +5 Ta 180.948 -32-11-2	74 +6 W 183.85 -32-12-2	75 +4 +6 +7 Re 186.207 32 13 2	76 +3 +4 Os 190.2 -32 14 2	77 +3 +4 Ir 192.9 32 15 2	78 +2 +4 Pt 195.09 -32-16-2	79 +1 +3 Au 196.9665 -32-18-1	80 +1 +2 Hg 200.59 -32-18-2	81 +1 +3 Tl 204.37 -32-19-3	82 +2 +4 Pb 207.19 -32-18-4	83 +3 +5 Bi 208.980 -32-18-5	84 +2 +4 Po (209) -32-18-6	85 At (210) -32-18-7	86 0 Rn (222) -32-18-8	N-O-P
87 +1 Fr (223) -18-8-1	88 +2 Ra 226.0254 -18-8-2	89** +3 Ac (227) -18-9-2	104 +4 Rf (261) -32-10-2	105 Ha (262) -32-11-2	106 (263) -32-12-2													-O-P-Q

*Lanthanides	58 +3 +4 Ce 140.12 -20-8-2	59 +3 Pr 140.9077 -21-8-2	60 +3 Nd 144.24 -22-8-2	61 +3 Pm 147 -23-8-2	62 +2 +3 Sm 150.4 -24-8-2	63 +2 +3 Eu 151.96 -25-8-2	64 +3 Gd 157.25 -25-9-2	65 +3 Tb 158.925 -27-8-2	66 +3 Dy 162.50 -28-8-2	67 +3 Ho 164.9304 -29-8-2	68 +3 Er 167.26 -30-8-2	69 +3 Tm 168.9342 -31-8-2	70 +2 +3 Yb 173.04 -32-8-2	71 +3 Lu 174.967 -32-9-2	-N-O-P
**Actinides	90 +4 Th 232.038 -18-10-2	91 +5 +4 Pa 231.0359 -20-9-2	92 +3 +4 +5 +6 U 238.029 -21-9-2	93 +3 +4 +5 +6 Np 237.0482 -22-9-2	94 +3 +4 +5 +6 Pu 239.052 -24-8-2	95 +3 +4 +5 +6 Am (243) -25-8-2	96 +3 Cm (247) -25-9-2	97 +3 +4 Bk (247) -27-8-2	98 +3 Cf (251) -28-8-2	99 +3 Es (254) -29-8-2	100 +3 Fm (257) -30-8-2	101 +2 +3 Md (258) -31-8-2	102 +2 +3 No (259) -32-8-2	103 +3 Lr (260) -32-9-2	-O-P-Q

Numbers in parentheses are mass numbers of most stable isotope of that element

Guide to General Information Sources

THIS ARTICLE is a directory of information sources on composite materials technology, including processing, properties, and government/industry standardization documents. All technical documents described here are unclassified, although some are distribution limited or subject to export control. Organizations that do provide documents that are classified, subject to export control, or otherwise restricted in distribution have been included in the following listings; the reader is responsible for complying with applicable laws and regulations in order to obtain restricted documents. The guide does not include the numerous scientific and engineering journals and books that deal with composites technology.

American Carbon Society (ACS)
The Stackpole Corporation
St. Marys, PA 15857
(814) 781-8410

The ACS offers conferences and publications dealing with the chemistry, physics, and scientific aspects of materials ranging from carbons and graphites to organic crystals and polymers.

American Ceramic Society (ACS)
65 Ceramic Drive
Columbus, Oh 43214
(614) 268-8645

The ACS publishes two monthly periodicals and a bimonthly abstract, and provides scientific and technical information to scientists and engineers involved in the glass, cements, refractories, ceramic-metal systems, nuclear ceramics, electronics, white wares, and structural clay products industries.

American Society for Testing and Materials (ASTM)
1916 Race Street
Philadelphia, PA 19103
(215) 299-5585

ASTM Standards. ASTM Committee D-30 on High Modulus Fibers and Their Composites and ASTM Committee D-20 on Plastics both prepare standards pertaining to composite materials and test methods. The D-30 standards encompass both polymeric and metallic reinforced composite materials. The D-30 and D-20 standards are published in the Annual Book of ASTM standards.

ASTM Special Technical Publications (STPs). Approximately 30 STPs covering various aspects of composites technology have been published thus far.

ASM International
Metals Park, OH 44073
(216) 338-5151

Information on composites is available from ASM International, a technical society for materials, through its reference books, conferences, and videocourses. It also produces a monthly publication, in both print and database form, which abstracts and indexes articles on composites. In addition, ASM is assembling, in conjunction with Materials Sciences Corporation, a composites properties database for Military Handbook 17B. The database also will be available from ASM as personal computer software.

Chemical Propulsion Information Agency (CPIA)
The Johns Hopkins University
Applied Physics Laboratory
Johns Hopkins Road
Laural, MD 20707
(301) 953-5000

The CPIA is a Department of Defense Information Analysis Center operated by The Johns Hopkins University Applied Physics Laboratory in accordance with DoD Instruction 5100.45, "Centers for Analysis of Scientific and Technical Information." The CPIA provides technical and administrative support to the Joint Army-Navy-NASA-Air Force (JANNAF) Interagency Propulsion Committee. It maintains a very broad data base on composites for military and aerospace applications.

Department of Defense (DoD)
Commanding Officer, Naval Publications and Forms Center
5801 Tabor Avenue
Philadelphia, PA 19120
(215) 697-3321

Government/Industry Standardization Documents. As defined by DoD 4120 3-M standardization documents include specifications, standards, handbooks, and related engineering documents used in engineering design, development, manufacturing, maintenance, and supply management. Standardization documents pertaining to composite materials, processes, and tests methods are prepared by three primary standardization organizations; miliary standardization (SD-1) activities; ASTM Committees D-30 and D-20, and the Society of Automotive Engineers. Aerospace Materials Division, Composites Committee. The DoD *Index of Specifications and Standards* (DoDISS) is the primary information source for these three groups of standardization documents. It consists of three volumes: an alphabetical index, a numerical index, and a Federal Supply Class (FSC) listing. Standardization documents related to composites can be found under the Composites Technology (CMPS) standardization area listing in the FSC volume. The establishment of a new CMPS area has been approved by DoD, but at this time the DoDISS does not contain a CMPS listing of documents.

Military Handbook 17 is the primary military standardization document that encompasses engineering properties of polymer matrix composites. It is updated continuously by a government-industry working group led by the U.S. Army Materials Technology Laboratory (AMTL) in Watertown, MA.

Military Standard 1944 contains military and industry specifications and standards for polymer matrix composite materials, processes, and test methods.

Guide to General Information Sources (continued)

Engineering Information, Inc. (EI)
345 E. 47th Street
New York, NY 10017-2387
(800) 221-1044

The EI, a monthy publication, compiles bibliographic citations and abstracts covering the technological literature in all engineering disciplines. The literature covered is found in journals, technical reports, books, and conference proceedings.

Fiber Society (FS)
Box 625
Princeton, NJ 08540
(609) 924-3150

The FS serves scientists and engineers involved in the research of fibers, fibrous materials, and fiber products.

Fiberglass Fabrication Association (FFA)
1010 Wisconsin Avenue N.W.
Suite 630
Washington, D.C. 20007
(202) 544-0262

The FFA disseminates information to companies that fabricate fiberglass products.

International Aerospace Abstracts (IAA)
American Institute of Aeronautics and Astronautics, Inc.
555 W. 57th Street
New York, NY 10019
(212) 247-6500

The IAA covers the literature in the field of aeronautics and space science and technology. Periodicals, books, conference proceedings, and translations are abstracted, indexed, and published twice monthly. A cumulative index is prepared annually. IAA complements NASA's STAR abstract journal (see the section on NASA in this article).

Materials Research Society (MRS)
110 Materials Research Laboratory
University Park, PA 16802
(814) 865-3424

The MRS promotes interaction among researchers at universities and in industry. It provides short courses, conferences, and tutorial lectures.

Metal Matrix Composites Information Analysis Center (MMCIAC)
Kaman Tempo
816 State Street, P.O. Drawer QQ
Santa Barbara, CA 93102-1479
(805) 963-6497

The MMCIAC was established by the DoD in 1980 to provide scientific and technical information analysis services on metal matrix composite (MMC) materials technology to government agencies and the private sector. This center establishes and maintains an MMC properties data base for designers concerned with MMC applications.

National Aeronautics and Space Administration (NASA)
NASA Scientific and Technical Information Facility
ATTN: Registration Activity
P.O. Box 8757
Baltimore-Washington International Airport, MD 21240
(301) 859-5300

The Scientific and Technical Aerospace Reports (STAR) is a major component of the comprehensive NASA information system. The STAR abstract journal, published twice monthly, includes abstracts on composite materials.

National Institute of Ceramic Engineers (NICE)
65 Ceramic Drive
Columbus, OH 43214
(614) 268-8645

The NICE is a professional organization for ceramics engineers. It confers accreditation on educational programs.

Plastics Institute of America (PIA)
Stevens Institute of Technology
Castle Point Station
Hoboken, NJ 07030
(201) 420-5553

The PIA is a cooperative venture of companies in the plastics industry to support education and research in the plastic fields. It conducts a graduate-level program of education in plastics in cooperation with over 100 universities and colleges.

The Refractories Institute (TRI)
3760 One Oliver Plaza
Pittsburgh, PA 15222
(412) 281-6787

TRI serves producers and suppliers in the refractory industry and publishes a product directory. It supports research and awards scholarships.

Guide to General Information Sources (continued)

National Technical Information Service (NTIS)
Springfield, VA 22161
(703) 487-4600

The NTIS serves as a clearinghouse for reports funded and issued by government agencies. NTIS publishes *Government Reports Announcements & Index* twice monthly, which abstracts and provides acquisition information for reports on advanced materials and technologies. NTIS also publishes weekly newsletters in 27 subject categories, one newsletter being *Materials Science*. The newsletters summarize unclassified federally funded research as it is made available to the public and provide abstracts of reports. NTIS also offers Selected Research in Microfiche, (SRIM) on an automatic biweekly basis and a master catalog of published searches, which lists more than 3000 bibliographies.

Suppliers of Advanced Composite Materials Association (SACMA)
1600 Wilson Boulevard, Suite 1008
Arlington, VA 22209
(703) 841-1556

SACMA is a trade association formed in 1985 to address technical problems in the development of composite materials, such as standardization of test methods, specifications, certification procedures, and export control. It disseminates relevant business, marketing, and technical information on advanced composite materials.

Society of Automotive Engineers (SAE)
400 Commonwealth Drive
Warrendale, PA 15096
(412) 776-4841

The Composites Committee of the SAE Aerospace Materials Division prepares Aerospace Materials Specifications (AMS) for composite materials. These specifications are listed in the annual SAE/AMS index under the AMS series 3000 documents for.nonmetallic materials.

Society for the Advancement of Material and
Process Engineering (SAMPE)
843 West Glentana
P.O. Box 2459
Covina, CA 91722
(818) 331-0616

SAMPE is a technical society that publishes a journal, a quarterly, and symposia and technical conference proceedings on a variety of processes and materials, including composites.

Society of Plastics Engineers (SPE)
14 Fairfield Drive
Brookfield, CT 06805
(203) 775-0471

SPE is a technical society and an information source for a wide variety of publications related to composites technology.

Society of the Plastics Industry, Inc. (SPI)
The Reinforced Plastics/Composites Institute
355 Lexington Avenue
New York, NY 10017
(212) 503-0600

The Reinforced Plastics/Composites Institute of SPI is an information source for reinforced plastics and composites technology. The proceedings of its annual meetings have a wealth of technological and commercial information.

Other Sources

Periodic conferences, symposia, and seminars related to composites technology include:

- DoD/NASA conferences on fibrous composites in structural design
- High-temperature plastic laminate evaluation (high temple) workshops, coordinated by a DoD/NASA steering group. Contact: Wright-Patterson Air Force Base, OH 45433. (513) 257-1110
- Technology Transfer Society (TTS), advanced composites seminars. Contact: TTS Conferences, P.O. Box 3608, 3420 Kashiwa Street, Suite 2000, Torrance, CA 90510-3608. (213) 534-3922
- Advanced composites working groups of the American Ceramic Society, Ceramic-Metal Systems Division, which is coordinated by DoD/NASA. Contact: NASA Langley Research Center, Mail Stop 387, Hampton, VA 23665. (804) 865-3131.

Automated Literature Searching

The storage of technical information in computer data banks is increasing. In this practice, computer terminals, linked to a central database by telephone, are used to search the stored information and retrieve relevant data.

ASM INTERNATIONAL maintains Metadex, a database on metals and materials. The contents of this and other databases relevant to metals and materials are described in the following section on DIALOG information services. Because the formats and search techniques for most databases are similar, the Metadex database will be used to illustrate briefly automated literature searching.

Each month ASM editors abstract documents for the Metadex database. These documents include:

- Seminar proceedings
- Journals
- Magazines
- Books
- Technical reports
- Patents
- Dissertations
- Translations

The document, as keyed for input to the system by the editor, consists of this appropriate information:

Distribution tape field No.	Contents
1	Document number
2	Title
3	Abstract, number of references, abstract, author's initials
4	Authors, corporate authors
5a	Conference name
5b	Conference place
5c	Conference date
5d	Publisher name
5e	Publisher address
5f	Publishing date
5g	Journal title
5h	Report number
5i	Date
5j	Number
5k	Volume reference
5l	Pages
5m	Languages
5n	Patent number
5o	Patent country
5p	Patent application date
5q	ISSN or ISBN
6	Indexing terms
7	Alloy indexing terms
8	Cross references, security code

The document information is also compiled into an abstract, similar to the example given here, which is entered into the database file.

Abstracts are arranged in logical subject groupings for publication in the bound volumes of *Metals Abstracts*. The two digit categories are included in the document number and may be used in retrieval. The subject codes and categories for grouping abstracts are:

Subject codes	Categories
11	Constitution
12	Crystal properties
13	Lattice defects
14	Structural hardening
15	Physics of metals
16	Irradiation effects
21	Metallography
22	Testing and control
23	Analysis
31	Mechanical properties
32	Physical properties
33	Electrical and magnetic
34	Chemical and electrochemical properties
35	Corrosion
41	Ores and raw materials
42	Extraction and smelting
43	Refining and purification
44	Physical chemistry of extraction and refining
45	Ferrous alloy production

Subject codes	Categories
46	Nonferrous alloy production
51	Foundry
52	Working
53	Machining
54	Powder technology
55	Joining
56	Thermal treatment
57	Finishing
58	Metallic coating
61	Engineering components and structures
62	Composites
63	Electronic devices
71	General and nonclassified
72	Special publications

Subject codes, standard indexing terms (descriptors), and key words can be used to search for abstracts in the database. For example, the computer can be used to select every abstract which:

- Is categorized in subject code 51 (Foundry)
- Contains the standard indexing term "casting" in the descriptor field
- Contains the phrase "improvements in quality" somewhere in the abstract text

Any of these three commands, which are circled on the sample abstract, would retrieve the abstract and display it on the computer terminal.

Sample abstract

771587 81-510668
Trends in Casting Technology.
Chandler, H E ; Baxter Jr, D F
Met. Prog. , Jan. 1981, 119, (1), 96-100
Language: ENGLISH
Document Type: ARTICLE
New die-casting applications for Al in the U.S. auto industry will include: engine blocks, cylinder heads, intake manifolds and pistons; the use of cold chamber cast Mg for transmission cases is being considered. Improvements in casting technology include: use of hot isostatic pressing to cast complex Ti components for the aerospace industry and to improve performance of Al castings; investment casting of stainless steels and Ti; spin casting of iron rolls and Al. Improvements in quality will result from automation, use of the AOD (Ar O decarburization) process, metal stream inoculation, in-mold measurement of gas pressure and ceramic foam filter for liquid metal filtration. --M.G.S.
Descriptors: Ferrous alloys, Casting; Nonferrous alloys, Casting; Automotive components, Casting; Foundry practice
Section Heading: 51, (FOUNDRY) Journal Announcement: 8107

Select Databases Available Through DIALOG Information Services, Inc.

The DIALOG Information Retrieval Service, from DIALOG Information Services, Inc., has been operating since 1972. There are 160 databases available on the system.

The databases on the DIALOG system contain in excess of 50,000,000 records. Records, or units of information, can range from a directory-type listing of specific manufacturing plants to a citation with bibliographic information and an abstract referencing a journal, conference paper, or other original source. For more information on DIALOG write:

DIALOG Information Service, Inc.
Marketing Department
3460 Hillview Avenue
Palo Alto, California 94304

The DIALOG service may also be called at these numbers:

- (800) 227-1927, marketing and training
- (800) 227-1960, customer services. Toll-free in the continental U.S., except California
- (800) 982-5838, California
- (415) 858-3785, all other locations

The following databases of interest to materials professionals are available on the DIALOG system:

AEROSPACE DATABASE **File 108**

Coverage: 1962 to the present
File Size: 1,536,984 records
Updates: Twice a month
Provider: American Institute of Aeronautics and Astronautics/Technial Information Service (AIAA/TIS), New York, NY

The AEROSPACE DATABASE provides references, abstracts, and controlled-vocabulary indexing of key scientific and technical documents, as well as books, reports, and conferences, covering aerospace research and development in over 40 countries, including Japan and Communist-bloc nations. This database supports basic and applied research in aeronautics, astronauts, and space sciences, as well as technology development and applications in complementary and supporting fields such as chemistry, geosciences, physics, communications, and electronics. The AEROSPACE DATABASE combines in one database two publications: *Scientific and Technical Aerospace Reports (STAR)*, produced by the National Aeronautics and Space Administration (NASA), and *International Aerospace Abstracts (IAA)*, produced by AIAA under contract to NASA.

The AEROSPACE DATABASE is available only in the United States. Access by non-U.S. governments, organizations, or persons acting on their behalf is not allowed without written approval of the Amerian Institute of Aeronautics and Astronautics, Technical Information Service.

CURRENT TECHNOLOGY INDEX **File 142**

Coverage: 1981 to the present
File Size: 115,662 records
Updates: Monthly
Provider: Library Association Publishing Ltd., The Library Association, London, United Kingdom.

CURRENT TECHNOLOGY INDEX (CTI) provides an index to current periodicals from all fields of modern technology. All journals indexed are published in the United Kingdom. The journal articles include aspects of technology on a worldwide basis. Subjects covered include, but are not limited to, the following: acoustic engineering, architecture, aircraft engineering, building, chemical engineering, architecture, aircraft engineering, building, chemical engineering, fishing, mining, food industry, packaging, printing, papermaking, space science, and more.

ENGINEERED MATERIALS ABSTRACTS™ **File 293**

Coverage: January 1986 to the present
File Size: 23,785 records
Updates: Monthly
Provider: ASM International, Metals Park, OH and Institute of Metals, London

ENGINEERED MATERIALS ABSTRACTS™ (EMA) is a joint production of ASM International (U.S.) and Institute of Metals (U.K.). The database corresponds to the publication of the same name, which began publication in January 1986. Informative abstracts are included for most records.

EMA provides comprehensive coverage of the world's published literature concerning 1) the science of polymers, ceramics, and composite materials intended for use in the design, construction, and operation of structures, equipment and systems; and 2) the practices of materials science and engineering as they relate to these materials. Composite materials represented in the coverage are primarily non-metallic, with ceramic or polymeric matrices. Laminated and composite structures, such as honeycombs, essentially consist of non-metallic materials, although metal flakes, fibers, or other forms may occur as reinforcement. Subjects covered include materials, properties, processes, products, and forms of these engineered materials.

HEILBRON **File 303**

Coverage: Current
File Size: 196,370 records
Updates: Semiannually
Provider: Chapman & Hall Ltd., London, England

HEILBRON, the chemical properties database, represents the complete text of two major chemical dictionaries from Chapman & Hall, Ltd.: *Dictionary of Organic Compounds* (Fifth edition), and *Dictionary of Organometallic Compounds.* Also included are other source books including: *Carbohydrates, Amino Acids and Peptides, The Dictionary of Antibiotics and Related Compounds,* and *the Dictionary of Organophophorus Compounds.* Other publications will be added to the file in the future.

HEILBRON is a source database of chemical identification, physical-chemical properties, use, hazard, and key reference data to the world's more important chemical substances, as selected by a panel of experts. HEILBRON provides chemical substance identification through searching physical and/or chemical properties, compound variants, derivative names, synonyms, CAS Registry numbers, molecular formulae and molecular weight, source statements, use/importance data, melting point, freezing point, boiling point, solubility, relative density, optical rotation, and dissociation constants.

Select Databases Available Through DIALOG (continued)

JAPAN TECHNOLOGY **File 582**

Coverage: January 1986 to the present
File Size: 108,374 records
Updates: Monthly
Provider: University Microfilms International, -Ann Arnor, MI

JAPAN TECHNOLOGY corresponds to the printed *Japanese Technical Abstracts*. It contains abstracts from the major Japanese journals in technology, applied sciences, engineering, and business management, as well as articles by Japanese authors published in journals outside Japan. Coverage includes journals as well as governmental, commercial, and society research reports. Major subject areas covered include: aerodynamic engineering, automotive engineering, biological sciences, energy and power, chemistry, medicine, physics, metallurgy, transportation, and more.

Subscribers may contact UMI for authorization to search JAPAN TECHNOLOGY at a reduced rate (File 972).

MATERIALS BUSINESS FILE **File 269**

Coverage: 1985 to the present
File Size: 29,780 records
Updates: Monthly
Provider: ASM International, Metals Park, OH

MATERIALS BUSINESS FILE covers technical and commercial developments in iron and steel, nonferrous metals, composites, plastics, etc. Over 1,300 publications including magazines, trade publications, financial reports, dissertations, and conference proceedings are reviewed for inclusion. Subjects covered are grouped into 9 catagories: 1) Fuel, Energy Usage, Raw Materials, Recycling; 2) Plant Developments and Descriptions; 3) Engineering, Control and Testing, Machinery; 4) Environmental Issues, Waste Treatment, Health and Safety; 5) Product and Process Development; 6) Applications, Competitive Materials, Substitution; 7) Management, Training, Regulations, Marketing; 8) Economics, Statistics, Resources, and Reserves, and 9) World Industry News, Company Information, and General Issues.

METADEX **File 32**

Coverage: 1966 to the present (*Alloys Index*, 1974-present; *Steels Supplement*, 1983-present)
File Size: 701,921 records
Updates: Monthly
Provider: ASM International, Metals Park, OH and The Metals Society, London, England

The METADEX database, produced by ASM International and The Metals Society (London), provides the most comprehensive coverage of international literature on the science and practice of metallurgy. Included in this database are *Review of Metal Literature* (1966-67), *Metals Abstracts* (1968 to present), and *Alloys Index* (1974 to present). The *Steels Supplement* to *Metals Abstracts* was added in 1983. *Metals Abstracts* includes about 30,000 citations each year from about 1,100 primary journal sources. *Alloys Index* supplements *Metals Abstracts* by providing access to the citations through commercial, numerical, and compositional alloy designations; specific metallic systems; and intermetallic compounds found within these systems. The *Steels Supplement* covers information on steel production, fabrication, and use, and developments of value to steel producing and steel-using industries. In addition to specialized topics (including specific alloy designations, intermetallic compounds, and metallurgical systems), six basic categories of metallurgy are covered: materials, processes, properties, products, forms, and influencing factors. Each month about 3,500 new documents related to metals technologies are scanned and abstracted for the ASM database, with intensive coverge of appropriate conference papers, reviews, technical reports, and books. These sources are international in scope, including the U.S.S.R. and Eastern European nations among the 43 countries covered.

NONFERROUS METALS ABSTRACTS **File 118**

Coverage: 1961 through 1983
File Size: 120,924 records
Updates: Closed file
Provider: British Non-Ferrous Metals Technology Centre, Wantage, Oxfordshire, England

NONFERROUS METALS ABSTRACTS covers all aspects of nonferrous metallurgy and technology. Sources include journals, monographs, British patents, reports, standards, and conference papers. The majority of the publications indexed are English language, but a large number of German and French publications are cited as well. NONFERROUS METALS ABSTRACTS corresponds to the printed publication, *BNF Abstracts*.

NTIS **File 6**

Coverage: 1964 to the present
File Size: 1,285,216 records
Updates: Biweekly
Provider: National Technical Information Service (NTIS): U.S. Department of Commerce, Springfield, VA

NTIS is available from DIALOG as an online database and in compact-disc format, DIALOG® NTIS OnDisc™. The NTIS database consists of government-sponsored research, development, and engineering plus analyses prepared by federal agencies, their contractors, or grantees. It is the means through which unclassified, publicly available, unlimited distribution reports are made available for sale from agencies such as NASA, DDC, DOE, HUD, DOT, Department of Commerce, and some 240 other agencies. In addition, some state and local government agencies now contribute their reports to the database.

Truly multi-disciplinary, this database covers a wide spectrum of subjects including: administration and management, agriculture and food, behavior and society, building, business and economics, chemistry, civil engineering, energy, health planning, library and information science, materials science, medicine and biology, military science, transportation, and much more.

SOVIET SCIENCE AND TECHNOLOGY **File 270**

Coverage: 1975 to the present
File Size: 120,363 records
Updates: Monthly
Provider: IFI/Plenum Data Corporation, Alexandria, VA

SOVIET SCIENCE AND TECHNOLOGY provides access to scientific and technological data which is published in Soviet-bloc countries. Information contained in the database covers: aerospace, aeronautics, communications, computer technology, cybernetics, earth science, energy use and conservation, environmental sciences, fuels and petroleum products, geosciences, metallurgy, meteorology, pollution abatement and control, robotics, and solar energy. Information is indexed from journal articles, patents, technical reports, and conference papers.

SSIE CURRENT RESEARCH **File 65**

Coverage: 1978 through February 1982
File Size: 439,265 records
Updates: Closed file
Provider: National Technical Information Service (NTIS), Springfield, VA

SSIE (Smithsonian Science Information Exchange) CURRENT RESEARCH is a database containing reports of both government and privately funded scientific research projects, either in progress or initiated and completed during 1978-1982. SSIE CURRENT RESEARCH covers projects funded from over 1,300 federal, state, and local government agencies; nonprofit associations and foundations; and colleges and universities. Some material is provided from private industry and foreign research organizations; 90% of the information in the database is provided by agencies of the federal government.

SSIE CURRENT RESEARCH is a closed file; however, updated information on federally-funded research projects may be obtained from Files 265 and 266, FEDERAL RESEARCH IN PROGRESS. SSIE CURRENT RESEARCH is an excellent source of historical background on many federally-funded projects currently in progress.

STANDARDS & SPECIFICATIONS **File 113**

Coverage: 1950 to the present
File Size: 116,239 records
Updates: Monthly
Provider: National Standards Association, Inc., Bethesda, MD

STANDARDS & SPECIFICATIONS provides access to all government and industry standards, specifications, and related documents which specify terminology, performance testing, safety, materials, products or other requirements and characteristics of interest to a particular technology or industry.

To identify standards and specifications the database contains the following information: issuing organization, Federal Supply Classification code, whether documents have been cancelled or superseded, if adopted by a U.S. government agency, designated an American National Standard, and, for international standards, whether approved for use by an agency of the U.S. Government. Suppliers of products and services conforming to standards and specifications are also provided. While the bulk of these standards and specifications documents have been issued since 1950, some go back as early as 1920.

WELDASEARCH **File 99**

Coverage: 1967 to the present
File Size: 93,581 records
Updates: Monthly
Provider: The Welding Institute, Cambridge, England

The WELDASEARCH database provides primary coverage of the international literature on all aspects of the joining of metals and plastics and related areas such as metals spraying and thermal cutting. WELDASEARCH includes materials on welded design, welding metallurgy, fatigue and fracture mechanics as well as welding and joining equipment, corrosion, thermal cutting, and quality control. Approximately 5,000 new records are added to WELDASEARCH each year from several thousand journals as well as research reports, books, standards, patents, theses, and special publications.

WORLD ALUMINUM ABSTRACTS **File 33**

Coverage: 1968 to the present
File Size: 126,697 records
Updates: Monthly
Provider: ASM International, Metals Park, OH

WORLD ALUMINUM ABSTRACTS provides coverage of the world's technical literature on aluminum, ranging from ore processing (exclusive of mining) through end uses. The WORLD ALUMINUM ABSTRACTS database includes information abstracted from approximately 1,600 scientific and technical patents, government reports, conference proceedings, dissertations, books and journals. All aspects of the aluminum industry, aside from mining, are covered, including the following major subject areas: aluminum industry general, ores, alumina production and extraction, melting, casting, and foundry, metalworking, fabrication, finishing, physical and mechanical metallurgy, engineering properties and tests, quality control and tests, and end uses.

Directory of Composites Laboratories and Information Centers

Ames Laboratory
Iowa State University
Ames, IA 50011
(515) 294-4037

Director: R.S. Hansen
Contact: D.K. Finnemore, Associate Director
Founded: 1949

Research work includes the study of dendritic growth in metal-metal composites and synthesis of ceramic powders. R&D services are offered for industry, but they must have consent of the U.S. Department of Energy. The DOE also determines patent and licensing policies. Ninety-eight percent of the lab's funding is provided by the government.

Brookhaven National Laboratory
Upton, NY 11973
(516) 282-2123

Associate Director: Dr. Martin Blume
Contacts: Dr. Victor Emery, Assoc. Chairman, Physics Dept.
Dr. Allen Goland, Assoc. Chairman, Dept. of Applied Science
Dr. William Marcuse, Head, Office of Research and Technology Applications
Founded: 1947

The laboratory provides basic and applied research on materials, including polymer/aggregate composites and polymer concrete development. Collaborative and proprietary research is conducted at BNL with engineers and scientists from the private sector. There are 40 senior researchers at the laboratory.

Carnegie-Mellon University
Polymer Science Program
Department of Chemistry
4400 Fifth Avenue
Pittsburgh, PA 15213
(412) 578-3131

Head: Professor G.C. Berry
Founded: 1960

Research is under way involving suspensions in polymeric matrices (e.g., magnetic particles) and interfacial polymerization, and major work has been accomplished in elucidation of interparticle organization in suspensions. Individual research contracts are awarded and faculty consulting is available. Patents and licenses are primarily state and federally owned. The library contains conference proceedings and key journals.

California Polytechnic State University
Metallurgical & Welding Engineering Department
San Luis Obispo, CA 93407
(805) 546-2568

Acting Head: George T. Murray

Composites are part of the current research program. Individual research contracts are available and patent and licensing policies vary. Faculty consulting is also available.

Colorado School of Mines
Advanced Materials Institute
Golden, CO 80401
(303) 273-3830

Acting Director: Jerome G. Morse, Ph.D.
Founded: 1984

Amorphous materials, ceramics, coatings, composites, particulates, polymers, magnetic materials, metals, and semiconductors are part of the current research areas. Graphite/epoxy composites are also currently under investigation. Industrial affiliates have access to long and short range problem-solving assistance, seminars and workshops, an expanding materials library, computer programming development, access to National Standard Reference Data and Information Systems in related fields, and faculty consulting. Individual research contracts are available. Although funded primarily by the state legislature, research orientation is also determined by industry.

Georgia Institute of Technology
School of Ceramic Engineering
Atlanta, GA 30332
(404) 894-2850

Director: Joseph L. Pentecost
Founded: 1924

Current research includes in-situ composite and crystal growth. Research involvement extends to industry on a formal contract basis and the facility is available on a fee basis. Seminars are offered periodically and other services include the use or purchase of some computer programs. Some proprietary research for ultimate patenting or licensing is available, but not preferred.

Composites Laboratories and Information Centers (continued)

Jet Propulsion Laboratory
Applied Mechanics Technology Section (354)
California Institute of Technology
4800 Oak Grove Drive
Pasadena, CA 91109
(818) 354-6580

Section Manager: Charles E. Lifer
Founded: Mid-1930's

Advanced composite technology has been developed and applied to the space program. Current space research includes work on durable materials and coatings and ceramic composites for armor and space shielding. JPL transfers and disseminates much of its research through published reports. Research contracts and faculty consulting are available to industry but the lab does not engage in proprietary R&D. Computer programs and space flight and science data are available through NASA.

Massachusetts Institute of Technology
Materials Processing Center
77 Massachusetts Ave., Bldg. 12-007
Cambridge, MA 02139
(617) 253-3217

Director: Professor H. Kent Bowen
Contact: Dr. George B. Kenney, Assistant Director
Founded: 1980

Metal- and polymer-matrix composites are under current research. Workshops and seminars are sponsored each year, as well as continuing education courses. Research contracts to industry are single and multiclient. Although somewhat flexible, patents generated from sponsored research programs are the property of MIT. The sponsors are granted nonexclusive royalty-free licenses.

Metals and Ceramics Information Center (MCIC)
Battelle-Columbus Laboratories
505 King Avenue
Columbus, OH 43201
(614) 424-5000

Director: Dr. Harold Mindlin
Contact: Frank Jelinck, Associate Director

This center covers the following subjects. Materials: type or base. Metals: titanium; aluminum; magnesium; beryllium; refractory metals; high-strength steels; superalloys. Ceramics: borides; carbides; carbonographite; nitrides; oxides; sulfides; silicides; selection glass; glass ceramics. Coverage: composites of these materials; coatings; environmental effects; mechanical and physical properties; materials applications; test methods; sources/suppliers; specifications; design characteristics. Processes: basic materials production; primary fabrication (forging, casting, rolling, extrusion, etc.); joining; powder processes; surface treatment; quality control and inspection.

Metal Matrix Composites Information Analysis Center (MMCIAC)
Kaman-TEMPO (formerly General Electric-TEMPO)
816 State Street
P. O. Drawer QQ
Santa Barbara, CA 93102
(805) 963-6497

Director: Louis Gonzales
Contact: Jacques E. Schoutens, Manager Data Analysis
William E. Rogers, Manager Information Services

Metal-matrix composite materials technology includes: continuous fibers; wires; discontinuous whiskers with L/D 10; directionally solidified eutectics. Fibers: boron; graphite; silicon carbide; borsic; nitride; alumina; boron carbide; titanium diboride. Wires: stainless steel; tungsten; molybdenum; beryllium; titanium; niobium alloys; compounds. Whiskers: alumina; silicon carbide; silicon nitride. MMC systems: alumina/magnesium; beryllium/titanium; boron/stainless steel/aluminum; boron/titanium/aluminum; borsic/aluminum; borsic/titanium; copper/graphite; graphite/aluminum; graphite/lead; tungsten/nickel. MMC properties: physical properties (thermal, chemical, optical, electromagnetic); specific modulus; strength; fatigue; environmental response; creep and wear resistance. Technical areas cover manufacturing; fabrication process development; defense system applications; performance computations; cost; testing and evaluation techniques and methods; properties data; operational; serviceability/repair; environmental protection; other MMC-related areas.

Michigan State University
Center for Composite Materials and Structures
Room 330, Engineering Building
East Lansing, MI 48824
(517) 353-5466

Interim Director: David L. Sikarski
Founded: 1983

Research areas include stress field determination, mechanical characterization and testing, and mechanical fastening of joints in composites. Studies of fatigue in composites and polymers and polymer processing are also under way. The center works closely with the Michigan Molecular Institute of Midland, Michigan, which provides expertise in the area of molecular-oriented research. The center offers short courses, workshops, contract research, consultation, industrial intern programs, and referral services. Future plans include an advanced materials/composites database containing pertinent materials selection information. The database will be available to industry.

Composites Laboratories and Information Centers (continued)

Mississippi State University
Raspet Flight Research Laboratory
Department of Aerospace Engineering
Drawer A
Mississippi State, MS 39762
(601) 325-3623

Director: Dr. George Bennett
Founded: 1948

The lab specializes in fabrication of aircraft composite structures. Facilities also include flight test evaluation. Services to industry range from consulting to research contracts, and patent and licensing arrangements are liberal. The lab also offers courses on applied design and fabrication of composite structures with hands-on experience in prepreg fabrication. A wide range of composite structural design computer programs are available.

North Carolina A & T State University
Composite Materials Research Laboratory
Department of Mechanical Engineering
108 Graham Hall
Greensboro, NC 27411
(919) 379-7620

Head: Dr. V. Sarma Avva

Composite studies include the frequency and stacking sequence effects on composites, the effect of thermal loads and fatigue on graphite-glass and SiC-glass, and the buckling of composite materials under hot and wet environments. SEM, LM, and X-radiographic studies are also being conducted on composite materials. Cooperative research and development projects, as well as individual contracts, are available. Faculty consulting is provided, and seminars and workshops can be arranged. Patent and licensing policies are negotiable.

North Carolina State University
Department of Materials Engineering
Box 7907
Raleigh, NC 27695
(919) 737-2377

Department Head: Dr. Hans Conrad

Composite materials research is part of the engineering department's program. Services offered to industry include testing, consulting, and individual and affiliated research contracts.

North Carolina State University
School of Textiles
Raleigh, NC 27650
(919) 737-3057

Head: William K. Walsh, Associate Dean

Besides the research and production of fibers, there is a complete polymer analytical lab. Polymer research areas include physical chemistry, synthesis, physics, electrical properties, and diffusion and radiation curing methods. The school awards research contracts, offers faculty consulting, service agreements, and rental equipment to industries. The patent and licensing policies may be negotiated, with the institution retaining the rights to use the invention free of royalty fees.

Ohio State University
Materials Research Laboratory
174 W. 18th Avenue
Columbus, OH 43210
(614) 422-5190

Director: James C. Garland
Contact: Richard L. McGeery, Associate Director
Founded: 1982

Particulate composites are currently being researched. Individual research contracts are awarded and faculty consulting and use of the facilities are also available. The lab is funded primarily by the National Science Foundation.

Plastics Technical Evaluation Center (PLASTEC)
U.S. Army Armament Research and Development Command
Dover, NJ 07801
(201) 724-4222

Director: Harry E. Pebly
Contact: Lee Ann Chervnisk, Publications

The center is responsible for the generation, evaluation, and exchange of technical information related to plastics, adhesives, and organic matrix composites. Covers technology from applied research through fabrication with emphasis on properties and performance. Subject areas include structural, electrical, electronic, and packaging applications. Includes molded, formed, foamed, and laminated materials. Maintains computerized data file on compatibility of polymers with propellants and explosives (COMPAT); also HAZARD-FAILURE file. Maintains complete file of standards, specifications, and handbooks in subject areas. Provides following services on a fee basis: technical inquiries, state-of-the-art studies, data compilations, handbooks, consultant, analysis and evaluations, background studies, bibliographic, and literature searches.

Composites Laboratories and Information Centers (continued)

Polytechnic Institute of New York
Polymer Research Institute
333 Jay Street
Brooklyn, NY 11201
(212) 643-5235

Director: Professor Eli M. Pearce
Founded: 1942

The study of thermosetting and thermoplastic composites are part of the polymer research at the institute. All phases of the polymer disciplines are currently under investigation.

Rensselaer Polytechnic Institute
Center for Manufacturing Productivity and Technology Transfer
JEC 5009
Troy, NY 12181
(518) 266-6950

Director: Dr. Leo E. Hanifin
Contact: Dr. Robert W. Messler, Jr., Associate Director

Current research of advanced composites is being performed at the institute. The work includes the study of high-modulus and high-strength fibers and chemical vapor deposition (CVD). The center publishes reports on individual projects and technical articles. Use of computer programs is proprietary to project sponsors and member companies. The center also employs project managers from industry, who manage staff members working on R&D programs.

Rensselaer Polytechnic Institute
Materials Engineering Department
Troy, NY 12181
(518) 266-6372

Chairman: Martin E. Glicksman
Contact: Roger N. Wright, Executive Officer

Current research areas include polymer-matrix composites. Services to industry include contract research, use of student talent, and consulting services. Patent and licensing policies are available upon request.

Rutgers, The State University of New Jersey
Center for Ceramics Research
Box 909
Piscataway, NJ 08854
(201) 932-2724

Director: John B. Wachtman, Jr.
Founded: 1982

Composite bonding and joining, as well as high-strength and high-temperature ceramic composites, are under research. The center provides cooperative research with industry and maintains confidential results for 1 year. Patents and licenses are usually held by the university with royalty-free, non-exclusive licenses given to sponsors. A company may obtain ownership of a patent or license at additional cost under single-client contract agreements.

Southern Illinois University
Materials Technology Center
Carbondale, IL 62901
(618) 536-2129

Director: Kenneth E. Tempelmeyer
Founded: 1983

The technology center specializes in the research of carbon and graphite composites and fibers. Major accomplishments include the development of applications of composite materials (particularly graphite) in commercial and construction areas. The center holds annual meetings on materials technology, and working relations with industry include an industrial advisory board, group and single client research projects, and consulting. Under patent and licensing agreements the center, inventors, and sponsors share benefits. Quarterly newsletters and research reports are published.

Southwest Research Institute
Materials Science Department
6220 Culebra Road
P.O. Drawer 28510
San Antonio, TX 78284
(512) 684-5111, Ext. 2500

Director: Dr. U. S. Lindholm
Contact: Dr. G. R. Leverant, Assistant Director

The study of polymer composites is included in the department's current research. R&D in the areas of materials development, processing, characterization, application, and service performance are available to industry, and both laboratory and field studies are conducted. Industrial subscription programs, consulting, and single-client research contracts are offered. Clients retain all rights to patents developed.

Composites Laboratories and Information Centers (continued)

Textile Research Institute
P.O. Box 625
Princeton, NJ 08542
(609) 924-3150

President and Director: Dr. Ludwig Rebenfeld
Founded: 1930

The institute is currently researching adhesion- and fiber-reinforced composites. Also, special emphasis is on fiber structure and dye transport processes. Services offered to industry include conferences, symposiums, and individual and multi-client research projects. Sponsored projects are given licensing and patent rights. Along with the *Textile Research Journal,* the institute also publishes many technical reports.

U.S. Army Material Development and Readiness Command
Army Materials and Mechanics Research Center
Director, AMMRC, Attn: ORXMR-PP
Watertown, MA 02172
(617) 925-5527

Contact: David W. Seitz, Technology Transfer Coordinator
Founded: 1966

Composites are currently under research at the AMMRC. Contact the Technology Transfer Coordinator for questions about research contracts. U.S. Government policies apply to all patents and licenses.

University of California
Lawrence Livermore National Laboratory
Box 808
Livermore, CA 94550
(415) 422-6416

Director: R. E. Batzel
Contact: Charles Miller, Interim Technology Transfer & Exchange Officer

Fundamental and applied composite research.

University of Cincinnati
Polymer Research Center
Mail Location 172
Cincinnati, OH 45221
(513) 475-2453

Director: Professor James E. Mark

Reinforced polymer composites are part of the center's current research.

University of Connecticut
Institute of Materials Science
Storrs, CT 06268
(203) 486-4623/4

Director: Leonid V. Azaroff
Founded: 1966

Current research areas include composite materials studies. The institute also offers lectures and workshops in metallurgy and polymer science. Working arrangements with industry include faculty consulting and single- or multi-client research contracts. The institute has liberal licensing policies, including exclusive licenses to sponsors.

University of Delaware
Center for Composite Materials
201 Spencer Laboratory
Newark, DE 19716
(302) 451-8149

Director: R. Byron Pipes, Ph.D.
Contact: William A. Dick, Assistant Director
Founded: 1974

The center deals exclusively with the research of composite materials. An encyclopedia of over 3,000 pages has been authored and distributed by the center. R&D services to industry are directed towards the transfer of composite technology to the workplace. The center also conducts annual workshops and symposiums. Nonexclusive royalty bearing agreements are offered on patents and licenses, but this policy is negotiable. Center publications include *Composite Update,* a quarterly newsletter, and over 20 technical reports a year. Computer programs and a materials micromechanics database, which includes fiber and matrix properties, are other services available to industry.

University of Detroit
Polymer Institute
400 West McNichols Road
Detroit, MI 48221
(313) 927-1270

Director: Dr. Kurt C. Frisch
Contact: Dr. Daniel Klempner, Associate Director
Founded: 1968

Fiber-reinforced and laminate composites are part of the current polymer research. The institute conducts applied as well as basic research and development, on new and improved materials. Polymer coatings are the subject of periodic workshops and conferences. Services offered to industry include faculty consulting, training programs, and individual research contracts. Grants and contracts are generally lengthy and run 1 to 2 years in duration. The institute requires no licensing and all patents are the property of the sponsors.

Composites Laboratories and Information Centers (continued)

University of Illinois at Urbana-Champaign
Materials Engineering Research Laboratory
100 Talbot Laboratory
Urbana, IL 61801
(217) 333-3751

Head: Dr. Frederick V. Lawrence, Jr.
Founded: 1978

The study of the fatigue behavior of composite structures is part of the lab's current research. The lab sponsors The Fracture Control Program and The Materials Processing Consortium as part of their university-industry cooperative program. Along with the usual grants, the lab also offers engineering testing agreements and consulting arrangements. The patent and licensing policies vary. "Material Engineering — Mechanical Behavior," a report series, is distributed to sponsors of industrial programs.

University of Florida
Program in New Materials (Composites)
Department of Engineering Sciences
Gainesville, FL 32611
(904) 392-0961

Head: Dr. L. E. Malvern

Research on filament-reinforced laminated plates and other fiber-reinforced composites includes the dynamic response of composite materials and structures, characterization of damping of composite materials, and the response of composite materials under high-strain-rate load. Individual research contracts are available, as well as consulting services. Patent and licensing policies are flexible.

University of Michigan
Materials and Metallurgical Engineering Department
Dow Building
Ann Arbor, MI 48109
(313) 764-7489

Head: Professor Robert D. Pehlke

Metal, ceramic, polymer, and elastomer composites are under research at the university. Along with individual research contracts, faculty consulting and fellowships are available.

University of Notre Dame
Metallurgy and Materials Science Department
P.O. Box E
Notre Dame, IN 46556
(219) 239-5330

Chairman: Dr. Gordon A. Sargent

Composites are part of the current research at the university. The department offers fundamental research support to industry, along with workshops and seminars. Research contracts and consulting are available and patent/licensing policies are negotiable.

University of Southern California
Polymer Program
Department of Chemical Engineering
University Park, MC 1211
Los Angeles, CA 90089-1211
(213) 743-7051

Departmental Chairman: Professor J. D. Goddard
Contact: Professor R. Salovey, Director Polymer Program
Founded: 1975

The department specializes in the study of polymers and rubberlike materials. Composites, filled copolymers, and blends are also under research. Included in the seminar and workshop programs are instructional television and material science courses. Research contracts are available for specific work, and contracts with other industrial groups are through advisory boards. Faculty consulting is also available. The university usually retains patent and license rights, although this is sometimes negotiated. Some computer programs are under development concerning the prediction of material properties. Industrial support and advisory groups are: Urethane Group; TLARGI (The Los Angeles Rubber Group Inc.) Foundation, 1975-; SCARAB (Sealants, Composites, Adhesives Research Advisory Board), 1984-; Participating University in Adhesive and Sealants Council Fellowship Program, 1984-.

University of Texas at Austin
Center for Materials Science and Engineering
ETC 9.102
Austin, TX 78712
(512) 471-1504

Director: Harris L. Marcus, Ph.D.
Founded: 1983

Basic research is conducted on metals, polymers, ceramics, and semiconductors. Polymer and metal composites are also under study. The center offers short courses, single-client research, faculty consulting, and cooperative teaching/research programs with industry. Patents are owned by the university, but exclusive licenses with or without a royalty to the university are granted, depending on the balance of equity.

Directory of Composites Manufacturers, Suppliers, and Services

Following is a select listing of manufacturers, suppliers, and services pertaining to the various composites industries. Space limitations preclude representing the entire field of composites-related organizations, or companies' entire product or service lines.

Company Name	Address	Phone	Description
A & M Engineered Composites	3 Hayes Memorial Dr. Marlboro, MA 01752	(617) 485-8000	Manufactures Kevlar, fiberglass, and graphite/epoxy structures for military and aerospace applications
Accudyne Engineering & Equipment Co.	Bell Gardens, CA		Supplies molding (compression) presses for laminates and other composite materials
ACI Fiberglass	Box 327 Frankston Rd. Victoria 3175 Australia	Telex No.: AA32275	Manufacturer of fiberglass products, including continuous-filament yarns, rovings, woven rovings, muffler fiber, chopped strands, chopped strand materials
Acoustic Emission Technology Corporation	1812 J Tribute Rd. Sacramento, CA 95815	(916) 927-3861 Telex: 171356	Testing and inspection equipment for composite materials and structures. Instruments for testing composite strengths within ASTM-SPI/CARP-ASME requirements
Acurex Corp., Aerotherm Division	555 Clyde Ave. Mt. View, CA 94039	(415) 964-3200 Telex: 34-6391 CA:FHX TWX: 910-379-6593	Manufactures composite structures and materials for extreme environmental conditions. Produces Chemceram, a family of ceramic composites currently under research, which includes graphite fiber, silica fiber, sapphire filament, and glass-reinforced composites. Also produces ceramic foams
Advanced Composite Products & Technology, Inc.	7415 Mount Jay Huntington Beach, CA 92648	(714) 848-3123	Complete composite fabrication capability, from hand lay-up through commercial rolling to filament winding. A team of aerospace experts supplies technological support. Additional services include design and analysis and marketing functions
Advanced Composite Products, Inc.	37 Washington Ave. East Haven, CT 06512	(203) 469-4647	Product development center for composites and composite structures. Manufactures components for the aerospace industry and provides technology and composite structures for governmental and OEM applications
Aerospace Technologies, Inc.	P.O. Box 50727-T Fort Worth, TX 76105	(817) 451-0620	Manufactures composite structures: honeycomb bonded panels, glass-reinforced plastics, and aircraft parts. Also supplies design engineering services
Airco Carbon, Division of the BOC Group, Inc.	800-0 Theresia St. St. Marys, PA 15857	(814) 781-2611 TWX: 510-693-4515 TELEX: 914513	Manufactures carbon and graphite tooling products. Features tooling used for fabrication of reinforced plastic composites for aerospace applications
Airtech International	Carson, CA		Features a line of layup materials for ovens and autoclaves used in the manufacture of composites. Operations include tooling and bonding shops
Aluminum Company of America, Chemicals Division	1501 Alcoa Building Pittsburgh, PA 15219		Produces sintered tubular alumina in crushed, graded, and ground sizes. Applications include use as base material in refractories. Also produces aluminum trihydrate, a filler used in plastic systems

Directory of Composites Manufacturers (continued)

Company Name	Address	Phone	Description
Amalga Corporation	10600 W. Mitchell St. West Allis, WI 53214	(414) 453-9555 Telex: 26-9503	Produces Amalgon, a filament-wound, glass fiber-reinforced epoxy resin tubing for hydraulic cylinder applications. Black Amalgon tubing has self-lubricating features and is used for low-pressure air and water applications
Amercom, Inc.	8949 Fullbright Ave. Chatsworth, CA 91311	(213) 882-4821	Fabricates metal and ceramic matrix composites. Products include aluminum, titanium, magnesium, and copper matrix materials reinforced with boron, silicon carbide, and graphite fibers, and filaments
American Cyanamid Co.	Engineered Materials Department One Cyanamid Plaza Wayne, NJ 07470	(201) 831-2000	Manufactures Dura-Core aluminum honeycomb and Cycom structural resins and prepregs with glass, quartz, ceramic, graphite, and aramid fibers
American Klegecell Corp.	204 N. Dooley Grapevine, TX 76051	(817) 481-3547 Telex: 73318	Produces high strength-to-weight ratio polyvinyl foam cores for composite uses
American Tempering, Inc.	33200 Western Ave. Union City, CA 94587	(415) 471-6811 Telex: 336302	Manufactures full range of glass products including laminates. Also produces polycarbonate laminates
Amoco Chemicals Corporation	200 East Randolph Dr. Chicago, IL 60601	(800) 621-4557 312/856-3414	Manufactures Torlon, poly (amide-imide) resins in glass, carbon, or graphite fiber-reinforced grades. Other grades are reinforced or filled with graphite powders and fluorocarbons for wear-resistant properties
The Andersons	P.O. Box 119 Maumee, OH 43537	(800) 472-3220 (800) 532-3370	Produces corn cob granule extenders and fillers for plastics as well as other industrial, chemical and agricultural applications
Andus Corporation	Canoga Park, CA		Develops and manufactures thin-film coatings on polymer substrates. Products are used in hi-tech applications
Applied Plastics Company, Inc.	612 E. Franklin St. El Segundo, CA 90245	(213) 322-8050	Features epoxy resins and hardeners for use in composite construction
Applied Polymer Technology, Inc.	6078-B Corte Del Cedro Carlsbad, CA 92008	(619) 438-8977 Telex: 821426	Manufactures composite fabrication control systems
Arco Chemical Company — Advanced Materials	Route 6, Box A Greer, SC 29651	(803) 877-0123	Manufactures high-performance silicon carbide-reinforced aluminum composite products
Artech Corp.	2901 Telestar Ct. Falls Church, VA 22042	(703) 560-3292	Products include: ceramic-to-metal seals and composites, thin-film resistors, ceramic adhesives and coatings, radiation detectors, and consumer protection equipment
Arvin Industries, Inc., Arvinyl Division	Department T Columbus, IN 47201	(812) 379-3000	Manufactures vinyl-to-metal laminates and glass-reinforced nylon and polypropylene
Asahi Nippon Carbon Fiber Co. Ltd.	Chiyoda-Ku Tokyo, Japan		Manufactures chopped carbon fiber and high-strength, high-strain continuous carbon fiber. Also produces Nicalon, a silicon carbide fiber
Astro Met Associates, Inc.	9974 Springfield Pike Cincinnati, OH 45215	(513) 772-1242 Telex: 23295526	Supplies research, development, and production of ceramics, including cermets

Company Name	Address	Phone	Description
Atacs Products, Inc.	Renton, WA		Features composite repair systems
Atlantic Research Corporation	5390 Cherokee Ave. Alexandria, VA 22314	(703) 642-4530	Supplies engineering, application analysis, testing, and fabrication services for advanced composite materials
Automation/Sperry	20327 Nordhoff Street Chatsworth, CA 91311	(818) 882-2600	Manufactures state-of-the-art ultrasonic inspection systems for composites and bond testing
Avco, Specialty Materials Division	2 Industrial Ave. Lowell, MA 01851	(617) 452-8961	Produces boron, graphite and silicon carbide reinforced aluminum and titanium. Also manufactures epoxy prepregs, fire protection, and graphite materials
Bentley-Harris Mfg. Co.	241 Welsh Pool Rd. Lionville, PA 19353	(215) 363-2600	Manufactures braided and woven reinforcements and materials of graphite, Kevlar, glass, ceramic, thermoset prepregs, PEEK/graphite, PPS/graphite, nylon/glass, and metal strip. Also provides design technology and is active in R&D with major industries and universities
Bisco Products, Inc.	Park Ridge, IL		Manufactures specialty silicone products, including a number of silicone coated fiberglass products
Boeing Technology Services	P.O. Box 3707 M/S 9R-28 Seattle, WA 98124	(206) 237-4490 Telex: 32-9430	Features advanced, automated manufacturing, and fabrication of composite structures. Also serves the sporting goods industry with analysis and consultation for use of composites
Boeing Aerospace Company	P.O. Box 3999 M/S 3E-79 Seattle, WA 98124	(206) 773-5889	Composite fabrication technology includes layup, filament winding, injection and compression molding, foams and bonding capabilities. Facilities also include testing and inspection equipment and technology
Bonded Technology, Inc.	One Alcap Ridge Rd. P.O. Box 160 Cromwell, CT 06416	(203) 635-1150	Custom designing, engineering, and manufacturing of bonded honeycomb and fiber composites to aerospace and governmental specifications. Production technology includes two autoclaves, two ovens, platen presses, FPL etch line, and phosphoric acid anodizing. Extensive use of Kevlar graphite and S-glass in composite applications
Boots Company PLC	Nottingham N62 3AA England	0602 506255 Telex: 377811	Manufactures thermosetting bismaleimide resins used primarily for fiber-reinforced composites. Also produces structural laminates and parts for aerospace, electrical, and thermal industries
Borg-Warner Chemicals, Inc.	International Center Parkersburg, WV 26102	(304) 424-5411	Manufactures phenylene ether copolymer alloys and Prevex in reinforced and nonreinforced grades
Bridon Composites Limited	Fairoak Lane Whitehouse Industrial Runcorn Estate Runcorn, Cheshire WA7 3DU England	Telex: 629765 701515 STD Code 0928	Manufactures and distributes high-performance woven, braided, and knitted fabrics, pultrusions, and thermoplastic products
Briskheat	Columbus, OH		Features line of curing and hot debulking composite equipment

Directory of Composites Manufacturers (continued)

Company Name	Address	Phone	Description
Bristol Composite Materials, Inc.	P.O. Box 789 1363 North Gaffey St. San Pedro, CA 90733-0789	(213) 514-3755 Telex: 691-549	Manufactures advanced composite components for the aerospace industry. Also provides designing and testing services
L. J. Broutman & Associates Ltd.	3424 S. State St. Chicago, IL 60616	(312) 842-4100	Mechanical testing, chemical/thermal analysis, failure analysis, stress analysis, and design and R & D in polymers and composites
Burnham Products, Inc.	4203-T W. Harry St. P.O. Box 12950 Wichita, KS 67277	(316) 942-3208	Features Fiber-Lok composite tooling reinforcements. Also manufactures honeycomb and laminate composites of fiberglass and graphite
C-E Refractories, Division of Combustion Engineering, Inc.	P.O. Box 828 Valley Forge, PA 19482	(215) 337-1100	Manufactures ceramic fibers
The Carborundum Co.	P.O. Box 156 Niagara Falls, NY 14302	(716) 278-2000	Advanced materials division produces high-performance ceramics, and composite materials and structures
Celion Carbon Fibers, Division of Celanese Corporation	Cherry Road Station Rock Hill, SC 29730	(803) 366-5656	Manufactures ultrahigh- and intermediate-modulus carbon and graphite fibers for applications in the aerospace aircraft, recreation, marine, and industrial markets
Chem-Tronics Inc.	1150 W. Bradley P.O. Box 1604 El Cajon, CA 92022	(619) 448-2320	Produces polyimide foams, adhesives, and laminating resins. Also manufactures Unistructure, a thermoplastic, titanium-reinforced composite, and other lightweight metal structures of aluminum, stainless steel, and nickel-base alloys for aerospace industries
Ciba-Geigy Corporation	Composite Materials 10910 Talbert Ave. Fountain Valley, CA 92708	(714) 964-2731	Manufactures impact- and damage-tolerant resin systems. Also features custom Kevlar, and graphite weavings, hybrid constructions, woven and unidirectional materials of Kevlar, graphite, Nomex, and high-temperature fiberglass, honeycomb core materials, fabricated sandwich panels, laminates, and preimpregnated fiberglass
Cincinnati Milacron Industries Inc.	Electronic Circuit Materials Division Route 28 Blanchester, OH 45107	Telex: 214-472	Manufactures glass-reinforced polyester and copper-clad laminates. Also produces aircraft parts of composite materials with the aid of the Milacron 7-axis, a numerically controlled computer
Composition Materials Co. Inc.	26 Sixth St. Stamford, CT 06905	(203) 324-0000 Telex: 131-454	Features a broad range of fillers, extenders, and abrasives, including cotton and cellulose fiber fillers for plastics
Comp-Tite SPS Aerospace & Industrial Products Division	Highland Ave. Jenkintown, PA 19046	(215) 572-3000	Manufactures blind fastener systems for use with advanced composites. Compatible with graphite-fiber composites
Cosby Newsom Associates	Norwalk, CA		Designers and manufacturers of composite debulking and curing equipment

Company Name	Address	Phone	Description
Custom Coating and Laminating Corp.	77 Goddard Industrial Park 717 Plantation St. Worcester, MA 01605	(617) 852-3072 Telex: 955329	Provides custom coatings and laminations
Cyro Industries	P.O. Box 8588 Woodcliff Lake, NJ 07015	(201) 285-1544 (800) 631-5384	Manufactures polymethacrylimide foam core materials for sandwich construction
DWA Composite Specialties, Inc.	21119 Superior St. Chatsworth, CA 91311	(818) 998-1504	Specialists in metal matrix composites and hardware. Manufactures DWG, the only thin-ply, zero CTE uniaxial graphite aluminum products available
Delsen Testing Laboratories, Inc.	1024 Grand Central Ave. Glendale, CA 91201	(213) 245-8517	Mechanical testing of advanced composites. Other testing includes thermal analysis, flammability testing, HPLC-GPC, electrical, metallographic, and environmental testing
Diab-Barracuda, Inc.	Grand Prairie, TX		Manufactures polyvinyl foam core for use at the high-temperature ranges required in the aerospace industry
Dow Corning Corporation	Department A0021 P.O. Box 0994 Midland, MI 48640	(800) 447-4700	Through various subsidiaries, manufactures thermoset resin matrix systems and high-technology silicone materials for composite applications. Also manufactures ceramic specialty products for the ceramics industry and Nicalon SiC Fibers
Dresser Industries	Dresser Manufacturing Division Nil Cor Operations P.O. Box 2058 Alliance, OH 44601	(216) 823-0500 Telex: 98-3461	Produces Nil-Cor ball and butterfly valves from composite materials. Valves are constructed from graphite fiber-reinforced vinyl ester, glass fiber-reinforced vinyl ester, and graphite fiber-reinforced ryton polyphenylene sulfide
E.I. du Pont de Nemours & Co.	Textile Fibers Department Kevlar Special Products Centre Road Building Wilmington, DE 19898	(302) 999-3728	Manufactures Kevlar (aromatic polyamide) fibers, fabrics, and preimpregnated thermoplastic resins, and Nomex aramid honeycomb. Ceramic (fiber FP) and pitch carbon fibers are currently under development
Duramic Products Inc.	426 Commercial Ave. Palisades Park, NJ	(201) 947-8313 Telex: 710-991-9632	Custom fabricators of ceramic composites and other high-temperature ceramics and materials
Eurocarbon Tilburg Bu	Tilburg, Holland		Produces tubular and flat fiber braidings, woven unidirectional, and bidirectional tapes and woven cloth, of carbon fiber, aramid, E of S2 glass, silicon carbide, and ceramics. Also features hybrid constructions and high-temperature thermoplastic strips
Ed Fagan, Inc.	Component Materials Division 33 Whitney Rd. Mahwah, NJ 07430	(201) 891-4003	Manufactures metal-to-metal laminates. Also produces ceramic and glass sealing alloys
Emser Industries	P.O. Box 1717 Industrial Park & Corporate Way Sumter, SC 29151	(803) 481-3172 Telex: 805077	Specializes in general purpose and specialty nylon engineering thermoplastics. Produces Grilon(R), a nylon 6 resin and Grilamid(R), a nylon 12 resin in reinforced and unreinforced grades

Directory of Composites Manufacturers (continued)

Company Name	Address	Phone	Description
Emerson & Cuming Inc.	Canton, MA 02021	(617) 828-3300	Features a line of ceramic microsphere products, Eccospheres, for plastic, ceramic, glass, plaster and concrete applications
Ferro Corporation, Composites Division	8790 National Blvd. Culver City, CA 90232	(213) 870-7873	Fabricates epoxy, polyimide, phenolic, and polyester resin prepregs with graphite, glass, silica, quartz, aramid, silicon carbide, alumina, and other ceramic reinforcements. Also manufactures broad goods, unidirectional tapes, rovings, molding compounds, and adhesive films
Fiber Composites, Inc.	101-T S. Highland Greens Port Ludlow, WA 98365	(206) 437-2152 Telex: 215406	Provides filament winding services
Fiber Innovations, Inc.	Norwood, MA		Manufactures triaxial braided fiber reinforcements. Designs and develops fibrous structures for aerospace, industrial, and sporting goods industries
Fiber Materials, Inc.	Biddeford Industrial Park Biddeford, ME 04005	(207) 282-5911 Telex: 944480	Produces advanced composite materials, including carbon/carbon and resin and metal matrix composites. Also available are carbon and graphite felts and Microfil high-modulus graphite fibers
Fiber Science, Inc./ Edo Corp.	506 N. Billy Mitchell Rd. Salt Lake City, UT 84116	(801) 539-0747	Manufactures composite structures
Fiberite Corporation	501 West 3rd St. Winona, MN 55987	(507) 454-3611	Supplies advanced composite materials for aerospace and recreational industries. Also provides design assistance, materials selection, product development, analysis and testing, fabrication processes and production troubleshooting
Frekote, Inc., A subsidiary of The Dexter Corporation, Hysol Division	1701 Spanish River Blvd. West Boca Raton, FL 33431	(305) 395-3083 Telex: 51-8918	Produces wide range of mold-release additives for plastics, advanced aerospace polymer composites, and rubber/elastomers
GAF Corporation	1361 Alps Road Wayne, NJ 07470	(201) 628-3000	Extensive line of reinforced and unreinforced engineering plastics. Products include Gafite, a thermoplastic polyester, and Gaflex, a thermoplastic polyester elastomer
General Electric, Plastics Operations	One Plastics Ave. Pittsfield, MA 01201	(413) 494-1110	Manufactures Valox, Noryl, and Xenoy thermoplastic glass fiber-reinforced and nonreinforced resins. Other resin systems include thermoplastic foams, polycarbonates, and polyetherimides, marketed under the tradenames Lexan and Ultrem
GGT	Aerospace Division 55 Gerber Rd. South Windsor, CT 06074	(203) 644-2401 Telex: 643-771	Produces advanced composite manufacturing systems for the aerospace industry
Genstar Stone Products Co.	Executive Plaza IV Hunt Valley, MD 21031	(301) 628-4000	Produces various grades of calcium carbonate for plastics, paint and paper industries

Company Name	Address	Phone	Description
D.A. Gordon Co.	Downy, CA		Supplies composite and polymer testing equipment
Gordon Plastics, Inc.	2872 S. Santa Fe Vista, CA 92083	(619) 727-2008	Provides custom manufacturing and design of FRP laminates and structural products
Great Lakes Carbon Corp.	360 Rainbow Blvd. S. P.O. Box 727 Niagara Falls, NY 14302	(800) 828-6601 (716) 285-5200	Produces Fortafil carbon fibers and prepreg composites. Fibers available in continuous filament tows and chopped strands. Also manufactures graphite and coke particles and graphite structural rods and shapes
Greene, Tweed & Co.	North Wales, PA 19454	(215) 256-9521 Telex: 6851164	Produces Arlon 1000, a high temperature semi-crystalline thermoplastic in reinforced & non-reinforced grades, useful for seal applications, high-temperature connectors and insulators, valve slats and compressor rings
Greenleaf Technical Ceramics	25019 Viking St. Hayward, CA 94545		Offers a complete line of full-density, high-purity ceramic composite materials, pressed and ground to specifications
Grumman Aerospace Corp.	S. Oyster Bay Rd. Bethpage, NY 11714	(516) 575-0574 Telex: 961-440	Produces advanced composites for aerospace applications
Hartford Steam Boiler, AE International Division	One State St. Hartford, CT 06102	(203) 722-1866 Telex: 99354	Provides acoustic emission testing services and equipment for composites
Heatcon, Inc.	Seattle, WA		Supplies advanced composite repair equipment
Heath Tecna Precision Structures, Inc.	19819 84th Ave. S Kent, WA 98031	(206) 872-7500	Plastic and fiberglass fabricators
Hercules, Inc.	Hercules Aerospace Division Hercules Plaza Wilmington, DE 19899	(302) 594-5000	Manufactures Magnamine graphite fibers and products. Provides composite structures for aircraft, missiles, and satellites. Also produces resins and esters and epoxy prepreg tapes
Hexcel Trevarno	P.O. Box 888 Lancaster, OH 43130	(614) 653-1528 Telex: 24-5391	Produces bonded honeycomb sandwich constructed composite materials, preimpregnated fabrics, resin systems, and fiberglass textiles
Hi-Tech Composites, Inc.	5447 Equity Ave. Reno, NV 89502	(702) 786-8666	Manufactures fabrics of polyester, glass, carbon, and Kevlar for the automotive, aerospace, marine and recreational markets
HITCO Materials Division	P.O. Box 1097 Gardena, CA 90249	(800) 243-8160	Manufactures epoxy and polyimide Kevlar-reinforced resin materials. Also produces carbon/carbon, ablative, honeycomb, and bonded structures and materials
J.M. Huber Corporation	P.O. Box 2831 Borger, TX 79008		Specializes in reinforcing additives. A new amorphous silica fiber/whisker product is under development and will be available in mid-1986
Hysol Grafil Co.	2850 Willow Pass Rd. Pittsburg, CA 94565	(415) 938-5533	Manufactures resins including SynCore, a group of syntactic thermosetting films and pastes used in advanced composite systems. Also produces carbon/graphite fibers through its Courtaulds' Carbon Fibers Division

Directory of Composites Manufacturers (continued)

Company Name	Address	Phone	Description
ICD Group Inc.	641 Lexington Ave. New York, NY 10022	(212) 644-1500	Specialists in raw materials supply to the ceramic industry including refractories, abrasives, composites, cutting tools, electronics and glass. Also distributes Tateho SiC and Si_3N_4 whiskers in U.S.
ICI Americas, Inc.	Concord Pike & New Murphy Rd. Wilmington, DE 19897	(302) 575-4466	Manufactures alumina fiber in bulk and mat forms for ceramic, glass, and metal reinforcement. Also produces polymer resins
Instron Corp.	100 TR Royall St. Canton, MA 02021	(617) 828-2500 Telex: 92-4434	Supplies instruments and systems for testing and evaluating advanced materials
KDI Composite Technology, Inc.	6881 Eighth St. Buena Park, CA 90620	(714) 739-8045	Manufactures fiberglass-reinforced plastics and composites demonstrating laminating, compression molding and metal bonding
Kaiser Aerotech	P.O. Box 1678 San Leandro, CA 94577	(415) 562-2456	Manufactures carbon/carbon composite structures
Koppers Company, Inc.	1250 Koppers Bldg. Pittsburgh, PA 15219	(412) 227-2000	Produces extensive line of reinforced and unreinforced polyester resin systems. Continued study of polyester resins is conducted at a centralized research facility
LNP Corporation	412 King St. Malvern, PA 19355		Produces Thermocomp, carbon, and glass-reinforced thermoplastic resins, LNP internally lubricated reinforced-thermoplastics and filled fluoropolymer composites. Also manufactures statically conductive and EMI attenuating composites
Lamination Technology, Inc.	2720 S. Main St. Santa Ana, CA 92707	(714) 556-1460	Products include: mass lamination for multi-layers, laminated copper-clad sheets, and epoxy fiberglass prepregs
Laminations	P.O. Box 1874 Auburn, WA 98002	(206) 833-8200	Features custom fiber-resin laminating and metal bonding. Also manufactures parts and assemblies
Laserage Technological Group	4201 Grove Ave. Gurnee, IL 60031	(312) 249-5900	Provides laser machining of all materials including composites, plastics, most metals, ceramics, sapphire, and quartz
Leesona Corp.	Regional Center Hwy. 521 Fort Hill, SC 29715	(401) 739-7100 Telex: 927715	Manufactures fiber producing machinery for textile and plastics industries
Lewcott Chemicals & Plastics	P.O. Box 319 Millbury, MA 01527	(617) 865-4466	Produces composite prepreg materials and phenolic foam
Lipton Steel & Metal Products Inc.	454 South St. P.O. Box 1159 Pittsfield, MA 01202	(413) 499-1661 Telex: 955317	Supplies bonding and laminating autoclave systems
C.A. Litzler Co. Inc.	4802 W. 160 St. Cleveland, OH 44135	(216) 267-8020 Telex: 98-0234	Manufactures and designs continuous process ovens for composite materials production
Lockheed Missiles & Space Co. Inc.	P.O. Box 504 Sunnyvale, CA 94086	(804) 742-4321	Produces advanced structural composites for the aerospace industry

Company Name	Address	Phone	Description
Lord Corporation	Chemical Products Group 2000 West Grandview Blvd. Box 10038 Erie, PA 16514	(814) 868-3611 Telex: 91-4445	Produces structural adhesives, coatings, binders, mold-release agents, and specialty chemicals for polymers and elastomers
Lunn Industries, Inc.	Lunn Bldg. Wyandanch, NY	(516) 643-8900	Features molding and fabrication of reinforced plastics
Lydall, Inc.	Composite Materials Division 615 Parker St. Manchester, CT 06040	(203) 646-1233	Manufactures fiber board, fiber composites, and fiber/polymer alloys
MFG, Molded Fiber Glass Companies	P.O. Box 675 Ashtabula, OH 44004		Over 80% of manufacturing is devoted to custom-molding of FRP. The remaining capacity is applied to proprietary products including flat sheet and ribbed panels
MTS Systems Corp.	P.O. Box 24012 TR Minneapolis, MN 55424	(612) 937-4000	Supplies systems for material testing
Manville	P.O. Box 5108 Denver, CO 80217	(303) 978-4900	Manufactures fiberglass fiber reinforcements and products. Also produces plastic and plastic material fillers and refractory products
Marshall Consulting, Inc.	720 Appaloosa Dr. Walnut Creek, CA 94596	(415) 945-6051	Offers assistance in the construction, design, production, tooling, and selection of materials for sandwich construction. Also offers in-plant seminars
Martin-Marietta Corporation	6801 Rockledge Dr. Bethesda, MD 20817		Aerospace division manufactures advanced composite structures. Other divisions produce ceramic and cement industry-related supplies and products
Material Concept, Inc.	666 N. Hague Ave. Columbus, OH 43204	(614) 272-5785	Provides research and development in the field of metal matrix composites. Also produces graphite machined parts, tool and dies
Materials Sciences Corp.	Gwynedd Plaza II Spring House, PA 19477		Offers research, design, and analysis services for materials, structures and composites
Metglass Products, Division of Allied Corporation	6 Eastmans Rd. Parsippany, NJ 07054	(201) 581-7700 Telex: 136044	Supplies Metglas amorphous alloy ribbons, which are produced through rapid solidification of molten metals at cooling rates of about a million degrees centigrade per second. Composite applications are being explored with promising results being reported
Mobay Chemical Corporation	Mobay Rd. Pittsburgh, PA 15205	(412) 777-2000	Produces a wide range of plastics and polymer additives. Features reinforced and nonreinforced grades of polycarbonates, ABS blends, and polyesters. Products are marketed under Pocan, Merlon, Petlon, Makroblend tradenames

Directory of Composites Manufacturers (continued)

Company Name	Address	Phone	Description
Monsanto Polymer Products Co.	800 N. Lindbergh Blvd. St. Louis, MO 63167	(314) 694-1000	Produces a wide range of reinforced and nonreinforced thermoplastic resins. Also manufactures Santoprene, a thermoplastic rubber compound. Engineering resins include Nyrim, a rubber modified nylon copolymer, Lustran ABS resins, polystyrene, Lustrex resins, Cadon resins, and Vydyne glass and mineral-reinforced nylon and polyamide resins
Mossberg Industries, Inc.	160 Bear Hill Road Cumberland RI 02864	(401) 333-3000	Manufactures the New England Butt composite maypole braider for continuous braiding of advanced composite fibers. Machines can be custom designed for specific requirements
Mutual Industries	1400 Goldmine Rd. Monroe, NC 28110	(704) 283-2147	Manufactures knit and woven graphite, Kevlar, carbon, and fiberglass tapes
NYCO	Box 368 Willsboro, NY 12996	(518) 963-4262	Produces mineral fillers including mica particulate, alumina trihydrate and wollastonite, calcium metasilicate for the plastics industry. Also produces glass fillers including hollow microspheres
Narmco Materials, Inc.	1440 N. Kraemer Blvd. Anaheim, CA 92806	(714) 630-9400 Telex: 18307	Features structural adhesives and advanced composites. High-modulus Rigidite resin systems include epoxies and bismaleimides reinforced with carbon, glass, and Kevlar fibers. Structural Narmco plastics include reinforced modified epoxies and phenolics. Also provides extensive R & D in the composite area
National Beryllia Corporation	Greenwood Ave. Haskell, NJ 07420	(201) 839-1600	Manufactures ceramic composite structures including ceramic-to-metal and glass-to-metal assemblies. Also produces pure oxide body composites
National-Standard	Woven Products Division P.O. Box 1620 Department IT Corbin, KY 40701	(606) 528-2141 Telex: 218486	Produces fine nickel fibers in loose or in sinter bonded mat forms. Also manufactures conductive fiber fabrics
National Starch & Chemical Corporation	Finderne Ave. Bridgewater, NJ 08807	(201) 685-5122	Manufactures Thermid, polyimides, and polyisoimide resins for high-performance structural composites, adhesives, and coatings
Newport Plastics, Subsidiary of Kidde, Inc.	Box 466 Derby Rd. Newport, VT 05855	(802) 334-7941	Specializes in low- to medium-volume production of fiberglass-reinforced plastic housings. Offers complete engineering and manufacturing services
Ontario Die Company of America	2735 20th St. P.O. Box 376 Port Huron, MI 48060	(313) 987-5060	Supplies cutting dies for composites. Includes cutting dies for graphite, Kevlar, fiberglass, and honeycomb core materials
PPG Industries	Fiber Glass Products One Gateway Center Pittsburgh, PA 15222	(412) 434-3131	Manufactures fiberglass reinforcements for thermoset and thermoplastic resins. Fibers available in rovings, chopped strands, yarns, mat and filament winding systems. Also produces Azdel, a polypropylene glass-reinforced laminate moldable sheet
PQ Corporation	P.O. Box 840 Valley Forge, PA 19482	(215) 293-7200	Produces Q-Cel, siliceous and ceramic hollow microsphere fillers for plastics and cements. Features wide range of particle sizes and densities

Company Name	Address	Phone	Description
Perkin-Elmer Corporation	Main Ave. Norwalk, CT 06856	(203) 762-1000 Telex: 965-954	Manufactures and designs scientific instruments, including testing equipment for composites, metals, and polymer applications
Pfizer, Inc.	235 E. 42nd St. New York, NY 10017	(800) 421-6330	Produces a line of talc and calcium carbonate fillers for use as extenders and reinforcements in plastics
Phillips Petroleum Co.	91-G Research Center Bartlesville, OK 74004	(918) 661-6600	Produces thermoplastic composite materials reinforced with carbon or glass fibers. Is the only manufacturer of glass- and mineral-reinforced polyphenylene sulfide, also known as Rytron
Physical Acoustics Corporation	819 Alexander Rd. P.O. Box 3135 Princeton, NJ 08540	(609) 452-2510 Telex: 642236	Designs and manufactures acoustic emission systems for testing composites and metals
Pierce & Stevens Chemical Corp.	710 Ohio Street Buffalo, NY 14240	(716) 631-8991 Telex: 91-202	Features Miralite, an ultra-low density PVDC microsphere in preexpanded form. Unexpanded micropheres are also available
Pollux Corporation	8280 Patuxent Range Road Jessup, MD 20794	(301) 953-2008	Manufactures aluminum honeycomb core for composite construction
Potters Industries, Inc.	377 Route 17 Hasbrouck Hts., NJ 07604	(201) 288-4700 Telex: 133447	Manufactures solid glass microsphere particles for additives and reinforcements of plastics
Quantum Composites, Inc.	4702 James Savage Rd. Midland, MI 48640	(517) 496-2884	Manufactures Lytex, an epoxy SMC reinforced with chopped glass fibers. Also produces Moulage, an epoxy composite, single component extruded sheet tooling compound, which eliminates layup method of creating mold tooling
Refractory Composites, Inc.	12220-A Rivera Rd. Whittier, CA 90606	(213) 696-8061	Specialists in the design, engineering, and manufacturing of ceramic composite structures, from inception to finished product. All aspects of customer needs and product applications are reviewed. Resources include consultants throughout industry who can assist in any phase of ceramic composites development
Reichhold Chemicals Inc.	Resins & Binders Division P.O. Box 1433 Pensacola, FL 32596	(904) 433-7621 Telex: 702-424	Features epoxy and phenolic resins, as well as glass cloth and fiber materials
Resinoid Engineering Corp.	7557 N. St. Louis Ave. Skokie, IL 60076	(312) 673-1050	Manufactures materials for engineering applications, including asbestos, glass, fabric, and mineral reinforcements
Rheometrics, Inc.	2438 U.S. Highway No. 22 Union, NJ 07038	(201) 687-4838 Telex: 138-816	Manufactures rheological and impact testing instrumentation
Ribbon Technology Corp.	Box 30758 Gahanna, OH 43230	(614) 864-5444 Telex: 246-518	Markets carbon and stainless steel fibers as reinforcements for Portland Cement concretes and refractory castables. R&D in flakes, fibers, foils, and powder materials for composite applications

Directory of Composites Manufacturers (continued)

Company Name	Address	Phone	Description
Rockwell International	North American Aircraft Operations 100 N. Sepulveda Bl. P.O. Box 92098 Los Angeles, CA 90009	(213) 647-1000 Telex: 664363	Manufactures composite and metallic structures for the aerospace industries
Rogers Corporation, Molding Materials Division	Box 550 Manchester, CT 06040	(203) 646-5500	Produces fiber-reinforced thermoset molding compounds. The University of Delaware Center for Composite Materials assists with R & D efforts
RTP Co.	580 East Front St. P.O. Box 439 Winona, MN 55987	(507) 454-6900 Telex: 910-565-2276	Manufactures reinforced thermoplastics from a variety of over 24 resins and a large selection of fillers and fibers
SEP	Tour Roussel Nobel Cedex No. 3 F-92080 Paris France		Develops and manufactures carbon/carbon and ceramic materials for aircraft structures
Shur-Tronics	2541 White Rd. Irvine, CA 92714	(714) 474-6000	Manufacturer of Compositest, a composite testing system with user friendly software
Stackpole Fibers Co., Inc.	Foundry Industrial Park Lowell, MA 01852	(617) 454-0409	Produces carbon fibers and carbon fiber products. Panex products include carbon fiber fabrics, spun yarns, chopped carbon fibers, continuous carbon fiber filament, and tow, and carbon fiber paper. Also produces carbon fiber-reinforced injection-molded parts for industrial applications
A.E. Staley Mfg. Co., Polymerizable Products Department	2200 East Eldorado St. Decatur, IL 62525	(217) 423-4411	Features Stalink polymerizable cellulosics for FRP resin binders and tie coats for FRP fillers and fibers
Stevens Products, Inc.	128 North Park St. East Orange, NJ 07019	(201) 672-2140	Manufactures ready-to-use shapes of high-modulus graphite epoxy, glass epoxy, and fiberglass laminates
Tateho Chemical Industries Co., Ltd.	974 kariya. Ako-shi Hyogo-ken, Japan	07914-5-2041 Telex: 5778625	Manufactures SiC and Si_3N_4 single crystal ceramic whiskers. Also produces other ceramic and refractory materials
Thermal Equipment Corp.	1301 W. 228th St. Torrance, CA 90501	(213) 775-6745 Telex: 182577	Designers and manufacturers of autoclave systems for composites, as well as fabricators of hydroclaves and composite presses
3-M Ceramic Fiber Products	225-4N 3-M Center St. Paul, MN 55144	(612) 733-1558	Produces continuous polycrystalline metal oxide ceramic fibers, Nextel. Fibers are available in rovings, ply-twisted yarns, tapes, sleevings, fabrics, and special constructions
Tiodize Co., Inc.	15701 Industry Lane Huntington Beach, CA 92649	(714) 898-4377	Features graphite composites, lubricants, and coatings
Unicel Corporation	Department TR 1520 Industrial Ave. Escondido, CA 92025	(619) 741-3912	Manufactures aluminum, low-carbon steel, and stainless steel honeycomb cores

Company Name	Address	Phone	Description
Union Carbide Corporation	Old Ridgebury Rd. Danbury, CT 06817	(213) 794-5300	Produces carbon fibers and carbon fiber composite products
Versar Manufacturing Inc.	Specialty Products Division 6850 Versar Center P.O. Box 1549 Springfield, VA 22151	(703) 750-3000	Manufactures conductive or nonconductive composite filler particles of carbon
WSF Industries, Inc.	Box 400, Kenwood Dr. Buffalo, NY 14217	(800) 874-8265 (716) 692-4930	Engineers and manufactures composite bonding equipment
Washington Penn Plastic Co., Inc.	2080 North Main St. P.O. Box 236 Washington, PA 15301	(412) 228-1260	Features calcium carbonate, talc, mica, glass, and combination filled polypropylene and polyethylene resins
Westinghouse Electric Corp., Insulating Materials Div.	West Mifflin, PA 15122	(412) 256-3975	Manufacturers fiberglass-reinforced polyester composite shapes. Polyglass structures are produced through the pultrusion process and are available in general and flame-retardant grades. Over 100 shapes are available as stock items
Westlake Plastics Co.	P.O. Box 127 161 W. Lenni Rd. Lenni, PA 19052	(215) 459-1000 Telex: 83-5406	Features extrusion of engineering thermoplastics and thermosets. Extrudes thermosetting plastics with high-performance fibers/fillers
Wilson-Fiberfil International	P.O. Box 3333 2267 W. Mill Rd. Evansville, IN 47732	(812) 424-3831 Telex: 752708	Supplies an extensive and complete line of electrically conductive thermoplastic compounds. Electrofil(R) resins are reinforced with carbon fibers, carbon black, aluminum flakes, nickel coated carbon fibers or stainless steel fibers
Zircar Products, Inc.	110 North Main St. Florida, NY 10921	(914) 651-4481 Telex: 996-608	Produces zirconia, alumina, and alumina-silica fiber products. Also supplies alumina and zirconia bulk fibers, refractory sheets, rigidizers and cements and textiles of yttrium oxide, hafnium oxide, tantalum oxide, titanium oxide, and cerium oxide

International Standards-Issuing Organizations

Australia

Standards Association of Australia
Standards House
8085 Arthur Street
North Sydney, NSW 2060
Telex: 26514 astan
Abbreviation: SAA

SAA standards are used by firms engaged in business in Australia and the southwest Pacific area. Australian standards appear as 1- to 4-digit numerical codes preceded by the *upper case* letters AS. Designations may also appear with the standard and should be separated by a space.

Example: AS 1446
AS 1565 80 A
AS 1867 1050

Austria

Osterreichisches Normungsinstitut
Leopoldsgasse 4, Postlach 130
A-1021, Wein
Telephone: (0222) 33 55 19-0
Telex: 75960
Abbreviation: ON

The Austrian Standards Institute was founded in 1920 and creates and publishes Austrian standards. The organization also recommends foreign standards for use in Austria. Report the Austrian standards and designations in the following way:

Example: ONORM M 3430
ONORM M 2429
ONORM A199.98
ONORM S-A199.8
ONORM AlMgSiO,5

Belgium

Institut Belge de Normalisation
Avenue de la Brabanconne 29
1040 Bruxelles
Telephone: (02) 734 92 05
Abbreviation: IBN

Created in 1946, the IBN consists of approximately 600 members. Both individuals and businesses prepare and propose standards for Belgium. Examples of IBM standards and their reporting methods are as follows. All letters should be in *upper case.*

Example: NBN D 02-002
NBN T 52-074
NBN N 22-101

Brazil

Associacao Brasileira de Normas Technicas
Av. 13 de Malo no 13-28° andar
Caixa Postal 1680, CEP 20 003
Rio de Janeiro
Abbreviation: ABNT

The Standards and designations of Brazil begin with the upper case letters ABNT and are followed by a numerical or alphanumerical code.

Example: ABNT 03.049
ABNT b4
ABNT CaSn4Pb4Zn4P

Bulgaria

State Committee for Standardization
at the Council of Ministers
21 6th September Street
Sofia
Telephone: 85 91
Telex: 22570 dks bg
Abbreviation: DKC

Bulgarian Standards designations begin with the *upper case* letters BDS and are followed by the standard's numerical code or the designations alphanumeric code, or both. If a designation is included, leave a space between the designation and the standard.

Example: BDS 7938
BDS 7938 9G2F
BDS 6751 U8A

Canada

Canadian Standards Association
178 Rexdale Blvd., Rexdale
Toronto, Ontario M9W 1R3
Telephone: (416) 747-4082
Telex: 06-989344
Abbreviation: CSA

Canadian standards are used heavily for commerce between Canada and the United States and provide more than 1000 standards in eight major programs. CSA is a membership organization and a national panel of more than 6000 experts develops the various standards. All Canadian standards are preceded by the *upper case* letters CSA. The standard or designation then follows.

Aluminum and Aluminum Alloy standards appear with the letters HA in *upper case* followed by a period, then a digit followed by another period and finally a series of digits. *No spaces* occur in this format.

Example: CSA HA.4.1100

Magnesium and Magnesium Alloys are reported with the letters HG in *upper case* followed by a period and then a digit. If more numuers follow, they are also separated by periods. The standard does not end with a period and there are *no* spaces between the digits or letters.

Example: CSA HG.1.7.3

Zinc and Zinc Alloy standards are reported with the letters HZ in *upper case* followed by a period and a digit. There are *no* spaces between the digits or letters.

International Standards-Issuing Organizations (continued)

Example: CSA HZ.20
CSA HZ.121
CSA HZ.5

Czechoslovakia

Urad pro Normalizacia Mereni
Office for Standards and Measurements
Vactavske Namesti 19
113 47 Praha 1
Telephone: 22 68 45
Telex: 12 19 48
Abbreviation: CSN

The Czechoslovakian Office for Standards and Measurements is a government agency concerned with standardization, metrology, and quality assurance. The CSN was founded in 1952 and is also a member of the ISO. Czechoslovakian standards are arranged according to classes and subgroups by a 6-digit reference number. All standards are preceded by CSN or ON (sectional standards) in *upper case* letters.

Example: CSN 01 0010
ON 41 0451

Denmark

Dansk Standardiseringsraad
(Danish Standards Association)
Aurehojvej 12, Postbox 77
DK-2900 Hellerup
Telex: 15615 Dansta DK
Abbreviation: DS

The Danish Standards Association was founded in 1926 and is involved in the standardization of all fields except electrical technology. The organization is composed of 50 members and a chairman, which is appointed by the Ministry of Industry. Danish standards are reported as follows. (If the code or serial number contains a letter, it appears in upper case letters and there are *no* spaces between the digits or letters.)

Example: DS 12012A
DS 10010

Europe

Association Eurpoeenne des Constructeurs
de Materiel Aerospatial
(European Association of Manufacturers
of Aerospace Material)
88 Bd Malesherbes
75008 Paris, France
Telephone: 563 82 85
Telex: 642701 F AECMA
Abbreviation: AECMA

The Standardization Committee (CN) within AECMA is responsible for the work of standardization. The committees that carry out this work are represented by the aerospace industry users, the processing industries, public bodies and authorities in commerce and science. The number of members is not limited. AECMA standards and designations begin with the prefix AECMA. The standard's numerical code is preceded with the lower *and* upper case letters prEN. Designations are alphanumeric and include hyphens and spaces.

Example: AECMA prEN2002-03
AECMA prEN2389
AECMA Co-P 92-HT
AECMA A-1-P13 PI-T3

Commission of the European Communities
Rue de la Loi 200
B-1049 Brussels, Belgium
Telephone: 235 11 11
Telex: COMEU B 21877
Abbreviation: EURONORM

The standardization of European steel by the Commission of the European Communities has ended and is now under the auspices of the new European Committee for Iron and Steel Standardization (ECISS), in close cooperation with the European Committee for Standardization (CEN), which produces European Standards (EN). The address of ECISS is

ECISS
Rue Brederode 2, Bte 5
B-1000 Bruxelles
Belgium

Euronorm standards, however, that still appear are prefaced by the *upper case* letters EURONORM and are followed by the numerical code. Designations appear as alphanumeric codes and are sometimes attached to the standard number. Note the embedded spaces and hyphens.

Example: EURONORM 80
EURONORM 80 Fe B 400
EURONORM 039-2

Finland

Suomen Standarisolimisliitto r.y.
Finnish Standards Association
Bulevardi 5 A 7
P.O. Box 205
SF-00121, Helsinki 12
Telephone: (90) 645 601
Telex: 122303 stand sf
Abbreviation: SFS

The Finnish Standards Association consists of approximately 37 organizations and develops about 300 standards per year. The Association originated in 1924. Finnish standards and designations are preceded by the letters SFS.

Example: SFS 251
SFS 701
SFS 255 Fe 3550
SFS 455 C 35-07
SFS 368 G-34CrMo4

France

Association Francaise de Normalisation
Tour Europe Cedex 7
92080 Paris La Defense
Telex:s 611974 AFNOR F
Abbreviation: AFNOR

AFNOR is a nonprofit organization created in 1926, with a membership of approximately 5200 members. Of nearly 4000 standards, over 700 relate to metallurgy and are used widely in Europe, Africa, Asia, the Middle East, and the Caribbean. The correct format for reporting AFNOR standards is as follows. Place an *upper case* NF to the left of the alpha numerical code. The alpha numerical code consists of an uppercase letter, followed by a space and a series of digits. The digits consist of two numerals followed by a hyphen and then another series of digits.

Example: NF A 32-050
NF L 09-001

Ministere de la Defense
Delegation Generale Pour L'Armement
Direction Technique des Constructions
Services Technique Aeronautique
4 Avenue de la Porte-d'Issy
75753 Paris CEDEX 15, France
Abbreviation: AIR

Report AIR standards with the prefix AIR in *upper case* letters. The numerical code is often hyphened, but does *not* include spaces.

Example: AIR 9165-001
AIR 9165-211

Germany

Deutsches Institut fur Normung e.V.
Burggrafenstrasse 4-10, Postfach 1107
D-1000 Berlin 30
Telephone: (030) 26 01 329
Telex: 184 273 din d
Abbreviation: DIN

DIN standards are developed by a nonprofit organization of approximately 130 standards committees with representatives from all technical circles. Over 20,000 standards by 40,000 honorary collaborators have been created. Membership is voluntary and open to both German and foreign companies. All German standards are preceded by the upper case letters DIN and followed by a numerical or alphanumerical code. An *upper case* letter sometimes precedes this code. German standards are now reported in one of two methods. One method uses a descriptive code number with chemical symbols and numbers in the designation; the second, known as the *Werkstoff number*, uses numbers only with a decimal point after the first digit. The latter method was devised to be more compatible with computerization. If you are reporting the designation along with the standard, leave a space between the two, but *do not* place any other spaces in the designation. All letters are in upper case *except* lower case letters that are part of a chemical symbol. Hyphens or parentheses may also appear in the descriptive method and, again, if so *do not* include spaces.

Example: DIN 17442 G-X20CrMo13
DIN E17440 X5CrNi1810
DIN 17745
DIN 1.4120
DIN 1.4301

Hungary

Magyar Szabvarnyugyi Hivatal
The Hungarian Office for Standardization
Postafiok 24
1450 Budapest, Hungary
Telephone: 183 011
Telex: 225723 norm h
Abbreviation: MSZH

Hungarian standards are developed by the Hungarian Office for Standardization founded in 1921. The agency is also a member of the ISO. The following are examples of how MSZH Standards are reported.

Example: MSZ 1300
MSZ NI 499/2
MSZ 5744
MSZ KGST 483

India

Indian Standards Institution
Manak Bhavan
6 Bahadur Shah Zafar Marg
New Delhi 110002
Telex: 031-2970
Abbreviation: ISI

Indian standards begin with the prefix IS and are followed by a numerical code.

Example: IS 1471

Israel

Standards Institution of Israel
42 University Street
Tel Aviv 09977
Telephone: 422811
Telex: 35508 SIIT IL
Abbreviation: SII

The Standards Institution of Israel is involved in the development and writing of standards for all areas of industry, ranging from metallurgy to pesticides. Some standards are declared Official Government Standards in the interest of public safety, or for the protection of public health. Such standards become obligatory and no commodity may be manufactured, sold, used, exported or imported unless it complies with the standard. Report the standards in the following way.

Example: SI Gde18
SI Gde22

Italy

Ente Nazionale Italiano di Unificazione
Piazza A. Diaz, 2
20123 Milano
Telephone: 87 69 14
Telex: 312481 UNI
Abbreviation: UNI

International Standards-Issuing Organizations (continued)

The Ente Nazionale Italiano di Unificazione was founded in 1921, and is a member of both the ISO and the European Committee for Standardization. Italian standards are preceded by the upper case letters UNI and followed by a 4-digit code. A designation will consist of an alphanumerical code.

Example: UNI 3159
UNI AT30
UNI GD-AlSi5Fe

Japan

Japanese Industrial Standards Committee
Ministry of International Trade & Industry
33rd Mori Bldg., 3-8-21, Torakomon
Minato-ku, Tokyo 105
Abbreviation: JISC

JIS standards cover industrial or mineral products with the exception of those regulated by their own special standards organizations. The standards are divided into 17 divisions and are used both by commercial and government organizations involved in design engineering, quality assurance, research and development, construction, testing and maintenance. JIS standards begin with the upper case letteers JIS and are followed by an *upper case* letter which designates the standard's division. This is then followed by a space and a series of digits.

Example: JIS G 3311
JIS G 3311
JIS S 20CK

Mexico

Direccion General de Normas
Tuxpan No. 2
Mexico 7, D.F.
Abbreviation: DGN

Mexican Standards begin with the *upper case* letters DGN. The code that follows consists of an upper case letter which denotes the standard's classification, followed by a hyphen and a number with no spaces.

Example: DGN C-189

Netherlands

Nederlands Normalisatie-Instituut
Postbus 5059
2600 GB Delft
Telephone: 90 68 00
Telex: 32123
Abbreviation: NNI

The NNI is composed of approximately 3000 firms and companies, and 200 various organizations. The Association helps prepare Dutch standards and cooperates in the development of international standardization. Dutch standards are prefixed by the letters NEN and followed by a numerical code.

Example: NEN 213
NEN 3077

Norway

Norges Standardiseringsforbund
(Norwegian Standards Association)
Haakon VII's Gate 2
Oslo 1
Telephone: (02) 46 60 94
Telex: 19050 nsf n
Abbreviation: NSF

The Norwegian Standards Association consists of approximately 80 firms, 64 organizations, and 35 government agencies and local authorities. The Association was founded in 1923 and is a member of both the ISO and the European Committee for Standardization. The organization is concerned with the preparation of standards, certification of goods in conformity with standards and publishing and marketing of national and international standards. Norwegian Standards appear as follows.

Example: NS 824
NS 6097
NS 17570

Pan America

Pan American Standards Commission
Lima 629
1073 Buenos Aires, Argentina
Cable: COPANTEC
Abbreviation: COPANT

The Pan American Standards Commission, COPANT, is comprised of national standards bodies of the United States and many Latin American countries. To report the COPANT standards precede the numerical code with COPANT in upper case letters. In some cases, the code number may begin with a letter. Leave a space between the letter and the number.

Example: COPANT R 12
COPANT 1188

Poland

Polish Committee for Standardization
Measures and Quality Control
ul. Elektoraina 2
00-139 Warszawa
Telephone: 20 02 41; 20 54 34
Telex: 813642 pkn pl
Abbreviation: PKNiM

Polish standards are prefixed with the upper case letters PN. The designations or standards may appear in a number of ways. Many are similar to the IS and GOST formats.

Example: PN 0800-02
PN H-88026
PN PA29N
PN AM4

Portugal

Direccao-Geral da Qualidade
Rua Jose Estevao 83-A
1199 Lisboa Codex

Telephone: 53 98 91
Telex: 13042 qualit
Abbreviation: DGQ

The Direccao-Geral de Qualidade is the official standards agency of Portugal and is also a member of the European Committee for Standardization. Portugese standards begin with the prefix DGQ and are followed by an alphanumeric code.

Example: DGQ R 001 FC 10
DGQ NP 968 C 720

South Africa

South African Bureau of Standards
Private Bag X191
Pretoria 0001
Republic of South Africa
Telex: 3-626 SA
Abbreviation: SABS

The SABS was officially established by the South African government in 1945, although unofficial work in the area of standardization by other organizations began in the early 1900's. The letters SABS precede the South African Standards and are followed by a numerical code. Designations are alphanumeric and refer to the chemical composition of the metal or alloy.

Example: SABS 712
SABS Al-CU4MgS
SABS 460
SABS Cu-ETP
SABS Cu-Zn28Sn1

Sweden

Standardiseringskommissionen i Sverige
Swedish Standards Institution
Tegnergatan 11, Box 3295
103 66 Stockholm
Telephone: (08) 23 04 00
Telex: 17453 sis s
Abbreviation: SIS

The Swedish Standards Institution was founded in 1922 and is a membership organization of approximately 29 organizations. The institution also consists of 500 technical committees and is a member of the ISO and European Committee for Standardization. All Standards begin with the prefix SS, or SIS if the standard was written prior to 1978.

Example: SS 11 21 19
SIS 14 01 00

Switzerland

International Organization for Standardization
Case Postale 56
CH-1211 Geneve 20
Telex: 23887 iso ch
Abbreviation: ISO

ISO is the specialized international agency for standardization. It is a non-governmental organization established in 1947 for the purpose of developing worldwide standards to improve international communication and collaboration, and to promote the smooth and equitable growth of international trade. Its members are standards bodies of some 90 countries representing more than 95% of the world's industrial production. ISO standards and designations are prefixed with the letters ISO. Standards appear as a numerical code and the designations as an alphanumeric code relating to the composition of the metal or alloy.

Example: ISO 3522
ISO AlMn1Cu
ISO AlZn6MgCu

Turkey

Turk Standardlari Enstitusu
Turkish Standards Institution
Necatibey Caddesi 112
Bakanliklar, Ankara
Telephone: 18 72 40
Telex: 42047
Abbreviation: TSE

Founded in 1960, the Turkish Standards Institution is a government agency dedicated to the preparation and publication of standards. The TSE is also a member of the ISO. The prefix for Turkish Standards is the letters TS. The code number follows, or in the case of a designation, the alphanumeric code.

Example: TS 2276
TS Mg-Al6Zz1

United Kingdom

British Standards Institution
2 Park Street
London W1A 2BS
United Kingdom
Telephone: 01-629 9000
Telex: 26693 BSILON G
Abbreviation: BSI

The British Standards Institute develops and publishes standards that are used extensively by exporters and importers. They are used both in government and commercial industry by those who are involved in engineering, designing, production, testing and construction. The letters BS precede the Standards numerical code, and may also include the alloy's designation. If so, leave a space between the designation and the specification.

Example: BS 3100
BS 2874 CZ101

United States of America

Aluminum Association
900 19th Street N.W., Suite 300
Washington, DC 20006
Telephone: (202) 862-5100
Abbreviation: AA

The Aluminum Association is an industry-wide trade organization representing all the primary producers of aluminum in the United States, leading manufacturers of semi-fabricated aluminum products and principal foundries and smelters. For the aluminum industry and those industries that use alumi-

International Standards-Issuing Organizations (continued)

num, the association helps develop standards and designation systems. To report an Aluminum Association standard designation place an upper case AA before the code number. Wrought alloy designations differ from the cast alloys by the inclusion of a decimal point between the third and fourth digits of the code number and occasionally by an upper case letter immediately preceding the digits. Alloy temper and subdivisions of temper designations are separated from the main alloy designation by a hyphen. The temper code letter is in upper case letters. Both the letter and numbers follow the hyphen immediately with no space in between. If more than one temper is included, separate it from the first with a comma and then a space.

Example: AA 6060-T4
AA 201.0-T6
AA C355.0
AA 6066-T4, T451

Alloy Casting Institute
c/o Steel Founders' Society of America
Cast Metals Federation Bldg.
455 State St.
Des Plaines, IL 60016
Telephone: (312) 299-9160
Abbreviation: ACI

The ACI has established a standards designating system for cast high-alloy stainless steels. Originally a separate organization, it was absorbed by the Steel Founders' Society of America in 1970. The correct format for reporting ACI designations is to prefix the alphanumeric code with the letters ACI. The code should be reported with *all upper case letters* and contain *no* spaces, including between the hyphens.

Example: ACI CF-8M
ACI CF-3MA
ACI CB-7Cu-1
ACI CA-6NM-B

American Iron and Steel Institute
1000 16th Street, N.W.
Washington, DC 20036
Telephone: (202) 452-7100
Abbreviation: AISI

AISI is not a material specification writing body, although many times steels are referred to as AISI Standard Steels. This designation refers only to the chemical composition ranges and limits of such grades. AISI designations are reported in the same manner as the SAE steel designations, except AISI is placed in front of the code. Refer to the example for steel designations listed under SAE for the correct format.

American National Standards Institute
1430 Broadway
New York, NY 10018
Telephone: (212) 354-3300
Abbreviation: ANSI

The American National Standards Institute standards are used widely throughout industry, and cover a tremendous variety of items, from architectural products to consumer goods to nuclear safety standards. The Institute is also the coordinator of the United States federal national standards system and acts by assisting participants in the voluntary system to reach agreement on standards needs and priorities, arranging for competent organizations to undertake standards development work, providing fair and effective procedures for standards development, and resolving conflicts and preventing duplication of efforts. The ANSI alphanumeric code begins with an *upper case letter* that is followed by 1-3 digits, which are then followed by additional digits that are separated by decimal points. If the code ends with a letter, it is reported in *upper case* letters. All standards begin with the prefix ANSI.

Example: ANSI H35.2
ANSI A156.2
ANSI B18.2.3.6M

American Society of Mechanical Engineers
345 East 47th Street
New York, NY 10017
Telephone: (212) 705-7722
Abbreviation: ASME

ASME standards are used by personnel in research, testing and design of power-producing machines such as internal combustion engines, steam and gas turbines, and jet and rocket engines. They are also used to design and develop power-using machines such as refrigeration and air-conditioning equipment, elevators, machine tools, printing presses and steel rolling mills. ASME standards should be reported in the following way. Place the *upper case* letters ASME to the left of the specification. The standards alphanumeric code is then reported in the same manner as the ANSI standard that is mentioned above. Some codes, however, are slightly different, as will be shown in the next example. Follow the format, however, of *no* embedded spaces and *all* letters in upper case.

Example: ASME B30.2 (same as ANSI)
ASME HST-6M-1N

American Petroleum Institute
2101 L. St. N.W.
Washington, DC 20037
Telephone: (202) 457-7000
Abbreviation: API

The American Petroleum Institute fosters the development of standards, codes and safe practices in the petroleum industries. These standards and codes are used by persons involved in the engineering, production, transportation, handling and use of petroleum products. API standards appear with the letters API before the specification.

Example: API Spec 5AC
API Spec 5L

ASTM
1916 Race Street
Philadelphia, PA 19103
Telephone: (215) 299-5400
Abbreviation: ASTM

ASTM, founded in 1898, is a scientific and technical organization formed for "the development of standards on characteristics and performance of materials, products, systems, and services; and the promotion of related knowledge." It is also the world's largest source of voluntary consensus standards. All standards begin with the prefix ASTM, followed by the actual standard code number. These standards are preceded with an *upper case* letter, which is then followed by a space and then the 1-4 digit numerical code. Sometimes this code number is immediately followed with *an upper case* letter (*no space*). If the standard or specification has a Group, Grade, Class or Type attached to the code number, they should be reported in this order with a space separating it from the digits. Also, the first *letter of each word* should be in upper case and if the Group, Grade, Class or Type is followed by a letter or number there *should* be a space in between. All letters should be in upper case thereafter, and all numerals are in arabic unless otherwise reported. If space in the field becomes limited, abbreviate 'Type' as Ty, and 'Grade' as Gr.

Example: ASTM A311
ASTM A372 Class V Type B
ASTM A723 Grade 1 Class 1
ASTM A336 Grade F 31
ASTM B209
ASTM B308M

Many metals have been given ASTM designations that may also be used in the specifications field. Magnesium alloys, for instance, are prefaced by *upper case* letters, one or two, which are then followed by 1-3 digits, and with an *upper case* A.

Example: ASTM AZ63A
ASTM EZ33A
ASTM ZH62A

American Welding Society
550 N.W. LeJeune Road
Miami, FL 33126
Telephone: (305) 433-9353
Abbreviation: AWS

AWS standards are used in the design, fabrication, testing, quality assurance and related functions in shipbuilding design/construction and heavy construction industries. Report the standard beginning with the *upper case* letters AWS.

Example: AWS A5.24
AWS C5.7
AWS B4.0

Copper Development Association
57th Floor Chrysler Bldg.
405 Lexington Ave.
New York, NY 10174
Telephone: (212) 953-7300
Abbreviation: CDA

The Copper Development Association was founded in 1963 to help expand the uses and applications of copper and copper products, and to broaden the markets for such products. The organization has also developed a standards system for wrought and cast copper and copper alloy products. The CDA standards are identical to the UNS numbers for copper alloys. All standards begin with an *upper case* C and are followed by 5 digits. There are *no* spaces in the standard.

Example: CDA C52400

FED & MIL Standards
Naval Publications & Forms Center
5801 Tabor Ave.
Philadelphia, PA 19120
Telephone: (215) 697-2000
Abbreviation: FED and MIL

Both military and federal standards and specifications can be obtained through the above address, but a DD Form 1425 requisition is necessary. These are obtained through the Government Printing Office, Superintendent of Documents, Washington, DC 20402.

MIL

Military specifications are issued by DOD to define materials, products, or services used only or predominately by military entities. All military specifications begin with the *upper case* letters MIL. The actual specification that follows begins with an *upper case* code letter that represents the first letter of the title for the item, followed immediately by a hyphen and then the serial number or digits. If the serial number is followed by another letter it is in upper case and there is *no* space in between. Additional information is in parentheses with no space between it and the last letter or digit.

Example: MIL S-9041B(AER)
MIL C-11866

Military standards provide procedures for design, manufacturing, and testing, rather than giving a particular material description. Report them in the following way. The standard reference organization letters, MIL, are again placed to the left of the standard and then followed by the upper case letters STD, hyphen, no space, and finally the serial number or digits.

Example: MIL STD-45662

FED

Federal specifications and standards are similar to the military, except they are issued by the General Services Administration, and are primarily for federal agencies. Their use, however, is now acceptable to the military when there is *no* separate MIL specification available. Federal specifications begin with the upper case letters QQ followed by the code numbers and letters. The beginning letter of the code should be in upper case and separated from the subsequent digits by a hyphen, and contain *no* spaces. If any slashes are embedded in the code, again, no spaces.

Example: QQ A-430
QQ A-200/10

National Bureau of Standards
U.S. Department of Commerce
Washington, DC 20234
Abbreviation: NBS

The National Bureau of Standards is the centralized U.S. repository, for reference purposes, of the National Standards of the World.

International Standards-Issuing Organizations (continued)

Society of Automotive Engineers
400 Commonwealth Drive
Warrendale, PA 15096
Telephone: (412) 776-4841
Telex: 866 355
Abbreviation: SAE

SAE standards are used primarily by designers, manufacturers and maintenance personnel in the automotive and aerospace industries. These standards are also useful and effective series for the metals, plastics, rubber, chemical and fastener industries in their standardization efforts. SAE standards begin with the upper case letters SAE. The SAE numerical code which follows is immediately preceded by an upper case J. In some cases the digits are followed by another letter which should appear in lower case and contain no space between the numbers and the letter.

Example: SAE J450
SAE J993b

SAE's designation or uniform numbering system may sometimes also appear in the specifications field. Below are the various categories of ferrous and nonferrous metals and examples of SAE's designations. Note the use of upper case vs lower case letters and the spaces for reporting these designations.

STEELS

Example: SAE 1020
SAE 950X
SAE 10B62
SAE 950A
SAE 950B
SAE 51420F Se
SAE EX 10
SAE 1080 +Mn
SAE 1005-1009

FERROUS MATERIALS

The designations for the different ferrous cast metals begin with the prefix letters G, M or D which represent iron, malleable iron and nodular iron, respectively. These upper case previxes are followed immediately by a 4-digit number.

Example: SAE D5506
SAE 6061
SAE 323

COPPER

Example: SAE CA670

TOOL STEELS

All tool steel designations are prefaced by an upper case letter that represents the different categories of the metals and is immediately followed by one to three digits (no space).

Example: SAE L6
SAE M3
SAE T1
SAE W108
SAE S1
SAE O1
SAE A2
SAE D5

AEROSPACE MATERIAL SPECIFICATIONS

SAE also produces the Aerospace Material Specifications. These are reported by placing the upper case letters AMS to the left of the 4-digit material code.

Example: AMS 5063

USSR

Gosudarstvennyi Komitet Standartov
USSR State Committee for Standards
Leninsky, Prospekt 9
Moskva 117049
Telephone: 236 40 44
Telex: 411378 gost su
Abbreviation: GOST

USSR State Standards number over 23,000 and cover most areas of commerce, industry, agriculture and public health. The standards are defined within groups, i.e., mining minerals; petroleum products; metals and metallic products, etc. The standards are prefaced with the upper case letters GOST and are followed by a numerical code.

Example: GOST 13819
GOST 5.1491
GOST 22974.9

Yugoslavia

Jugoslovenski zavodza Standardizeciju
Sonodana Penozica Krcunabr. 35
Post. Pregr. 933
11000 Beograd
Telephone: 644 066
Telex: 12089 yu
Abbreviation: SZS

The SZS was founded in 1946 and concerned with the adoption and application of standards, technical norms of quality of products and services and the regulations covered by legislation. Yugoslavian standards begin with the prefix JUS and are followed by an alphanumeric code. The first letter of the code denotes the section under which the standard is classified. Most standards relating to metallurgy are in section C.

Example: JUS C.AO.003
JUS C.K6.150
JUS C.T3.005

Universities with Faculties in Polymer Science

University of Akron
Department of Polymer Science
Whitby Hall
(216-375-7542)
Institute of Polymer Science
Auburn Science and Engineering Center
(216-375-7500)
Akron, OH 44325

Case Western Reserve University
Case Institute of Technology
Department of Macromolecular Science
University Circle
Cleveland, OH 44106
(216-368-4284)
(B.S., M.S., Ph.D.)

University of Connecticut
Polymer Science Program (Institute of Materials Science)
Storrs, CT 06268
(203-486-3582)
(M.S., Ph.D.)

University of Illinois at Urbana-Champaign
Polymer Group
Urbana, IL 61801
(217-333-1440)
(M.S., Ph.D., Polymer Science)

Lehigh University
Polymer Science and Engineering Program
Coxe Laboratory No. 32
Bethlehem, PA 18015
(215-758-3844)
(M.S., Ph.D., Polymer Science and Engineering

University of Lowell
Department of Plastics Engineering
Lowell, MA 01854
(617-452-5000, ext. 2324)

University of Massachusetts
Department of Polymer Science and Engineering
Amherst, MA 01003
(413-545-0433)

Massachusetts Institute of Technology
Polymer Division
Cambridge, MA 02139
(617-253-3300)

University of Michigan
Macromolecular Science and Engineering Program
Ann Arbor, MI 48109
(313-763-2316)
(M.S., Ph.D.)

University of Southern Mississippi
Department of Polymer Science
Hattiesburg, MS 39406-0076
(601-266-4868)

Pennsylvania State University
Materials Science and Engineering Department
Polymer Science Program
University Park, PA 16802
(814-865-1288)
(B.S., M.S., Ph.D. Polymer Science)

Stevens Institute of Technology
Department of Chemistry and Chemical Engineering
Hoboken, NJ 07030
(201-420-5546)
(B.E., B.S., M.E., M.S., Ph.D. Chemical Engineering)

The University of Utah
Department of Materials Science and Engineering
Salt Lake City, UT 84112
(801-581-6863)
(B.S., [Polymer Science option], M.S., Ph.D. Materials Science and Engineering)

Universities with Faculties in Ceramics

Alfred University
New York State College of Ceramics
Division of Engineering and Science
Alfred, NY 14802
(607-871-2448)
(B.S., M.S. Ceramic Engineering, Ceramic Science, Glass Science; Ph.D. Ceramics)

University of California-Berkeley
Department of Materials Science and Mineral Engineering
Berkeley, CA 94720
(415-642-3801)

University of California-Los Angeles
Los Angeles, CA 90024
(213-825-5473)

Clemson University
Department of Ceramic Engineering
Clemson, SC 29634-0907
(803-656-3093 or 803-656-3038)
(B.S., M.S. Ceramic Engineering)

University of Florida
Ceramics Division
Gainesville, FL 32611
(904-392-1451)

University of Illinois
Department of Ceramic Engineering
Urbana, IL 61801
(217-333-1770)
(B.S., M.S., Ph.D. Ceramic Engineering; M.S., Ph.D. Ceramic Science)

Iowa State University of Science and Technology
Department of Materials Science and Engineering
Ames, IA 50011
(515-294-1214)
(B.S., M.S., Ph.D. Ceramic Engineering)

Lehigh University
Department of Materials Science and Engineering
Whitaker Laboratory No. 5
Bethlehem, PA 18015
(215-861-4220)

Massachusetts Institute of Technology
Ceramics Division
Cambridge, MA 02139
(617-253-3300)

The University of Missouri-Rolla
School of Mines and Metallurgy
Department of Ceramic Engineering
Rolla, MO 65401
(314-341-4401)
(B.S., M.S., Ph.D. Ceramic Engineering)

The Ohio State University
Columbus, OH 43210
(614-422-2960)
(B.S., M.S., Ph.D. Ceramic Engineering)

The Pennsylvania State University
Materials Science and Engineering Department
Ceramic Science and Engineering Program
University Park, PA 16802
(814-865-4992)
(B.S., M.S., Ph.D. Ceramic Science)

Rutgers, The State University
Department of Ceramics
Piscataway, NJ 08854
(201-932-2220)
(B.S., M.S., Ph.D. Ceramics and Ceramic Engineering)

The University of Utah
College of Engineering
Department of Materials Science and Engineering
Salt Lake City, UT 84112
(801-581-6863)
(B.S., M.S., Ph.D. Materials Science and Engineering)

Virginia Polytechnic Institute and State University
Blacksburg, VA 24061
(703-961-6640)
(B.S., M.S., Ph.D. Materials Engineering Science)

University of Washington
Department of Materials Science and Engineering
Seattle, WA 98195
(206-543-2600)
(B.S., M.S., Ceramic Engineering; M.S. Engineering; Ph.D.)